P9-CCW-725

THE THERMAL ENVIRONMENT

SERIES IN MECHANICAL ENGINEERING

BURGESS H. JENNINGS
Northwestern University

THE THERMAL ENVIRONMENT

Conditioning and Control

HARPER & ROW, Publishers
New York Hagerstown San Francisco London

Sponsoring Editor: *Charlie Dresser*
Project Editor: *Penelope Schmukler*
Production Manager: *Marion Palen*
Compositor: *Composition House Limited*
Printer and Binder: *The Maple Press Company*
Art Studio: *J & R Technical Services, Inc.*

Library of Congress Cataloging in Publication Data

Jennings, Burgess Hill
The thermal environment.

(Series in mechanical engineering)
Includes index.
1. Air conditioning. 2. Heating. 3. Refrigeration and refrigerating machinery. I. Title.
TH687.J46 697 78-17824
ISBN 0-06-043311-6

CONTENTS

10 Physiological Reactions to the Environment 350

11 Warm-Air and Panel Heating Systems 370

12 Fluid Flow, Duct Design, and Air-Distribution Methods 397

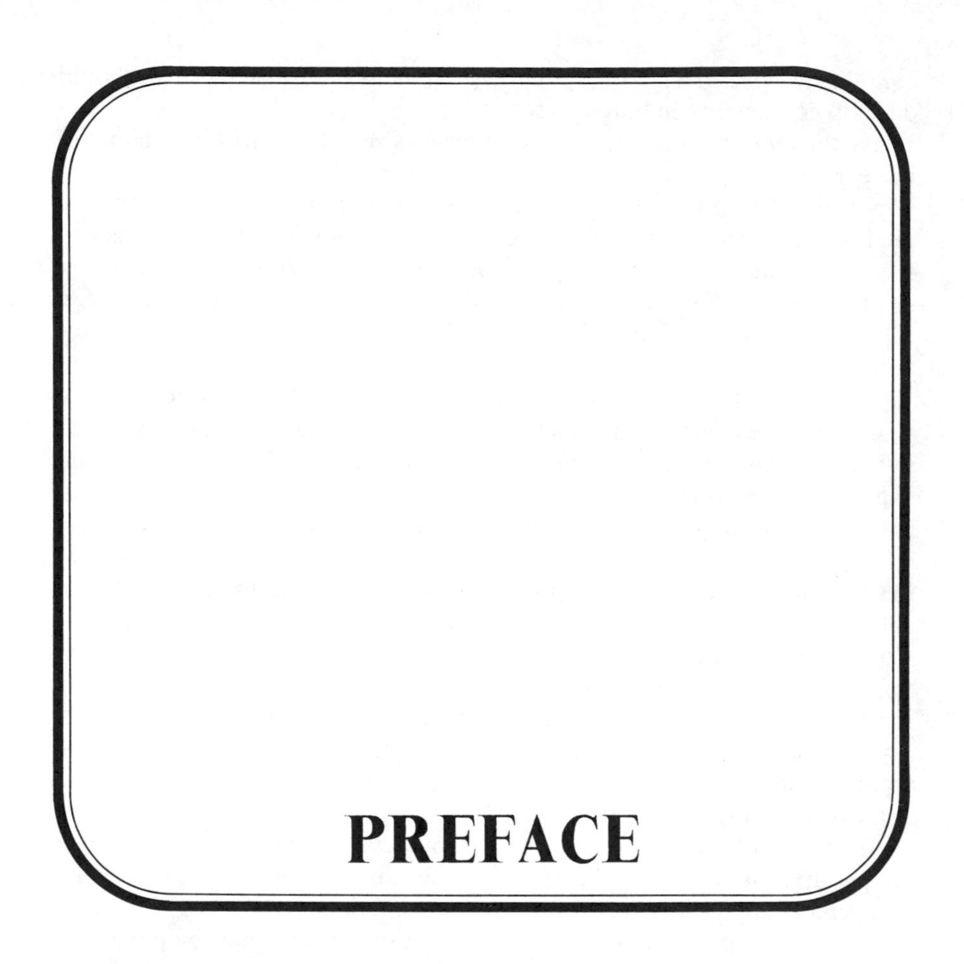

PREFACE

The Thermal Environment represents a new approach to the subject matter of *Environmental Engineering*. It is, however, based on the latter and includes excerpts from it. Naturally, the major reason for developing another book was to cover the many technological changes which have taken place since *Environmental Engineering* first appeared in 1970. Indeed, technical change has been most rapid because of the more stringent need for energy conservation in the light of diminishing energy resources.

During this period also, rapid progress has been made toward converting engineering practice to a greater understanding and usage of SI (metric) units. The new text presents SI in depth and makes much use of it, but not to the exclusion of current engineering nomenclature and units. A complete transition to SI units at this time would be ill advised because now and for some time in the future most engineering practice will be conducted using conventional units. It is also true that almost all the fundamental reference data used in calculations are still expressed in conventional engineering units and, until the changeover is completed, it is preferable to use source data in original form. In this text many illustrative examples are given which are completely worked in both systems, so that any user of the text should develop competency in both systems.

Another goal for this text was to reduce its size in comparison with that of the earlier book in order to give greater emphasis to items of maximum importance to the thermal environment by eliminating topics of peripheral interest. For example, the older book is designed to give complete coverage to the whole range of refrigeration, included chapters or sections in cryogenics, gas liquefaction, food preservation, and thermoelectric refrigeration. These topics, along with others of limited interest, do not appear in this new volume.

Although the new text has been shortened in some areas, it has been significantly expanded in others. More illustrative material has been introduced, and many of the more difficult topics have been rewritten and expanded to provide for ease of comprehension. Much attention is devoted in the text to energy conservation and to ways and means of using alternative energy sources when the need arises. A discussion of the earth-solar system has been presented in depth to provide a basis for understanding how to use solar energy to supplement other heating and cooling sources of energy. Cooling for the environment is now presented in three chapters covering refrigeration and refrigerating equipment. A psychrometric chart completely in SI units has been added, along with illustrative examples.

The author has drawn freely from source material published by the technical societies. The American Society of Heating, Refrigerating, and Air-Conditioning Engineers in particular has provided data in many areas, each specifically noted in the text. Manufacturers have also been most cooperative in providing illustrative material. The author acknowledges and is most appreciative of all such source material.

The text is designed for use by college-level students and by practicing engineers. There is more material provided in the text than can be easily covered in a year of study, but by judicious selection of material, a teacher can easily arrange a program to suit the needs of his class. The author has provided many illustrative examples throughout the text as well as problems for solution at the end of each chapter. With most of these problems, answers are given; a few, however, are left unanswered to satisfy the wishes of teachers who perfer to have such problems.

BURGESS H. JENNINGS

ABBREVIATIONS AND SYMBOLS

ASHRAE—American Society of Heating, Refrigerating, and Air-Conditioning Engineers. Applicable also for its predecessor societies, ASHAE, ASRE, and ASHVE.

ASME—American Society of Mechanical Engineers

Btu—British thermal unit (1.0543 kJ)

°C—symbol for degrees Celsius (centigrade)

C—specific heat capacity (kJ/kg · K in SI; also Btu/lb · R or kcal/kg · K)

C_p—specific heat capacity at constant pressure

C_v—specific heat capacity at constant volume

C_m—mean specific heat capacity

cal—calorie; 1 kcal = 4184 J = 3.9683 Btu

cfm—volume flow in ft^3/min

d—density (kg/m^3 or lb/ft^3)

f—surface coefficient of heat transfer (usually in Btu/h · ft² · F or W/m² · °C)

F—force (N or lbf)

F—symbol for degrees Fahrenheit

ft—feet; 1 ft = 0.3048 m

g_c—ratio (unity in SI; 32.174 for the slug to pound-mass)

gpm—liquid flow in gallons/min

h—enthalpy (Btu/lb, J/kg, or kJ/kg)

Hg—element mercury

hp—horsepower (0.746 kW)

J—joule (newton · meter)

K—symbol for degrees Kelvin

k—thermal conductivity (W/m · K or Btu/h · ft · R and Btu · in./h · ft² · R)

kcal—kilocalorie

$\mathbf{K}_e$—kinetic energy

kg—kilogram (SI unit of mass = 2.2046 lb)

km—kilometer

kPa—kilopascal (1000 Pa = 0.145038 psi)

kW—kilowatt (1000 W)

kWh—kilowatthour

lb—pound (0.45359 kg)

lbf—pound force (4.4482216 N)

m—mass (sometimes weight, in kg or lb)

m—meter (SI unit of length = 3.28084 ft)

mm—millimeter (0.001 m)

m^2—meter squared (area)

m^3—meter cubed (volume)

M—molecular weight, also mol

mlb—millipounds (0.001 lb)

N—newton (SI unit of force = 0.224809 lbf)

Pa—pascals (N/m^2)

p, P—pressure in various units (N/m^2 = Pa, psi, psf, psia, psfa)

$\mathbf{P}_e$—potential energy (J or ft lbf)

psi—lbf/in.^{2}

psia—lbf/in.^{2} absolute

q, Q—heat exchanged (usually Btuh, W, kW, Btu/lb, kJ/kg, etc.)

q_s—sensible-heat load

q_L—latent-heat load

q_T—total heat load

R—degrees Rankine, i.e., degrees Fahrenheit absolute

R—gas constant (usually ft · lbf/lb · R or J/kg · K)

R—Reynolds number

R—resistance to heat flow; reciprocal of conductance or of U, the overall heat transfer coefficient (usually h · ft^2 · F/Btu or m^2 · °C/W)

R—universal gas constant (8309.5 J/kg · mol · K or 1545.3 ft · lbf/lb · mol · R)

s—entropy (usually Btu/lb · R or kJ/kg · K)

s—second

SI—Système International d'Unités, name given to the new metrically oriented international standard units

t—temperature (°C or F)

T—temperature (K or R)

u, U—internal energy (usually Btu/lb or kJ/kg)

U—overall coefficient of heat transfer (Btu/h · ft^2 · F or W/m^2 · °C)

v—specific volume (m^3/kg or ft^3/lb)

V—velocity (m/s or ft/s: also in m/min or fpm)

W—Watts (J/s)

W—humidity ratio (usually lb or mlb/lb dry air or kg or g/kg dry air)

W_s—humidity ratio at saturation

W—work (in appropriate energy units, i.e., Btu, ft · lbf, J, kWh)

α—coefficient of linear expansion (1/F or 1/°C)

ρ—density (kg/m^3 or lb/ft^3)

μ—symbol for degree of saturation (W/W_s); also symbol for dynamic viscosity (usually $N \cdot s/m^2 = Pa \cdot s$ or $lb/ft \cdot s$)

ν—symbol for kinematic viscosity (m^2/s or ft^2/s)

φ—relative humidity expressed as a decimal

THE THERMAL ENVIRONMENT

CHAPTER 1

BASIC CONCEPTS, INSTRUMENTATION, AND ENERGY

1-1. HEATING AND AIR CONDITIONING

The practice of heating and ventilating has made it possible for man to exist under forbidding climatic conditions. (The term *heating*, as used here, means the maintenance of a space at a temperature above that of its surroundings, while *ventilation* means the supplying of atmospheric air, and the removal of inside air, in sufficient amounts to provide satisfactory living conditions.) Ever since early man in his cave huddled close to a fire, and worried about the removal of smoke, he has used the combustion of fuel to aid him in adjusting to a harsh environment. Heating and ventilating methods have changed greatly since those early days of prehistoric man, but the fundamental problem remains; in temperate climates, heating and ventilating in winter are necessary to life as we know it.

Early civilizations had their origin in tropical areas, where heating requirements were slight or even nonexistent; but residents of tropical areas were (and still are) faced with the opposite problem—how to devise satisfactory cooling methods in order to keep body temperature at sufficiently low levels. Under both warm and cold climatic conditions, a balance must be maintained between the individual and his environment. The objective in heating or cooling for comfort is to provide an atmosphere having such characteristics that the occupants of a space can effectively lose enough heat to permit proper functioning of the metabolic processes in their bodies and yet not lose this heat at so rapid a rate that the body lowers in temperature. The regulatory mechanisms of the human body endeavor to keep body temperature at or about 37.0°C, which is the normal temperature for human beings. In hot summer weather, dissipation of body heat may be difficult, whereas in winter the heat dissipation is not difficult and in fact must be controlled so as not to be excessive. Outdoors, the customary heavier clothing of winter reduces the rate of heat loss. However, it is also desirable and usually necessary to keep the indoor temperature in winter within a suitable comfort range so that body functions are properly carried out without taxing the heat regulatory system.

Heating. Early heating involved the use of fireplaces or room stoves. These were not efficient in the use of fuel, since much of the heat from combustion went directly up the chimney. Also, the space being heated was drafty, because of the excessive amount of cool air brought in and wasted up the chimney, and the occupants were therefore frequently overheated on one side and too cool on the other. The individual stoves in each room were an improvement over open fires, but in addition to the inconvenience of operating such separate heaters, the stoves had certain of the disadvantages associated with fireplaces.

Toward the middle of the nineteenth century, *central heating systems* began to come into use. In such systems, a furnace, which is placed in a convenient location, frequently in the basement of a building, burns fuel and the resultant heat generated is carried by a suitable medium to other parts of the building. The transfer mediums usually employed are air, steam, or hot water. *Warm-air systems* are usually of the forced-air type, using fans and ductwork to circulate the warm air mechanically. Such systems make it possible to clean and humidify the air as well to provide controlled amounts of air for respective areas. *Steam systems* are of numerous types from the simple one-pipe system to complex vapor and vacuum systems. *Hot-water* or *hydronic* systems use mechanical circulation to force the hot water to radiators or convectors located in the rooms to be heated.

The ever-rising cost of fuels has made the usage of electric heating increasingly significant. This is particularly the case where electric power is produced largely from coal or nuclear energy and where gas is becoming difficult to obtain at the low prices for which it was formerly sold. Electric energy used in resistance units is more costly than energy provided from combustion of gas or oil. However, in regions where summer cooling is also important, heat-pump systems will often be cheaper to own and operate than year-round oil- or gas-fired furnace systems with conventional air-conditioning compressors. A similar condition also exists in large buildings where interior spaces require cooling year around. Here also the heat pump can provide cooling for the interior spaces, aid in heating the exposed building areas, and accomplish both functions at lower operating cost than would be involved in using a furnace and air-conditioning compressor. Electrical resistance heating is also being extensively employed in large new apartment structures where heat loss can be minimized because of adjacent heated space, the cost of installation is reasonably low, and each area can have its space temperature accurately controlled independent of other areas.

Air conditioning. The term air conditioning, although it has been used in many different concepts, implies the creation and maintenance of an atmosphere having such conditions of temperature, humidity, air circulation, and purity as to produce desired effects upon the occupants of that space, or upon the materials that are handled or stored there. In relation to temperature, it should be noted that comfort for the individual depends not only on the ambient air temperature but also on radiation to and from surrounding surfaces. The simultaneous control of these four factors within required limits, when directed toward human comfort and health or when industrially directed toward conditions permitting the best product yield during manufacture and storage, can rightly be called air conditioning. Air conditioning is independent of time or season and can function effectively under all extremes of weather.

Complete air conditioning for comfort, and for industrial control of product, developed during the last 60 years. The term *air conditioning* was first employed in connection with the practice of humidifying the air in textile mills to control static-electricity effects and to reduce the breaking of fibers. This was very necessary in winter because, when the cold outside air with its low moisture content was heated, the static electric charges produced by the moving threads could not leak off under the resulting low relative-humidity conditions and the looms operated with difficulty. The very dry threads were also brittle and had a greater tendency to break. Humidifying (that is, adding moisture to the air) reduced or eliminated these difficulties and gave significant impetus to the development of air conditioning in industry.

Moisture can easily be introduced into the air for humidifying by evaporating moisture into the air, but the problem of removing surplus moisture (dehumidification) is more difficult. Dehumidification can be accomplished by the use of desiccant materials which can be periodically reactivated. Most frequently, however, dehumidification is brought about by using refrigeration to chill the air to a sufficiently low temperature that the excess water vapor can be removed by condensation. The resulting water can then be withdrawn from the system. Water removal by condensation from air is illustrated by the familiar formation of moisture on the cold surfaces of a glass of ice water. For comfort, the lowering of temperature is a most important function of the air-conditioning system whether or not dehumidification is also necessary. A refrigeration system is thus an adjunct of any air-conditioning system whenever temperatures below those of the surrounding atmosphere are required.

1-2. REFRIGERATION

Control of the environment, whever cooling is required, involves the use of refrigeration in one form or another. Comfort air conditioning does not require extremely low temperatures for the cooling process, and the equipment seldom has to operate at temperatures lower than 40 F (4.4°C). However, in food processing and storage, temperatures down to −40 F (−40°C) may be required. Industrial processes may employ temperatures as low as −150 F (−101°C). Temperatures in these ranges can be reached by conventional refrigeration equipment.

Refrigeration is thus intimately associated with man and his environment and needs to be thoroughly understood. It is significant that every aspect of heating, air conditioning, and refrigeration involves the application of fundamental principles drawn from the concepts of thermodynamics, fluid flow, and heat transfer coordinated to adapt mediums to serve the environmental requirements for occupants or materials in a space, whether for the comfort of individuals, the preservation and creation of products, or the production of power. The development of interrelated principles toward these desired ends will constitute the subject matter of much of this text.

1-3. RELATION OF PROPERTIES OF MATERIALS TO INSTRUMENTATION

The reproducible response of materials to changes in temperature and pressure can often furnish a basis on which to design measuring devices.

Thermal expansion with temperature increase is a useful property of matter which is used in instrumentation. Most materials increase in size (length and volume) as the temperature increases. Over small temperature ranges (not exceeding some 150°C) the change in length is essentially a linear function of temperature and can be expressed by the relation

$$L_t = L_o[1 + \alpha(t - t_o)] \tag{1-1}$$

$$\Delta L = L_t - L_o = L_o\alpha(t - t_o) \tag{1-2}$$

where L_o is the original length of the object at temperature t_o, in meters, feet, or inches; L_t is the length of the object at temperature t in consistent units of meters, feet, or inches; t is temperature in degrees Celsius or Fahrenheit; t_o is temperature in degrees Celsius or Fahrenheit; ΔL is the change in length, in meters, feet, or inches; α is the linear coefficient of expansion in meters per meter °Fahrenheit, feet per foot °Fahrenheit or inches per inch °Fahrenheit. Values of α from Table 1-1 when used with Celsius temperature changes must be multiplied by the factor $\frac{9}{5}$ or 1.8, for meters per meter °Celsius or feet per foot °Celsius. Values of linear expansion coefficients are tabulated in Table 1-1 for a variety of substances. The volume coefficient of expansion is closely equal to 3α over moderate temperature changes. Volume coefficients of expansion for some liquids are listed in Table 1-2. For gases, which undergo relatively large volume change under temperature variation, computation methods are discussed in Section 2-5.

Table 1-1 Linear Coefficients of Thermal Expansion per Degree Fahrenheit

Material	Coefficient α [1/(deg F)]	Temperature Range Applicable (deg F)
Aluminum	0.0000123	32–212
	0.0000143	68–570
Brass	0.0000106	32–212
Brick	0.0000053	68
Fireclay	0.0000061	600
Bronze	0.0000102	32–212
Carbon, graphite	0.0000044	104
Concrete	0.0000056	68
Copper	0.0000093	32–212
	0.0000098	77–570
Glass		
Jena	0.0000045	32–212
Plate	0.0000049	32–212
Pyrex	0.0000020	70–880
Quartz	0.0000003	60–1800
Gold	0.0000082	32–212
Granite	0.0000046	68
Ice	0.0000283	0–30
Inconel (0.8 Ni, 0.14 Cr)	0.0000089	100–1400
Invar	0.0000005	68
Iron, cast	0.0000059	104
Iron, wrought	0.0000063	0–212
Lead	0.0000139	32–212
Marble	0.0000065	60–212
Masonry	0.000002	68
	0.000004	68
Monel metal (.67 Ni, .33 Cu)	0.0000078	70–212
	0.0000089	70–1100
Paraffin	0.0000723	60–100
Silver	0.0000105	68
Steel	0.0000064	32–212
	0.0000068	32–392
	0.0000072	32–572
Stainless	0.0000054	68–392
Tin	0.0000149	64–212
Wood		
Oak	0.0000027	36–92
Across fiber	0.0000302	36–92
Pine	0.0000030	36–92
Across fiber	0.0000189	36–92
Zinc	0.0000165	32–212

Table 1-2 Volume Coefficients of Thermal Expansion and Specific Volumes of Liquids

Material	Volume Coefficient β_v [1/(deg F)]	Applicable Temperature Range (deg F)	Specific Volume and Density at Temperature Indicated			
			Cu Ft per Lb	Cu Cm per Gram	Lb per Cu Ft	Deg F
Alcohol, ethyl..........	0.000562	80–115	0.02029	1.267	49.27	68
50% water (weight)...	0.000413	32–102	0.01753	1.094	57.05	68
Alcohol, methyl.........	0.000630	32–142	0.02153	1.2566	49.62	60
50% water (weight)...	....	....	0.01746	1.0887	57.27	60
Benzene................	0.000650	52–176	0.01782	1.1123	56.12	32
Calcium chloride, 40% solution in water....	0.000235	68	0.01148	0.7164	87.13	68
Mercury..............	0.000101	32–212	0.00118	0.0735	848.7	32
			0.00118	0.0738	845.6	68
			0.00119	0.0741	842.9	100
			0.00120	0.0749	833.5	212
Petroleum (sp gr 0.899)..	0.00044	75–248	0.01784	1.1138	56.05	60
Sea water	....	....	0.01565	0.9756	63.90	60
Sodium chloride 20% solution in water....	0.00020	32–85	0.01396	0.8712	71.65	68
Water................	....	....	0.01602	1.0001	62.422	32
	....	....	0.01602	1.0000	62.426	39.6
	....	....	0.01604	1.0009	62.35	60
	....	....	0.01605	1.0018	62.31	68
	0.00022	60–160	0.01613	1.0070	61.99	100

Example 1-1. A steam pipe of ordinary steel is 80 ft long at 60 F when installed. Find the increase in the length of this pipe when it is carrying steam at 215 F.

Solution: From Table 1-1, α is read as 0.0000064. Then, by Eq. (1-2),

$$\Delta L = (80)(0.0000064)(215 - 60) = 0.079 \text{ ft}$$
$$= 0.95 \text{ in.} \qquad \textit{Ans.}$$

Example 1-2. A steam pipe of ordinary steel is 24.38 m long at 15.56°C when installed. Find the increase in length of this pipe when it is carrying steam at 101.67°C.

Solution: From Table 1-1, α is 0.0000064 m/(m · F). Multiply by 1.8 to find α = 0.00001152 m/(m · °C). By Eq. (1-2),

$$\Delta L = (24.38)(0.00001152)(101.67 - 15.56) = 0.0241 \text{ m} \qquad \textit{Ans.}$$

The familiar mercury-in-glass thermometer consists of a small glass bulb, and a stem in which there is a passage of capillary size. Liquid mercury in the bulb expands when the bulb warms, and the expansion is in evidence as the mercury thread moves upward in the capillary. Proper calibration of thermometers from standardized temperature points is accomplished by accurately etching and marking the stem between these points to indicate temperature values on a selected scale.

Another example of expansion effect in instrumentation is the use of the bimetal strip, which consists of two dissimilar metals fused lengthwise to each other. Under temperature increase, the side of the strip made from the metal with the greater coefficient of expansion elongates more than the other side of the strip, which causes the strip to bend out of a straight-line position. The resultant movement of the free end of the strip can be used to close electric contacts for relay operation, or in pneumatic systems can control the branch-line air flow to actuate air-operated damper or valve motors.

Figure 1-1 shows a thermostat which makes use of a bimetal strip in its operation. Note the bimetal strip (2) in the operating diagram. This elongates on its upper side with rise in ambient temperature and in so doing presses on the flapper (1) with increasing force as temperature increases. Pressure from the air line, reduced after passage through a restrictor, acts on the opposite side of the flapper, tending to open it and thereby reduce the pressure to the branch line and its controlled device, except to the extent that this action is prevented by the bimetal strip. A number of adjustments for the operating range exist. Near the end of the bimetal is a throttling-range adjustment and at the fixed end is the calibration screw. The setpoint cam, under control of the user, sets the desired temperature. At control conditions, depending on the settings of the setpoint cam, the throttling-range button, and the calibration screw, there is a fixed branch-line pressure for each temperature. The forces within the assembly always move toward a fixed branch-line pressure for each temperature setting irrespective of fluctuations in mainline air pressure. In the external view of the thermostat one scale indicates space temperature and the other shows the temperature selected by the setpoint cam.

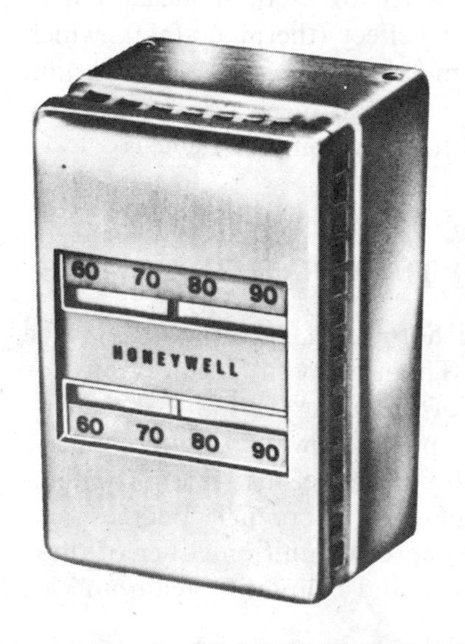

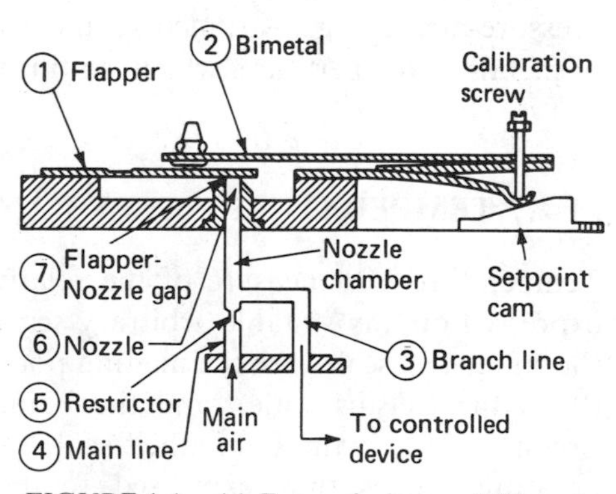

FIGURE 1-1. (a) External view and (b) operating section of a bimetal-element pneumatic thermostat. (Courtesy of Honeywell, Inc.)

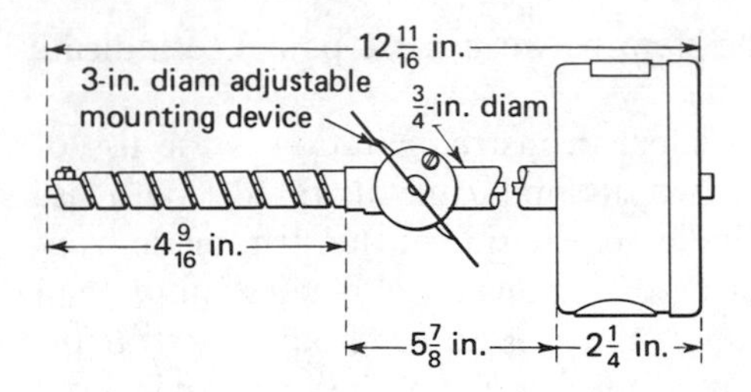

FIGURE 1-2. Warm-air limit control for insertion in furnace duct or plenum chamber. (Courtesy Mercoid Corporation.)

Another example of a bimetal control device is the limit control, illustrated in Fig. 1-2. The particular control shown is used to prevent overheating of a warm-air furnace operated by an automatic burner. It is designed for insertion in the hot-air outlet duct leading from the furnace. As the temperature in the duct rises, the coiled bimetal element reacts by twisting about its central axis until, when the predetermined limit temperature is reached, the coil has twisted sufficiently to trip open the circuit. This action stops the burner. The limit setting of the control can easily be adjusted by pressing the knurled knob on the front of the control either to the left (for a lower temperature) or to the right (for a higher temperature). The adjustable range for this control is from 70 to 310 F, and with its mercury trip switch it can be used for either line-voltage (110 V) or reduced-voltage applications. Devices similar to this but with the bimetal coil mounted in a leakproof separable well or shield are also applicable for limit control of water temperatures in systems or in boilers alone.

Invar and brass have been used in some bimetallic designs. Invar, which is an alloy of iron containing 36% nickel, has an extremely low coefficient of thermal expansion, and when used with brass in a bimetallic strip the high coefficient of expansion of the brass in contrast to that of the invar makes possible a relatively large travel for small temperature change.

Among other characteristics of matter which are useful in measurement and instrumentation are the thermoelectric effect (thermocouple), which will be described later; the elasticity of metals, which is used in certain pressure-measuring instruments; and the variation in electrical resistance of metals to temperature, which is utilized for thermometric devices.

1-4. TEMPERATURE SCALES AND TEMPERATURE DEVICES

Temperature is a measure of the relative hotness of a body and may be expressed on any suitable arbitrary scale. The most used datum points for thermometric scales are the melting point of ice, 32 on the Fahrenheit scale (0 on the Celsius scale), and the boiling point of water at atmospheric pressure, 212 on the Fahrenheit scale (100 on the Celsius scale). The temperature ranges thus expressed represent 180 Fahrenheit degrees and 100 Celsius degrees, and having defined a degree or unit on either of these arbitrary scales they may be extended above and below the steam and ice points. To change from one scale to the other the following obvious relations can be used:

$$t_f = \tfrac{9}{5}t_c + 32 = 1.8t_c + 32 \tag{1-3}$$

$$t_c = \tfrac{5}{9}(t_f - 32) = \frac{t_f - 32}{1.8} \tag{1-4}$$

where t_f and t_c represent temperatures expressed in degrees Fahrenheit and Celsius, respectively.

Consideration of the laws which govern the behavior of gases and the thermodynamic work scale render the concept of an absolute zero of energy and temperature a reasonable conclusion. It can be shown by calculation that the absolute zero of temperature occurs at 459.69 deg below zero on the Fahrenheit scale and 273.16 deg below zero on the Celsius scale. Absolute temperature can be found on the Fahrenheit and Celsius scales by the relationships

$$T_f = t_f + 459.69 \text{ R (degrees Rankine)} \tag{1-5}$$

or

$$T_f = t_f + 460 \text{ (approx) R} \tag{1-6}$$

and

$$T_c = t_c + 273.16 \text{ K}$$

where T_f represents degrees Fahrenheit absolute (degrees Rankine), and T_c degrees Celsius absolute (degrees Kelvin). In most calculations the value 460 is employed instead of the more exact 459.69.

The main use that will be made of the absolute-temperature scale in this work will be in relationships dealing with gases such as air and low-pressure superheated steam.

Mention has already been made of the mercury-in-glass thermometer, which has wide utility. Mercury thermometers can be used to temperatures which approach 1000 F (540°C), particularly when the stem is filled with nitrogen or another inert gas. However, the temperature range on the low side is limited by the freezing point of mercury, which is −39.6 F (−39.2°C). For ranges below this point it is customary to use thermometers filled with colored alcohol or pentane.

Usually mercury thermometers are calibrated for complete immersion in the medium whose temperature is being measured, but they are often used under conditions of partial immersion. Partial immersion means that only the bulb and a portion of the stem are in the medium being investigated and that the rest of the stem is in a colder or warmer atmosphere. Partial immersion leads to inaccurate indications of the thermometer and it is therefore customary to make a *stem correction*. Stem-correction data can be determined by observing the number of degrees of mercury thread exposed outside of the medium being measured and by finding the temperature of the stem itself. The temperature of the stem can be determined approximately by tying an auxiliary thermometer to the exposed stem and insulating the bulb and stem at the point of attachment. The temperature t of the medium being measured to a close approximation is then

$$t = t_1 + EK(t_1 - t_s) \tag{1-7}$$

where t_1 is the temperature indicated by the thermometer, in degrees Celsius or Fahrenheit; t_s is the temperature of the stem as indicated by the auxiliary thermometer, in degrees Celsius or Fahrenheit; E is the number of Celsius of Fahrenheit degrees of emergent mercury thread outside the fluid medium under measurement; and K is the difference between the coefficients of expansion of mercury and the glass of the thermometer. Use $K = 0.000158$ when Celsius temperatures are employed and $K = 0.000088$ when Fahrenheit temperatures are employed.

Thermometric elements are also made by filling a temperature-actuated bulb with a suitable fluid such as mercury, methane gas, aniline, or the like, and connecting this bulb through a capillary tube to a pressure-responsive gage. Under proper calibration the pressure readings of the gage can be related directly to the increase in the volume of the fluid, and thus to the temperature of the bulb and of the medium being measured. Such instruments have an essentially uniform scale for temperature indication. Vapor-pressure instruments are built in a similar manner but depend on the vapor pressure of the fluid in the bulb being transmitted through the capillary tube to an indicating pressure gage calibrated to read in temperature. The relation of temperature to the vapor pressure of a fluid at saturation is not a linear function, so these instruments have nonuniform divisions on their temperature scales. Remote-reading and recording-type gages are frequently one or the other of these two types. The bulb of such an instrument is immersed in the hot medium whose temperature is being controlled, and under variations in temperature of the medium the contents of the bulb expand or contract. The resulting pressure changes then actuate a control element or switch.

Thermocouples. The simplest thermocouple circuit consists of two wires of dissimilar metals, with juntions made at both ends. If each junction is maintained at a different temperature, it is found that a measurable electric current will flow in the completed circuit, or if one of the wires is broken a potential difference (small voltage) of measurable amount will exist at the point of breakage in the circuit. Figure 1-3a is a diagram of such a circuit. A thermoelectric pyrometer makes use of this effect, and serves as a temperature-indicating device.

In Figure 1-3b, a simple thermocouple pyrometer circuit is shown. Here one junction in this case the hot one, is indicated at H, and the other junction, the cold one, is indicated at C. The thermocouple element is the portion CHB, and from C and B lead wires complete a circuit to an indicating instrument, which is most cases is essentially a high-resistant voltmeter. One of the temperature-responsive junctions is frequently enclosed in a protective

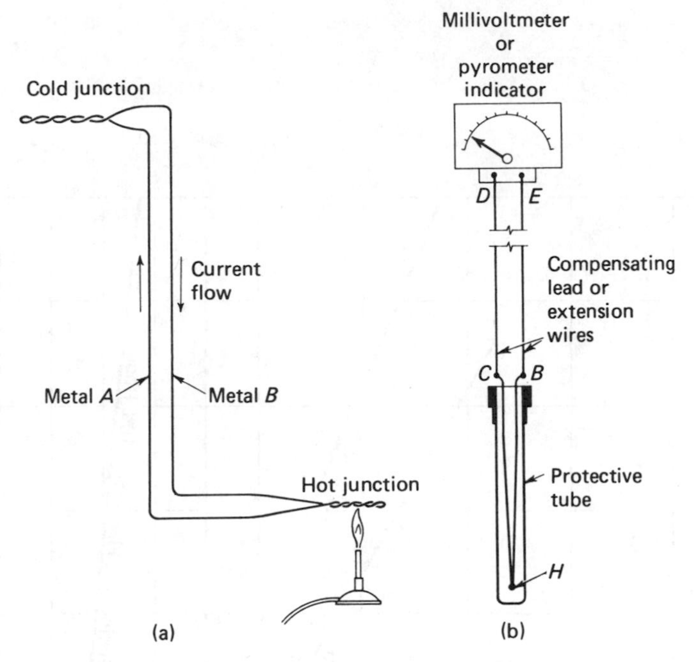

FIGURE 1-3. (a) Basic thermocouple circuit. (b) Thermocouple circuit with temperature-indicating device.

tube. When C and H are at different temperatures a current flows through this circuit, and the reading of the instrument can then be used to indicate the temperature difference between the hot and cold junction. In this simple circuit the wire CH is of one metal and BH of a dissimilar metal. The cold junction is located at C, although by use of compensating lead wires it can be transferred to the instrument location at D or E if this is desired. In a simple instrument of this type, it is very desirable that the temperature of the cold junction remain essentially constant, as the reading of the instrument is always a measure of the temperature differential above that of the cold junction. The compensating lead wires are made of one of the metals used in the couple, or of a metal with similar thermoelectric characteristics.

Various metal combinations are used in thermocouples. In the temperature range to 400°C, and in refrigeration work, copper and constantan are frequently used. Constantan is an alloy containing approximately 60% copper and 40% nickel, and it develops approximately 0.045 mV/°C of temperature difference, existing between the two junctions, in the lower part of the working range. For temperatures reaching to 900°C, iron and constantan can be employed; and for the range to 1100°C, chromel and alumel. Chromel is an alloy of 10% chromium and 90% nickel; alumel is an alloy of 2% aluminum and 98% nickel. For higher temperature ranges, and also in certain scientific work, "noble" metals are often used, of which platinum and platinum-rhodium represent a frequently employed combination.

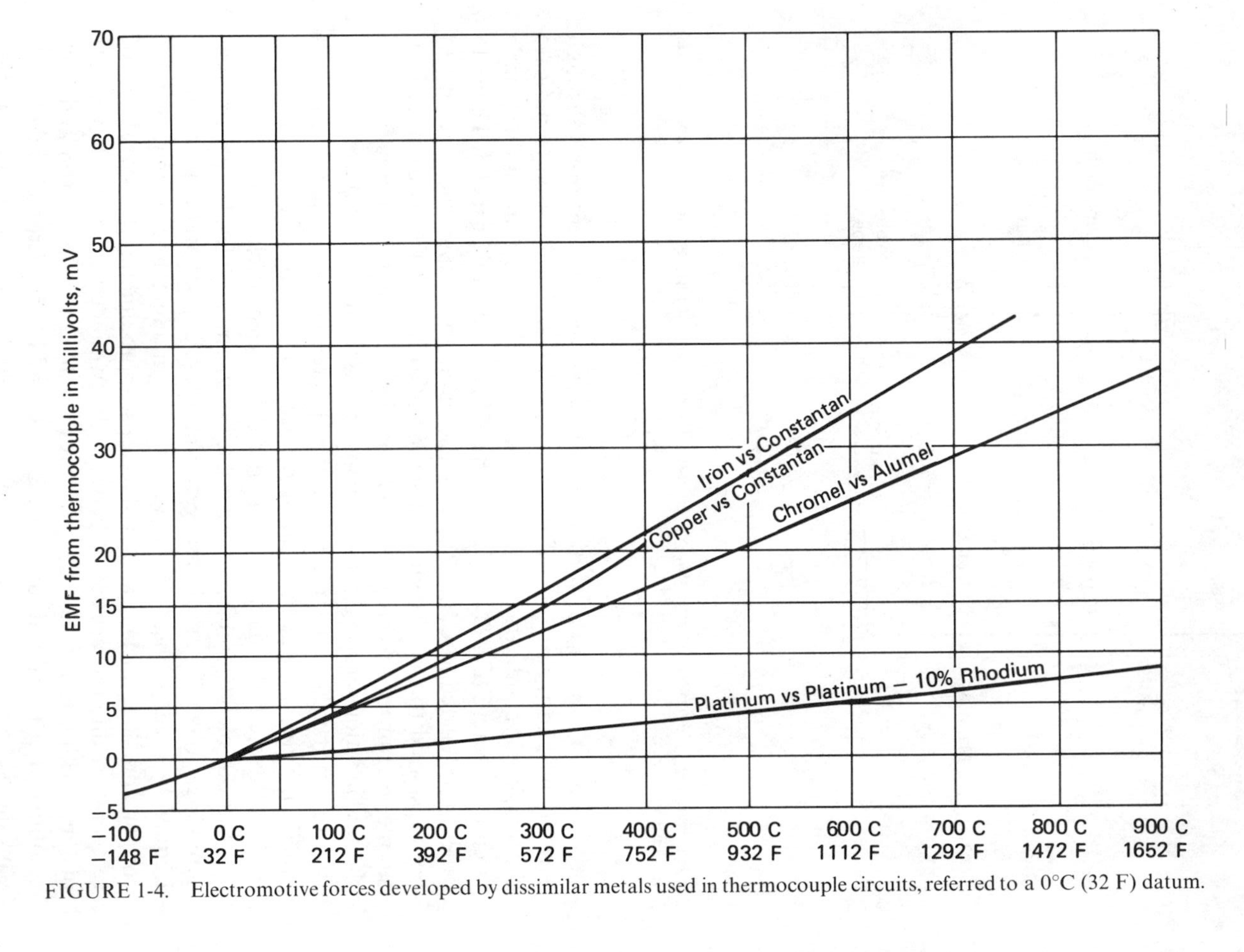

FIGURE 1-4. Electromotive forces developed by dissimilar metals used in thermocouple circuits, referred to a 0°C (32 F) datum.

Couple junctions can be joined by welding, brazing, soldering, or, for very temporary work, sometimes even by twisting, provided the surfaces are clean and corrosion is not a factor.

The electromotive effects for various thermocouple combinations are plotted in Fig. 1-4. Notice that values of emf are not constant but vary slightly in different temperature regimes.

Thermocouple circuits, in addition to serving as temperature indicators, have also been employed in control devices. One such device is illustrated in Fig. 1-5. Here the hot junction and the cold junction are shown, and the complete electric circuit from the hot and cold junction includes an operating electric coil. When the junctions are at different temperatures a current flows in the circuit, and under sufficient temperature difference the current is strong enough to operate switches, control valves, or other devices in the circuit. One use which is made of such a circuit is in connection with gas burners, where the hot junction is placed adjacent to the pilot light. As long as the pilot is burning and the junction is hot, the electric circuit holds open the safety trip in the main gas-supply circuit, so that when the thermostat calls for gas flow and active combustion, the gas valve can open and combustion can proceed following ignition by the pilot flame. On the other hand, if the pilot goes out, the gas-valve safety trip releases, and this prevents the gas valve from opening and loading the furnace with unignited gas. The same thermocouple principle is applicable to many other uses.

In scientific work, and in other types of work requiring precision measurement, it is often desirable to eliminate the electrical resistance of the circuits, and for this a potentiometer-type of instrument is often employed. In Fig. 1-6, the circuit of such an instrument is shown. An electromotive force is generated from the junctions C and H. By means of the battery, a current is caused to flow through the circuit $RGABDE$, and by adjusting the position of the movable slide B the potential drop along the slide wire from D to B can be changed. When the potential between B and D is exactly the same as the thermocouple voltage, the galvanometer will not indicate current flow if the key at K is closed; on the other hand, if the voltages are unbalanced the galvanometer will indicate current flow, and the slide position on the wire DBA is then adjusted until current flow ceases. Because the voltage of a dry- or wet-cell battery is not constant, most instruments are provided with a standard (cadmium) cell. Standard cells are precision-made to produce a constant voltage if used so that a minimal current of short

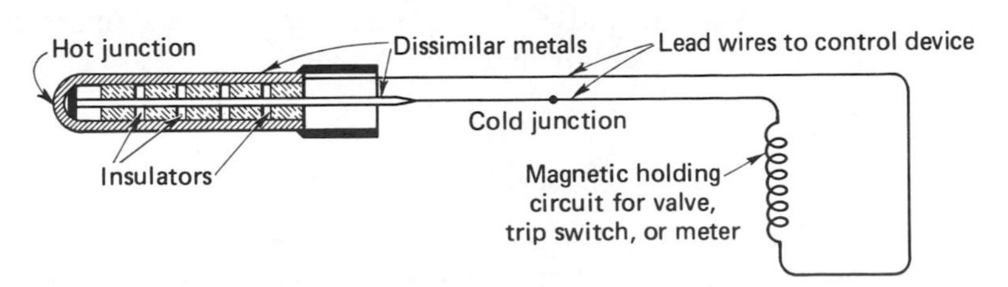

FIGURE 1-5. Thermocouple circuit used as control device.

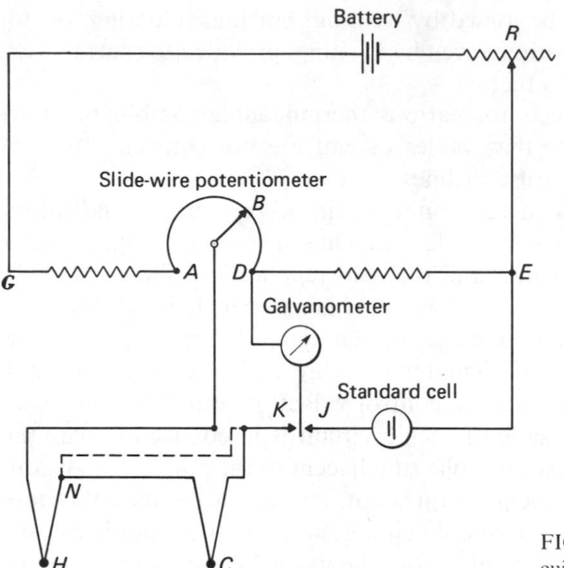

FIGURE 1-6. Potentiometer circuit for thermocouple.

duration is taken from them. To check the battery circuit, the key J is momentarily closed to indicate whether the proper current flows through DE and also through the slide wire DBA. If the current is not as desired, the galvanometer deflects and the resistance R in the battery circuit is then adjusted until the desired current flows through the slide wire. Thus readings of the slide wire are then automatically reevaluated in terms of a standard current flow, and from the basic calibration of the instrument the slide-wire contact setting B can be used to indicate the electromotive force, or temperature difference of the external thermocouple circuit.

In some cases, the junction C is immersed in an ice and water bath and enclosed in an insulated container so that a fixed datum of 0°C (32.0 F) is maintained by the melting ice. In this way a constant reference point is obtained, in terms of which the other couple can be evaluated. More frequently, however, in industrial instrumentation, a datum point which is essentially room temperature is used as the reference point, and the partially dotted circuit HNK is substituted for the circuit $HNCK$. Under these conditions, the second junction of the thermocouple circuit is either at N or K. If the same metal runs from H to K, the second junction will be at the instrument point K, whereas if the lead wires are of a different, noncompensating metal, the second junction point would be at N.

1-5. FUNDAMENTAL UNITS

The engineer or scientist in making computations must have at his disposal a system of units which is completely consistent and one which is readily understood by all those who may wish to make use of his final results. The

English-speaking peoples, with the passage of time, have developed consistent but complex systems of units in which we encounter such familiar units as inch, foot, pound, slug, ton, gallon, and horsepower, to name a few. These consistent systems go by such names as the foot-pound-second system, the English system, or the engineering system. All have the disadvantage of not being decimal in nature. It is also true that most of the non-English-speaking countries of the world make use of a decimal-oriented metric system. Even in the English-speaking countries, pure-science areas such as chemistry and physics have used metric units and terminology to the point that the gram, milligram, liter, meter, kilowatt, and such terms are familiar in daily activities.

International discussions over a period of years have led to agreement on a carefully developed decimal system known as "Système International d'Unités," most usually referred to as SI. This system closely resembles the metric system, on which it was based, but it differs in regard to precise definition of standards, recommended practices of notation, and the elimination of certain terminology.

The United Kingdom has made the decision to convert to SI and the transition is now in progress. In the United States many units are already legally defined in terms of SI standards, and it is merely a question of time before full authorization to convert to SI can be expected. The conversion will involve great expense even if the change is not sudden, and some 10 to 15 years might elapse before full transition could occur. Thus now and during the transition period it will be necessary to understand and to be able to use both systems.

Each system is employed in this text. Most of the sources of data appear in engineering units and most computations involve such units. However, a parallel SI approach is usually given and every effort should be made to develop equal familiarity with SI usage.

In SI the *second* continues as the basic time unit. Definitions of the basic units of *mass*, *length*, and *force* follow.

1. *Kilogram*. This unit of mass is equal to the mass of the international prototype kilogram.
2. *Meter*. This measure of length, formerly defined as the distance between two fine lines on a standard platinum-iridium bar held at 0°C, is now defined as a length equal to 1 650 763.73 wavelengths in vacuum corresponding to the transition between two levels of the krypton-86 atom.
3. *Newton*. The unit of force, the *newton*, is a force of such magnitude that when applied to a body having a mass of 1 kilogram (kg) it will give it an acceleration of 1 meter per second per second (m/s^2).

The basic units in the engineering (nonmetric) system are the *pound* (lb), *foot* (ft), and *second* (s) for mass, length, and time. The pound has multiple uses in this system because it is used as a measure of mass as well as of force and weight. This can be confusing and the usage is further complicated by the fact that there may be variations in the pound itself. The *avoirdupois* pound is usually understood when the pound is mentioned, but the *troy* pound is used in coinage and its equivalent, the *apothecary* pound, was formerly used in the drug industry. In the avoirdupois pound there are 16 ounces (oz), whereas in the troy and apothecary pound there are 12 oz. In fact, the only common unit in all three pounds is the *grain*, and 7000 grains equal 1 lb avoirdupois, while 5760 grains equal 1 lb troy or apothecary.

Engineering practice is concerned only with the avoirdupois pound of mass, and this is defined in terms of the prototype SI kilogram with this equivalence:

$$1 \text{ lb} = 0.4535924 \text{ kg} = 4.535924 \text{ E}-01 \text{ kg} = 453.5924 \text{ g}$$

The foot unit of length is expressed in terms of the SI meter as:

$$1 \text{ ft} = 0.3048000 \text{ m} = 3.048000 \text{ E}-01 \text{ m}$$

$$1 \text{ m} = 3.28084 \text{ ft} = 3.28084 \text{ E}+00 \text{ ft} = 39.3670 \text{ in.}$$

Force can be related to mass by making use of Newton's second law, expressed $F \sim ma$, or, in equational form,

$$F = \frac{m}{g_c} a \tag{1-8}$$

where, in most instances, g_c can be considered merely a constant of proportionality. In SI, the units have been selected so that g_c is unity and the SI force unit, called the *newton* (N), will give a mass of 1 kg an acceleration of 1 m/s^2, or

$$1 \text{ N} = 1 \text{ kg} \times 1 \text{ m/s}^2 \tag{1-9}$$

The weight of a body is the force which the earth's gravitational pull exerts on the body at any point on the earth. It varies locally with altitude and with latitude. Referred to sea level, it is greatest at the poles and least at the equator. The standard or reference value is taken as $g = 9.80665$ m/s^2 and 32.1740 ft/s^2. However, the variations in g are small and 1 kg of mass exerts almost the same gravitational force everywhere. Since weighing is usually done by comparing the mass to be weighed with other masses (weights), calibrated in standard kilograms or pounds, the weighing process produces an exact measure of the mass of a body, except when spring balances are used.

Former use was made in the metric system of the kilogram force unit. If we use Newton's law, Eq. (1-8), under conditions of standard gravity with g_c taken first as unity, it follows that

$$\text{kilogram force} = \text{kilogram} \times 9.80665 \text{ m/s}^2 \tag{1-10}$$

Comparing Eqs. (1-9) and (1-10), it can be seen that

$$1 \text{ kgf} = 9.80665 \text{ N} \tag{1-11}$$

This development gives a physical indication of the magnitude of the newton and it should be noted that it is approximately one-tenth of the weight (force) of 1 kg. The kilogram of force is not used in SI.

In the foot-pound-second system of units some use was made of a force unit called the *poundal*. With g_c taken as unity in Eq. (1-8), the poundal was a force unit of such magnitude that it could accelerate 1 lb of mass at 1 ft/s^2.

$$1 \text{ poundal} = 1 \text{ lb}_{\text{mass}} \times 1 \text{ ft/s}^2 \tag{1-12}$$

The mass, which a 1 lb force (lbf) could accelerate at 1 ft/s^2, was named the *slug*, and it follows that

$$1 \text{ slug} = 32.1740 \text{ lb}_{\text{mass}} \tag{1-13}$$

This numerical value arises from the fact that a force of 1 lb is defined as the gravitational pull on a 1 lb mass at a location where the acceleration of gravity is standard at 32.1740 ft/s^2, or expressed in equational form,

$$1 \text{ lbf} = 1 \text{ lb} \times 32.174 \text{ ft/s}^2 \tag{1-14}$$

$$1 \text{ lbf} = 1 \text{ slug} \times 1 \text{ ft/s}^2 \tag{1-15}$$

If the pound force and pound mass are used together, then the value of g_c is 32.174 in Eq. (1-8), and it follows dimensionally that:

$$g_c = \frac{m}{F} a = 32.174 \frac{\text{lb}}{\text{lbf}} \frac{\text{ft}}{\text{s}^2} = \frac{\text{slug}}{\text{lbf}} \frac{\text{ft}}{\text{s}^2} \tag{1-16}$$

However g_c can also be considered dimensionless as the ratio of two mass units

$$g_c = \frac{\text{slug}}{\text{lb}} = 32.1740 \tag{1-17}$$

There is no inconsistency in g_c having a dual character since its non-dimensional aspect arises when Newton's second law is used as a dimensional definition for force, $F = M^1L^1T^{-2}$, and in this connection g_c is merely a numerical ratio.

To summarize for the two systems:

1. a. When the newton force unit and the kilogram mass unit are employed together, g_c has unit value and needs no further consideration. This is the SI approach.
   b. When the kilogram of force is used with the kilogram for mass, the mass value must be divided by $g_c = 9.80665$.
2. a. When the poundal force unit and the pound mass unit are employed together, g_c has unit value and needs no further consideration.
   b. When the pound of force is used with the pound for mass, the mass value must be divided by $g_c = 32.174$.

The following conversion factors are applicable:

$$1 \text{ N} = 0.10197 \text{ kgf} = 0.224809 \text{ lbf} = 7.2330 \text{ poundals} \tag{1-18}$$

$$1 \text{ kgf} = 9.80665 \text{ N} = 2.20462 \text{ lbf} = 70.93152 \text{ poundals} \tag{1-19}$$

$$1 \text{ lbf} = 4.44822 \text{ N} = 0.453592 \text{ kgf} = 32.1740 \text{ poundals} \tag{1-20}$$

$$1 \text{ dyne} = 1 \times 10^{-5} \text{ N} = 2.24809 \times 10^{-6} \text{ lbf} \tag{1-21}$$

The dyne was a common force unit in the metric system.

Example 1-3. A metric ton (1000 kg) of raw rubber is shipped from Singapore to Cleveland, Ohio. The value of g at Singapore, near the equator, is 9.7808 m/s^2, while at Cleveland g is 9.8024. Compute the force exerted by the rubber on its supporting skid (a) at Singapore and (b) at Cleveland. (c) Compute its weight at both places.

Solution: By Newton's second law, the force exerted,

$$F = \frac{ma}{g_c}$$

becomes

$$F = \frac{mg}{g_c}$$

Here g_c is unity if force, measured in newtons, is desired and

(a) $$F = (1000)(9.7808) = 9780.0 \text{ N} \qquad \textit{Ans.}$$

or

$$F = \frac{(1000)(g)}{g_c} = \frac{(1000)(9.7808)}{9.80665} = 997.4 \text{ kgf}$$

(b) $$F = (1000)(9.8024) = 9802.4 \text{ N} \qquad \textit{Ans.}$$

or

$$F = (1000)\frac{9.8024}{9.80665} = 999.6 \text{ kgf}$$

(c) Its weight at both places is 1000 kg, assuming it was weighed on platform scales using standard scale weights.

Example 1-4. A ton (2000 lb_{mass}) of supplies loaded on a plane in Chicago is delivered to north-central Greenland. The supplies are reweighed on arrival in Greenland on a conventional platform scales using standard weights on its beam. The local acceleration of gravity in Chicago $g = 32.16$ ft/s^2 and $g = 32.24$ ft/s^2 in north-central Greenland. (a) What did the supplies weigh in Greenland compared to their prior weight in Chicago? (b) What force did these supplies exert on their loading platforms when in Chicago and when in Greenland?

Solution: (a) The supplies weighed exactly the same at both places provided they were weighed on scales using standard weights.

(b) To find the force exerted, use Newton's second-law equation

$$F = \frac{ma}{g_c} = \frac{mg}{g_c}$$

$$F = \frac{(2000)(32.16)}{32.174} = 1999.5 \text{ lbf in Chicago}$$

$$F = \frac{(2000)(32.24)}{32.174} = 2004.4 \text{ lbf in Greenland}$$

1-6. PRESSURE

Unit pressure is defined as force per unit area. The total pressure acting on an area represents the force acting over the whole area and is equal to the product of the unit pressure and the area. Pressure is nearly always measured relative to some datum of pressure, usually that of the atmosphere, and the measurement is really an indication of how much the pressure is greater or less than that of the atmosphere. Pressures measured in this way are known as *gage pressures*. Thus the gage on a steam boiler indicating 150 psi signifies that the pressure in the boiler is 150 psi higher than atmospheric (barometric) pressure. If the barometric pressure is 14.7 psi, the pressure exerted by the steam in the boiler is really 150 + 14.7 = 164.7 psi. This total, or real, pressure is known as the *absolute pressure*.

When a pressure measured is less than atmospheric it is usually called a *vacuum*, and the absolute pressure is found by subtracting the vacuum pressure from the barometric (atmospheric) pressure.

Figure 1-7 is a diagram illustrating gage and absolute pressures, making use of the engineering (foot-pound-second) system. The barometric pressure is shown as standard at 14.696 psi corresponding to 29.9212 in. Hg (0.76 m Hg) measured at 0°C.

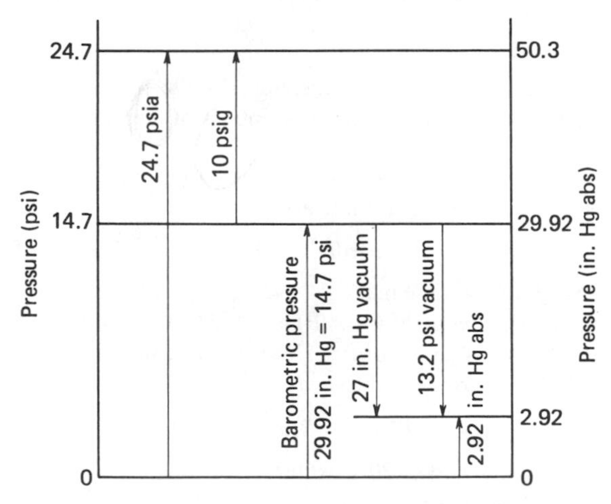

FIGURE 1-7. Diagrammatic representation of barometric, gage, and absolute pressures.

In SI, the pressure unit, the newton per square meter, is named the pascal (Pa). The pascal is of such small magnitude that frequent use is made of the kilopascal (kPa) because of its more convenient size,

$$1 \text{ kPa} = 0.145038 \text{ psi}$$
$$1 \text{ psi} = 6.8948 \text{ kPa} \tag{1-22}$$

This means that a pound per square inch is about seven times larger than a kilopascal. Use is also made of a pressure unit called the *bar*, defined as 100 000 N/m^2. Since the standard atmosphere is 101 325 Pa (101.325 kPa), the bar can be seen to be 0.98066 times as large as the standard atmosphere, at 14.5038 psi, and of course is 100 kPa.

A simple method of measuring pressure is to balance a column of liquid against the pressure and then measure the height of the column. It is well-known from the principles of hydrostatics that the unit pressure exerted by a column of liquid is a function of its height and its density and is independent of the cross section of the column. For example, a cubic foot of 68-F water weights 62.33 lb. If this water is in a cubical container the pressure on the bottom of the container is 62.33 psf, or 0.433 psi. The same unit pressure applies to every vertical section in the cube, and this would be true whether the cube were cut in half to make two containers or reduced any number of times. Water itself can be used as a measuring material, but mercury, with its greater density, is a more common measuring medium. Mercury has a density of 0.4912 lb/in.^{3} (1.3596×10^4 kg/m^3) measured at 0°C. At 20°C (68 F) the corresponding values are 0.4894 lb/in.^{3} (1.3546×10^4 kg/m^3).

To change fluid pressure units, this obvious relation is applicable:

$$h_1 d_1 = h_2 d_2 \tag{1-23}$$

where h_1 and h_2 are the respective heights of the two fluids in question, measured in consistent units (meters, feet, inches, etc.) and where d_1 and d_2 are the densities or relative densities of the two fluids in question, measured in consistent units (kilograms per cubic meter, pounds per cubic foot, specific gravity, etc.).

Example 1-5. Express one standard atmosphere 14.696 psi as (a) feet of water and meters of water and (b) meters of mercury and inches of mercury at 20°C.

Solution: When in Eq. (1-23), h and d are expressed in units of feet, and pounds per cubic foot, respectively, the resultant units are

$$\text{ft} \times \text{lb/ft}^3 = \text{lbf/ft}^2$$

because of the gravity force on the mass.

(a) The density of water at 20°C (68 F) is 62.32 lb/ft^3, and a column of this water 1 ft high exerts a pressure of 62.32 lbf/ft^2. Here set atmospheric pressure expressed in pounds of force per square foot equal to $h_1 d_1$.

$$(144)(14.696) = h_1 d_1 = h_1(62.32)$$

$$h_1 = 33.96 \text{ ft of 20°C water} \qquad \textit{Ans.}$$

$$h_1 = 33.96 \times 0.3048 = 10.35 \text{ m of 20°C water} \qquad \textit{Ans.}$$

(b) The density of mercury at 20°C is 1.3548×10^4 kg/m^3 and a 1-m column of mercury exerts a pressure of 1.3548×10^4 kgf/m^2,

$$1.3548 \times 10^4 \times 9.80665 = 1.32860 \times 10^5 \text{ N/m}^2$$

and h meters of column would exert h times this pressure,

$$14.696 \text{ lbf/in.}^2 \times 4.44822 \text{ N/lbf} \times 1549.99 \text{ in.}^2/\text{m}^2 = 101\,325 \text{ N/m}^2$$

$$101\,325 = 1.32860 \times 10^5 \times h$$

$$h = 0.7626 \text{ m Hg} \qquad \textit{Ans.}$$

$$h = 0.7626 \times 39.37008 \text{ in./m} = 30.02 \text{ in. Hg} \qquad \textit{Ans.}$$

Manometers, shown at the top of Fig. 1-8, consist of a tube bent in the shape of the letter U and about half-filled with mercury or other suitable indicating liquid—water, carbon tetrachloride, or colored kerosene. When the ends of the tube are connected to different pressure regions, the liquid surfaces will stand at different levels in the two legs, and the difference in height of liquid will represent the difference in pressure of the two regions, expressed in inches of mercury or inches of the particular indicating fluid.

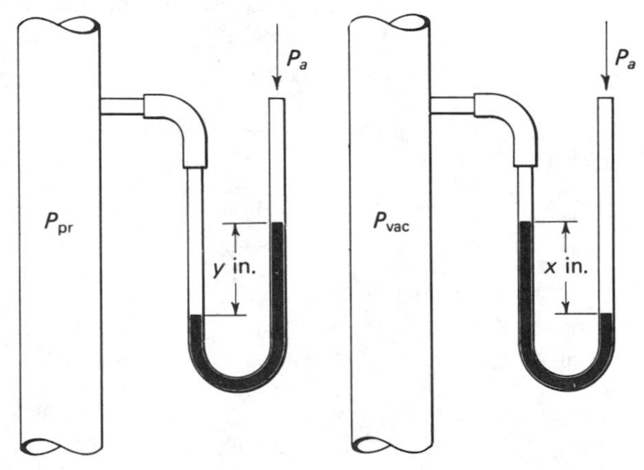

U-tube manometer for pressure and vacuum measurement

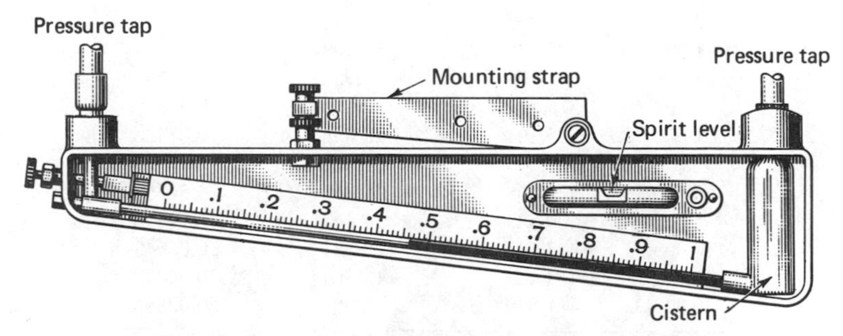

Inclined draft gage for measuring small pressure differences

FIGURE 1-8. Manometric pressure-measuring devices.

Frequently the atmospheric (barometric) pressure is one of the pressure regions employed.

At the bottom of Fig. 1-8 is shown a draft-gage type of manometer. Draft gages are used to measure small pressure differences, such as those that exist in duct systems or in combustion chambers and stacks of furnaces.

A barometer measures atmospheric pressure. Mercury barometers are made by filling with mercury a tube sealed at one end and, after removing extraneous gases and sealing the other end, inserting the mercury tube upside down in a mercury cistern. Atmospheric pressure supports the column and prevents the mercury from falling out of the tube. The height of the supported mercury column is a measure of the atmospheric pressure. The space above the mercury in the sealed top of the tube is almost a perfect vacuum except for the slight vapor pressure of mercury existing at ordinary temperatures. Barometric (atmospheric) pressure in a given locality varies from day to day but only to a slight extent. It decreases markedly with altitude, as can be seen in Table 1-3.

Table 1-3 Variations in Standard Atmosphere with Altitude

Altitude	Temperature	Pressure	Air Density
	Using SI (Metric) Units		
(m)	(K)	(Pa or N/m²)	(kg/m³)
−1 000	294.651	11 3931	1.3470
0	288.150	10 1325	1.2250
300	286.200	9 7773	1.1901
600	284.250	9 4343	1.1560
850	282.625	9 1522	1.1281
1 000	281.651	8 9874	1.1117
1 500	278.402	8 4555	1.0581
2 000	275.154	7 9501	1.0066
4 000	262.166	6 1661	0.8194
8 000	236.215	3 5652	0.5258
10 000	223.252	2 6500	0.4135
(ft)	(R)	(in. Hg at 32 F)	(lb/ft³)
−3 000	529.370	33.311	0.08342
−2 000	525.803	32.148	0.08105
−1 000	522.236	31.109	0.07874
0	518.670	29.921	0.07647
500	516.887	29.385	0.07536
1 000	515.104	28.856	0.07426
2 000	511.538	27.821	0.07210
4 000	504.408	25.843	0.06792
5 000	500.843	24.897	0.06590
10 000	483.025	20.581	0.05648
20 000	447.415	13.761	0.04077
30 000	411.839	8.903	0.02866

From *U.S. Standard Atmosphere, 1962*, and *Supplement, 1966*, National Aeronautics and Space Administration, Washington, D.C.

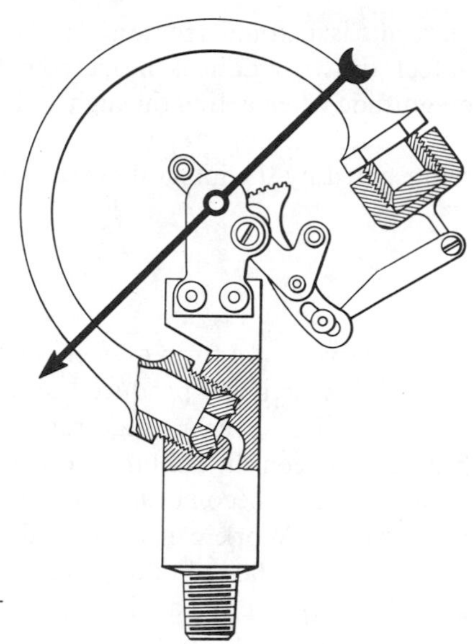

FIGURE 1-9. Bourdon-tube pressure-gage element.

The Bourdon-tube gage represents a very common type of pressure-measuring instrument used in engineering practice. The Bourdon tube, one form of which is illustrated in Fig. 1-9, is usually made of brass or steel and has a flattened cross section resembling an ellipse or, in some cases, a rectangle with rounded corners. The tube is bent in the form of a circular arc covering from 100° to some 300°. The open end of the tube is attached to the case of the instrument and connected to the source of pressure. The other end is sealed and free to move. When pressure is applied to a bent tube of this type, it tends to increase in volume, and in so doing its radius of curvature also increases so that the circular arc tends to straighten. The movement of the free end is attached through a suitable linkage to operate an indicating pointer. Bourdon tubes can measure partial vacuums as well as positive (gage) pressures. In the case of a partial vacuum, the tube tends to contract in depth and its radius of curvature decreases.

Bourdon-tube types of control instruments are manufactured which are responsive to the changes in liquid or vapor volume in a closed system. The motion of the free end of the tube actuates a switch or other control mechanism upon changes in pressure (temperature) of the medium in the control bulb. Temperature-responsive bulbs, when filled with liquid or gas, cause the tube to move in a linear path under changes in temperature.

1-7. WORK, POTENTIAL ENERGY, KINETIC ENERGY, POWER

Work (**W**) is a form of energy in transition, and we define work as the product of force (F) acting through distance (x) in the direction of the force. In equational form, if F is constant,

$$\mathbf{W}_{(1-2)} = F(x_2 - x_1) \qquad (1\text{-}24)$$

Here, if F is in pounds force and $(x_2 - x_1)$ is the distance traversed, measured in feet, the work done is measured in foot-pounds. If F is a variable force, its magnitude when acting through a differential distance (dx) must be summed up for each variable step $(F\,dx)$ to give the total work, and this summation is represented by the integral expression

$$\mathbf{W}_{(1-2)} = \int_1^2 F\,dx \tag{1-25}$$

Since work is a form of energy, we may very properly ask what happens to it while it is being produced or afterwards. If the force is being used to push a boulder over a rough roadway, the work is dissipated as frictional heating between the rubbing surfaces; or when a variable force compresses a spring, energy becomes stored in the elastic coils of the spring and can easily be retrieved. Work can also produce gravitational *potential energy*. If a pulley hoists a 100 lb building block to the roof ledge of a house 30 ft above ground, work is done as the rope exerts a 100 lb force or pull on the block to overcome the effect of gravity. In this case the work has not been lost but resides in the block in relation to the earth. The elevated block in turn could be made to produce work by means of a simple mechanism as it is lowered to the ground. The amount of the work required to elevate the block is simply:

constant force $\times$ distance (or elevation), $100 \times 30 = 3000$ ft $\cdot$ lbf.

More specifically, that form of energy existing through positional configuration such as the position of a body at an elevation above a selected datum plane on the earth, or by virtue of a spring in restraint, is called *potential energy*.

For the gravitational case, potential energy $(\mathbf{P}_e)$ can be expressed in foot-pounds as

$$\mathbf{P}_e = \text{weight of body (lbf)} \times \text{elevation above datum (ft)} \tag{1-26}$$

In SI, each kilogram of mass exerts a weight force of 9.8066 N; thus gravitational potential energy becomes:

$$\mathbf{P}_e = 9.8066 \times \text{weight (kg)} \times \text{elevation above datum (m)} \tag{1-27}$$

with the energy measured in newton meters (N $\cdot$ m).

That form of energy associated with a body by virtue of its motion is known as *kinetic energy*. For example, the energy associated with a moving bullet is kinetic energy. When a bullet is suddenly brought to rest, as in hitting a wall, the kinetic energy changes form. The resultant energy, which in this case would be of thermal form, appears in heating the bullet and surroundings. It may be intense enough to melt the bullet. In any event, *thermal energy*, equivalent to the kinetic energy originally possessed by the bullet, would appear. The expression for kinetic energy ($\mathbf{K}_e$) which can be developed from elementary physics, is

$$\mathbf{K}_e = \frac{mV^2}{2g_c} \tag{1-28}$$

where, in engineering (foot-pound-second), units, m is the mass (weight) of the moving body in pounds; V is the velocity in feet per second (fps); g_c is equal to 32.174; and $\mathbf{K}_e$ is in foot-pounds.

In SI, $\mathbf{P}_e$ and $\mathbf{K}_e$ are most conveniently expressed in meter × newton units, that is, in joule units, since the *joule* (J) be definition is 1 m · N. Thus Eq. (1-28) should use the following consistent units: $\mathbf{K}_e$ in joules (i.e., meter × newton, m · N); m, the mass (weight), in kilograms (kg); V, velocity, in meters per second (m/s); and g_c, unity or 1.000.

Example 1-8. A 30-lb ball drops to the ground from a height of 20 ft. What velocity will the ball have attained at the instant of striking the ground? Work the problem in engineering units and in SI (metric) units.

Solution: The potential energy in the ball-earth system converts directly to kinetic energy. By Eq. (1-26),

$$\mathbf{P}_e = 30 \times 20 = 600 \text{ ft} \cdot \text{lbf}$$

This is equivalent to the $\mathbf{K}_e$ as expressed in Eq. (1-28)

$$\mathbf{K}_e = 600 = \frac{mV^2}{2g_c} = \frac{(30)V^2}{(2)(32.174)}$$

$$V = \sqrt{\frac{(600)(2)(32.174)}{30}} = 35.8 \text{ fps} \qquad \textit{Ans.}$$

$$20 \text{ ft} \times 0.3048 = 6.096 \text{ m}$$

$$30 \text{ lb} \times 0.45359 = 13.608 \text{ kg}$$

30 lb exerts a weight force of 30 lbf; thus

$$30 \text{ lbf} \times 4.4482 = 133.45 \text{ N}$$

$$\mathbf{P}_e = (6.096)(133.45) = 813.5 \text{ J} = 813.5 \text{ m} \cdot \text{N}$$

$$\mathbf{P}_e = \mathbf{K}_e = \frac{mV^2}{2g_c} = \frac{(13.608)V^2}{(2)(1)} = 813.5$$

$$V = 10.93 \text{ m/s} \qquad \textit{Ans.}$$

Power is defined as the rate at which work is performed. The *horse-power* (hp) is a common power unit in engineering practice. By definition each horsepower signifies work performed at the rate of 33 000 ft · lbf/min, 550 ft · lbf/s.

The *watt*, or 1000 watts called the *kilowatt* (kW), the most common of all power units, is used extensively both in SI and in engineering practice.

The watt signifies work performed at the rate of 1 joule per second. The joule represents 1 meter newton of work. From the fundamental units presented in Section 1-5 it is easy to show that

$$1 \text{ joule (J)} = 1 \text{ m} \cdot \text{N} = 0.73757 \text{ ft} \cdot \text{lbf}$$

$$1 \text{ watt (W)} = 1 \text{ J/s} = 0.73757 \text{ ft} \cdot \text{lbf/s}$$

$$1 \text{ kilowatt (kW)} = 737.57 \text{ ft} \cdot \text{lbf/s}$$

$$1 \text{ horsepower (hp)} = 0.746 \text{ kW}$$

$$1 \text{ kilowatt (kW)} = 1.340 \text{ hp}$$

The product of power multiplied by time is an energy unit. One kilowatt acting for one hour is the very common energy unit of a kilowatt-hour (kWh).

$$1 \text{ kWh} = 1.340 \text{ hph} = 2\,655\,300 \text{ ft} \cdot \text{lbf}$$

Other forms of energy and power and their conventional units are developed in the next chapter.

PROBLEMS

1-1. Compare the relative increase in length of a 10 ft length of copper pipe imbedded in concrete at 70 F, when the pipe, while carrying water, gradually warms to 140 F. (Obviously the increase in length is not sufficient to break or disturb the relative bond between the tube and the concrete.)

Ans. 0.031 in.

1-2. Write Eq. (1-1) in the form $L_t = L_o[1 + \alpha(\Delta t)]$ and then cube both sides of the equation. Define the volume coefficient of expansion (β_v) in an equation of the form $L_t^3 = V_t = V_o(1 + \beta_v \Delta t)^3$ and show that for moderate temperature changes and representative coefficients of linear expansion, $\beta_v = 3\alpha$ for all practical purposes.

1-3. Change 20°C to degrees Fahrenheit and to degrees Fahrenheit absolute.

Ans. 68 F, 528 F abs

1-4. Change 14 F and 200 F to degrees Celsius.

Ans. −10°C, 93.3°C

1-5. A mercury-in-glass thermometer is used to measure the temperature of the gases flowing through an insulated metal duct. The thermometer indicates a temperature of 640 F and has an emergent mercury thread of 100 deg showing. The ambient temperature at the stem of the thermometer is 110 F. Find the probable true temperature that would be indicated by this thermometer if completely immersed.

Ans. 644.7 F

1-6. A mercury-in-glass thermometer inserted in a thermometer well is being used to measure the temperature in a steam line. The thermometer reading is 400 F and the emergent thread of mercury outside the well is 70 deg. If the ambient temperature is 90 F at the thermometer stem, what is the probable true temperature in the steam line?

Ans. 401.9 F

1-7. An underground steam pipe, made of steel, runs 2000 ft underground. In summer, with no steam flowing, the pipe can be considered to be at a ground temperature of 60 F. In winter the pipe carries saturated steam at 115 psia at a temperature of 338.1 F. Compute the increase in pipe length in winter, which must be absorbed by expansion devices, over the summer length.

Ans. 3.8 ft

1-8. Compare the millivoltage generated by thermocouples employed to measure the temperature of hot gases at 260°C (500 F), when the cold junction of the thermocouple is at 37.8°C (100 F) and when the couples are (a) copper-constantan, (b) iron-constantan, and (c) chromel-alumel.

Ans. (a) 11.1; (b) 12; (c) 9.2

1-9. Thermopiles consist of several thermocouple circuits connected in series. The cold junctions are brought out to a common region of low or standard temperature, while the hot junctions, in a separate compartment, are exposed to an unknown temperature, to a source of radiant energy, or the like. A particular thermopile has 50 iron-constantan thermocouple circuits in series. This thermopile shows 250.0 mV on its potentiometer when exposed to a source of radiation with the cold junctions held at 32 F. Find (a) the emf per couple and (b) the temperature in the hot compartment of the thermopile.

Ans. (a) 5 mV; (b) 200 F

1-10. (a) For the thermocouple control device in Fig. 1-5, using an iron-constantan couple, find the emf developed when the hot junction is 315.6°C (600 F) and the cold junction is 37.8°C (100 F). (b) If the electrical resistance of the circuit is 0.1 ohm, what is the probable current flowing?

Ans. (a) 15 mV; (b) 0.15 A

1-11. Transform a pressure of 12 psig into inches of mercury at (a) 32 F and (b) 70 F.

Ans. (a) 24.4; (b) 24.5

1-12. What is the pressure, in inches of water, of an air column 900 ft high? Assume constant air density at 0.078 lb/ft^3 and water density at 62.3 lb/ft^3.

Ans. 13.5 in.

1-13. Convert a mercury column 20 in. high, measured at 70 F, into pounds per square inch, meters of mercury, and feet of water, at 68 F (20°C).

Ans. 9.78 psi, 0.508 m Hg, 22.6 ft water

1-14. Assume that in Problem 1-13 the barometric pressure is 29.92 in. Hg at 32 F. Find answers expressed as absolute pressures.

Ans. 24.47 psia, 56.5 ft water

1-15. A less commonly used method of changing Celsius to Fahrenheit temperatures employs the following sequence of operations: to the temperature in degrees Celsius, 40 is added and the result is multiplied by 1.8. From this product, 40 is subtracted and the resulting number represents the temperature in degrees Fahrenheit. By simple proof, show that this method is the equivalent of Eq. (1-3).

1-16. Refer to the ton (2000 lb) of supplies discussed in Example 1-4 and assume that they are destined for use on the surface of the moon where the moon's acceleration of gravity is 5.34 ft/s^2. What force will these supplies exert on the surface of the moon? Express your answer in pounds and in newtons.

Ans. 332 lbf, 1477 N

1-17. The U.S. gallon is a volume, used in measuring liquids, of exactly 231 in.^{3} and the quart is one-fourth of a gallon. The liter is a volume of 1000 cm^3. Using only the basic equivalent that 1 in. = 2.54 cm, find the precise relationship between the quart and the liter. Compare this with the approximation that a liter is 5% larger than a quart. What is the exact percentage?

1-18. Refer to Problem 1-17 and compare the magnitudes of one-fifth of a gallon (U.S.) and 1 liter in terms of their respective volumes.

1-19. The four jet engines of a commercial plane by producing a constant thrust of 120 000 lbf accelerate the 600 000-lb plane to a speed of 160 mph before the plane becomes airborne in 4500 ft of runway. (a) How much work is done by the engines in bringing the plane to lift-off speed? (b) How much kinetic energy is stored in the plane at the instant of lift-off? (c) Account for the disposition of the energy difference in parts a and b.

Ans. (a) 5.4×10^8 ft lbf; (b) 5.15×10^8 ft lbf; (c) rolling friction and windage drag

1-20. The engine of a 3200 lb automobile, running at 60 mph, stops while climbing a hill that rises 10 ft in each 100 ft of roadway. If the car is immediately put into neutral and permitted to continue its forward progress, how much further can it travel before stopping? Disregard rolling friction and windage losses.

Ans. 1200 ft

1-21. Consider an automobile in a situation similar to that described in Problem 1-20 except that the car weighs 1450 kg, is traveling at 96.6 km/h before the engine stops. It is traveling up a hill with a 10 m rise in every 100 m of roadway. For conditions of no friction, how far up the roadway can this car travel without engine power? Carry out the solution completely in SI units.

Ans. 367 m

REFERENCES

ASTM/IEEE Standard Metric Practice, ASTM E380-76, IEEE Std 268-1976, also ANSI Z 210.1-1976. A very complete jointly sponsored publication covering SI units that can be obtained from IEEE, 345 E. 47th St., New York, N.Y. 10017, for $4.

C. H. Page and P. Vigoureux, eds., *The International System of Units* (*SI*), National Bureau of Standards special publication 330. Available from Supt. of Documents, U.S. Government Printing Office, Washington, D.C. 20402, for 40 cents.

E. A. Mechtly, *SI Units—Physical Constants and Conversion Factors*, Publication SP-7012. Available from Supt. of Documents, U.S. Government Printing Office, Washington, D.C. 20402.

ASME, *Orientation and Guide for Use of SI* (*Metric*) *Units*, 4th ed., 1974. Available from ASME, 345 E. 47th St., New York, N.Y. 10017.

E. A. Mechtly, *The International System of Units, Physical Constants and Conversion Factors*, 2nd rev. ed., 1973. NASA, SP-7012. Available from Supt. of Documents, U.S. Government Printing Office, Washington, D.C. 20402, for 50 cents.

CHAPTER 2

THERMODYNAMICS, STEAM, AND GAS PROPERTIES

2-1. THE FIRST LAW OF THERMODYNAMICS

Thermodynamics (that branch of science which deals with energy and its transformations), heat transfer, and fluid flow are the basic disciplines underlying air conditioning and refrigeration. The first law of thermodynamics is fundamentally equivalent to the law of conservation of energy, which states that energy can neither be created nor destroyed but can change in form. When energy does change from one form to another, it always transforms in definite fixed ratios.

The first law, when written in a form applicable to a nonflow process, appears

$$Q_{(1-2)} = U_2 - U_1 + \mathbf{W}_{(1-2)} \tag{2-1}$$

This states the balance that must exist for the energy relationships in the system during a process. Here $Q_{(1-2)}$, which represents the heat added as the process takes place, is shown reappearing on the right side of the equation as an increase in internal energy of the system, with U_2 the final and U_1 the original values; or the heat can transform directly to work produced by the system. Since this is an equation, any of these terms can be positive, zero, or negative, and it is customary to give Q a positive sign for heat added to a system, and **W** is positive when work is being delivered from or produced by the system. The internal energy, U, is a thermodynamic property really defined by the equation and measures the thermal energy stored in the medium of the system. Heat, Q, is not a thermodynamic property but is energy in transition flowing under the impetus of a temperature gradient. Likewise work, **W**, is energy in transition.

The fundamental character of work as one form of energy has already been discussed in Chapter 1 and its relation to potential and kinetic energy was shown at that time. Here we need to extend our thinking to recognize that work also relates to other aspects of energy and more particularly to those of a thermal nature. To this end let us consider the situation of a gas trapped in a cylinder by means of a movable piston. If the piston is moved inward to compress the gas into a smaller volume, work is required to overcome the resistance to compression. Let P represent the unit pressure in pounds force per square foot (psf) exerted by the gas on the piston face. The force in pounds that the piston exerts on the gas and that the gas exerts on the piston is obviously

$$F = PA$$

and from Eq. (1-25),

$$\mathbf{W}_{(1-2)} = \int_1^2 F\,dx = \int_1^2 PA\,dx = \int_1^2 P\,dV \tag{2-2}$$

The term dV evolves since the piston area A sweeping through distance dx necessarily produces the volume change dV. This concept is perfectly general and applies equally well to an expending gas or to situations other than those associated with a cylinder and piston. Thus it is proper in many applications to use this expression for work, and Eq. (2-1) can be written

$$Q_{(1-2)} = U_2 - U_1 + \int_1^2 P\,dV \tag{2-3}$$

During compression since the gas volume is decreasing, dV is minus in sense, indicating that work is being done on the gas in the system and not by the system.

In broader terms we should recognize that work is far reaching in its character, relating not only to physical phenomena such as raising or lowering weights, or compressing springs, but also to thermal effects. For example if during this compression of the gas we give the work of Eq. (2-3) a minus sign and rearrange the equation

$$\mathbf{W}_{(1-2)} = \int_1^2 P\,dV = U_2 - U_1 - Q_{(1-2)} \tag{2-4}$$

it appears that the work done will either be stored in the gas to increase its internal energy $(U_2 - U_1)$ or pass out of the system as heat flow $Q_{(1-2)}$

through the walls of the compressor cylinder. The internal energy increase shows up as a rise in temperature of the gas.

Let us consider another situation, with the cylinder being heated from outside while the piston is locked in fixed position. With the volume unable to change, work is necessarily zero and the heat additions to the gaseous medium must all reappear as an increase in internal energy, and Eq. (2-4) would appear

$$Q_{(1-2)} = U_2 - U_1 + \int_1^2 P\,dV = U_2 - U_1 + 0 \tag{2-5}$$

When matter, either gaseous, liquid, or solid, does not change in phase, heat addition to it would be evidenced by a rise in its temperature as energy is stored in increased activity of its constituent molecules. For example, a bar of iron heated to redness differs from the cold bar only in that its molecular activity is greater and the greater internal energy that the bar possesses is evidenced by its greater temperature.

When a solid is heated it gradually rises in temperature until a point is reached when the molecular activity becomes so great that the substance can no longer exist as a solid and gradually begins to change over to liquid form. During this process a large amount of energy is required to bring about the change in state from solid to liquid, and the temperature remains constant during the process. This energy is known as the *heat of fusion.* If more heat is then added to the resulting liquid, the temperature will continue to rise until a temperature is reached at which the liquid begins to change into the vapor or gaseous state. Here again the temperature remains constant during vaporization, and much energy, called the *heat of vaporization*, must be added in order to effect this change of state.

Consider the familiar ice, water, steam process as an example. If ice is heated from below 0°C (32 F), when its temperature reaches 0°C (32 F), it starts to change to liquid and the temperature will stay constant at 0°C until all the ice in contact with the water has melted. To accomplish this change the heat of fusion for ice, amounting to 333.2 kJ/kg or 143.35 Btu/lb must be supplied. (The Btu (British thermal unit), discussed later, is an energy unit equal to 1054.35 J and to 777.65 ft lbf.) If the water were now further heated, at standard atmospheric pressure, it would not boil (vaporize) until a temperature of 100°C (212 F) were reached, and to effect this change from liquid to vapor, 2255.4 kJ/kg (970.3 Btu/lb) of water would have to be supplied. This large amount of energy is required to break up the molecular-bond conditions in the liquid state as compared to the vapor state, and to supply the energy required as the increased volume of the steam over that of the liquid makes room for itself in expanding against the surrounding atmospheric pressure. The temperature remains constant during this vaporization. If the resultant vapor is further heated out of contact with the liquid, there will be a continuous temperature rise above the saturation or boiling temperature, and such vapor is called *superheated*. If the whole process is reversed and heat is removed, exactly the same changes will occur, but in reversed order.

During condensation, which will take place at a constant temperature of 100°C, 2255.4 kJ/kg of vapor condensed must be removed; and during freezing, which will take place when the temperature lowers to 0°C, 333.2 kJ/kg of water frozen must be removed. Similarly, in the foot-pound-second system, 970.3 Btu/lb of vapor condensed must be removed at 212 F, and during freezing, which will take place at 32 F, 143.35 Btu/lb of water must be removed.

In the last century much experimental work was conducted to determine the relationship of mechanical work to heat and internal energy. The historic work of James Prescott Joule in 1843 to 1850, and later that of Henry A. Rowland, showed that an invariant relation held between these forms of energy. Both Joule and Rowland demonstrated by doing work on water, such as by intensively paddling or churning it, that exactly the same results could be obtained as if the water were heated by fire. From such experiments it was found that almost exactly 778 ft lbf are equivalent to 1 British thermal unit (Btu). The Btu was defined as the amount of heat required to raise the temperature of 1 lb of water through 1 Fahrenheit degree (from 62 F to 63 F). Although the specific heat of water is approximately unity, it is not exactly unity over a range of temperature, so the Btu became defined as $\frac{1}{180}$ of the amount of energy required to raise 1 lb of water from 32 F to 212 F. In similar fashion, for countries using the metric system, there came into use the *calorie*, which is the amount of energy required to raise the temperature of 1 g of water at 15°C by 1 Celsius (centigrade) degree, or the kilocalorie applicable for 1 kg raised by 1°C.

It is obvious that definition of an important unit in terms of the response of a medium to energy addition is not an ideal way of setting its magnitude, so other approaches were followed to set a supposedly equivalent value. In 1956, at the Fifth International Conference on the Properties of Steam, agreement was reached that by definition

$$1 \text{ kcal} = 4186.8 \text{ J (exactly)}$$

and it followed that:

$$1 \text{ kWh} = 859.845 \text{ kcal} = 3412.1 \text{ Btu}$$

$$1 \text{ Btu} = 1055.06 \text{ J} = 778.17 \text{ ft} \cdot \text{lbf}$$

$$1 \text{ kcal} = 3.9690 \text{ Btu} = 3088.0 \text{ ft} \cdot \text{lbf}$$

However, another approach to interrelating thermal energy equivalents led to the so-called thermochemical calorie of SI, defined as

$$1 \text{ kcal} = 4.184 \text{ J} = 1.16222 \times 10^{-3} \text{ kWh}$$

and

$$1 \text{ kWh} = 860.42 \text{ J} = 3414.4 \text{ Btu}$$

$$1 \text{ Btu} = 1054.35 \text{ J} = 777.65 \text{ ft} \cdot \text{lbf}$$

$$= 251.99 \text{ cal} \equiv 0.252 \text{ kcal}$$

$$1 \text{ kcal} = 3.9683 \text{ Btu} = 3085.9 \text{ ft} \cdot \text{lbf}$$

It is unfortunate that there are differences in defining equivalents of the calorie to the joule because this leads to confusion in values of conversion factors. The difference is, of course, insignificant when the values are rounded off to 3 or 4 places for slide-rule usage. In this text, the values are given to 5 and 6 places, not only to show the real equivalents but also to make it possible to make precise checks using desk-type electronic calculators when SI and foot-pound-second computations are carried out simultaneously. For most purposes a general value of 778 ft · lbf/Btu is satisfactory in the foot-pound-second system.

It is most important to make a definite distinction between internal energy and heat. Internal energy is stored in the molecules or atoms of a system, but because internal-energy changes are often associated with the heat flows into or from a system, internal energy itself is often inaccurately called heat. It is preferable and less confusing if we call the energy, which is associated with the substance itself, *internal energy* and restrict the term *heat* to the energy transfer occurring whenever differences in temperature exists. *Heat* will thus be used in this text to describe that energy flow, or energy in transition, which results from the driving action of a temperature difference. Note that internal energy, work, and heat can be expressed in any consistent energy units, and we find use made of Btu (British thermal units), foot-pounds, calories, kilowatt-hours, gram-centimeters, joules, and others, all as possible units for expressing energy in any of its forms.

2-2. THE OPEN SYSTEM AND THE STEADY-FLOW ENERGY EQUATION

Our considerations, up to this point, have tacitly dealt with the first law in relation to a closed system, that is, one in which a medium participated in energy interchanges but was of itself invariant in mass. Of much greater interest for thermodynamic application is a so-called open system, which is one into which mass can flow at a given time rate or from which, except for periods of variable mass storage in the system, mass at an equivalent time rate of flow leaves the system. The medium flowing into the system has associated with it various forms of energy, as is similarly the case for the medium leaving the system. Energy also can be delivered to, or abstracted from, the flowing medium as it passes through the system, with this interchange most commonly occurring as heat or as work (electricity) through shafts or wires leading to the external environment. For purposes of this discussion, we can define a system as any convenient region that we wish to isolate for purposes of analysis and then consider the mass and energy interchanges which take place at its boundaries.

Fortunately, many real processes involve essentially steady flow, that is, the situation existing when the time rate of mass flow into and the time rate of mass flow from a system are equal, and we can expect the law of continuity of mass to be applicable for the flow patterns of the system. Many illustrations of steady-flow systems exist. For example, in a water pump, as the fluid flows through the pump at an essentially constant rate the water absorbs energy from the work transmitted into the system by the pump shaft. A steam boiler, after it has reached equilibrium operation, is essentially a steady-flow device receiving a constant supply of feed water per hour and sending out an equivalent weight of steam per hour. Heat from the burning fuel flows into the water, changing the condition of the water to that of the delivered steam at a greater energy level. Many devices in engineering employ processes which are steady-flow in character, or which approach this condition so closely as to permit treatment by steady-flow methods. Other examples of such equipment are steam turbines, nozzles, centrifugal compressors, and even reciprocating machines, such as steam engines and piston-type compressors.

When the laws of the conservation of energy and continuity of mass flow are applied to a steady-flow system or process, a useful equation can be developed which is known as the steady-flow energy equation. To develop the equation, imagine a device of any kind (boiler, pump, compressor, etc.) to which a fluid is supplied. In the device (Fig. 2-1), work can be added or removed, as through a power shaft. Heat can be added, as from a fire or from steam coils, or removed, as by a refrigeration evaporator. Finally the fluid departs, with energy different from that which is possessed when it entered the device.

Each unit mass of fluid entering the system has associated with it at entry:

1. *Potential energy* in amount $(1)Z_1$ where Z_1 is the elevation above a convenient datum plane.
2. *Kinetic energy* in amount $(1)V_1^2/2g_c$, where V_1 is the representative velocity at point 1.
3. *Internal energy* in amount $(1)(u_1)$.
4. *Flow work*, in amount P_1v_1.

Flow work, which may also be called the *work of intrusion*, or the *work of extrusion*, is the energy required to force unit mass of the fluid, having a specific volume v_1, into the system at the constant pressure P_1 existing under the steady-flow conditions. Also at exit, flow work is required to force unit

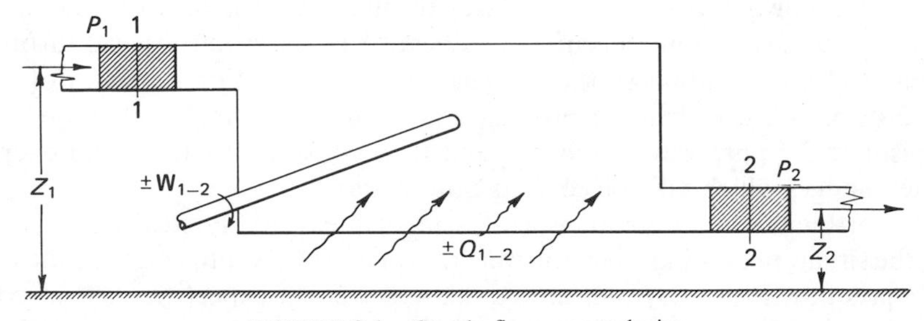

FIGURE 2-1. Steady-flow energy device.

mass from the system under conditions P_2 and v_2. *Proof*:

$$\text{work} = \text{force} \times \text{distance}$$

$$\text{work} = \text{pressure} \times \text{area} \times \text{distance}$$

$$\text{area} \times \text{distance} = \text{volume}$$

Therefore

$$\text{flow work} = Pv$$

This proof is true under the steady pressure P existing when the specific volume of unit mass of fluid is being forced into or from the device by the succeeding portions of fluid.

This flow work has little significance except for steady-flow processes. Its use will be illustrated in some later examples.

The fluid, having entered the device, bringing with it the forms of energy heretofore enumerated, can now do *mechanical work* in amount $-\mathbf{W}_{(1-2)}$ or mechanical work can be added to the fluid in amount $+\mathbf{W}_{(1-2)}$. The subscripts 1–2 under $\mathbf{W}$ simplify signify that work occurs between the fluid inlet 1 and outlet 2 points. In the case of a turbine or engine, work energy leaves the fluid through the driven shaft in amount $-\mathbf{W}_{(1-2)}$. In the case of a compressor, work is supplied to the fluid system through the driven shaft in amount $+\mathbf{W}_{(1-2)}$. In the engine example, some of the work $-\mathbf{W}_{(1-2)}$ generated in the cylinder may be dissipated as friction in the bearings; not all of it will appear as delivered energy from the shaft. Such a condition also exists in an opposite sense for a compressor.

Heat in amount $+Q_{(1-2)}$ can be added in certain devices, as from a fire in a boiler, or perhaps may leave, in amount $-Q_{(1-2)}$, by radiation or conduction to cooling coils, colder surroundings, or the like.

From the law of conservation of energy, the energy in the leaving fluid must equal the energy in the entering fluid plus any energy supplied, or minus any energy diminution, in the device itself. Thus, in mathematical form, for unit mass of fluid the energy equation appears as follows:

$$Z_1 + \frac{V_1^2}{2g_c} + u_1 + P_1 v_1 \pm \mathbf{W}_{(1-2)} \pm Q_{(1-2)} = Z_2 + \frac{V_2^2}{2g_c} + u_2 + P_2 v_2 \tag{2-6}$$

Let us now write Eq. (2-6) in appropriate format for use with SI units. Unit mass is the kilogram and energy will be expressed in terms of the newton meter under its name the joule (J). The elevation Z is in meters; velocity, V, is in meters per second; g_c is unity; P is in pascals (newtons per square meter); and v is specific volume in cubic meters per kilogram. We should note that the standard weight force of 1 kg of mass is 9.807 N and potential energy requires this factor. For SI units, Eq. (2-6) then becomes:

$$9.807\, Z_1 + \frac{V_1^2}{2g_c} + u_1 + P_1 v_1 \pm \mathbf{W}_{(1-2)} \pm Q_{(1-2)}$$

$$= 9.807\, Z_2 + \frac{V_2^2}{2g_c} + u_2 + P_2 v_2 \tag{2-7}$$

In the foot-pound-second (engineering) system the unit of mass is the pound, elevation Z is expressed in feet, V is in feet per second, and g_c is 32.174. Since it is customary to express internal energy and Q in Btu per pound mass, the factor 778 is required for these two terms whenever the equation is arranged for energy in foot-pounds; thus

$$Z_1 + \frac{V_1^2}{2(32.174)} + 778\, u_1 + P_1 v_1 \pm \mathbf{W}_{(1-2)} \pm 778\, Q_{(1-2)}$$

$$= Z_2 + \frac{V_2^2}{2(32.174)} + 778\, u_2 + P_2 v_2 \quad (2\text{-}8)$$

The foot-pound is an inconveniently small unit so that thermal calculations are usually carried out using units of Btu per pound mass for each term of the equation. This necessitates dividing each term by the factor 778. In addition it is customary to introduce a thermodynamic property called *enthalpy*. Enthalpy, a composite term, is the sum of internal energy and flow work. It is the term which is tabulated in most tables of vapor properties and uses **h** as its conventional symbol.

$$\mathbf{h} = u + \frac{Pv}{778} \text{ Btu/lb}$$

where u is the internal energy, in Btu per pound; P is the absolute pressure, in pounds per square foot; and v is the specific volume, in cubic feet. Enthalpy in SI, $u + Pv$, is expressed basically as joules per kilogram. However, frequent use is made of kilojoules per kilogram and also of watthours per kilogram.

Dividing the preceding steady-flow energy equation by 778 and substituting enthalpy terms, there results

$$\frac{Z_1}{778} + \frac{V_1^2}{2g_c(778)} + \mathbf{h}_1 \pm \frac{\mathbf{W}_{(1-2)}}{778} \pm Q_{(1-2)} = \frac{Z_2}{778} + \frac{V_2^2}{2g_c(778)} + \mathbf{h}_2 \quad (2\text{-}9)$$

Equation (2-9) is more convenient than Eq. (2-8) for steady-flow processes involving steam or a refrigerant vapor.

The general steady-flow energy equation, Eq. (2-7) or Eq. (2-9), is applied to a given process by eliminating negligibly small or irrelevant terms to obtain the simplest possible form of equation.

Apply Eq. (2-9) first to a steam boiler. Consider the $Z/778$ terms, it can be shown that where steam is concerned, small changes in potential energy are negligible, since the enthalpy term values have magnitudes about 1000 Btu/lb (778 000 ft lbf), and thus differences in elevation up to 100 ft (100 ft lbf/lb) are relatively negligible. Considering the kinetic-energy terms, it will be realized that they are small and that entering and leaving values will practically cancel out in well-designed feed and discharge pipes. The work term also does not appear, as no shafts supply or remove energy from a boiler. There remains the term $+Q_{(1-2)}$, for heat added, and the equation appears as follows:

$$\mathbf{h}_1 + Q_{(1-2)} = \mathbf{h}_2$$

Thus, for a boiler,

$$Q_{(1-2)} = \mathbf{h}_2 - \mathbf{h}_1 \text{ Btu/lb} \tag{2-10}$$

is the heat added and is equal to the difference in enthalpies of the leaving steam ($\mathbf{h}_2$) and the feed water entering ($\mathbf{h}_1$).

It should be noted that this equation does not apply to a nonsteaming boiler, such as would be encountered when a cold boiler is being warmed up for service. It happens that for this case the heat added equals the difference in the u's based on each composite pound of water-steam at the end of the heating period and each composite pound of water-steam at the beginning of the heating period.

Let us apply the energy equation, Eq. (2-8), to a fan. The Z terms are relatively negligible for slight changes in elevation of air. Work is supplied in amount $+\mathbf{W}_{(1-2)}$. The heat flow $Q_{(1-2)}$ is usually negligible unless the surrounding air is much hotter or colder than the circulated air. Evaluating, we have

$$\mathbf{W}_{(1-2)} = \frac{V_2^2 - V_1^2}{2g_c} + P_2 v_2 - P_1 v_1 + 778(u_2 - u_1) \text{ ft lbf}$$

The work applied to a fan appears primarily as the change in kinetic energy of the leaving and entering air at the points of measurement and as the increase in the flow-work term. The frictional losses, and a portion of the compression work, appear as an increase in the u, or internal energy, term. The slight increase in temperature of the air because of a change in u is difficult to measure or even detect, and in fan testing the other two terms, which might be called the useful output of the fan, are the ones considered. Thus, per pound of air,

$$\mathbf{W}_{(1-2)} = \frac{V_2^2 - V_1^2}{2g_c} + (P_2 - P_1)v_1 \text{ ft lbf} \tag{2-11}$$

As the specific volume of the air, v_1, does not change appreciably in going through a fan, v_1 may be called equal to v_2 and the equation appears as above. The $(P_2 - P_1)$ term must be expressed in consistent units, usually pounds per square foot of pressure change, and represents what is called the static-pressure change in the fan. Actually, in fan work it is customary to measure this change in pressure units of inches of water. The equation given, with the kinetic-energy term present, is called the *total fan work*. If the kinetic-energy term is removed, because the practical application of the fan prevents any possible utilization of the kinetic energy in the delivery, the equation represents the *static fan work*.

These two illustrations show how the steady-flow energy equation is used. For any practical problem, the method of use is to eliminate irrelevant or negligible terms and then to analyze what remains. Further use of this equation will be made in other parts of the book.

Example 2-1. A fan delivers 20 200 cfm of air, having a density of 0.075 lb/ft^3, at an outlet velocity of 1800 fpm and at a static pressure of 2.25 in. water. Air enters the fan at 1500 fpm at a negative static pressure of −0.25 in. water. Calculate the work per pound of air, disregarding turbulence and friction, for delivery under the conditions indicated. Compute similarly the work per minute or power requirement.

Solution: For the small pressure rise involved, [2.25 − (−0.25)] or 2.50 in. water, there is no significant change in air density, which remains at 0.075 lb/ft^3.

$$\text{specific volume} = \frac{1}{0.075} = 13.33 \text{ ft}^3/\text{lb}$$

$$\text{air flow} = \frac{20\,200}{13.33} = 1515 \text{ lb/min}$$

To use Eq. (2-11) for the $(P_2 - P_1)$ term, first convert the 2.5 in. water to pounds per square foot (psf) units. Refer to Table A-3 for a conversion factor, or recall that a cubic foot of water weighing 62.32 lb would exert a pressure of 62.32 psf if its height were 12 in. so a 1-in. column exerts a pressure one-twelfth as great.

$$P_2 - P_1 = 2.5 \times \frac{62.32}{12} = 2.5 \times 5.19 = 12.98 \text{ psf}$$

Using Eq. (2-11), per pound of air, we note that:

$$\mathbf{W}_{(1-2)} = \left[\left(\frac{1800}{60}\right)^2 - \left(\frac{1500}{60}\right)^2\right]\frac{1}{(2)(32.174)} + (12.98)(13.33) = 177.3 \text{ ft lbf}$$

$$\text{work per minute} = (1515)(177.3) = 288\,660 \text{ ft lbf/min}$$

$$\text{horsepower} = \frac{288\,660}{33\,000} = 8.14$$

$$\text{kilowatts} = (8.14)(0.746) = 6.07$$

This result is the so-called *fan air horsepower* or *fan air kilowatts* and as such represents only the power directly needed to move and pressurize the air. The motor shaft necessarily delivers more power to offset turbulences in the stream passages, overcome bearing friction, and contribute to the slight increase in internal energy of the air.

Example 2-2. A fan delivers 572 m^3/min of air of density 1.20 kg/m^3 at an outlet velocity of 9.14 mps and at a static pressure of 5.72 cm water. Air enters the fan at 7.62 mps at a negative static pressure of −0.63 cm water. Calculate for each kilogram the work associated with the air under these conditions. Calculate the air power (work per minute) required for the specified conditions. Express in joules per second, kilowatts, and horsepower.

Solution: For the small pressure rise of [5.72 − (−0.63)] or 6.35 cm water, there is no significant change in the density (or specific volume) of the air. The air flow is 572 m^3/min × 1.20 kg/m^3 = 687.2 kg/min or 11.44 kg/s. To convert centimeters of water at 20°C to newtons per square meter (pascals), multiply by 97.9 to yield 6.35 × 97.9 = 621 N/m^2. Equation (2-11) is applicable for this problem when proper and consistent SI units are used with it.

$$\mathbf{W}_{(1-2)} = (V_2^2 - V_1^2)\frac{1}{2g_c} + (P_2 - P_1)v_2$$

$$= [(9.14)^2 - (7.62)^2]\tfrac{1}{2} + 97.9 \times 6.35\left(\frac{1}{1.20}\right) = 530.8 \text{ J/kg}$$

work per second or power = 11.44 × 530.8 = 6079 N · m/s

$$6079 \text{ N} \cdot \text{m/s} = 6079 \text{ J/s} = 6079 \text{ W} = 6.08 \text{ kW}$$

$$\text{horsepower} = 6.08 \div 0.746 = 8.15$$

This result is the so-called fan air kilowatts or fan air horsepower.

In preceding pages unusual attention has been given to illustrating the relationships between the types of units that are employed for energy and other physical measures. Care is required for all numerical solutions whether these are made in SI, conventional metric, foot-pound-second, or mixed units. However, in the derivation and development of equations and other relationships units are usually disregarded until the final formulation is made, following which consistency in the use of units again is imperative.

2-3. HEAT CAPACITY, SPECIFIC HEAT

Consider a process during which heat in amount $Q_{(1-2)}$ is added to a system while its temperature increases from T_1 to T_2. For this situation one can write the equation

$$C'_m = \frac{Q_{(1-2)}}{T_2 - T_1} \tag{2-12}$$

where C'_m represents the average heat capacity of the system over the temperature range $T_2 - T_1$. For a small increment of heat addition (Q) and a resulting small temperature change (ΔT), we can arrive at the concept of an instantaneous heat capacity

$$C' = \frac{Q}{\Delta T}$$

Instantaneous heat capacities for a system can vary greatly at different portions of a temperature range and differ significantly from a mean or average value. Since it may be desirable to know the heat capacity per unit mass in any system of mass m, during heat addition or rejection, it is convenient to write the two prior equations in the form

$$C_m = \frac{Q_{(1-2)}}{m(T_2 - T_1)} \tag{2-13}$$

$$C = \frac{Q}{m\Delta T} \tag{2-14}$$

Each of these C values is called *specific heat capacity*, or the name may be shortened to *specific heat*. In SI specific heat would be expressed in units of kilojoules per kilogram degree Kelvin (or degree Celsius).

Specific heat is more usually expressed as the kilocalories of energy required to raise 1 kilogram of mass through 1 degree Kelvin (or Celsius). In the foot-pound-second notation, specific heat is expressed as the Btu of energy required to raise 1 pound of mass through 1 degree Rankine (or Fahrenheit). These two, last-defined specific heats are equivalent numerically.

There are two kinds of specific heat that are of importance when dealing with gases: that at constant volume, C_v, and that at constant pressure, C_p. If, for example, 1 pound of air is heated in a closed container in which no change in volume is possible, 0.171 Btu will be required to cause a temperature rise of 1 degree F. If, however, the air is heated through the same temperature range but is free to expand against the surrounding constant pressure, 0.240 Btu will be required. This extra energy in the specific-heat value really represents the work that must be done by the air as it expands against the surrounding pressure. For liquids the volume change is so small that the distinction between C_p and C_v is negligible. Note that 0.171 Btu/lb · F has the equivalent value of 0.171 kcal/kg · °C. In SI, to convert to the required form we must multiply 0.171 by 4.184 to yield kJ/kg · °C.

It can easily be shown that during a constant pressure process the heat addition (or rejection) is equal to the change in enthalpy of the medium. Thus for such a process Eq. (2-13) would appear

$$C_p = \left.\frac{Q}{m\Delta t}\right)_p = \left.\frac{\Delta \mathbf{h}}{\Delta t}\right)_p \tag{2-15}$$

Similarly, in a constant-volume process

$$C_v = \left.\frac{Q}{m\Delta t}\right)_p = \left.\frac{\Delta u}{\Delta t}\right)_v \tag{2-16}$$

Table 2-1 gives values of specific heat for selected gases and vapors, and Table 4-1 provides similar information for representative solids and liquids. It should be noticed that the value for water at moderate temperatures is unity and also that the value for most other substances is less than unity. Specific-heat values of materials such as meats and vegetables containing varying amounts of water are variable. The specific heat of a frozen or

Table 2-1 Properties of Gases and Vapors

Gases and Vapors	Chemical Symbol	Molecular Weight	Gas Constant, R ($PV = mRT$)		Properties at 1 atm = 14.696 psia = 101 325 N/m² = 760 mm Hg: Melting Point		Boiling Point		Specific Volume of Vapor at 32 F (0°C)		Latent Heat h_{fg}		Specific Heat of Vapor at 68 F (20°C) (Btu/lb · R or cal/g · K)		k
			(ft · lbf/lb · R)	(J/kg · K)	(R)	(K)	(F)	(°C)	(ft³/lb)	(m³/kg)	(Btu/lb)	(J/g)	C_p	C_v	C_p/C_v
GASES															
Air (R-729)		28.97	53.35	286.9			−317.8	−194.4	12.40	0.773	88.2	205.0	0.240	0.171	1.40
Argon (R-740)	Ar	39.95	38.68	208.0	150.9	83.8	−302.6	−185.9	8.99	0.561	69.7	161.9	0.125	0.075	1.67
Helium (R-704)	He	4.00	386.0	2075	—	—	−452.1	−268.9	89.68	5.60	8.85	20.6	1.230	0.737	1.67
Hydrogen (R-702p)	H_2	2.02	765.0	4114	25.2	14.0	−423.0	−252.8	178.09	11.118	191.7	445.6	3.39	2.404	1.41
Nitrogen (R-728)	N_2	28.02	55.15	296.6	113.7	63.2	−320.4	−195.8	12.81	0.800	85.9	199.6	0.248	0.177	1.40
Oxygen (R-732)	O_2	32.00	48.29	259.6	97.9	54.4	−297.3	−183.0	11.22	0.700	91.6	212.9	0.218	0.156	1.40
VAPORS															
Ammonia (R-717)	NH_3	17.03	90.73	487.8	351.8	195.4	−28.0	−33.3	21.08	1.316	589.4	1370.0	0.500	0.382	1.31
Azeotrope (R-502)		111.6	13.85	74.5			−49.8	−45.4	3.126	0.195	74.2	172.4	0.162	0.145	1.117
Carbon dioxide (R-744)	CO_2	44.00	35.12	188.8	350.4_s	194.7_s	—	—	8.16	0.509			0.206	0.158	1.304
Carbon monoxide	CO	28.00	55.18	296.7	118.9	66.1	−312.7	−191.5	12.82	0.800	90.7	210.8	0.250	0.179	1.40
Carbontetrafluoride (R-14)	CF_4	88.01	17.56	94.4	160.7	89.3	−198.3	−127.9	4.08	0.254	99.8	232.0	0.166	0.143	1.16
Chlorodifluoromethane (R-22)	$CHClF_2$	86.48	17.87	96.1	203.7	113.2	−41.4	−40.8	4.068	0.254	100.4	233.4	0.156	0.131	1.195
Dichlorodifluoromethane (R-12)	CCl_2F_2	120.9	12.78	68.7	207.7	115.4	−21.6	−29.8	2.886	0.180	71.0	165.0	0.146	0.128	1.143
Dichlorofluoromethane (R-21)	$CHCl_2F$	102.9	15.01	80.7	248.7	138.2	48.1	8.9			104.2	242.2	0.137	0.116	1.186
Ethane (R-170)	C_2H_6	30.04	51.44	276.6	181.7	100.9	127.5	53.2			209.6	487.2	0.386	0.316	1.22
Ethylene	C_2H_4	28.03	55.13	296.4	187.7	104.3	155.0	68.3			206.7	480.5	0.375	0.299	1.255
Isobutane (R-600a)	C_4H_{10}	58.12	26.59	143.0	230.7	128.2	10.3	−12.0	6.18	0.385	156.6	364.0	0.376	0.338	1.11
Methane (R-50)	CH_4	16.03	96.40	518.4	162.7	90.4	−258.9	−161.6	22.40	1.398	219.7	510.7	0.520	0.397	1.31
Nitrous oxide (R-744a)	N_2O	44.02	35.10	188.8	307.7	170.9	127.0	52.8			162.3	377.3	0.203	0.156	1.303
Octane	C_8H_{18}	114.2	13.53	72.8	389.7	198.7	258.0	125.5			128.0	297.5			
Propane (R-290)	C_3H_8	44.06	35.07	188.6	149.9	83.3	−44.0	−42.2	8.15	0.509	182.2	423.5	0.375	0.332	1.13
Sulfur dioxide (R-764)	SO_2	64.06	24.12	129.7	355.8	197.7	14.0	−10.0	5.60	0.350	167.0	388.2	0.152	0.118	1.29
Trichlorofluoromethane (R-11)	CCl_3F	137.4	11.25	60.5	291.7	162.1	74.9	23.8			77.5	180.1	0.140	0.123	1.136
Trichlorofluoroethane (R-113)	$C_2Cl_3F_3$	187.4	8.25	44.3	428.7	238.2	117.6	47.6			63.1	146.7	0.158	0.150	1.050
Water (R-718)	H_2O	18.02	85.8	461.1	491.7	273.2	212.0	100.0			970.3	2255.4	0.485	0.367	1.32

solidified substance is different from that of the same substance in liquid form; for example, C_p for ice is 0.487, C_p for water is about 1.0. Specific-heat values for most substances increase with temperature. This is particularly true of gases, for which the specific heat shows a large increase in temperature ranges near and above 550°C. However, it is fortunately true that in the range below 250°C, variations in the specific heat values are not excessive.

2-4. THE SECOND LAW AND ENTROPY

The second law of thermodynamics is an expression of man's experiences and the conclusions he has reached from his dealings with nature. One wording of the second law states: it is impossible to develop a machine that can operate continuously at the expense of energy from a single reservoir at one temperature. The significance of this statement is not at first apparent, unless we realize that if it were false, boats could travel over the ocean by merely taking energy from the sea and using this energy to power the vessel. It is indeed possible to develop work from a single reservoir at a given temperature but to do so continuously leads to an impossible situation, in terms of our experience. We find that at least two reservoirs at different temperatures are needed if an engine is to produce work continuously, with the engine taking energy from the higher temperature source, producing some work from this energy, and then discharging the remainder of the energy, received by its working medium, to a sink reservoir at lower temperature. We should note that temperature is a peculiar property of matter which can indicate the direction in which heat will flow of itself between two bodies at different temperatures. In terms of temperature an alternate wording of the second law is sometimes employed: it is impossible to form a self-acting machine which can make heat flow continuously from a region of lower to one of higher temperature.

Analyses based on the second law, in addition to considerations regarding temperature, bring into play the concept of reservoirs and the concept of reversibility. Reversibility implies, for a thermal process, that merely changing the direction of the driving force by an infinitesimal amount will cause a process to reverse itself and return to essentially its initial condition. Reversibility is an idealized or limiting condition that can never be fully realized because of the natural phenomena which inhibit its realization, one of the most obvious of which is friction. Heat flow to be capable of reversibility must take place under infinitesimally small temperature differences and would never be possible with the finite and often large temperature differences that naturally occur. Fluid friction and turbulence negate reversibility as is also the case for a fluid suddenly expanding into a region of lower pressure or even into a vacuum. Nevertheless, as a measure of optimum attainment or as a base from which to make comparisons, reversibility is a valuable concept.

From the idea of reversibility, we can arrive at two extremely important conclusions. First, no engine operating between two reservoirs at different temperatures can have a higher efficiency that a reversible engine; second, if this is true, all reversible engines operating between the two reservoirs will have the same efficiency. On the basis of this premise it is possible to develop the so-called absolute thermodynamic temperature scale, which was first envisaged by Carnot and later developed by Lord Kelvin.

The efficiency of any heat engine working in a cycle with the medium returning to its original state can be expressed as

$$\eta = \frac{\mathbf{W}}{Q_a} = \frac{Q_a - Q_R}{Q_a} = 1 - \frac{Q_R}{Q_a} \tag{2-17}$$

since by the first law, if a cycle exists, the work, **W**, must be equivalent to the difference between the heat added, Q_a, and the heat rejected Q_R. We can envisage that for each reservoir a temperature scale can be found that bears a direct relationship to the heat delivered from the reservoir to a reversible engine, Q_a, and to the heat rejected by the reversible engine, Q_R, to the sink reservoir. Such a temperature scale, which we could justifiably call the thermodynamic scale of temperature, would define efficiency in the same way as Eq. (2-17) and could be written:

$$\eta = \frac{\mathbf{W}}{Q_a} = \frac{T_a - T_R}{T_a} = 1 - \frac{T_R}{T_a} \tag{2-18}$$

It follows that

$$\frac{Q_R}{Q_a} = \frac{T_R}{T_a}$$

and

$$\frac{Q_a}{T_a} = \frac{Q_R}{T_R} \tag{2-19}$$

This last relationship for a reversible process brings together the ratio of the heat received (or rejected) to the thermodynamic temperature at which the heat exchange occurred and tacitly indicates that each exchange occurred at a constant temperature. Since many processes occur under conditions for which the temperature is changing, let us introduce the symbol s and use infinitesimals for representation of varying-temperature heat addition or rejection. Then:

$$ds = \left.\frac{dQ}{T}\right)_{\text{rev}} \tag{2-20}$$

$$s_2 - s_1 = \int_1^2 \left.\frac{dQ}{T}\right)_{\text{rev}} \tag{2-21}$$

The name entropy has been given to the above defined ratio, with the symbol s, and it should be noted that by its definition and derivation it is applicable only for reversible conditions and for this reason the restriction, rev, has been noted in Eqs. (2-20) and (2-21). The same restriction also holds for Eq. (2-19) even though not written there.

Entropy has been developed at this point primarily as a mathematical concept yet it can be used in many ways. Some of these will be shown, and at the same time its physical significance will be developed. However, at this point we make only two comments: (1) entropy, s, is a thermodynamic property in the same sense that internal energy, u, pressure, P, temperature, T, and volume, v, also are properties; (2) entropy can be used for analyzing irreversible processes by replacing the irreversible process between two state points by reversible paths which can lead to the same end point.

Establishment of quantitative values for the thermodynamic temperature scale is possible since computations can be carried out by which the efficiency of a reversible engine working between any chosen temperature state points can be found. Such computations lead to the result that on the thermodynamic (or absolute) temperature scale, the freezing point of water, the so-called ice point 32 F or 0°C, occurs at

$$491.69 \text{ R} = 459.69 \text{ F} + 32 \text{ F}$$

or

$$273.16 \text{ K} = 273.16°\text{C} + 0°\text{C}$$

In thermodynamic computations absolute temperatures on any chosen scale, Rankine or Kelvin, should always be employed except where the computation involves merely temperature differences, in which case the additive term 459.7 or 273.16 is automatically eliminated.

2-5. THE PERFECT GAS

A gas which fully satisfies the equation of state represented by

$$Pv = RT \tag{2-22}$$

is called a perfect or ideal gas. While no actual gas has an equation of state as simple as that represented by Eq. (2-22), nevertheless many gases in the range where their temperatures are far from their critical temperature and where their pressures are low in relation to their critical pressure closely obey this relationship. Equation (2-22) can be derived, if a derivation is considered necessary, by making use of Boyle's law which states that at constant temperature the volume of a fixed mass of gas varies inversely with its pressure and by use of Charles' law which states that at constant pressure the volume of a gas varies in direct relation to its absolute temperature. These laws readily show that

$$\frac{Pv}{T} = \frac{P_1 v_1}{T_1} = \frac{P_2 v_2}{T_2} = R \tag{2-23}$$

R is a constant for a particular gas with v appearing as specific volume. The expression can be generalized for any volume V by introducing the mass of gas, m,

$$\frac{PV}{Tm} = \frac{PV_1}{T_1 m} = \frac{PV_2}{T_2 m} = R$$

$$PV = mRT \tag{2-24}$$

In consistent SI units, P is the pressure in pascals (Pa), that is, newtons per square meter; T is the temperature in degrees Kelvin (K); V is the volume in cubic meters; m is the mass of the gas in kilograms; v is the specific volume in cubic meters per kilogram; and R is the gas constant in joules per kilogram degree Kelvin.

In the foot-pound-second system, P is the pressure in pounds per square foot, or 144 p, when p is in pounds per square inch; T is the temperature, in degrees Rankine; V is the volume in cubic feet; m is the pounds of gas; v is the specific volume, in cubic feet per pound; and R is the gas constant in foot-pounds per pound degree Rankine.

The gas constant R is different for each gas when based on unit mass of that gas. However, for a mol (molecular weight) of a gas, the value of molal **R**, the universal gas constant, is invariant and is the same for each and every gas. It has the following numerical values:

$$\begin{aligned} \mathbf{R} &= 8309.5 \text{ J/kg-mol} \cdot \text{K} \\ &= 1545.3 \text{ ft} \cdot \text{lbf/lb-mol} \cdot \text{R} \\ &= 1.986 \text{ kcal/kg-mol} \cdot \text{K} \end{aligned}$$

or

$$= 1.986 \text{ Btu/lb-mol} \cdot \text{R}$$

The mol values (molecular weights) of many gases are tabulated in Table 2-1 and to find the value of R to use in Eq. (2-24), divide by the molecular weight M, as follows:

$$R = \frac{\mathbf{R}}{M} = \frac{8309.5}{M} \text{ J/kg} \cdot \text{K} \tag{2-25}$$

or

$$R = \frac{\mathbf{R}}{M} = \frac{1545.3}{M} \text{ ft} \cdot \text{lbf/lb} \cdot \text{R} \tag{2-26}$$

At low pressures approaching zero, Eq. (2-24) is extremely precise for any gas, and the accuracy is almost independent of the temperature range. The equation is also accurate for most gases and vapors at atmospheric pressure, and even up to some 150 psi. However, when pressures exceed this range, deviations from the equation can become appreciable as other compressibility deviations take place. When gases (vapors) are close to their condensation temperature for the pressure in question, the perfect-gas equation may be far from reliable.

To illustrate the magnitude of deviation which might be expected in a real vapor when close to its condensation temperature or when under significant pressure, we can compute for steam that

$$R = \frac{1545.3}{M} = \frac{1545.3}{18.016} = 85.77 \text{ ft} \cdot \text{lbf/lb} \cdot \text{R}$$

This value gives accurate results for superheated steam (water vapor) at pressures of 1.0 psia and lower. However, for steam at 250 F (709.7 R) at a pressure as low as 14.7 psia, a value of 84.7 would be needed to provide accurate values for the specific volume of the steam, and at higher pressures for steam it is inadvisable to use Eq. (2-24).

Example 2-3. Dry air is a mixture of diatomic gas and has a composite molecular weight (M) of 28.966. Considering air as a perfect gas, find (a) the volume occupied by 1 kg of air at standard atmospheric pressure 101 325 Pa (101.325 kPa) and 20°C (14.696 psia and 68 F) and (b) the volume of 1 lb of air at the same conditions.

Solution: (a) R can be found by making use of the universal gas constant in appropriate units; here, by use of Eq. (2-25),

$$R = \frac{8309.5}{M} = \frac{8309.5}{28.966} = 286.9 \text{ J/kg} \cdot \text{K}$$

Substitute in Eq. (2-24)

$$(101\,325)(v) = (1)(286.9)(273.16 + 20)$$

$$v = 0.830 \text{ m}^3\text{/kg} \qquad \textit{Ans.}$$

(b) The answer to this part can be found most easily by converting cubic meters per kilogram to cubic feet per pound.

$$0.8300 \text{ m}^3\text{/kg} \times 35.314 \text{ ft}^3\text{/m}^3 \times 0.45359 \text{ kg/lb} = 13.30 \text{ ft}^3\text{/lb}$$

However, if a complete solution is desired, first find R by use of Eq. (2-26) and substitute appropriate values in Eq. (2-24).

$$R = \frac{1545.3}{28.966} = 53.35 \text{ ft lb/lb} \cdot \text{F}$$

$$(144)(14.696)(v) = (1)(53.35)(459.7 + 68)$$

$$v = 13.3 \text{ ft}^3\text{/lb} \qquad \textit{Ans.}$$

2-6. PERFECT-GAS RELATIONSHIPS

The specific internal energy of a perfect gas is a function of temperature only, with the value of u not influenced by pressure. Joule demonstrated this in 1843 when he connected two small pressure tanks together by a pipe in which a valve has been installed. When closed, the valve isolated the gaseous contents of the two tanks. For his experiment, he filled one tank with air under high pressure and he evacuated the other tank. Both tanks were then placed in a water bath surrounded by an insulated shell. The tanks were allowed to reach thermal equilibrium and the temperature of the water was precisely recorded. The valve was then opened and air rushed from the high-pressure tank into the evacuated one, quickly reaching pressure equalization. Although this highly irreversible process raised the temperature in the evacuated tank and lowered the temperature in the pressure tank, Joule was unable to detect any temperature change in the water bath after equilibrium was reached. He thus had to conclude that no net heat transferred to or from the total air mass. With such the case, $dQ = 0$, and with no work being done, $d\mathbf{W} = 0$; thus by Eq. (2-1), $du = 0$. Thus there is no change in internal energy as long as the temperature is constant and the property, internal energy, is independent of pressure and volume. Because of this fact, the constant-volume and constant-pressure restrictions placed on C_v and C_p in Eqs. (2-15) and (2-16) are no longer required when these relate to a perfect gas, and it follows that:

$$C_v = \frac{\Delta u}{\Delta T} = \frac{du}{dT}$$

$$C_p = \frac{\Delta \mathbf{h}}{\Delta T} = \frac{d\mathbf{h}}{dT}$$

If Joule had possessed more precise instrumentation, he would have found that a slight change in temperature resulted in the bath following expansion, which would have indicated that u is not a unique function of temperature. Other evidence shows that this perfect-gas relationship is fully honored only at pressures approaching zeros under which condition the molecules of a gas are extremely far apart. However, the approximation is sufficiently close under conditions of low and moderate pressure to be of great value in making computations for real gases. Thus it follows that for a perfect gas, since internal energy U, is essentially independent of volume and pressure, changes in U can be expressed in terms of temperature alone. Thus

$$U_2 - U_1 = mC_v(T_2 - T_1) \tag{2-27}$$

or for unit mass

$$u_2 - u_1 = C_v(T_2 - T_1) \tag{2-28}$$

This expression is true only if C_v is the average value over the range in question or if C_v is essentially constant in magnitude. Following similar reasoning, we can show that changes in enthalpy, **h**, for a perfect gas can be written

$$\mathbf{h}_2 - \mathbf{h}_1 = C_p(T_2 - T_1) \tag{2-29}$$

Frequently it is desirable to list values of **h** in tabular form for a gas, in which case an arbitrary datum temperature such as 0 F or 0°C or even 0 K or 0 R can be chosen. If, for example, we choose 0 F and the foot-pound-second system is used,

$$\mathbf{h} - \mathbf{h}_0 = C_p(T - T_0) = C_p[459.7 + t - (459.7 + 0)] = C_p t \tag{2-30}$$

and if $\mathbf{h}_0$ is also set at zero,

$$\mathbf{h} = C_p t \tag{2-31}$$

In computing such tabular values, these would be correct only if a proper mean value were used for C_p or if in constructing the table small temperature steps with the proper C_p value for each step were used in a summation or integration process.

Two other useful generalizations can be shown for the perfect gas. The first is that the difference between the specific heats for any gas is essentially a constant which happens to be its R value, namely,

$$C_p - C_v = R \tag{2-32}$$

Second, it is also true that the ratio of these specific heats is also a constant, namely,

$$\frac{C_p}{C_v} = k \tag{2-33}$$

The value of k can be found from computed or experimentally-determined values of C_p and C_v. Values of k can also be found by generalizations developed from the kinetic theory of gases. These relations show that for diatomic gases, such as oxygen, nitrogen, air, and hydrogen, the value of k is in the neighborhood of 1.4. For triatomic gases such as CO_2, H_2O, and SO_2, the value of k is in the range of 1.3. Monatomic gases, of which helium is one, have a value of 1.66 for k. For real gases, since both C_p and C_v vary significantly with temperature increase, it follows that k also varies. However, over temperature ranges from -40 F to some 300 F, these variations are small for gases such as air, oxygen, nitrogen, and hydrogen, and this makes it possible to simplify computations with them in this useful range of temperature.

Example 2-4. A small residence of loose frame construction is 15 m long by 10 m deep, and is 6 m high in its two stories. The windows are not weatherstripped. It was determined, on a windy day, that outside air equivalent to 4 complete changes of inside air entered and left the house per hour. When an average temperature of 23.3°C (73.94 F) is maintained in the house, all of the outside infiltration air must warm to this temperature before it leaves by exfiltration. Compute the extra load on the heating system caused by infiltration when it is 0°C (32 F) outside and the barometric pressure is 101 325 Pa (101.325 kPa). Work (a) in SI units and (b) in foot-pound-second units.

Solution: (a) This process takes place at constant pressure so the specific heat at constant pressure represents the heat addition needed for each degree rise in temperature. From Table 2-1 note that the specific heat of dry air is given as 0.240 kcal/kg · K. For SI, joules must be used and

$$C_p = 0.240 \times 4.184 = 1.0042 \text{ kJ/kg} \cdot \text{K}$$

Find the specific volume of the inside air by use of Eq. (2-24) with R for air, 286.9 J/kg · K from Table 2-1:

$$(101\ 325)(v) = (1)(286.9)(273.16 + 23.3)$$

$$v = 0.8394\ \text{m}^3/\text{kg}$$

For 4 air changes per hour, the volume involved is

$$4[(15\ \text{m})(10\ \text{m})(6\ \text{m})] = 3600\ \text{m}^3/\text{h}$$

and

$$\frac{3600}{v} = \frac{3600}{0.8394} = 4289\ \text{kg/h}$$

$$q = (\text{kg/h})(C_p)(\Delta t) = (4289)(1.0042)(23.3 - 0) = 100\ 350\ \text{kJ/h}$$

$$q = \frac{100\ 350}{3600} = 27.9\ \text{kW}$$ *Ans.*

(b) Use Eq. (2-24) to find the specific volume, noting that

$$101\ 325\ \text{Pa} \times 1.4504 \times 10^{-4} = 14.696\ \text{psia}$$

$$(144)(14.696)(v) = (1)(53.35)(459.7 + 73.94)$$

$$v = 13.45\ \text{ft}^3/\text{lb}$$

For the 4 air changes per hour, the building volume must be converted to cubic feet:

$$4[(15)(10)(6) \times 3.2808^3] = 127\ 130\ \text{ft}^3/\text{h}$$

$$\text{pounds of air per hour} = \frac{127\ 130}{13.45} = 9452$$

$$q = (\text{lb/h})(C_p)(\Delta t) = (9452)(0.240)(73.94 - 32) = 95\ 140\ \text{Btuh}$$ *Ans.*

or

$$q = \frac{95\ 140}{3414} = 27.9\ \text{kW}$$ *Ans.*

2-7. PERFECT-GAS WORK PROCESSES

Anyone who is in the vicinity of an air compressor at the time of its operation could almost certainly observe that the action of compression produces a substantial increase in the temperature of the compressed air. Work is required to compress (squeeze) air into a smaller volume, and this work energy necessarily becomes stored in the air with a resulting increase in temperature, except for any heat which transfers from the hot gas to outside during the compression. It should be noted that this significant temperature rise is not the result of friction since frictional effects contribute only trivially to the whole process. It is also true that in most instances compression is essentially an *adiabatic* process. An adiabatic process is one in which no heat transfer takes place across the boundary walls of the system. Here, if the walls of the compressor were carefully insulated so that almost no heat loss occurred during compression or if insufficient time were available during compression for significant heat transfer (cooling) to occur, an adiabatic condition would almost exist. It is convenient to treat certain processes as being adiabatic even while we recognize that an adiabatic process is a limiting condition, never fully available.

Let us now consider that compression takes place in a noninsulated cylinder so slowly that heat flow from the compressed air takes place in adequate amount to keep the temperature from the gas from rising more than a trivial amount. Such a process carried out at essentially constant temperature is called *isothermal.* In such an isothermal process with T constant, Eq. (2-24) shows that values of PV at different steps during the compression would necessarily be equal, and for air or any perfect gas

$$P_1V_1 = P_2V_2 = P_3V_3 \equiv PV \equiv \text{constant} \qquad (2\text{-}34)$$

On a PV plane a plot of such values would represent the path of state points followed during the compression.

The preceding discussions relative to a gas being compressed are equally applicable to a previously compressed gas as it expands or drops in pressure. Figure 2-2 is a plot of an isothermally expanding gas starting at point 1, dropping in pressure and increasing in volume to point 3 as the expansion takes place. During such an expansion heat would have to be added to the gas to keep the temperature from falling as work is delivered.

Consider now what happens if the gas expands adiabatically; here the work energy must be provided by the gas, its temperature will drop substantially to point 2, and the colder gas will have a smaller volume (2) than

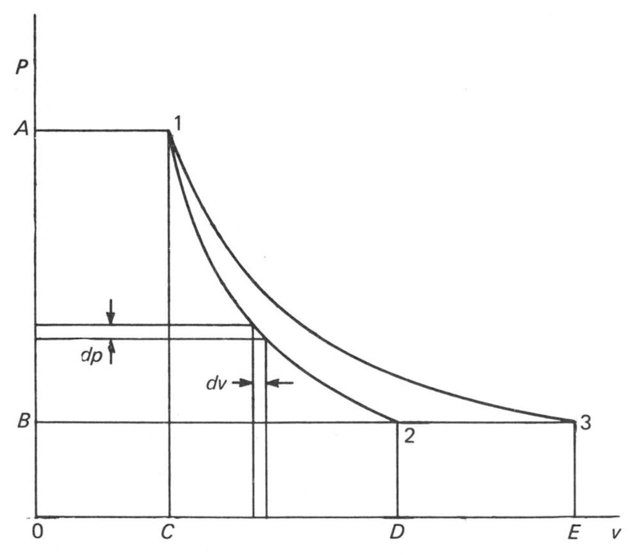

FIGURE 2-2. *Pv* diagram to illustrate work under isentropic (1–2) and isothermal (1–3) conditions.

the isothermal gas (3) when both expansions reach the same pressure P_B. If an adiabatic expansion is carried out reversibly—that is, with no friction or turbulence—it will also be isentropic, and this is the condition which is shown in Fig. 2-2.

A mathematical approach to show the shape of the paths of processes on the PV plane would provide an exponent to the V term in each equation. For example, for the isothermal case in Eq. (2-34), the exponent for each V term could be (1) unity without altering the expression in any way. Standard thermodynamic texts develop the fact that for a reversible-adiabatic expansion or compression.

$$PV^k = \text{constant} \equiv P_1 V_1^k = P_2 V_2^k = P_3 V_3^k, \text{etc.} \qquad (2\text{-}35)$$

where k is the ratio C_p/C_v. PV^k = constant thus represents the equation of the curve (path), and if the small pressure changes (dp) and the corresponding volume changes (dv) are integrated for such a process, it is found that the work during compression or expansion, and represented by the area $12DC$, is

$$\mathbf{W}_{(1-2)} = \frac{P_2 v_2 - P_1 v_1}{1 - k} = \frac{P_1 v_1 - P v_2}{k - 1}$$

$$= \frac{P_1 V_1}{k - 1}\left[1 - \left(\frac{P_2}{P_1}\right)^{(k-1)/k}\right] \qquad (2\text{-}36)$$

These equations are developed for nonflow conditions; that is, they measure the work delivered (or received) as the gas expands (or is compressed) but do not include the work of admission and the work of delivery from the compressor. These additional amounts of work can be found as follows. Referring to Fig. 2-2, we can readily see that for an expander engine (turbine), the work done by the gas in merely entering (flow work) is represented by the area $A1C0$ and equals $P_1 v_1$ in magnitude. The flow work of delivery is $B2D0$ and equals $P_2 v_2$. If we add these values to Eq. (2-36), we find that for a steady-flow machine

$$\mathbf{W} = P_1 v_1 + \frac{P_2 v_2 - P_1 v_1}{1 - k} - P_2 v_2 = \frac{k}{k - 1}(P_1 v_1 - P_2 v_2) \qquad (2\text{-}37)$$

or

$$\mathbf{W} = \frac{k}{k - 1} P_1 v_1 \left[1 - \left(\frac{P_2}{P_1}\right)^{(k-1)/k}\right] \qquad (2\text{-}38)$$

In similar manner we could integrate the work production under isothermal conditions making use of Eq. (2-34) and would find that the work of expansion or compression is

$$\mathbf{W}_{(1-3)} = P_1 v_1 \log \frac{v_3}{v_1} = P_1 v_1 \log_e \frac{P_1}{P_3} = RT \log_e \frac{P_1}{P_3} \qquad (2\text{-}39)$$

Let us now consider an isothermal expansion in the light of Eq. (2-1), which we can write as

$$Q_{(1-3)} = u_3 - u_1 + \mathbf{W}_{(1-3)}$$

Since for an isothermal process both the temperature and internal energy are constant, it follows that

$$u_3 = u_1$$

and

$$Q_{(1-3)} = \mathbf{W}_{(1-3)} = P_1 v_1 \log_e \frac{v_3}{v_1} = RT\ 2.3 \log_{10} \frac{P_1}{P_3} \tag{2-40}$$

This equation shows the obvious fact that since work is not created at the expense of internal energy, all of it must be supplied by heat addition during an expansion process. During isothermal compression, conversely, all of the work supplied is simultaneously delivered as heat since no additional internal energy can reside in the gas. Note also that the work during a reversible isothermal expansion, represented by the area under the line 1–3 in Fig. 2–2, is appreciably greater than the work produced under a reversible-adiabatic expansion for the same pressure range and initial conditions.

The steady-flow work under isothermal compression (or expansion) of a perfect gas by happenstance is exactly the same as nonflow work since the work of admission ($P_1 v_1$) and of delivery ($P_3 v_3$) are the same except for opposite sign and thus, of course, offset each other.

The reversible adiabatic process can be used as an idealized model by which to gage the performance of real systems. Temperature changes are most convenient for making such comparisons, and the following relationships can often be advantageously employed. Let us combine equations and let

$$P_1 v_1^k = P_2 v_2^k$$

$$P_1 v_1 = RT_1$$

$$P_2 v_2 = RT_2$$

By simple algebra there results

$$\frac{T_2}{T_1} = \left(\frac{P_2}{P_1}\right)^{(k-1)/k} \tag{2-41}$$

$$\frac{T_2}{T_1} = \left(\frac{v_1}{v_2}\right)^{k-1} \tag{2-42}$$

It is thus easy to find temperature in terms of the pressure ratio or in terms of the volume ratio in which a reversible-adiabatic process has been involved.

Temperature has here been singled out because both enthalpy and internal energy can be expressed in terms of it. This was shown in Eqs. (2-28) and (2-29). To make use of enthalpy and temperature changes in relation to work, refer to Eq. (2-9) where if we consider an adiabatic condition ($Q_{(1-2)} = 0$) and eliminate the irrelevant potential and kinetic energies, we get

$$\mathbf{W}_{(1-2)} = \mathbf{h}_1 - \mathbf{h}_2 \tag{2-43}$$

for an engine or turbine using air or gas, and

$$\mathbf{W}_{(1-2)} = \mathbf{h}_2 - \mathbf{h}_1 \tag{2-44}$$

for a compressor using air or gas. It of course follows from Eq. (2-29) that

$$\mathbf{W}_{(1-2)} = \mathbf{h}_2 - \mathbf{h}_1 = C_p(T_2 - T_1) \tag{2-45}$$

provided C_p is a true average value for the temperature range in question.

Example 2-5. Air at standard atmospheric pressure (14.7 psia) at 68 F is compressed, through a pressure compression ratio of 8, to 117.6 psia in a steady-flow compressor. Find the work required to compress each pound of air isentropically (reversibly adiabatic) by (a) the pressure-volume method and (b) the enthalpy-difference (temperature-difference) method. (c) If because of turbulence, friction, and internal leakage, it is found that 33% more work is required to compress each pound of air than is indicated by the reversible-adiabatic computation, find the actual work per pound of air. (d) Find the actual air horsepower needed to deliver air at a rate of 50 lb/h.

Solution: The specific volume of the supply air can be found by use of Eq. (2-24). Here we will make use of the value found in Example 2-3, 13.30 ft^3/lb.

(a) Use will be made of Eq. (2-38) after finding k from Table 2-1 as 1.4.

$$\mathbf{W}_{(1-2)} = \left(\frac{1.4}{1.4 - 1}\right)(144)(14.7)(13.30)[1 - (8)^{(1.4-1)/1.4}]$$

$$= 98\,540[1 - 1.811] = -79\,910 \text{ ft lbf}$$

The work is thus 79 910 ft lbf/lb of air compressed and delivered. The minus sign merely indicates work being done on the air during compression.

(b) Here, use Eq. (2-41) to find the ideal temperature resulting from isentropic compression.

$$\frac{T_2}{T_1} = \left(\frac{P_2}{P_1}\right)^{(k-1)/k} = \left(\frac{117.6}{14.7}\right)^{(1.4-1)/1.4} = (8)^{0.286} = 1.811$$

$$T_2 = (T_1)(1.811) = (459.7 + 68)(1.811) = 955.7$$

The mean specific heat of air $C_p = 0.24$ Btu/lb F from Table 2-1. Using Eq. (2-29),

$$\mathbf{W}_{(1-2)} = (1)(0.24)[955.7 - (459.7 + 68)] = 102.72 \text{ Btu/lb}$$

$$102.72 \text{ Btu/lb} \times 778 = 79\,910 \text{ ft lbf}$$

(c) actual work = (79 910)(1.33) = 106 280 ft lbf/lb air
(d) horsepower required for 50 lb/h = (50)(106 280)/(60)(33 000) = 2.68

where 33 000 ft lbf/min are in each horsepower, and multiplying by 0.746 gives the equivalent kW:

$$(2.68)(0.746) = 2.0 \text{ kW}$$

In addition to the reversible adiabatic and the isothermal processes, which have been described, other compressions (or expansions) can be envisaged, in which less (or more) heat is exchanged than during the isothermal. For such a case, if reversible, we could find an appropriate exponent and set up an equation of the form

$$PV^n = P_1 V_1^n = P_2 V_2^n = \text{constant} \tag{2-46}$$

and a plot of such an equation on the PV plane would indicate the path followed during the process. For example, if n were 1.25, the path would rest between the paths 1–2 and 1–3 shown in Fig. 2-2. Such generalized paths are called polytropics, and n can have varying values ranging from zero to infinity. Real processes are not reversible, but nevertheless appropriate values of n can be found which will approximate idealized (reversible) conditions and equations of the form of (2-38), (2-41), and (2-42) can be used with n substituted for k.

Note that these expressions are not applicable for values of $n = 1$, $n = 0$, $n = \pm\infty$, but the proper forms are easily derived as was done for $n = 1$ earlier in this chapter. Work is, of course, zero for $n = \pm\infty$, where $v = \text{constant}$, since $dv = 0$.

Example 2-6. Air at standard atmospheric pressure 760 mm Hg (101 325 N/m^2) at 20°C is compressed through a compression ratio of 8 to 810 600 Pa (N/m^2). Find (a) the work required to compress each kilogram of air isothermally and (b) the heat rejected during the compression. (c) Do the answers to parts a and b change if the process is considered to involve steady flow into and from the compressor?

Solution: (a) Make use of Eq. (2-39), realizing that a compression would start at point 3 and move to 1 as this equation is set up in relation to Fig. 2-2. Also note that RT can be used in place of $P_1 V_1$ where

$$R = \frac{8309.5}{28.966} = 286.9 \text{ J/kg} \cdot \text{K}$$

$$\begin{aligned} \mathbf{W}_{(1-3)} &= (286.9)(273.16 + 20)\, 2.3 \log_{10} \tfrac{1}{8} \\ &= -(286.9)(293.16)\, 2.3 \log_{10} 8 \\ &= -(286.9)(293.16)(2.3)(0.9031) = -174\,600 \text{ J/kg} \end{aligned}$$

The minus sign merely implies that the work is done on and not by the air.

(b) By Eq. (2-40) it is obvious that the work supplied is all dissipated as heat and $\mathbf{W}_{(1-3)} = Q_{(1-3)}$.

(c) The nonflow work and the steady-flow work in an isothermal process are equivalent. To compare the work done (and equivalent heat dissipated) in the isothermal compression with the corresponding work done on the air in an equivalent reversible-adiabatic compression (part *b* of Example 2-5 at 102.72 Btu/lb), convert 174 000 J/kg, to Btu/lb and find:

$$174\,000 \times 0.948 \times 10^{-3} \times \frac{1}{2.2046} = 74.8 \text{ Btu/lb}$$

Thus the isothermal compression requires about 27% less work supply than does the reversible adiabatic for this case of an 8 to 1 compression.

2-8. WATER-STEAM

Earlier material in this chapter has made use of the perfect-gas relationships to illustrate thermodynamic principles. This served a useful purpose because the simplicity of the perfect-gas equation of state makes it easy to use and the results obtained from its use are sufficiently accurate for many engineering calculations. However, thermodynamic substances serve not only in their vapor (gaseous) phase but in the liquid, liquid-vapor, and solid phases. For such a spectrum, simple equations of state are not applicable and formulations based on experimental data and generalized thermodynamic relationships are required. However, these are usually so complex that for convenience it is customary to prepare tables of properties for widely-used materials.

Water in its three states, solid (ice), liquid (water), and vapor (steam) has been thoroughly investigated and elaborate tables of its complete properties are available. In this text, abbreviated tables for water-steam appear as Tables 2-2, 2-3, and 2-4 and also in connection with air as Table 3-1. Water thus serves as a good medium to illustrate the range of properties for a representative multiphase substance.

The thermodynamic properties of state already introduced apply to the solid and liquid as well as to the gaseous phase, but further explanation is necessary in some instances. Specific volume (v) has been mentioned earlier as the volume occupied by unit mass of a medium. It has conventional units of cubic feet per pound or in SI cubic meters per kilogram. Density (d) is the reciprocal of specific volume and appears in conventional units of pounds per cubic foot, or kilograms per cubic meter. Tables of properties may record either specific volume or density.

The density or specific volume of a liquid is very slightly affected by changes in pressure but alters appreciably under changes in temperature. There is no simple relationship to describe the change in specific volume of a liquid with temperature and so it is customary to give tabular values. The water-steam Tables 2-2 and 2-3 employ v_f, the specific volume, instead of density. For example, in Table 2-2, water at 100 F shows v_f equal to 0.01613 ft^3/lb. The corresponding density of water at 100 F is then $1/0.01613 = 61.99\ lb/ft^3$.

For each liquid upon which a given pressure is being exerted there exists a certain definite temperature for that pressure at which boiling or vaporization will take place. This state at which a liquid and vapor are in temperature-pressure equilibrium is known as the *saturation state.* Tables of the properties of steam (Tables 2-2, 2-3, and 2-4) all list saturation pressures and corresponding temperatures. For example, at 14.696-psia pressure (atmospheric), the saturation or boiling temperature is 212 F; at 100 psia water will not boil until its temperature reaches 327.81 F.

The steam tables take water at 32 F as a datum point from which to start enthalpy values. This means that the magnitude of the **h** of water at 32 F is arbitrarily taken as zero. If 1 lb of water were heated at atmospheric pressure from 32 F to 212 F, then (since the mean specific heat of water over

TABLE 2-2 Saturated Steam: Temperature Table

Temperature F t	Absolute Pressure Lb per sq in. p	Specific Volume		Enthalpy			Entropy	
		Sat. Liquid v_f	Sat. Vapor v_g	Sat. Liquid h_f	Evap. h_{fg}	Sat. Vapor h_g	Sat. Liquid s_f	Sat. Vapor s_g
32°	0.08854	0.01602	3306	0.00	1075.8	1075.8	0.0000	2.1877
35	0.09995	0.01602	2947	3.02	1074.1	1077.1	0.0061	2.1770
40	0.12170	0.01602	2444	8.05	1071.3	1079.3	0.0162	2.1597
45	0.14752	0.01602	2036.4	13.06	1068.4	1081.5	0.0262	2.1429
50	0.17811	0.01603	1703.2	18.07	1065.6	1083.7	0.0361	2.1264
60°	0.2563	0.01604	1206.7	28.06	1059.9	1088.0	0.0555	2.0948
70	0.3631	0.01606	867.9	38.04	1054.3	1092.3	0.0745	2.0647
80	0.5069	0.01608	633.1	48.02	1048.6	1096.6	0.0932	2.0360
90	0.6982	0.01610	468.0	57.99	1042.9	1100.9	0.1115	2.0087
100	0.9492	0.01613	350.4	67.97	1037.2	1105.2	0.1295	1.9826
110°	1.2748	0.01617	265.4	77.94	1031.6	1109.5	0.1471	1.9577
120	1.6924	0.01620	203.27	87.92	1025.8	1113.7	0.1645	1.9339
130	2.2225	0.01625	157.34	97.90	1020.0	1117.9	0.1816	1.9112
140	2.8886	0.01629	123.01	107.89	1014.1	1122.0	0.1984	1.8894
150	3.718	0.01634	97.07	117.89	1008.2	1126.1	0.2149	1.8685
160°	4.741	0.01639	77.29	127.89	1002.3	1130.2	0.2311	1.8485
170	5.992	0.01645	62.06	137.90	996.3	1134.2	0.2472	1.8293
180	7.510	0.01651	50.23	147.92	990.2	1138.1	0.2630	1.8109
190	9.339	0.01657	40.96	157.95	984.1	1142.0	0.2785	1.7932
200	11.526	0.01663	33.64	167.99	977.9	1145.9	0·2938	1.7762
210°	14.123	0.01670	27.82	178.05	971.6	1149.7	0.3090	1.7598
212	14.696	0.01672	26.80	180.07	970.3	1150.4	0.3120	1.7566
220	17.186	0.01677	23.15	188.13	965.2	1153.4	0.3239	1.7440
230	20.780	0.01684	19.382	198.23	958.8	1157.0	0.3387	1.7288
240	24.969	0.01692	16.323	208.34	952.2	1160.5	0.3531	1.7140
250°	29.825	0.01700	13.821	218.48	945.5	1164.0	0.3675	1.6998
260	35.429	0.01709	11.763	228.64	938.7	1167.3	0.3817	1.6860
270	41.858	0.01717	10.061	238.84	931.8	1170.6	0.3958	1.6727
280	49.203	0.01726	8.645	249.06	924.7	1173.8	0.4096	1.6597
290	57.556	0.01735	7.461	259.31	917.5	1176.8	0.4234	1.6472
300°	67.013	0.01745	6.466	269.59	910.1	1179.7	0.4369	1.6350
310	77.68	0.01755	5.626	279.92	902.6	1182.5	0.4504	1.6231
320	89.66	0.01765	4.914	290.28	894.9	1185.2	0.4637	1.6115
330	103.06	0.01776	4.307	300.68	887.0	1187.7	0.4769	1.6002
340	118.01	0.01787	3.788	311.13	879.0	1190.1	0.4900	1.5891
350°	134.63	0.01799	3.342	321.63	870.7	1192.3	0.5029	1.5783
360	153.04	0.01811	2.957	332.18	862.2	1194.4	0.5158	1.5677
370	173.37	0.01823	2.625	342.79	853.5	1196.3	0.5286	1.5573
380	195.77	0.01836	2.335	353.45	844.6	1198.1	0.5413	1.5471
390	220.37	0.01850	2.0836	364.17	835.4	1199.6	0.5539	1.5371
400°	247.31	0.01864	1.8633	374.97	826.0	1201.0	0.5664	1.5272
410	276.75	0.01878	1.6700	385.83	816.3	1202.1	0.5788	1.5174
420	308.83	0.01894	1.5000	396.77	806.3	1203.1	0.5912	1.5078
430	343.72	0.01910	1.3499	407.79	796.0	1203.8	0.6035	1.4982
440	381.59	0.01926	1.2171	418.90	785.4	1204.3	0.6158	1.4887
450°	422.6	0.0194	1.0993	430.1	774.5	1204.6	0.6280	1.4793
460	466.9	0.0196	0.9944	441.4	763.2	1204.6	0.6402	1.4700
470	514.7	0.0198	0.9009	452.8	751.5	1204.3	0.6523	1.4606
480	566.1	0.0200	0.8172	464.4	739.4	1203.7	0.6645	1.4513
490	621.4	0.0202	0.7423	476.0	726.8	1202.8	0.6766	1.4419
500°	680.8	0.0204	0.6749	487.8	713.9	1201.7	0.6887	1.4325
520	812.4	0.0209	0.5594	511.9	686.4	1198.2	0.7130	1.4136
540	962.5	0.0215	0.4649	536.6	656.6	1193.2	0.7374	1.3942
560	1133.1	0.0221	0.3868	562.2	624.2	1186.4	0.7621	1.3742
580	1325.8	0.0228	0.3217	588.9	588.4	1177.3	0.7872	1.3532
600°	1542.9	0.0236	0.2668	617.0	548.5	1165.5	0.8131	1.3307
620	1786.6	0.0247	0.2201	646.7	503.6	1150.3	0.8398	1.3062
640	2059.7	0.0260	0.1798	678.6	452.0	1130.5	0.8679	1.2789
660	2365.4	0.0278	0.1442	714.2	390.2	1104.4	0.8987	1.2472
680	2708.1	0.0305	0.1115	757.3	309.9	1067.2	0.9351	1.2071
700°	3093.7	0.0369	0.0761	823.3	172.1	995.4	0.9905	1.1389
705.4	3206.2	0.0503	0.0503	902.7	0	902.7	1.0580	1.0580

Table 2-3 Saturated Steam: Pressure Table

Absolute Press. Lb per Sq in. p	Temperature F t	Specific Volume Sat. Liquid v_f	Specific Volume Sat. Vapor v_g	Enthalpy Sat. Liquid h_f	Enthalpy Evap. h_{fg}	Enthalpy Sat. Vapor h_g	Entropy Sat. Liquid s_f	Entropy Sat. Vapor s_g
1.0	101.74	0.01614	333.6	69.70	1036.3	1106.0	0.1326	1.9782
2.0	126.08	0.01623	173.73	93.99	1022.2	1116.2	0.1749	1.9200
3.0	141.48	0.01630	118.71	109.37	1013.2	1122.6	0.2008	1.8863
4.0	152.97	0.01636	90.63	120.86	1006.4	1127.3	0.2198	1.8625
5.0	162.24	0.01640	73.52	130.13	1001.0	1131.1	0.2347	1.8441
6.0	170.06	0.01645	61.98	137.96	996.2	1134.2	0.2472	1.8292
7.0	176.85	0.01649	53.64	144.76	992.1	1136.9	0.2581	1.8167
8.0	182.86	0.01653	47.34	150.79	988.5	1139.3	0.2674	1.8057
9.0	188.28	0.01656	42.40	156.22	985.2	1141.4	0.2759	1.7962
10	193.21	0.01659	38.42	161.17	982.1	1143.3	0.2835	1.7876
11	197.75	0.01662	35.14	165.73	979.3	1145.0	0.2903	1.7800
12	201.96	0.01665	32.40	169.96	976.6	1146.6	0.2967	1.7730
13	205.88	0.01667	30.06	173.91	974.2	1148.1	0.3027	1.7665
14	209.56	0.01670	28.04	177.61	971.9	1149.5	0.3083	1.7605
14.696	212.00	0.01672	26.80	180.07	970.3	1150.4	0.3120	1.7566
15	213.03	0.01672	26.29	181.11	969.7	1150.8	0.3135	1.7549
16	216.32	0.01674	24.75	184.42	967.6	1152.0	0.3184	1.7497
17	219.44	0.01677	23.39	187.56	965.5	1153.1	0.3231	1.7449
18	222.41	0.01679	22.17	190.56	963.6	1154.2	0.3275	1.7403
19	225.24	0.01681	21.08	193.42	961.9	1155.3	0.3317	1.7360
20	227.96	0.01683	20.089	196.16	960.1	1156.3	0.3356	1.7319
21	230.57	0.01685	19.192	198.79	958.4	1157.2	0.3395	1.7280
22	233.07	0.01687	18.375	201.33	956.8	1158.1	0.3431	1.7242
23	235.49	0.01689	17.627	203.78	955.2	1159.0	0.3466	1.7206
24	237.82	0.01691	16.938	206.14	953.7	1159.8	0.3500	1.7172
25	240.07	0.01692	16.303	208.42	952.1	1160.6	0.3533	1.7139
26	242.25	0.01694	15.715	210.62	950.7	1161.3	0.3564	1.7108
27	244.36	0.01696	15.170	212.75	949.3	1162.0	0.3594	1.7078
28	246.41	0.01698	14.663	214.83	947.9	1162.7	0.3623	1.7048
29	248.40	0.01699	14.189	216.86	946.5	1163.4	0.3652	1.7020
30	250.33	0.01701	13.746	218.82	945.3	1164.1	0.3680	1.6993
35	259.28	0.01708	11.898	227.91	939.2	1167.1	0.3807	1.6870
40	267.25	0.01715	10.498	236.03	933.7	1169.7	0.3919	1.6763
45	274.44	0.01721	9.401	243.36	928.6	1172.0	0.4019	1.6669
50	281.01	0.01727	8.515	250.09	924.0	1174.1	0.4110	1.6585
55	287.07	0.01732	7.787	256.30	919.6	1175.9	0.4193	1.6509
60	292.71	0.01738	7.175	262.09	915.5	1177.6	0.4270	1.6438
65	297.97	0.01743	6.655	267.50	911.6	1179.1	0.4342	1.6374
70	302.92	0.01748	6.206	272.61	907.9	1180.6	0.4409	1.6315
75	307.60	0.01753	5.816	277.43	904.5	1181.9	0.4472	1.6259
80	312.03	0.01757	5.472	282.02	901.1	1183.1	0.4531	1.6207
85	316.25	0.01761	5.168	285.39	897.8	1184.2	0.4587	1.6158
90	320.27	0.01766	4.896	290.56	894.7	1185.3	0.4641	1.6112
95	324.12	0.01770	4.652	294.56	891.7	1186.2	0.4692	1.6068
100	327.81	0.01774	4.432	298.40	888.8	1187.2	0.4740	1.6026
110	334.77	0.01782	4.049	305.66	883.2	1188.9	0.4832	1.5948
120	341.25	0.01789	3.728	312.44	877.9	1190.4	0.4916	1.5878
130	347.32	0.01796	3.455	318.81	872.9	1191.7	0.4995	1.5812
140	353.02	0.01802	3.220	324.82	868.2	1193.0	0.5069	1.5751
150	358.42	0.01809	3.015	330.51	863.6	1194.1	0.5138	1.5694
160	363.53	0.01815	2.834	335.93	859.2	1195.1	0.5204	1.5640
170	368.41	0.01822	2.675	341.09	854.9	1196.0	0.5266	1.5590
180	373.06	0.01827	2.532	346.03	850.8	1196.9	0.5325	1.5542
190	377.51	0.01833	2.404	350.79	846.8	1197.6	0.5381	1.5497
200	381.79	0.01839	2.288	355.36	843.0	1198.4	0.5435	1.5453
250	400.95	0.01865	1.8438	376.00	825.1	1201.1	0.5675	1.5263
300	417.33	0.01890	1.5433	393.84	809.0	1202.8	0.5879	1.5104
400	444.59	0.0193	1.1613	424.0	780.5	1204.5	0.6214	1.4844
500	467.01	0.0197	0.9278	449.4	755.0	1204.4	0.6487	1.4634
600	486.21	0.0201	0.7698	471.6	731.6	1203.2	0.6720	1.4454
700	503.10	0.0205	0.6554	491.5	709.7	2101.2	0.6925	1.4296
800	518.23	0.0209	0.5687	509.7	688.9	1198.6	0.7108	1.4153
1000	544.61	0.0216	0.4456	542.4	649.4	1191.8	0.7430	1.3897
2000	635.82	0.0257	0.1878	671.7	463.4	1135.1	0.8619	1.2849
3000	695.36	0.0346	0.0858	802.5	217.8	1020.3	0.9731	1.1615
3206.2	705.40	0.0503	0.0503	902.7	0	902.7	1.0580	1.0580

Reprinted, by permission, from J. H. Keenan and F. G. Keyes, *Thermodynamic Properties of Steam* (New York: John Wiley & Sons, Inc., 1936).

Table 2-4 Superheated Steam: Temperature

Pressure (psia) Sat. Temp.		Sat. Vapor	140 F	180 F	220 F	260 F	300 F	400 F	500 F	600 F	700 F	800 F	1000 F
1.0	v	333.6	356.5	380.6	404.5	428.4	452.3	512.0	571.6	631.2	690.8	750.4	869.5
101.74 F	**h**	1106.0	1123.3	1141.4	1159.5	1177.6	1195.8	1241.7	1288.3	1335.7	1383.8	1432.8	1533.5
	s	1.9782	2.0081	2.0373	2.0647	2.0907	2.1153	2.1720	2.2233	2.2702	2.3137	2.3542	2.4283
2.0	v	173.7	177.96	190.04	2.201	214.1	226.0	255.9	285.7	315.5	345.4	375.1	434.7
126.08 F	**h**	1116.2	1122.6	1140.9	1159.1	1177.3	1195.6	1241.6	1288.2	1335.6	1383.8	1432.8	1533.5
	s	1.9200	1.9308	1.9603	1.9879	2.0140	2.0387	2.0955	2.1468	2.1938	2.2372	2.2778	2.3519
5.0	v	73.52		75.71	80.59	85.43	90.25	102.26	114.22	126.16	138.10	150.03	173.87
162.24 F	**h**	1131.1		1139.4	1158.1	1176.5	1195.0	1241.2	1288.0	1335.4	1383.6	1432.7	1533.4
	s	1.8441		1.8574	1.8857	1.9121	1.9370	1.9942	2.0456	2.0927	2.1361	2.1767	2.2509
10	v	38.42			40.09	42.56	45.00	51.04	57.05	63.03	69.01	83.32	96.58
193.21 F	**h**	1143.3			1156.2	1175.1	1193.9	1240.6	1287.5	1335.1	1383.4	1432.5	1533.3
	s	1.7876			1.8017	1.8341	1.8595	1.9172	1.9689	2.0160	2.0595	2.1118	2.1860
14.696	v	26.80			27.15	28.85	30.53	34.68	38.78	42.86	46.94	51.00	59.13
212.00 F	**h**	1150.4			1154.4	1173.8	1192.8	1239.9	1287.1	1334.8	1383.2	1432.3	1533.1
	s	1.7566			1.7624	1.7902	1.8160	1.8743	1.9261	1.9734	2.0170	2.0576	2.1319
20	v	20.09				21.11	22.36	25.43	28.46	31.47	34.47	37.46	43.44
227.96 F	**h**	1156.3				1172.2	1191.6	1239.2	1286.6	1334.4	1382.9	1432.1	1533.0
	s	1.7319				1.7545	1.7808	1.8396	1.8918	1.9392	1.9829	2.0235	2.0978
60	v	7.175					7.259	8.357	9.403	10.427	11.441	12.449	14.454
292.71 F	**h**	1177.6					1181.6	1233.6	1283.0	1331.8	1380.9	1430.5	1531.9
	s	1.6438					1.6492	1.7135	1.7678	1.8162	1.8605	1.9015	1.9762
100	v	4.432						4.937	5.589	6.218	6.835	7.446	8.656
327.81 F	**h**	1187.2						1227.6	1279.1	1329.1	1378.9	1428.9	1530.8
	s	1.6026						1.6518	1.7085	1.7581	1.8029	1.8443	1.9193
300	v	1.5433							1.7675	2.005	2.227	2.442	2.859
417.33 F	**h**	1202.8							1257.6	1314.7	1368.3	1420.6	1525.2
	s	1.5104							1.5701	1.6268	1.6751	1.7184	1.7954
500	v	0.9278							0.9927	1.1591	1.3044	1.4405	1.6996
467.01 F	**h**	1204.4							1231.3	1298.6	1357.0	1412.1	1519.6
	s	1.4634							1.4919	1.5588	1.6115	1.6571	1.7363

this range is unity), the heat added, equal to the increase in enthalpy, would be

$$mC_p(t_2 - t_1) = (1)(1)(212 - 32) = 180.0 \text{ Btu}$$

and this is the value for $\mathbf{h}_f$ at 212 F. The value 180.07 that appears in the tables is based on the Btu arbitrarily defined in terms of the pre-1956 International Steam Table Calorie.

As the specific heat of water is not unity over a large range, $\mathbf{h}_f$ should not be calculated but selected from the tables. For example, at 100 psia for saturated liquid, $\mathbf{h}_f = 298.33$ Btu/lb. For water at temperatures below 212 F, at any temperature t the enthalpy is very closely

$$\mathbf{h}_f = (1)(t - 32) \text{ Btu/lb} \tag{2-47}$$

If a pound of water at 212 F is heated at atmospheric pressure it will be found that the temperature remains constant, but that a large amount of energy from the heating source is utilized in changing this pound of water into vapor (steam). This energy required to transform 1 lb of water into 1 lb of steam is known as the *heat of vaporization*, or *latent heat*. At 212 F (14.696 psi pressure) it amounts to 970.3 Btu and is listed under the heading $\mathbf{h}_{fg}$ (Table 2-2). The enthalpy of each pound of dry saturated vapor referred to a 32 F datum is the sum of $\mathbf{h}_f + \mathbf{h}_{fg}$ and is called $\mathbf{h}_g$. At 212 F (14.696 psia), $\mathbf{h}_g = 180.07 + 970.3 = 1150.4$ Btu/lb.

During rapid boiling it often happens that some of the water is entrained in the steam delivered, in the form of fine droplets or mist. If it should happen that the steam carries 90% by weight of moisture it is obvious that there is 90% of dry steam present. This weight percentage of dry steam present is known as the *steam quality* and the customary symbol for it is x.

The enthalpy of wet steam at any quality x (expressed as a decimal) is

$$\mathbf{h}_x = \mathbf{h}_f + x\mathbf{h}_{fg} \text{ Btu/lb} \tag{2-48}$$

The quality x can range from 0.0 to 1.0.

The specific volume of a vapor or gas is affected by both pressure and temperature. In the case of a vapor or imperfect gas, tables of properties of the vapor must be used to find all values. In the case of the steam properties given in Tables 2-2 and 2-3, the column v_g represents the specific volume in cubic feet per pound of saturated steam at the corresponding pressure-temperature indicated. The symbol v_{fg} indicates the amount by which the volume of 1 lb of the medium increases in changing from liquid to dry saturated steam. The reciprocal of v_g, namely, $1/v_g$, represents the density (the weight, in pounds of 1 ft^3 of steam).

The specific volume of wet steam is

$$v_x = v_f + xv_{fg} = (1 - x)v_f + xv_g \text{ ft}^3\text{/lb} \tag{2-49}$$

or

$$v_x = xv_g \text{ ft}^3\text{/lb} \tag{2-50}$$

is used as a close approximation.

Example 2-7. (a) Find the enthalpy $\mathbf{h}$ of steam delivered from a boiler operating at 85.3 psig pressure if its quality is 95%. The barometric pressure is 14.7 psi. (b) Find also the specific volume of the steam delivered. (c) At what temperature does the steam and water mixture leave the boiler, and in what way does the 5% moisture affect the temperature?

Solution: (a) Boiler pressure = 85.3 + 14.7 = 100.0 psia. At this pressure, from Table 2-3,

$$\mathbf{h}_f = 298.4 \text{ Btu}$$

and

$$\mathbf{h}_{fg} = 888.8 \text{ Btu}$$

In this case all of 298.4 Btu has been supplied to bring the liquid up to saturation conditions above the datum of the steam tables, but only 95% of the heat of vaporization has been supplied. Consequently the enthalpy of this wet steam, $\mathbf{h}_x$, must be

$$\mathbf{h}_x = \mathbf{h}_f + x\mathbf{h}_{fg} = 298.4 + (0.95)(888.8) = 1142.8 \text{ Btu/lb} \qquad \textit{Ans.}$$

NOTE. Multiplying $x\mathbf{h}_g$ to find $\mathbf{h}_x$ is incorrect and must never be done.

(b) Similarly, the specific volume of the wet steam v_x can be found, using $v_x = v_f + xv_{fg}$ or an equivalent form:

$$v_x = v_f(1 - x) + xv_g = 0.01774(0.05) + 0.95(4.432) = 4.21 \text{ ft}^3\text{/lb} \qquad \textit{Ans.}$$

It is theoretically wrong to say $v_x = xv_g$, but for low-pressure steam, and at relatively high qualities, this approximation is often made.

(c) The temperature of the saturated steam–water mixture at 100 psia is 327.81 F (column 2) and is not affected by the quality variations. *Ans.*

If steam is heated until all traces of moisture are gone, its temperature will rise above the saturation temperature corresponding to the pressure. Such steam, at a temperature higher than saturation temperature, is called *superheated*. Superheated steam resembles a gas in its behavior and in some cases can be treated by using the simple gas laws, but for most calculations, except those at very low pressures, recourse should be had to complete tables of the properties of steam in the superheated state. Table 2-4 is an abridged table of superheated-steam properties and lists, for each pressure at various superheat temperatures, specific volume v, enthalpy $\mathbf{h}$, and entropy s.

2-9. TEMPERATURE-ENTROPY (*Ts*) DIAGRAM FOR WATER-STEAM

It will be noticed that the last two columns of Tables 2-2 and 2-3 list the entropy of saturated liquid and the entropy of saturated vapor. Similarly in Table 2-4, the entropy of superheated steam is included. The property entropy is extremely valuable for analyzing processes involving work. It has already been shown that processes taking place under isentropic (reversible and adiabatic) conditions produce optimum work and the entropy function when plotted on a *Ts* plane can show many important relationships to advantage. One important usage relates to the fact that areas on a *Ts* plane can represent the heat added or rejected during a process.

A temperature-entropy diagram for steam appears in Fig. 2-3 and because it is similar to *Ts* diagrams for most fluids, it will be described in detail. For areas to be in scale, absolute temperatures are required. Temperature is usually the ordinate with entropy the abscissa. The zero of entropy for plotting such diagrams can be arbitrarily selected, and in the case of water-steam, the zero value of entropy is taken at 32 F under a pressure of 0.08854 psia. The line from this datum point (A) up to the critical pressure point (C) is known as the saturated-liquid line while the line from C to F represents the saturated-vapor line. Along the saturated-liquid and saturated-vapor lines at any temperature there is a corresponding pressure of saturation. For example, at 212 F under a pressure of 14.696 psia, liquid will change into vapor (steam) and increase in volume when heat is added with the temperature remaining constant at 212 F until all of the liquid is converted to steam. If further heat is added to the water-free vapor, it will increase in temperature and become superheated to a higher temperature such as E. The saturated liquid-vapor line for 14.696 psia thus appears as BD and the area under BD—namely, $BDdb$—is a measure of the amount of heat required to change water to steam under steady-flow conditions, the $\mathbf{h}_{fg}$ value of Table 2-3. If the steam were under a higher pressure—say, 200 psia—vaporization of the liquid would not start until the temperature reached 381.7 F, as indicated on the chart.

It is a characteristic of most substances that as the saturation temperature and pressure are increased, the value of the heat of vaporization (latent heat) diminishes until at a point C, no latent heat is needed and the liquid and vapor phase become indistinguishable. This point is known as the critical point and represents the transition (maximum) point of the liquid-vapor line. For steam the critical pressure is 3206 psia and the temperature 705.4 F. Above its critical temperature, it is not possible to liquefy a vapor. This is true although at the usually high pressures and temperatures involved, it may not be possible to distinguish the liquid phase from the vapor phase so the statement may be of theoretical importance only. The critical temperature and pressure are most significant in the analysis of any vapor.

It is easily possible to take a saturated liquid, such as water at 212 F and 14.696 psia, and compress it to a higher pressure, such as 200 psia or even 5000 psia. Such a liquid is classified as a compressed liquid and although its specific volume is not significantly altered by the higher pressure, the state point of such a compressed liquid does not lie precisely on the saturation line but is to the left of the saturated liquid line and the region to the left of the line ABC is known as the compressed-liquid region. A representative line has been drawn in for 4000 psia in Fig. 2-3. Below the critical temperature it is customary to speak of a liquid as being compressed or saturated. Above the critical temperature, it is customary to speak of the medium as compressed vapor or superheated vapor (superheated steam in the case of water).

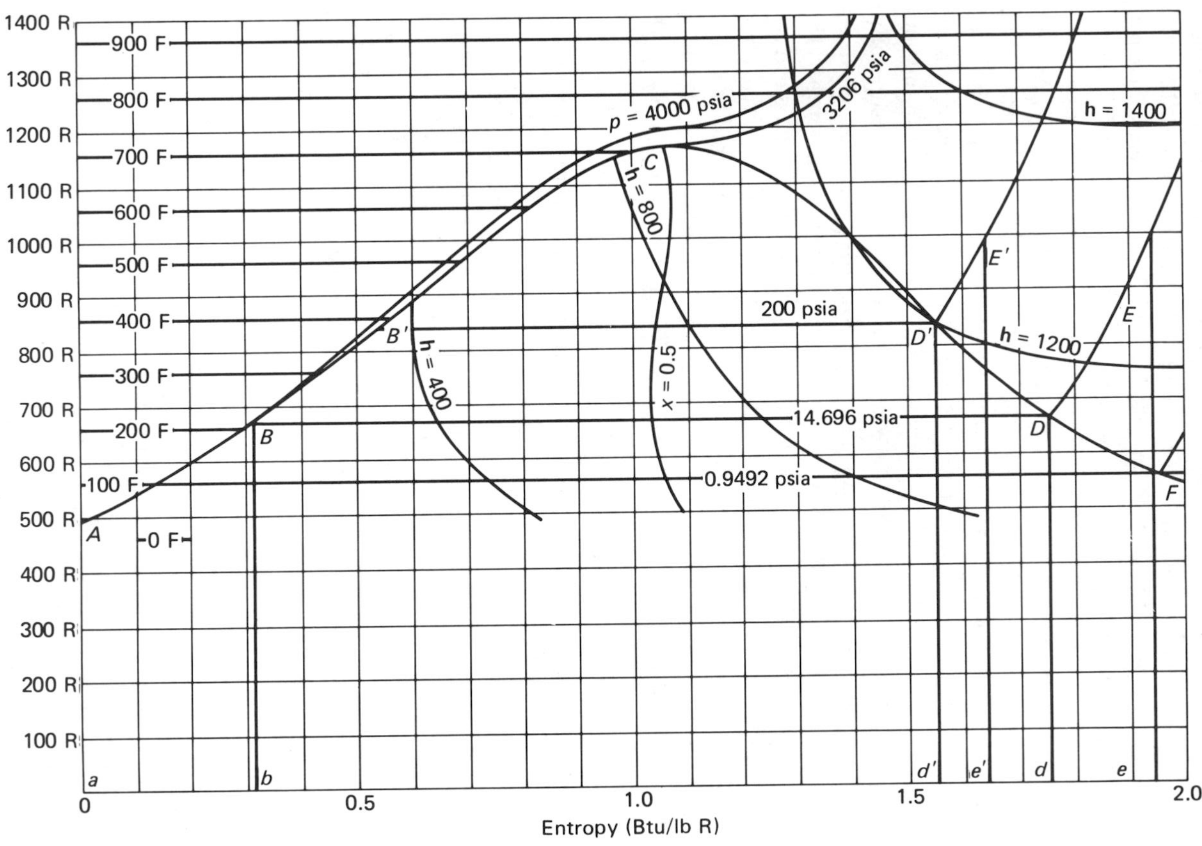

Figure 2-3. Temperature-entropy (Ts) diagram for steam.

The lines of constant enthalpy in the superheat region swing down and to the right, tending to level off as the steam becomes far removed from its saturated-vapor state.

The area under the curve *DE* (i.e., *EedD*) naturally represents the heat required to superheat steam from the temperature *D* to the temperature *E*. The area under the curve *AB* represents the heat required to warm liquid (water) along the saturation curve from 32 F to 212 F. Similar areas are applicable for other pressures such as those under *AB'*, *B'D'*, and *D'E'* for a pressure of 200 psia.

2-10. GAS MIXTURES

Mixtures of gases are so common that methods for computing the properties of gaseous mixtures are most necessary. The mol concept, which has been mentioned earlier, is particularly useful. A *mol* of a chemical compound or element is a mass (weight) of that compound equal to its molecular weight expressed in pounds (grams in the metric system). Further, the volume occupied by a mol of any essentially perfect gas is the same at any specified temperature and pressure (379.6 ft^3 at 60 F and 14.7 psia). Thus it follows that the density of a gas is closely proportional to the molecular weight of that particular gas. The constancy of mol volume should be recognized as only a close approximation, for except at low pressures no real gas satisfies either Avogadro's law or the perfect-gas equation. It should also be noted that the gas constant, 1545.3 ft lbf/lb-mol · F, used earlier in this chapter, is a mol value and applies to a gas in any range where the gas approaches perfect-gas behavior.

In SI, the mol volume at one standard atmosphere = 1.01325 bar = 101 325 *Pa* (N/m^2) and at 0°C (273.16 K) is 22.40 $m^3/kg \cdot mol$. It is easily possible to determine the mol volume at other temperatures and pressure by simple multiplication of the basic value by the proper temperature and pressure ratios. For example, the mol volume at 20°C and 1 atm_{std} pressure is

$$22.40 \times \frac{273.16 + 20}{273.16} \times \frac{1 \text{ atm}_{std}}{1 \text{ atm}} = 24.04 \text{ m}^3$$

In gaseous mixtures at low and moderate pressures, the respective gases in the mixture frequently resemble perfect gases in their behavior and also follow a further generalization known as the Gibbs–Dalton law of partial pressures. This law states: *in a given mixture of gases or vapors each gas or vapor exerts the same pressure it would exert if it occurred alone in the same space and at the same temperature as exists in the mixture.* Partial pressures, volumetric percentages, and molal concentrations are directly related to each other in gaseous mixtures which obey the Gibbs–Dalton law.

Example 2-8. A natural gas consists of 95.6% methane (CH_4), 3.8% nitrogen (N_2), and 0.6% water vapor (H_2O), with each constituent measured by volume. The gas is under a pressure of 15 psia and at 60 F. Compute the partial pressure of each constituent in the mixture, the mol concentrations, the weight analysis, the molecular weight, and the density.

Solution: It is most convenient for this type of problem to use a solution in tabular form as shown on the following page.

In the table, note that volumetric percentages (column 2) and mol concentrations (column 3) are the same. To change to a weight analysis, multiply each volumetric or mol fraction by its respective molecular weight and find what fraction each resulting product is of the total. At the bottom of column 5 is indicated a composite molecular weight of the gas, and 1 mole of gas can be said to weigh 16.46 lb. This value is found from the representative total of column 6. Use this molecular weight with Eq. (2-24) find the density:

Constituent	Volumetric Percentages	mols/100 mols	Partial Pressure $p_T \times \frac{\text{col. 2}}{100}$ (psi)	Molecular Weight	Col. 2 × Col. 5	Weight Percentage
1	2	3	4	5	6	7
Methane	95.6	95.6	14.34	16	1529	92.89
Nitrogen	3.8	3.8	0.57	28	106	6.44
Water vapor	0.6	0.6	0.09	18	11	0.67
TOTAL	100.0	100.0	15.00	(16.46)	1646	100.00

$$(144)(15)(1) = m\left(\frac{1545.3}{16.46}\right)(460 + 60)$$

$$m = 0.0442 \text{ lb/ft}^3 \qquad \textit{Ans.}$$

Alternate Solution: An alternate solution would employ the mol volume. Thus at 14.7 psia,

$$v = \frac{379.6}{16.46} = 23.06 \text{ ft}^3\text{/lb}$$

and

$$\text{density} = \frac{1}{v} = \frac{1}{23.06} = 0.0435 \text{ lb/ft}^3$$

At a constant temperature the density of a gas is directly proportional to pressure. Thus at 60 F and 15 psia,

$$\text{density} = 0.0435\left(\frac{15}{14.7}\right) = 0.0442 \text{ lb/ft}^3 \qquad \textit{Ans.}$$

In the event that the analysis of a gas is given with the constituents expressed in weight percentages, a transformation can be made to a volumetric (mol) concentration basis by dividing each constituent by its appropriate molecular weight. After adding the resulting quotients, the volumetric analysis can be found by using each constituent quotient as a fraction of the total.

PROBLEMS

2-1. In SI, the newton is the unit of force, the meter is the unit of length, and 1 newton meter is a joule. The watt, in turn, is defined as 1 joule per second. (a) Show that 1 kWh is the equivalent of 3.6 MJ, i.e., 3.6×10^6 J. (b) Find the number of foot-pounds in a killowatt-hour making use of the equivalents 1 N = 0.22481 lbf and 1 m = 3.28084 ft.

Ans. (b) 2.6552×10^6 ft lbf

2-2. During an extended period from 1936 to 1956 use was made of a unit known as the International Steam Table kilocalorie, defined as $\frac{1}{860}$ kWh. On the basis of this early definition find the value of this original International Steam Table kilocalorie in joules.

Ans. 4186.1 J

2-3. A kilocalorie can be related to 1 kg of mass and 1 K (or 1°C) while the Btu can be related to 1 lb of mass and 1 F. Since 1 kg = 2.2046 lb and 1 K = 1.8 F, find the relation of the kilocalorie to the Btu.

Ans. 3.9683

2-4. A blower fan takes standard atmospheric air from an open inlet, raises it to a pressure of 1.5 in. Hg (1.5 inches of mercury) at 68 F, and delivers it into the discharge duct at 2000 fpm (feet per minute). If 10 000 cfm (cubic feet per minute) measured at inlet conditions are delivered, compute the minimum air horsepower required to compress and deliver the air. Refer to Table 1-2 or A-3 for data on mercury density or pressure equivalents. The specific volume of standard dry air at 68 F is 13.3 ft^3/lb.

Ans. 32.4 hp

2-5. A blower fan takes 20°C standard atmospheric air from an open inlet, raises it to a pressure of 38.1 mm Hg (millimeters of mercury measured at 20°C), and delivers it into a discharge duct at 10.16 m/s (meters per second). If 283.2 m^3/min of air measured at inlet conditions are delivered, compute the minimum air kilowatts required to compress and deliver the air. Refer to Table A-3 for pressure equivalents. The specific volume of standard dry air at 20°C is 0.8299 m^3/kg.

Ans. 24.2 kW

2-6. In an industrial refrigeration plant 1000 lb of brine per minute are cooled from 25 F (−3.9°C) to 14 F (−10°C). The specific heat of the brine is 0.748 Btu/lb · F (0.748 kcal/kg · °C). (a) Compute the heat removal expressed as tons of refrigeration. (b) Compute the heat removal in kilowatts (c) Express the specific heat of this brine in units of kilojoules per kilogram degree Kelvin.

Ans. (a) 41.1; (b) 144 kW; (c) 3.12

2-7. A heavy steel tank weighing 300 lbs contains 3 ft^3 of hydrogen at 40 F (4.44°C) at a pressure of 500 psia. The tank warms in the ambient air to a temperature of 77 F (25°C) with no significant change in its volume. (a) Compute the hydrogen pressure in the tank at 77 F. (b) Find out how much heat flow from the atmosphere took place in warming the tank and its contents from 40 F to 77 F. This should be considered a constant-volume process.

Ans. (a) 537 psia; (b) 1383 Btu

2-8. Work Problem 2-7 in SI units. Express the answers in pascals (newtons per square meter) and in kilowatt-hours.

2-9. A certain dwelling has inside dimensions of 56 × 31 ft with a height of 9.2 ft. It was found by test that on a certain windy day 2 complete air changes per hour took place in the dwelling, measured at inside conditions of 77 F (25°C). (a) Considering the house as a rectangular box and disregarding furniture and other fillers, find the mass of outside air which moves into and out of the house each hour. (b) How much heat is required to warm this air from an outside temperature of 32 F (0°C) to 77 F?

Ans. (a) 2360 lb/h; (b) 25 920 Btuh

2-10. Work Problem 2-9 completely in SI (metric) units if the dimensions of the bouse are 17.06 × 9.45 × 2.80 m.

Ans. (a) 1070 kg; (b) 7.59 kW

2-11. Air at 147.0 psia rapidly expands in a power turbine to atmospheric pressure of 14.7 psia. If the air is initially at 70.3 F (530 R), at what temperature will it leave the turbine if the expansion is considered to be reversibly adiabatic?

Ans. 274.4 R

2-12. Air at 1 013 250 N/m^2 rapidly expands in a power turbine to atmospheric pressure of 101 325 N/m^2 (Pa). If the air is initially at 21.28°C (294.44 K), at what temperature will it leave the turbine if the expansion is considered to be reversibly adiabatic.

Ans. 152.4 K

2-13. Dry air is compressed isothermally at 530 R from 14.7 to 147.0 psia. (a) Compute the work required and the heat rejected. (b) Compute the change in entropy which takes place.

Ans. (a) 83.5 Btu/lb; (b) −0.1576 Btu/lb · R

2-14. Dry air expands isothermally from 105 to 15 psia at 530 R. (a) Compute the work produced per pound of air and the heat added during the expansion. (b) Compute the entropy change and find the heat addition by making a $T\Delta s$ computation.

2-15. (a) How much energy must be added to 10 lb of ice at 25 F to change it to water and warm it to 70 F? (b) What is the enthalpy of the 70 F water? (c) What additional amount of energy is required to warm this 70 F water and change it into dry saturated steam at 212 F and 14.7 psi?

Ans. (a) 1847.1 Btu; (b) 38.0 Btu/lb; (c) 11 124 Btu

2-16. Find the specific volume and enthalpy of a pound of steam at a pressure of 200 psia and 95% quality.

Ans. 2.17, 1156.2

2-17. Twelve tons of ice from an indoor skating rink are disposed of by using an atmospheric-pressure steam blast. Assume that the ice is at 26 F at the start of the process and that the disposed water reaches a temperature of 40 F before disposal to the sewer drains. Dry saturated steam at 14.7 psia condenses and cools to mix with the drain water at 40 F. How many pounds of steam are required to warm, melt, and dispose of the ice water, if 20% of the useful enthalpy associated with the steam and its subcooled water is lost to the atmosphere?

Ans. 4080 lb

2-18. Water at 300 psia in a saturated state enters a drain trap serving a steam line, and it drops to atmospheric pressure by the time it enters the trap discharge pipe. (a) Find the final temperature of the discharged water-steam and (b) compute what fraction of each original pound of water flashes to steam. Note that in a process of this type the enthalpy stays constant, so the enthalpy of the original water equals the enthalpy [Eq. (2-48)] of the discharged mixture of 14.7 psia.

Ans. (a) 212 F; (b) 0.22

2-19. Water at 100 psia in a saturated state enters a drain trap serving a steam line and drops to atmospheric pressure by the time it enters the trap discharge pipe. (a) Find the final temperature of the discharged water-steam and (b) compute what fraction of each original pound of water flashes to steam. Note that in a process of this type the enthalpy stays constant, so that the enthalpy of the original water equals the enthalpy [Eq. (2-48)] of the discharged mixture at 14.7 psia.

Ans. (a) 212 F; (b) 0.122

2-20. By volume analysis, the dry flue gases from a furnace consist, in percentages, of CO_2, 11; O_2, 8; CO, 1; and N_2, 80. Compute (a) the weight analysis of the flue gas, (b) its molecular weight, and (c) its density measured at 68 F and 14.7 psia.

Ans. (a) 14.9, 8.6, 0.9, 75.6; (b) 29.64; (c) 0.0768 lb/ft^3

2-21. The wet flue gases from the combustion of a gas show a volumetric analysis, in percentages, of CO_2, 8.5; O_2, 0.5; CO, 0.9; H_2O, 18.9; and N_2, 71.2. Compute (a) the weight analysis of the wet flue gas, (b) its molecular weight, and (c) its density at 300 F and 14.7 psia.

Ans. (a) 13.6, 0.6, 0.9, 12.4, 72.5; (b) 27.49; (c) 0.0496

2-22. If the flue gas of Problem 2-21 is cooled to 70 F, most of the water vapor condenses and the resulting dry analysis by volume is, in percentages, CO_2, 10.5; O_2, 0.6; CO, 1.1; and N_2, 87.8. Assume that this dry gas is reheated to 300 F at 14.7 psia and compute (a) the weight analysis, (b) the molecular weight, and (c) the density.

Ans. (a) 15.6, 0.6, 1.1, 82.7; (b) 29.7; (c) 0.0536 lb/ft^3

2-23. Nitrogen gas in an ammonia synthesizing plant is compressed from a storage tank at 100.3 psig at 70 F to a pressure of 1135.3 psig. Assume that the actual compression is approximated by the conditions of a reversible polytropic for which the exponent n is 1.3. Compute (a) specific volume of the gas at the start of compression, (b) temperature of the gas leaving the compressor, and (c) work required to carry out the compression of each pound of gas. Note that equations of the form (2-41) and (2-38) are applicable with use being made of the appropriate n value.

Ans. (a) 1.696; 902 R; 109.7 Btu

2-24. Dry air is supplied an expander turbine at 20 atm (294 psia) and expands to 2 atm (29.4 psia). Assume that the expansion is reversibly adiabatic. (a) Compute the temperature at exit from the expander turbine if air enters it at 70 F (530 R). (b) Compute the power produced if 15 lb/min flow through the expander. (c) If 75% of this power is actually delivered at the turbine shaft, make use of an energy balance to find the actual temperature of the air leaving the turbine.

Ans. (a) 274.6 R, (b) 920 Btu/min; (c) 338.3 R

REFERENCES

J. H. Keenan and J. Kaye, *Gas Tables* (New York: John Wiley and Sons, 1948).

F. Din, *Thermodynamic Functions of Gases, Vol. 2 for Air* (London: Butterworth's Scientific Publications, 1956).

E. F. Obert and R. A. Gaggioli, *Thermodynamics*, 2d ed. (New York: McGraw-Hill Book Co., 1963), pp. 219–227.

B. F. Dodge, *Chemical Engineering Thermodynamics* (New York: McGraw-Hill Book Co., 1944), pp. 159–187.

CHAPTER

3

AIR AND HUMIDITY CALCULATIONS

3-1. ATMOSPHERIC AIR

Air as found in the earth's atmosphere is a mixture of constituent gases. Even when uncontaminated, its composition varies slightly at different points of the globe and at different altitudes. However, for scientific purposes it became necessary to define a composition that can be called standard air. The composition of standard air which is commonly accepted is listed in the publication entitled *U.S. Standard Atmosphere* (Ref. 1). It is reproduced here in the following tabulation with the values given in percentage composition by volume or equivalently as mols per 100 mols. In addition to the gases listed by name, there are trace amounts of other gases such as methane, sulfur dioxide, hydrogen, krypton, and xenon. These vary in composition and are simply listed in the table as other gases.

The apparent molecular weight of standard dry air is 28.966.

The items listed in the table are the constituents of dry air. However, most air is moist and has water vapor associated with it. Moist air is defined as a binary mixture of dry air and water vapor (steam) with the amount of steam varying from zero to the condition of saturation, that is,

Standard Air

Nitrogen	78.084000
Oxygen	20.947600
Argon	0.934000
Carbon dioxide	0.031400
Neon	0.001818
Helium	0.000524
Other gases	0.000658
	100.000000

with the water vapor in the mixture capable of coexisting in neutral equilibrium with a flat surface of liquid or solid water. Saturation denotes the maximum amount of steam that can exist in a cubic foot of space at any given temperature and is essentially independent of the weight and pressure of the air which may simultaneously exist in the same space. For example, if water at saturation temperature were sprayed into a saturated air mixture, the water would remain in liquid phase and not evaporate. The water would either fall in the space or remain suspended as a mist or fog. Further, if the steam-saturated air were cooled in the slightest degree, it would be impossible for this existing amount of moisture to remain as steam, and the excess would condense out in the form of fog or dew. At the new, lower temperature, a different and lower saturation pressure must exist.

Saturation conditions for steam, and the corresponding maximum weight of steam that can exist in a space, are not directly predictable and must therefore be found from tabulations of the properties of steam or of the properties of steam and air mixtures. Table 3-1 in this text can be employed. It should also be mentioned that a more extensive and complete tabulation of moist-air properties has been prepared by J. A. Goff and S. Gratch (Ref. 5). Notice that moist air tabulations are based on a fixed total (barometric) pressure, usually that of the standard atmosphere and for other pressures, use must be made of appropriate correction factors.

3-2. MOIST-AIR TERMINOLOGY

Steam-air mixtures do not exactly follow the perfect-gas laws but for total pressures up to some three atmospheres, sufficient accuracy is possible to permit the use of these laws in engineering computations for such mixtures. This approach will first be developed, following which a more scientific analysis of the subject will be indicated.

In atmospheric air-steam mixtures the Gibbs–Dalton law(Section 2-10) is closely obeyed. This means that in any gas mixture the total pressure exerted is the summation of the partial pressures exerted independently by each of the constituent gases. Atmospheric air exists at a total pressure equal to barometric pressure (p_B), and this pressure in turn is made up of the partial pressures exerted by all the gases, mainly the nitrogen, p_{N2}; the oxygen, p_{O2}; and the water vapor (steam), p. Or, in mathematical terms,

$$p_B = p_{N2} + p_{O2} + p = p_a + p \tag{3-1}$$

In this equation, since there is no need of separating the pressures of the nitrogen and oxygen, it is customary to list the total barometric pressure as the sum of the pressures of the air portion (p_a) and the steam portion (p). The nitrogen and oxygen together are frequently called "dry air," even though some steam is usually mixed with the air.

By referring to Table 3-1 it will be seen that at 80 F, saturated steam exerts a pressure of 1.0323 in. Hg and weighs 1/633.0, or 0.001579 lb/ft^3. These values are essentially correct whether air exists in the same cubic foot or steam alone is present. For this low-pressure steam, even when it is saturated, the gas equation $pV = mRT$ holds closely. For steam $R = 1545.3/18.016 = 85.77$ ft · lbf/lb · R (461.2 J/kg · K), and by using this value and the figures just given, it can be seen that the equivalence is close. Thus

$$(144)(0.4912)(1.0323)(1) = (0.001579)(85.77)(459.7 + 80)$$

or

$$73.02 \equiv 73.09$$

It happens that over the range used for air-steam mixtures, values of $R = 85.6$ (460.3 for SI units) give the most consistent results and therefore should be used in the solution of problems.

The water vapor (steam) mixed with dry air in the atmosphere is known as *humidity*. The weight of water vapor per unit volume is really *vapor density* (d) and in this text will be called by no other name although in the technical literature various names are used. Conventional units for vapor density are pounds per cubic foot, or in SI, grams per cubic meter. The weight of water vapor associated with unit weight of dry-air constituents is called *humidity ratio* or *specific humidity*. Conventional units are pounds (or millipounds) of water vapor per pound of dry air constituents or, in SI, grams of water vapor per kilogram of dry-air constituents. Humidity ratio employs the symbol W, or W_s for air saturated with water vapor.

Relative humidity (ϕ) by definition is the mol fraction of water vapor (x_w) in moist air to the mol fraction (x_{ws}) of water vapor in saturated air at the same temperature (t) and total pressure (p_T)

$$\phi = \left.\frac{x_w}{x_{ws}}\right)_{t,\,p_T} \tag{3-2}$$

Under the conditions where the perfect gas laws and the Gibbs–Dalton relationships hold, the above definition is equivalent to one based on partial pressures and also on densities, namely: *relative humidity* is the ratio of the

Table 3-1 Thermodynamic Properties of Air, Water, and Steam (from 32 to 200 F and from 32 to −120 F)

	Properties of Water and Steam				Properties of Dry Air at a Pressure of 29.921 in. Hg abs		Properties of Mixture of Dry Air and Sat. Steam at a Total Pressure of 29.921 in. Hg abs		
		Enthalpy							
Temperature (F) t	Saturation Pressure of Water and Steam (in. Hg) p_s	Saturated Water (Btu/lb) $\mathbf{h}_f$	Saturated Steam (Btu/lb) $\mathbf{h}_g$	Specific Volume of Sat. Steam (ft³/lb) v_g	True Specific Volume (ft³/lb) v_a	Enthalpy (Btu/lb) $\mathbf{h}_a$	Volume of Mixture per lb of Dry Air (ft³) v_s	Enthalpy of Mixture per lb of Dry Air (Btu) $\mathbf{h}_s$	Humidity Ratio (mlb/lb dry air, or g/kg dry air) W_s
32	0.1803	0.0	1075.2	3305	12.389	7.69	12.46	11.75	3.771
33	0.1878	1.0	1075.6	3180	12.414	7.93	12.49	12.16	3.927
34	0.1955	2.0	1076.0	3062	12.439	8.17	12.52	12.57	4.090
35	0.2034	3.0	1076.5	2948	12.464	8.41	12.55	13.00	4.257
36	0.2117	4.0	1076.9	2839	12.490	8.65	12.58	13.42	4.431
37	0.2202	5.0	1077.4	2734	12.515	8.89	12.61	13.86	4.611
38	0.2290	6.0	1077.8	2634	12.540	9.13	12.64	14.30	4.797
39	0.2382	7.0	1078.2	2538	12.565	9.37	12.67	14.75	4.991
40	0.2477	8.0	1078.7	2445	12.591	9.61	12.70	15.21	5.191
41	0.2575	9.0	1079.1	2357	12.616	9.85	12.73	15.68	5.400
42	0.2676	10.1	1079.5	2272	12.641	10.09	12.76	16.16	5.614
43	0.2781	11.1	1080.0	2190	12.667	10.34	12.79	16.64	5.837
44	0.2890	12.1	1080.4	2112	12.692	10.58	12.82	17.13	6.067
45	0.3002	13.1	1080.9	2037	12.717	10.82	12.85	17.63	6.306
46	0.3119	14.1	1081.3	1965	12.742	11.06	12.88	18.13	6.551
47	0.3239	15.1	1081.7	1896	12.768	11.30	12.91	18.66	6.807
48	0.3363	16.1	1082.2	1829	12.793	11.54	12.94	19.19	7.073
49	0.3491	17.1	1082.6	1766	12.818	11.78	12.97	19.73	7.346
50	0.3624	18.1	1083.1	1704	12.844	12.02	13.00	20.28	7.628
51	0.3761	19.1	1083.5	1645	12.869	12.26	13.03	20.84	7.920
52	0.3903	20.1	1083.9	1589	12.894	12.50	13.06	21.41	8.223
53	0.4049	21.1	1084.4	1534	12.919	12.74	13.10	21.99	8.536
54	0.4200	22.1	1084.8	1482	12.945	12.98	13.13	22.59	8.859
55	0.4356	23.1	1085.2	1431	12.970	13.22	13.16	23.20	9.194

56	0.4518	24.1	1085.7	1383	12.995	13.46	13.19	23.82	9.540
57	0.4684	25.1	1086.1	1336	13.020	13.70	13.23	24.45	9.897
58	0.4856	26.1	1086.5	1292	13.046	13.94	13.26	25.10	10.27
59	0.5033	27.1	1087.0	1249	13.071	14.18	13.29	25.76	10.65
60	0.5216	28.1	1087.4	1207	13.096	14.42	13.33	26.43	11.04
61	0.5405	29.1	1087.9	1167	13.122	14.66	13.36	27.11	11.45
62	0.5599	30.1	1088.3	1129	13.147	14.90	13.40	27.82	11.87
63	0.5800	31.1	1088.7	1092	13.172	15.14	13.43	28.54	12.30
64	0.6007	32.1	1089.2	1056	13.197	15.38	13.47	29.27	12.75
65	0.6221	33.1	1089.6	1022	13.223	15.62	13.50	30.03	13.22
66	0.6441	34.1	1090.0	988.6	13.248	15.86	13.54	30.79	13.69
67	0.6668	35.1	1090.5	956.8	13.273	16.10	13.58	31.58	14.19
68	0.6902	36.1	1090.9	926.1	13.298	16.35	13.61	32.38	14.70
69	0.7143	37.1	1091.3	896.5	13.324	16.59	13.65	33.20	15.23
70	0.7392	38.1	1091.8	868.0	13.349	16.83	13.69	34.04	15.77
71	0.7648	39.1	1092.2	840.5	13.374	17.07	13.72	34.90	16.33
72	0.7911	40.1	1092.6	814.0	13.399	17.31	13.76	35.79	16.91
73	0.8183	41.1	1093.1	788.4	13.425	17.55	13.80	36.69	17.51
74	0.8463	42.1	1093.5	763.8	13.450	17.79	13.84	37.61	18.13
75	0.8751	43.1	1093.9	740.0	13.475	18.03	13.88	38.55	18.76
76	0.9047	44.1	1094.4	717.0	13.501	18.27	13.92	39.52	19.41
77	0.9352	45.1	1094.8	694.9	13.526	18.51	13.96	40.51	20.09
78	0.9667	46.1	1095.2	673.5	13.551	18.75	14.00	41.52	20.79
79	0.9990	47.1	1095.7	652.9	13.576	18.99	14.04	42.56	21.51
80	1.0323	48.1	1096.1	633.0	13.602	19.23	14.09	43.63	22.26
81	1.0665	49.1	1096.6	613.8	13.627	19.47	14.13	44.72	23.02
82	1.1017	50.1	1097.0	595.3	13.652	19.71	14.17	45.84	23.81
83	1.1380	51.1	1097.4	577.4	13.678	19.95	14.22	46.98	24.63
84	1.1752	52.1	1097.8	560.1	13.703	20.19	14.26	48.16	25.47
85	1.2136	53.1	1098.3	543.3	13.738	20.43	14.31	49.36	26.34
86	1.2530	54.0	1098.7	527.2	13.753	20.67	14.35	50.59	27.23
87	1.2935	55.0	1099.1	511.6	13.778	20.91	14.40	51.86	28.14
88	1.3351	56.0	1099.6	496.5	13.804	21.15	14.45	53.14	29.10
89	1.3779	57.0	1100.0	482.0	13.829	21.39	14.50	54.48	30.09
90	1.4219	58.0	1100.4	467.9	13.854	21.64	14.55	55.85	31.09

Table 3-1 (*Continued*)

	Properties of Water and Steam				Properties of Dry Air at a Pressure of 29.921 in. Hg abs		Properties of Mixture of Dry Air and Sat. Steam at a Total Pressure of 29.921 in. Hg abs		
		Enthalpy							
Temperature (F) t	Saturation Pressure of Water and Steam (in. Hg) p_s	Saturated Water (Btu/lb) $\mathbf{h}_f$	Saturated Steam (Btu/lb) $\mathbf{h}_g$	Specific Volume of Sat. Steam (ft³/lb) v_g	True Specific Volume (ft³/lb) v_a	Enthalpy (Btu/lb) $\mathbf{h}_a$	Volume of Mixture per lb of Dry Air (ft³) v_s	Enthalpy of Mixture per lb of Dry Air (Btu) $\mathbf{h}_s$	Humidity Ratio (mlb/lb dry air, or g/kg dry air) W_s
91	1.4671	59.0	1100.9	454.3	13.880	21.88	14.60	57.25	32.13
92	1.5136	60.0	1101.3	441.1	13.905	22.12	14.65	58.69	33.20
93	1.5613	61.0	1101.7	428.4	13.930	22.36	14.70	60.16	34.30
94	1.6103	62.0	1102.2	416.1	13.955	22.60	14.75	61.67	35.44
95	1.6607	63.0	1102.6	404.2	13.981	22.84	14.80	63.22	36.63
96	1.7124	64.0	1103.0	392.7	14.006	23.08	14.86	64.81	37.83
97	1.7655	65.0	1103.4	381.5	14.031	23.32	14.91	66.45	39.09
98	1.8200	66.0	1103.9	370.7	14.057	23.56	14.97	68.13	40.37
99	1.8759	67.0	1104.3	360.3	14.082	23.80	15.02	69.86	41.70
100	1.9334	68.0	1104.7	350.2	14.107	24.04	15.08	71.62	43.07
101	1.9923	69.0	1105.2	340.4	14.132	24.28	15.14	73.44	44.47
102	2.0529	70.0	1105.6	331.0	14.157	24.52	15.20	75.31	45.93
103	2.1149	71.0	1106.0	321.8	14.183	24.76	15.26	77.22	47.43
104	2.1786	72.0	1106.4	313.0	14.208	25.00	15.32	79.19	48.97
105	2.2440	73.0	1106.9	304.4	14.233	25.24	15.39	81.21	50.56
106	2.3110	74.0	1107.3	296.0	14.259	25.48	15.45	83.29	52.20
107	2.3798	75.0	1107.7	288.0	14.284	25.72	15.52	85.42	53.89
108	2.4503	76.0	1108.2	280.2	14.309	25.96	15.59	87.62	55.63
109	2.5226	77.0	1108.6	272.6	14.334	26.20	15.65	89.87	57.43
110	2.5968	78.0	1109.0	265.3	14.360	26.45	15.72	92.19	59.27
111	2.6728	79.0	1109.4	258.2	14.385	26.69	15.80	94.58	61.19
112	2.7507	80.0	1109.0	251.3	14.410	26.93	15.87	97.03	63.16
113	2.8306	81.0	1110.3	244.6	14.435	27.17	15.94	99.55	65.19
114	2.9125	82.0	1110.7	238.1	14.461	27.41	16.02	102.16	67.29
115	2.9963	83.0	1111.1	231.8	14.486	27.65	16.10	104.81	69.44

116	3.0823	84.0	1111.6	225.8	14.511	27.89	16.18	107.55	71.66
117	3.1703	85.0	1112.0	219.9	14.537	28.13	16.26	110.38	73.46
118	3.2606	86.0	1112.4	214.1	14.562	28.37	16.34	113.29	76.33
119	3.3530	87.0	1112.8	208.6	14.587	28.61	16.43	116.28	78.77
120	3.4477	88.0	1113.3	203.2	14.612	28.85	16.51	119.36	81.29
121	3.5446	89.0	1113.7	197.9	14.637	29.09	16.60	122.52	83.89
122	3.6439	90.0	1114.1	192.9	14.663	29.33	16.70	125.79	86.57
123	3.7455	91.0	1114.5	188.0	14.688	29.57	16.79	129.15	89.33
124	3.8496	92.0	1114.9	183.2	14.713	29.82	16.89	132.61	92.19
125	3.9561	93.0	1115.4	178.5	14.739	30.06	16.98	136.17	95.13
126	4.0651	94.0	1115.8	174.0	14.764	30.30	17.08	139.88	98.17
127	4.1768	95.0	1116.2	169.6	14.789	30.54	17.19	143.64	101.3
128	4.2910	96.0	1116.6	165.4	14.814	30.78	17.29	147.54	104.6
129	4.4078	97.0	1117.0	161.3	14.839	31.02	17.40	151.57	107.9
130	4.5274	98.0	1117.5	157.3	14.865	31.26	17.52	155.72	111.4
131	4.6498	99.0	1117.9	153.4	14.890	31.50	17.63	160.00	114.9
132	4.7750	100.0	1118.3	149.6	14.915	31.74	17.75	164.43	118.6
133	4.9030	101.0	1118.7	145.9	14.941	31.98	17.87	168.98	122.5
134	5.0340	102.0	1119.2	142.4	14.966	32.22	17.99	173.69	126.4
135	5.1679	103.0	1119.6	138.9	14.991	32.46	18.12	178.54	130.5
136	5.3049	104.0	1120.0	135.5	15.016	32.70	18.25	183.57	134.7
137	5.4450	105.0	1120.4	132.2	15.043	32.94	18.39	188.75	139.1
138	5.5881	106.0	1120.8	129.1	15.067	33.18	18.53	194.09	142.9
139	5.7345	107.0	1121.2	126.0	15.092	33.43	18.67	199.64	148.3
140	5.8842	108.0	1121.7	123.0	15.117	33.67	18.82	205.34	153.0
141	6.0371	109.0	1122.1	120.0	15.143	33.91	18.97	211.27	158.0
142	6.1934	110.0	1122.5	117.2	15.168	34.15	19.13	217.39	163.3
143	6.3532	111.0	1122.9	114.4	15.193	34.39	19.29	223.70	168.6
144	6.5164	112.0	1123.3	111.7	15.218	34.63	19.45	230.28	174.1
145	6.6832	113.0	1123.7	109.1	15.244	34.87	19.62	236.94	179.5
146	6.8536	114.0	1124.1	106.6	15.269	35.11	19.81	244.06	185.9
147	7.0277	115.0	1124.6	104.1	15.294	35.35	19.99	251.34	192.0
148	7.2056	116.0	1125.0	101.7	15.319	35.59	20.18	258.88	198.4
149	7.3872	117.0	1125.4	99.32	15.345	35.83	20.37	266.71	205.1
150	7.5727	118.0	1125.8	97.04	15.370	36.07	20.58	274.84	212.1

Table 3-1 *(Continued)*

	Properties of Water and Steam				Properties of Dry Air at a Pressure of 29.921 in. Hg abs		Properties of Mixture of Dry Air and Sat. Steam at a Total Pressure of 29.921 in. Hg abs		
		Enthalpy							
Temperature (F) t	Saturation Pressure of Water and Steam (in. Hg) p_s	Saturated Water (Btu/lb) $\mathbf{h}_f$	Saturated Steam (Btu/lb) $\mathbf{h}_g$	Specific Volume of Sat. Steam (ft³/lb) v_g	True Specific Volume (ft³/lb) v_a	Enthalpy (Btu/lb) $\mathbf{h}_a$	Volume of Mixture per lb of Dry Air (ft³) v_s	Enthalpy of Mixture per lb of Dry Air (Btu) $\mathbf{h}_s$	Humidity Ratio (mlb/lb dry air, or g/kg dry air) W_s
151	7.7622	119.0	1126.2	94.81	15.395	36.31	20.79	283.25	219.3
152	7.9556	120.0	1126.6	92.65	15.420	36.56	21.01	292.00	226.7
153	8.1532	121.0	1127.0	90.54	15.446	36.80	21.23	301.07	234.4
154	8.3548	122.0	1127.4	88.49	15.471	37.04	21.46	310.53	242.6
155	8.5607	123.0	1127.8	86.50	15.496	37.28	21.71	320.34	251.0
156	8.7708	124.0	1128.3	84.55	15.521	37.52	21.96	330.57	259.7
157	8.9853	125.0	1128.7	82.66	15.547	37.76	22.22	341.18	268.9
158	9.2042	126.0	1129.1	80.81	15.572	38.00	22.49	352.24	278.3
159	9.4276	127.0	1129.5	79.02	15.597	38.24	22.77	363.29	288.3
160	9.6556	128.0	1129.9	77.27	15.622	38.48	23.07	375.81	298.6
161	9.8882	129.0	1130.3	75.56	15.648	38.72	23.37	388.34	309.3
162	10.126	130.0	1130.7	73.90	15.673	38.96	23.69	401.45	320.6
163	10.368	131.0	1131.1	72.28	15.698	39.21	24.02	415.14	332.3
164	10.615	132.0	1131.5	70.71	15.724	39.45	24.37	429.44	344.6
165	10.867	133.0	1131.9	69.17	15.749	39.69	24.73	444.41	357.6
166	11.124	134.0	1132.3	67.67	15.774	39.93	25.11	460.10	371.0
167	11.386	135.0	1132.7	66.21	15.799	40.17	25.50	476.53	385.1
168	11.653	136.0	1133.1	64.79	15.824	40.41	25.92	493.77	411.4
169	11.925	137.0	1133.5	63.40	15.850	40.65	26.35	511.83	415.7
170	12.203	138.0	1133.9	62.04	15.875	40.89	26.81	530.86	432.0

171	12.487	139.0	1134.4	60.73	15.900	41.13	27.29	550.89	449.3
172	12.775	140.0	1134.8	59.44	15.925	41.37	27.79	571.97	467.6
173	13.080	141.0	1135.2	58.18	15.951	41.61	28.32	594.19	486.7
174	13.370	142.0	1135.6	56.96	15.976	41.85	28.88	617.65	507.0
175	13.676	143.0	1136.0	55.77	16.001	42.10	29.47	642.48	528.4
176	13.987	144.0	1136.4	54.60	16.026	42.34	30.09	668.67	551.1
177	14.305	145.0	1136.8	53.47	16.052	42.58	30.75	696.49	575.1
178	14.029	146.0	1137.2	52.36	16.077	42.82	31.46	726.04	600.7
179	14.959	147.0	1137.6	51.28	16.103	43.06	32.21	757.46	628.0
180	15.295	148.0	1137.9	50.22	16.128	43.30	32.99	790.88	656.8
181	15.637	149.0	1138.3	49.19	16.153	43.54	33.83	826.46	687.9
182	15.986	150.0	1138.7	48.19	16.178	43.78	34.74	864.74	720.9
183	16.341	151.0	1139.1	47.20	16.203	44.02	35.70	905.58	756.3
184	16.703	152.0	1139.5	46.25	16.229	44.26	36.74	949.49	794.3
185	17.071	153.0	1139.9	45.31	16.254	44.51	37.85	996.86	895.3
186	17.446	154.0	1140.3	44.40	16.279	44.75	39.04	1047.7	879.4
187	17.829	155.0	1140.7	43.51	16.304	44.99	40.34	1102.8	927.3
188	18.218	156.0	1141.1	42.64	16.329	45.23	41.75	1162.6	979.1
189	18.614	157.0	1141.5	41.79	16.354	45.47	43.28	1227.8	1035.7
190	19.017	158.0	1141.9	40.96	16.380	45.71	44.94	1298.9	1083.1
191	19.428	159.1	1142.3	40.14	16.405	45.94	46.78	1376.9	1165.1
192	19.846	160.1	1142.7	39.35	16.430	46.19	48.79	1463.0	1242.7
193	20.271	161.1	1143.1	38.58	16.456	46.43	51.02	1558.2	1322.4
194	20.704	162.1	1143.5	37.82	16.481	46.68	53.50	1664.0	1414.1
195	21.145	163.1	1143.8	37.09	16.506	46.92	56.27	1782.6	1517.1
196	21.594	164.1	1144.2	36.36	16.531	47.16	59.40	1916.2	1633.4
197	22.050	165.1	1144.6	35.66	16.556	47.40	62.93	2067.6	1764.9
198	22.515	166.1	1145.0	34.97	16.582	47.64	66.99	2240.9	1915.6
199	22.987	167.1	1145.4	34.30	16.607	47.88	71.66	2440.9	2089.1
200	23.468	168.1	1145.8	33.64	16.632	48.12	77.14	2675.6	2293.1

Table 3-1 (*Continued*)

	Properties of Ice and Steam				Properties of Dry Air at a Pressure of 29.921 in. Hg abs		Properties of Mixture of Dry Air and Sat. Steam at a Total Pressure of 29.921 in. Hg abs		
		Enthalpy							
Temperature (F) t	Saturation Pressure of Ice and Steam (in. Hg) p_s	Saturated Ice (Btu/lb) $\mathbf{h}_f$	Saturated Steam (Btu/lb) $\mathbf{h}_g$	Specific Volume of Sat. Steam (ft³/lb) v_g	True Specific Volume (ft³/lb) v_a	Enthalpy (Btu/lb) $\mathbf{h}_a$	Volume of Mixture per lb of Dry Air (ft³) v_s	Enthalpy of Mixture per lb of Dry Air (Btu) $\mathbf{h}_s$	Humidity Ratio (mlb/lb dry air, or g/kg dry air) W_s
32	0.1803	−143.4	1075.2	3305	12.389	7.69	12.46	11.75	3.771
31	0.1723	−143.9	1074.7	3453	12.363	7.45	12.43	11.32	3.601
30	0.1645	−144.4	1074.3	3608	12.338	7.21	12.41	10.90	3.439
29	0.1571	−144.9	1073.8	3771	12.313	6.97	12.38	10.49	3.283
28	0.1500	−145.4	1073.4	3943	12.287	6.73	12.35	10.09	3.131
27	0.1431	−145.9	1073.0	4122	12.262	6.49	12.32	9.70	2.989
26	0.1366	−146.4	1072.5	4311	12.237	6.25	12.29	9.31	2.851
25	0.1303	−146.9	1072.1	4509	12.211	6.01	12.27	8.92	2.720
24	0.1243	−147.4	1071.7	4717	12.186	5.77	12.24	8.55	2.594
23	0.1186	−147.9	1071.2	4936	12.161	5.53	12.21	8.18	2.474
22	0.1130	−148.4	1070.8	5166	12.136	5.29	12.18	7.81	2.359
21	0.1078	−148.9	1070.3	5408	12.110	5.05	12.15	7.45	2.247
20	0.1027	−149.4	1069.9	5662	12.085	4.81	12.13	7.10	2.141
19	$9.789\,(10)^{-2}$	−149.8	1069.5	5929	12.060	4.56	12.10	6.75	2.040
18	$9.326\,(10)^{-2}$	−150.3	1069.0	6210	12.035	4.32	12.07	6.40	1.944
17	$8.884\,(10)^{-2}$	−150.8	1068.6	6505	12.009	4.08	12.05	6.06	1.851
16	$8.461\,(10)^{-2}$	−151.3	1068.1	6817	11.984	3.84	12.02	5.73	1.763
15	$8.056\,(10)^{-2}$	−151.8	1067.7	7144	11.959	3.60	11.99	5.40	1.678
14	$7.669\,(10)^{-2}$	−152.3	1067.3	7489	11.933	3.36	11.96	5.07	1.597
13	$7.300\,(10)^{-2}$	−152.8	1066.8	7851	11.918	3.12	11.94	4.75	1.580
12	$6.946\,(10)^{-2}$	−153.3	1066.4	8234	11.883	2.88	11.91	4.43	1.447
11	$6.608\,(10)^{-2}$	−153.7	1065.9	8636	11.857	2.64	11.88	4.11	1.376

10	$6.286 (10)^{-2}$	−154.2	1065.5	9060	11.832	2.40	11.86	3.80	1.309
9	$5.977 (10)^{-2}$	−154.7	1065.1	9507	11.807	2.16	11.83	3.49	1.244
8	$5.683 (10)^{-2}$	−155.2	1064.6	9979	11.782	1.92	11.80	3.18	1.183
7	$5.402 (10)^{-2}$	−155.7	1064.2	$1.048 (10)^{4}$	11.756	1.68	11.78	2.88	1.124
6	$5.134 (10)^{-2}$	−156.1	1063.7	$1.100 (10)^{4}$	11.731	1.44	11.75	2.58	1.070
5	$4.878 (10)^{-2}$	−156.6	1063.3	$1.155 (10)^{4}$	11.706	1.20	11.72	2.28	1.015
4	$4.633 (10)^{-2}$	−157.1	1062.8	$1.214 (10)^{4}$	11.680	0.96	11.70	1.98	0.9636
3	$4.400 (10)^{-2}$	−157.6	1062.4	$1.275 (10)^{4}$	11.655	0.72	11.67	1.68	0.9069
2	$4.178 (10)^{-2}$	−158.0	1062.0	$1.340 (10)^{4}$	11.630	0.48	11.65	1.40	0.8691
1	$3.966 (10)^{-2}$	−158.5	1061.5	$1.408 (10)^{4}$	11.604	+ 0.24	11.62	1.12	0.8252
0	$3.764 (10)^{-2}$	−159.0	1061.1	$1.481 (10)^{4}$	11.579	0.00	11.59	+ 0.83	0.7829
− 5	$2.888 (10)^{-2}$	−161.3	1058.9	$1.909 (10)^{4}$	11.453	− 1.20	11.46	− 0.57	0.6006
− 10	$2.203 (10)^{-2}$	−163.6	1056.7	$2.475 (10)^{4}$	11.326	− 2.40	11.33	− 1.92	0.4580
− 15	$1.670 (10)^{-2}$	−165.9	1054.5	$3.228 (10)^{4}$	11.200	− 3.61	11.21	− 3.24	0.3471
− 20	$1.259 (10)^{-2}$	−168.2	1052.3	$4.237 (10)^{4}$	11.073	− 4.81	11.08	− 4.53	0.2614
− 25	$9.420 (10)^{-3}$	−170.5	1050.1	$5.596 (10)^{4}$	10.947	− 6.01	10.95	− 5.80	0.1957
− 30	$7.003 (10)^{-3}$	−172.7	1047.8	$7.441 (10)^{4}$	10.820	− 7.21	10.82	− 7.06	0.1454
− 35	$5.170 (10)^{-3}$	−174.9	1045.6	$9.961 (10)^{4}$	10.693	− 8.41	10.70	− 8.30	0.1074
− 40	$3.790 (10)^{-3}$	−177.1	1043.4	$1.343 (10)^{5}$	10.567	− 9.61	10.57	− 9.53	0.07867
− 45	$2.757 (10)^{-3}$	−179.2	1041.2	$1.824 (10)^{5}$	10.440	−10.82	10.44	−10.76	0.05724
− 50	$1.990 (10)^{-3}$	−181.3	1039.0	$2.496 (10)^{5}$	10.314	−12.01	10.31	−11.97	0.04132
− 55	$1.426 (10)^{-3}$	−183.4	1036.8	$3.443 (10)^{5}$	10.187	−13.22	10.19	−13.19	0.02959
− 60	$1.012 (10)^{-3}$	−185.5	1034.6	$4.738 (10)^{5}$	10.060	−14.42	10.06	−14.40	0.02101
− 70	$4.974 (10)^{-4}$	−189.6	1030.2	$9.501 (10)^{5}$	9.807	−16.82	9.807	−16.81	0.01032
− 80	$2.355 (10)^{-4}$	−193.6	1025.7	$1.955 (10)^{6}$	9.553	−19.23	9.553	−19.23	0.00488
− 90	$1.071 (10)^{-4}$	−197.5	1021.3	$4.186 (10)^{6}$	9.299	−21.64	9.300	−21.64	0.00221
−100	$4.664 (10)^{-5}$	−201.3	1016.9	$9.352 (10)^{6}$	9.046	−24.04	9.046	−24.04	0.00097
−120	$7.649 (10)^{-6}$	−208.6	1008.0	$5.386 (10)^{7}$	8.538	−28.86	8.538	−28.86	0.00016

Reproduced by permission of The American Society of Refrigerating Engineers from the A.S.R.E. Data Book. Compiled by C. O. Mackey, using basic Goff and Gratch data.

partial pressure of water vapor in the air to the pressure that saturated steam exerts at the temperature of the air. In equational form

$$\phi = \left(\frac{p}{p_s}\right)_{td} \tag{3-3}$$

or

$$\phi = \left(\frac{d}{d_s}\right)_{td} \tag{3-4}$$

where ϕ is the relative humidity, expressed as a decimal; p is the partial pressure of the water vapor (steam) in the air; p_s is the pressure of saturated steam at the air temperature (dry-bulb temperature, t); d is the density of the water vapor in the air; and d_s is the density of saturated water vapor at the air temperature (dry-bulb).

The pressures p and p_s must be expressed in consistent units. In the case of space saturated with steam (water vapor) at 80 F, Table 3-1 shows the specific volume of steam to be 633.0 ft^3/lb; and the density of the steam is, of course, the reciprocal of specific volume, or $d = 1/v = 1/633 = 0.001579$ lb/ft^3. The saturation pressure, from Table 3-1, is 1.0323 in. Hg. If it happened that, with the temperature 80 F at a given time, only 0.001579/2 lb of steam existed in each cubic foot, then the characteristic gas equation would show that the pressure exerted by the steam would be only half as great as the value given in Table 3-1 for 80 F. Using $pV = mRT$, we may write:

$$(144)(0.491)(p)(1) = \frac{0.001579}{2}(85.6)(460 + 80)$$

$$p_s = 0.5162 \text{ in. Hg}$$

Thus the relative humidity is

$$\phi = \frac{p}{p_s} = \frac{d}{d_s} = \frac{0.5162}{1.0323} = \frac{0.001579/2}{0.001579} = 0.50, \text{ or } 50\%$$

It can be seen that Table 3-1 is not expressed in SI units and a similar situation exists for much of the tabular information appearing in this text and in fact throughout most of the technical literature written in the English language. Conversion factors are well known and it is easy to convert to SI terminology. However, until much more of our reference literature is converted to the SI basis it is more convenient to use the foot-pound-second system in calculations. This practice will be followed in this text but parallel SI solutions will be given whenever necessary for clarity.

Example 3-1. The temperature in a certain room is 70 F and the relative humidity is 30%. The barometric pressure p_B is 29.2 in. Hg. Find (a) the partial pressure of the steam in the air, (b) the weight of steam per cubic foot (vapor density), and (c) the weight of steam associated with each pound of dry air (humidity ratio or specific humidity).

Solution: (a) From Table 3-1, the pressure of saturated steam at 70 F is 0.7392 in. Hg. At 30% relative humidity,

$$p = \phi p_s = (0.30)(0.7392) = 0.2218 \text{ in. Hg} \qquad \textit{Ans.}$$

(b) The actual partial pressure of the steam in the air being known, d can be found from $pV = mRT$. Thus

$$(144)(0.491)(0.2218)(1) = (70.7)(0.2218) = d(85.6)(460 + 70)$$

and

$$d = 0.000345 \text{ lb/ft}^3 \qquad \textit{Ans.}$$

The value of d can also be found from the steam tables for saturated steam at 70 F and a relative humidity $\phi = 0.30$. Thus

$$d = \phi d_s = (0.30)\left(\frac{1}{868.0}\right) = 0.000345 \text{ lb/ft}^3$$

(c) From Dalton's law of partial pressures, expressed in Eq. (3-1),

$$p_B = p_a + p$$

Therefore

$$p_a = p_B - p = 29.2 - 0.2218 = 28.98 \text{ in. Hg}$$

where 29.2 is the barometric pressure and p_a is the partial pressure of the dry air.

Using $pV = mRT$, with R for air $= 53.3$, we find the volume occupied by 1 lb of the dry air ($m = 1$):

$$(144)(0.491)(28.98)(v) = (70.7)(28.98)v = (1)(53.3)(460 + 70)$$

$$v = 13.79 \text{ ft}^3$$

The weight of steam per cubic foot of space = 0.000345 lb (cf. part b). Therefore the weight of steam associated with 1 lb of dry air, that is, with 13.79 ft^3 of space, is

$$(13.79)(0.000345) = 0.00476 \text{ lb} \qquad \textit{Ans.}$$

Humidity ratio (*specific humidity*) can be calculated in one step from a simple relationship derived from $pV = mRT$. The volume occupied by 1 lb of air ($m = 1$) at the partial pressure of the air, $p_a = p_B - p$, is, in cubic feet,

$$v = \frac{mRT}{p_a} = \frac{(1)(53.3)(T)}{p_B - p} \tag{A}$$

The weight of steam in 1 lb of dry air (v ft^3) is, from $pv = mRT$,

$$m = W = \frac{pv}{RT} = \frac{pv}{(85.6)T} \tag{B}$$

Substitution of Eq. (A) in Eq. (B) gives

$$W = \frac{p(53.3)T}{85.6T(p_B - p)} = 0.622 \frac{p}{p_B - p} \tag{3-5}$$

where W is the humidity ratio (specific humidity) in pounds of steam (water vapor) per pound of dry-air constituents, or in kilograms of water vapor per kilogram of dry-air constituents; p is the partial pressure of steam (water vapor) in air; and p_B is the barometric pressure in the same units as p.

Equation (3-5) can also be expressed as

$$W = 622 \frac{p}{p_B - p} \qquad (3\text{-}6)$$

and for this case, W is expressed in millipounds (mlb) of steam per pound of dry-air constituents, or in grams (g) of steam per kilogram (kg) of dry-air constituents.

Example 3-2. A room is at 21.11°C (70 F) at a relative humidity of 30%. The barometric pressure is 98 884 Pa or 29.2 in. Hg. Outside air is at −12.22°C (10 F) at 50% relative humidity. (a) Find the humidity ratios of the indoor air and of the outside air. (b) How much water has to be added by means of a humidifier for each kilogram of outside air supplied to this room? (c) Solve part b using foot-pound-second system units.

Solution: (a) Refer to Table 3-1 and read at 70 F that $p_{si} = 0.7392$ in. Hg, and at 10 F, $p_{so} = 0.06286$ in Hg.

$$p_{si} = 0.7392 \times 3386.45 = 2503.3 \text{ Pa}$$
$$p_{so} = 0.06286 \times 3386.45 = 212.9 \text{ Pa}$$

By Eq. (3-6),

$$W_i = 622 \frac{(0.3)(2503.3)}{98\,884 - (0.3)(2503.3)} = 4.760 \text{ g/kg} \qquad \textit{Ans.}$$

$$W_o = 622 \frac{(0.5)(212.9)}{98\,884 - (0.5)(212.9)} = 0.670 \text{ g/kg} \qquad \textit{Ans.}$$

(b) Water added = 4.760 − 0.670 = 4.09 g/kg of outside dry-air constituents. *Ans.*

(c) Read saturation water-vapor values from Table 3-1 at 70 F and 10 F and make use of Eq. (3-6).

$$W_i = 622 \frac{(0.3)(0.7392)}{29.2 - (0.3)(0.7392)} = 4.760 \text{ mlb/lb}$$

$$W_o = 622 \frac{(0.5)(0.06282)}{29.2 - (0.5)(0.06286)} = 0.670 \text{ mlb/lb}$$

Water added = 4.760 − 0.670 = 4.09 millipounds of steam per pound of outside dry-air constituents. *Ans.*

Example 3-3. Outside air at 10 F and 60% relative humidity, after passing through a heater and humidifier, enters an auditorium at 76 F and 50% relative humidity. Find how much water vapor has to be added to each pound of dry outside air to bring it to inside conditions. The barometer reads 29.8 in. Hg.

Solution: Using subscript 1 to denote outside conditions, and subscript 2 to denote inside conditions,

$$p_1 = \phi_1 p_{s1} = (0.6)(0.06286) = 0.03772 \text{ in. Hg}$$

where 0.06286 is taken from Table 3-1 at 10 F.

Since $p_B = 29.8$ in. Hg, the specific humidity of the outside air is, by Eq. (3-5),

$$W_1 = 0.622 \frac{0.03772}{29.8 - 0.03772} = 0.00079 \text{ lb}$$

Also,

$$p_2 = \phi_2 p_{s2} = (0.5)(0.9047) = 0.4523 \text{ in. Hg}$$

where 0.9047 is taken from Table 3-1 at 76 F.

The specific humidity of the inside air is

$$W_2 = 0.622 \frac{0.4523}{29.8 - 0.4523} = 0.00958 \text{ lb}$$

The weight of water vapor (steam) added to each pound of air supplied to the auditorium is therefore

$$W_2 - W_1 = 0.00958 - 0.00079 = 0.00879 \text{ lb}$$

Formerly in psychrometric studies use was made of the *grain*, where 7000 grains = 1 lb. Thus in this example,

$$0.00879 \times 7000 = 61.53 \text{ grains of water vapor added} \qquad \textit{Ans.}$$

Example 3-4. The temperature in a room is 21.1°C, the barometric pressure is 98 066 Pa, and the relative humidity is 30%. Find (a) the partial pressure of the steam in the air, (b) the density of the steam in the air in kilograms per cubic meter, and (c) the humidity ratio.

Solution: (a) Since 21.1°C is also 70 F, refer to Table 3-1 and read the pressure of saturated steam as 0.7392 in. Hg. Find the conversion factor from Table A-3 and compute the pressure as

$$0.7392 \times 3386 = 2503 \text{ Pa.}$$

Partial pressure of the steam from Eq. (3-3) is thus

$$2503 \times 0.3 = 750.9 \text{ Pa.} \qquad \textit{Ans.}$$

(b) The steam density can be found from Eq. (2-24),

$$pV = mRT$$

where R for steam is $8309.5/18.016 = 461.2$ J/kg · K

$$(750.9)(1) = (d)(461.2)(273.16 + 21.1)$$

$$d = 0.00553 \text{ kg/m}^3 \qquad \textit{Ans.}$$

(c) Use Eq. (3-6) and employ consistent pressure units.

$$W = 622 \frac{750.9}{98\,066 - 750.9} = 4.80 \text{ g/kg dry air} \qquad \textit{Ans.}$$

and also 4.80 mlb/lb of dry-air constituents.

Degree of saturation (μ), also known as *saturation ratio*, is the humidity ratio, W, of the moist air to the humidity ratio that air could possess at saturation (W_s) at the same temperature and pressure. In equational form,

$$\mu = \left.\frac{W}{W_s}\right)_{t,\,p_T} \tag{3-7}$$

In the technical literature, degree of saturation is called by various other names, such as percentage humidity and percentage saturation.

Relationship between the relative humidity ϕ and the degree of saturation μ can be found. By definition,

$$\phi = \frac{p}{p_s}$$

and

$$\mu = \frac{W}{W_s} = \frac{0.622[p/(p_B - p)]}{0.622[p_s/(p_B - p_s)]}$$

By transformations and substitutions, we get

$$\mu = \frac{p}{p_s}\frac{p_B - p_s}{p_B - p} = \phi\frac{1 - (p_s/p_B)}{1 - (p/p_B)(p_s/p_s)} = \phi\frac{1 - (p_s/p_B)}{1 - \phi(p_s/p_B)} \tag{3-8}$$

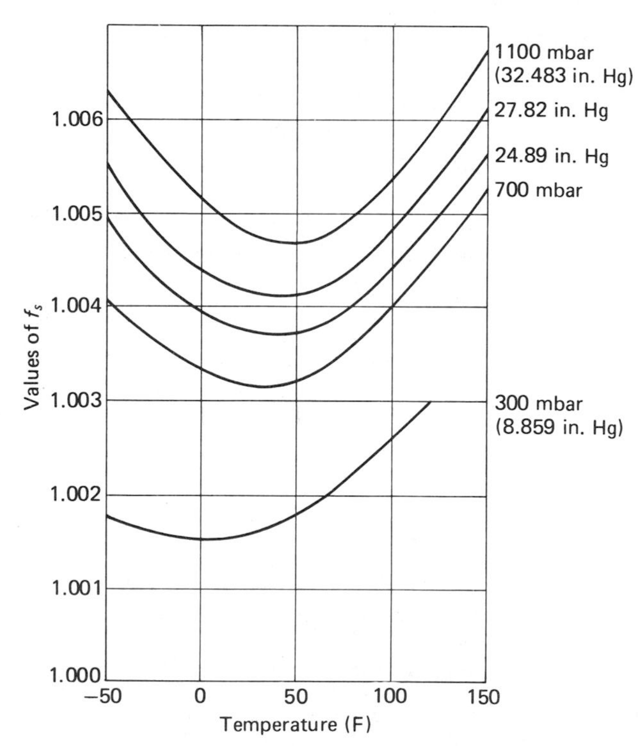

FIGURE 3-1. Plot of the factor f_s for use in accurate determination of humidity ratio. (From Ref. 5.)

Similarly,

$$\phi = \mu \frac{p_B - p}{p_B - p_s} = \mu \frac{1 - \phi(p_s/p_B)}{1 - (p_s/p_B)} = \frac{\mu}{1 - (1 - \mu)(p_s/p_B)} \tag{3-9}$$

It has been mentioned that use of the Gibbs–Dalton law and the perfect-gas equation can give close approximations for determination of values in connection with air-steam mixtures. However, real gaseous molecules do not behave ideally, since there are interactions between various types of molecules, and mutual solubilities between gaseous and liquid phases. More rigorous equations covering air-steam mixtures can be developed by the methods of statistical mechanics. These are developed in some detail in Refs. 5 and 6, listed at the end of this chapter. In order to give some idea of the magnitude of the deviations which exist, Fig. 3-1 has been prepared. The factor f_s applies for humidity ratio when consideration is given to the more precise methods of analysis. It is employed as follows:

$$W = 0.622 \frac{f_s(p/p_B)}{1 - f_s(p/p_B)} \tag{3-10}$$

At 14.7 psia and 212 F, 100% relative humidity has no significance in terms of air, since at this condition no air can be present. However, values

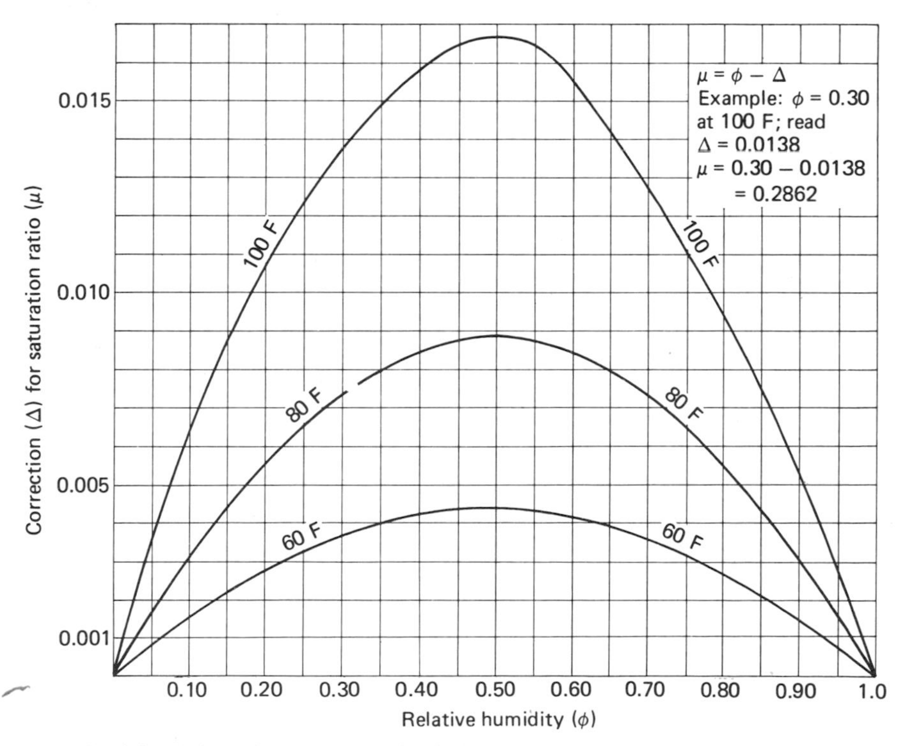

FIGURE 3-2. Subtractive corrections for finding values of saturation ratio μ when relative humidity ϕ is given. Drawn for $p_B = 29.92$ in. Hg.

less than 100% are significant. Relative humidities of air mixtures are sometimes used at temperatures above 212 F but are low in numerical magnitude because the denominator of the expression for relative humidity is the saturation pressure of steam at the temperature actually existing, whereas the term p_s cannot exceed the atmospheric pressure (or the total pressure for a compressed mixture). The degree of saturation has no significance at temperatures of 212 F and above unless the mixture is compressed.

Figure 3-2 is a graph of Eq. (3-8) for a barometric pressure of 29.92 in. Hg and selected temperatures. At a given value of ϕ (relative humidity) it is possible to read the correction Δ to be subtracted to find the corresponding value of μ (saturation ratio); that is, $\mu = \phi - \Delta$. For example, if ϕ is 0.30 at 100 F, the value of Δ is 0.0138 and $\mu = 0.30 - 0.0138 = 0.2862$. It should be noted that ϕ and μ are numerically close at low and high values but differ appreciably in the middle range, the value of Δ increasing as the temperature becomes higher.

3-3. ENTHALPY OF AIR

The enthalpy of humid air, when the air is considered to act as a perfect gas, is found by adding the enthalpy per pound of dry air and the enthalpy of the water vapor (steam) associated with the pound of dry air.

The enthalpy $\mathbf{h}$ of a gas was given in Eq. (2-30) as $\mathbf{h} = C_p(T - T_0)$. In most air-conditioning processes *enthalpy changes* alone are important. Thus the temperature t can be referred to any convenient datum, with 0 F customarily taken as the reference temperature, and

$$\mathbf{h}_a = C_p(t - 0) + W\mathbf{h}_v \tag{3-11}$$

where $\mathbf{h}_a$ is the enthalpy of 1 lb of "dry" air, in Btu; C_p is the specific heat of air at constant pressure, usually 0.24 Btu/lb F; 0.2403 in range 0 F to 200 F; t is the dry-bulb temperature, in degrees Fahrenheit; W is the humidity ratio or specific humidity (pounds of steam associated with 1 lb of dry air); and $\mathbf{h}_v$ is the enthalpy of steam at the dry-bulb temperature, in Btu per pound (usually referred to a 32 F datum).

The value of $\mathbf{h}_v$ for steam can in some cases be taken directly from the tables compiled for superheated or saturated steam, but for the usual air-conditioning ranges of temperature and pressure the steam-table values are incomplete. It is usually more convenient to use relations derived directly from the steam-table values. For temperatures in the range from 70 F to 150 F,

$$\mathbf{h}_v = 1060.5 + 0.45t \tag{3-12}$$

For temperatures below 70 F, a more precise relation is

$$\mathbf{h}_v = 1061.4 + 0.44t \tag{3-13}$$

In Eqs. (3-12) and (3-13), $\mathbf{h}_v$ and t have the same meanings as in Eq. (3-11). Where the temperature is above 70 F, Eq. (3-11) becomes

$$\mathbf{h}_a = C_p(t - 0) + W(1060.5 + 0.45t) \tag{3-14}$$

The enthalpy of moist air can be determined to a slightly greater degree of precision by making use of moist-air tables such as Table 3-1. This approach is not required for most engineering calculations, but the method involves reading the property values at any given dry-bulb temperature (t) for air saturated with water vapor ($\mathbf{h}_{s,t}$), and the corresponding value for completely dry air ($\mathbf{h}_{a,t}$). The degree of saturation, μ, must then be applied to the difference between saturated and dry air. In equation form the enthalpy of moist air, $\mathbf{h}_a$, is thus

$$\mathbf{h}_a = \mathbf{h}_{a,t} + \mu(\mathbf{h}_{s,t} - \mathbf{h}_{a,t}) \tag{3-15}$$

Even this method starts to lose precision at temperatures above 150 F (66°C) but most air-conditioning calculations are fortunately in a lower temperature regime.

3-4. AIR-HUMIDITY PROCESS, THERMODYNAMIC WET-BULB TEMPERATURE

The application of the general energy equation [Eq. (2-9)] to any continuous process in which air and water-vapor interactions are taking place is as follows: Consider a device, such as a humidifier, a dehumidifier, or a drier, to which air is supplied and in which water may be added or removed from the air, and in which heat Q also may be added or rejected. Refer to Fig. 3-3 and imagine that 1 lb of dry air at temperature t_1 and carrying W_1 lb of steam is supplied. In the device, steam can be condensed from the air, or water can be evaporated into the air, the action depending on the function of the device. Heat (Q Btu, based on each pound of dry air entering) may be added or removed. Finally, 1 lb of dry air, at temperature t_2, with steam

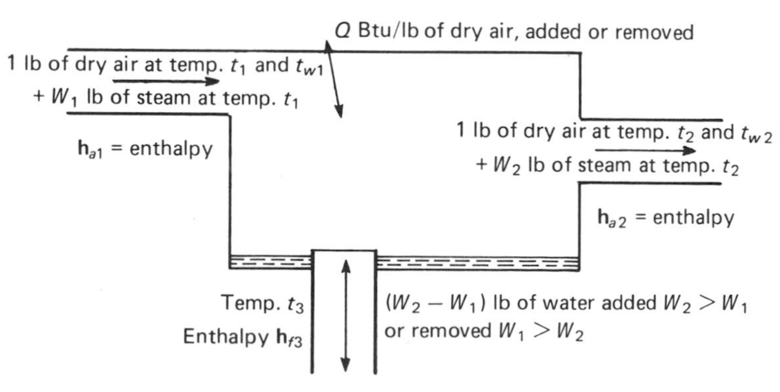

FIGURE 3-3. Air-humidity device.

content W_2, leaves. Applying the energy Eq. (2-9) and removing nonessential terms, we obtain:

$$\mathbf{h}_{a1} \pm Q_{(1-2)} + (W_2 - W_1)\mathbf{h}_{f3} = \mathbf{h}_{a2} \tag{3-16}$$

If heat is added, a plus sign is used with $Q_{(1-2)}$ in Eq. (3-16), and the Btu per pound of dry air is

$$Q_{(1-2)} = \mathbf{h}_{a2} - \mathbf{h}_{a1} - (W_2 - W_1)\mathbf{h}_{f3} \tag{3-17}$$

Substituting values from Eq. (3-14) for $\mathbf{h}_{a2}$ and $\mathbf{h}_{a1}$, and algebraically rearranging, we may write:

$$\begin{aligned} Q_{(1-2)} = {} & (0.24)(t_2 - t_1) \\ & + (W_2 - W_1)(1060.5 + 0.45t_2 - \mathbf{h}_{f3}) + 0.45W_1(t_2 - t_1) \end{aligned}$$

Since $\mathbf{h}_{f3} = (1)(t_3 - 32)$ approximately,

$$\begin{aligned} Q_{(1-2)} = {} & (0.24 + 0.45W_1)(t_2 - t_1) \\ & + (W_2 - W_1)(1092.5 - t_3 + 0.45t_2) \end{aligned} \tag{3-18}$$

If heat is removed, a minus sign is used for $Q_{(1-2)}$ in Eq. (3-16) and the Btu per pound of dry air is

$$Q_{(1-2)} = \mathbf{h}_{a1} - \mathbf{h}_{a2} + (W_2 - W_1)\mathbf{h}_{f3} \tag{3-19}$$

From this, there results:

$$\begin{aligned} Q_{(1-2)} = {} & (0.24 + 0.45W_1)(t_1 - t_2) \\ & + (W_1 - W_2)(1092.5 - t_3 + 0.45t_2) \end{aligned} \tag{3-20}$$

Let us imagine that the device of Fig. 3-3 is provided with an insulating jacket so effective that it eliminates heat flow into or from the system to the point that it is adiabatic. Figure 3-4 shows the modified device provided with a water reservoir of large surface area. Air is shown entering it at a dry-bulb temperature t_1 and leaving in saturated condition at a dry-bulb temperature t_2. Analysis, confirmed by experimentation, will show that when such a device is provided with a continuous stream of air at a constant inlet state, the air will be brought to its "temperature of adiabatic saturation," commonly known as the "thermodynamic wet-bulb temperature." Water in the reservoir, if not at this temperature initially, will also reach the thermodynamic wet-bulb temperature. The process is one in which the air gradually cools down in contact with the water surface as water evaporates into it and the air

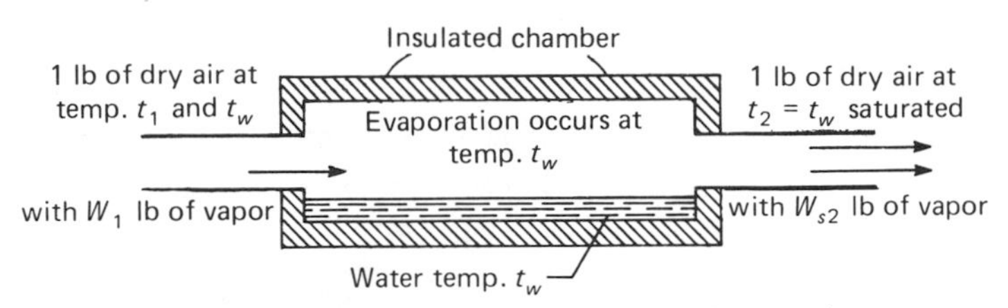

FIGURE 3-4. Diagram illustrating adiabatic saturation.

finally becomes saturated at the thermodynamic wet-bulb temperature. This phenomenon has broad implications, one of which yields a definition, namely: *thermodynamic wet-bulb temperature* is that temperature, for any state of moist air, at which water or ice may be evaporated (sublimed) into the air to bring it to saturation at the same temperature.

Thus "thermodynamic wet-bulb temperature," usually shortened to "wet-bulb temperature," is a characteristic property for air at a given state and air can be thermodynamically and definitively described by giving its dry-bulb and wet-bulb temperatures at a given barometric (total) pressure. Air can, of course, also be described in several other ways, as, for example, by stating dry-bulb temperature and humidity ratio and pressure.

Apply Eq. (3-16) to the process of adiabatic saturation for water-air. Heat $Q_{(1-2)}$ equals zero by definition; and to make adiabatic saturation a steady-flow process, enough water $(W_{s2} - W_1)$ must be added at the wet-bulb temperature t_w to make up for evaporation. The resulting equation is

$$\mathbf{h}_{a1} + (W_{s2} - W_1)\mathbf{h}_{fw} = \mathbf{h}_{a2} \tag{3-21}$$

Computation will show that at moderate temperatures (below 100 F), the term $(W_{s2} - W_1)\mathbf{h}_{fw}$ is quite small. If this term is neglected, it can be seen that in adiabatic saturation,

$$\mathbf{h}_{a1} = \mathbf{h}_{a2} \tag{3-22}$$

That is, the enthalpy remains constant. Also, since adiabatic saturation is a process at constant wet-bulb temperature, it immediately appears that the enthalpy of air at the same wet-bulb temperature is practically constant even though the dry-bulb temperature varies. (This last statement, although not precisely true, is accurate enough for many engineering computations, and some psychometric charts are constructed on this basis.)

Equation (3-21) may be rewritten in the form

$$\mathbf{h}_{a1} - W_1\mathbf{h}_{fw} = \mathbf{h}_{a2} - W_2\mathbf{h}_{fw} \tag{3-23}$$

Notice that inlet conditions to the process represented by this equation are restricted only in regard to the wet-bulb temperature. Therefore the inlet term can be generalized and the expression can be written as follows:

$$\Sigma = \mathbf{h}_{a2} - W_2\mathbf{h}_{fw} = \mathbf{h}_{a1} - W_1\mathbf{h}_{fw} = \mathbf{h}_{ax} - W_x\mathbf{h}_{fw} \tag{3-24}$$

This means that the *sigma function* (Σ), defined by Eq. (3-24), is constant for any given wet-bulb temperature.

The significance of Eq. (3-21) is further exemplified when it is written as follows:

$$\mathbf{h}_{a1} = \mathbf{h}_{a2} - (W_{s2} - W_1)\mathbf{h}_{fw}$$

or

$$\mathbf{h}_{a1} = \mathbf{h}_{aw} - (W_{sw} - W_1)\mathbf{h}_{fw} = \mathbf{h}_{aw} \pm D \tag{3-25}$$

This means that the enthalpy of an air-stream mixture is equal to the enthalpy of saturated air at the same wet-bulb temperature, less the small corrective term $(W_{s2} - W_1)\mathbf{h}_{fw}$. This corrective term is called the deviation and carries the symbol D.

From the adiabatic-saturation Eq. (3-21), it is possible to derive a valuable formula for finding the weight of water vapor in the atmosphere in terms of the wet-bulb and dry-bulb temperatures. Starting with Eq. (3-21), we use values for the enthalpy $\mathbf{h}_{a1}$ and $\mathbf{h}_{a2}$ from Eqs. (3-11) and (3-14). The result is

$$C_p t_1 + W_1 \mathbf{h}_{v1} + (W_{s2} - W_1)\mathbf{h}_{fw} = C_p t_w + W_{s2}\mathbf{h}_{v2}$$

From this,

$$W_1 = \frac{C_p(t_w - t_1) + W_{s2}(\mathbf{h}_{v2} - \mathbf{h}_{fw})}{\mathbf{h}_{v1} - \mathbf{h}_{fw}} \tag{3-26}$$

At exit condition 2, the dry-bulb temperature equals the wet-bulb temperature. Therefore $\mathbf{h}_{v2} - \mathbf{h}_{fw} = \mathbf{h}_{vw} - \mathbf{h}_{fw}$. But this latter expression equals $\mathbf{h}_{fgw}$, since $\mathbf{h}_{vw} - \mathbf{h}_{fw}$ is the change in enthalpy when liquid changes to vapor at the wet-bulb temperature, or it is the latent heat of vaporization $(\mathbf{h}_{fgw})$ at the wet-bulb temperature. By Eq. (3-12), the term $(\mathbf{h}_{v1} - \mathbf{h}_{fw})$ is equal to $(1060.5 + 0.45t_1 - \mathbf{h}_{fw})$. Let us now algebraically add the identity $(0.45t_w - 0.45t_w)$ to the latter expression, giving

$$\begin{aligned}(1060.5 + 0.45t_w) - \mathbf{h}_{fw} + 0.45t_1 - 0.45t_w \\ = \mathbf{h}_{vw} - \mathbf{h}_{fw} + 0.45(t_1 - t_w) = \mathbf{h}_{fgw} + 0.45(t_1 - t_w)\end{aligned}$$

If these identities are applied to Eq. (3-26), there results

$$W_1 = \frac{C_p(t_w - t_1) + W_{sw}\mathbf{h}_{fgw}}{\mathbf{h}_{fgw} + 0.45(t_1 - t_w)}$$

or

$$W = \frac{W_{sw}\mathbf{h}_{fgw} - 0.24(t - t_w)}{\mathbf{h}_{fgw} + 0.45(t - t_w)} \tag{3-27}$$

where W is the weight of water vapor carried with each pound of dry air at dry-bulb temperature t and wet-bulb temperature t_w, in pounds, humidity ratio; $\mathbf{h}_{fgw}$ is the heat of vaporization of 1 lb of steam at the wet-bulb temperature; and W_{sw} is the weight of water vapor associated with each pound of dry air when saturated at the wet-bulb temperature t_w, in pounds. The value of W_{sw} can be found by Eq. (3-5), or from Table 3-1, for standard barometric pressure.

The concept of wet-bulb temperature as a characteristic property of moist air has lead to the reason for calling conventional air temperature by the name *dry-bulb temperature*. In this text when unspecified air temperature is mentioned, dry-bulb temperature is meant. Its symbol is the letter t.

Specifically, air temperature or dry-bulb air temperature represents air temperature in a closed or open space when the temperature is independent of radiation effects from surroundings and when air motion relative to the measuring device is not significant.

When moist air is cooled at constant pressure with its humidity ratio remaining fixed, a temperature is reached at which the air becomes saturated and further cooling results in the appearance of fog or deposition of moisture on adjacent surfaces. This saturation or condensation temperature for a given humidity ratio of air is called the *dew-point temperature*. Instrumentation has been developed using the temperature of the dew point as a means of determining the exact moisture condition of humid air. The wet-bulb temperature can serve a similar purpose. Its instrument, the wet-bulb psychrometer which serves as a supplement to the saturator, is described in a subsequent section.

Example 3-5. Air leaves a well-insulated (adiabatic) saturator at 66 F, at a pressure of 29.92 in. Hg. It enters the saturator at 80 F and water is supplied as needed at 66 F. Find the humidity ratio, degree of saturation, enthalpy, and specific volume of the entering air.

Solution: Note that the saturated air at exit is at the thermodynamic wet-bulb temperature, t_w, and this is of course also the dry-bulb temperature, t_2, at that point. Read appropriate values from Table 3-1 and use with Eq. (3-27) or substitute values directly in Eq. (3-21).

At 66 F on exit

$$\mathbf{h}_{sw} = \mathbf{h}_{a2} = 30.79 \text{ Btu/lb air}$$
$$W_{sw} = 13.69 \text{ mlb} = 0.01369 \text{ lb/lb air}$$

By Eqs. (3-11) and (3-12),

$$\mathbf{h}_{a1} = (0.2403)(80 - 0) + W_1[1060.5 + (0.45)(80)] = 19.23 + W_1(1096.5)$$

$$\mathbf{h}_{fw} = 34.1 \text{ Btu/lb for the water added (Table 3-1 at 66 F)}$$

Substitute values in Eq. (3-21) and solve for W_1.

$$19.23 + W_1(1096.5) + (0.01369 - W_1)\,34.1 = 30.79$$

$$W_1 = 0.01043 \text{ lb/lb air} \qquad \textit{Ans.}$$

at 80 F, $W_s = 0.02226$ lb $= 22.26$ mlb

$$\mu = \frac{0.01043}{0.02226} = 0.47 \qquad \textit{Ans.}$$

$$\mathbf{h}_{a1} = 19.23 + 0.01043\,(1096.5) = 30.68 \text{ Btu/lb at inlet} \qquad \textit{Ans.}$$

Specific volume can be found by use of Table 3-1 and use of degree of saturation. At inlet temperature, 80 F,

$$v = v_a + \mu(v_s - v_a) \qquad (3\text{-}28)$$
$$= 13.60 + 0.47\,(14.09 - 13.60) = 13.83 \text{ ft}^3\text{/lb} \qquad \textit{Ans.}$$

Table 3-2 Corrections to Enthalpy of Moist Air in Btu/lb Dry Air for Atmospheric Pressures Other Than 29.92 in. Hg

Thermodynamic Wet-Bulb Temp, t_w(F)	Atmospheric Pressure, in. Hg						Thermodynamic Wet-Bulb Temp, t_w(F)	Atmospheric Pressure, in. Hg					
	24.92	25.92	26.92	27.92	28.92	30.92		24.92	25.92	26.92	27.92	28.92	30.92
33	0.87	0.67	0.49	0.31	0.15	−0.14	62	2.67	2.05	1.49	0.95	0.46	−0.43
34	0.91	0.70	0.51	0.32	0.16	−0.15	63	2.77	2.13	1.54	0.99	0.48	−0.45
35	0.95	0.73	0.53	0.34	0.16	−0.15	64	2.88	2.21	1.60	1.03	0.50	−0.46
36	0.99	0.76	0.55	0.35	0.17	−0.16	65	2.98	2.29	1.66	1.06	0.51	−0.48
37	1.02	0.79	0.57	0.37	0.18	−0.17	66	3.09	2.38	1.71	1.10	0.53	−0.50
38	1.06	0.82	0.59	0.38	0.18	−0.17	67	3.21	2.47	1.78	1.14	0.55	−0.51
39	1.11	0.85	0.61	0.39	0.19	−0.18	68	3.33	2.56	1.85	1.19	0.57	−0.53
40	1.15	0.89	0.64	0.41	0.20	−0.19	69	3.45	2.65	1.91	1.23	0.59	−0.55
41	1.20	0.92	0.67	0.43	0.21	−0.19	70	3.58	2.75	1.98	1.27	0.62	−0.57
42	1.25	0.97	0.70	0.45	0.22	−0.20	71	3.70	2.85	2.06	1.32	0.64	−0.59
43	1.30	1.00	0.72	0.46	0.22	−0.21	72	3.85	2.96	2.13	1.37	0.66	−0.62
44	1.35	1.04	0.75	0.48	0.23	−0.22	73	3.99	3.06	2.21	1.42	0.68	−0.64
45	1.41	1.08	0.78	0.50	0.24	−0.23	74	4.14	3.18	2.29	1.47	0.71	−0.66
46	1.46	1.12	0.81	0.52	0.24	−0.24	75	4.28	3.29	2.38	1.53	0.74	−0.68
47	1.52	1.17	0.84	0.54	0.26	−0.24	76	4.44	3.41	2.46	1.58	0.76	−0.71
48	1.58	1.21	0.88	0.56	0.27	−0.26	77	4.60	3.54	2.55	1.64	0.79	−0.73
49	1.64	1.26	0.91	0.58	0.28	−0.27	78	4.77	3.66	2.64	1.70	0.82	−0.77
50	1.68	1.31	0.95	0.61	0.29	−0.27	79	4.94	3.80	2.74	1.76	0.85	−0.79
51	1.77	1.36	0.99	0.63	0.31	−0.29	80	5.12	3.93	2.84	1.82	0.88	−0.82
52	1.84	1.41	1.02	0.66	0.32	−0.30	81	5.31	4.08	2.94	1.89	0.91	−0.85
53	1.91	1.47	1.06	0.68	0.33	−0.31	82	5.50	4.22	3.04	1.95	0.94	−0.88
54	1.98	1.52	1.10	0.71	0.34	−0.32	83	5.70	4.38	3.16	2.02	0.97	−0.91
55	2.06	1.58	1.14	0.74	0.36	−0.33	84	5.90	4.53	3.27	2.09	1.01	−0.95
56	2.13	1.64	1.19	0.76	0.37	−0.34	85	6.12	4.70	3.39	2.17	1.05	−0.98
57	2.22	1.71	1.23	0.79	0.38	−0.35	86	6.35	4.86	3.50	2.25	1.08	−1.01
58	2.30	1.77	1.28	0.82	0.40	−0.37	87	6.56	5.04	3.63	2.33	1.12	−1.05
59	2.39	1.84	1.33	0.85	0.41	−0.38	88	6.81	5.23	3.77	2.42	1.16	−1.08
60	2.48	1.90	1.38	0.88	0.43	−0.40	89	7.04	5.41	3.89	2.50	1.20	−1.12
61	2.58	1.98	1.43	0.92	0.44	−0.41	90	7.30	5.60	4.04	2.59	1.25	−1.16

Reproduced by permission from *ASHRAE Handbook of Fundamentals, 1967*. Compilation prepared by J. C. Davis, National Bureau of Standards.

Let us compute the middle term of Eq. (3-21) and compare with the values found: $(W_{sw} - W_1)\mathbf{h}_{fw}$, and we find,

$$(0.01369 - 0.01043)(34.1) = 0.11 \text{ Btu/lb air}$$

Notice that this is the amount by which the inlet enthalpy of the air $\mathbf{h}_{a1}$ is less than the enthalpy of saturated air at the thermodynamic wet-bulb temperature of the entering air.

$$\mathbf{h}_{aw} - \mathbf{h}_{a1} = 30.79 - 30.68 = 0.11 \text{ Btu/lb air}$$

Observe how small this term is in relative magnitude.

The derivations and developments in this chapter have largely been based on the perfect-gas relationships for the tabular approaches to the properties of moist air. It is unfortunately true that each table is constructed for a single barometric pressure and the difficulty of adjusting to other barometric pressures can produce great inaccuracies.

Because of the problems which arise from variations in barometric pressure, Table 3-2 has been included. This gives corrections to enthalpy at standard barometric pressure for a number of other pressures. To generalize for changes in barometric pressure, it should be mentioned that for a given dry-bulb and wet-bulb temperature, the values of humidity ratio and enthalpy increase with altitude, that is, for lower barometric pressures. The greatest change occurs with specific volume which for a given dry-bulb temperature and humidity ratio varies in an almost inversely proportional ratio to barometric pressure. Relative humidity changes little with changes in barometric pressure.

3-5. THE PSYCHROMETER

The phenomena of adiabatic saturation and thermodynamic wet-bulb temperature have been explained. However, it is obvious that an adiabatic saturator is not a workable instrument for measuring moist-air conditions and for many years an instrument known as a psychrometer has been used in its stead. The psychrometer measures the dry-bulb temperature and also a wet-bulb temperature. The wet-bulb temperature measured by this instrument is not precisely the thermodynamic wet-bulb temperature. Fortuitously however, it gives a reading which is very close to the thermodynamic wet-bulb temperature.

The *sling psychrometer* (Fig. 3-5) consists of two thermometers mounted side by side on a holder, with provision for whirling the whole device through the air. The dry-bulb thermometer is bare, and the wet bulb is covered by a wick which is kept wetted with clean water. After being whirled for a sufficient time the wet-bulb thermometer reaches its equilibrium point, and both the wet-bulb and dry-bulb thermometers are then quickly read. Rapid relative movement of the air past the wet-bulb thermometer is necessary to get dependable readings.

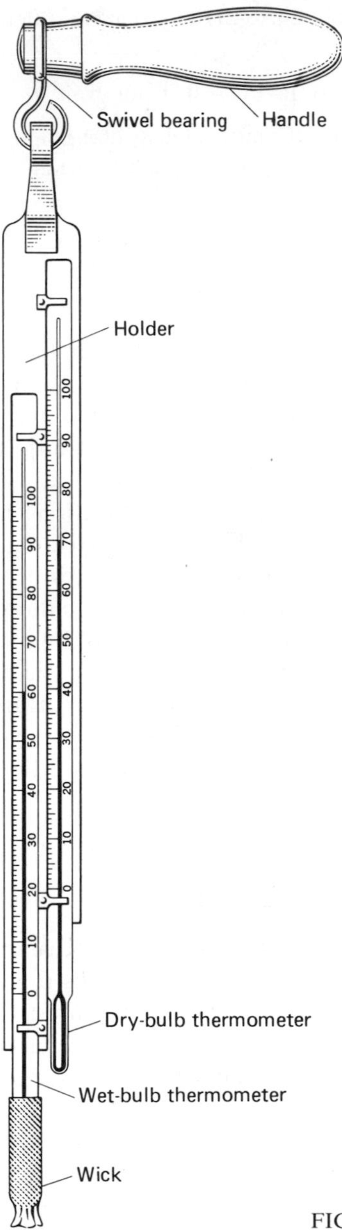

FIGURE 3-5. Sling psychrometer.

In the *aspiration psychrometer* (Fig. 3-6) a small fan is used to pull the air past the dry-bulb and wet-bulb thermometers to bring about wet-bulb equilibrium. If the water-supply temperature for the wick is much higher or lower than the wet-bulb temperature, readings should not be taken until it is certain that equilibrium has been reached.

To explain the phenomenon of wet-bulb temperature depression, imagine a free surface of water existing in unsaturated air. If the temperature

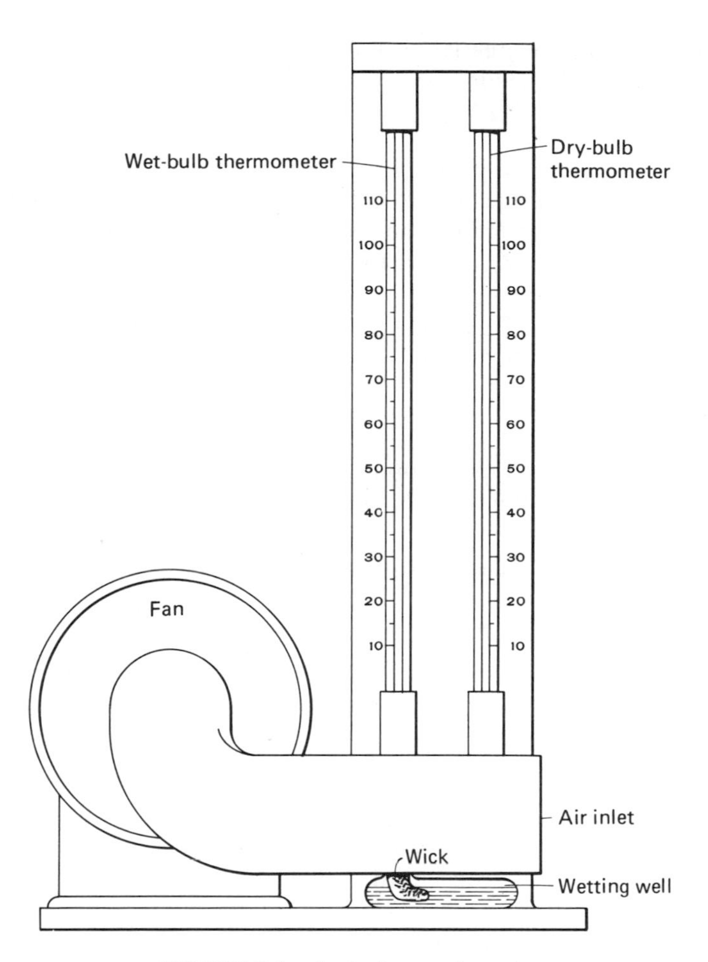

FIGURE 3-6. Aspiration psychrometer.

of the water is above the dew-point temperature, evaporation will take place from this surface into the air. The vapor leaves this surface of area, S, by diffusion through the contiguous air-vapor film. The rate of evaporation is proportional to the difference between the pressure p_w of the vapor at the liquid surface and the partial pressure p of the vapor in the air. Calling dW the weight of water evaporated in a differential of time, dT, we may write an equation as follows:

$$\frac{dW}{dT} = kS(p_w - p) \tag{A}$$

where k represents the coefficient of diffusion through a unit surface of film contiguous to the liquid.

Since the temperature t of the air is higher than the temperature t_w of the water, heat will flow from the air to the water. Calling dQ the amount of heat flowing in a differential of time dT, and calling f the coefficient of heat transfer through the film, we obtain the equation

$$\frac{dQ}{dT} = fS(t - t_w) \tag{B}$$

As heat flows from the air into the water, the temperature of the water will tend to rise and to reduce the heat inflow, but evaporation tends to increase as the temperature rises. Eventually, a condition of equilibrium will be reached when the heat dQ flowing into the water in a differential of time is exactly balanced by the heat used in evaporating the water, or $(dW/dT)(\mathbf{h}_{fg})$. Thus

$$\frac{dQ}{dT} = \frac{dW}{dT}(\mathbf{h}_{fg}) \tag{C}$$

where $\mathbf{h}_{fg}$ is the heat of vaporization of the water and dW/dT is the weight of water evaporated in time dT from the surface of area S.

Eliminating dW/dT from Eqs (A) and (C), and using the resulting equation with Eq. (B) to eliminate dQ/dT, we get

$$p_w - p = \frac{f}{\mathbf{h}_{fg}k}(t - t_w)$$

or

$$p = p_w - \frac{f}{\mathbf{h}_{fg}k}(t - t_w) \tag{3-29}$$

Equation (3-29) gives a means of finding the water-vapor pressure in the air if the term $f/\mathbf{h}_{fg}k$ can be evaluated either experimentally or otherwise. This term can almost be considered a constant, since the value of $\mathbf{h}_{fg}$ for steam does not greatly vary over a 20- to 30-deg temperature range, and f and k depend on the gas-film thickness contiguous to the water surface. If f and k are determined for a normal set of operating conditions, as with a sling psychrometer, they, also, may be considered practically constant. Heat radiating to the wetted bulb from surrounding warmer air and surfaces tends to raise the temperature t_w slightly, and so a psychrometer should be operated with such velocity as to reduce the ratio f/k to a minimum and essentially constant value, in order to make such radiation effects of small importance.

The value t_w in the equilibrium condition is the so-called wet-bulb temperature. To make Eq. (3-29) usable, the essentially constant term $f/\mathbf{h}_{fg}k$ must be evaluated, and it will be realized that the barometric pressure p_B will also affect the equation. Various forms of this equation, with the constants evaluated, are now in use. The most common of these equations are as follows:

1. The modified Apjohn equation, proposed in 1837:

$$p = p_w - \frac{p_B}{30}\frac{t - t_w}{90} \tag{3-30}$$

2. The modified Ferrel equation, proposed in 1886:

$$p = p_w - 0.000367 p_B(t - t_w)\left(1 + \frac{t_w - 32}{1571}\right) \tag{3-31}$$

3. The Carrier equation, proposed in 1911:

$$p = p_w - \frac{(p_B - p_w)(t - t_w)}{2800 - 1.3t_w} \tag{3-32}$$

where p is the partial pressure of the water vapor in the atmosphere; p_w is the pressure of saturated water vapor at the wet-bulb temperature; p_B is the barometric pressure; t is the dry-bulb temperature, degrees Fahrenheit; and t_w is the wet-bulb temperature, degrees Fahrenheit.

The pressures p, p_w, and p_B must be expressed in the same units, as inches or mercury or pounds per square inch.

Equations (3-31) and (3-32) have more correction terms and give results more accurate than the Apjohn equation [Eq. (3-30)]. However, in many cases Eq. (3-30) is sufficiently accurate, particularly when the experimental wet-bulb and dry-bulb readings have not been found to a precision closer than ± 0.3 deg. The Carrier equation has been used extensively in air-conditioning work.

In actual practice, however, these equations are not used to any great extent; graphical charts made from plots of them are employed instead. Charts of this type are known as psychrometric charts, several of which are included in this book.

It must be realized that the wet-bulb temperature and the dew-point temperature are different entities. However, in the one case of saturated air, the dry-bulb, wet-bulb, and dew-point temperatures must, of course, all be the same. In the case of nonsaturated air, the dry-bulb temperature represents the actual temperature of the air as measured by an ordinary thermometer; the dew-point temperature is that temperature to which the air, with its existing moisture content, would have to be cooled before saturation and any condensation could occur; and the wet-bulb temperature represents that temperature which a thermometer having a bulb covered with a wetted wick would reach if whirled through the air.

Example 3-6. Air has a dry-bulb temperature of 70 F and a wet-bulb temperature of 64 F, and the barometer indicates 29.9 in. Hg. Without the use of a psychrometric chart, find (a) the relative humidity of the air, (b) the (water) vapor density in the air, (c) the humidity ratio, and (d) the dew point.

Solution: (a) Equation (3-30), (3-31), or (3-32) may be applied to find p, but Eq. (3-32) will be used here. Thus

$$p = p_w - \frac{(p_B - p_w)(t - t_w)}{2800 - 1.3t_w}$$

From Table 3-1, $p_w = 0.6007$ in. Hg for the wet-bulb temperature of 64 F. Also, $p_B = 29.9$ in. Hg; $t = 70$ F; $t_w = 64$ F. Then, the partial pressure of the vapor in the air is

$$p = 0.6007 - \frac{(29.9 - 0.6007)(70 - 64)}{2800 - (1.3)(64)} = 0.5360 \text{ in. Hg}$$

The relative humidity may now be found by Eq. (3-3). From Table 3-1, $p_s = 0.7392$ in. Hg for the dry-bulb temperature of 70 F. Hence

$$\phi = \frac{0.5360}{0.7392} = 0.725, \text{ or } 72.5\% \qquad \textit{Ans.}$$

(b) From Eq. (3-4),

$$d = \phi d_s$$

and from Table 3-1, the specific volume of saturated steam at $t = 70$ F is $v_g = 868.0$ ft^3/lb; and $d_s = 1/868 = 0.001152$. Therefore, the weight of vapor in the air is

$$d = (0.725)(0.001152) = 0.000835 \text{ lb/ft}^3 \qquad \textit{Ans.}$$

(c) By Eq. (3-5)

$$W = 622 \frac{0.536}{29.9 - 0.536} = 11.35 \text{ mlb/lb dry air}$$

$$= 11.35 \text{ g/kg dry air} \qquad \textit{Ans.}$$

(d) The dew-point temperature exists for saturated air having the above humidity ratio (specific humidity). Refer to Table 3-1 and for $W_s = 11.35$ mlb read by interpolation the dew-point temperature as 60.8 F.

The question naturally arises as to how closely the thermodynamic wet-bulb temperature (temperature of adiabatic saturation) and the wet-bulb temperature as determined by a psychrometer are in agreement. Fortunately experimentation has shown that the agreement is very close, well within the range of experimental error, and results determined by either method can be considered reliable. Analytic comparisons of Eqs. (3-27) and (3-29) also show that when the various parameters are evaluated and compared the agreement is very close. It is even possible to obtain reliable data from a psychrometer at temperatures below 32 F (0°C). In this region a thin layer of ice is frozen onto the wet bulb and when the psychrometer is operated, sublimation from the iced bulb causes a depression in temperature similar to that which occurs at temperatures above 32 F in the nonfrost regime.

Dew-point instrumentation is also available for determining the moisture characteristics of air. In such instruments a polished surface, the temperature of which can be carefully controlled and measured, is chilled until moisture (dew or frost) is observed to form as the air sample is brought in contact with the surface. At the instant of dew formation the temperature is read. The surface is then warmed and the temperature at which the frost disappears is noted. The two temperatures should be very close and, under proper operation, the instrument furnishes very accurate dew-point readings. Liquid nitrogen or carbon dioxide can be employed for controlling the surface temperature.

3-6. THE PSYCHROMETRIC CHART

It has now been shown how all the important properties of air-stream mixtures are interrelated and how each of them can be calculated. To simplify the labor of making calculations and to illustrate processes, charts representing air-stream properties, drawn for a given barometric (or total) pressure, are of inestimable value. Such charts have taken many forms, and each form may have special advantages.

In this text the psychrometric chart which was developed by the American Society of Heating, Refrigerating, and Air-Conditioning Engineers was selected for inclusion. This chart uses as basic coordinates dry-bulb temperature as abscissa and humidity ratio as ordinate. However, a diagonal axis presents enthalpy as a coordinate so the chart has at least one characteristic of a Mollier chart, so-called because Richard Mollier was the first to use a psychrometric chart with enthalpy as a coordinate. An abridgement of the ASHRAE chart for normal temperature is shown in Fig. 3-7. The horizontal axis represents dry-bulb temperature by degree intervals. The dry-bulb temperature lines are straight but are not precisely parallel to each other and incline slightly from the vertical, usually to the left. Humidity ratio in pounds of water per pound of dry air appears at the right side of the chart using a uniform scale with lines horizontal. The saturation curve, which includes the wet-bulb and dew-point temperature, swings upward to the right. On this curve it follows that at saturation, wet-bulb, dry-bulb, and dew-point temperatures are equivalent and equal. Relative-humidity lines of similar shape are shown on 20% intervals. The enthalpy lines appear drawn obliquely down the chart expressed in units of Btu per pound of dry air. The enthalpy lines are parallel to each other and close readings can be made on the scale to which the enthalpy lines are extended. The wet-bulb temperature lines are straight but are not parallel to each other, and since enthalpy and wet-bulb temperature are not in a fixed equivalent ratio to each other, the enthalpy and wet-bulb temperature lines, although coincident at the saturation curve, diverge from each other in the body of the chart.

A portion of the fog region has been drawn with enthalpy and wet-bulb temperature lines extended into it. This two-phase region is a mechanical mixture of saturated moist air and liquid water in suspension. Specific volume lines are obliquely drawn at intervals of 0.5 ft^3/lb dry air. These lines are straight but adjacent lines are not precisely parallel to each other.

The chart has been constructed, using the thermodynamic data from the tables of Goff and Gratch (Ref. 5). The basic construction of a psychrometric chart of this type is not difficult since the dry-bulb scale and the humidity-ratio scale are arbitrarily chosen, and it requires merely plotting values

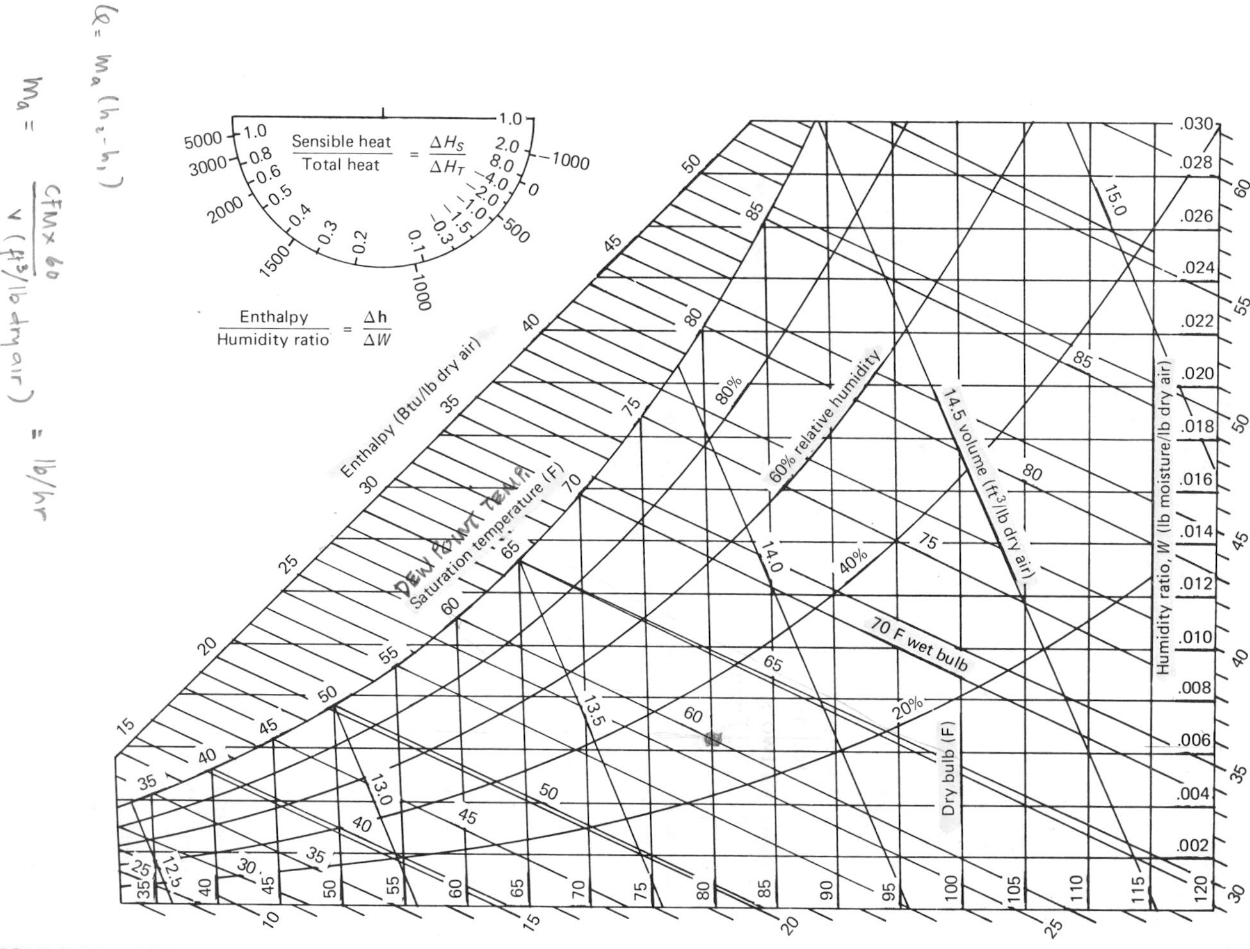

FIGURE 3-7. ASHRAE skeleton psychrometric chart for normal temperatures. (Reproduced by permission from *ASHRAE Guide and Data Book 1966*.)

$Q = m_a (h_2 - h_1)$

$m_a = \dfrac{CFM \times 60}{v \ (ft^3/lb \text{ dry air})} = lb/hr$

from the moist-air tables to locate wet-bulb and dew-point temperatures on the saturation line of the chart. Enthalpy values can be added for saturation conditions. A more difficult problem is to locate enthalpy and wet-bulb temperatures in the body of the chart, but since both enthalpy and wet-bulb temperatures are defined in terms of dry-bulb and humidity ratio, each point can be uniquely located and it is possible to find the whole body of the chart. Better accuracy can be obtained if the dry-bulb lines are shifted from their orthogonal position relative to humidity ratio.

Since it is awkward to cover a large range of dry-bulb temperatures on one chart, it is customary to provide three charts for the working temperature range, Plate I for normal (medium) temperatures 32 F to 120 F, Plate II for low temperatures −40 F to 50 F, and Plate III for high temperatures 60 F to 250 F. Mention has already been made that a basic psychrometric chart is applicable for one specific pressure and in this case, the three charts are for standard barometric pressure 29.92 in. Hg. If psychrometric information is needed at another pressure, either a separate chart has to be constructed for that pressure or corrections to a basic chart are required. In this text an additional chart has been provided for 5000 ft elevation representing a standard atmospheric pressure of 24.89 in. Hg and is included as Plate IV.

A protractor and nomograph appear at the left of the chart. The protractor shows two scales; one involves the ratio of enthalpy difference to humidity-ratio difference, the other the ratio of sensible to total heat. The use of the protractor nomograph will be illustrated in subsequent problems.

Example 3-7. Read from the psychrometric chart the properties of moist air at 80 F dry-bulb, 60 F wet-bulb at 29.92 in. Hg barometric pressure.

Solution: Refer to the skeleton chart of Fig. 3-7 for guidance and then read from Plate I the desired results with precision.

Humidity ratio (specific humidity), W: Move up the 80 F dry-bulb line to its intersection with the 60 F wet-bulb line. Follow the W line to the right and read $W = 0.0066$ lb steam/lb dry air.

Dew-point temperature: From the same intersection follow the W line to the saturation temperature curve and read $t_d = 45.9$ F.

Enthalpy: Method 1. Make use of two triangles and from the $t = 80$ F, $t_w = 60$ F intersection locate a line parallel to an adjacent enthalpy line and extend this to the edge scale to read $\mathbf{h} = 26.35$ Btu/lb dry air.

Enthalpy: Method 2. The magnitude of D in Eq. (3-25) shows the amount by which the enthalpy of unsaturated moist air differs from the enthalpy of saturated air at the same wet-bulb temperature. This difference is called the deviation. At $t_w = 60$ F on the saturation line, read $\mathbf{h}_{aw} = 26.46$. To find D, read the nomograph at the upper left of Plate I for $W = 0.0066$ and $t_w = 60$ F as -0.11. Thus the enthalpy is $\mathbf{h} = \mathbf{h}_{aw} - D = 26.46 - 0.11 = 26.35$ Btu/lb dry air.

Relative humidity: Read by linear interpolation at the $t = 80$ F, $t_w = 60$ F intersection point, $\phi = 30\%$.

Specific volume: At the $t = 80$ F, $t_w = 60$ F intersection point read by linear interpolation between adjacent volume lines $v = 13.72$ ft^3/lb dry air.

Later in this chapter another chart will be presented. That chart is in SI units and uses the "deviation" approach to correct enthalpy values for any air which is not at saturation.

3-7. AIR-CONDITIONING PROCESSES USING PSYCHROMETRIC CHARTS AND MOIST-AIR TABLES

Heating of Air: If air is heated or cooled without the addition of moisture, the humidity ratio (specific humidity) remains constant and this process appears as a straight horizontal line on the psychrometric chart. On the skeleton chart of Fig. 3-8 heating is shown taking place from 1 to 2 or cooling taking place from 2 to 1. In equational form, we would read

$$q_{1-2} = m_a(\mathbf{h}_2 - \mathbf{h}_1) \tag{3-33}$$

The enthalpy values can be read from the chart as shown in the diagram.

The heat added or removed during such a process which takes place at constant humidity can also be computed by use of the following equation since the process is a sensible-heat change:

$$\begin{aligned} q_{1-2} = q_s &= m_a[C_{pa}(t_2 - t_1) + C_{ps}W(t_2 - t_1)] \\ &= m_a(C_{pa} + C_{ps}W)(t_2 - t_1) \\ &= m_a C_p(t_2 - t_1) \end{aligned} \tag{3-34}$$

where q_s is the heat added or removed with no moisture change, in Btu per m_a pounds of dry air; C_{pa} is the specific heat of dry air, equal to 0.24 (approximately); C_{ps} is the specific heat of steam, equal to 0.45 (approximately);

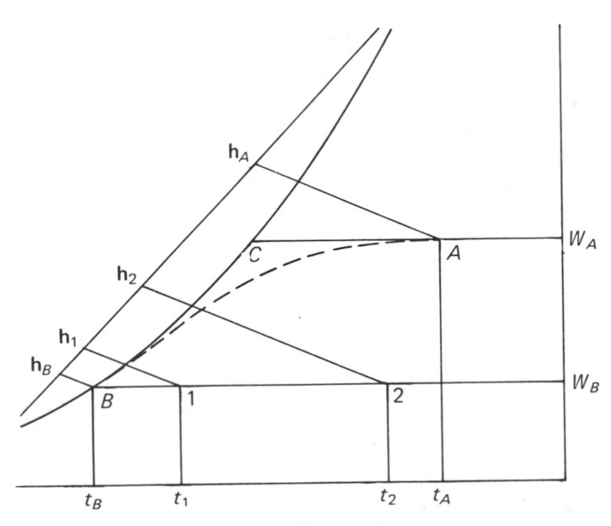

FIGURE 3-8. Processes on psychrometric chart. Heating with humidity ratio constant, path 1 to 2; cooling with humidity ratio constant, path 2 to 1; cooling and dehumidification to a saturated condition, path A to B.

W is the humidity ratio, in pounds per pound of dry air; and $t_2 - t_1$ is the dry-bulb temperature change, in degrees Fahrenheit.

In the technical literature, the composite term $C_{pa} + C_{ps}W = C_p$ has been called *humid heat*. At the low temperatures of the air-conditioning range, W is so small that humid heat differs very little from the value of C_p for air. At higher temperatures, in particular those above 100 F, the second term of this expression for humid heat assumes much greater significance. For general values of humid heat to use in the air-conditioning range, 0.242 to 0.245 are representative.

Example 3-8. Find the heat required to warm 5000 cfm of outside air at 39 F and 80% relative humidity to 90 F without addition of moisture. Barometric pressure = 29.92 in. Hg.

Solution: Use the psychrometric chart Plate I and read that at $t_1 = 39$ F, $\phi = 80\%$, that $\mathbf{h}_1 = 13.68$ Btu/lb air, $W_1 = 0.0040$ lb/lb dry air. At the same value of W_1, read at $t_2 = 90$ F that $\mathbf{h}_2 = 26.02$ Btu/lb, $\phi = 13.93\%$. At $t_1 = 39$ F, $\phi = 80\%$, read on the chart that $v = 12.64$ ft^3/lb dry air; then

$$m_a = \frac{5000 \text{ cfm} \times 60}{12.64} = 23\,700 \text{ lb/h}$$

Substituting in Eq. (3-33),

$$q_{1-2} = 23\,700\,(26.02 - 13.68) = 292\,500 \text{ Btu/h}$$

or by use of Eq. (3-34),

$$q_{1-2} = (23\,700)(0.242)(90 - 39) = 292\,500 \text{ Btu/h}$$

Example 3-8 shows how the relative humidity decreases when air is heated with no addition of moisture. The relative-humidity decrease represents a usual winter condition. When air is cooled with no moisture change, the relative humidity, of course, increases. This condition is common in summer.

Cooling and dehumidifying moist air. Whenever air is cooled to a temperature lower than its original dew point, some of the water vapor or steam in the original air necessarily condenses out. This condensation does not take place at a fixed temperature but continues over a variable path related to the surface temperature and placement of the cooling coils or the temperature and manner by which chilled water is sprayed into the air in a direct-contact cooler. The path of cooling and dehumidification to a saturated state is indicated in Fig. 3-8 by the dotted line A to B. However, an actual path might be quite different, perhaps moving close to C without any condensation and then dropping to A on a path much closer to the saturation curve. For computing the energy interchange the actual path is immaterial since the total energy and mass transfer from initial to final states alone are of significance, a fact which follows from a steady-flow energy analysis [Eq. (2-9)]. Thus we can write an equation for the heat removal $-q_{A-B}$ as follows:

$$m_a \mathbf{h}_A - q_{A-B} = m_a \mathbf{h}_B + m_a(W_A - W_B)\mathbf{h}_{fB} \tag{3-35}$$

m_w, the water condensed from m_a lb dry air, is

$$m_a(W_A - W_B) = m_w \tag{3-36}$$

$$q_{A-B} = m_a[(\mathbf{h}_A - \mathbf{h}_B) - (W_A - W_B)\mathbf{h}_{fB}] \tag{3-37}$$

For m_a lb of dry air flowing per hour, q_{A-B} represents the heat removal required for the process in Btu per hour. In the equation above it is assumed that all the condensate is removed at the saturation temperature, t_B. If some is removed, at any other temperature, the enthalpy of the liquid, $\mathbf{h}_f$, must also be selected for the other temperature and used with the appropriate mass involved, to show its contribution to the total.

In connection with refrigeration and cooling processes, use is made of a unit called the *ton of refrigeration*, defined as an energy interchange of 12 000 Btu per hour. In spite of its name, the ton of refrigeration is a power unit and has no relation to mass or weight. Historically the unit came into existence as representing the cooling created when 1 ton (2000 lb) of ice melted during a 24-hour period (approximately 288 000 Btu).

$$\begin{aligned} \text{1 ton of refrigeration} &= 12\,000 \text{ Btu/h} = 200 \text{ Btu/min} \\ &= 3024 \text{ kcal/h} = 3.51 \text{ kW} \end{aligned} \tag{3-38}$$

$$\begin{aligned} 1 \text{ kW} &= 0.2845 \text{ ton of refrigeration} = 860.4 \text{ kcal/h} \\ &= 56.9 \text{ Btu/min} \end{aligned} \tag{3-39}$$

Example 3-9. Find the refrigeration required to cool 10 000 cfm of outside air at 90 Fdb, 80 Fwb to a condition of saturation at 56 F. Condensate is removed at 56 F and barometer is at 29.92 in. Hg.

Solution: Using Plate I, at $t_A = 90$ F, $t_{wa} = 80$ F, read $\mathbf{h}_A = 43.52$ Btu/lb dry air, $\phi = 65\%$. $W_A = 0.0195$ lb/lb dry air, $v = 14.3$ ft^3/lb dry air. Also at $t_B = 56$ F $= t_{wB} = t_{dB}$, read $\mathbf{h}_B = 23.84$ Btu/lb dry air, $W_B = 0.0096$ lb/lb dry air, $\phi = 100\%$ at saturation. Read from Table 3-1 at 56 F that $\mathbf{h}_{fB} = 24.1$ Btu/lb water.

$$m_a = \frac{10\,000 \times 60}{14.3} = 41\,950 \text{ lb dry air/h, being cooled}$$

By use of Eq. (3-37) the refrigeration required is

$$q_{A-B} = 41\,950\,[43.52 - 23.84 - (0.0195 - 0.0096)\,24.1] = 815\,500 \text{ Btu/h}$$

Expressed in tons of refrigeration,

$$q_{A-B} = \frac{815\,500}{12\,000} = 67.9 \text{ tons}$$

Note that the condensate removed per hour is

$$m_a(W_A - W_B) = 41\,950\,(0.0195 - 0.0096) = 415 \text{ lb water/h}$$

Air mixing. When two quantities of air having different enthalpies and different specific humidities are mixed, the final condition of the air mixture depends on the masses (weights) involved, and on the enthalpy and humidity ratio of each of the constituent masses which enters the mixture. If m_A pounds of air at enthalpy $\mathbf{h}_A$ and specific humidity W_A are mixed with m_B pounds of air at enthalpy $\mathbf{h}_B$ and specific humidity W_B, the following equations will obviously apply:

$$m_A \mathbf{h}_A + m_B \mathbf{h}_B = (m_A + m_B)\mathbf{h}_m \tag{3-40}$$

or

$$\mathbf{h}_m = \mathbf{h}_A + \frac{m_B}{m_A + m_B}(\mathbf{h}_B - \mathbf{h}_A) \tag{3-41}$$

and

$$m_A W_A + m_B W_B = (m_A + m_B)W_m \tag{3-42}$$

or

$$W_m = W_A + \frac{m_B}{m_A + m_B}(W_B - W_A) \tag{3-43}$$

where m_A and m_B are the weights of air mixed, in pounds or pounds per unit time; $\mathbf{h}_A$ and $\mathbf{h}_B$ are the enthalpies per pound associated with each of the weights of dry air mixed; and W_A and W_B are the humidity ratios (specific humidities) associated with each pound of dry air being mixed, in grams or pounds.

If, on a psychrometric chart like the skeleton chart shown in Fig. 3-9, a straight line is drawn to connect the state points A and B of two weights of air which enter into a mixing process, the resultant mixture will be found to have a state point which falls on the line or lies very close to the line.

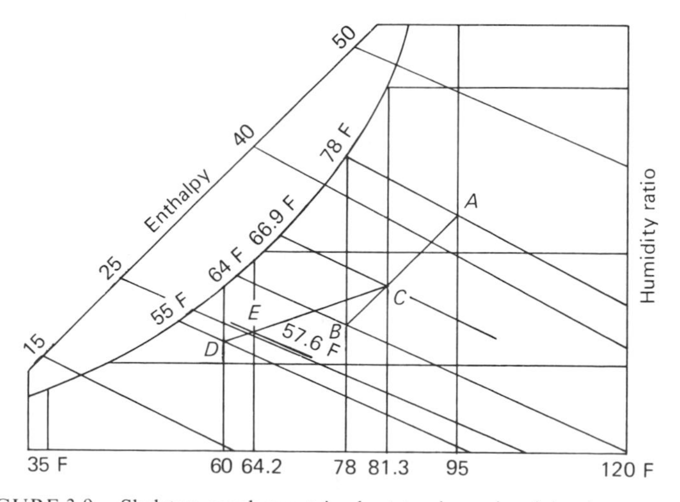

FIGURE 3-9. Skeleton psychrometric chart to show air mixing in two steps.

The location of the mixing point on the line is determined by the relative weights of the materials entering into the mixing. For example, if equal weights were mixed, the point would be midway between the two end points. Similarly, if 2 parts by weight were mixed with 1 part by weight, the point would lie one-third of the distance along the line from the point corresponding to the larger quantity entering into the process. This straight-line relationship which occurs in mixing is extremely convenient, and sufficiently accurate for most types of calculation.

It is easy to rearrange Eqs. (3-40) and (3-42) to show that

$$\frac{\mathbf{h}_B - \mathbf{h}_m}{\mathbf{h}_m - \mathbf{h}_A} = \frac{W_B - W_m}{W_m - W_A} = \frac{m_A}{m_B} \tag{3-44}$$

In this form it is obvious that the line segments bear a direct ratio to the masses of dry air (m_A and m_B) in the mixing supply streams.

It is also possible to show, to a very close approximation, that the dry-bulb temperature resulting from two weights of air being mixed is:

$$t = t_A + \frac{m_B}{m_A + m_B}(t_B - t_A) \tag{3-45}$$

For Eqs. (3-41) and (3-43), a further approximation can be made—if the points are not too far apart—to the effect that the terms in the ratio $m_B/(m_A + m_B)$ may be expressed in cubic feet of air (cfm) instead of in pounds or pounds per hour.

Example 3-10. From a conditioned space, 8000 cfm of air at 78 Fdb and 64 Fwb are recirculated and mix with 2000 cfm of outside air at 95 F and 78 Fwb. (a) Find the condition of the resultant mixture. This mixture then enters the conditioner where 80% of the mixture, by weight, is cooled and dehumidified to 60 F and 55 Fwb while the rest by-passes the coils and remains unchanged. The cooled and the by-passed air then mix to become the supply air for the conditioned space. (b) Find the temperature and relative humidity of this supply air.

Solution: Refer to Fig. 3-9 which illustrates this problem on a skeleton psychrometric chart. Then on Plate I locate points A (95 F, 78 F) and B (78 F, 64 F) and connect these with a straight line. Read $\mathbf{h}_A = 41.32$ Btu/lb dry air, $\mathbf{h}_B = 29.21$, $v_A = 14.37$ ft^3/lb dry air, and $v_B = 13.73$. $m_A = 2000/14.37 = 139.2$ lb/min, $m_B = 8000/13.73 = 583$ lb/min. By Eq. (3-41),

$$\mathbf{h}_M = 41.32 + \frac{583}{722.2}(29.21 - 41.32) = 31.52 \text{ Btu/lb dry air}$$

From this value on the diagonal enthalpy scale, use two triangles and run the enthalpy line to intersect the connecting line at point C, which is seen to be at 81.3 F, 66.9 Fwb. The same result can be found by making use of the teaching of Eq. (3-44) which shows that lengths on the line are directly proportional to the supply mass flow rates m_A and m_B. Here

$$\frac{m_A}{m_B} = \frac{139.2}{583} = \frac{1}{4.18}$$

and the split is in the ratio of 1 to 4.18 so that the line split is 1/5.18 of the length and 4.18/5.18 of the length. The length can be measured in inches or millimeters and the point, C, can then be located after multiplying the total length by either ratio. Note that the shorter length lies next to the larger air-flow rate.

Of the 722.2 lb/min entering the conditioner, 80% is cooled to 60 F, 55 F and the remaining 20% is by-passed. Locate point D at 60 F, 55 Fwb on Plate I and draw a line to the previous mixture point, C, just determined at 81.3 F, 66.9 F. Measure the length of this line and then lay off 0.20 of that length from D to locate the resultant leaving air at E having temperatures of 64.2 F and 57.6 Fwb.

Humidification of moist air. Humidification can be accomplished when steam is supplied either from direct source or by vaporization from heated or unheated water surfaces or from a water spray in the air supply. Equation (3-16) is applicable to a humidification process, but some modification in interpretation of the terms is necessary. In that equation, $\mathbf{h}_{f3}$ represents the enthalpy of water added or removed and it is tacitly assumed that the water is in a liquid state; however, this energy balance is equally applicable to water in any form (liquid, steam, or solid), provided the proper value of enthalpy is assigned. Therefore $\mathbf{h}_{f3}$ will be written $\mathbf{h}_3$ and will indicate merely enthalpy, in Btu per pound, of the actual water, water vapor, or ice added or removed. Thus the equation takes the following form:

$$\mathbf{h}_{a1} \pm Q_{(1-2)} + (W_2 - W_1)\mathbf{h}_3 = \mathbf{h}_{a2} \tag{3-46}$$

If the process is also adiabatic, the equation becomes

$$\mathbf{h}_{a1} + (W_2 - W_1)\mathbf{h}_3 = \mathbf{h}_{a2} \tag{3-47}$$

Equation (3-47) can be written in the following form:

$$\frac{\mathbf{h}_{a2} - \mathbf{h}_{a1}}{W_2 - W_1} = \mathbf{h}_3 \tag{3-48}$$

From this, for adiabatic mixing of steam and air, it can be seen that the ratio of enthalpy difference ($\Delta\mathbf{h}$) to humidity ratio (ΔW) sets the slope of a line on the psychrometric chart, the direction of which is fixed by the enthalpy of the water (steam) supply. The line in question passes through the initial and final state points.

Example 3-11. General supply air in a certain industrial plant is available at 70 F and 60% relative humidity. Air is required at 100 F for a particular process, and a higher specific humidity is not objectionable. If steam from the plant boilers at 600 psia, superheated to 1000 F (for which $\mathbf{h}$ = 1516.7 Btu/lb), is throttled into the supply air (70 F and $\phi = 60\%$) and the final temperature is 100 F, what characteristics will the air have after mixing with the steam?

Solution: This process is adiabatic and therefore Eq. (3-47) is applicable. From Plate I and given values, we read

$$\mathbf{h}_{a1} = 27.02 \text{ Btu/lb dry air}$$
$$W_1 = 0.0094 \text{ lb/lb dry air}$$

and

$$\mathbf{h}_3 = 1516.7 \text{ Btu/lb steam}$$

With the values of $\mathbf{h}_{a2}$ and W_2 at 100 F unknown, to solve this problem locate $\mathbf{h}_{a1}$ at 70 F and $\phi = 60\%$ on Plate I and through this point draw a line parallel to a line slope of 1516.7, located on the protractor at the left of Plate I. The intersection of a line having this slope when carried to 100 F shows that $\mathbf{h}_{a2} = 52.68$ Btu/lb dry air, $W_2 = 0.0260$ lb/lb dry air. The parallel line can be drawn with the help of two triangles.

Figure (3-10) is a skeleton chart showing this process for the slope 1516.7 and points 1 and 2.

If use is not made of the protractor, Eq. (3-47) can be written as follows:

$$27.02 + (W_2 - 0.0094)\,1516.7 = \mathbf{h}_{a2}$$

and a solution to accurately match the two unknown values is sought.

The problem involves finding the respective values of W_2 and $\mathbf{h}_{a2}$ which constitute the solution to this equation. A trial-and-error solution usually can produce a close answer after about three or four substitutions of appropriate humidity ratio, enthalpy values. This approach can also be used to check the accuracy of the graphical method and this should be done since the slope of the line cannot be determined with a high degree of precision. If necessary, slightly modified values of $\mathbf{h}_{a2}$ and W_2 should be selected to make the equation reach the degree of accuracy desired.

It should be observed that, for the heating of air, direct addition of steam is not particularly effective, although such mixing is effective in increasing the humidity ratio. Relative humidity may increase or decrease in the process.

Evaporative cooling. In Section 3-4 it was shown that an adiabatic saturator, using recirculated water, will bring the dry-bulb temperature of the entering air to its original wet-bulb temperature whenever the air is brought to complete saturation. In the process the dry-bulb temperature is lowered

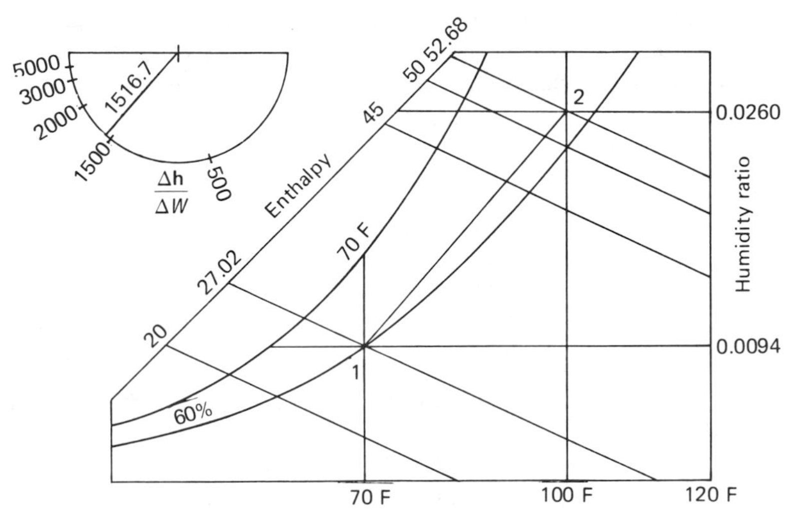

FIGURE 3-10. Skeleton psychrometric chart showing use of $\Delta\mathbf{h}/\Delta W$ slope, employed for enthalpy of steam added.

toward the wet-bulb temperature because its sensible heat is absorbed in evaporating water. If the moist air leaves the saturator before complete saturation occurs, the air will be at a temperature higher than that corresponding to its wet-bulb temperature at saturation. Even so, the wet-bulb temperature of the air has not changed from its original value and the process of adiabatic evaporative cooling still takes place at the wet-bulb temperature of the entering air. On a psychrometric chart then, a line of constant wet-bulb temperature represents the process. Equation (3-21) is applicable and for the process to be truly consistent, the make-up water to the cooler would have to be provided at the wet-bulb temperature. If it is not, there is a trivial deviation from the wet-bulb line. So-called desert coolers depend on evaporative cooling to lower air temperature.

Example 3-12. On a given day, air in the southwestern United States was supplied, at 105 Fdb and 20% relative humidity, to an air-washer type of humidifier using recirculated water. If the humidifier had adequate spray-nozzle capacity to bring the air to within 70% of the original wet-bulb depression, what were the conditions of the air leaving the humidifier?

Solution: From Plate I, original air at 105 Fdb and 20% relative humidity has a wet-bulb temperature of 72.3 F. The original wet-bulb depression is 105 − 72.3 = 32.7 deg. The final air temperature required to reach 70% of the original wet-bulb depression is 105 − 0.7 (32.6) = 82.1 Fdb.

The final condition is at 72.3 Fwb and 82.1 Fdb and $\phi = 62\%$.

This process, which is adiabatic, follows a constant wet-bulb line and therefore Eq. (3-21) applies. The recirculated water with a small make-up reached essentially the wet-bulb temperature of the entering air, 72.3 F, at which temperature $\mathbf{h}_{fw} = 40.4$ Btu/lb (see Table 3-1). From Plate I, $\mathbf{h}_{a1} = 35.78$ Btu/lb dry air, $W_1 = 0.00945$ lb/lb dry air, $\mathbf{h}_{a2} = 36.00$ Btu/lb dry air, and $W_2 = 0.0148$ lb/lb dry air when these are read at 82.1 F and 72.3 Fwb. Substitution in Eq. (3-21) shows a very close check.

$$35.78 + (0.0148 - 0.00945)(40.4) = 36.00$$

The relative humidity on leaving the humidifier has reached 62%. This is not objectionable for comfort and is so preferable to a 105 F temperature that in dry atmospheres evaporative cooling can be employed to advantage, when mechanically refrigerated cooling is not available.

Spray-type equipment can also be used to dehumidify air, in which case the water supply must be externally cooled below the dew-point temperature of the supply air and the process is no longer an adiabatic one at constant wet-bulb temperature. Humidifying with externally heated hot water is also possible with spray-type equipment. This process is not an adiabatic constant wet-bulb process.

Dew point and relative humidity. An important relationship holds approximately true between relative humidity and the dew-point differential.

For example, for a dew-point differential 10 deg below dry-bulb temperature the relative humidity of the air is approximately 70% over a moderate range of dry-bulb temperatures. Again, for a dew-point differential of 20 deg, the relative humidity is approximately 50% over a moderate range of dry-bulb temperatures. Similar relative-humidity values for other dew-point differentials can be read from the psychrometric chart. This comparative constancy of relative humidity for constant dew-point differentials is utilized in some air-conditioning control systems.

3-8. AIR-SUPPLY CONDITIONS, SPACE AIR CONDITIONING

When a space is heated by warm air, the air must be supplied at a dry-bulb temperature which is sufficiently high that in cooling to the desired room temperature it offsets the heat losses from the space. The cooling load presents a similar problem, in a reverse sense. For example, the cooling load on an auditorium filled with people involves the problem of supplying air at a dry-bulb temperature low enough to keep the air in the auditorium from rising above a desired maximum temperature, and also the problem of supplying air that is sufficiently low in moisture content to keep the humidity of the air in the auditorium from rising above a desired value.

A major part of the heat load (really cooling load) in a space arises from heat transfers through boundaries and from heat sources in a space such as lights, machinery, and from a portion of the heat added by occupants in the space. This type of heat addition, if not offset, produces solely a temperature rise in the space which would be evident to the senses; consequently it is called sensible heat (q_s). However, occupants, through their metabolic processes, also deliver moisture into the air, as do many cooking and drying processes so that steam may also be added to the air of a space. The supply air must thus be sufficiently dry to keep the moisture load from exceeding a predetermined value. The energy associated with steam additions is obviously the various masses of steam Σm_v multiplied by appropriate enthalpy values, $\mathbf{h}_v$, for the steam. Let us then call $\Sigma m_v \mathbf{h}_v$ a summation term for all of the steam additions made at respective enthalpy values $\mathbf{h}_v$. An energy equation for a place provided with an air stream of m_a pounds of dry air per hour, with entering enthalpy $\mathbf{h}_{a1}$ and with sensible energy (q_s), and steam additions being added in the space would appear:

$$m_a \mathbf{h}_{a1} + q_s + \Sigma m_v \mathbf{h}_v = m_a \mathbf{h}_{a2} \tag{3-49}$$

This is a true equation since the same amount of "dry" air (m_a) must leave the space as entered and the leaving enthalpy must carry out the total entering enthalpy and added energy from the space. The moisture that leaves the space in the air stream must equal both the steam added in the space and that brought in by the original supply stream; thus

$$\Sigma m_v = m_a(W_2 - W_1) \tag{3-50}$$

where W_2 and W_1 are the respective humidity ratios in the leaving and entering air streams in pounds per pound of dry air. Combine Eq. (3-49) and Eq. (3-50) by division:

$$\frac{m_a(\mathbf{h}_{a2} - \mathbf{h}_{a1})}{m_a(W_2 - W_1)} = \frac{q_s + \Sigma m_v \mathbf{h}_v}{\Sigma m_v}$$

$$\frac{\mathbf{h}_{a2} - \mathbf{h}_{a1}}{W_2 - W_1} = \frac{\Delta \mathbf{h}_a}{\Delta W} = \frac{q_s + \Sigma m_v \mathbf{h}_v}{\Sigma m_v} \tag{3-51}$$

In this form it can be seen that the rate of inlet air flow has disappeared and the ratio $\Delta \mathbf{h}_a/\Delta W$ is a slope related only to the ratio of the total space cooling load, $q_s + \Sigma m_v \mathbf{h}_v$, and to the moisture added in the space, Σm_v. If the final conditions in and leaving the space, $\mathbf{h}_{a2}$ and W_2, are set, the values of $\mathbf{h}_{a1}$ and W_1 may be allowed to vary but they must always lie on the line set by the slope $\Delta \mathbf{h}_a/\Delta W$.

Values of $\Delta \mathbf{h}_a/\Delta W$ appear on the protractor at the left of Plate I.

The water-vapor (steam) produced by occupants for most calculations can be considered as though produced in the range 85 F to 95 F and for a round value, $\mathbf{h}_v = \mathbf{h}_g = 1100$ Btu/lb can be used.

It should be recognized that in a cooling process the water vapor removed from air follows a similar pattern and the heat removal in condensing out water vapor from air, involving as it does removal of latent heat and subcooling of liquid, can be expressed as

$$Q_L = m_a(W_1 - W_2)\mathbf{h}_v = \Sigma m_v \mathbf{h}_v \tag{3-52}$$

where Q_L is the latent (water-vapor) load, Btuh; m_a is the pounds of dry air per hour air flow through the conditioner; W_1 is the humidity ratio (specific humidity) of air into and W_2 is the air at exit from the conditioner, in pounds per pound of dry air; $\mathbf{h}_v$ is the enthalpy of water vapor in Btu per lb and, $\Sigma m_v = m_a(W_1 - W_2)$ is the pounds per hour of water vapor condensed out, or alternatively created in a space.

Example 3-13. (*Cooling Load*). An auditorium is to be maintained at a temperature not to exceed 76 Fdb and 66 Fwb. The air supplied to the auditorium should not be lower than 64 F. It is found that the sensible heat load on the auditorium is 300 000 Btuh and 130 lbs of water are added to the air per hour. (a) Compute the ratio of enthalpy difference to moisture added for the auditorium. (b) Using this value, determine the supply air conditions to the auditorium for minimum air flow. (c) Making use of the enthalpy of the entering and leaving air, compute the air flow. (d) Find the air flow and the other data needed without making use of the line slope from the protractor.

Solution: (a) For this problem $q_s = 300\,000$ Btuh and $\Sigma m_v \mathbf{h}_v = (130)(1100) = 143\,000$ Btuh. By Eq. (3-50)

$$\frac{\Delta \mathbf{h}_a}{\Delta W} = \frac{300\,000 + 143\,000}{130} = 3408 \text{ Btu/lb steam}$$

(b) Locate this ratio on the protractor of Plate I to find the slope and then on the chart body of Plate I draw the line parallel to it which passes through the design leaving air conditions 76 F and 66 F ($\mathbf{h}_{a2} = 30.76$). For air supplied at 64 F and lying on this line the wet-bulb temperature is 60.0 F and the enthalpy is $\mathbf{h}_{a1} = 26.42$ Btu/lb dry air.

(c) The air-flow rate can be found by using the cooling load and enthalpy change; thus

$$m_a = \frac{q_s + \Sigma m_v \mathbf{h}_v}{\mathbf{h}_{a2} - \mathbf{h}_{a1}} = \frac{443\,000}{30.76 - 26.42} = 102\,100 \text{ lb/h air flow}$$

The humidity ratio of the supply air is 0.0102 lb/lb dry air (10.2 mlb) and the dew-point temperature is 57.5 F.

(d) An alternate solution to this problem would be to make use of Eq. (3-34) and apply it to the sensible heat load q_s; here then

$$300\,000 = m_a(0.244)(76 - 64)$$
$$m_a = 102\,100 \text{ lb dry air/h}$$

The total moisture load of 130 lb must be absorbed by the supply air and the pickup per pound of dry air is thus

$$W_2 - W_1 = \frac{130}{102\,100} = 0.00127 \text{ lb/lb dry air}$$

Since W_2 for the leaving air at 76 F, 66 F is 0.0114 lb/lb dry air from Plate I, then $W_1 = 0.0114 - 0.00127 = 0.01013$. For $t = 64$ F and $W_1 = 0.01013$, the wet-bulb temperature is found from Plate I as 59.9 F. The slight difference in this answer from the slope method arises from the difficulty in reading the slope with a high degree of accuracy and the alternate method may be a preferable method under some conditions.

Sensible to total-heat ratio. Instead of using the water added in a space as one of the variables in establishing the condition line, the ratio of sensible to total heat may serve equally well. Here

$$\frac{q_s}{q_T} = \frac{q_s}{q_s + \Sigma m_v \mathbf{h}_v} = \frac{\Delta H_s}{\Delta H_T} \tag{3-53}$$

Under the symbolism used at the right of Eq. (3-53) the sensible-to total-heat ratio ($\Delta H_s/\Delta H_T$) is plotted on the inside curve of the protractor which is provided on the psychrometric charts, Plates I through IV. $\Delta H_s/\Delta H_T$ is a true dimensionless ratio. Its numerical values are usually decimal fractions and thus are easier to work with on the protractor than the large numerical values involved in the use of $\Delta\mathbf{h}/\Delta W$ with its customary dimensions of Btu per lb of water. Both of these ratio-type parameters are used in the same way, to find a slope on the protractor, and then this slope has to be transferred to the body of Plates I, II, III, or IV.

3-9. PSYCHROMETRIC PROPERTIES UNDER VARYING BAROMETRIC PRESSURE

Attention is again called to the fact that Table 3-1 and the psychrometric charts (Plates I, II, and III) have been constructed for standard barometric pressure, essentially at sea-level elevation, 29.92 in. Hg. When these charts or Table 3-1 are used for computations involving other than standard pressure, some error in the result arises. The error is not serious for small variations of less than an inch of mercury pressure but are significant for higher deviations. Plate IV, constructed for 24.90 in. Hg., essentially 5000-ft altitude, is available for mountainous or high-plateau regions but also needs correction for sizeable deviations from its basic pressure or elevation.

A number of methods have been developed for making corrections with a correction table for enthalpy included as Table 3-2. For other properties recourse must be had to use of the basic equations of this chapter, in nearly all cases with help necessary from generalizations based on the perfect-gas laws. With barometric pressure as well as dry-bulb and dew-point temperatures known, it is easy to get the other properties. However, if dry-bulb and wet-bulb are the temperatures known, Eq. (3-27) is awkward to use and the humidity ratio, W_{sw}, for saturation at the thermodynamic wet-bulb temperature must be computed for the barometric pressure before the equation can be used.

Example 3-14. Compute the properties of moist air in a region where the barometric pressure is 26.92 in. Hg and the measured temperatures are $t = 80$ Fdb, $t_w = 70$ Fwb.

Solution: The humidity ratio for saturated air at the wet-bulb temperature can be found by Eq. (3-5), or by Eq. (3-10) if slightly more accuracy is desired. The pressure of saturated steam at 70 F is independent of barometric pressure and thus can be read from Table 3-1 or Table 2-2 as 0.7392 in. Hg. If Eq. (3-10) is used, $f_s = 1.0046$ from Fig. 3-1.

$$W_{sw} = 0.622 \frac{(1.0046)(0.7392/26.92)}{1 - (1.0046)(0.7392/26.92)} = 0.0176 = 0.0176 \text{ lb/lb dry air}$$

Comparison of this value with the saturation humidity ratio from Table 3-1 at standard barometer, namely, 15.8 mlb or 0.0158 lb, shows that W_s is significantly greater at a reduced barometric pressure.

Employing Eq. (3-27) and reading $\mathbf{h}_{fgw}$ at the 70 Fwb in Table 3-1 as $1091.8 - 38.1 = 1053.7$,

$$W = \frac{(0.0176)(1053.7) - 0.24(80 - 70)}{1053.7 + 0.45(10)} = 0.01525 \text{ lb/lb dry air}$$

is the humidity ratio.

Enthalpy can be found from Eq. (3-14):

$$\begin{aligned} \mathbf{h}_a &= 0.24(80 - 0) + 0.01525(1060.5 + 0.45 \times 80) \\ &= 35.92 \text{ Btu/lb air} \end{aligned}$$

Enthalpy can also be found by reading the enthalpy at standard barometer 29.92 from Plate I and adding the corrective term from Table 3-2; thus

$$\mathbf{h}_a = 34.00 + 1.98 = 35.98 \text{ Btu/lb dry air}$$

Enthalpy by the two methods shows reasonable agreement.

The partial pressure of the water vapor in the air can be found by solving for the partial pressure of the water vapor using Eq. (3-5) or Eq. (3-10). Using Eq. (3-5),

$$0.01525 = 0.622 \frac{p}{26.92 - p}$$

$$p = 0.645 \text{ in. Hg}$$

The dew-point temperature corresponds to this pressure and can be read from Table 3-1, closely 66 F.

Specific volume of the moist air can also be found by using $p_a V = mR_a T$ with $p_a = p_B - p = 26.92 - 0.645 = 26.275$ in. Hg.

$$(144)(0.491)(26.275)v = 1(53.34)(460 + 80)$$

$$v = 15.5 \text{ ft}^3\text{/lb dry air}$$

Comparison with values at standard barometric pressure will show that the relative humidity is little changed by pressure deviation but that great changes arise with v, **h**, and W.

The labor in the computation of psychrometric properties at a pressure varying from standard is very great if an appropriate chart is not available, so much so that it may be advisable to construct such a chart, at least in skeleton form, to minimize labor. For this purpose the layout of the standard chart can be used as a base on which to construct a new chart.

Based on the perfect-gas and Gibbs-Dalton relations it is easily possible to develop the following working equations:

$$v = \frac{0.754(t + 460)}{p_B}\left(1 + \frac{W}{0.622}\right) \tag{3-54}$$

$$p = \frac{W p_B}{0.622 + W} \tag{3-55}$$

where v is the specific volume in cubic feet per pound of dry air with associated water vapor; t is the dry-bulb temperature, in degrees Fahrenheit; p_B is the barometric pressure in inches of mercury; and W is the humidity ratio (specific humidity) in pounds per pound of dry air.

3-10. PSYCHROMETRIC CHART IN SI UNITS

The psychrometric chart presented earlier in this chapter, Fig. 3-7 and I, II, III and IV, is difficult to use because enthalpy and wet-bulb lines both appear in the body of the chart, making it extremely crowded. This crowding sometimes causes confusion. Of course, the same situation would exist if the charts had been drawn using SI units. Figure 3-11 is presented in the text, to show a different design approach and to make available a chart in SI (metric) units.

A superficial comparison of Fig. 3-11 with Fig. 3-7 (or Plate I) shows that, in general character, the two charts are very much alike. The major

PSYCHROMETRIC CHART

NORMAL TEMPERATURES

SI METRIC UNITS

Barometric Pressure 101.325 kPa

SEA LEVEL

Reproduced by permission of Carrier Corporation.

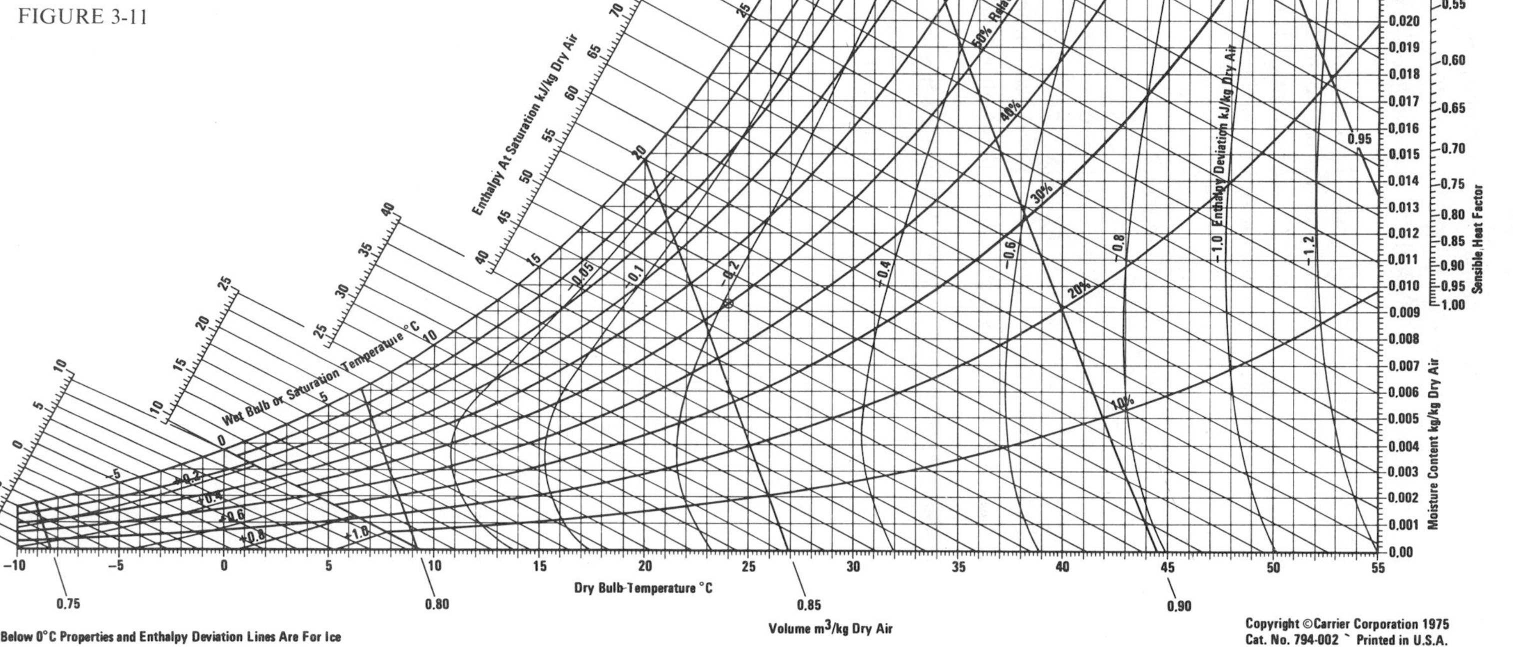

FIGURE 3-11

difference lies in the fact that the wet-bulb lines which start at the saturation line and run downward to the right also serve as enthalpy lines. However, as previously explained, enthalpy deviates from its value at saturation whenever the dry-bulb temperature exceeds the wet-bulb temperature. The magnitude of the enthalpy deviation is plotted on the chart and by use of the deviation lines the enthalpy can easily be corrected in any region of the chart. For example, saturated air at 20°C wet bulb and 20°C dry bulb is read as 55.51 kJ/kg dry air on the diagonal scale. Air at the same wet-bulb temperature but at 31.6°C db falls near the −0.4 deviation line, so the enthalpy is 55.51 − 0.4 = 55.11 kJ/kg. Between deviation lines, interpolation by eye is sufficiently accurate; for example, 27°C db and 20°C wb would show an **h** of 55.51 − 0.24 or 55.27 kJ/kg.

Note that for heat-load calculations the sensible-heat factor (ratio) approach is employed, making use of the small circle shown near the lower middle of the chart and the right-hand scale lines. This chart is developed from a pattern that is explained in detail in Refs. 10 and 7.

Example 3-15. (*Cooling Load*). An auditorium is to be maintained at a temperature not to exceed 24.44°C db and 18.89°C wb. The air supply to the auditorium should not be colder than 17.77°C. The sensible load of the auditorium is 87.86 kW and 58.97 kg of water transfer into the moving air per hour from occupants in the space. (a) Compute the total space load in kilowatts. (b) Find the sensible-heat ratio (factor) for this load condition. (c) Compute the minimum air flow to serve this load based on the sensible-heat requirement. (d) Apply the sensible-heat factor from the chart to find the appropriate supply-air conditions.

Solution: (a)

$$q_s = 87.86 \text{ kW}$$

$$q_L = \Sigma m_v \mathbf{h}_v = (58.97)(0.710) = 41.87 \text{ kW}$$

where $\mathbf{h}_v = 0.710$ kWh/kg

$$q_T = 87.86 + 41.87 = 129.73 \text{ kW} \qquad \textit{Ans.}$$

(b) By Eq. (3-53),

$$\text{SHF} = \frac{q_s}{q_T} = \frac{87.86}{129.73} = 0.677 \qquad \textit{Ans.}$$

(c) Use Eq. (3-36) in connection with the lowest allowable supply-air temperature. Note that 0.244 Btu/lb · F × 4.184 = 1.02 kJ/kg · °C = 1.02 kW · s/kg · °C.

$$q_s = mC_p\Delta t$$

$$(87.86) = m(1.02)(24.44 - 17.77)$$

$$m = 12.91 \text{ kg/s or } 46\,490 \text{ kg/h} \qquad \textit{Ans.}$$

(d) On Fig. 3-11, locate the SHF of 0.677 on the right axis and draw a line from this through the index circle. Parallel to this line draw a second line passing through the space dry-bulb and wet-bulb temperatures, 24.4°C and 18.9°C. Where this second line intersects 17.8°C, read the humidity ratio and wet-bulb temperature as 0.01019 kg/kg and 15.5°C. *Ans.*

This condition of supply air will serve the latent load as the supply air with 0.01019 kg/kg picks up moisture and reaches the 0.01146 humidity ratio of the space. We can check the calculation by multiplying the air flow by the moisture pickup:

$$(46\,490)(0.01146 - 0.01019) = 59.04 \text{ kg/h}.$$

This can be seen to check, within chart accuracy, the indicated moisture load of 58.97 kg/h.

PROBLEMS

3-1. The temperature in a certain room is 72 F (22.22°C) and the relative humidity is 50%. The barometric pressure is 29.92 in. Hg. (760 mm Hg). Find (a) the partial pressures of the air and water vapor, (b) the vapor density, and (c) the humidity ratio of the mixture.
Ans. (a) 14.496 psi, 0.1941 psi; (b) 0.000613 lb/ft^3; (c) 0.0083 lb/lb

3-2. The temperature in a certain room is 22.22°C and the relative humidity is 50%. The barometric pressure is 101 325 N/m^2 (101.325 kPa). Find (a) the partial pressures of the air and water vapor; (b) the vapor density in grams per cubic meter; and (c) the humidity ratio of the mixture in grams per kilogram.
Ans. (a) 99 985 and 1340 N/m^2; (b) 9.84 g/m^3; (c) 8.34 g/kg

3-3. Certain air has a dry-bulb temperature of 75 F and a relative humidity of 50%, and the barometric pressure is 29.8 in. Hg. Calculate (a) the weight of 1 ft^3 of the mixture of air and moisture and (b) the weight of moisture per pound of dry air.
Ans. (a) 0.0734 lb/ft^3; (b) 0.0093 lb/lb

3-4. Air with a dry-bulb temperature of 70 F and a wet-bulb temperatue of 65 F is at a barometric pressure of 29.92 in. Hg. Without making use of the psychrometric chart, find (a) the relative humidity of the air; (b) the vapor density; (c) the dew-point temperature; (d) the humidity ratio; (e) and the volume occupied by the mixture associated with a pound of dry air.
Ans. (a) 76.8%; (b) 0.000883 lb/ft^3; (c) 62.4 F; (d) 0.01204 lb; (e) v = 13.62

3-5. Rework Problem 3-4, but assume that the barometric pressure is 28.0 in. Hg.
Ans. (a) 77.3%; (b) 0.000891 lb/ft^3; (c) 62.5 F; (d) 0.0129 lb; (e) v = 14.56

3-6. For a barometric pressure of 29.5 in. Hg, compute the saturation ratio at 80 F for relative humidities of 30, 50, and 90%. Compare these values with the 80 F line of Fig. 3-2.

3-7. Air is supplied to a room at a 72 Fdb temperature and a 68 Fwb temperature from outside air at 40 Fdb and 37 Fwb. The barometric pressure is 29.92 in. Hg. Find (a) the dew-point temperatures of the inside and outside air, (b) the moisture added to each pound of dry air, (c) the enthalpy of the outside air, (d) the enthalpy of the inside air.
Ans. (a) 66.1 F, 33.6 F; (b) 0.0098 lb; (c) 13.86 Btu/lb; (d) 32.39 Btu/lb

3-8. (a) Compute the humidity ratio of air at 90 F and 60% relative humidity, at 29.92 in. Hg, by use of Eq. (3-5). (b) Then recompute, making use of Eq. (3-10) and Fig. 3-1. (c) Find the percentage error involved in using Eq. (3-5).
Ans. (a) 0.018251; (b) 0.018346; (c) 0.5%

3-9. (a) Find from the psychrometric chart the dew point and humidity ratio of 1 lb of dry air at 29.92in. Hg if the dry-bulb temperature is 80 F and the wet-bulb temperature is 70 F. (b) Find also the enthalpy and specific volume.

Ans. (a) 65.5 F, $W = 0.0135$; (b) $\mathbf{h} = 34.0$, $v = 13.9\ \text{ft}^3$

3-10. Air is heated to 80 F, without the addition of water, from 60 Fdb and 50 Fwb. By use of the chart find (a) the relative humidity of the original mixture; (b) the dew-point temperature, (c) the humidity ratio; (d) the original enthalpy; (e) the final enthalpy; (f) the heat added; and (g) the final relative humidity.

Ans. (a) 49%; (b) 40.7 F; (c) 0.00533 lb; (d) 20.28; (e) 25.11; (f) 4.83; (g) 24%

3-11. Air is cooled from 75 Fdb and 70 Fwb to 55 F. Find (a) the moisture removed per pound of dry air; (b) the heat removed to condense the moisture; (c) the sensible heat removed, and (d) the total amount of heat removed.

Ans. (a) 0.0054 lb; (b) 5.8; (c) 5.0; (d) 10.8

3-12. Air is supplied to a certain room from the outside, where the temperature is 20 F and the relative humidity is 70%. It is desired to keep the room at 70 F and 60% relative humidity. How many pounds of water must be supplied to each pound of air entering the room if these conditions are to be met? The barometric pressure is 29.5 in Hg.

Ans. 0.00797 lb/lb

3-13. Air at 100 Fdb and 65 Fwb is humidified adiabatically with steam. The steam supplied contains 20% of moisture saturated at 16 psia. When sufficient steam is added to humidify the air to 60% relative humidity, what is the dry-bulb temperature of the humidified air? The barometer is at 29.92 in. Hg.

Ans. 92 F

3-14. Air at 84 Fdb and 60 Fwb and at 29.92 in. Hg is humidified with the dry-bulb temperature remaining constant. Saturated steam is supplied for humidification at 14.7 psia in wet condition. What quality must the steam have (a) to provide saturated air and (b) to provide air at 70% relative humidity?

Ans. (a) (b) 94.2%, essentially constant

3-15. In the text Example 3-11, air at 70 F and 60% relative humidity was adiabatically humidified by adding superheated steam at 600 psia and 1000 F ($\mathbf{h} = 1516.7$ Btu/lb). (a) What is the final dry-bulb temperature if enough steam is added to raise the relative humidity to 65%? (b) if the final temperature is 180 F, what relative humidity can be attained by the addition of an adequate amount of this steam?

Ans. (a) 75 F and $W = 0.0121$; (b) $W = 0.077$ and 22%

3-16. Air is humidified by the addition of dry saturated steam at 200 psia ($\mathbf{h} = 1198$). The air is initially at 60 Fdb and 50 Fwb and at 29.92 in. Hg. By varying the weight of steam supplied to each pound of dry air, the air temperature is raised to 63 F and 62 F, respectively. For each of these final temperatures, find the corresponding values of humidity ratio, relative humidity, and enthalpy.

Ans. For 63 F: 0.00111, 89.5%, 27.2

3-17. Air at 90 Fdb and 60 Fwb and at 29.92 in. Hg is humidified to a final dew point of 70 F by use of dry saturated steam at 120 psia ($\mathbf{h} = 1190.4$). If the process takes place under adiabatic conditions, what is the final dry-bulb temperature?

Ans. 95 F

3-18. Calculate the wet-bulb temperature of dry air at 90 F. The barometric pressure is 29.92 in. Hg. Base the solution on Tables 3-1 and 2-2, not on the psychrometric chart.

Ans. 52.7 F

3-19. Air at 100 Fdb and 70 Fwb and 29.92 in. Hg. is adiabatically mixed with water supplied at 140 F, in such proportions that the mixture has a relative humidity of 80%. Find the dry-bulb temperature of the mixture.

Ans. 75 F

3-20. Air at 40 Fdb and 35 Fwb is mixed with warm air at 100 Fdb and 77 Fwb in the ratio of 2 lb of cool air to 1 lb of warm air. Compute the resultant humidity ratio and enthalpy of the mixed air by Eqs. (3-43) and (3-41). On the psychrometric chart of Plate I, connect by a straight line the points representing the two kinds of air, and locate a point on this line at a distance of one-third of its length from the cooler point. Read at this point the humidity ratio and enthalpy of the mixed air, as shown on the straight line, and compare the readings with the computed values.

Ans. $W = 0.007$ lb, $\mathbf{h} = 22.1$

3-21. Assume that 3000 ft^3 of air at 50 F and 100% relative humidity are mixed with 2500 ft^3 of air at 75 F and 50% relative humidity. Compute the temperature, relative humidity, and humidity ratio of the resulting mixture.

Ans. 61.1 F, 72%, 0.0083

3-22. In an auditorium maintained at a temperature not to exceed 75 F, and at a relative humidity not to exceed 60%, a sensible-heat load of 450 000 Btuh and 171.4 lb of moisture per hour must be removed. Air is supplied to the auditorium at 65 F. (a) How many pounds of air per hour must be supplied? (b) What is the dew-point temperature of the entering air, and what is its relative humidity? (c) How much latent-heat load is picked up in the auditorium? (d) What is the sensible-heat ratio?

Ans. (a) 184 400 lb/h; (b) 57.6 F, 0.78; (c) 188 500 Btuh; (d) 0.705

3-23. A meeting hall is to be maintained at 80 Fdb and 68 Fwb. The barometric pressure is 29.92 in. Hg. The space has a load of 200 000 Btuh, sensible, and 200 000 Btuh, latent. The temperature of the supply air to the space cannot be lower than 66 Fdb. (a) How many pounds of air per hour must be supplied? (b) What is the required wet-bulb temperature of the supply air? (c) What is the sensible-heat ratio?

Ans. (a) 58 500 lb/h; (b) 58.4 F; (c) 0.5

3-24. A building with a heat loss of 200 000 Btuh is heated by warm air which is supplied at 135 F and which has a humidity ratio of 0.006 lb/lb dry air. Air returns to the furnaces at 65 F, with no significant change in humidity ratio. Find (a) the number of pounds of air which must be circulated per hour for heating and (b) the number of cubic feet per minute measured at inlet conditions.

Ans. (a) 11 720 lb/h; (b) 2930 cfm

3-25. If pipes carrying water at 50 F run through a room which has an air temperature of 70 F, what is the maximum relative humidity that can be held in the room without any water condensing on the pipes?

Ans. 49%

3-26. Outside air at 95 Fdb and 80 F dew point at 27.92 in. Hg barometric pressure is cooled to saturated state at 64 F. Without using the psychrometric chart find (a) the humidity-ratio of the moist air before and after cooling; (b) the weight of moisture condensed; (c) the enthalpy of the moist air before and after cooling, and (d) the heat removed in Btu per hour when 300 lb/min of air pass through the air cooler.

3-28. Outdoor air with a temperature of 40 Fdb and 35 Fwb, and with a barometric pressure of 29 in. Hg, is heated and humidified under steady-flow conditions to a final dry-bulb temperature of 70 F and 40% relative humidity. (a) Find the weight of water vapor added to each pound of dry air. (b) If the water is supplied at 50 F, how much heat is added per pound of dry air?

Ans. (a) 3.13 mlb; (b) 10.5 Btu/lb

3-29. Outdoor air at 95 Fdb and 79 Fwb, and at a barometric pressure of 28.92 in. Hg, is cooled and dehumidified under steady-flow conditions until it becomes saturated at 60 F. (a) Find the weight of water condensed per pound of dry air. (b) If the condensate is removed at 60 F, what quantity of heat is removed per pound of dry air?

Ans. (a) 0.00707 lb; (b) 15.4 Btu/lb

3-30. If the air in Problem 3-29, after cooling, were reheated to 70 F, what would be its relative humidity?

Ans. 70.4%

3-31. Making use of the SI (metric) psychrometric chart for air at 26.7°Cdb and 21.1°Cwb at a pressure of 101.325 kPa, read (a) the dew point and humidity ratio; and (b) the enthalpy and specific volume.

Ans. 18.5°C, 0.0135; 61.3 kJ, 0.865 m^3/kg

3-32. Make use of the SI (metric) psychrometric chart at standard barometric pressure, 101.325 kPa. (a) Find the dew point and humidity ratio for air at 28°Cdb and 22°Cwb. (b) Find also the enthalpy and specific volume.

Ans. (a) 19.4°C, 0.0143; (b) 64.4 kJ, 0.87 m^3/kg

3-33. Air is warmed from 15.6°Cdb and 10°Cwb without the addition of moisture to 26.7°Cdb. Make use of the SI (metric) chart and find (a) the relative humidity, humidity ratio, and dew point of the original air; (b) the relative humidity, humidity ratio, and dew point of the heated air; (c) the original and final enthalpy of the air; and (d) the heat added. (e) Check the answer to part d by making use of the specific heat of air as 1.017 kJ/kg · K and the temperature change.

Ans. (a) 49%, 0.0054, $t = 4.4$°C; (b) 5%, 0.0054, 4.4°C; (c) 29.2 and 40.5 kJ/kg; (d) 11.3

3-34. Make use of the SI (metric) chart and find (a) the moisture which must be removed in cooling air from 24°Cdb, 21°Cwb to 13°C. (b) Find the total heat removal for such cooling. (c) Find the sensible heat removal and by difference the latent heat removal.

Ans. 0.0051 kg/kg; 24.2 kJ/kg; 11.2, 13.0 latent

3-35. Air at 4°Cdb, 2°Cwb is mixed with an equal weight of air at 38°Cdb, 25°Cwb. Compute the humidity ratio and enthalpy of the resultant air mixture. Make use of Eqs. (3-43) and (3-41) to obtain your answers. Next locate the two state points on Fig. 3-11 and draw a straight line on the chart connecting them. At the midpoint of this line, read the desired result and compare with the result obtained by computation.

Ans. 0.0091 kg/kg, 44.3 kJ/kg; close agreement

3-36. Rework Problem 3-35 on the basis of mixing 2 parts of cooler air with 1 part of warmer air. Note that in this case the mixture point will lie one-third of the line length from the cool-air state point.

Ans. 0.0072, 33.8 kJ/kg dry air

3-37. A meeting hall at full occupancy in summer has a computed sensible load of 132 kW and a latent load of 53 kW. The space should not exceed 25°Cdb, 19°Cwb and the supply air should not be colder than 18°Cdb to prevent discomfort to the occupants. (a) Compute the minimum acceptable air flow in kilograms per hour to maintain desired conditions, based on sensible-load conditions. (b) Compute the sensible-heat ratio (factor) on the chart and make use of this to find satisfactory supply-air conditions. (c) What weight of moisture per hour is actually picked up by the supply air under the indicated load conditions?

Ans. (a) 66 751 kg/h; (b) 0.717 SHR, 15.6°Cwb, $W = 0.0103$; (c) 73.3 kg/h

3-38. Rework Problem 3-37 if the space is held at 24°Cdb, 19°Cwb and the latent load is 70 kW with all other conditions unchanged.

Ans. (a) 77 800 kg/h; (b) 0.653, $W = 0.0103$; (c) 99 kg/h

REFERENCES

NASA, *U.S. Standard Atmosphere, 1962*, and *Supplement, 1966* (Washington, D.C.: National Aeronautics and Space Administration).

W. H. Carrier, "Rational Psychrometric Formulas," *Trans. ASME*, vol. 23 (1911), p. 1005.

J. H. Arnold, "The Theory of the Psychrometer," *Physics*, vol. 4 (1933), pp. 255, 334.

B. H. Jennings and A. Torloni, "Psychrometric Charts for Use at Altitudes above Sea Level," *Refrigerating Engineering*, vol. 62 (June 1954), pp. 71–76, 118–26.

J. A. Goff and S. Gratch, "Thermodynamic Properties of Moist Air," *Trans. ASHVE*, vol. 51 (1945), pp. 31–36.

E. P. Palmatier and D. D. Wile, "A New Psychrometric Chart," *Refrigerating Engineering*, vol. 52 (July 1946), pp. 31–36.

E. P. Palmatier, "Construction of the Normal Temperature Psychrometric Chart," *ASHRAE Journal*, vol. 5 (May 1963), pp. 55–60.

ASHRAE, *Handbook of Fundamentals* (1972), chap. 5.

M. Costantino et al., "ASHRAE Psychrometric Chart Converted to Metric System," *ASHRAE Journal*, vol. 8 (April 1966), pp. 68–69.

C. E. Bullock and J. H. Carpenter, "New Psychrometric Charts in SI (Metric) Units," *ASHRAE Journal*, vol. 17 (Dec. 1975), pp. 30–33.

CHAPTER 4

HEAT TRANSFER AND TRANSMISSION COEFFICIENTS

4-1. MODES OF HEAT TRANSFER

The necessity of calculating heat flow occurs in so many engineering problems that the need of a thorough understanding of heat flow cannot be overestimated. The design of every heating or cooling system is based primarily on the heat-transfer characteristics of the building structure. Also, when both the heat flow to or from the building and the internal heat load have been calculated, the heat-transfer problems again appears in finding the size (surface) of heaters, cooling coils, or other appliances to carry the load.

Heat is gained or lost through the walls and structure of a building in two general ways: first, by transmission through the wall from the air on one side to the air on the other side, and second, by actual leakage of warmer or colder air into the building. Thus to reduce heat transfer, the insulating quality of the walls must be improved by use of building insulation or by use of insulating air spaces in walls and between roofs and ceilings. Leakage is reduced by the installation of weather strips, by use of double windows and doors, and by caulking or otherwise reducing air flow through cracks.

Transfer of heat takes place by conduction, convection, radiation, or by some combination of these processes, whenever a temperature difference exists.

Conduction is a process in which heat is transmitted from and to adjacent molecules along the path of flow, whereby some of the thermal agitation of the hotter molecules is passed on to the adjacent cooler molecules. An example of heat transfer by conduction is the passage of heat along an iron bar, one end of which is being heated in a fire. In conduction of heat there is no appreciable displacement of the particles of the material.

Convection is (1) the transfer of heat between a *moving* fluid medium (liquid or gas) and a surface or (2) the transfer of heat from one point to another within a fluid by movements within the fluid, which movements intermix different portions of the fluid. The final method of heat transfer in convection is eventually some form of conduction or radiation. In convection, if the fluid moves because of differences in density resulting from temperature changes, the process is called *natural convection*, or *free convection*; if the fluid is moved by mechanical means (pumps or fans) the process is called *forced convection*.

Radiation is the name given to the transfer of heat by means of wave or quanta action. Two bodies at different temperatures both emit and absorb impinging radiation but the hotter the body emits more energy than it receives, resulting in a net transfer of heat. The radiation which falls on a body may be absorbed, reflected, or transmitted. Media, which transmit radiant energy without themselves being affected, are *diathermous*. Glass, air, and many gases are relatively diathermous but only a vacuum could be considered completely diathermous. Carbon dioxide and water-vapor molecules in air react to radiant enrgy while the nitrogen and oxygen diatomic molecules are little affected. The radiant energy absorbed by a body produces an increase in its internal energy, usually evidenced by a rise in temperature.

4-2. HEAT-TRANSFER EQUATIONS

The theory of heat conduction was first mathematically developed by the French mathematician J. B. Fourier, although Sir Isaac Newton had long before started work on the subject. Fourier's equation for unidirectional heat flow, based on experimental evidence, is

$$\frac{dQ}{d\theta} = -kA\frac{dt}{dx} \tag{4-1}$$

where $dQ/d\theta$ is the heat transfer per unit of time (θ); A is the area of the section through which heat is flowing; dt is the temperature difference causing the heat flow; dx is the length of the path through the material, in the direction of the heat flow; and k is the proportionality factor called thermal conductivity.

In the terms dQ, $d\theta$, dt, and dx the differential notation is used to indicate infinitesimal changes in these quantities. It will be noted that the rate of flow of heat is inversely proportional to the thickness of the insulation; that is, less heat is transferred as the thickness of the insulation is increased. Equation (4-1), although of little more than academic interest for the purposes of this book, forms a convenient starting point for the development of more important equations.

When the temperature t varies with both time θ and position x, as when a substance is warming or cooling, the flow is called *unsteady flow* and solutions of the differential equation [Eq. (4-1)] are usually very complex. When, however, equilibrium in heat transfer is reached and temperature depends only on position, the flow is called *steady flow*. For steady flow, which is a very common case, the heat transferred is constant, $dQ/d\theta = q$, and thus

$$q = -kA \frac{dt}{dx} \tag{4-2}$$

Since the minus sign appears in this equation merely as a mathematical indication that positive heat flow occurs in the direction of a falling temperature gradient, it serves no purpose here and will be dropped. The units of k in SI are:

$$k = \frac{q}{A} \times \frac{dx}{dt} = \frac{(\text{W})(\text{m})}{(\text{m}^2)(^\circ\text{C or K})} = \frac{\text{W}}{(\text{m})(^\circ\text{C})} \tag{4-3}$$

In the foot-pound-second system, both the foot and the inch are used for length of path in the direction of heat flow and conventional units for k are:

$$k = \frac{q}{A} \times \frac{dx}{dt} = \frac{(\text{Btu})(\text{ft})}{(\text{h})(\text{ft}^2)(\text{F})} \quad \text{or} \quad \frac{(\text{Btu})(\text{in.})}{(\text{h})(\text{ft}^2)(\text{F})} \tag{4-4}$$

$$\frac{(\text{Btu})(\text{in.})}{(\text{h})(\text{ft}^2)(\text{F})} = \frac{\text{Btu}}{(\text{h})(\text{ft}^2)(\text{F/in.})} = \frac{12(\text{Btu})(\text{ft})}{(\text{h})(\text{ft}^2)(\text{F})} \tag{4-5}$$

At the present time so much reference data are expressed in the latter system that it appears advisable to employ mainly this system and convert final results to SI units whenever this is desired. The conversion factors are readily available for the two systems and it should also be noted that earlier metric-system units also appear in heat-transfer literature. Note for reference that:

$$\frac{(\text{W})(\text{m})}{(\text{m}^2)(^\circ\text{C})} = 6.938 \frac{(\text{Btu})(\text{in.})}{(\text{h})(\text{ft}^2)(\text{F})} = 0.5782 \frac{(\text{Btu})(\text{ft})}{(\text{h})(\text{ft}^2)(\text{F})} \tag{4-6}$$

$$\begin{aligned} \frac{(\text{Btu})(\text{ft})}{(\text{h})(\text{ft}^2)(\text{F})} &= 1.73 \frac{\text{W}}{(\text{m})^\circ\text{C})} = 0.0173 \frac{\text{W}}{(\text{cm})(^\circ\text{C})} \\ &= 1.489 \times 10^{-2} \frac{\text{kcal}}{(\text{h})(\text{cm})(^\circ\text{C})} = 4.13 \times 10^{-6} \frac{\text{kcal}}{(\text{s})(\text{cm})(^\circ\text{C})} \end{aligned} \tag{4-7}$$

$$\frac{(\text{Btu})(\text{in.})}{(\text{h})(\text{ft}^2)(\text{F})} = 0.1441 \frac{\text{W}}{(\text{m})(^\circ\text{C})} = 0.124 \frac{\text{kcal}}{(\text{h})(\text{m})(^\circ\text{C})} \tag{4-8}$$

The value of k for different materials varies over wide limits; and k even varies with the same material at different temperatures and under varying densities of packing. Typical values of k for many substances and for selected building materials appear in Table 4-1 and Table 4-3.

Conduction through a plain wall leads to the following simplification of Fourier's equation:

$$q = k\frac{A}{x}(t_1 - t_2) = k\frac{A}{x}\Delta t \tag{4-9}$$

where q is the heat transferred per unit of time (hour), in Btu; A is the area of the wall, in square feet; x is the thickness of the wall, in feet or inches (depending on the units of k); k is the thermal conductivity, in units of Btu · ft (or in.)/h · ft^2 · F; and $t_1 - t_2 = \Delta t$ is the temperature difference on the two sides of the wall which causes heat flow, in degrees Fahrenheit.

Conduction through a composite wall, shown in Fig. 4-1, can be treated by using Eq. (4-9) in connection with the so-called resistance concept. Writing Eq. (4-9) in more general form and algebraically rearranging the terms, there results for a plain wall,

$$q = k\frac{A}{x}(t_1 - t_2) = \frac{t_1 - t_2}{x/kA} = \frac{t_1 - t_2}{R} \tag{4-10}$$

where $R = x/kA$ is called the thermal resistance.

In the case of the composite wall shown in Fig. 4-1, the heat flow under steady conditions through each section of the wall is the same; that is, in Btu per hour through the whole wall,

$$q = q_a = q_b = q_c$$

and

$$q = \frac{t_1 - t_2}{R_a} = \frac{t_2 - t_3}{R_b} = \frac{t_3 - t_4}{R_c} = \frac{t_1 - t_4}{R_t} \tag{4-11}$$

Table 4-1 Thermal Conductivity and Other Properties of Miscellaneous Substances

Material	Specific Heat C_p (Btu per lb deg F)	Density at 68 F (lb per cu ft)	Conductivity k $\left[\frac{\text{(Btu)(in.)}}{\text{(hr)(sq ft)(deg F)}}\right]$	Temperature Range (F)
Air, still	**0.24**	.....	**0.169–0.215**	**32–200**
Aluminum	**0.21**	**168.0**	**1404–1429**	**32–600**
Ammonia				
Liquid	**1.128**	**38.0**	**3.48**	**5–86**
Vapor	**0.52**	**0.67**	**0.144**	**32**
Asbestos board with cement	**0.20**	**123**	**2.7**	**85**
Asbestos, wool	**0.20**	**25.0**	**0.62**	**32**
Bagasse	**0.32**	**13.5**	**0.336**	**68**
Benzol	**0.34**	**55.5**	**1.18**	**68**
Brass				
Red	**0.090**	**536.0**	**715.0**	**32**
Yellow	**0.088**	**534.0**	**592.0**	**32**
Brick				
Common	**0.22**	**112.0**	**5.0**	...
Face	**0.22**	**125.0**	**9.2**	...
Fire	**0.20**	**115.0**	**6.96**	**392**

Table 4-1 (*Continued*)

Material	Specific Heat C_p (Btu per lb deg F)	Density at 68 F (lb per cu ft)	Conductivity k $\left[\frac{(\text{Btu})(\text{in.})}{(\text{hr})(\text{sq ft})(\text{deg F})}\right]$	Temperature Range (F)
Bronze	0.10	509.0	522.0	32
Cellulose, dry	0.37	94.0	1.66	59
Celotex	0.32	13.2	0.34	...
Cement mortar	0.19	118.0	12.0	...
Chalk	0.21	142.0	6.35	...
Cinders	0.18	40–45	1.1	...
Clay				
Dry	0.22	63.0	3.5–4.0	68–212
Wet	0.60	110.0	4.5–9.5	...
Concrete				
Cinder	0.18	97.0	4.9	75
Stone	0.19	140.0	12.0	75
Cork, granulated	0.42	8.1	0.31	90
Corkboard	0.42	8.3	0.28	60
Cornstalk, insulating board	0.32	15.0	0.33	71
Cotton	0.32	5.06	0.39	32
Foamglas	0.16–0.19	10.5	0.40	50
Gasoline	0.53	42.0	0.94	86
Glass wool	0.22	1.5	0.27	75
Glass				
Common thermometer	0.20	164.0	5.5	68–212
Flint	0.12	247.0	5.1	50–122
Pyrex	0.20	140.0	7.56	...
Gold	0.031	1205.0	2028	64–212
Granite	0.20	159.0	15.4	...
Gypsum, solid	0.26	78.0	3.0	68
Hair felt	0.33	13.0	0.26	90
Ice	0.50	57.5*	15.6	32
Iron				
Cast	0.13	442.0	326.0	129–216
Wrought	0.11	485.0	417.0	64–212
Iron oxide	0.17	306–330	3.63	68
Lampblack	....	10.0	0.45	104
Lead	0.030	710.0	240.0	64–212
Leather, sole	0.36	54.0	1.10	86
Lime				
Mortar	0.22	106.0	2.42	...
Slaked	0.13	81.0–87.0	....	...
Limestone	0.22	132.0	10.8	75
Marble	0.21	162.0	20.6	32–212
Mineral wool				
Board	0.25	15.0	0.33	75
Fill-type	0.20	9.4	0.27	103

*At 32 F.

Table 4-1 (*Continued*)

Material	Specific Heat C_p (Btu per lb deg F)	Density at 68 F (lb per cu ft)	Conductivity k $\left[\frac{(Btu)(in.)}{(hr)(sq\ ft)(deg\ F)}\right]$	Temperature Range (F)
Nickel	0.10	537.0	406.5	64–212
Paper	0.32	58.0	0.90	...
Paraffin	0.69	55.6	1.68	32–68
Plaster				
Cement and sand	0.20	73.8	5.00	68
Gypsum	0.20	46.2	5.60	73
Redwood bark	....	5.0	0.26	75
Rock wool	0.20	10.0	0.27	90
Rubber				
Hard	0.40	74.3	11.0	100
India	0.48	59.0	1.302	68–212
Sand, dry	0.19	94.6	2.28	68
Sandstone	0.22	143.0	12.6	68
Silver	0.056	656.0	2905.0	64–212
Soil				
Crushed quartz (4% moisture)	[0.16–0.19	100 (dry)	11.5	40
Crushed quartz (4% moisture)	dry reaching	110 (dry)	16.0	40
Fairbanks sand	to			
Moisture, 4%	0.3	100 (dry)	8.5	40
Moisture, 10%	wet]	110 (dry)	15.0	40
Dakota sandy loam	↓			
Moisture, 4%	↓	110 (dry)	6.5	40
Moisture, 10%	↓	110 (dry)	13.0	40
Healy clay	↓			
Moisture, 10%	↓	90 (dry)	5.5	40
Moisture, 20%	↓	100 (dry)	10.0	40
Steam	0.48	0.037*	0.151	212
Steel				
1% C	0.12	487.0	310.0	64–212
Stainless	0.12	515.0	200.0	...
Tar, bituminous	0.35	75.0	....	86
Water				
Fresh	1.00	62.4	4.10	70
Sea	0.94	64.0	3.93	64
Wood				
Fir	0.65	34.0	0.80	75
Maple	....	40.0	1.2	75
Red oak	0.57	48.0	1.10	86
White pine	0.67	31.2	0.780	86
Wood fiber board	0.34	16.9	0.34	90
Wool	0.33	4.99	0.264	86

*At 212 F and 14.7 psi.

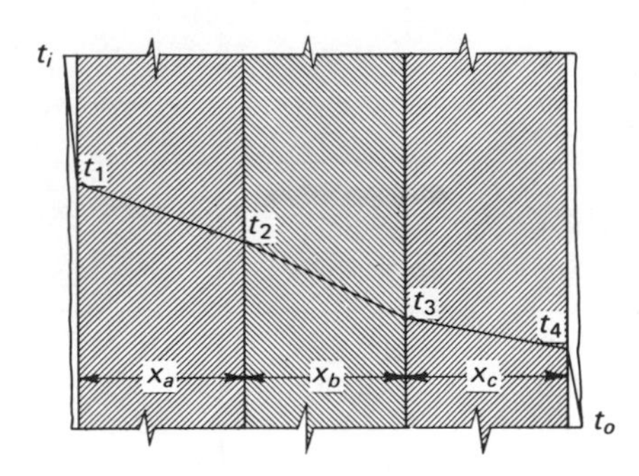

FIGURE 4-1. Sketch of composite wall.

And, in the general case,

$$q = \frac{t_1 - t_n}{R_t} = \frac{t_1 - t_n}{\dfrac{x_a}{k_a A_a} + \dfrac{x_b}{k_b A_b} + \dfrac{x_c}{k_c A_c} + \cdots + \dfrac{x_n}{k_n A_n}} \tag{4-12}$$

In the case of flat walls per square foot of surface, in Btu per hour,

$$q = \frac{t_1 - t_n}{x_a/k_a + x_b/k_b + x_c/k_c} = \frac{t_1 - t_n}{R_t} \tag{4-13}$$

In heat flow between a fluid (gas or liquid) and a solid, there always exists a thin fluid film which tends to cling to the surface as a relatively stagnant layer and which acts as an additional resistance to heat flow. The thickness of this film is greatly influenced by the convection conditions of the system. For example, in building construction, a wind blowing on the outside walls of a building greatly changes the film condition over that which exists on a quiet day with little wind. The films, and the temperature drops through them, are indicated in Fig. 4-1.

In the case of heat-transfer surfaces, such as those for heat exchangers, boiler tubes, and evaporators, the film conditions are affected by (1) the fluid velocity, (2) the shape and kind of surface, (3) whether natural or forced convection exists, and (4) whether boiling or condensing is occurring. The resistance of gas films to heat flow is usually many times greater than that of similar liquid films. Because of the many variables involved, exact calculation of heat transfer through such films is in many cases impossible, and recourse must be had to overall heat-transfer coefficients based on experimental determinations and set up for use in the form of empirical equations or tables. In heat transfer through metal surfaces, the films offer the greatest resistance to heat flow and must be considered. In transfer through effective insulation, the film resistance may be relatively unimportant.

In this text the letter f is used as the symbol for this *film coefficient*, or *surface coefficient* as it is sometimes called. The letter h is also used for this coefficient in the technical literature. The units of f are $\text{Btu/h} \cdot \text{ft}^2 \cdot \text{F}$ or, in SI, $\text{W/m}^2 \cdot {}^\circ\text{C}$. The magnitude of the surface coefficient is affected by many variables, including the inherent thermal conductivity of the air or gas film and the thickness of the film. The latter, in turn, is strongly affected by velocity, as well as by the character of the surface underlying the film. Temperature is significant when considering films adjacent to air spaces. In terms of surface placement, data on f values for building construction appear as the first topic of Table 4-3. Film thickness is indefinite but may vary from a few to some 20-thousandths of an inch. Thus

$$\frac{1}{f} = \text{film resistance to heat flow per square foot} \tag{4-14}$$

or

$$\frac{1}{fA} = \text{film resistance to heat flow for an area } A \tag{4-15}$$

Considering films f_0 and f_i on two sides of a material, Eq. (4-9) becomes, in units of Btu/h · ft^2,

$$q = \frac{t_i - t_o}{\dfrac{1}{f_i(1)} + \dfrac{x_a}{k_a(1)} + \dfrac{x_b}{k_b(1)} + \dfrac{x_c}{k_c(1)} + \cdots + \dfrac{1}{f_o(1)}} \tag{4-16}$$

The use of air spaces in building construction is a common and effective way of improving the insulation capability of walls and partitions. Table 4-2 shows how the conductance (a) of an air space in vertical wall rapidly

Table 4-2 Thermal Conductances of Vertical Air Spaces at Various Mean Temperatures

Mean Temperature (F)	Width of Air Space (in.)						
	0.128	0.250	0.364	0.493	0.713	1.00	1.50
	Value of Conductance a (Btu per hr sq ft deg F)						
20	2.300	1.370	1.180	1.100	1.040	1.030	1.022
30	2.385	1.425	1.234	1.148	1.080	1.070	1.065
40	2.470	1.480	1.288	1.193	1.125	1.112	1.105
50	2.560	1.535	1.340	1.242	1.168	1.152	1.149
60	2.650	1.590	1.390	1.295	1.210	1.195	1.188
70	2.730	1.648	1.440	1.340	1.250	1.240	1.228
80	2.819	1.702	1.492	1.390	1.295	1.280	1.270
90	2.908	1.757	1.547	1.433	1.340	1.320	1.310
100	2.990	1.813	1.600	1.486	1.380	1.362	1.350
110	3.078	1.870	1.650	1.534	1.425	1.402	1.392
120	3.167	1.928	1.700	1.580	1.467	1.445	1.435
130	3.250	1.980	1.750	1.630	1.510	1.485	1.475
140	3.340	2.035	1.800	1.680	1.550	1.530	1.519
150	3.425	2.090	1.852	1.728	1.592	1.569	1.559

Trans. ASHVE, vol. 35 (1929). p. 165.

decreases up to some 0.7 in. of width, decreasing at a slower rate thereafter. Because of convection currents (and radiation), air-space conductance also increases with temperature. Radiation heat transfer, which is discussed in Section 4-10, is dependent not only on temperature difference but also on the character of the surfaces in relation to their emissivities, and on the geometric configuration of the radiating surfaces. In Section 4-10 it is shown how to compute the configuration factor, F_a, for parallel surfaces such as are commonly met in building construction. A low value of F_a means highly reflective walls, while a high value applies for ordinary walls of wood, plaster, masonry, or glass and latex-painted and oil-painted surfaces. Table 4-3 gives the values for a number of air-space arrangements. Note that for the case of horizontally lying air spaces with heat leakage upward, the conductance is higher than for vertical spaces because of increased convection currents. For heat flow downward the values for vertical spaces can be applied with little error. With the passage of time, reflective surfaces collect dust and undergo chemical action. Therefore emissivity values (Section 4-10) of wall surfaces adjacent to air spaces in building construction are not always constant, and thus the rate of heat flow across an air space may change.

It is often impracticable to calculate the heat-flow conditions through the various subsections of a heat-transfer barrier, but an overall heat-transfer coefficient can be found, either experimentally or from tabulations for surfaces built up in a similar manner. For a composite wall, Eq. (4-12) becomes

$$q = UA(t_i - t_o)\text{Btuh} \tag{4-17}$$

where U is the overall coefficient of heat transfer expressed in units of $\text{Btu/h} \cdot \text{ft}^2 \cdot \text{F}$, found either by direct experiment or by calculation from various items in Eq. (4-16), or selected, when possible, from suitable tabulated values of heat-transfer data.

For composite walls of standard construction, actual tests or computations to determine values of the overall heat-transfer coefficient U, the heat conductivity C, or the resistance R have been made by various investigators, and where these values can be found they should be used in preference to making detailed calculations. Tables 4-4 through 4-17 give some of the values, including those for film resistance on each side of a wall, with a 15-mph wind assumed on the outside wall.

SYMBOLS USED IN HEAT-TRANSFER EQUATIONS AND TABLES.

U = overall coefficient of heat transfer, in units of $\text{Btu/h} \cdot \text{ft}^2 \cdot \text{F}$, existing between the air or other fluid on the two sides of a wall, floor, ceiling, roof, or heat-transfer surface under consideration.

k = thermal conductivity, in units of $\text{Btu} \cdot \text{in./h} \cdot \text{ft}^2 \cdot \text{F}$.

f = film or surface coefficient, for computing heat transfer through the gas or liquid film by conduction, radiation, and convection to the adjacent medium, in units of $\text{Btu/h} \cdot \text{ft}^2 \cdot \text{F}$.

C = thermal conductance (the heat transmitted through a nonhomogeneous or composite material of the thickness and type for which C is given), expressed in units of Btu/h · ft^2 · F. Film effects not usually considered.

a = thermal conductance of an air space, in units of Btu/h · ft^2 · F.

R = resistance to heat flow, usually for 1 ft^2 of area. A general term for various kinds of resistances.

= $1/U$ = overall resistance, in units of h · ft^2 · F/Btu.

= $1/C$ = resistivity of a composite section. Film effects not considered.

= $1/f$ = film resistance.

= $1/a$ = air-space resistance.

For a composite wall, general expressions for resistance and heat transfer are as follows:

$$R = \frac{1}{f_i} + \frac{x_1}{k_1} + \frac{1}{a_1} + \frac{x_2}{k_2} + \frac{1}{a_2} + \cdots + \frac{1}{f_o} = \frac{1}{U} \tag{4-18}$$

$$= \frac{1}{f_i} + \frac{1}{C} + \cdots + \frac{1}{f_o} = \frac{1}{U} \tag{4-19}$$

$$q = UA(t_i - t_o) \text{ Btuh} \tag{4-20}$$

$$q = \frac{A}{R}(t_i - t_o) \text{ Btuh} \tag{4-21}$$

$$q = CA(t_i - t_2) \text{ Btuh (no films considered)} \tag{4-22}$$

$$q = \frac{k}{x} A(t_1 - t_2) \text{ Btuh through homogeneous substance of thickness } x \text{ inches (no films considered)} \tag{4-9}$$

where t_i and t_o are temperatures of the air on two sides of barrier, in degrees Fahrenheit, and A is the area in square feet.

In SI, wall areas are expressed in square meters and wall thickness in centimeters. To employ these units and also SI energy units, the following conversions are in order, for U, C, f, and a.

$$\frac{\text{Btu}}{(\text{h})(\text{ft}^2)(\text{F})} = 5.68 \frac{\text{W}}{(\text{m}^2)(\text{K})} = 0.00568 \frac{\text{kW}}{(\text{m}^2)(\text{K})} = 4.88 \frac{\text{kcal}}{(\text{h})(\text{m}^2)(\text{K})} \tag{4-23}$$

$$\frac{\text{kcal}}{(\text{h})(\text{m}^2)(\text{K})} = 0.2048 \frac{\text{Btu}}{(\text{h})(\text{ft}^2)(\text{F})} = 1.162 \frac{\text{W}}{(\text{m}^2)(\text{K})} \tag{4-24}$$

$$\frac{k\text{W}}{(\text{m}^2)(\text{K})} = 176 \frac{\text{Btu}}{(\text{h})(\text{ft}^2)(\text{F})} = 860.4 \frac{\text{kcal}}{(\text{h})(\text{m}^2)(\text{K})} \tag{4-25}$$

For q in Btu per hour,

$$\frac{\text{Btu}}{\text{h}} = 2.93 \times 10^{-4} \text{ kW} = 0.252 \frac{\text{kcal}}{\text{h}} \tag{4-26}$$

Table 4-3 Thermal Conductivities (k) and Conductances (C) of Building and Insulating Materials

k in $\frac{(\text{Btu})(\text{in.})}{(\text{h})(\text{ft}^2)(\text{F})}$; C in $\frac{\text{Btu}}{(\text{h})(\text{ft}^2)(\text{F})}$, for thickness of material in question

Material	Type and Condition	Density (lb/ft^3)	Mean Temp. (F)	Thermal Conductivity k	Resistance per inch $1/k$	Thermal Conductance C	Resistance $1/C$
Air film (surface):							
Still air (f_i)	General value	—	50–90	—	—	1.65	0.61
Heat flow, up	Horizontal	—	50–90	—	—	1.63	0.61
	Sloping, 45°	—	50–90	—	—	1.60	0.62
Heat flow, down	Horizontal	—	50–90	—	—	1.08	0.92
	Sloping, 45°	—	50–90	—	—	1.32	0.76
Heat flow, horizontal	Vertical	—	50–90	—	—	1.46	0.68
Wind, 15 mph; (f_o)	Any position	—	—	—	—	6.00	0.17
Wind, 7.5 mph; (f_o)	Any position	—	—	—	—	4.00	0.25
Air spaces:	*Bounded by:* Structural materials						
Horizontal space	$E_a = 0.82$, $\Delta t = 10°$	—	90	—	—	1.32	0.76
$\frac{3}{4}$- to 4-in. deep	$E_a = 0.82$, $\Delta t = 10°$	—	50	—	—	1.15	0.87
Heat flow up	$E_a = 0.82$, $\Delta t = 20°$	—	−50	—	—	0.95	1.05
Vertical space	$E_a = 0.82$, $\Delta t = 10°$	—	90	—	—	1.15	0.87
$\frac{3}{4}$- to 4-in. deep	$E_a = 0.82$, $\Delta t = 10°$	—	50	—	—	0.99	1.01
Heat flow across	$E_a = 0.82$, $\Delta t = 20°$	—	−50	—	—	0.71	1.41
	Both sides bright aluminum foil						
Horizontal space	$E_a = 0.03$, $\Delta t = 10°$	—	90	—	—	0.42	2.38
$\frac{3}{4}$- to 4-in. deep	$E_a = 0.03$, $\Delta t = 10°$	—	50	—	—	0.43	2.33
Heat flow up	$E_a = 0.03$, $\Delta t = 20°$	—	−50	—	—	0.58	1.72

Vertical space $\frac{3}{4}$- to 4-in. deep Heat flow across	$E_a = 0.03$ $\Delta t = 10°$	—	90	—	—	0.27	3.70
	$\Delta t = 10°$	—	50	—	—	0.27	3.70
	$\Delta t = 20°$	—	−50	—	—	0.34	2.94
	Structural surface on one side, aluminum-coated paper on other						
Vertical space $\frac{3}{4}$- to 4-in. deep Heat flow across	$E_a = 0.2$ $\Delta t = 10°$	—	90	—	—	0.47	2.13
	$\Delta t = 10°$	—	50	—	—	0.43	2.34
	$\Delta t = 20°$	—	−50	—	—	0.41	2.44
	Structural surface with space divided in two by single aluminum curtain						
Vertical space $\frac{3}{4}$- to 4-in. deep Heat flow across	$\Delta t = 10°$	—	50	—	—	0.23	4.35
	Structural surface on one side, aluminum foil on other						
Vertical space $\frac{3}{8}$-in. deep Heat flow across	$\Delta t = 10°$	—	50	—	—	0.62	1.61
Building boards:							
Asbestos-cement	Compressed, $\frac{1}{8}$ in.	120	75	—	—	33.00	0.03
	Compressed, $\frac{1}{4}$ in.	120	75	3.6	0.27	16.50	0.07
Gypsum plaster, paper covered	$\frac{3}{8}$ in.	50	75	—	—	3.10	0.32
	$\frac{1}{2}$ in.	50	75	—	—	2.25	0.45

Table 4-3 *(Continued)*

Material	Type and Condition	Density (lb/ft^3)	Mean Temp. (F)	Thermal Conductivity k	Resistance per inch $1/k$ = R	Thermal Conductance C	Resistance $1/C$ = R
Insulating							
Fiberboard		15	75	0.33	3.03	—	—
Hardboard		40–50	75	0.73	1.37	—	—
Plywood	$\frac{1}{4}$ in.	34	75	—	—	3.21	0.31
Plywood	$\frac{3}{8}$ in.	34	75	—	—	2.13	0.47
Plywood	$\frac{1}{2}$ in.	34	75	—	—	1.60	0.63
Plywood	$\frac{3}{4}$ in.	34	75	—	—	1.07	0.93
Repulped paper		30	75	0.5	2.00	—	—
Sheathing							
Asphalt—impregnated insulating	$\frac{1}{2}$ in.	—	75	—	—	0.69	1.46
Nail base	$\frac{1}{2}$ in.	25	75	—	—	0.88	1.14
Regular density	$\frac{1}{2}$ in.	18	75	—	—	0.76	1.32
Shingle backer	$\frac{5}{16}$ in.	18	75	—	—	1.28	0.78
Wood, sub floor	$\frac{3}{4}$ in.	—	75	—	—	1.06	0.94
Building paper							
Vapor seal 2 layers	Mopped 15-lb felt	—	75	—	—	8.35	0.12
Exterior finishes	Frame walls						
Brick veneer	4 in.	—	—	—	—	2.27	0.44
Shingles	Asbestos-cement	120	75	—	—	4.76	0.21
Shingles	Wood	—	75	—	—	1.28	0.78

Siding							
Aluminum or steel	Oversheathing	—	75	—	—	1.61	0.61
Asbestos cement	$\frac{1}{4}$ in., lapped	—	75	—	—	4.76	0.21
Insulating board	$\frac{3}{8}$ in., foil back	—	75	—	—	0.55	1.82
Wood bevel	$\frac{1}{2} \times 8$ in., lapped	—	75	—	—	1.23	0.81
Wood plywood	$\frac{3}{8}$ in., lapped	—	75	—	—	1.59	0.59
Stucco	1 in.	—	—	12.50	0.08	—	—
Flooring construction:							
	Asphalt tile, $\frac{1}{8}$ in.	120	—	—	—	24.8	0.04
	California redwood, dry	28.0	75	0.70	1.43	—	—
	Ceramic tile, 1 in.	—	—	—	—	12.50	0.08
	Composition felt	—	—	—	—	16.70	0.06
	Maple across grain	40.0	75	1.20	0.83	—	—
	Rubber tile, $\frac{1}{8}$ in.	110.0	—	—	—	42.4	0.02
	White pine	31.2	86	0.78	1.28	—	—
	Yellow pine, dry	36.0	75	0.91	1.10	—	—
	Linoleum, $\frac{1}{8}$ in.	—	—	—	—	12.00	0.08
Insulation:							
Bat- or blanket-type	Made from mineral or vegetable fiber, or animal hair; closed or open	—	—	0.27	3.70	—	—
Bat-type	Corkboard, no added binder	—	—	0.30	3.33	—	—
	Mineral wool						
	Made from rock slag or glass	—	—	0.29	3.50	—	—
	2–$2\frac{3}{4}$ in. Paperbacked	—	—	—	—	—	7.0
	3–$3\frac{3}{4}$ in. Paperbacked	—	—	—	—	—	11.0

Table 4-3 *(Continued)*

Material	Type and Condition	Density (lb/ft^3)	Mean Temp. (F)	Thermal Conductivity k	Resistance per inch $1/k$ = R	Thermal Conductance C	Resistance $1/C$ = R
Loose-fill type	Chemically treated wood fiber	4.0	75	0.28	3.45	—	—
	Fibrous material made from dolomite and silica	1.50	75	0.27	3.70	—	—
	Fibrous material made from slag	9.4	103	0.27	3.70	—	—
	Glass wool fibers	1.50	75	0.27	3.70	—	—
	Redwood bark	3.0	90	0.31	3.22	—	—
		5.0	75	0.26	3.84	—	—
	Regranulated cork in particles	8.1	90	0.31	3.22	—	—
	Rock wool	10.0	90	0.27	3.70	—	—
	Sawdust	12.0	90	0.41	2.44	—	—
	Vermiculite, expanded	7.0	70	0.48	2.08	—	—
	Wood planer shavings	8.8	90	0.41	2.44	—	—
Slab-type	Polystyrene, expanded	3.5	75	0.19	5.26	—	—
	Polyurethane, expanded	1.5	75	0.16	6.25	—	—
	Wood, shredded, cemented, and preformed	22.0	75	0.60	1.67	—	—
Interior finishes and plasters:							
Composition wallboard	$\frac{3}{16}$ to $\frac{3}{8}$ in. thick	—	—	0.50	2.00	—	—
Gypsum board	Plain or decorated ($\frac{3}{8}$ in.)	—	—	—	—	3.7	0.27
Gypsum lath, $\frac{3}{8}$ in. (and plaster)	Plaster thickness assumed $\frac{1}{2}$ in.	—	—	—	—	2.4	0.42
Insulating board, $\frac{1}{2}$ in.	Plain or decorated	—	—	—	—	0.66	1.52

Insulating board lath, $\frac{1}{2}$ in. (and plaster)	Plaster thickness $\frac{1}{2}$ in.	—	—	—	—	0.60	1.67
Insulating board lath, 1 in. (and plaster)	Plaster thickness $\frac{1}{2}$ in.	—	—	—	—	0.31	3.18
Metal lath and plaster	Cement-sand, $\frac{3}{4}$ in.	—	—	—	—	6.66	0.15
Conventional	Gypsum-sand, $\frac{3}{4}$ in.	—	—	—	—	4.40	0.23
High sand	Gypsum, $\frac{3}{4}$ in.	—	—	—	—	7.70	0.10
Plaster							
Cement and sand	—	—	—	5.0	0.20	—	—
High cement mix	—	—	—	8.0	0.13	—	—
Gypsum and sand	—	—	—	5.6	0.18	—	—
Plaster, vermiculite-gypsum	—	—	—	1.7	0.59	—	—
Wood lath and plaster	Total thickness $\frac{3}{4}$ in.	—	—	—	—	2.50	0.40
Masonry materials:							
Brick	Adobe, 4 in. thick	—	—	—	—	0.89	1.12
	Common, 4 in. thick	—	—	5.0	0.20	1.25	0.80
	Face, 4 in. thick	—	—	9.0	0.11	2.27	0.44
	One tier common clay brick, one tier face brick, approximately 8 in. thick	—	—	—	—	0.77	1.30
Cement mortar	Mortar or plaster	—	—	5.0	0.20	—	—
Clay tile, hollow sections	3 in. thick, 1 cell	—	—	—	—	1.25	0.80
	4 in. thick, 1 cell	—	—	—	—	0.90	1.11
	6 in. thick, 2 cells	—	—	—	—	0.66	1.52
	8 in. thick, 2 cells	—	—	—	—	0.54	1.85
	10 in, thick, 2 cells	—	—	—	—	0.45	2.22
	12 in. thick, 3 cells	—	—	—	—	0.40	2.50
	16 in. thick, 3 cells	—	—	—	—	0.31	3.23
Concrete	Light aggregate of expanded slag, clay, cinders, or pumice	100	—	3.6	0.28	—	—
		80	—	2.5	0.40	—	—
		40	—	1.15	0.86	—	—

Table 4-3 (*Continued*)

Material	Type and Condition	Density (lb/ft^3)	Mean Temp. (F)	Thermal Conductivity k	Resistance per inch $1/k$	Thermal Conductance C	Resistance $1/C$
Concrete	Sand and gravel aggregate	140	—	12.0	0.08	—	—
3-in. concrete blocks	Hollow; cinder aggregate	—	—	—	—	1.16	0.86
4-in. concrete blocks	Hollow; cinder aggregate	—	—	—	—	0.90	1.11
8-in. concrete blocks	Hollow; cinder	—	—	—	—	0.58	1.72
	aggregate	—	—	—	—	0.58	1.72
12-in. concrete blocks	Hollow; cinder aggregate	—	—	—	—	0.53	1.89
8-in. concrete blocks	Hollow; sand and gravel aggregate	—	—	—	—	0.90	1.11
12-in. concrete blocks	Hollow; sand and gravel aggregate	—	—	—	—	0.78	1.28
8-in. concrete blocks	Hollow; lightweight aggregate	—	—	—	—	0.50	2.00
12-in. concrete blocks	Hollow; lightweight aggregate	—	—	—	—	0.44	2.27
3-in. gypsum tile	3-cell partition	—	—	—	—	0.74	1.35
4-in. gypsum tile	3-cell partition	—	—	—	—	0.60	1.67
3-in. gypsum tile	Hollow	—	—	—	—	0.61	1.64
4-in. gypsum tile	Hollow	—	—	—	—	0.46	2.18
Stone, typical	—	—	—	12.50	0.08	—	—
Stucco	—	—	—	5.0	0.20	—	—
Roofing:							
Asbestos-cement shingles	—	120	75	—	—	4.76	0.21
Asphalt singles	—	70.0	75	—	—	2.27	0.44
Built-up roofing	Assumed thickness $\frac{3}{8}$ in.	—	—	—	—	3.00	0.33

Heavy rolled roof material	—	—	—	—	—	1.10	0.91
Slate shingles	Assumed thickness $\frac{1}{2}$ in.	—	—	10.0	0.10	20.0	0.05
Wood shingles	—	—	—	—	—	1.06	0.94
Structural insulation:							
Cellular glass	—	9.0	75	0.42	2.38	—	—
(Foamglas)	—	9.0	60	0.38	2.62	—	—
		9.0	300	0.55	1.82	—	—
Corkboard	No added binder	10.6	90	0.30	3.33	—	—
		7.0	90	0.27	3.70	—	—
Corkboard	Asphaltic binder	14.5	90	0.32	3.12	—	—
Mineral (rock) wool	—	15.7	90	0.32	3.12	—	—
		15.7	30	0.29	3.44	—	—
Shredded wood and cement	—	29.8	—	0.77	1.30	—	—
Styrofoam	—	1.3–2.0	40	0.23–0.30	4.35–3.33	—	—
Sugar cane, fiber insulation	Encased in asphalt membrane	13.8	70	0.30	3.33	—	—
Thermoflex, refractory	Felted blanket	6.0	500	0.47	2.12	—	—
fiber		6.0	1000	0.92	1.09	—	—
Woods:	Balsa	8.8	90	0.38	2.63	—	—
	California redwood, dry	28.0	75	0.70	1.43	—	—
	Long-leaf yellow pine, dry	40.0	75	0.86	1.16	—	—
	Red oak, dry	48.0	75	1.18	0.85	—	—
	Short-leaf yellow pine, dry	36.0	75	0.91	1.10	—	—
	Fir or pine, average	—	—	0.80	1.25	—	—
	Maple or oak, average	—	—	1.10	0.91	—	—
Sheathing	1-in. fir sheathing, building paper, and pine-lap siding	—	—	—	—	0.50	2.00

The data for Table 4-3 are from many sources but largely from the *ASHRAE Handbook of Fundamentals 1972*, by permission.

In making heat-loss calculations for a building for which the U factor must be determined in advance, it is convenient to search the tables to determine whether a wall is described which exactly fits the situation. If one cannot be found in this way, it is then necessary to make a computation for the actual wall under consideration. This is done by summing up the individual resistances of the components of the wall, selecting appropriate values from Table 4-3. Illustrative examples of how this is done follow in the text. The first example is set up to illustrate fundamentals; the second shows how to make quick use of the tabular material.

Example 4-1. Assume that the composite wall of Fig. 4-1 is made up of 4 in. of common brick against 6 in. of concrete, with $\frac{1}{2}$ in. of cement plaster on the inside (concrete) wall. Assume still air at 76 F outside, outside air at 20 F, and a 15-mph wind velocity. Find (a) the thermal resistance of the wall, (b) the overall conductivity of the wall, (c) the heat transferred per hour, and (d) the heat transfer if the film resistances are disregarded.

Solution: (a) Referring to Table 4-3, we find the following thermal conductivities, in units of Btu in./h · ft² · F:

$$k_b = 5.00 \text{ for common brick}$$

$$k_c = 12.00 \text{ for concrete}$$

$$k_p = 5.00 \text{ for cement plaster}$$

We also find from Table 4-3 the following film coefficients for the inside and outside walls, in units of Btu/h · ft² · F:

$$f_o = 6.00 \text{ for 15-mph wind} \qquad R_i = 0.17$$
$$f_i = 1.46 \text{ for still air, vertical wall} \quad R_o = 0.68$$

Therefore, from Eq. (4-18) or (4-19), the thermal resistance is, in units of h · ft² · F/Btu,

$$R = \frac{1}{f_o} + \frac{x_b}{k_b} + \frac{x_c}{k_c} + \frac{x_p}{k_p} + \frac{1}{f_i} = 0.17 + \frac{4.0}{5.0} + \frac{6.0}{12.0} + \frac{0.5}{5.0} + 0.68 = 2.25 \qquad \textit{Ans.}$$

(b) The overall conductivity of the wall is, then, in units of Btu/h · ft² · F,

$$U = \frac{1}{R} = \frac{1}{2.25} = 0.444 \qquad \textit{Ans.}$$

(c) The heat transferred is, therefore, in units of Btu/h · ft²

$$q = UA(t_i - t_o) = (0.444)(1)(76 - 20) = 24.8 \qquad \textit{Ans.}$$

(d) If, in part a, $1/f_o$ and $1/f_i$ are disregarded, R becomes 1.400 and the heat transferred is, in units of Btu/h · ft²,

$$q = \frac{1}{R} A(t_i - t_o) = \frac{1}{1.400}(1)(76 - 20) = 40.0 \qquad \textit{Ans.}$$

NOTE. Although the fact is not mentioned in the foregoing example, usually $\frac{1}{4}$ to $\frac{1}{2}$ in. of bonding plaster would be used between masonry divisions, as here between the brick and concrete, or at least an air space would exist which would also contribute to the total thermal resistance.

Table 4-4 Coefficients of Transmission (U) of Frame Walls
[No Insulation between Studs* (see Table 4-5)]

Exterior Finish	Interior Finish	Type of Sheathing: Gypsum (1/2 In. Thick)	Plywood (5/16 In. Thick)	Wood¶ (25/32 In. Thick) Bldg. Paper	Insulating Board (25/32 In. Thick)	Wall Number
		A	B	C	D	
Wood siding (clapboard)	Metal lath and plaster†	0.33	0.32	0.26	0.20	1
	Gypsum board (3/8 in.) decorated	0.32	0.32	0.26	0.20	2
	Wood lath and plaster	0.31	0.31	0.25	0.19	3
	Gypsum lath (3/8 in.) plastered‡	0.31	0.30	0.25	0.19	4
	Plywood (3/8 in.) plain or decorated	0.30	0.30	0.24	0.19	5
	Insulating board (1/2 in.) plain or decorated	0.23	0.23	0.19	0.16	6
	Insulating board lath (1/2 in.) plastered‡	0.22	0.22	0.19	0.15	7
	Insulating board lath (1 in.) plastered‡	0.17	0.17	0.15	0.12	8
Wood shingles§	Metal lath and plaster†	0.25	0.25	0.26	0.17	9
	Gypsum board (3/8 in.) decorated	0.25	0.25	0.26	0.17	10
	Wood lath and plaster	0.24	0.24	0.25	0.16	11
	Gypsum lath (3/8 in.) plastered‡	0.24	0.24	0.25	0.16	12
	Plywood (3/8 in.) plain or decorated	0.24	0.24	0.24	0.16	13
	Insulating board (1/2 in.) plain or decorated	0.19	0.19	0.19	0.14	14
	Insulating board lath (1/2 in.) plastered‡	0.19	0.18	0.19	0.13	15
	Insulating board lath (1 in.) plastered‡	0.14	0.14	0.15	0.11	16
Stucco	Metal lath and plaster†	.43	0.42	0.32	0.23	17
	Gypsum board (3/8 in.) decorated	0.42	0.41	0.31	0.23	18
	Wood lath and plaster	0.40	0.39	0.30	0.22	19
	Gypsum lath (3/8 in.) plastered‡	0.39	0.39	0.30	0.22	20
	Plywood (3/8 in.) plain or decorated	0.39	0.38	0.29	0.22	21
	Insulating board (1/2 in.) plain or decorated	0.27	0.27	0.22	0.18	22
	Insulating board lath (1/2 in.) plastered‡	0.26	0.26	0.22	0.17	23
	Insulating board lath (1 in.) plastered‡	0.19	0.19	0.16	0.14	24
Brick veneer‖	Metal lath and plaster†	0.37	0.36	0.28	0.21	25
	Gypsum board (3/8 in.) decorated	0.36	0.36	0.28	0.21	26
	Wood lath and plaster	0.35	0.34	0.27	0.20	27
	Gypsum lath (3/8 in.) plastered‡	0.34	0.34	0.27	0.20	28
	Plywood (3/8 in.) plain or decorated	0.34	0.33	0.27	0.20	29
	Insulating board (1/2 in.) plain or decorated	0.25	0.25	0.21	0.17	30
	Insulating board lath (1/2 in.) plastered‡	0.24	0.24	0.20	0.16	31
	Insulating board lath (1 in.) plastered‡	0.18	0.18	0.15	0.13	32

Note. Coefficients are expressed in Btu per (hour) (square foot) (Fahrenheit degree difference in temperature between the air on both sides), and are based on an outside wind velocity of 15 mph.

* Coefficients not weighted; effect of studding neglected.
† Plaster assumed 3/4 in. thick.
‡ Plaster assumed 1/2 in. thick.
§ Furring strips (1 in. nominal thickness) between wood shingles and all sheathings except wood.
‖ Small air space and mortar between building paper and brick veneer neglected.
¶ Nominal thickness, 1 in.

Table 4-5 Coefficients of Transmission (U) of Frame Walls and Roofs with Insulation between Framing*

Coefficient with *No* Insulation between Framing	Coefficient with Insulation between Framing				Number
	Mineral Wool or Vegetable Fibers in Blanket or Bat·Form†			$3\frac{5}{8}$ In. Mineral Wool between Framing‡	
	1 In. Thick	2 In. Thick	3 In. Thick		
	A	B	C	D	
0.11	0.078	0.063	0.054	0.051	33
0.13	0.088	0.070	0.058	0.055	35
0.15	0.097	0.075	0.062	0.059	37
0.17	0.10	0.080	0.066	0.062	39
0.19	0.11	0.084	0.069	0.065	41
0.21	0.12	0.088	0.072	0.067	43
0.23	0.12	0.091	0.074	0.069	45
0.25	0.13	0.094	0.076	0.071	47
0.27	0.14	0.097	0.078	0.073	49
0.29	0.14	0.10	0.080	0.075	51
0.31	0.14	0.10	0.081	0.076	53
0.33	0.15	0.10	0.083	0.077	55
0.35	0.15	0.11	0.084	0.078	57
0.37	0.16	0.11	0.085	0.080	59
0.39	0.16	0.11	0.086	0.081	61
0.41	0.16	0.11	0.087	0.082	63
0.43	0.17	0.11	0.088	0.082	65

*** Coefficients corrected for 2 × 4 framing, 16 in. on centers—15 per cent of surface area.**
† Based on one air space between framing.
‡ No air space.

NOTE. In Tables 4-4 to 4-18 inclusive, computed values of U are given for composite wall structures. For outside walls and roofs, a film factor based on a 15-mph wind has been considered for the outside surface, with a still-air film factor of 1.65 used for all inside wall surfaces. If attention is given to the exact orientation of each wall in relation to film factors (see Table 4-3), slightly different results will appear. New materials and new manufacturing processes also generate different values of thermal conductivities which in some cases could lead to deviations from these reported U values. In case of doubt, a user should compute the overall U value using data from Table 4-3. Table 4-5 provides a means of quickly correcting a U value for the addition of inorganic insulation between framing, and Table 4-18 provides a similar means for correcting for varying outside wind velocities. Tables 4-4 through 4-18 are reprinted from Chapter 9 of the *ASHRAE Heating, Ventilating, Air-Conditioning Guide* of 1955, by permission.

Table 4-6 Coefficients of Transmission (U) of Masonry Walls

Type of Masonry	Thickness of Masonry (in.)	Interior Finish (Plus Insulation Where Indicated): Plain Walls—No Interior Finish	Plaster (½ in.) on Walls	Metal Lath and Plaster¶ Furred ††	Gypsum Board (⅜ in.) Decorated—Furred ††	Gypsum Lath (⅜ in.) Plastered **—Furred ††	Insulating Board (½ in.) Plain or Decorated—Furred ††	Insulating Board Lath (½ in.) Plastered **—Furred ††	Insulating Board Lath (1 in.) Plastered **—Furred ††	Gypsum Lath ** Plastered Plus 1 in. Blanket Insulation—Furred ††	Wall Number
		A	B	C	D	E	F	G	H	I	
Solid Brick*	8	0.50	0.46	0.32	0.31	0.30	0.22	0.22	0.16	0.14	67
	12	0.36	0.34	0.25	0.25	0.24	0.19	0.19	0.14	0.13	68
	16	0.28	0.27	0.21	0.21	0.20	0.17	0.16	0.13	0.12	69
Hollow Tile† (Stucco Exterior Finish)	8	0.40	0.37	0.27	0.27	0.26	0.20	0.20	0.15	0.13	70
	10	0.39	0.37	0.27	0.27	0.26	0.20	0.19	0.15	0.13	71
	12	0.30	0.28	0.22	0.22	0.21	0.17	0.17	0.13	0.12	72
	16	0.24	0.24	0.19	0.19	0.18	0.15	0.15	0.12	0.11	73
Stone ‡	8	0.70	0.64	0.39	0.38	0.36	0.26	0.25	0.18	0.16	74
	12	0.57	0.53	0.35	0.34	0.33	0.24	0.23	0.17	0.15	75
	16	0.49	0.45	0.31	0.31	0.29	0.22	0.22	0.16	0.14	76
	24	0.37	0.35	0.26	0.26	0.25	0.19	0.19	0.15	0.13	77
Poured Concrete §	6	0.79	0.71	0.42	0.41	0.39	0.27	0.26	0.19	0.16	78
	8	0.70	0.64	0.39	0.38	0.36	0.26	0.25	0.18	0.16	79
	10	0.63	0.58	0.37	0.36	0.34	0.25	0.24	0.18	0.15	80
	12	0.57	0.53	0.35	0.34	0.33	0.24	0.23	0.17	0.15	81
Hollow Concrete Blocks	Gravel Aggregate										
	8	0.56	0.52	0.34	0.34	0.32	0.24	0.23	0.17	0.15	82
	12	0.49	0.46	0.32	0.31	0.30	0.22	0.22	0.16	0.14	83
	Cinder Aggregate										
	8	0.41	0.39	0.28	0.28	0.27	0.21	0.20	0.15	0.13	84
	12	0.38	0.36	0.26	0.26	0.25	0.20	0.19	0.15	0.13	85
	Lightweight Aggregate‖										
	8	0.36	0.34	0.26	0.25	0.24	0.19	0.19	0.15	0.13	86
	12	0.34	0.33	0.25	0.24	0.24	0.19	0.18	0.14	0.13	87

* Based on 4-in. hard brick and remainder common brick.
† The 8-in. and 10-in. tile figures are based on two cells in the direction of heat flow. The 12-in. tile is based on three cells in the direction of heat flow. The 16-in. tile consists of one 10-in. and one 6-in. tile, each having two cells in the direction of heat flow.
‡ Limestone or sandstone.
§ These figures may be used with sufficient accuracy for concrete walls with stucco exterior finish.
‖ Expanded slag, burned clay or pumice.
¶ Thickness of plaster assumed ¾ in.
** Thickness of plaster assumed ½ in.
†† Based on 2-in. furring strips; one air space.

Table 4-7 Coefficients of Transmission (U) of Brick- and Stone-Veneer Masonry Walls

Typical Construction	Facing	Backing	Interior Finish (Plus Insulation Where Indicated)									Wall Number
			Plain Walls—No Interior Finish	Plaster (½ in.) on Walls	Metal Lath and Plaster ‖—Furred **	Gypsum Board (⅜ in.) Decorated—Furred **	Gypsum Lath (⅜ in.) Plastered ¶—Furred **	Insulating Board (½ in.) Plain or Decorated—Furred **	Insulating Board Lath (½ in.) Plastered ¶—Furred **	Insulating Board Lath (1 in.) Plastered ¶—Furred **	Gypsum Lath Plastered ¶ Plus 1-in. Blanket Insulation—Furred **	
			A	B	C	D	E	F	G	H	I	
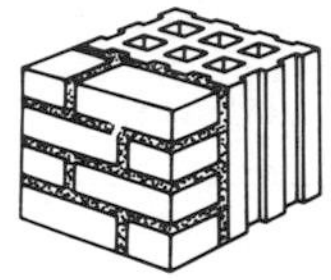	4-in. Brick Veneer*	6-in. hollow tile†	0.35	0.34	0.25	0.25	0.24	0.19	0.18	0.14	0.13	88
		8-in. hollow tile†	0.34	0.32	0.25	0.24	0.23	0.19	0.18	0.14	0.13	89
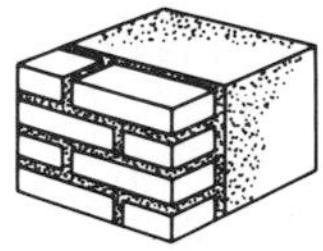		6-in. concrete	0.59	0.54	0.35	0.35	0.33	0.24	0.23	0.17	0.15	90
		8-in. concrete	0.54	0.50	0.33	0.33	0.31	0.23	0.23	0.17	0.15	91
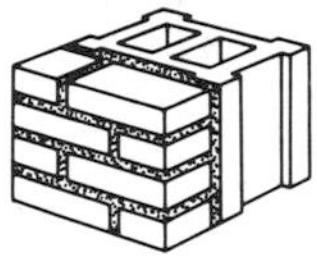		8-in. concrete blocks‡ (gravel aggregate)	0.44	0.41	0.29	0.29	0.28	0.21	0.21	0.16	0.14	92
		8-in. concrete blocks‡ (cinder aggregate)	0.34	0.33	0.25	0.24	0.24	0.19	0.18	0.14	0.13	93
		8-in. concrete blocks‡ (lightweight aggregate)§	0.31	0.29	0.23	0.23	0.22	0.18	0.17	0.14	0.12	94
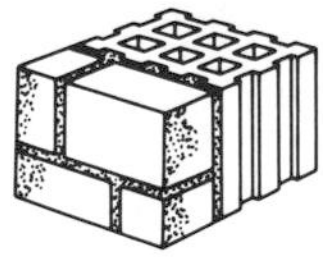	4-in. Cut-Stone Veneer*	6-in. hollow tile†	0.37	0.35	0.26	0.26	0.25	0.19	0.19	0.15	0.13	95
		8-in. hollow tile†	0.36	0.34	0.25	0.25	0.24	0.19	0.19	0.14	0.13	96
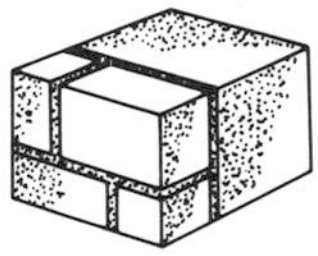		6-in. concrete	0.63	0.58	0.37	0.36	0.34	0.25	0.24	0.18	0.15	97
		8-in. concrete	0.57	0.53	0.35	0.34	0.33	0.24	0.23	0.17	0.15	98
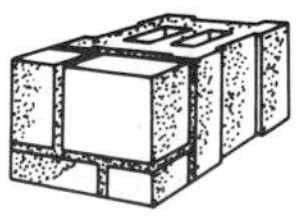		8-in. concrete blocks‡ (gravel aggregate)	0.47	0.44	0.30	0.30	0.29	0.22	0.21	0.16	0.14	99
		8-in. concrete blocks‡ (cinder aggregate)	0.36	0.34	0.25	0.25	0.24	0.19	0.19	0.15	0.13	100
		8-in. concrete blocks‡ (lightweight aggregate)§	0.32	0.30	0.23	0.23	0.22	0.18	0.17	0.14	0.12	101

* Calculation based on ½-in. cement mortar between backing and facing, except in the case of the concrete backing, which is assumed to be poured in place.
† The hollow tile figures are based on two air cells in the direction of heat flow.
‡ Hollow concrete blocks.
§ Expanded slag, burned clay or pumice.
‖ Thickness of plaster assumed ¾ in.
¶ Thickness of plaster assumed ½ in.
** Based on 2-in. furring strips; one air space.

Table 4-8 Coefficients of Transmission (U) of Frame Partitions or Interior Walls*

Type of interior finish	Single Partition (Finish on One Side Only of Studs)	Double Partition (Finish on Both Sides of Studs)		Partition Number
		No Insulation between Studs	1-in. Blanket § between Studs. One Air Space	
	A	B	C	
Metal lath and plaster†	0.69	0.39	0.16	1
Gypsum board (3/8 in.) decorated	0.67	0.37	0.16	2
Wood lath and plaster	0.62	0.34	0.15	3
Gypsum lath (3/8 in.) plastered‡	0.61	0.34	0.15	4
Plywood (3/8 in.) plain or decorated	0.59	0.33	0.15	5
Insulating board (1/2 in.) plain or decorated	0.36	0.19	0.11	6
Insulating board lath (1/2 in.) plastered‡	0.35	0.18	0.11	7
Insulating board lath (1 in.) plastered‡	0.23	0.12	0.082	8

* Coefficients not weighted; effect of studding neglected.

† Plaster assumed 3/4 in. thick.

‡ Plaster assumed 1/2 in. thick.

§ For partitions with other insulations between studs refer to Table 4–5, using values in column B of above table, in left-hand column of Table 4–5. *Example:* What is the coefficient of transmission (U) of a partition consisting of gypsum lath and plaster on both sides of studs with 2-in. blanket between studs? *Solution:* According to above table, this partition with no insulation between studs (No. 4B) has a coefficient of 0.34. Interpolating from Table 4–5, it will be found that a wall having a coefficient of 0.34 with no insulation between studs, will have a coefficient of 0.10 with 2 in. of blanket insulation between studs (No. 55B).

Table 4-9 Coefficients of Transmission (U) of Masonry Partitions

Type of partition		Thickness of Masonry (in.)	Type of Finish			Partition Number
			No Finish (Plain Walls)	Plaster One Side	Plaster Both Sides*	
			A	B	C	
Hollow clay tile		3	0.50	0.47	0.43	9
		4	0.45	0.42	0.40	10
Hollow gypsum tile		3	0.35	0.33	0.32	11
		4	0.29	0.28	0.27	12
Hollow Concrete Tile or Blocks	Cinder aggregate	3	0.50	0.47	0.43	13
		4	0.45	0.42	0.40	14
	Lightweight aggregate†	3	0.41	0.39	0.37	15
		4	0.35	0.34	0.32	16
Common brick		4	0.50	0.46	0.43	17

* 2-in. solid plaster partition, $U = 0.53$.

† Expanded slag, burned clay or pumice.

Table 4-10 Coefficients of Transmission (U) of Frame-Construction Ceilings and Floors

Type of ceiling	INSULATION BETWEEN, OR ON TOP OF, JOISTS (NO FLOORING ABOVE)												WITH FLOORING** (ON TOP OF CEILING JOISTS)		NUMBER
	None	Insulating Board on Top of Joists		Blanket or Bat Insulation¶ between Joists*			Vermiculite Insulation between Joists*			Mineral Wool Insulation between Joists*			Single Wood Floor†	Double Wood Floor‡	
		½ In.	1 In.	1 In.	2 In.	3 In.	2 In.	3 In.	4 In.	2 In.	3 In.	4 In.			
	A	B	C	D	E	F	G	H	I	J	K	L	M	N	
No ceiling		0.37	0.24										0.45	0.34	1
Metal lath and plaster§	0.69	0.26	0.19	0.19	0.12	0.093	0.18	0.14	0.11	0.12	0.093	0.077	0.30	0.25	2
Gypsum board (⅜ in.) plain or decorated	0.67	0.26	0.18	0.19	0.12	0.092	0.18	0.13	0.10	0.12	0.092	0.077	0.30	0.24	3
Wood lath and plaster	0.62	0.25	0.18	0.19	0.12	0.091	0.17	0.13	0.10	0.12	0.091	0.076	0.28	0.24	4
Gypsum lath (⅜ in.) plastered‖	0.61	0.25	0.18	0.19	0.12	0.091	0.17	0.13	0.10	0.12	0.091	0.076	0.28	0.24	5
Plywood (⅜ in.) plain or decorated	0.59	0.24	0.18	0.19	0.12	0.091	0.17	0.13	0.10	0.12	0.091	0.076	0.28	0.23	6
Insulating board (½ in.) plain or decorated	0.36	0.19	0.15	0.16	0.10	0.082	0.14	0.12	0.097	0.10	0.082	0.069	0.22††	0.19††	7
Insulating board lath (½ in.) plastered‖	0.35	0.19	0.15	0.15	0.10	0.081	0.14	0.11	0.096	0.10	0.081	0.068	0.21	0.18	8
Insulating board lath (1 in.) plastered‖	0.23	0.15	0.12	0.12	0.089	0.072	0.12	0.097	0.084	0.089	0.072	0.061	0.16	0.14	9

* Coefficients corrected for framing on basis of 15 per cent area, 2 in. by 4 in. (nominal) framing, 16 in. on centers.

† 25/32-in. yellow pine or fir.

‡ 25/32-in. pine or fir subflooring plus 13/16-in. hardwood-finish flooring.

§ Plaster assumed ¾ in. thick.

‖ Plaster assumed ½ in. thick.

¶ Based on insulation in contact with ceiling, and consequently no air space between.

** For coefficients for constructions in columns M and N (except No. 1) with insulation between joists, refer to Table 4–5. *Example:* The coefficient for No. 3-N of Table 4–10 is 0.24. With 2-in. blanket insulation between joists, the coefficient will be 0.093. (See Table 4–5.) (Column D of Table 4–5 applicable only for 3⅝-in. joists.)

†† For 25/32-in. insulating board sheathing applied to the under side of the joists, the coefficient for *single* wood floor (col. M) is 0.18 and for *double* wood floor (col. N) is 0.16. For coefficients with insulation between joists, see Table 4–5.

Table 4-11 Coefficients of Transmission (U) of Concrete-Construction Floors and Ceilings

Flooring / Ceiling — Type of ceiling	Thickness of Concrete** (in.)	Type of Flooring: No Flooring (Concrete Bare)	Tile* or Terrazzo Flooring on Concrete	⅛-In. Asphalt Tile† Directly on Concrete	Parquet‡ Flooring in Mastic on Concrete	Double Wood Floor on Sleepers§	Number
		A	B	C	D	E	
No ceiling	3	0.68	0.65	0.66	0.45	0.25	1
	6	0.59	0.56	0.58	0.41	0.23	2
	10	0.50	0.48	0.49	0.36	0.22	3
½-in. plaster applied to underside of concrete	3	0.62	0.59	0.60	0.43	0.24	4
	6	0.54	0.52	0.53	0.39	0.22	5
	10	0.46	0.44	0.45	0.34	0.21	6
Metal lath and plaster‖—suspended or furred	3	0.38	0.37	0.37	0.30	0.19	7
	6	0.35	0.34	0.35	0.28	0.18	8
	10	0.32	0.31	0.32	0.26	0.17	9
Gypsum board (⅜ in.) and plaster¶—suspended or furred	3	0.36	0.35	0.35	0.28	0.19	10
	6	0.33	0.32	0.33	0.27	0.18	11
	10	0.30	0.29	0.30	0.24	0.17	12
Insulating board lath (½ in.) and plaster¶ suspended or furred	3	0.25	0.24	0.25	0.21	0.15	13
	6	0.23	0.23	0.23	0.20	0.15	14
	10	0.22	0.21	0.22	0.19	0.14	15

*Thickness of tile assumed to be 1 in.

† Conductivity of asphalt tile assumed to be 3.1.

‡ Thickness of wood assumed to be 13/16 in.; thickness of mastic, ⅛ in. ($k = 4.5$). Column D may also be used for concrete covered with carpet.

§ Based on 25/32-in. yellow pine or fir subflooring and 13/16-in. hardwood-finish flooring with an air space between subfloor and concrete.

‖ Thickness of plaster assumed to be ¾ in.

¶ Thickness of plaster assumed to be ½ in.

** For other thickness of concrete, interpolate.

Table 4-12 Coefficients of Transmission (U) for Wood and Metal Doors

Nominal Thickness and Type	Actual Thickness (in.)	U for Door; No Storm Door	U with Storm Door Considered as 50% glass: Wood	Metal
1-in. wood	0.78	0.64	0.30	0.39
$1\frac{1}{4}$-in. wood[a]	1.06	0.55	0.28	0.34
$1\frac{1}{2}$-in. wood	1.31	0.49	0.27	0.33
$1\frac{3}{4}$-in. wood	1.38	0.48	0.26	0.32
2-in. wood	1.63	0.43	0.24	0.29
$2\frac{1}{2}$-in. wood	2.13	0.36	0.22	0.26
$1\frac{3}{4}$-in. steel				
mineral fiber core		0.59		
polystyrene core		0.47		
urethane foam core		0.40		

[a] Use $U = 0.85$ for hollow-construction, thin-wood-panel doors.

Table 4-13 Coefficients of Transmission (U) of Flat Roofs Covered with Built-up Roofing. No Ceiling—Underside of Roof Exposed

Type of Roof Deck	Thickness of Roof Deck (in.)	No Insulation	Insulation on Top of Deck (Covered with Built-up Roofing)							Number
			Insulating Board Thickness				Corkboard Thickness			
			½ In.	1 In.	1½ In.	2 In.	1 In.	1½ In.	2 In.	
		A	B	C	D	E	F	G	H	
Flat metal roof deck* (Roofing, Insulation, Metal deck)		0.94	0.39	0.24	0.18	0.14	0.23	0.17	0.13	1
Precast cement tile (Roofing, Cast tile, Supports)	1⅝ in.	0.84	0.3	0.24	0.17	0.14	0.22	0.16	0.13	2
Concrete (Roofing, Insulation, Concrete)	2 in.	0.82	0.36	0.24	0.17	0.14	0.22	0.16	0.13	3
	4 in.	0.72	0.34	0.23	0.17	0.13	0.21	0.16	0.12	4
	6 in.	0.65	0.33	0.22	0.16	0.13	0.21	0.15	0.12	5
Gypsum fiber concrete† on ½-in. gypsum board (Roofing, Insulation, Gypsum, Gypsum board)	2½ in.	0.38	0.24	0.18	0.14	0.12	0.17	0.13	0.11	6
	3½ in.	0.31	0.21	0.16	0.13	0.11	0.15	0.12	0.10	7
Wood‡ (Roofing, Insulation, Wood)	1 in.	0.49	0.28	0.20	0.15	0.12	0.19	0.14	0.12	8
	1½ in.	0.37	0.24	0.17	0.14	0.11	0.17	0.13	0.11	9
	2 in.	0.32	0.22	0.16	0.13	0.11	0.16	0.12	0.10	10
	3 in.	0.23	0.17	0.14	0.11	0.096	0.13	0.11	0.091	11

* Coefficient of transmission of bare corrugated iron (no roofing) is 1.50 Btu per (hr) (sq ft of projected area) (F deg difference in temperature) based on an outside wind velocity of 15 mph.

† 87½ per cent gypsum, 12½ per cent wood fiber. Thickness indicated includes ½ in. gypsum board.

‡ Nominal thicknesses specified—actual thicknesses used in calculations.

Table 4-14 Coefficients of Transmission (U) of Flat Roofs Covered with Built-up Roofing. With Lath and Plaster Ceilings[a]

Type of Roof Deck	Thickness of Roof Deck (in.)	No Insulation	Insulation on Top of Deck (Covered with Built-up Roofing): Insulating Board Thickness				Corkboard Thickness			Number
			½ In.	1 In.	1½ In.	2 In.	1 In.	1½ In.	2 In.	
		A	B	C	D	E	F	G	H	
Flat metal roof deck Roofing, Insulation, Metal deck, Ceiling		0.46	0.27	0.19	0.15	0.12	0.18	0.14	0.11	12
Precast cement tile Roofing, Cast tile, Supports, Ceiling	1⅝ in.	0.43	0.26	0.19	0.15	0.12	0.18	0.14	0.11	13
Concrete Roofing, Insulation, Concrete, Ceiling	2 in.	0.42	0.26	0.19	0.14	0.12	0.18	0.14	0.11	14
	4 in.	0.40	0.25	0.18	0.14	0.12	0.17	0.13	0.11	15
	6 in.	0.37	0.24	0.18	0.14	0.11	0.17	0.13	0.11	16
Gypsum fiber concrete† on ½-in. gypsum board Roofing, Insulation, Gypsum, Gypsum board, Ceiling	2½ in.	0.27	0.19	0.15	0.12	0.10	0.14	0.12	0.097	17
	3½ in.	0.23	0.17	0.14	0.11	0.097	0.13	0.11	0.091	18
Wood‡ Roofing, Insulation, Wood, Ceiling	1 in.	0.31	0.21	0.16	0.13	0.11	0.15	0.12	0.10	19
	1½ in.	0.26	0.19	0.15	0.12	0.10	0.14	0.11	0.095	20
	2 in.	0.24	0.17	0.14	0.11	0.097	0.13	0.11	0.092	21
	3 in.	0.18	0.14	0.12	0.10	0.087	0.11	0.095	0.082	22

*** Calculations based on metal lath and plaster ceilings, but coefficients may be used with sufficient accuracy for gypsum lath or wood lath and plaster ceilings. It is assumed that there is an air space between the under side of the roof deck and the upper side of the ceiling.**

† 87½ per cent gypsum, 12½ per cent wood fiber. Thickness indicated includes ½ in. gypsum board.

‡ Nominal thicknesses specified—actual thicknesses used in calculations.

Table 4-15 Coefficients of Transmission (U) of Pitched Roofs

Roof sheathing, Shingles, Rafters, Plaster, Plaster base — Type of ceiling (applied directly to roof rafters)	Wood Shingles (On 1 × 4 Wood Strips* Spaced 2 In. Apart)				Asphalt Shingles or Roll Roofing (On Solid Wood Sheathing)*				Slate or Tile† (On Solid Wood Sheathing)*				Number
	Insulation Between Rafters												
	None	Blanket or Bat (Thickness Below)			None	Blanket or Bat (Thickness Below)			None	Blanket or Bat (Thickness Below)			
		1 In.	2 In.	3 In.		1 In.	2 In.	3 In.		1 In.	2 In.	3 In.	
	A	B	C‡	D‡	E	F	G‡	H‡	I	J	K‡	L‡	
No ceiling applied to rafters	0.48§	0.15	0.10	0.081	0.52§	0.15	0.11	0.084	0.55§	0.16	0.11	0.085	1
Metal lath and plaster‖	0.31	0.14	0.10	0.081	0.33	0.15	0.10	0.083	0.34	0.15	0.10	0.083	2
Gypsum board (3/8 in.) decorated	0.30	0.14	0.10	0.080	0.32	0.15	0.10	0.082	0.33	0.15	0.10	0.083	3
Wood lath and plaster	0.29	0.14	0.10	0.080	0.31	0.14	0.10	0.081	0.32	0.15	0.10	0.082	4
Gypsum lath (3/8 in.) plastered¶	0.29	0.14	0.10	0.079	0.31	0.14	0.10	0.081	0.32	0.15	0.10	0.082	5
Plywood (3/8 in.) plain or decorated	0.29	0.14	0.099	0.079	0.30	0.14	0.10	0.081	0.31	0.15	0.10	0.081	6
Insulating board (1/2 in.) plain or decorated	0.22	0.12	0.090	0.072	0.23	0.12	0.091	0.074	0.24	0.13	0.092	0.074	7
Insulating board lath (1/2 in.) plastered¶	0.22	0.12	0.088	0.072	0.22	0.12	0.090	0.073	0.23	0.12	0.091	0.074	8
Insulating board lath (1 in.) plastered¶	0.16	0.10	0.078	0.064	0.17	0.10	0.079	0.065	0.17	0.10	0.080	0.066	9

* Sheathing and wood strips assumed 25/32 in. thick.
† Figures in columns I, J, K, and L may be used with sufficient accuracy for rigid asbestos shingles on wood sheathing. Layer of slater's felt neglected.
‡ Coefficients corrected for framing on basis of 15 per cent area, 2 in. by 4 in. (nominal), 16 in. on centers.
§ No air space included; all other coefficients based on one air space.
‖ Plaster assumed 3/4 in. thick.
¶ Plaster assumed 1/2 in. thick.

Table 4-16 Combined Coefficients of Transmission (U) of Unvented Pitched Roofs* and Horizontal Ceilings—Based on Ceiling Area†

CEILING COEFFICIENTS (FROM TABLE 4-10)	TYPE OF ROOFING AND ROOF SHEATHING						NUMBER
	Wood Shingles on Wood Strips ‡			Asphalt Shingles § or Roll Roofing on Wood Sheathing ‖			
	No Roof Insulation (Rafters Exposed) ($U_r = 0.48$)	½-In. Insulating Board on Underside of Rafters ($U_r = 0.22$)	1-In. Insulating Board on Underside of Rafters ($U_r = 0.16$)	No Roof Insulation (Rafters Exposed) ($U_r = 0.53$)	½-In. Insulating Board on Underside of Rafters ($U_r = 0.23$)	1-In. Insulating Board on Underside of Rafters ($U_r = 0.17$)	
	A	B	C	D	E	F	
0.10	0.085	0.073	0.066	0.087	0.074	0.067	19
0.11	0.092	0.078	0.07	0.094	0.079	0.071	20
0.12	0.099	0.082	0.074	0.10	0.083	0.075	21
0.13	0.11	0.087	0.078	0.11	0.088	0.079	22
0.14	0.11	0.091	0.081	0.11	0.093	0.083	23
0.15	0.12	0.096	0.084	0.12	0.097	0.086	24
0.16	0.13	0.10	0.087	0.13	0.10	0.089	25
0.17	0.13	0.10	0.090	0.13	0.10	0.092	26
0.18	0.14	0.11	0.093	0.14	0.11	0.095	27
0.19	0.14	0.11	0.095	0.15	0.11	0.098	28
0.20	0.15	0.11	0.098	0.15	0.12	0.10	29
0.21	0.15	0.12	0.10	0.16	0.12	0.10	30
0.22	0.16	0.12	0.10	0.17	0.12	0.11	31
0.23	0.16	0.12	0.10	0.17	0.12	0.11	32
0.24	0.17	0.13	0.11	0.18	0.12	0.11	33
0.25	0.17	0.13	0.11	0.18	0.13	0.11	34
0.26	0.18	0.13	0.11	0.19	0.13	0.11	35
0.27	0.18	0.13	0.11	0.19	0.13	0.12	36
0.28	0.19	0.14	0.12	0.19	0.14	0.12	37
0.29	0.19	0.14	0.12	0.20	0.14	0.12	38
0.30	0.20	0.14	0.12	0.20	0.14	0.12	39
0.34	0.21	0.15	0.12	0.22	0.15	0.13	40
0.35	0.22	0.15	0.13	0.22	0.15	0.13	41
0.36	0.22	0.15	0.13	0.23	0.15	0.13	42
0.37	0.23	0.15	0.13	0.23	0.16	0.13	43
0.45	0.25	0.17	0.13	0.26	0.17	0.14	44
0.59	0.29	0.18	0.14	0.30	0.19	0.15	45
0.61	0.29	0.18	0.15	0.31	0.19	0.15	46
0.62	0.30	0.19	0.15	0.31	0.19	0.15	47
0.67	0.31	0.19	0.15	0.33	0.20	0.16	48
0.69	0.31	0.19	0.15	0.33	0.20	0.16	49

* Calculations based on 1:3-pitch roof ($n = 1.2$), using the following formula:

$$U = \frac{U_r \times U_c}{U_r + (U_c/n)}$$

where U = combined coefficient to be used with ceiling area; U_r = coefficient of transmission of the roof; U_c = coefficient of transmission of the ceiling; n = the ratio of the area of the roof to the area of the ceiling.

† Use ceiling area (not roof area) with these coefficients.

‡ Based on 1-in. by 4-in. strips spaced 2 in. apart.

§ Coefficients in columns D, E, and F may be used with sufficient accuracy for tile, slate, and rigid asbestos shingles on wood sheathing.

‖ Sheathing assumed 25/32 in. thick.

Table 4-17 Coefficients of Transmission (U) of Windows, Skylights, and Glass-Block Walls

Section A—Vertical Glass Sheets

Number of Sheets	One	Two			Three		
Air space (in.)	None	¼	½	1*	¼	½	1*
Outdoor exposure	1.13	0.61	0.55	0.53	0.41	0.36	0.34
Indoor partition	0.75	0.50	0.46	0.45	0.38	0.33	0.32

Section B—Horizontal Glass Sheets (Heat Flow Up)

Number of Sheets	One	Two		
Air space (in.)	None	¼	½	1*
Outdoor exposure	1.40	0.70	0.66	0.63
Indoor partition	0.96	0.59	0.56	0.56

Section C—Walls of Hollow Glass Block

Description	U	
	Outdoor Exposure	Indoor Partition
5¾ by 5¾ by 3⅞ in. thick	0.60	0.46
7¾ by 7¾ by 3⅞ in. thick	0.56	0.44
11¾ by 11¾ by 3⅞ in. thick	0.52	0.40
7¾ by 3¾ by 3⅞ in. thick with glass fiber screen dividing the cavity	0.48	0.38
11¾ by 11¾ by 3⅞ in. thick with glass fiber screen dividing the cavity	0.44	0.36

Section D—Approximate Application Factors for Windows (Multiply Flat-Glass U Values by These Factors)

Window Description	Single Glass		Double Glass †		Windows with Storm Sash ‡	
	Per Cent § Glass	Factor	Per Cent § Glass	Factor	Per Cent § Glass	Factor
Sheets	100	1.00	100	1.00	......	....
Wood sash	80	0.90	80	0.95	80	0.90
	60	0.80	60	0.85	60	0.80
Metal sash	80	1.00	80	1.20	80	1.00
Aluminum	80	1.10	80	1.30	80	1.10‖

* For 1 in. or greater.

† Unit-type double glazing (two lights or panes in same opening).

‡ Use with U values for two sheets with 1 in. air space.

§ Based on area of exposed portion of sash; does not include frame or portions of sash concealed by frame.

‖ For metal storm sash, or metal sash with attached storm pane.

Table 4-18 Conversion Table for Wall Coefficient U for Various Wind Velocities

U for 15 MPH*	U for 0 to 30 MPH Wind Velocities					
	0	5	10	20	25	30
0.050	0.049	0.050	0.050	0.050	0.050	0.050
0.060	0.059	0.059	0.060	0.060	0.060	0.060
0.070	0.068	0.069	0.070	0.070	0.070	0.070
0.080	0.078	0.079	0.080	0.080	0.080	0.080
0.090	0.087	0.089	0.090	0.090	0.091	0.091
0.100	0.096	0.099	0.100	0.100	0.101	0.101
0.110	0.105	0.108	0.109	0.110	0.111	0.111
0.130	0.123	0.127	0.129	0.131	0.131	0.131
0.150	0.141	0.147	0.149	0.151	0.151	0.152
0.170	0.158	0.166	0.169	0.171	0.172	0.172
0.190	0.175	0.184	0.188	0.191	0.192	0.193
0.210	0.192	0.203	0.208	0.212	0.213	0.213
0.230	0.209	0.222	0.227	0.232	0.233	0.234
0.250	0.226	0.241	0.247	0.252	0.253	0.254
0.270	0.241	0.259	0.266	0.273	0.274	0.275
0.290	0.257	0.278	0.286	0.293	0.295	0.296
0.310	0.273	0.296	0.305	0.313	0.315	0.317
0.330	0.288	0.314	0.324	0.333	0.336	0.338
0.350	0.303	0.332	0.344	0.354	0.357	0.359
0.370	0.318	0.350	0.363	0.375	0.378	0.380
0.390	0.333	0.368	0.382	0.395	0.399	0.401
0.410	0.347	0.385	0.402	0.416	0.420	0.422
0.430	0.362	0.403	0.421	0.436	0.441	0.444
0.450	0.376	0.420	0.439	0.457	0.462	0.465
0.500	0.410	0.464	0.487	0.509	0.514	0.518
0.600	0.474	0.548	0.581	0.612	0.620	0.626
0.700	0.535	0.631	0.675	0.716	0.728	0.736
0.800	0.592	0.711	0.766	0.821	0.836	0.847
0.900	0.645	0.789	0.858	0.927	0.946	0.960
1.000	0.695	0.865	0.949	1.034	1.058	1.075
1.100	0.742	0.939	1.039	1.142	1.170	1.192
1.200	0.786	1.010	1.129	1.250	1.285	1.318
1.300	0.828	1.080	1.217	1.359	1.400	1.430

* In first column, U is from previous tables or as calculated for 15-mph wind velocity.

Example 4-2. Consider the composite wall of Table 4-7 and compute its resistance (and U factor) when 6-in. hollow tile is used followed by a 2-in. air space provided by furring strips, with $\frac{3}{4}$-in. metal lath and plaster inside. Select basic data from Table 4-3.

Solution: Total the resistances:

		R
1.	Outside air film (15 mph)	0.17
2.	4-in. face brick	0.44
3.	6-in. hollow tile	1.52
4.	2-in. air space, chosen at 50 F mean temp.	1.01
5.	Metal lath and plaster	0.23
6.	Inside air film	0.68
		$R_T = 4.05$

$$U = \frac{1}{R_T} = \frac{1}{4.05} = 0.25 \text{ Btu/h} \cdot \text{ft}^2 \cdot \text{F}$$

Note that this corresponds to wall 88C.

Find the U value for the same wall except that the 2-in. furred air space is removed and a $\frac{1}{2}$-in. gypsum plaster board having reflective insulation on its inner side is installed using $\frac{3}{4}$-in. nailing strips for mounting.

Original value total resistance	4.05
Deduct 2-in. air space	1.01
Deduct metal lath and plaster	0.23
Difference	2.81
Add $\frac{3}{4}$-in. air-space reflective insulation on one side	2.36
Add $\frac{1}{2}$-in. plaster board	0.45
New resistance total, R_T	5.62

$$U = \frac{1}{R_T} = \frac{1}{5.62} = 0.18 \text{ Btu/h} \cdot \text{ft}^2 \cdot \text{F}$$

Example 4-3. An inside room of a house faces on an enclosed porch. The room wall adjacent to the porch is 20 ft long by 8 ft high and contains one glazed door 3 ft by 7 ft. Glass forms 80% of the total area of the door, which is built of $1\frac{1}{2}$-in. wood. The wall is constructed of 2- by 4-in. studding, with $\frac{3}{4}$ in. of plaster and wood lath on each side. There is an air space between the studding. At a certain time the temperature on the enclosed porch is 40 F and the inside room is 72 F. Find the heat loss from the room to the porch.

Solution: Wall resistance:

Still air film, vertical wall	0.68
Wood lath and plaster, $\frac{3}{4}$ in.	0.40
Air space, nonreflective	1.01
Wood lath and plaster, $\frac{3}{4}$ in.	0.40
Still air film, vertical wall	0.68
	$R = 3.17$

$$U = \frac{1}{R} = 0.32 \text{ Btu/h} \cdot \text{ft}^2 \cdot \text{F for the wall}$$

(Note that this wall is No. 3 of Table 4-8 but differs slightly because the general f_i value of 1.65 was used in Table 4-8.)

$$\text{wall area} = 20 \times 8 - 3 \times 7 = 139 \text{ ft}^2$$

By Eq. (4-17), heat loss through the wall is

$$q_w = 0.32(139)(72 - 40) = 1423 \text{ Btuh}$$

Door wood area = (21 ft^2)(0.20) = 4.2 ft^2, for which the U value is 0.49 (Table 4-12). Door glass area = (21 ft^2)(0.80) = 16.8 ft^2, for which the indoor (no-wind) U value is 0.75 (Table 4-17). Heat loss through door is thus

$$q_d = 0.49(4.2)(72 - 40) + (0.75)(16.8)(72 - 40) = 469 \text{ Btuh}$$

The total heat transfer loss from the room to the porch is thus

$$1423 + 469 = 1892 \text{ Btuh} \qquad \textit{Ans.}$$

Example 4-4. Let us assume that except for the wall itself all of the data for Example 4-3 are given in SI units; namely, the room wall is 6.10 m long by 2.44 m high and the door, which is 80% glass, is 213 by 91.4 cm. The temperatures are 22.22°C and 4.44°C in the room and porch spaces, respectively. Convert the U values to SI units and compute the heat loss in this system.

Solution: Tabular data on wall construction in SI are not presented in this text and, in fact, are not extensively developed in English-language references. Thus for a wall given in SI units, it is most convenient to convert its constituent dimensions to inches and then compute the resistance (R) and overall coefficient (U) in foot-pound-second units followed by converting the U value to SI units; or, alternatively, k or C values in the foot-pound-second system can be directly converted to SI units with the computation following through from that point.

For the wall $U = 0.32$ Btu/h · ft^2 · F, by Eq. (4-23),

$$U_m = (0.32)(0.00568) = 0.00182 \text{ kW/m}^2 \cdot \text{K}$$

For the wood and glass in the door,

$$U_w = (0.49)(0.00568) = 0.00278 \text{ kW/m}^2 \cdot \text{K}$$

$$U_g = (0.75)(0.00568) = 0.00426 \text{ kW/m}^2 \cdot \text{K}$$

Net wall area = 6.1 × 2.44 − 2.13 × 0.914 = 14.88 − 1.95 = 12.93 m^2. Heat loss through wall = (0.00182)(12.93)(22.24 − 4.44) = 0.418 kW. Heat loss through door = (0.00278)(195 × 0.2)(17.78 deg) + (0.00426)(1.95 × 0.8)(17.78) = 0.137 kW. Total heat transfer loss from room is thus

$$0.418 + 0.137 = 0.555 \text{ kW} \qquad \textit{Ans.}$$

or

$$0.555 \text{ kW} \times 860.6 = 477.6 \text{ kcal/h} \qquad \textit{Ans.}$$

4-3. COMMENTS ON INSULATING MATERIALS, TABULAR VALUES

In general, the materials used for insulation and those used for construction purposes exhibit similar characteristics. Most nonmetallic materials have a basic structure in which there are numerous cells containing air or other gas. As the temperature increases, the thermal conductivity nearly always increases, and in many cases this increase can be attributed largely to increased molecular activity within the cells of the material. If the cells are extremely small, convection effects are not significant. Carbon black (lamp black), although difficult to use, is a good insulator; the particle size is extremely small and the resulting air cells are infinitesimally small. On the other hand, there appears to be an upper limit of cell size which if exceeded leads to a rapid decrease in insulating effectiveness. For example, carbon, charcoal, or graphite—say, in pieces of $\frac{1}{2}$-in. size and packed into a space—are not effective for insulating because of greater gaseous activity in the interstices, with gas actually moving from one space to another.

The insulation problem is also related to the density characteristics of the material. A compressible substance, such a glass wool, if loosely packed, is a better insulator than is the same material if closely packed (compressed) to a higher density. Thus the generalization can be made that as the density of the material decreases, its insulating effectiveness usually increases. On the other hand, this premise cannot be carried too far, since it is possible to pack a material so loosely that insulation effectiveness is poor.

In general, materials which are moist have higher thermal conductivity than dry materials. For example, soil with associated moistures of (say) 25 to 15% has better thermal conductivity than the same soil when it is dry. It is also of interest to note that if a moist soil exists at a temperature low enough to freeze the entrained moisture, the conductivity exceeds that of nonfrozen moist soil.

Tables 4-4 through 4-18 give either computed or test values for composite building structures. For both the computed and test values, the effect of surface films is considered. In addition to this, attention has been given to parallel heat-flow paths within the structure. For example, in a frame structure with wood studding on 16-in. centers, a representative space would consist partly of a free air space, with or without insulation, and at 16-in. intervals a solid piece of wood. The air space has a different resistance to heat flow than does the wood studding. Consequently, if an accurately representative value is desired for such a wall section, the proportional area associated with each type of heat-flow path should be considered in connection with its respective U factor. In the case of parallel paths for heat flow, it should be realized that the conductances or overall coefficients of each path in relation to respective areas are additive in determining the composite U value, whereas, with uniform walls involving different types of insulations in series, the respective thermal resistances of the elements constituting the thermal circuit are additive. Little inaccuracy arises if the parallel paths involve materials which do not have widely different characteristics, but it should be recognized that a wall of high thermal resistance will be greatly altered if metal parts used either for support or for fastening purposes pass through the wall.

The wall transmission-coefficient tables in this text make use of parallel-circuit corrections in instances where framing increases the U value, but not where the correction would show a decrease. Note that the finished (planed) size of lumber differs significantly from the nominal size. For example, a 2- by 4-in. stud, which was formerly finished to $1\frac{5}{8}$ by $3\frac{5}{8}$ in., since 1971 is conventionally sized to $1\frac{1}{2}$ by $3\frac{1}{2}$ in. Nominal 1-in. finished timber is $\frac{25}{32}$ in. or slightly less in thickness.

Example 4-5. A controlled-temperature test room, designed to be held at 110 F, was built in the storage space of an industrial plant where the temperature is held at 50 F in winter. Using material at hand in the plant, the wall of the room was constructed of 1-in. smooth pine boards ($\frac{25}{32}$ in. thick), 4 in. of rigid mineral-wool (rock-wool) insulation blocks, and an inside cover of pine boarding similar to the outside layer. The boards were held in place over the insulation by $\frac{3}{4}$-in. steel through-bolts, in such a way that one bolt was used for each 2 ft^2 of wall area. In addition, a lattice of 2- by 4-in. studs and cross-ties on 6-ft centers was attached to give the wall rigidity. Disregard the thermal effect of the bracing and compute (a) the U factor of a section of the wall, not considering the presence of the steel bolts, and (b) the same, but considering the effect of the steel bolts. Consider film surface factors to apply at the outer surfaces.

Solution: (a) Reference to Table 4-3 shows that the value of f_i on inside walls is 1.65, and that k for mineral-wool block at 90 F is 0.32, and 0.86 for pine wood. Thus the wall resistance is, by Eq. (4-18),

$$R = \frac{1}{1.65} + \left(\frac{25}{32}\right)\frac{1}{0.86} + \frac{4}{0.32} + \left(\frac{25}{32}\right)\frac{1}{0.86} + \frac{1}{1.65}$$

$$= 0.606 + 0.908 + 12.50 + 0.908 + 0.606 = 15.528 \text{ h} \cdot \text{ft}^2 \cdot \text{F/Btu}$$

Therefore the U factor is, in units of Btu/h · ft^2 · F,

$$U = \frac{1}{R} = 0.0644 \qquad \textit{Ans.}$$

(b) The thermal conductivity k for steel is 310 (from Table 4-1), and the resistance of the metallic-bolt circuit through the 5.56 in. in effective length ($\frac{25}{32} + 4 + \frac{25}{32}$) is, allowing for films at each side,

$$R = \frac{1}{1.65} + \frac{5.56}{310} + \frac{1}{1.65} = 1.229$$

The overall coefficient of heat transfer is therefore

$$U = \frac{1}{R} = \frac{1}{1.229} = 0.814 \text{ Btu/h} \cdot \text{ft}^2 \cdot \text{F}$$

The bolt area is $(0.75)^2(\pi/4)(1/144) = 0.00307$ ft^2, so that the total conductance through each bolt is

$$C_m = UA_m$$
$$= (0.814)(0.00307) = 0.0025 \text{ Btu/h} \cdot \text{F}$$

Since one bolt is associated with each 2 ft^2 of wall surface, the total conductance of the nonmetal basic wall section of 2 ft^2 gross area is

$$C_w = UA_w = (0.0644)(2 - 0.00307) = 0.1286 \text{ Btu/h} \cdot \text{F}$$

The total conductance of each 2 ft^2 section is thus

$$0.1286 + 0.0025 = 0.1311 \text{ Btu/h} \cdot \text{F}$$

Thus the composite-wall U value is

$$0.1311 \div 2 = 0.0655 \text{ Btu/h} \cdot \text{ft}^2 \cdot \text{F} \qquad \textit{Ans.}$$

and the composite wall resistance = 1/0.0655 = 15.26. Thus if the metal bolts are disregarded because of their relatively small area, the error would be

$$\frac{15.528 - 15.26}{15.26} \times 100 = 1.76\%$$

This error is not great, but is shows that whenever a low-thermal-resistance circuit is used in parallel with a high-thermal-resistance circuit, an error can be created if both circuits are not considered. In particular, metal ties in insulating material can seriously reduce insulation effectiveness.

4-4. WALL-SURFACE TEMPERATURE

The temperature on the inside-wall surface or ceiling surface of a building cannot be considered to be the same as the temperature inside the building. This surface temperature depends on the convection (film) conditions in the building, the insulating ability of the wall, and the outside conditions of temperature and wind. If the surface temperature is lower than the inside dew-point temperature, moisture will condense. A wet wall or ceiling results, and the moisture may cause serious damage to plaster and woodwork, besides constituting a nuisance. In winter, if the insulation effectiveness of the wall cannot be increased, this moisture formation necessitates lowering the inside relative humidity, or decreasing the inside film resistance by increasing the air circulation over the inside surfaces. Even if the dew problem does not exist, a wall of inadequate insulating capacity may chill the occupants in winter, by radiation to the cold wall, and cause discomfort in summer, by permitting too much heat inflow.

The calculation of inside surface temperature for a given wall at certain inside and outside temperatures can most easily be made by making use of the ratio of the surface film resistance to that of the whole wall. The following example will show the method.

Example 4-6. A building wall consists of 10-in. concrete with $\frac{3}{4}$-in. plaster on metal lath on the inside surface. (a) With a 15-mph wind outside at 0 F, what is the temperature on the inside-wall surface when the room is held at 74 Fdb and 62 Fwb? (b) Will moisture condense on this wall?

Solution: (a) Select resistance values for the wall from Table 4-3:

Film on outside wall, 15 mph	0.17
Concrete at $k = 12$, for 10 in.	0.833
Metal lath, cement-sand plaster, $\frac{3}{4}$ in.	0.15
Film, still air, vertical wall	0.68

$$R_{\text{total}} = 1.833 \text{ h} \cdot \text{ft}^2 \cdot \text{F/Btu}$$

The total temperature drop of 74 deg (74 F − 0 F) is used in sending heat through this resistance of 1.833. The temperature drop through the inside film resistance of 0.68 can then be found by the proportion

$$\frac{R_{\text{film}}}{R_{\text{total}}} = \frac{\Delta t_{\text{film}}}{\Delta t_{\text{total}}} \tag{4-27}$$

Substituting,

$$\frac{0.68}{1.833} = \frac{\Delta t_{\text{film}}}{74}$$

Computing, the temperature drop in the film $= \Delta t_{\text{film}} = 27.4$ deg. Therefore the inside-wall temperature $= 74 - 27.4 = 46.6$ F. *Ans.*

(b) Moisture will condense if the inside dew-point temperature is higher than the inside-wall temperature. For $t_d = 74$ F and $t_w = 62$ F, the psychrometric chart (Plate I) shows a dew point of 54.7 F. Thus moisture will form on the 46.6 F wall. *Ans.*

Example 4-7. How many layers of typical insulating fiberboard $\frac{1}{2}$ in. thick with a thermal conductivity k of 0.33 must be fastened to the inside-wall surface in Example 4-4 to prevent moisture deposition? Temperatures are assumed the same.

Solution: Since the dew-point temperature is 54.7 F, the allowable temperature drop in the film cannot exceed 74 F − 54.7 F, or 19.3 deg. The film resistance will not appreciably change from its value of 0.68, but the total wall resistance must change. By Eq. (4-27),

$$\frac{R_{\text{film}}}{R_{\text{total}}} = \frac{\Delta t_{\text{film}}}{\Delta t_{\text{total}}}$$

or

$$\frac{0.68}{R_{\text{total}}} = \frac{19.3}{74}$$

Therefore $R_{\text{total}} = 2.61$, the minimum total wall resistance needed to prevent condensation, and $2.61 - 1.833 = 0.777$, the additional resistance needed over that for the original wall.

For fiberboard, $R = 1/k = 1/0.33 = 3.03$, the resistance per inch of thickness, and $3.03/2 = 1.515$, the resistance per $\frac{1}{2}$ in. of thickness, or for one board.

As only 0.777 additional resistance is needed, one board at $R = 1.515$ is adequate to prevent condensation. *Ans.*

In winter, condensation of humidity from the inside air on windows often presents a serious problem. The use of double-glass windows or storm windows probably furnishes the best means of alleviating such a nuisance. Lowering inside humidity to reduce condensation can seldom be recommended, since the relative humidity maintained in most homes and many public buildings is in general too low. That more trouble is not experienced from moisture condensation in winter often corroborates the fact that the inside humidity is too low. In some places where high humidity is maintained it is customary to mount drip pans on the lower window sash to catch the condensation if it should become excessive. Data on transmission coefficients of glass are given in Table 4-17. A great increase in thermal resistance occurs when double glazing (two glass sheets) is used instead of single glazing. For example, with two glass sheets, one or more inches apart, the U factor reduces from 1.13 for single glass to 0.53 for double glass.

4-5. INSIDE AIR-TEMPERATURES

The temperatures that should be maintained in the rooms of a building during the heating season vary over a wide range, the exact temperatures depending primarily upon the use for which the rooms are planned. Table 5-1 gives temperatures considered representative of good practice for winter. These temperatures are dry-bulb temperatures and do not necessarily imply comfort conditions, since comfort is also influenced by relative humidity, air motion, activity of the occupants, and radiation effects. For spaces which are heated by radiation methods, slightly lower dry-bulb temperatures may be acceptable.

When inside temperatures for a given type of installation are selected for design purposes, the temperature at the *breathing line* of 5 ft (or less frequently at the 30-in. *comfort level*) is considered. However, in making calculations for heat-transfer losses, the breathing-line temperature may be far from the average or mean room temperature because of the tendency for air to stratify as the warm air rises to the top. For ceilings not over 20 ft high it has been found that the temperature rises approximately 2% for each foot of height above the breathing line.

Example 4-8. Consider a 20-ft room in which the temperature is 70 F at the 5-ft line and find the temperature at the ceiling (15 ft above the breathing line) and at the floor (5 ft below the breathing line).

Solution: The ceiling temperature $= [1.00 + (0.02)(15)]70 = 91$ F, and the floor temperature $= [1.00 - (0.02)(5)]70 = 63$ F. The average room temperature is thus taken as $(91 + 63)/2 = 77$ F for heat-transfer calculations. *Ans.*

For calculation of the average room temperature (t_{avg}) in a moderately high room (height H feet), the following formula is applicable when the temperature (t_b) is given at the 5-ft breathing line and, with less accuracy, at the 30-in. comfort line:

$$t_{avg} = t_b\left[1.0 + 0.02\left(\frac{H}{2} - 5\right)\right] \text{F} \tag{4-28}$$

For the previously worked case (Example 4-7) this formula becomes

$$t_{avg} = 70\left[1.0 + 0.02\left(\frac{20}{2} - 5\right)\right] = 77 \text{ F (as before)}$$

The foregoing calculation applies closely when a room is heated by direct radiation. With certain types of heating systems having positive (fan) air circulation, as when the air is distributed into the space in such a manner as to oppose the tendency of warm air to rise, the temperature difference between floor and ceiling may be quite small and the foregoing rule should not be followed. For this condition a temperature-rise factor of 1 %/ft of height above the breathing-line temperature should be used. Where ceiling heights exceed 15 ft, allow 1 %/ft of height up to 15 ft, and then allow 0.1 deg for each foot of height in excess of 15 ft. Very often a temperature at the floor 5 deg less than at the breathing line is assumed.

Inside design temperatures for summer are considered in a later chapter. These should be set slightly lower than outside temperatures but not so far below as to cause temperature shock to an occupant entering the conditioned space.

4-6. DESIGN TEMPERATURES FOR UNHEATED INSIDE SPACES

If there is an unheated or uncooled room adjacent to the room for which the heat transfer is being computed, some intermediate temperature (higher than that out of doors but lower than that of the room when heating, and lower than that out of doors but higher than that of the room when cooling) must be assumed.

Reasonable accuracy for ordinary rooms may be attained if the following rules are used in determining the design temperatures:

1. *Cooling with unconditioned room adjacent.* Select for computation a temperature equal to $t_i + (t_o - t_i) \times 0.667$. In other words, add to the room temperature two-thirds of the difference between the indoor and outdoor temperatures.
2. *Adjacent room having unusual heat sources.* Here, as would be the case with (say) a kitchen or boiler room, add 10 to 20 deg to the usually taken outside cooling-design temperature in computing the heat gain.
3. *Heating season, with adjacent room unheated.* Take $\Delta t = (t_i - t_o) \times 0.5$. That is, use one-half the temperature difference between inside and outside in computing the heat loss through the wall to the adjacent room.

Example 4-9. The outside temperature is 95 F, and that of a conditioned room is 80 F, and an unconditioned room is adjacent to the conditioned room. Find the temperature to be used for the inside in computing the heat gain.

Solution: By rule 1 preceding, the inside temperature for computation = 80 + (95 − 80) × 0.667 = 90 F. *Ans.*

Example 4-10. In heating, the outdoor temperature is 0 F and that of a heated room is 70 F, and an unheated room is adjacent to the heated room. Find the temperature difference to be used for finding the heat loss from the heated room.

Solution: By rule 3 preceding, $\Delta t = (70 - 0) \times 0.5 = 35$ deg. *Ans.*

4. *Ground floors.* For floors directly on the ground, 50 F to 55 F may be assumed for the ground temperature, although some people prefer merely to assume an arbitrary temperature difference existing between the inside air and the ground. See also Section 4-7.
5. *Attics.* In certain cases it is possible to compute the temperature of an adjacent space from the conductivity characteristics of the building walls and partitions. This computation is easily possible in the case of spaces which are rather simply connected to the heated (or cooled) space, as in the case of an attic adjacent only to the ceiling of the top story. The heat passing through the ceiling to the attic equals the heat passing to the outside. In equational form, this becomes

$$U_r A_r(t - t_o) = U_c A_c(t_i - t) = U_c \frac{A_r}{n}(t_i - t)$$

$$t = \frac{U_c t_i + nU_r t_o}{U_c + nU_r} \tag{4-29}$$

where t is the attic air temperature in degrees Fahrenheit; t_i is the inside air temperature in degrees Fahrenheit; t_o is the outside air temperature in degrees Fahrenheit; U_r is the coefficient of heat transfer of roof; U_c is the coefficient of heat transfer of ceiling; and n is the ratio of roof area A_r to ceiling area A_c.

If it is desired to find an overall heat-transfer coefficient U for a ceiling, attic, and roof, the heat transferred through the composite structure is equated to the heat transfer through either the ceiling or the roof. When a value of U, based on the ceiling area, is desired, equate

$$UA_c(t_i - t_o) = U_c A_c(t_i - t)$$

then substitute for t from Eq. (4-29), which leads to

$$U = \frac{U_c U_r}{U_r + U_c/n} \tag{4-30}$$

where U is the coefficient of heat transfer in British thermal units per square foot of ceiling area, per degree difference between inside-room and outside temperatures, when applied to a ceiling, attic, and roof. Other symbols are as for Eq. (4-29).

Equations (4-29) and (4-30) may not be exact, since convection conditions in the attic affected by air circulation may change the surface factors normally used in computing the values of U_c and U_r, and also, heat transfer by radiation within the attic has some effect. However, for most cases of closed attics in heating seasons the formulas can be used. If the attic is ventilated in winter it is more accurate to assume the attic temperature as practically that of the outside and to use only the ceiling in the heat-transfer computations.

In the case of summer air conditioning, attics should be well ventilated, since the sun, shining on the roof, will raise the inside air temperature greatly above the outside temperature and thereby impose an additional load on the conditioning system. Equations (4-29) and (4-30) should generally not be used for summer air conditioning.

In the event that none of the previously mentioned methods seems applicable, or where only a rough approximation is desired for a heating calculation, the following values may be used when outside temperatures vary between 0 and 20 F:

Basements and unheated rooms 32 F
Vestibules, frequently opened 25 F
Attic under slate or metal roof, insulating capacity poor 25 F
Attic under a roof of good insulating capacity 40 F

4-7. GROUND FLOOR AND BASEMENT SLABS

The heat loss through basement floors and through building-wall surface below ground level is difficult to compute with accuracy, since the ground temperature varies with depth and with the amount of heat flowing into it from the structure above. The ground temperature near the surface varies with the season of the year and the climate; in the United States, however, frost seldom penetrates more than 4 ft, and below 3 or 4 ft the ground temperature undergoes only moderate swings the year round. In fact, it is customary to consider the temperature of ground water as indicative of subsurface earth temperature. Ground-water temperature, even in the most northerly sections of the United States, is seldom below 40 F and in warm sections of the country seldom exceeds 60 F. Using these temperatures as a range, and considering the overall U of a representative slab as 0.10, Table 4-19 has been constructed.

In one type of building construction which has found frequent use, the structure is placed on a concrete slab laid directly on the ground over a well-drained gravel or ash fill. For heating, in this case, the effect of loss into the ground and air near the outside edge of the slab is extremely important, and it is almost essential to provide for edge insulation. It is also desirable to put a waterproof membrane over the gravel subfill to keep water from seeping through the floor by capillary action. This seepage can happen even though the ground under the floor is well-drained.

Table 4-19 Below-Grade Heat Losses for Basement Walls and Floors*

Ground-Water Temperature	Basement-Floor Loss † (Btuh per sq ft)	Below-Grade Wall Loss † (Btuh per sq ft)
40	3.0	6.0
50	2.0	4.0
60	1.0	2.0

* Reprinted, by permission, from *Heating Ventilating Air Conditioning Guide 1955*, Chapter 12.
† Based on basement temperature of 70 F and *U* of 0.10.

Table 4-20 Heat Loss of Concrete Floors at or Near Grade Level*

Outdoor Design Temperature (F)	Heat Loss per Foot of Exposed Edge (Btuh)			
	Recommended 2-in. edge insulation	1-in. edge insulation	1-in. edge insulation	No. edge† insulation
−20 to −30	50	55	60	75
−10 to −20	45	50	55	65
0 to −10	40	45	50	60

* Reprinted, by permission, from *Heating Ventilating Air Conditioning Guide 1955*, Chapter 12.
† This construction not recommended; shown for comparison only.

Table 4-20 gives representative heat-loss factors for concrete floors placed at or near grade level. This loss is based on the perimeter of the floor edge, but the values are compensated to take care of total loss from the whole slab.

4-8. HEAT-TRANSFER COMPUTATION FOR BUILDINGS

Outside temperatures for building-design purposes are based on the geographical location of the building. Table 5-2 gives winter design temperatures for many places. Since temperatures approaching the lowest ever recorded are usually rare and of short duration, the recommended outside design temperatures in Table 5-2 are higher than the recorded minimum. After selecting a design outside temperature, and the inside temperature from Table 5-1 (usually 70 F) to be maintained in a structure for a given purpose, heat-transfer losses through walls of a given type can readily be obtained by using the heat-transmission values from Tables 4-4 to 4-18. Detailed methods of calculating heating and cooling loads for whole buildings, when infiltration, ventilation, exposure, sun, and internal loads may act together, are considered in Chapters 5 and 6.

Example 4-11. The top-story ceiling of a building is 27 by 34 ft. An unventilated attic above the ceiling is surmounted by a double-sloping roof which has an area 1.8 times greater than that of the ceiling. The ceiling is made of 2- by 4-in. joists ($1\frac{5}{8}$ in. by $3\frac{5}{8}$ in. finished size) on 16-in. centers. The $\frac{3}{4}$-in. wood lath and plaster is attached to the joists, and on top of the joists 1-in. yellow-pine flooring is laid ($\frac{25}{32}$ in. finished size). The roof is made of exposed 2- by 4-in. rafters ($1\frac{5}{8}$ in. by $3\frac{5}{8}$ in. actual size) on 2-ft centers, and the rafters are surmounted by 1-in. wood sheathing ($\frac{25}{32}$ in. thick) and covered with slate shingles. The temperature just under the ceiling is 84 F, and the outside temperature is 10 F. Find (a) the coefficients of transmission for the ceiling and the roof, (b) the probable attic temperature, (c) the composite coefficient of heat transfer per square foot of roof area for the ceiling, attic, and roof combination, and (d) the heat loss from the building through the roof.

Solution: (a) Ceiling: no wind or forced circulation. Take the value of U as equal to 0.28 (from Table 4-10) for such a ceiling. Roof: wind on one side, still air on other. Take the value of U as equal to 0.55 (from Table 4-15) for such a roof. *Ans.*

(b) Use Eq. (4-29) to find the probable attic temperature:

$$t = \frac{U_c t_i + nU_r t_o}{U_c + nU_r} = \frac{(0.28)(84) + (1.8)(0.55)(10)}{0.28 + 1.8(0.55)} = 26.3 \text{ F} \qquad \textit{Ans.}$$

(c) Use Eq. (4-30) to find U. Thus, in Btu per square foot of roof surface per degree Fahrenheit temperature difference between inside and outside,

$$U = \frac{U_c U_r}{U_c + nU_r} = \frac{(0.28)(0.55)}{0.28 + 1.8(0.55)} = 0.12 \qquad \textit{Ans.}$$

(d) Using in Eq. (4-17) the value of U from (part c), $Q = UA_r(t_i - t_o) = 0.12(1.8 \times 27 \times 34)(84 - 10) = 14\,800$ Btuh, the heat loss through the top-story ceiling.

This result can also be found by using U_c for the ceiling, along with the ceiling area and the difference between the room and attic temperatures. Thus the heat loss through the top-story ceiling is $Q = U_c A_c(t_i - t_a) = (0.28)(27)(34)(84 - 26.3) = 14\,800$ Btuh. *Ans.*

4-9. HEAT TRANSFER THROUGH PIPES, PIPE COVERINGS

In the case of pipe lagging, and similar annular coverings, the cross section of the path through which heat must flow varies in proportion to the linear surface through the section.

Figure 4-2 shows a pipe, indicated by the dark circle, which is covered by insulation of thickness $(r_o - r_i)$. If we assume the pipe is hot and heat is being lost through the insulation it is obvious that the heat-flow area at the outer circumference, $2\pi r_o \times L$ (where L is the length of pipe), is much greater than the heat-flow area at the pipe, $2\pi r_i \times L$. Nevertheless, under steady-flow conditions the heat flow through both of these areas is exactly the same. A simple derivation (Ref. 6, p. 178), shows that for pipe lagging

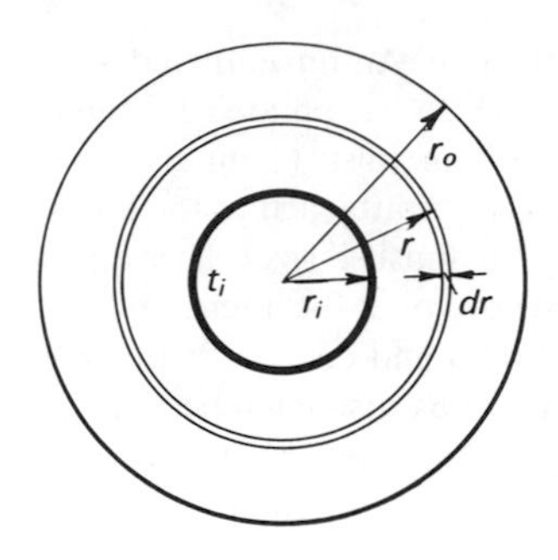

FIGURE 4-2. Section through cylindrical pipe covered with lagging.

such as this, the following equation is correct and takes into account the variations in area in the direction of heat flow:

$$q_L = \frac{2\pi k L(t_i - t_o)}{2.3 \log_{10}(r_o/r_i)} \text{ Btuh} \tag{4-31}$$

where q_L are the Btuh transferred through the lagging; t_i, t_o are the temperatures on each side of the lagging, in degrees Fahrenheit; L is the length of the lagging, measured along the axis of the pipe, preferably in feet; and k is the specific conductivity. If L is in feet, express k in units of

$$\frac{(\text{Btu})(\text{ft})}{(\text{h})(\text{ft}^2)(\text{F})} = \frac{1}{12} \times \frac{(\text{Btu})(\text{in.})}{(\text{h})(\text{ft}^2)(\text{F})}$$

r_i, r_o are radii to innermost and outermost sections of lagging, in any consistent length units (inches, feet, etc.).

To find the average area of a pipe lagging, of variable cross section in the direction of heat flow, that could be accurately used in an equation of the same form as Eq. (4-9), equate the two expressions (4-9) and (4-31) for q and for unit length of pipe, $L = 1$:

$$k \frac{A_{\text{avg}}}{x} (t_i - t_o) = \frac{2\pi k(t_i - t_o)}{2.3 \log_{10}(r_o/r_i)}$$

$$A_{\text{avg}} = \frac{2\pi x}{2.3 \log_{10}(r_o/r_i)} = \frac{2\pi(r_o - r_i)}{2.3 \log_{10}(r_o/r_i)} \tag{4-32}$$

where area A_{avg} per unit length of pipe is either in square feet or in square inches, depending on whether r_o and r_i are in feet or inches.

This equation can also be expressed, in square feet or square inches per unit length of pipe, as

$$A_{\text{avg}} = \frac{A_o - A_i}{2.3 \log_{10}(A_o/A_i)} \tag{4-33}$$

In Eqs. (4-32) and (4-33), A_{avg} is that average area per unit length which can be used for the lagging, as though this area were of constant cross section in the direction of heat flow. Where thin-wall tubing or insulation exists, there is little difference between an arithmetic average of outside and inside of the tubing compared with the average area as found by the accurate relation—Eq. (4-32) or (4-33).

Example 4-12. A 6-in. steel pipe (OD = 6.625 in.) carries saturated steam at 100 psig (337.9 F), and the surrounding air is at 80 F. Under these conditions, when the pipe was covered with 2 in. of insulation of the so-called 85% magnesia type, for which $k = 0.5$ Btu · in./h · ft² · F, the temperature at the outer surface was found to be 100 F. Assuming that the outer surface of the steel pipe is at the same temperature as the steam, an assumption that is approximately correct, find the heat loss per hour through the insulation, per 100 ft of pipe.

Solution: By Eq. (4-32), the average area per unit length is expressed by the relation

$$A_{avg} = \frac{2\pi(r_o - r_i)}{2.3 \log_{10} (r_o/r_i)}$$

Therefore, since

$$r_i = \frac{6.625}{2} \text{ in.,} \quad \text{or} \quad \frac{3.312}{12} \text{ ft}$$

and

$$r_o = \left(\frac{6.625}{2} + 2\right) \text{ in.,} \quad \text{or} \quad \frac{5.312}{12} \text{ ft}$$

it follows that the average area per foot of pipe length is

$$A_{avg} = \frac{2\pi[(5.312 - 3.312)/12]}{2.3 \log_{10} (5.312/3.312)} = \frac{2\pi(2/12)}{2.3(0.2052)} = 2.219 \text{ ft}^2$$

Therefore, the heat lost per 1-ft length of pipe is

$$q = k \frac{A_{avg}}{x}(t_i - t_o) = (0.5)\left(\frac{2.219}{2}\right)(337.9 - 100) = 131.97 \text{ Btuh}$$

And for 100 ft of length, the heat loss is

$$q_{100} = (100)(q) = 13\,200 \text{ Btuh} \qquad \textit{Ans.}$$

Alternate Solution An alternate solution employs Eq. (4-31):

$$q_L = \frac{2\pi k(L)(t_i - t_o)}{2.3 \log_{10} (r_o/r_i)}$$

It is necessary to change the units of k from

$$k = 0.5 \frac{(\text{Btu})(\text{in.})}{(\text{h})(\text{ft}^2)(\text{F})}$$

to

$$\frac{0.5}{12} \frac{(\text{Btu})(\text{ft})}{(\text{h})(\text{ft}^2)(\text{F})}$$

Then the heat loss, for the length given, is

$$q_L = \frac{2\pi(0.5/12)(100)(337.9 - 100)}{2.3 \log_{10} (5.312/3.312)} = 13\,200 \text{ Btuh} \qquad \textit{Ans.}$$

Heat loss from the outside of a hot bare pipe, or even from one with some insulation, is affected not only by convection conditions around the pipe but also by radiation from the hot pipe to the cooler surroundings. To make possible the use of ordinary conduction equations for pipe, when radiation loss from the surface of the pipe is also appreciable, the so-called film factor is sometimes determined experimentally to include the effect of both convection and radiation and is called $f_c + f_r$. For this condition the factors are dependent on convection conditions and temperature difference, and also on size of pipe. Some of these values, based on the experimental work of Heilman, are found in Fig. 4-3. Although the surface factors in Fig. 4-3 are based on bare pipe, they can also be used with sufficient accuracy for the surface of insulation if the proper temperature difference between the surface of the insulation and an ambient air temperature is used in connection with the extreme outside diameter of the insulation.

Example 4-13. A bare 3-in. standard pipe in a building at 70 F carries saturated steam at 100 psig (337.9 F). Find (a) the heat loss per 100 ft of pipe and (b) the weight of steam condensed as a result of this heat loss.

Solution: (a) Assuming that the temperature of the pipe surface is that of the steam (closely true), the temperature difference existing is 337.9 − 70 = 267.9 deg. Interpolating for 267.9 deg in Fig. 4-3 gives $(f_c + f_r) = 3.12$ Btu/h · ft^2 of outside pipe surface per degree temperature difference.

For 3-in. pipe the OD = 3.5 in. = 3.5/12 ft. The area of outside surface per foot of length is therefore

$$\pi \times \frac{3.5}{12} \times 1 = 0.917 \text{ ft}^2$$

The heat loss from the surface of the pipe is found as follows:

$$q = (f_c + f_r)(A)(t_i - t_o)$$
$$q = (3.12)(0.917)(337.9 - 70) = 765.6 \text{ Btuh/ft length}$$
$$q_L = (100)(765.6) = 76\,560 \text{ Btuh/100-ft length} \qquad \textit{Ans.}$$

(b) The latent heat of steam at 100 psig (100 + 14.7 psia) is 880.7 Btu/lb (from Table 2-3). Therefore,

$$\text{steam condensed} = \frac{76\,560}{880.7} = 86.9 \text{ lb/h} \qquad \textit{Ans.}$$

Example 4-14. Assume that the 3-in. pipe (3.5 OD) of Example 4-13 is covered with a 1-in. layer of 85% magnesia insulation, which has a value of k of 0.5 Btu · in./h · ft^2 · F. Find (a) the heat loss per 100 ft of pipe and (b) the weight of steam condensed as a result of the loss. (c) Compare results with those for bare pipe.

Solution: (a) Here $f_c + f_r$ must be selected for an estimated temperature difference between the outside of the insulation and the 70 F air. Estimate 50-deg difference, giving a value of about 1.97 (from Fig. 4-3) for an outside diameter of (3.5 + 1 + 1), or 5.5 in.

Only one logarithmic area is involved, that for the 1-in. insulation of radii:

$$\left(\frac{3.5}{2} + 1\right)\text{in.} = \frac{2.75}{12}\text{ ft} = r_o$$

and

$$\frac{3.5}{2}\text{ in.} = \frac{1.75}{12}\text{ ft} = r_i$$

The inside film resistance and the pipe resistance are negligible.

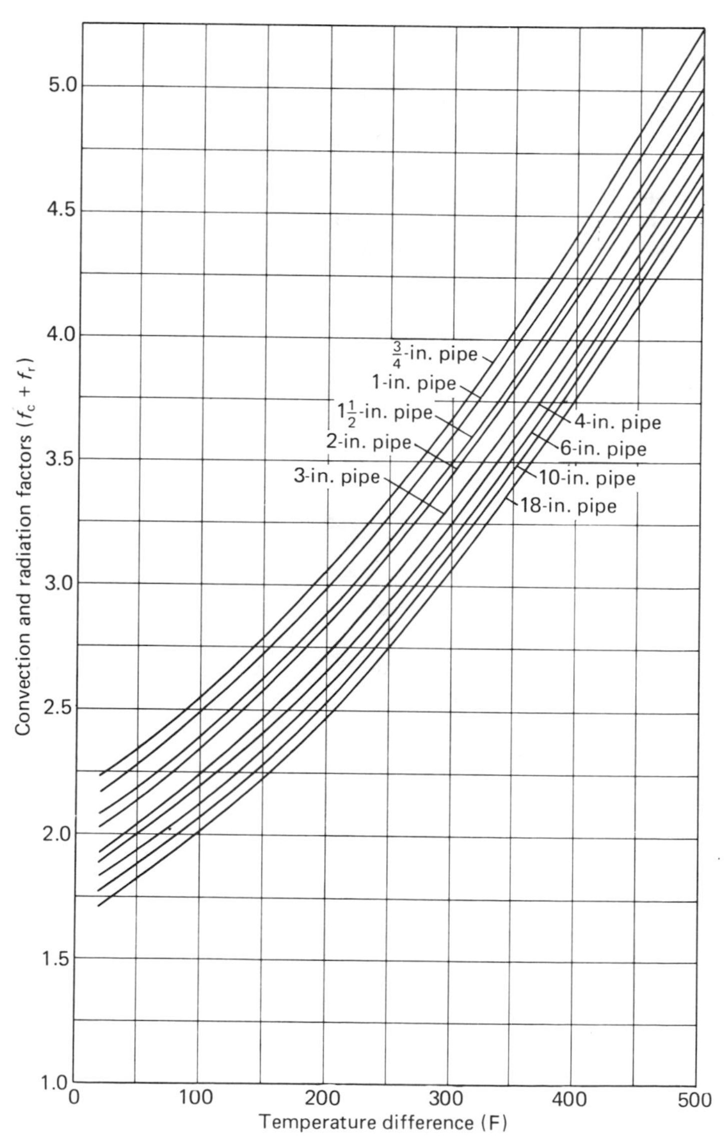

FIGURE 4-3. Values of $(f_c + f_r)$ for heat loss from horizontal bare iron pipes.

By Eq. (4-16) or (4-12), using the combined film factor and the proper logarithmic area, we find for unit length of pipe:

$$q = \frac{t_i - t_o}{\dfrac{1}{2\pi r_o(f_c + f_r)} + k\dfrac{r_o - r_i}{[2\pi(r_o - r_i)]/[2.3\log_{10}(r_o/r_i)]}} = \frac{2\pi(t_i - t_o)}{\dfrac{1}{(f_c + f_r)r_o} + \dfrac{2.3\log_{10}(r_o/r_i)}{k}}$$

Substituting, the heat loss per 1-ft length of pipe is

$$q = \frac{2\pi(337.9 - 70)}{\dfrac{1}{1.97(2.75/12)} + \dfrac{2.3\log_{10}(2.75/1.75)}{0.5/12}} = \frac{2\pi(337.9 - 70)}{2.21 + 10.83}$$

$$= 129.0 \text{ Btuh}$$

The heat loss per 100-ft length is therefore

$$q_{100} = (100)(129.0) = 12\,900 \text{ Btuh} \qquad \textit{Ans.}$$

The total temperature drop through the insulation and film occurs in direct ratio to the thermal resistances, which are proportional to the 2.21 + 10.83 above. The temperature drop through the surface film on the outer pipe insulation is thus (337.9 − 70) × (2.21)/(2.21 + 10.83) = 45 deg. This result is close enough to the 50 deg assumed to

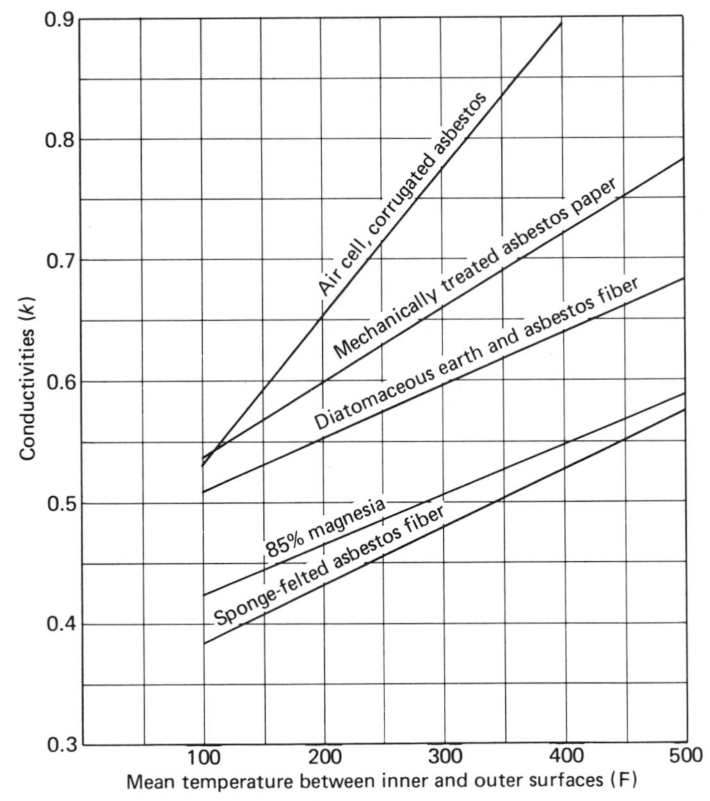

FIGURE 4-4. Conductivities per inch of thickness for insulation coverings for medium- and high-temperature pipes.

make a recalculation unnecessary, inasmuch as the film resistance is such a relatively small part of the total resistance.

(b) Latent heat of steam at 114.7 psia = 880.7 Btu/lb (from Table 2-3). Therefore, the steam condensed = 12 900/880.7 = 14.6 lb/h by heat loss from the 100-ft length of pipe. *Ans.*

(c) This 14.6 lb of steam condensed per hour in the insulated pipe compares with 86.9 lb condensed (lost) per hour from the uninsulated pipe. *Ans.*

For the case of two or more different kinds of insulation in series on a pipe, use of the resistance concept of heat transfer through insulation leads to an equation of the same form as Eq. (4-16), but requires the use of the logarithmic mean areas of each respective insulation.

Figure 4-4 shows conductivities of various types of insulating materials for covering medium- and high-temperature pipes.

4-10. RADIANT-HEAT TRANSFER

The heat transferred by radiant energy is proportional to the difference in the fourth powers of the absolute temperatures of the hot source and the more cool receiver of the radiation. Expressed as an equation,

$$q_r = 0.172A\left[\left(\frac{T_s}{100}\right)^4 - \left(\frac{T_r}{100}\right)^4\right] e\,F_a \tag{4-34}$$

where q_r is the Btu transferred per hour by radiation; A is the area of either the source or receiver, depending on the method of selecting F_a, in square feet; T_s, T_r are the temperatures of source and receiver, $(t_s + 460)$ and $(t_r + 460)$, in degrees Fahrenheit absolute; e is the effective absorptivity or emissivity factor, expressing the degree to which the source and receiver surfaces approach an "ideal black body" [an ideal black body is one which could absorb (or emit) all of the radiant energy falling on it]; and F_a is the factor to account for the geometric configuration between the radiating surfaces. This factor must be calculated mathematically, or found from curves or tables in complete heat-transfer treatises, for various geometrical arrangements. For the case of two infinite parallel plates ($F_a = 1$), use

$$e = \frac{1}{(1/e_s) + (1/e_r) - 1}$$

This is also reasonably applicable for the panel surfaces ordinarily found in radiant-heated rooms. For the case of a completely enclosed body, small compared with enclosure ($F_a = 1$), use e for the enclosed body and also use the enclosed body area, etc.

The shape, form, and relative position of radiating bodies greatly affect the amount of radiant-heat transfer and sometimes make it impossible to calculate mathematically the factor F_a.

The emissivity or absorptivity of no material is perfect ($e = 1$), although lamp black and certain oil paints may reach values of $e = 0.96$. The metals, particularly those highly polished, have low values: polished aluminum, $e = 0.039$ to 0.057; tinned sheet iron, 0.043 to 0.054; polished cast iron, 0.21; polished sheet steel, 0.55 to 0.60, Most nonmetallic substances, regardless of color, have emissivities in excess of 0.8 and the emissivities increase in value with increasing temperature. Aluminum paint has a rather low emissivity (0.3 to 0.5), but oil paints, regardless of color (including white), usually exceed 0.9. Thus painting of steam radiators or other heat-transfer surfaces has small effect on radiant-heat transfer unless an aluminum or other metal suspension paint is used. Oxidized metals usually show values of e greater than 0.8. For plaster, $e = 0.91$; and for asbestos board, e approaches 0.93.

It should be mentioned that emissivity (absorptivity) factors e are different for radiation from high-temperature sources, such as the sun (or incandescent bodies), than they are for radiation from low-temperature sources, such as steam radiators and radiant-heating panels. White clothes, whitewashed surfaces, plaster, and light cream paint, for example, have e values of 0.85 to 0.95 for low-temperature (infrared) radiation, whereas for solar radiation the corresponding e values range from 0.3 to 0.5. Window glass is transparent to solar radiation ($e \equiv 0$), whereas for low-temperature radiation glass is a good absorber ($e = 0.9$ to 0.95) and such radiation is trapped. This phenomenon is used in greenhouses as a means of heating on sunny days. For black nonmetallic surfaces, there is little difference between e values for solar and other radiant sources.

PROBLEMS

4-1. A solid stone wall 12 in. thick is finished on the inside with $\frac{1}{2}$ in. of gypsum plaster. Find the resistance and conductance of the wall both with and without film coefficients being considered.

Ans. $R = 1.822$ and 1.049

4-2. Find U for a wall that is made up of 10 in. of concrete, 2 in. of corkboard, and $\frac{1}{2}$ in. of plaster. Assume that there is a 15-mph wind on the concrete side of the wall.

Ans. $U = 0.119$

4-3. *Masonry Wall.* This wall consists of 4-in. face brick, $\frac{1}{2}$ in. of cement mortar, 6 in. of poured concerete, and $1\frac{1}{2}$ to 2 in. of furring strips on the concrete surface, on which metal lath with gypsum plaster is placed. Compute the resistance of this vertical wall and the U factor, allowing for 15-mph wind outside and still air inside.

Ans. $R = 2.98$, $U = 0.335$ Btu/h · ft^2 · F

4-4. *Masonry Wall.* This wall consists of sand-and-gravel concrete blocks (three-hole oval-cored) 8 in. deep, with 2-in. furring strips to which paper-coated $\frac{1}{2}$-in. gypsum wall board with reflective insulation on its inner surface is attached. Consider 15-mph wind outside, still air inside, and compute the wall resistance and U factor.

Ans. $R = 4.75$, $U = 0.211$ Btu/h · ft^2 · F

4-5. *Masonry Wall.* Modify the wall of Problem 4-4 by replacing the sand-and-gravel concrete blocks with similar blocks made with cinder aggregate.

Ans. $R = 5.36$, $U = 0.186$ Btu/h · ft^2 · F

4-6. *Frame Wall.* This vertical wall consists of lapped wood siding ($\frac{1}{2} \times 8$ in.), sheathing ($\frac{1}{2}$ in. regular density), a $3\frac{1}{2}$-in. air space, and gypsum wall board ($\frac{1}{2}$ in.). Consider 15-mph wind outside and still air inside.

Ans. $R = 4.44$, $U = 0.23$ Btu/h · ft² · F

4-7. *Frame Wall.* Modify the frame wall of Problem 4-6 to use 3-in. bat-type, paper-enclosed, mineral-wood insulation in the air space and compute R and U.

Ans. $R = 14.43$, $U = 0.069$ Btu/h · ft² · F

4-8. *Frame Ceiling or Floor.* This floor consists of asphalt tile on felt over $\frac{1}{2}$-in. plywood nailed to a $\frac{3}{4}$-in. pine subfloor, resting on 2- by 4-in. joists, to which $\frac{3}{4}$-in. metal lath and with lightweight aggregate plaster are attached. Compute R and U on the basis of heat flow from beneath, namely, unoccupied space above.

Ans. $R = 4.23$, $U = 0.235$ Btu/h · ft² · F

4-9. *Frame Ceiling or Floor.* Consider the construction of Problem 4-8 to be in use over an unheated basement area with the ceiling not plastered but consisting of $\frac{1}{4}$-in. plywood nailed to joists. Heat flow is downward. Compute R and U.

Ans. $R = 4.83$, $U = 0.21$ Btu/h · ft² · F

4-10. *Interior Wall.* This wall consists of $\frac{1}{2}$-in. gypsum boards nailed to both sides of studding set on 16-in. centers. An approximate $3\frac{5}{8}$-in. air space results and still air is considered applicable to both wall surfaces. Find the total resistance and the U factor.

Ans. $R = 3.27$, $U = 0.30$ Btu/h · ft² · F

4-11. *Sloping (Pitched) Roof.* This roof consists of asphalt shingles on double-layer, felt-mopped building paper laid on $\frac{1}{2}$-in. plywood, a $3\frac{5}{8}$-in. air space with both sides reflective surfaces, finished with $\frac{1}{2}$-in. gypsum wall board. Framing of 2- by 4-in. rafters on 20-in. centers will be disregarded in this computation.

Ans. 4.76 h · ft² · F/Btu, $U = 0.22$

4-12. *Sloping (Pitched) Roof.* Assume that ordinary building surface is used in the air space and compute R and U for Problem 4-11.

4-13. *Sloping Pitched Roof—Rafter Modification.* Rework Problem 4-11 making proper allowance for the 2- by 4-in. rafters of the framing if their finished size is $1\frac{5}{8}$ by $3\frac{5}{8}$ in.

Ans. $U = 0.21$ Btu/h · ft² · F

4-14. A solid stone wall 30.48 cm thick is finished on the inside with an additional 1.27 cm of gypsum plaster. Find the resistance and conductance of the wall both with and without film coefficients being considered. Use a general value for f_i, the inside film coefficient. Change the k, f, and C values to SI units and find the total resistance and U value for the wall.

Ans. $R = 321.0$ or 184.9 m² · K/kW, $U = 0.00311$ or 0.00541 kW/m² · K

4-15. The wall of Problem 4-14 constituting the side of one room in a factory is 10.0 m long and is 3.05 m high. Consider films acting on both sides of the wall. When it is 20°C inside and −10°C outside, compute the probable heat loss in kilowatts and in kilocalories per hour. Convert your answer to British thermal units per hour.

Ans. 2.86 kW, 2460 kcal/h, 9750 Btu/h.

4-16. The resistance of the wall of Example 4-3 was computed as $R = 3.17$ h · ft² · F/Btu without allowing for the fact that where wood framing on 16-in. centers is present a

different resistance exists than would be the case for 100% air space and no wood framing. Find the conductance or true resistance of this wall making allowance for the presence of the wood in the air space. The finished size of 2- by 4-in. lumber is $1\frac{5}{8}$ by $3\frac{5}{8}$ in. For the analysis it is convenient to consider a 1-ft^2 section of wall 16 in. wide, 9 in. high, and containing a $1\frac{5}{8}$-in. width of wood.
Ans. $R = 6.69$ at wood framing, $R = 3.17$ hr · ft^2 · F/Btu at air space; U for composite wall with framing, 0.302 Btu/h · ft^2 · F

4-17. Starting with the wall resistance as computed in Example 4-3, find the new resistance and U values if the air space is filled with 3 to $3\frac{3}{4}$ in. of paperbacked mineral-wool insulation. Then correct for the fact that wood framing on 16-in. centers is present. Refer to the preceding problem for a suggested method of solution.
Ans. $R = 13.16$ for mineral-wool wall with no correction for framing; $R = 6.69$ h · ft^2 · F/Btu at wood framing, U for composite wall of mineral wool and framing = 0.083 Btu/h · ft^2 · F

4-18. A building is constructed of 8-in. common-brick walls, finished with gypsum lath and plaster on furring strips (2-in. air space) on the inside surface. If the room is maintained at 70 F and the outside temperature is 20 F, what is the inside wall temperature?

Ans. 60.9 F

4-19. The outside wall of a building is constructed of 8 in. of common brick with $\frac{1}{2}$ in. of plaster on the inside surface. The inside of the building is maintained at 75 Fdb and 65 Fwb. How many layers of $\frac{1}{2}$-in. corkboard must be added to the wall to prevent moisture from condensing on the inside surface of the wall when the outside temperature is 0 F?

Ans. One layer

4-20. A certain room is 20 ft wide, 30 ft long, and 15 ft high. The temperature at the breathing line (5-ft level) is 70 F. In the room there is a door $3\frac{1}{2}$ by 7 ft and two windows 3 by 7 ft, with the lower edge of the windows 3 ft from the floor. What is the temperature (a) at the ceiling and (b) at the floor? (c) What average inside temperature might be used in calculating the heat loss through the walls, windows, and doors?

Ans. (a) 84 F; (b) 63 F; (c) 73.5 F

4-21. A ranch-type home is built on a concrete floor which is laid at ground level. Waterproof insulation 1 in. thick is placed at the edge of the slab where this joins the wall and footing of the building. The insulation also extends for 6 in. down along the side of the footing. The house is rectangular in shape, 60 by 26 ft. Making use of Table 4-20, compute the loss at the floor slab. Assume that the house is built in a locality where the outside design temperature is -10 F.

Ans. 7650 Btuh

4-22. Refer to Table 4-16. For a ceiling with a coefficient of $U_c = 0.24$ and for a roof coefficient of $U_r = 0.48$, check the value given in the table for overall (combined) coefficient. Take n equal to 1.2. Recompute on the basis of $n = 1.8$.

4-23. One side wall of a public building is made of hollow glass blocks, each $11\frac{3}{4}$ by $11\frac{3}{4}$ by $3\frac{7}{8}$ in. thick. Compute the heat loss through the 20-ft-long, 10-ft-high wall when the temperature is 70 F inside and 0 F outside.

Ans. 7280 Btuh

4-24. In front of the glass-block wall of Problem 4-23 it is usual at night to draw close-fitting drapes running from the ceiling to the floor. On the basis that the drapes produce the effect of introducing approximately a half-inch air space in additional insulating effectiveness (see Table 4-2), (a) compute the probable overall U of the glass-block wall with its drapes and (b) find the heat loss.

Ans. (a) $U = 0.37$ (0.52 without drapes); (b) 5190 Btuh

4-25. A heated basement is under a building of 26 by 30-ft floor-plan size. The wall height below grade is 4.5 ft. Estimate the heat loss through the basement floor and walls in contact with ground surface when the building is located in a region of 50 F groundwater temperature.

Ans. 3580 Btuh

4-26. Calculate the heat loss per 24 h from 100 lin ft of standard 2-in. steel pipe carrying saturated steam at 150 psig (365.8 F). The pipe is covered with 2-in. insulation of the 85% magnesia type. The temperature at the outer edge of the pipe covering is 90 F. (Neglect surface film and the thermal resistance of the pipe itself.)

Ans. 168 500 Btu/24 h

4-27. What is the heat loss from 100 ft of $\frac{1}{2}$-in. pipe carrying steam at 100 psig (a) if the pipe is bare and (b) if the pipe is covered with 1-in. insulation of 85% magnesia. The room temperature is 80 F.

Ans. (a) 19 600 Btuh; (b) 4640 Btuh

4-28. A bare 4-in. pipe carries steam at 150 psig through a room which is maintained at 80 F. Find (a) the heat loss per hour and (b) the weight of steam condensed per hour as a result of this heat loss, for 100 ft of pipe.

Ans. (a) 107 900 Btuh; (b) 125.8 lb/h · 100 ft

4-29. Determine the insulation heat loss through the four side walls of a building 40 by 60 by 20 ft. The walls are constructed of 4-in. face brick on the outside, and 6 in. of concrete, and with a $1\frac{1}{2}$- to 2-in, air space where the wall is furred for the metal lath and plaster on the inside. The building contains a total of 12 single-glazed windows $3\frac{1}{2}$ by 7 ft, and three 4- by 7-ft doors made of nominal 2-in. wood. The outside temperature $t_o = 10$ F, and the inside temperature $t_i = 80$ F. Disregard temperature variations in the room itself. This corresponds to a wall in Table 4-7 where a U factor of 0.35 Btu/h · ft^2 · F is reported. Compute the total glazed area and door area and subtract from the gross wall area to find (a) the net wall area and (b) the heat loss through the walls for the conditions indicated.

Ans. (a) 3622 ft^2; (b) 88 700 Btuh

4-30. For Problem 4-29 compute the heat loss; (a) through the windows and (b) through the doors.

Ans. 23 260 Btuh; (b) 2520 Btuh

4-31. Compute the emissivity-area factor for radiant heat transfer, E_a (also written $e\,F_a$), for an air space with essentially parallel walls, $3\frac{5}{8}$ in. deep, when one facing surface consists of plaster board, $e = 0.91$, and the other surface is painted plywood at $e = 0.9$. Note that the depth of the space is unimportant if the framing is not so close as to disturb the configuration. Use the relationship given in Section 4-10, realizing that $F_a = 1$ for parallel plates.

Ans. $E_a = 0.828$

4-32. Rework Problem 4-31 if the plaster board is surfaced with aluminum-coated paper, $e = 0.22$, and no other change.

Ans. $E_a = 0.21$

REFERENCES

ASHRAE Handbook of Fundamentals (New York: ASHRAE, 1972).

W. H. McAdams, *Heat Transmission* (New York: McGraw-Hill Book Co., 1954).

L. B. McMillan, "Heat Transfer through Insulation in Moderate and High Temperature Fields," *Trans. ASME*, vol. 48 (1926), pp. 1269–1317.

R. H. Heilman, "Heat Losses from Bare and Covered Wrought Iron Pipe," *Trans. ASME*, vol. 44 (1922), p. 299.

ASHRAE Heating and Ventilating Guide, chap. 9 (New York: ASHRAE, 1955).

B. H. Jennings, *Environmental Engineering* (New York: IEP, A Dun-Donnelley Publisher, 1970).

CHAPTER

5

THE HEATING LOAD

5-1. HEATING LOAD

The items entering into the heating load in a building or space are:

1. Heat loss through exposed wall area to outside, including that lost through outside walls, roofs, or ceiling to unheated attics, and through the floor to unheated spaces, but not including glass or door area.
2. Heat loss through glass surfaces and doors.
3. Heat required to warm the air entering by infiltration through outside windows and door cracks and through other points of leakage. In complete air-conditioning projects with sealed windows, the ventilation air brought in from outside may constitute this item.
4. Miscellaneous heat requirements, as for humidification of outside air, and safety factors to take care of contingencies.

Item 1. Item 1 largely constitutes the so-called insulation loss for the building. The general methods of computing this loss for various types of building construction were discussed in detail in Chapter 4. For a heat loss of this nature the following basic relationship for heat transfer applies:

$$Q_w = U_w A_w(t_i - t_o) \text{ Btu transmitted per hour} \tag{5-1}$$

In this equation, A_w is the square feet of net wall area for the space or room under consideration and equals the gross wall area less the area of any glass and doors in the space considered. In calculating gross wall area, use the length (feet) of exposed wall, measured on the inside, and use as height (feet) the distance from floor to floor when the heated space is above (otherwise use height to ceiling). The area of glass and doors, to be subtracted from the gross area, is the sum of the total sash and frame area of all windows and outside doors.

The inside temperature t_i to be used in the heated space should be modified for the height of the space or room. Values of design inside temperature should be selected from Table 5-1 for the particular type of space, although a value of 72 F can be arbitrarily taken. These temperatures are meant to apply at the breathing line, 5 ft above floor level (or at the 30-in. line). With an unusually high room heated with conventional radiation, the mean wall temperature will be different from the design inside temperature, and the proper value to use can be found with Eq. (4-28).

The winter design temperatures of living spaces, appearing in Table 5-1, indicate a range of temperatures for each type of space. In selecting a design temperature within one of these ranges, consideration should be given to the probable relative humidity and to the amount of possible exposed cold wall and glass surface in the space. Since relative humidity in heated spaces in winter may drop appreciably below 30%, a higher dry-bulb temperature may be required to give the same degree of comfort that would apply in the 45 to 50% relative-humidity range. This is particularly true during cold spells, when, if proper humidification is not provided,

Table 5-1 Inside Dry-bulb Temperatures Usually Specified for Heating Season (Winter)

Type of Space	Temperature Range (F)	(°C)	Type of Space	Temperature Range (F)	(°C)
Auditoriums	72–76	22–24	Hotel bedrooms and baths	72–76	22–24
Ballrooms	68–72	20–22	Kitchens and laundries	66	19
Bathrooms in general	74–80	23–27	Paint shops	80	27
Dining and lunch rooms	72–76	22–24	Public buildings	70–74	21–23
Factories			School classrooms	72–76	22–24
Light work	60–70	16–21	Steam baths	110	43
Heavy work	58–68	14–20	Stores	70–74	21–23
Gymnasiums	60–70	16–21	Swimming pools	75	24
Homes	72–76	22–24	Theater lounges	70	21
Hospitals			Toilets	70	21
Operating rooms	70–95	21–35			
Patients' rooms	74–76	23–24			

the dry outside air entering by infiltration may reduce relative humidities in an occupied space to values as low as 10 to 25%. The effect of such low humidities is to make the occupant feel as though he were in a space one or more degrees cooler. Also, an occupant in a space containing extensive cold glass area or poorly insulated exposed walls will radiate excessive heat to these areas and he will feel colder than he should in relation to the ambient dry-bulb temperature. Thus, if either of these cases is considered probable, a dry-bulb design temperature near the top of the range, or even above it, should be selected.

The design outside temperature t_o should usually be selected from Table 5-2. If the particular locality under consideration cannot be found in the table, use as a guide the United States Weather Bureau records, which give the lowest recorded temperatures for many places. Such minimum temperatures should not be used in design, since their occurrence is rare and usually of short duration; when only such minimum values are available, a figure 10 to 15 deg higher than minimum should be selected as the design temperature. One technique has been to employ an outside design temperature which is equaled or exceeded during 97.5% of the time during the months of December, January, and February.* The values in Table 5-2 are partly based on this premise, but more extensively on what engineers in given communities have found to be satisfactory in making actual heat-load calculations. Wind velocities and their direction are quite variable, so that the values indicated in Table 5-2 should be considered merely indicative.

U_w should be calculated by the methods shown in Chapter 4, using the prevailing wind velocity. If the factor U_w can be found in Tables 4-4 to 4-14, no correction for wind velocity need be made unless the actual velocity is extremely different from the 15-mph velocity on which the tables were computed. Table 4-18 can be of use in case of a different wind velocity.

Item 2. For computing glass and door loss, use a relationship of the form

$$Q_g = U_g A_g (t_i - t_o) \text{ Btuh} \tag{5-2}$$

Select U_g from Table 4-17. The window area is the whole area of all the window sashes and doors in question. For very high windows, or if extreme accuracy is desired, the temperature at the mean height of the window should be used for t_i.

Item 3. The amount of air which enters through cracks and clearances around windows and doors depends mainly on the tightness of the construction and on the wind velocity. Leakage is also increased by any chimney effect the building may exert by virtue of its height and the existing temperature differences inside and outside. The crack length is the perimeter of

* During a normal winter not more than 54 of the 2160 hours in these months is less than or equal to the indicated temperature.

Table 5-2 Outdoor Conditions for Cooling and Heating Design*

State and City	SUMMER			WINTER			
	Dry Bulb, °F		Wet Bulb, °F	Dry Bulb, °F		Wind Velocity, mph, and Direction	Degree, Days
	1%	Normal	Normal	97½%	Normal		
Alabama							
Birmingham	97	95	78	22	10	8N	2551
Mobile	95	95	80	29	15	10N	1560
Montgomery	98	95	78	26	10	7NW	2291
Alaska							
Anchorage	73	75	63	−20	−25		10,864
Fairbanks	82	80	64	−50	−50		14,279
Juneau	75	75	66	− 4	−10		9075
Arizona							
Flagstaff	84	90	65	5	−10	7SW	7152
Phoenix	108	105	76	34	25	5E	1765
Tucson	105	105	72	32	25	5NW	1800
Yuma	111	110	78	40	30	7N	974
Arkansas							
Fort Smith	101	95	76	19	10	8E	3292
Little Rock	99	95	78	23	5	10NW	3219
California							
Los Angeles	86	90	70	48	35	6NE	1391
Sacramento	100	100	72	32	30	8SE	2502
San Diego	86	85	68	44	35	5NW	1458
San Francisco	83	85	65	37	35	8N	3015
Colorado							
Colorado Springs	90	90	63	4	−25		6423
Denver	92	95	64	3	−10	8S	6283
Grand Junction	96	95	65	11	−15	5NW	5041
Connecticut							
Bridgeport	90	95	75	8	0		5617
Hartford	90	93	75	5	0	8NW	6235
New Haven	88	95	75	9	0	10N	5897
Delaware							
Wilmington	93	95	78	15	0	NW	4930
Dist. of Columbia							
Washington	94	95	78	19	0	7NW	4224
Florida							
Jacksonville	96	95	78	32	25	9	1239
Miami	92	91	79	47	35	13	214
Pensacola	92	95	78	32	20	11	1463
Tampa	92	95	78	39	30	8	683

*The data from which this table was developed came from many sources of which the most direct are the ASHRAE *Handbook of Fundamentals*, 1967, chapter 22 and the ASHRAE *Guide and Data Book*, applications Volume 1968, chapter 34. These volumes are highly recommended for the much more extensive weather data they obtain. *The Handbook of Air Conditioning Heating and Ventilating* (ref. 8) is of interest for the extensive tabulations of degree-day data and supporting information. The fundamental sources of weather information are found in references 1 through 5. Extensive data on design conditions used in many areas can be found in references 7 and 8.

Table 5-2 (*Continued*)

State and City	SUMMER Dry Bulb, °F		SUMMER Wet Bulb, °F	WINTER Dry Bulb, °F		WINTER Wind Velocity, mph, and Direction	Degree, Days
	1%	Normal	Normal	97½%	Normal		
Georgia							
Atlanta	95	95	76	23	10	11	2961
Augusta	98	98	76	23	10	6	2397
Macon	98	95	78	27	15	7	3326
Savannah	96	95	78	27	20	9	1819
Hawaii							
Honolulu	87	90	75	62			0
Idaho							
Boise	96	95	65	10	−10	9	5809
Lewiston	98	95	65	12	− 5	7	5542
Pocatello	94	95	65	− 2	− 5	10	7033
Illinois							
Chicago	95	95	75	1	−10	11	6282
Moline	94	96	76	− 3	−10	10	6408
Peoria	94	96	76	2	−10	9	6025
Springfield	95	98	77	4	−10	14	5429
Indiana							
Evansville	96	95	78	10	0	10	4435
Fort Wayne	93	95	75	5	−10	11	6205
Indianapolis	93	95	76	4	−10	13	5699
Iowa							
Des Moines	95	95	78	− 3	−15	10NW	6588
Dubuque	92	95	78	− 7	−20	7NW	7376
Sioux City	96	95	78	− 6	−20	12NW	6951
Kansas							
Dodge City	99	95	78	7	−10	15	4986
Topeka	99	100	78	0	−10	10	5182
Wichita	102	100	75	9	−10	12	4620
Kentucky							
Lexington	94	95	78	10	0	13SW	4683
Louisville	96	95	78	12	0	10SW	4660
Louisiana							
Baton Rouge	96	95	80	30	20	8	1560
New Orleans	93	95	80	35	20		1385
Shreveport	99	100	78	26	20	9	2184
Maine							
Eastport		90	70		−10	9	8445
Portland	88	90	73	0	− 5	9	7511
Maryland							
Baltimore	94	95	78	15	0	8NW	4487
Cumberland	94	95	75	9			
Massachusetts							
Boston	91	92	75	10	0	12W	5634
Springfield	91	93	75	2	−10	9SW	

Table 5-2 (*Continued*)

State and City	SUMMER			WINTER			
	Dry Bulb, °F		Wet Bulb, °F	Dry Bulb, °F		Wind Velocity, mph, and Direction	Degree, Days
	1%	Normal	Normal	97½%	Normal		
Michigan							
Detroit	92	95	75	8	−10	12SW	6232
Flint	89	95	75	3	−10	W	7377
Grand Rapids	91	95	75	6	−10	12NW	6894
Lansing	89	95	75	6	−10	10SW	6909
Marquette	88	93	73	− 4	−10	11NW	8398
Minnesota							
Duluth	85	93	73	−15	−25	13SW	10,000
Minneapolis	92	95	75	−10	−20	11NW	7966
Mississippi							
Jackson	98	95	78	24	15	8SE	2239
Meridian	97	95	79	24	10	6N	2289
Vicksburg	97	95	78	26	10	8SE	2041
Missouri							
Kansas City	100	100	76	8	−10	10NW	4711
St. Louis	98	95	78	8	0	12S	4900
Springfield	97	95	78	10	−10	11SE	4900
Montana							
Billings	94	90	66	− 6	−25	12W	7049
Butte	86	95	67	−16	−20	NW	
Helena	90	95	67	−13	−20	7SW	8129
Nebraska							
Lincoln	100	95	78	0	−10	11S	5864
North Platte	97	95	78	− 2	−10	8W	6684
Omaha	97	95	78	− 1	−10	10NW	6612
Nevada							
Las Vegas	108	110	70	26	20	S	2709
Reno	95	95	65	7	− 5	6W	6332
New Hampshire							
Concord	91	90	73	− 7	−15	6NW	7383
New Jersey							
Atlantic City	91	95	78	18	5	13SW	4812
Newark	94	95	75	15	0	11	4589
Trenton	92	95	78	16	0	9SW	4980
New Mexico							
Albuquerque	96	94	65	17	0	8SW	4348
Roswell	101	95	70	19	0	6S	3793
Santa Fe	90	90	65	11	0	6SE	6123
New York							
Albany	91	93	75	0	−10	11S	6875
Buffalo	88	93	73	6	− 5	17W	7062
New York	94	95	75	15	0	16NW	5280
Rochester	91	95	75	5	− 5	10W	6748
Syracuse	90	93	75	2	−10	11S	6756

Table 5-2 (*Continued*)

State and City	SUMMER Dry Bulb, °F 1%	SUMMER Dry Bulb, °F Normal	SUMMER Wet Bulb, °F Normal	WINTER Dry Bulb, °F 97½%	WINTER Dry Bulb, °F Normal	WINTER Wind Velocity, mph, and Direction	Degree, Days
North Carolina							
Asheville	91	93	75	17	0	9NW	4042
Charlotte	96	95	78	28	10	7SW	3191
Raleigh	95	95	78	20	10	8SW	3393
Wilmington	93	95	78	27	15	9SW	2347
Winston-Salem	94	95	78	17	10		3595
North Dakota							
Bismark	95	95	73	−19	−30	10	8851
Grand Forks	91	90	73	−23	−25		9870
Ohio							
Cincinnati	94	95	78	12	0	8SW	4410
Cleveland	91	95	75	7	0	15SW	6144
Columbus	92	95	76	7	− 5	12SW	5660
Toledo	92	95	75	5	−10	12SW	6494
Oklahoma							
Oklahoma City	100	100	77	15	0	11S	3725
Tulsa	102	100	77	16	0	N	3860
Oregon							
Eugene	91	90	68	26	−15		4726
Portland	89	90	68	24	10	7S	4635
Pennsylvania							
Bethlehem	92	95	75	5	− 5		5810
Erie	88	93	75	11	− 5	14SW	6451
Harrisburg	92	95	75	13	0	7NW	5251
Philadelphia	93	95	78	15	0	11NW	5144
Pittsburgh	90	95	75	9	0	11W	5987
Reading	92	95	75	9	0	9	4945
Scranton	89	95	75	0	− 5	8SW	6254
Rhode Island							
Providence	80	95	75	10	0	12NW	5954
South Carolina							
Charleston	95	95	78	30	15	10	2033
Columbia	98	95	75	23	10	8	2484
Greenville	95	95	75	23	10	8	2980
South Dakota							
Rapid City	96	95	70	− 6	−20	10	7345
Sioux Falls	95	95	75	−10	−20	11	7839
Tennessee							
Chattanooga	97	95	76	19	10	8	3384
Knoxville	95	95	75	17	0	7	3590
Memphis	98	95	78	21	0	12	3015
Nashville	97	95	78	16	0	10	3578

Table 5-2 (*Continued*)

State and City	SUMMER			WINTER			
	Dry Bulb, °F		Wet Bulb, °F	Dry Bulb, °F		Wind Velocity, mph, and Direction	Degree, Days
	1%	Normal	Normal	97½%	Normal		
Texas							
Amarillo	98	100	72	12	−10	13	3985
Austin	101	100	78	29	20	10	1711
Brownsville	94	100	80	40	30	12	600
Dallas	101	100	78	24	10	10	2363
El Paso	100	100	69	25	10	10	2700
Fort Worth	102	100	78	24	10	12	2405
Galveston	91	95	80	36	20	11	1274
Houston	96	95	80	32	20	11	1396
San Antonio	99	100	78	30	20	9	1546
Utah							
Salt Lake City	97	95	65	9	−10	8	6052
Vermont							
Burlington	88	90	73	− 7	−10	12	8269
Virginia							
Lynchburg	94	95	75	19	5	8	4166
Norfolk	94	95	78	23	15	12	3421
Richmond	96	95	78	18	15	8	3865
Roanoke	94	95	76	18	0	10	4150
Washington							
Seattle	81	85	65	32	15	10	4424
Spokane	93	90	64	4	−15	8	6655
Tacoma	85	85	64	24	15	8	5145
Walla Walla	98	95	65	16	− 5	5	4805
West Virginia							
Charleston	92	95	75	14	0	9	4476
Huntington	95	95	76	14	− 5	W	4446
Parkersburg	93	95	75	12	−10	7	4754
Wheeling	91	95	75	9	− 5		5218
Wisconsin							
La Crosse	90	95	75	− 8	−25	9	7589
Madison	92	95	75	− 5	−15	11	7863
Milwaukee	90	95	75	− 2	−15	13	7635
Wyoming							
Cheyenne	89	95	65	− 2	−15	14NW	7381
Sheridan	95	95	65	− 7	−30	5NW	7680
Argentina							
Buenos Aires	91		76	34			
Bahamas							
Nassau	90		80	63			
Brazil							
São Paulo	86		74	46			

Table 5-2 (*Continued*)

State and City	SUMMER Dry Bulb, °F 1%	Dry Bulb, °F Normal	Wet Bulb, °F Normal	WINTER Dry Bulb, °F 97½%	Dry Bulb, °F Normal	Wind Velocity, mph, and Direction	Degree, Days
Canada							
Calgary, Alta.	87	90	66	−25	−29		9703
Edmondton, Alta.	86	90	68	−26	−33		10,268
Fredericton, N.B.	89	90	75	−10	−10		8671
Halifax, N.S.	83	90	75	4	4		7361
Montreal, Que.	88	90	75	−10	−10		7899
Ottawa, Ont.	90	90	75	−13	−15		8735
Quebec, Que.	86	90	75	−13	−12		9372
Regina, Sask.	92	90	71	−29	−34		10,806
Toronto, Ont.	90	93	75	1	0		6827
Vancouver, B.C.	80	80	67	19	11		5515
Winnipeg, Man.	90	90	71	−25	−29		10,679
Yarmouth, N.S.	76	80	68	9	7		7340
Dominican Republic							
Santo Domingo	92		80	65			
Mexico							
Mexico City	83		60	39			
Monterrey	98		78	41			
Vera Cruz	91		83	62			
Panama							
Panama City	93		81	73			
Puerto Rico							
San Juan	89		80	68			
Venezuela							
Caracas	84		69	54			

each window sash, counting only once the crack at the meeting rail of double-hung sliding sash. Table 5-3 gives the infiltration, in cubic feet per hour per foot of crack, for different types of windows and doors, under different wind velocities.

The air entering a building by infiltration leaves at the same rate as that at which it enters. However, in the case of a building with many partitions and of tight construction, air may enter on the windward side in such quantity as to build up a slight positive pressure and thereby reduce infiltration. In general, it may be considered that the air which enters by infiltration on the windward side of a room leaves on the leeward side (or through vertical openings), so that only half of the total feet of crack of the openings in a room should be included in infiltration-loss calculations. In considering a single room with one exposed wall, all of the feet of crack should be used; for a room with two or more exposed walls, take the feet of crack in the wall which has the greatest amount, but in no case use less than half the total feet of crack in the room.

In the case of high buildings, chimney effect or stack effect acts, in addition to the effect of wind, in causing infiltration. That is, the warm column

Table 5-3 Infiltration through Window and Door Crack in Cubic Feet per Hour per Foot of Crack ($ft^3/h/ft$)
$ft^3/h/ft \times 0.0929 = m^3/h/m$

Type of Aperture	Remarks	Wind Velocity (mph) 5	10	15	20	25
		(km/h) 8.0	16.1	24.1	32.2	40.2
Double-hung wood-sash windows (unlocked)	Average; non-weather-stripped	7	21.4	39	59	80
	Average; weather-stripped	4	13	24	36	49
	Poorly fitted; non-weather-stripped	27	69	111	154	199
	Poorly fitted; weather-stripped	6	19	34	51	71
	Around window frame: masonry wall, uncalked	3	8	14	20	27
	Around window frame: masonry wall, calked	1	2	3	4	5
	Around window frame: wood frame structure	2	6	11	17	23
Double-hung metal windows	Non-weather-stripped; unlocked	20	47	74	104	137
	Non-weather-stripped; locked	20	45	70	96	125
	Weather-stripped; unlocked	6	19	32	46	60
Single-sash metal windows	Industrial; horizontally pivoted	52	108	176	244	304
	Residential casement	14	32	52	76	100
	Vertically pivoted	30	88	145	186	221
Doors	Well-fitted	27	69	110	154	199
	Poorly fitted	54	138	220	308	398

Data based on research papers in *Trans. ASHVE*, vols. 30, 34, 36, 37, and 39.

of air inside the building exerts less pressure at the lower part of the building than the corresponding column of cooler, denser air outside the building, and infiltration therefore occurs. This air which enters at the lower part of the building must leave at the upper part, and in the upper levels this exfiltration opposes the infiltration occurring from wind velocity. In the lower part of the building, however, wind effect and stack effect are cumulative. On the rough assumption that a neutral zone is located at the mid-height of a building, the following formulas can be used to determine wind velocity in connection with Table 5-3:

$$V_e = \sqrt{V^2 - 1.75a} \tag{5-3}$$

$$V_e = \sqrt{V^2 + 1.75b} \tag{5-4}$$

where V_e is the equivalent wind velocity in miles per hour; V is the wind velocity upon which infiltration is based if chimney effect is neglected, in miles per hour; a is the distance from windows under consideration to mid-height of the building when windows are above mid-height, in feet; and

b is the distance from windows under consideration to mid-height of the building when windows are below mid-height, in feet. To reduce chimney effect, the stairwells, elevator passages, and other vertical shafts in tall buildings should be sealed off as much as possible.

The effect of wind when it is stopped or forced through an opening is to create a resulting pressure, equivalent to the velocity head that would exist in an undisturbed moving stream. In Chapter 12 Eqs. (12-20) and (12-21) are derived, showing the resulting velocity head (or velocity pressure) for standard air (density 0.075 lb/ft^3) as

$$h_v = \left(\frac{V_m}{4005}\right)^2 \text{ in. water}$$

with V_m expressed in feet per minute (fpm). Also

$$h_v = 0.000483 V_w^2 \text{ in. water} \tag{5-5}$$

with V_w (wind velocity) expressed in miles per hour (mph). In SI (metric) units, the wind velocity, V_k, expressed in kilometers per hour, is

$$h_{vm} = 0.00472\ V_k^2 \text{ mm water} \tag{5-6}$$

Air infiltration through walls can be neglected in most cases if the walls are lathed and plastered, or if a good application of building paper is made. With poor wood construction and no plaster or building paper, infiltration loss may be extremely high. At 10-mph wind, the air passing through each square foot of wall is 4.20 ft^3/h for an $8\frac{1}{2}$-in brick wall unplastered, but only 0.037 ft^3/h for the same wall plastered. For a frame house with lath and plaster or sheet-rock-finished inner walls the infiltration at 10 mph approximates 0.09 ft^3/h ft^2 of surface. If in addition allowance is made for impervious wall boarding or heavy building paper it may be feasible to disregard wall leakage in present-day construction and compute the infiltration for the openings only.

An approximate method sometimes used for calculating air infiltration is the method of *assumed air changes*. For almost any type of room or space, a volume of air per hour equivalent to $\frac{1}{2}$ to 3 times the volume of the room or space enters by infiltration and an equivalent volume leaves. Typical values give $1\frac{1}{2}$ changes of air per hour for a room with two sides exposed, 1 to 3 changes per hour for stores, and 1 to 2 changes for living rooms. The method is so unreliable, however, that it cannot be recommended, but at least $\frac{1}{2}$ change per hour should be allowed for any space.

Expressed as an equation, the heat loss is

$$Q_{\text{inf}} = (0.244)(q_h)(d)(t_i - t_o) \text{ Btuh} \tag{5-7}$$

where Q_{inf} is the heat loss by infiltration in Btu per hour; q_h is the air entering by infiltration in cubic feet per hour, or 60 q; d is the density of air at conditions measured, in pounds per cubic foot; t_i is the inside temperature, in degrees Fahrenheit; and t_o is the outside temperature, in degrees Fahrenheit.

For ordinary conditions a density d of air of 0.075 lb/ft^3 is sufficiently accurate, so that $0.244 \times d$ in Eq. (5-7) reduces to 0.018 $Btu/ft^3 \cdot F$ as a representative specific heat of air, on a volume basis. Therefore, using the foregoing notation, there results

$$Q_{inf} = (0.018)(q_h)(t_i - t_o) \text{ Btuh} \tag{5-8}$$

Example 5-1. A small building has a total of 28 double-hung 5- by 3-ft wood-sash windows of average construction, evenly distributed on all four sides. They are not weather-stripped. There are also four poorly fitting doors 6.5 by 3.5 ft. The wind velocity is 9 mph NW and there are no obstructions around the building. What is the heat loss by infiltration when the temperature is 70 F inside and −5 F outside?

Solution: Total window crack = 28[2(5 + 3) + 3] = 532 ft. Total door crack = 4[2(6.5 + 3.5)] = 80 ft. Consider half the total window and door crack as contributing to infiltration. Interpolation from Table 5-3 to get infiltration for 9 mph gives 18.5 cfh for the window per foot of crack, and 121.2 cfh for the door per foot of crack. Hence the air entering by window infiltration is

$$q_w = (\tfrac{1}{2} \times 532)(18.5) = 4920 \text{ ft}^3/\text{h}$$

and the air entering by door infiltration is

$$q_d = (\tfrac{1}{2} \times 80)(121.2) = 4848 \text{ ft}^3/\text{h}$$

Therefore the total infiltration of air is 4920 + 4848 = 9768 ft^3/h.

Substituting in Eq. (5-8), the heat loss by infiltration is

$$Q_{inf} = (0.018)(9768)[70 - (-5)] = 13\,180 \text{ Btuh} \qquad \textit{Ans.}$$

Item 4. Specific computations for the miscellaneous items are often avoided, but in such cases a suitable factor for unforseen contingencies and for unusual exposure should be added. Thus if a building is in the open sweep of wind, from 10 to 20% should be added to the computed heat load. If the region where the building is located is subject to erratic and sudden temperature changes, or if the building is to be heated intermittently, the design capacity of the heating system should be increased materially.

With open fireplaces, loss from chimney effect is difficult to estimate but should be given some consideration. A figure of 2500 Btuh is sometimes arbitrarily used.

Often humidification for entering outside air is not adequately provided for, but when it is required a definite heat load results, as can be seen in the following example.

Example 5-2. If the infiltration air of Example 5-1 enters at −5 F, with a relative humidity of 50%, (a) how much water must be evaporated per hour to maintain the inside relative humidity at 35% at 70 F and (b) how much heat load is required for this evaporation?

Solution: (a) The humidity ratio and specific volume of the outside air at −5 F and 50% relative humidity can be found from Plate II as 0.0003 lb/lb dry air and v = 11.42 ft^3/lb. The humidity ratio of inside air can be read from Plate I, at 70 F and 35% relative humidity, as 0.0055 lb/lb dry air.

The water to be evaporated into each pound of infiltration air = 0.0055 − 0.0003 = 0.0052 lb/lb.

From Example 5-1, 9768 ft^3 of infiltration air, or 9768/11.42 = 855 lb dry-air constituents, enter per hour.

The water evaporated per hour = 0.0052 × 855 = 4.4 lb. *Ans.*

(b) The latent heat of evaporation of water at room temperatures = 1060 Btu/lb (a round figure; see Table 2-2). Therefore the heat required for evaporation (humidification) = 4.4 × 1060 = 4664 Btuh.

Many buildings, such as churches, are heated only intermittently. Heat absorption by a cold building structure when the building is warming up may amount to about one-half to twice as much as the calculated normal heat loss from the building. Consequently, where quick warming up is required, additional heating capacity should be installed. It is also true that in a building which has recently been cold, the walls for some time after heat is turned on remain much colder than the air, and the resulting excessive radiation loss from the occupants to the cold walls creates an uncomfortable sensation necessitating a higher temperature to offset this condition.

In contrast to the items constituting heating demand, there are some items which in themselves are heat producers. The heat generated by a large assemblage of people, for example, may be so excessive that the external heating supply can be reduced or even stopped. However, the full design heating capacity should be installed, since the building must be warmed before occupancy. Motors and lights also contribute largely to the heat produced in a space, particularly in manufacturing plants, and may reduce or even eliminate the need of heating capacity. Unless the work done by a motor is performed outside the building space itself, all the power supply to the motor reappears in the building. The heat output developed by a running motor can be computed as follows:

$$\frac{\text{motor horse power developed}}{\text{motor efficiency}} \times 2545 = \text{Btuh} \tag{5-9}$$

In the case of electric heaters or lights the heat output developed for each hour of operation can be computed thus:

$$\text{kilowatt capacity} \times 3414 = \text{Btuh}$$

or

$$\text{watt capacity} \times 3.414 = \text{Btuh}$$

Except for unusual layouts, it is not customary in designing heating systems to make deduction in capacity for power and electric lights, since these items are not usually in operation prior to occupancy. It is wise, however, to provide means of reducing the output from the heating apparatus in order to prevent overheating after these local heat supplies go into operation. In the case of well-insulated electrically-heated spaces, the lighting load constitutes a definite part of the heat supply. Also in the interior areas of large buildings the electrical loading may be so extensive that cooling is required to offset it.

5-2. HEATING-LOAD SUMMATION AND ESTIMATING

In computing the heat requirements of the rooms of a building, a systematic approach is required and it is often desirable to prepare in advance tabular forms for listing and summing the various heat losses. For an existing or proposed building, architectural drawings can furnish much of the basic information needed. Columns in a tabular form should be allowed for each room or space which is exposed to outside temperature or to the intermediate temperature of an unheated space such as an attic, cellar, or enclosed porch. The heat loss is computed by using the appropriate outside temperature or the temperature of the unheated space. It is desirable to consider together all the rooms involved on a given story before calculating the next story.

To illustrate how a tabular approach might be followed Fig. 5-1 has been worked up. The name of the room should be placed in the first column, as for example, "Living Room." No space has been left to record length, width, and height, since these dimensions are misleading when one is considering odd-shaped rooms and data should be obtained directly from the architectural drawings of the house. In the case of the house, used as an example in Fig. 5-1, calculations from architectural data show the exposed wall of the living room to be 200 ft^2 in area, and this is so recorded in the second column.

The glass-window and frame area and the door area in the exposed wall are next calculated as 45 ft^2 and are recorded in the third column, at the top of the row being filled in. The overall coefficient of heat transfer U for the glass is 1.13, from Table 4-17, and Δt is taken as 70 deg for this problem. These values have previously been entered in the column headings, and any changes from these values for other rooms must be noted. Temperature at the mean room height or mean glass height should be used. In the bottom half of the row can now be placed the computed heat load, or $Q = UA\Delta t = (1.13)(45)(70) = 3560$ Btuh.

The value for the fourth column, which is for net exposed wall area, equals the gross area minus the window area, or $200 - 45 = 155$ ft^2, and this value is placed in the top half of the row. Since $U = 0.35$ for the wall in question and since $\Delta t = 70$, Q is found as 3797 Btuh, and this value is placed in the bottom half of the row.

The floor in this specific case is over a partly heated cellar; therefore, using $\Delta t = 10$ and $U = 0.34$, and the computed area of 300 ft^2, which has been entered in the top half of the row, under column 5, we find that $Q =$ 1020 Btuh, and this value is entered in column 5, in the bottom of the row.

The ceiling area and other data are then filled in, but for these, since the bedroom above the living room is heated, it is better to use $\Delta t = 0$ and not consider slight interchanges of heat in the building except in cases where the loss to internal adjacent rooms is so great as to necessitate extra heating capacity in the space considered.

Similarly, loss to partitions is regarded as zero, but a fireplace is present, for which 2500 Btuh is allowed.

Room or Space	Gross Exposed Wall Area (sq ft)	Glass Area	Net Exposed Wall Area	Floor Area	Ceiling Area	Partitions, Fireplaces, Etc.	Crack $\left(\frac{Ft}{2}\right) q_h$ $\times$ (0.018)(Δt)	Btuh Sum	Exposure or Other Factor	Total Room or Space Load (Btuh)	Heating Unit Size or CFM
		$U = 1.13$ $\Delta t = 70$	$U = 0.35$ $\Delta t = 70$	$U = 0.34$ $\Delta t = 10$	$U = 0.28$ $\Delta t = 0$						
Living Room		45	155	300	300		½(57)				
	200	3560	3797	1020	0	2500	768	11,645	0.15	13,392	177 CFM

FIGURE 5-1. Heating data and estimate sheet filled in for one room.

HEAT LOAD = Q = UAΔt

Crack calculation brings up a decision—in regard to procedure—which must be adhered to throughout. If the living room is on the windward side of the building, then cold air will be entering all the crack space and leaving on the other side of the building as warm air. The living-room radiators should be sized to handle this total infiltration load instead of half of it. If they are sized in this way, the rooms on the leeward side can be regarded as exposed only to warm air exfiltration and little or no allowance for infiltration need be made. The trouble with this method of calculation is that the wind direction is not constant, and when the wind changes direction the rooms on the leeward side may become too cool. Thus it is better as a general (but not inflexible) rule to use only one half of the crack and, for exposed rooms, add to the Btuh sum an exposure (safety) factor. For the 57 ft of crack along the double-hung non-weather-stripped windows, from Table 5-3, at 10 mph, $q_h = 21.4$ ft^3/h. Thus

$$q = \tfrac{57}{2}(21.4)(0.018)(70) = 768 \text{ Btuh}$$

The sum of all the items is 3560 + 3797 + 1020 + 2500 + 768 = 11 645 Btuh.

Because this room is considered to be on the windward side of the building, an exposure factor of 15% is added.

This exposure factor, when used to modify the Btuh sum, gives the total room or space requirement as 11 645 × 1.15 = 13 392 Btuh.

Data on equipment and its selection are given in Chapters 7, 8, 9, and 11. By adding the Btuh requirements for each room, the total design hourly heat load for the building can be found.

5-3. SPACE HEATING

With knowledge of the heat requirements of a space available, it is possible to provide sufficient steam, hot water, electric energy, or warm air to offset the heat losses of that space and maintain desired conditions in it. Later chapters treat such distribution in detail but here computational patterns for energy distribution by means of air will be presented. With space heating by use of delivered air, the process is merely one involving sensible cooling of supply air and the basic equation is

$$Q_s = 60m_a(C_p)(t_2 - t_1) \text{ Btuh} \tag{5-10}$$

or

$$Q_s = 60\frac{q}{v}C_p(t_2 - t_1) \text{ Btuh} \tag{5-11}$$

$$= 60\frac{q}{v}(0.244)(t_2 - t_1)$$

$$= 14.6\frac{q}{v}(t_2 - t_1) \text{ Btuh} \tag{5-12}$$

where m_a is the pounds per minute air flow; C_p is the specific heat of moist air as delivered, about 0.244 Btu/lb · F; q is the cubic feet per minute air flow; v is the specific volume of the moist air delivered, in cubic feet per pound; t_2 and t_1 are the temperatures of the final air in the space and as delivered in degrees Fahrenheit; and Q_s is the heat load of the space, Btu per hour. For many cases it is sufficiently accurate to disregard the variation in v with temperature and using the value of $v = 13.51$ at 70 F and 50% relative humidity, Eq. (5-12) reduces to

$$Q_s = 1.08q(t_2 - t_1) \tag{5-13}$$

or with saturated air at 60 F,

$$v = 13.33 \text{ ft}^3/\text{lb}$$

and

$$Q_s = 1.10q(t_2 - t_1) \tag{5-14}$$

The latter two equations suffice for many computations.

5-4. SEASONAL HEAT AND FUEL REQUIREMENTS; DEGREE-DAYS

An extensive series of tests, conducted by the American Gas Association, showed that the fuel consumption in residences and public buildings varies almost directly as the difference between the outside temperature and 65 F. Other investigations have shown that 66 F is closer than 65 F in representing a datum point, but 66 F is so close to the 65 F that the earlier datum is still used. This datum 65 F (or 66 F) indicates that when the outside temperature is 65 F or above, practically no heat is required and the fuel consumption approaches zero; also, the fuel consumption would double if an outside temperature of 55 F (10 deg difference) changed to 45 F (20 deg difference).

The difference between 65 F and the average outside temperature is important as an index of heating requirements and gives the basis for the degree-day for specifying the nominal winter heating load. A *degree-day* accrues for every degree the average outside temperature is below 65 F during a 24-h period. Thus, in a given locality, if the outside temperature for 30 days averaged 50 F, the degree-days for the period would be (65 − 50) times (30), or 450. Degree-days vary greatly from place to place, and slightly from year to year.

In large cities, airport degree-day data may be significantly different from central-city data. Degree-days at a given location also vary from year to year and consequently average values are tabulated in most instances. The data in Table 5-2 are largely based on 30-yr averages but a number of shorter-period averages also appear in the table.

A normal heating season in the United States is considered to be from October 1 to May 1 (212 days, or 5088 h). This period varies throughout the country.

By making use of the degree-day values in a given location, it is possible to estimate with a fair degree of success the probable fuel or steam consumption required by a building during a heating season. The estimations, however, may be inaccurate for such reasons as minor variations in degree-days from year to year, local wind conditions, and unusual exposures. In making estimates, therefore, it is extremely helpful to have available an accurate heat-loss analysis of the building, made under known design conditions.

Many approaches to the estimation of fuel or steam consumption are in use, but only one or two will be indicated here. With the heat loss under design conditions known, it is desirable to find the heat loss per degree difference in temperature, which can be found by the relation

$$Q_D = \frac{Q_T}{t_i - t_o} \tag{5-15}$$

where Q_T is the total heat loss for a building or space, in Btu per hour, when based on the following: t_i is the inside design temperature, in degrees Fahrenheit; t_o is the outside design temperature (from Table 5-2), in degrees Fahrenheit; and Q_D is the heat loss per degree of temperature difference, in Btu per hour.

To find the steam consumption, use

$$S = \frac{(Q_D) \times (\text{degree-days}) \times (24)}{1000} \tag{5-16}$$

where Q_D is the calculated heat loss per degree of temperature difference, in Btu per hour; 1000 is the approximate number of Btu released for heating by each pound of steam; 24 is the number of hours per day; and S is the pounds of steam required for the degree-days in the period of estimation (heating season, month, etc.).

Example 5-3. For an office building in Cleveland, Ohio, having a calculated heat loss of 3 200 000 Btuh, estimate the weight of steam required during the heating season.

Solution: Degree-days in Cleveland are 6144 and the design temperatures are 0 F and 70 F. Thus by Eq. (5-15) and (5-16),

$$Q_D = \frac{3\,200\,000}{70 - 0} = 45\,700 \text{ Btuh/deg}$$

and

$$S = \frac{(45\,700)(6144)(24)}{1000} = 6\,740\,000 \text{ lb} \qquad \textit{Ans.}$$

The amount of fuel used per season depends on its heating value and on the efficiency of utilization. This latter factor is influenced by the kind of combustion equipment, and its condition, and by the manner in which it is operated. In the case of small domestic equipment, some of the otherwise wasted heat from the furnace or boiler room and chimney may be reabsorbed by the building itself. The manufacturer's published efficiency of a given unit may be used for making estimates, although such laboratory values are not usually maintained in actual practice. For rough estimation the values in Table 5.4 can be used.

Example 5-4. Consider that the office building of Example 5-3 is heated with its own boiler and can employ gas, oil, or coal. (a) Estimate for the heating season the cubic feet of gas, at a heating value of 800 Btu/ft^3, which would be required for this building. Also, convert the answer into therms of gas. (b) Estimate the gallons of fuel oil, at a heating value of 144 000 Btu/gal, required for the heating season. (c) Estimate the tons of coal required for the season. Assume that the coal has a heating value of 12 800 Btu/lb.

Solution: The season heat load is (45 700)(6144)(24) = 6 740 000 000 Btu.

(a) The furnace efficiency with gas should reach 80%, since gas can be burned efficiently although there is an unavoidably high loss resulting from hydrogen in the fuel (see Chapter 9). Thus

$$F_g = \frac{6\,740\,000\,000}{(0.80)(800)} = 105\,300\,000 \text{ ft}^3 \qquad \textit{Ans.}$$

Expressed in therms, the gas required is

$$F_t = \frac{6\,740\,000\,000}{(0.80)(100\,000)} = 84\,250 \text{ therms}$$

(b) Considering the oil-burning efficiency to be 78%,

$$F_o = \frac{6\,740\,000\,000}{(0.78)(144\,000)} = 60\,000 \text{ gal} \qquad \textit{Ans.}$$

(c) Considering stoker firing for the coal, with good controls on the boiler, an efficiency of 70% should be possible. Thus

$$F_c = \frac{6\,740\,000\,000}{(0.70)(12\,800)(2000)} = 376 \text{ tons} \qquad \textit{Ans.}$$

The decision as to which fuel is most desirable depends on (1) the cost of fuel delivered, ready to burn; (2) its availability; (3) the labor costs associated with the firing of the fuel; and (4) possible ash-removal and maintenance costs as related to the different types of combustion equipment.

Table 5-4 Performance of Representative Heating Equipment

	Efficiency (%)
Coal-fired small boiler or furnace (heating value of coal, about 13 000 Btu/lb)	45–70
Small oil-fired boiler or furnace (heating value of oil, about 144 000 Btu/gal)	60–85
Gas-fired furnace (heating value of manufactured gas, about 550 Btu/ft^3; of natural gas, about 900 Btu/ft^3 [a])	75–85

[a] Gas is also sold in therm units, where 1 therm = 100 000 Btu.

Table 5-4 gives data from which the steam-consumption requirements of various types of buildings can be estimated. The net volume of heated space in a building is approximately 80% of the gross volume of a building. Estimates made from data such as found in this table should be recognized as furnishing only approximations of heating demands.

Example 5-5. A residence has a total volume of 40 000 ft^3 and is located at a place where the degree-days are 4626. Find the probable steam consumption for a heating season.

Solution: The heated space in a building is usually only about 80% of the total cubage, or in this case,

$$40\,000 \times 0.8 = 32\,000 \text{ ft}^3$$

From Table 5-5, the steam consumption per degree-day is 0.962/1000 ft^3 of heated space. Therefore the probable steam consumption for a heating season is

$$0.962 \times \frac{32\,000}{1000} \times 4626 = 142\,000 \text{ lb} \qquad \textit{Ans.}$$

For monthly percentage estimates of fuel consumption during the heating season it has been found that the values given in Table 5-6 apply to average localities in the United States (that is, not localities in the extreme north or south).

Table 5-5 Steam Consumption for the Heating Season for Various Classes of Buildings

Building Classification	Steam Consumption for Heating (pounds per degree-day per 1000 cu ft of net heated space)	Average Hours of Occupancy
Apartment	0.962	22
Bank	0.786	12
Church	0.532	8
Club or hotel	0.990	22
Department store	0.385	11
Garage	0.202	22
Hospital	1.19	22
Lodge	0.390	12.5
Loft	0.588	10
Manufacturing	0.808	9.5
Municipal	0.587	16
Office	0.685	12
Office and shops	0.617	13
Printing	1.23	18
Residence	0.962	22
School	0.592	11
Stores	0.624	10.5
Theatre	0.482	13
Warehouse	0.459	9.5

Table 5-6 Estimated Monthly Fuel Consumption, in Percentage of Total Consumption for Winter Season

October	5	February	17
November	11	March	15
December	17	April	10
January	20	May	5

To furnish a better idea of degree-day variations throughout a season, Table 5-7 gives the monthly variation of degree-day values for several representative cites.

By use of a table such as Table 5-7, it is possible to compute the average temperature during a given month or during a winter season. For example, in Chicago during January, with 1218 degree-days accrued in 31 days, and using the 65 F base of the table,

$$1218 = 31(65 - t_{avg})$$

Therefore, for a typical January,

$$t_{avg} = 25.8 \text{ F}$$

To find the average temperature for a major part of the heating season in Chicago—say, from October 1 to May 1, or 212 days—first subtract the degree-day values for the unwanted months from the yearly total:

$$6282 - (95 + 73 + 259) = 5855 = 212(65 - t_{avg})$$

Then compute the temperature for these seven worst months of the heating season:

$$t_{avg} = 37.4 \text{ F}$$

In the operation of heating systems, much consideration has been given to the desirability of lowering the temperature maintained in the heated space during periods when the space is not in active use or is unoccupied. Some saving usually results if the space temperature is reduced during such periods. However, if the temperature is greatly reduced, as, for example,

Table 5-7 Average Monthly Degree-day Values for Representative Cities

City	Aug., Sept.	Oct.	Nov.	Dec.	Jan.	Feb.	Mar.	Apr.	May	June, July	Total
Baltimore	33	227	526	855	921	837	637	343	95	12	4487
Chicago	95	337	712	1116	1218	1080	861	531	259	73	6282
Cleveland	107	354	684	1045	1143	1067	876	553	252	63	6144
Los Angeles	5	43	110	225	272	235	212	158	103	28	1391
Minneapolis	190	481	942	1415	1587	1372	1072	577	260	70	7966
New York	54	272	594	940	1028	953	771	465	172	31	5280

over the night hours, the heating equipment may have to run overloaded, and inefficiently, to bring the space rapidly back to the desired temperature. If this is the case, the saving may be minimized or even disappear.

Example 5-6. Analyze the effect of lowered inside temperatures for portions of the time in relation to the building of Example 5-3. Consider that the daily temperature is lowered from 70 F to 64 F during the 8 h from 11 P.M. to 7 A.M. of any 24-h period. Find (a) the weighted average inside temperature and (b) the possible saving in steam over a season.

Solution: The weighted average inside temperature is

$$t_{avg} = \frac{(24 - 8)(70) + (8)(64)}{24} = 68 \text{ F} \qquad \textit{Ans.}$$

(b) The average outside temperature of the whole heating season for Cleveland can be found if the total days requiring heat are known. Refer to Table 5-7 and consider that heat is required during 6 days in September and during 4 days in June. When these are added to the 243 days from October 1 to May 31, the total is 253 days. To find the average outside temperature t_o for the season, use the total degree-days (6144) spread over 253 days:

$$6144 = (253)(65 - t_o)$$

$$t_o = 40.7 \text{ F}$$

The 6 740 000 lb of steam required per season with a 70 F inside temperature, when referred to a 68 F inside temperature, would reduce to

$$S = 6\,740\,000 \, \frac{68 - 40.7}{70 - 40.7} = 6\,280\,000$$

The saving is thus

$$6\,740\,000 - 6\,280\,000 = 460\,000 \text{ lb/season} \qquad \textit{Ans.}$$

The saving amounts to almost 7%. Although all of the saving is not realized, the possible magnitude of the saving requires giving serious consideration to inside temperature reduction at off hours.

Example 5-7. The manager of a large public building located in Baltimore was told by his fuel-gas supplier that, because of a system gas shortage, his gas allocation was to be cut by 10% from his prior year's usage. During the prior year, the building had been maintained at an average temperature of 71 F over a 24-h period; but the manager, realizing that he had to take some action, decided to decrease the average building temperature to 68 F even though he knew many tenants would complain. Make the necessary calculations to see whether this change will produce the necessary saving in fuel.

Solution: Refer to Table 5-7 and read 4487 degree-days for the Baltimore season, but note the manager probably used no heat in June and provided heat for 2 days only in September. The days of heat-system operation totaled 244. The degree-days involved are 4487 less 12 in June; thus

$$4475 = 244 \text{ days } (65 - t_{avg})$$

$$t_{avg} = 46.7 \text{ F average heating season temperature}$$

$$\% \text{ saving} = \frac{(71 - 46.7) - (68 - 46.7)}{(71 - 46.7)} \times 100 = \frac{300}{24.33} = 12.3$$

This indicates slightly more saving than the 10% required.

PROBLEMS

5-1. A room in a residence is 10 ft (3.05 m) long, 8 ft (2.44 m) wide, 9 ft (2.74 m) high, and has one double-hung window in each of the two outside walls. Each window measures 3 by 5 ft (0.91 by 1.52 m) and the bottom of the lower sash is 3 ft (0.91 m) from the floor. A temperature of 70 F (21.1°C) at the 5-ft (0.91-m) level is maintained when it is 0 F (−17.8°C) outside. (a) Find the temperature at each mean height of the single-glazed windows and find the heat loss through them. (b) The two walls of the room, which are exposed to outside, have an overall transmission U value of 0.35 Btu/h · ft^2 · F (1.985 W/m^2 · °C). Find the mean temperature of the two walls. Compute the net heat loss through them. (c) Compute the infiltration loss through the window crackage based on a 10-mph (16.1-km/h) wind. The long side of the room faces the prevailing wind and receives its full effect with this infiltration air then discharging elsewhere from the house.

Ans. (a) 70.7 F, 2397 Btuh; (b) 69.3 F at mean height, 3200 Btuh; 512 Btuh

5-2. A large business office in an old building is fitted with nine poorly fitted, double-hung, wood-sash windows, 3 ft wide by 6 ft high (0.91 by 1.83 m). Design conditions in the area are based on −10 F (−23.3°C) and a possible 15-mph (24.1-kmh) wind. Inside temperature is maintained at 70 F (21.1°C). It is desired to determine the possible maximum reduction in heating load that might be expected from weather stripping the windows. The solution should be based on the assumption that half of the crackage receives an infiltration of outside air while the other half is experiencing exfiltration. Compute the amount of outside-air infiltration, first without weather stripping and then with weather stripping. The difference represents the possible reduction. Note that, if specific heat on a volume basis is used, the value should take cognizance of the specific volume of outside air.

Ans. 7240 ft^3/h, probable reduction; 12 450 Btuh saving

5-3. A large business office is fitted with nine poorly fitted, double-hung, wood-sash windows, each 0.91 m wide by 1.83 m high. A prevailing wind of 24.1 km/h blows in such a manner that half of the windows receive infiltration while the remainder undergo exfiltration. To compute the probable net infiltration, use should be made of Table 5-3 where the results are expressed in engineering units. (a) From fundamental equivalents of the units involved, find the factor to convert a flow in cubic feet per hour per foot of opening to cubic meters per hour per meter of opening. (b) Similarly, find the factor to convert miles per hour to kilometers per hour. (c) Compute the net infiltration to the room in cubic meters per hour.

Ans. ft^3/h · ft × 0.0929 = m^3/h · m; (b) mph × 1.6093 = km/h; (c) 297 m^3/h

5-4. If the poorly fitted windows of the office of Problem 5-3 are weather stripped, compute the probable net infiltration in cubic meters per hour during 24.1-kmh wind conditions and, making use of the answer given for Problem 5-3, find the percentage reduction in infiltration air.

Ans. 90.8 m^3/h; 69.3%

5-5. The installation of weather stripping on the windows described in Problems 5-3 and 5-4 resulted in a reduction of outside infiltration air from 297 to 90.8 m^3 of air per hour. When it is −23.3°C outside and inside is maintained at 21.1°C, compute the reduction in heating load expressed in kilowatts that can be anticipated. *Suggestion*: find the specific heat of air on a volume basis. The density of −23.3°C air is 1.41 kg/m^3 and $C_p = 0.24$ kcal/kg of dry-air per degree Celsius.

Ans. 0.338 kcal/$m^3 \cdot$ °C at −23.3°C, reduction 3094 kcal/h and 3.6 kW

5-6. Outside air at 20 F and 60% relative humidity enters a small building by infiltration at a rate of 2000 ft^3/h. (a) How much moisture must be evaporated per hour to maintain an inside relative humidity of 40% at 65 F? (b) How much heat is required for humidification?

Ans. (a) 0.65 lb; (b) closely 700 Btuh

5-7. A room has four vertically pivoted, steel-sash windows 25 in. wide by 62 in. high. They are on the windward side of the room, which has two snugly fitting doors which open into the central part of a relatively open building. Assume that is it 10 F outside and 70 F inside, and that a prevailing 10-mph wind is blowing. Compute (a) the probable infiltration with both doors open; (b) the probable infiltration with both doors closed and locked; and (c) the heat loss under each of these conditions.

Ans. (a) 5100 ft^3/h; (b) 2550 ft^3/h; (c) 5500 and 2750 Btuh

5-8. What is the heat loss through the glass area of the room in Problem 5-7 (a) when no storm windows are used and (b) when complete-coverage storm windows are employed?

Ans. (a) 2920 Btuh; (b) 1370 Btuh

5-9. (a) How many pounds of steam are required during a winter season for an office building located in Chicago when the heat load of the building is 3 500 000 Btuh based on an inside temperature of 65 F and an outside temperature of −10 F? Use the degree-day method. (b) If this building is heated by fuel oil burned with an overall efficiency of 75%, how many gallons of No. 6 fuel oil are required for the heating season?

Ans. (a) 7 040 000 lb; (b) 65 200 gal

5-10. How many pounds of steam are required each winter for an office building in Philadelphia with a heat load of 1 000 000 Btuh based on an inside temperature of 70 F and an outside temperature of 0 F? Use the degree-day method.

Ans. 1 764 000 lb

5-11. What is the probable fuel consumption per season for the building in Problem 5-10 (a) if coal is used in the furnace and if the furnace has a 60% efficiency and (b) if oil is used and the furnace has an overall efficiency of 70%?

Ans. (a) 113 tons; (b) 17 500 gal

5-12. Estimate the district steam required per season to heat a department store having a net volume of 36 000 000 ft^3 if the store is located in (a) New York, (b) St. Louis, or (c) Minneapolis.

Ans. (a) 73 200 000 lb; (b) 63 500 000 lb; (c) 110 600 000 lb

5-13. The price of fuel and of energy is continually changing, but at one period the following quotations were applicable: 20 cents per therm (100 000 Btu) for gas, 40 cents per gallon of fuel oil (144 000 Btu), and 3 cents per kilowatt hour (kWh) for electric power. The gas and oil can develop 78% of their heating value into useful heating and electric energy is completely utilized. Compute the operating costs per hour for heating a large residence with a computed heat loss of 160 000 Btuh when operating at this maximum design demand.
Ans. \$0.41 gas, \$0.57 fuel oil, \$1.41 electricity. No conclusion should be drawn from these results except that electricity is substantially more costly unless the benefits of heat-pump operation are applicable.

5-14. Compute the average outside temperature throughout the Baltimore heating season, which extends from October 1 to May 1. Use data from Table 5-7.

Ans. 44.5 F

5-15. A certain building located in Baltimore has a heat loss of 150 000 Btuh based on 70 F inside and 0 F outside temperatures. Compute the probable fuel-oil consumption for the heating season, which extends from October 1 to May 1.

Ans. 2010 gal

5-16. It is decided to reduce the inside temperature of the building in Problem 5-10 from 70 F to 65 F from 10:30 P.M. to 7:30 A.M. Find the weighted average inside temperature for a 24-h period. Find the probable saving in fuel per season that would accrue from this night temperature reduction.

Ans. 68.1 F, 7.5%

5-17. What is the heat loss for the first floor of the residence shown in Fig. 11-7 if one assumes that the residence is located in Pittsburgh? Consider that the prevailing winter wind at 12 mph strikes the terrace and dining-room sides of the building. There is enough waste heat from the furnace to keep the basement warm, and this prevents any significant loss downward from the first-floor rooms. Heated space is above. Make computations for each room and tabulate them. Allow for the outside wall surface adjacent to the steps at the right of the entry door. Consider full infiltration on the windward sides of the building and one-half the maximum on the other sides. Consider the unheated garage to be at 20 deg above the outside design temperature. Consider the fireplace structure to be 7 ft wide, and of ceiling height, with its masonry having an overall U of 0.31 Btu/h · ft^2 · F. Also allow for excess exfiltration from the fireplace amounting to 2500 Btuh.

5-18. Compute the heat loss from the second floor of the residence of Fig. 11-8. Assume that the residence has the location and ambient conditions indicated in Problem 5-17, that there is no loss to the floor beneath, and that the attic temperature is 30 deg higher than the outside design temperature. Compute each room and also the stairwell, using the infiltration conditions stated in Problem 5-17. Consider the chimneys adjacent to room 23 merely as conventional wall surface.

REFERENCES

Engineering Weather Data, Army, Navy, and Air Force Manual TM 5-785, 1963.

Evaluated Weather Data for Cooling Equipment Design (Santa Rosa, Calif.: Fluor Products Co., 1958).

Evaluated Weather for Cooling Equipment Design, Addendum No. 1, Summer and Winter Data (Santa Rosa, Calif.: Fluor Products Co., 1964).

L. W. Crow, *Study of Weather Design Conditions*, ASHRAE Research Project No. 23, Bulletin, 1963.

D. W. Boyd, *Climatic Information for Building Design in Canada*, Supplement No. 1, Nat'l Building Code of Canada (Ottawa, Canada: Nat'l Research Council).

Monthly Normals of Temperature, Precipation and Heating Degree Days (Washington, D.C.: U.S. Weather Bureau, 1962).

Carrier Air Conditioning Co., *Handbook of Air-Conditioning System Design* (New York: McGraw Hill Book Company, Inc., 1965).

C. Strock and R. L. Koral, *Handbook of Air-Conditioning Heating and Ventilating* (New York: The Industrial Press, 1965).

CHAPTER

6

THE COOLING LOAD

6-1. BASIC AIR-CONDITIONING SYSTEM

The computations involved in heating load show that the heat losses occurring in a space have to be offset by heat supplied to the extent that a satisfactory thermal environment is set up for occupants in that space. In similar manner refrigeration or cooling of a space is required when environmental conditions are such that energy must be removed to provide a suitable thermal environment for occupants. Because of the complexities involved in computing cooling load, it will be helpful to understand at least one method of providing air conditioning to a space for which the cooling load is to be computed. Thus we will describe a basic system before developing details of the necessary computations.

Except for the rare usage of direct radiation for cooling a space, most cooling is brought about by supplying chilled air to a space, and many systemic arrangements are possible for this purpose (see Chapter 17). Similarities exist among all of them, and the basic features shown in Fig. 6-1 are broadly applicable. Conditioned (chilled) air is shown being provided to the space in adequate amount to maintain a desired space temperature. The

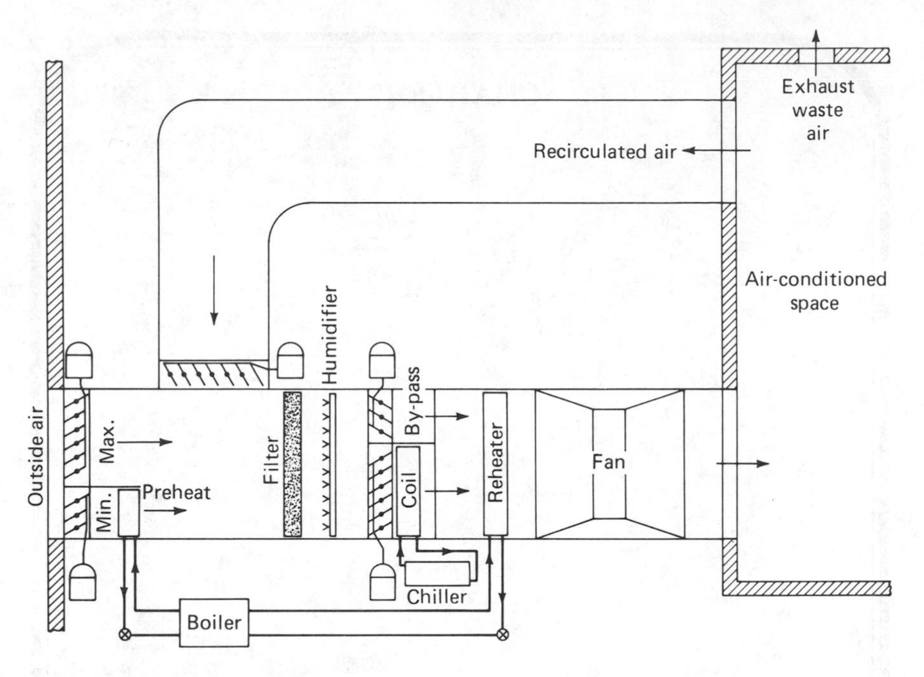

FIGURE 6-1. Basic year-around central air-conditioning system, constant volume, with face and by-pass damper control of conditioner.

supply air, in warming, offsets the heat gains (cooling load) of the space while at the same time the supply air must be sufficiently dry that it can pick up the moisture load produced by people or by other moisture sources in the space. Most of the air from the space is recirculated and, after intermixing with the outside air needed for ventilation, passes through filters and then to the air conditioner to be cooled and dehumidified. If the air has to be greatly chilled to make it sufficiently dry to absorb the moisture load it may then be too cold for comfort when it contacts the occupants of the space. In this instance, if enough air cannot by-pass the cooling coil to warm the dehumidified air, then some reheating is required. Preheating of the outside air may only be necessary under extremely cold outside conditions or sometimes under light-load conditions in the occupied space. With this type of system it may even be possible in intermediate seasons (fall and spring) to operate the system with largely outside air and impose minimum demand either on the boiler or on the refrigeration system. However, this arrangement is only possible if the conditioned space is provided with adequate air exhaust facilities. Operation of the detailed controls will not be described here but it should be mentioned that the return and outside air dampers must be interlocked and also the face and by-pass dampers. Thermostats and humidistats operate to control the operation of the refrigeration machine and boiler as well as the dampers.

6-2. COOLING LOAD

Heat sources for cooling load. In a space, the heat load which must be removed by means of cooling equipment arises from the following sources:

1. Heat transmission through barriers such as walls, doors, windows, ceilings, floors, and partitions, and caused by the different temperatures existing on the two sides of the barrier.
2. Heat from solar effects.
   a. Transmitted by radiation through glass and absorbed by inside surfaces and furnishings.
   b. Absorbed by walls or roofs exposed to rays of sun and transferred to the inside.
3. Heat and moisture introduced with infiltration air.
4. Heat load from occupants (sensible and latent).
5. Heat load from machinery, appliances, lights, and combustion equipment.

In addition, the cooling equipment must be adequate to bring ventilation air to the desired space temperature.

HEAT TRANSMISSION THROUGH BARRIERS. The transmission of heat through barriers is computed by the methods developed in Chapter 5 for computing the heating load. The basic equation, Eq. (5-1), using appropriate inside and outside design temperatures, is applicable. In the case of a cooling load, the outside temperature is normally higher than the inside temperature, and consequently the equation appears as

$$Q = UA(t_o - t_i) \tag{6-1}$$

The inside design dry-bulb temperature can be selected from Table 10-1, and for general usage might be considered as 80 F. For most localities, the outside design temperature can be selected by use of Table 5-2. This table gives the top temperature which experience has shown will not be exceeded more than 1% of the time. The normal design values are based on this along with the wet-bulb temperature which might be expected to occur simultaneously with outside design dry-bulb temperature.

Typical Outside Daily Temperature Variations in Summer

Time	Temperature (Fdb)	(°C db)	Temperature (Fwb)	(°C wb)
10 A.M.	89	31.7	74	23.3
12 M.	93	33.9	75	23.9
2 P.M.	95	35.0	76	24.4
4 P.M.	95	35.0	76	24.4
6 P.M.	94	34.4	76	24.4
8 P.M.	91	32.8	75	23.9
10 P.M.	88	31.1	74	23.3

The dry-bulb temperature throughout a day is variable, usually reaching its maximum about 4:00 P.M., as can be seen in the accompanying tabulation.

The simple relation of Eq. (6-1), while perfectly accurate, is not applicable in many instances because of the complicating effects of solar energy, which can so greatly increase the temperature of an outside wall surface that it greatly exceeds the outside air temperature, thus making the outside air temperature incorrect for use. Consequently other procedures should be used whenever solar effects are significant in influencing the heat flow through a roof or wall.

6-3. SOLAR CONTRIBUTION TO THE COOLING LOAD

Because of the numerous variables involved, solar effects are the most difficult to predict and compute of all the heat gains (cooling loads), listed in Section 6-2. The computations appear much less complex if the designer understands the interaction of the sun and the earth as elements of the solar system. Since this interaction may be unclear to some readers, a detailed explanation follows. Those who do fully understand earth-sun relationships may wish to skip this material and resume reading at the point where the computational procedures are discussed.

The celestial sphere, sun and earth. Except for the sun which is at some 150 million km (93×10^6 miles) from the earth, the nearest star is approximately 3.94×10^{13} km (24.5 million million) miles away. Because the stars are so distant, it is customary to think of them as being almost infinitely far away and, further, to imagine that they lie in a surrounding surface called the celestial sphere. To a person on earth watching the stars for several hours, it appears that these rise in the east and set in the west with their paths forming arcs of circles. If one faces to the north in the Northern Hemisphere, it will be found that the circles traced by the stars appear to be concentric about a certain point in the sky called the pole; that is, the whole celestial sphere appears to rotate about an axis. This, of course, is not true because the apparent rotation is caued by the rotation of the earth about its own axis from west to east, of course, in an opposite direction to that in which the stars appear to move. The stars closest to the pole trace out the smallest circular paths.

Stars are not fixed in position, but are moving in space. However, they change their positions so slowly relative to an observer on the earth that they appear to remain in fixed positions. Nevertheless, records over centuries of time show significant changes in the position of stars and nautical tables account for the changes taking place.

This permanence for the celestial sphere does not apply to the solar system in which rapid changes take place. The apparent motion of the sun is eastward among the stars at about 1° per day, while the earth makes its revolution around it in a year. The moon also appears to travel among the stars at an even faster rate since it moves an amount equal to its own diameter in about an hour and completes its revolution around the earth in approximately $27\frac{1}{3}$ days.

The orbit of the earth, as it moves eastward around the sun in 365.2422 days, constitutes an ellipse with the sun lying at one of the foci of the ellipse (Fig. 6-2). The axis of rotation of the earth is inclined to the plane of the orbit (called the ecliptic) at about a $66\frac{1}{2}°$ angle. The plane of the equator of the earth is inclined at an angle of about $23\frac{1}{2}°$ to the plane of the ecliptic. The direction of the earth's axis of rotation is nearly constant so that it points nearly to the same place in the sky (celestial sphere) year after year.

Changes of seasons result directly from the inclination of the earth's axis and the fact that the axis is essentially parallel to its mean position at all times. When the northern end is pointed away from the sun, it is winter in the Northern Hemisphere and, of course, summer in the Southern Hemisphere. In the Northern Hemisphere, the sun is farthest south of the equator about December 22, at which time the days are shortest and the nights longest. At this time the winter solstice occurs. This condition just precedes the situation where the earth is closest to the sun at the so-called perihelion. Even though the earth is closer to the sun in winter than in summer, winter is colder because the days are shorter and the rays of sunlight strike the surface of the ground more obliquely. At the summer solstice, when summer begins in the Northern Hemisphere, the days are longest and the nights shortest. The solstice points are at the extreme ends of the major axis of the ecliptic. When the earth passes either of the ends of the other axis of the ecliptic, the sun lies in the plane of the earth's equator. At these conjunctions, day and night are of equal length at all places on the earth, and the earth is said to be at an equinox. The vernal equinox occurs around March

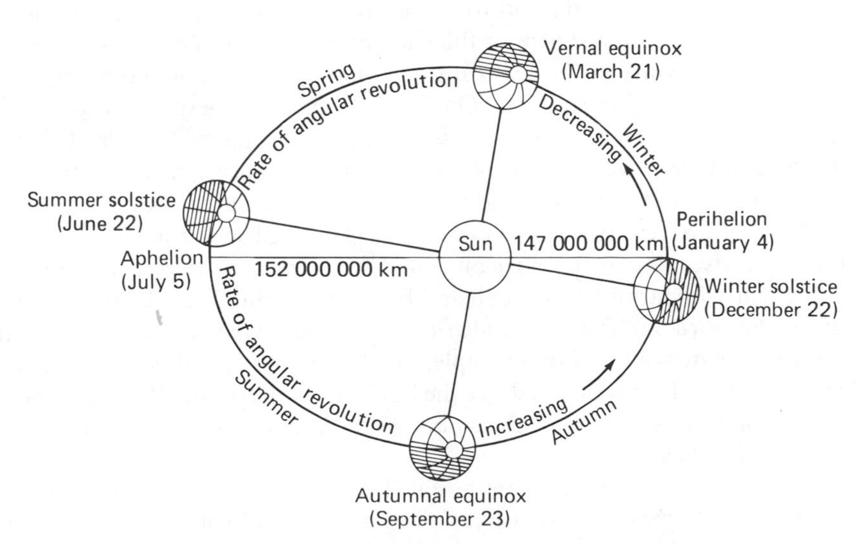

FIGURE 6-2. The seasons, with the usual starting dates shown.

21 and the autumnal equinox around September 23. There are approximately 186 days in spring and summer and 179 days in autumn and winter. Reference should be made to Fig. 6-2, which shows the relative positions of the earth to the sun at the various periods indicated.

The time of year in relation to the angle of declination (δ) alters the intensity of the solar radiation received on the earth. The angle of declination is that angle measured between the equatorial plane and a line from the earth to the sun. In the Northern Hemisphere the angle (δ) varies from $+23.45°$ on December 21–22 to 0° on March 21–22 (and September 21–22) to $-23.45°$ on June 21–22.

The relative position of the sun to the equator is continuously changing since half of the year the sun is north of the equator and the other half of the year it is south of the equator. On June 22 the sun is in its most northerly position and is visible continuously above the Arctic Circle and for more than half of the 24-h day elsewhere in the Northern Hemisphere. On December 22 it is at its most distant point south of the equator and is visible less than half of the day. The Tropic of Cancer at 23.45°N latitude is the northerly limit of places on the earth where the sun appears vertically overhead. A similar situation exists in the Southern Hemisphere at the Tropic of Capricorn. Above the Arctic Circle at 66.55°N latitude, daylight can exist for 24 h for periods of time during the summer. That is, there is no sunrise nor sunset. At the North Pole, there are 24 h of daylight from the period of the vernal equinox to the time of the autumnal equinox, and even at 70°N latitude, 24-h of daylight exist from May 20 to July 23.

The orbit of the moon is inclined about 5° to the ecliptic. This makes the inclination of its orbit to the plane of the earth's equator range between $18\frac{1}{2}°$ to $28\frac{1}{2}°$.

Figure 6-3 supplements Fig. 6-2 to show how the $23\frac{1}{2}°$ tilt of the axis of rotation affects solar radiation to the earth. The positions of the earth in relation to the sun are shown at the starting periods of the four seasons of the year. First, consider the earth at the period of the summer (June 21) and winter (December 22) solstices. On June 21 the Northern Hemisphere is starting summer and at noon on this day at the Tropic of Cancer (T of C) the sun is directly overhead. At 40°N latitude an observer would notice the sun at 73° above the horizon, and that portion of the earth above the Arctic Circle would have daylight 24 h each day. To an observer at point *X*, the sun never rises above the horizon. On December 22 the Arctic Circle is permanently in darkness although an observer at point *X* could glimpse the sun at the horizon. The sun is also lower in the heavens in the Northern Hemisphere in winter. For example, on December 22 at noon at 40°N latitude the sun is only at 27° above the horizon. In the Southern Hemisphere reverse conditions exist, with December 22 representing the start of summer, and June 21 the start of winter.

The horizon at any point on the earth coincides with the tangent line to the surface of the earth at that point. The direction of rotation of the earth is indicated on the N-S (north-south) line.

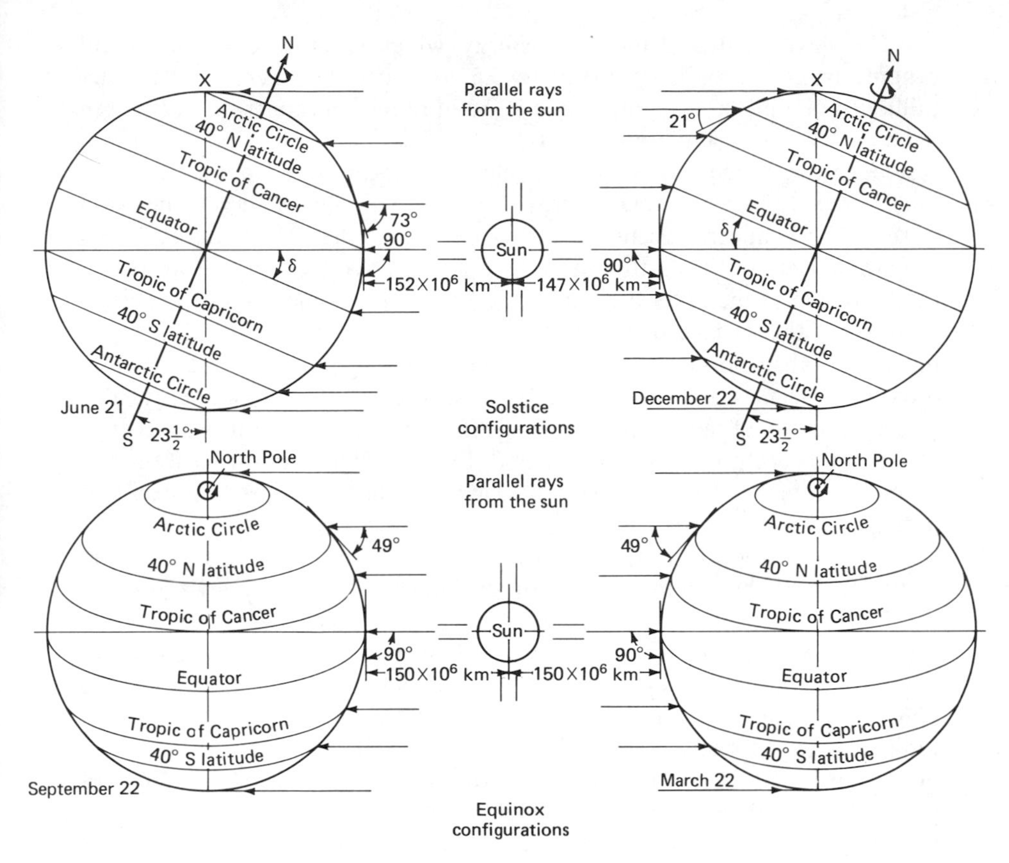

FIGURE 6-3. Upper diagrams depict the manner in which the rays of the sun strike the earth at the time of the summer and winter solstices. Lower figures depict the situation at the time of the equinoxes. The sun is so distant that its radiation impinges in parallel rays.

The lower part of Fig. 6-3 shows the position of the sun and earth at the time of the equinoxes, and when the earth is faced as shown, the N-S pole axis does not lie in the plane of the paper but is displaced forward at its northerly end by an angle of $23\frac{1}{2}°$. Consequently the lines of latitude do not appear as straight lines but are curved as shown. Day and night are of equal length.

Solar radiation. The intensity of solar radiation is greatly reduced in passing through the atmosphere before reaching the surface of the earth. Outside of the earth's atmosphere, rays of the sun falling on a surface normal to them can reach 1441 W/m^2 when the earth is closest to the sun in January and December (mid-winter) but decrease to a minimum value of around 428 W/m^2 on June 22. For reference purposes, the so-called solar constant or average value of the intensity of solar energy outside of the earth's atmosphere is usually taken as $I_{sc} = 1400\ W/m^2 = 444\ Btu/h \cdot ft^2 = 1204$ $kcal/hr \cdot m^2$.

The wave lengths of the solar energy, which is radiated, lie in a band ranging from approximately 0.29 to 3.5 μ, where the micron (μ) is one-millionth of a meter. The ultraviolet band of radiation ranges in wavelength from 0.29 to 0.40 μ and approximately 9% of the radiated energy of the sun resides in this range. The visible-wavelength band ranges from 0.4 to 0.7 μ and carries 40% of the solar energy. Infrared wavelengths lie on the band of 0.7 to 3.5 μ and represent 51% of the energy. When solar energy enters the atmosphere of the earth, nearly all of the ultraviolet energy is absorbed by the ozone in the outer atmosphere, so that only some 1% of the energy arriving at the surface of the earth lies in the ultraviolet. In addition to ultraviolet absorption at the outer ranges, water-vapor molecules, carbon dioxide molecules, and dust particles absorb and reflect this energy. The resulting scattered radiation, known as diffuse radiation, is transmitted even when clouds are present in the sky and direct sunlight cannot be seen. Appreciable diffuse radiation always exists at the surface of the earth in daylight hours.

In the NorthernHemisphere during spring and summer the sun appears to lie north of the equator, while for the other half of the year it appears south of the equator and then is visible for less than half the day. Figure 6-4 shows for an observer in the Northern Hemisphere the path that the sun appears to follow in its diurnal orbit at different times of the year. The diagram is not to scale, but to an observer at, say, 40°N latitude in June it indicates the sun rising in the northeast and traveling high in the sky to set in the northwest. In contrast in winter, the sun is low in the sky and moves southeast to southwest during the short hours of daylight. It can be observed in Fig. 6-3 that the Tropic of Cancer at 23.45° is the northernmost latitude at which the sun can be seen directly overhead.

Let us now refer to Fig. 6-5, which shows a portion of a vertical wall (or glass) at earth level with solar radiation impinging on it. As the earth rotates, the position of the wall relative to the sun changes. To describe accurately the position of the sun, relative to a surface on earth, two terms are extensively used, namely, the wall-solar azimuth angle ($\gamma = \angle ROH$) and the solar altitude angle ($\beta = \angle QOR$). The wall-solar azimuth angle

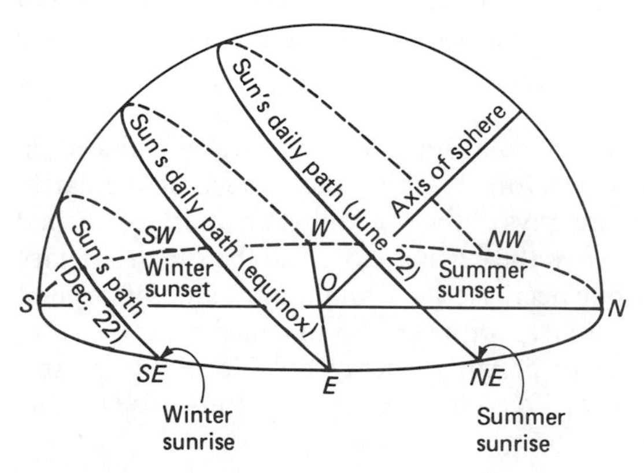

FIGURE 6-4. Apparent path of the sun at different seasons.

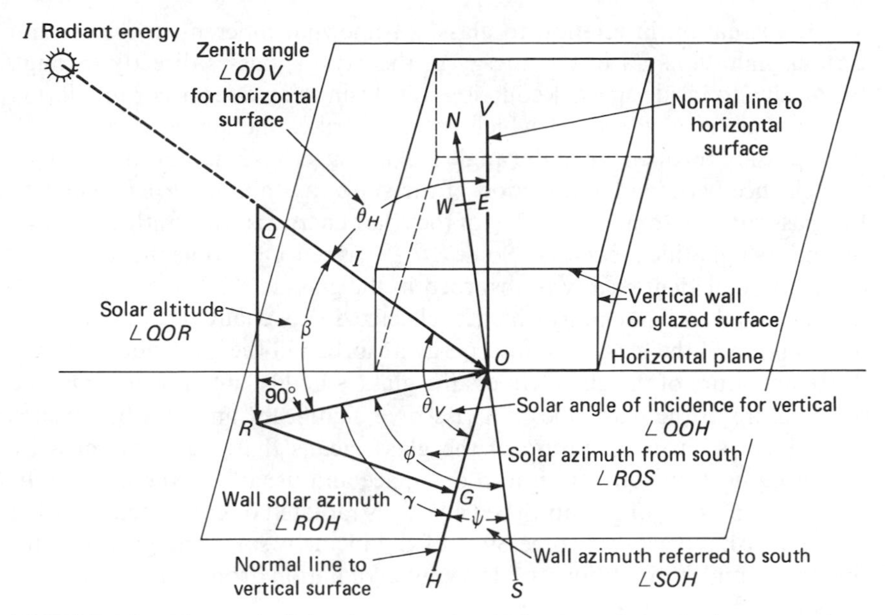

FIGURE 6-5. Diagram depicting the manner in which solar radiation impinges on surfaces.

necessarily changes with time of day. This angle lies in a horizontal plane between a perpendicular (*HO*) to the vertical (wall) and the vertical plane which passes through the sun and a point in the wall. In addition to the wall-solar azimuth angle, use is also made of the term solar-azimuth angle ($\phi = \angle ROS$). This is the angle between a south-north vertical plane and a vertical plane passing through the sun and the reference point in the wall.

The solar altitude angle (β) is the angle lying in a vertical plane between horizontal and a line passing through the point on earth and the sun. Solar altitude, in addition to changing with the time of day, varies with latitude and with season, the sun being higher in the sky in summer than in winter. For example, at 40°N latitude, solar-altitude angles at noon are as follows:

Jan. 21	Feb. 20	Mar. 20	Apr. 20	May 21	June 21
30°	39°	50°	61°	70°	73°
July 23	Aug. 24	Sept. 22	Oct. 23	Nov. 21	Dec. 22
70°	61°	50°	39°	30°	27°

In Fig. 6-5, let the length *QO* represent the direction and intensity, *I*, in watts per square meter or in Btu per hour per square foot, of the solar radiation impinging on the surface of a building or on its fenestration. The component radiation falling on a horizontal surface such as a roof is obviously

$$I_H = I \sin \beta \tag{6-2}$$

The component of this radiation on a vertical surface is the length *GO*, where its value is

$$\begin{aligned} I_v &= I \cos \theta_v = I \cos \beta \cos \gamma \\ I_v &= I \cos \beta \cos (\phi - \psi) \end{aligned} \tag{6-3}$$

The situation in relation to glass is somewhat different in that within certain limitations all of the energy in the rays, I, passes directly through the glazing to inside space. Reduction in transmission does occur in relation to the angle of incidence at which the sun strikes the glazed surface with most passing through when the angle (θ_v) in Fig. 6-5 is small. At a low angle of incidence between the direction of the sun's rays and a perpendicular to the glass surface, some 86 to 87% of the solar energy is transmitted through ordinary single-thickness glass. Some 8 to 9% is reflected while the remainder of the total incident energy is absorbed in the glass itself. The energy which is transmitted and the energy which is reflected do not change the temperature or condition of the glass, but any energy absorbed in the glass does increase the temperature of the glass. Unless the glass is highly tinted or opaque, the temperature of the glass does not rise by a significant amount, but even a slight increase in temperature of the glass means that the glass cools by delivering heat by convection into the space and usually to the outside. In general, some 40% of this absorbed energy is radiated or convected to inside and thus it happens that some 89% of the incident solar energy will enter the space under conditions of low-angle radiation from the sun on the glass.

When the altitude angle is steep, producing a high incident angle at, say, $\theta_v = 75$ to 80%, a much greater fraction of the solar energy is reflected, appreciably over half, and only some 40% is transmitted through the glass. This is a common situation, for example, on a south wall in the early morning or late afternoon and even around the noon period in summer.

The solar heat which passes through glass is largely absorbed by the interior furnishings, and by the inside walls and floors. Some of this heat is quickly given up and warms the inside air; but the heat which flows into heavy floors and walls may not be given up to the room air until hours later, so that the peak air temperature lags behind the peak of radiation from the sun. Glass, which is relatively diathermous to the radiant energy of the sun, acts as a screen preventing the outflow of low-temperature radiation from inside the building. Little sun energy which has entered a space can be reradiated out because of the "trap" effect of glass on low-temperature (long-wavelength) radiation.

COMPUTATION OF THE SOLAR LOAD. Extensive research and experimentation have been carried out on the relationships between solar phenomena and building response. It is obvious that with the many variables involved, extremely elaborate procedures for computing solar loading have resulted. To consider all of these variables brings in complexities that become too awkward for ordinary computation unless computer programs to handle them are brought into play. Even when this is done, the end result is still dependent upon the variability of solar effects, radiation, diffusivity, altered response in relation to adjacent buildings, trees, and the like. The ASHRAE *Handbook of Fundamentals* (Ref. 5) presents these detailed patterns most effectively and computer-program data are included. In this text a more simplified approach, which in itself is complex, is used for the solar computation. It has been found to give satisfactory results for design.

The heat delivery of sun radiation through glass may be materially reduced by proper use of awnings, Venetian blinds, or shades, thereby reducing the load for the air-conditioning equipment. When awnings or other barriers to direct solar rays are employed they are most effective when placed outside the glass. The effectiveness of various window-shading devices in reducing sun radiation can be observed from the factors F_S of Table 6-5.

Tables 6-1 and 6-2, with the help of Tables 6-3, 6-4, and 6-5, give a basis for finding the heat gain through glass under representative summer con-

Table 6-1 Instantaneous Rates of Heat Gain from Transmitted Direct and Diffuse or Sky Solar Radiation by Single-Sheet Unshaded Common Window Glass [For Clear Atmospheres and 18° Declination, North (August 1)]

Latitude	Sun Time		Instantaneous Heat Gain (Btu per hr sq ft)								
	A.M.→ ↓		N	NE	E	SE	S	SW	W	NW	Horiz.
30 Deg North	6 A.M.	6 P.M.	25	98	108	52	5	5	5	5	17
	7	5	23	155	190	110	10	10	10	10	71
	8	4	16	148	205	136	14	13	13	13	137
	9	3	16	106	180	136	21	15	15	15	195
	10	2	17	54	128	116	34	17	16	16	241
	11	1	18	20	59	78	45	19	18	18	267
	12		18	19	19	35	49	35	19	19	276
40 Deg North	5 A.M.	7 P.M.	3	7	6	2	0	0	0	0	1
	6	6	26	116	131	67	7	6	6	6	25
	7	5	16	149	195	124	11	10	10	10	77
	8	4	14	129	205	156	18	12	12	12	137
	9	3	15	79	180	162	42	14	14	14	188
	10	2	16	31	127	148	69	16	16	16	229
	11	1	17	18	58	113	90	23	17	17	252
	12		17	17	19	64	98	64	19	17	259
50 Deg North	5 A.M.	7 P.M.	20	54	54	20	3	3	3	3	6
	6	6	25	128	149	81	8	7	7	7	34
	7	5	12	139	197	136	12	10	10	10	80
	8	4	13	107	202	171	32	12	12	12	129
	9	3	14	54	176	183	72	14	14	14	173
	10	2	15	18	124	174	110	16	15	15	206
	11	1	16	16	57	143	136	42	16	16	227
	12		16	16	18	96	144	96	18	16	234
		↑ P.M.→	N	NW	W	SW	S	SE	E	NE	Horiz.

Note. For total instantaneous heat gain, add these values to the Table 6–2 values.

Table 6-2 Values of the Factor X: Instantaneous Rates of Heat Gain by Convection and Radiation from Single-Sheet Unshaded Common Window Glass [For Clear Atmospheres and 18° Declination, North (August 1); 80 F Indoor Temperature]

Sun Time	Dry-Bulb (deg F)	North Latitude Degrees	Instantaneous Heat Gain (Value X) (Btu per hr sq ft)								
			N	NE	E	SE	S	SW	W	NW	Horiz.
5 A.M.	74	↑	−6	−6	−6	−6	--6	−6	−6	−6	−6
6	74		−5	−4	−4	−5	−5	−6	−6	−6	−5
7	75		−5	−2	−2	−3	−5	−5	−5	−5	−3
8	77		−3	0	1	0	−2	−3	−3	−3	0
9	80		0	2	4	3	1	0	0	0	3
10	83		3	4	6	6	5	3	3	3	8
11	87		8	8	10	11	10	9	8	8	13
12	90	30, 40, 50	12	12	12	13	14	13	12	12	16
1 P.M.	93		15	15	15	16	17	17	17	15	20
2	94		16	16	16	16	18	19	19	17	21
3	95		17	17	17	17	19	21	21	19	21
4	94		16	16	16	16	17	20	20	19	19
5	93		15	15	15	15	15	18	19	18	17
6	91		13	13	13	13	13	14	15	15	13
7	87		8	8	8	8	8	8	8	8	8
8	85		6	6	6	6	6	6	6	6	6
9	83	↓	3	3	3	3	3	3	3	3	3

Note. For total instantaneous heat gain, add these values to the Table 6–1 values.

Table 6-3 Application Factors to Apply to Tables 6-1, 6-2, and 6-4 To Obtain Instantaneous Rates of Heat Gain for Various Types of Single Flat Glass and Combinations of Two Sheets of Flat Glass Spaced at $\frac{1}{4}$ in.

Glass *	Normal Incidence Transmittance	Factor to Apply to Table 6–1 (F_t)	Factors to Apply to Tables 6–2 and 6–3	
			C_1	C_2
Single common window	0.87	1.00	$1.0(X)$†	$+ 0.0(Y)$‡
Single regular plate	0.77	0.87	$1.0(X)$	$+ 0.25(Y)$
Single heat-absorbing plate	0.41	0.46§	$1.0(X)$	$+ 1.00(Y)$
Double common window	0 76	0.85	$0.6(X)$	$+ 0.10(Y)$
Double regular plate	0 60	0.66§	$0.6(X)$	$+ 0.55(Y)$
Heat-absorbing plate outdoors, Regular plate indoors	0.35	0.37§	$0.6(X)$	$+ 0.75(Y)$

*Common window glass $\frac{1}{8}$ in. thick. Plate glass $\frac{1}{4}$ in. thick.
†X values are Table 6–2 values.
‡Y Values are Table 6–4 values.
§For better precision, increase factors 10 per cent when glass is in the shade.

Table 6-4 Heat Absorbed in Glass. Values of Y To Be Used with Factors in Table 6-3 in the Determination of Instantaneous Rates in Heat Gain Due to Convection and Radiation for Various Types of Single Glass and Combinations of Two Sheets of Glass Spaced at $\frac{1}{4}$ in.
[For Clear Atmospheres and 18° Declination, North (August 1)]

Sun Time	Latitude	Values of Y (Btu per hr sq ft)*								
		N	NE	E	SE	S	SW	W	NW	Horiz.
5 A.M.	↑	0	0	1	0	0	0	0	0	0
6		4	16	18	9	1	1	1	1	3
7		2	24	30	20	2	2	2	2	11
8		2	22	33	25	2	2	2	2	21
9		2	16	30	29	8	3	3	3	32
10		3	5	25	27	14	3	3	3	37
11	40	3	3	12	21	18	3	3	3	42
12	Degrees	3	3	3	15	19	12	3	3	45
1 P.M.	North	3	3	3	3	19	22	10	3	44
2	Latitude	3	3	3	3	16	27	24	4	41
3		3	3	3	3	10	30	31	15	35
4		3	3	3	3	4	29	36	23	26
5		2	2	2	2	2	23	34	27	17
6		4	1	1	1	1	14	24	21	6
7	↓	0	0	0	0	0	2	3	3	1

Sun Time	Latitude	SE	S	SW	Sun Time	Latitude	SE	S	SW
5 A.M.	↑	0	0	0	5 A.M.	↑	2	0	0
6		7	1	1	6		13	1	1
7		18	2	2	7		22	2	2
8		22	2	2	8		28	3	2
9		24	3	3	9		·30	13	3
10	30†	22	5	3	10	50†	31	20	3
11	Degrees	16	7	3	11	Degrees	27	25	5
12	North	6	9	4	12	North	20	27	17
1 P.M.	Latitude	3	9	14	1 P.M.	Latitude	9	25	26
2		3	6	21	2		3	22	32
3		3	5	27	3		3	16	33
4		3	3	26	4		2	7	31
5		2	2	21	5		2	2	26
6		1	1	11	6		1	1	17
7	↓	0	0	0	7	↓	0	0	7

* Values of Y for 8 and 9 P.M. are zero.
† For N, NE, E, W, NW, and horizontal, use 40 deg north latitude values.

Table 6-5 Effect of Shading upon Instantaneous Solar Heat Gain through Single Thickness of Common Window Glass (F_S)

Type of Shading	Finish on Side Exposed to Sun	Fraction of Gain through Unshaded Window	
Canvas awning			
Sides open	Dark or medium	0.25	
Top and sides tight against building	Dark or medium	0.35	
Inside roller shade			
Fully drawn*	White, cream	0.41	
	Medium	0.62	
	Dark	0.81	
Half-drawn*	White, cream	0.71	
	Medium	0.81	
	Dark	0.91	
Inside Venetian blind, slats set at 45 deg†	White, cream	0.56	
	Diffuse-reflecting aluminum metal	0.45	
	Medium	0.65	
	Dark	0.75	
Outside Venetian blind			
Slats set at 45 deg†	White, cream	0.15	
Slats set at 45 deg,†‡ extended as awning fully covering window	White, cream	0.15	
Slats set at 45 deg, extended as awning covering $\frac{2}{3}$ of window‡	White, cream	0.43	
Outside shading screen			
Solar altitude		Dark§	Green Tint‖
10 deg	....	0.52	0.46
20 deg	....	0.40	0.35
30 deg	....	0.25	0.24
Above 40 deg	....	0.15	0.22

* Roller shades are assumed to be opaque. Some white shades may transmit considerable solar radiation. For white translucent shades fully drawn, use 0.55; when half-drawn, use 0.77.

† Venetian blinds are fully drawn and cover window. It is assumed that the occupant will adjust slats to prevent direct rays from passing between slats. If slats are fully closed (slats set at 90 deg), use same factors as used for roller shade fully drawn.

‡ Commercial shade with wide slats. The sun may shine on window through sides of shade. Estimate the exposed portion of glass as unshaded.

§ Commercial shade, bronze. Metal slats 0.05 in. wide, 17 per inch, and set at 17-deg angle with horizontal. At solar altitudes below 40 deg, some direct solar rays are allowed to pass between slats, and this amount becomes progressively greater at low solar altitudes.

‖ Commercial aluminum shade. Slats 0.057 in. wide, 17.5 per inch, set at 17-deg angle with horizontal. At solar altitudes below 40 deg, some direct solar rays are allowed to pass between slats and this amount becomes progressively greater at low solar altitude.

ditions.[1] Although the values indicated are for August, they can be used for other summer periods.

Example 6-1. Find the total instantaneous gain per square foot of single-sheet plate glass in a northwest wall at 3 P.M. in 40°N latitude location. Consider outside air at 95 F and inside conditions 80 Fdb.

Solution: Read, from Table 6-1, in the column labeled NW at the bottom, and opposite 40°N and 3 P.M., 79 Btu/h · ft² for the direct-transmitted and diffuse-radiation gain. Next, correct this value by a proper factor from Table 6-3—namely, 0.87—for single-plate glass: (79)(0.87) = 68.6. Read the convection and radiation gain from the glass surface, by use of Table 6-2, as 19. This reading is called the X value. For the second value, read from Table 6-4, in the NW column, and opposite 3 P.M. and 40°N latitude, the value 15; this is the Y value to use in the summation term $1.0X + 0.25Y$ of Table 6-3. Thus

$$(1.0)(19) + (0.25)(15) = 22.7$$

The total heat gain through the glass by direct transmittance, and by convection, and radiation from the glass surface, is thus

$$q_{\text{glass}} = (q_{\text{tr}})(F_t) + (C_1)(X) + (C_2)(Y) \tag{6-4}$$

Summing up, the total solar gain through the glass is

$$q_{\text{glass}} = (79)(0.87) + [(1.0)(19) + (0.25)(15)]$$

$$= 91.5 \text{ Btu/h} \cdot \text{ft}^2. \qquad \textit{Ans.}$$

This total in turn may have to be modified if shading is used, in which case an appropriate factor is selected from Table 6-5.

For walls and roofs the computations for solar heat gain are more complex, since the sun warms the surface and sets up a pulse flow of heat into the space as the solar warming rises to a peak and then falls off. The complex relationships associated with this problem can be somewhat simplified by use of the sol-air temperature concept, developed by Mackey and Wright (Refs. 1, 2). The sol-air temperature is that fictitious temperature of outside air which in the absence of all radiation effects would give the same rate of heat entry into the surface as would exist under the actual combination of incident solar radiation, radiant energy exchange with the sky and outdoor surroundings, and convective heat exchange with outdoor air. For example, for a dark-colored horizontal roof at 3 P.M., in full midsummer sun, the sol-air temperature might reach 132 F, in contrast with an air temperature of 95 F. The representative heat transfer through sun-activated walls and roofs is responsive to the character of the structure, to the daily range of temperature variation, to surface character (that is, dark or light), to the time of day, to latitude, and to other factors.

[1] Tables 6-1 through 6-8 have been reprinted, by permission, from *Heating Ventilating Air Conditioning Guide 1955*, chap. 13.

Table 6-6 Total Equivalent Temperature Differentials for Calculating Heat Gain through Sunlit and Shaded Walls

North Latitude Wall Facing:	Sun Time																		South Latitude Wall Facing:
	A.M.						P.M.												
	8		10		12		2		4		6		8		10		12		
	Exterior Color of Wall (D = Dark, L = Light)																		
	D	L	D	L	D	L	D	L	D	L	D	L	D	L	D	L	D	L	
	Frame																		
NE	22	10	24	12	14	10	12	10	14	14	14	14	10	10	6	4	2	2	SE
E	30	14	36	18	32	16	12	12	14	14	14	14	10	10	6	6	2	2	E
SE	13	6	26	16	28	18	24	16	16	14	14	14	10	10	6	4	2	2	NE
S	−4	−4	4	0	22	12	30	20	26	20	16	14	10	10	6	6	2	2	N
SW	−4	−4	0	−2	6	4	26	22	40	28	42	28	24	20	6	4	2	2	NW
W	−4	−4	0	0	6	6	20	12	40	28	48	34	22	22	8	8	2	2	W
NW	−4	−4	0	−2	6	4	12	10	24	20	40	26	34	24	6	4	2	2	SW
N (shade)	−4	−4	−2	−2	4	4	10	10	14	14	12	12	8	8	4	4	0	0	S (shade)
	4-In. Brick or Stone Veneer + Frame																		
NE	−2	−4	24	12	20	10	10	6	12	10	14	14	12	12	10	10	6	4	SE
E	2	0	30	14	31	17	14	14	12	12	14	14	12	12	10	8	6	6	E
SE	2	−2	20	10	28	16	26	16	18	14	14	14	12	12	10	8	6	6	NE
S	−4	−4	−2	−2	12	6	24	16	26	18	20	16	12	12	8	8	4	4	N
SW	0	−2	0	−2	2	2	12	8	32	22	36	26	34	24	10	8	6	6	NW
W	0	−2	0	0	4	2	10	8	26	18	40	28	42	28	16	14	6	6	W
NW	−4	−4	−2	−2	2	2	8	6	12	12	30	22	34	24	12	10	6	6	SW
N (shade)	−4	−4	−2	−2	0	0	6	6	10	10	12	12	12	12	8	8	4	4	S (shade)
	8-In. Hollow Tile or 8-In. Cinder Block																		
NE	0	0	0	0	20	10	16	10	10	6	12	10	14	12	12	10	8	8	SE
E	4	2	12	4	24	12	26	14	20	12	12	10	14	12	14	10	10	8	E
SE	2	0	2	0	16	8	20	12	20	14	14	12	14	12	12	10	8	6	NE
S	0	0	0	0	2	0	12	6	24	14	26	16	20	14	12	10	8	6	N
SW	2	0	2	0	2	0	6	4	12	10	26	18	30	20	26	18	8	6	NW
W	4	2	4	2	4	2	6	4	10	8	18	14	30	22	32	22	18	14	W
NW	0	0	0	0	2	0	4	2	8	6	12	10	22	18	30	22	10	8	SW
N (shade)	−2	−2	−2	−2	−2	−2	0	0	6	6	10	10	10	10	10	10	6	6	S (shade)
	8-In. Brick or 12-In. Hollow Tile or 12-In. Cinder Block																		
NE	2	2	2	2	10	2	16	8	14	8	10	6	10	8	10	10	10	8	SE
E	8	6	8	6	14	8	18	10	18	10	14	8	14	10	14	10	12	10	E
SE	8	4	6	4	6	4	14	10	18	12	16	12	12	10	12	10	12	10	NE
S	4	2	4	2	4	2	4	2	10	6	16	10	16	12	12	10	10	8	N
SW	8	4	6	4	6	4	8	4	10	6	12	8	20	12	24	16	20	14	NW
W	8	4	6	4	6	6	8	6	10	6	14	8	20	16	24	16	24	16	W
NW	2	2	2	2	2	2	4	2	6	4	8	6	10	8	16	14	18	14	SW
N (shade)	0	0	0	0	0	0	0	0	2	2	6	6	8	8	8	8	6	6	S (shade)
	12-In. Brick																		
NE	8	6	8	6	8	4	8	4	10	4	12	6	12	6	10	6	10	6	SE
E	12	8	12	8	12	8	10	6	12	8	14	10	14	10	14	8	14	8	E
SE	10	6	10	6	10	6	10	6	10	6	12	8	14	10	14	10	12	8	NE
S	8	6	8	6	6	4	6	4	6	4	8	4	10	6	12	8	12	8	N
SW	10	6	10	6	10	6	10	6	10	6	10	8	10	8	12	8	14	10	NW
W	12	8	12	8	12	8	10	6	10	6	10	6	10	6	12	8	16	10	W
NW	8	6	8	6	8	4	8	4	8	4	8	4	8	6	10	6	10	6	SW
N (shade)	4	4	2	2	2	2	2	2	2	2	2	2	2	2	4	4	6	6	S (shade)

Table 6-6 (*Continued*)

North Latitude Wall Facing:	Sun Time																		South Latitude Wall Facing:
	A.M.						P.M.												
	8		10		12		2		4		6		8		10		12		
	Exterior Color of Wall (D = Dark, L = Light)																		
	D	L	D	L	D	L	D	L	D	L	D	L	D	L	D	L	D	L	
8-In. Concrete or Stone or 6-In. or 8-In. Concrete Block																			
NE	4	2	4	0	16	8	14	8	10	6	12	8	12	10	10	8	8	6	SE
E	6	4	14	8	24	12	24	12	18	10	14	10	14	10	12	10	10	8	E
SE	6	2	6	4	16	10	18	12	18	12	14	12	12	10	12	10	10	8	NE
S	2	1	2	1	4	1	12	6	16	12	18	12	14	12	10	8	8	6	N
SW	6	2	4	2	6	2	8	4	14	10	22	16	24	16	22	16	10	8	NW
W	6	4	6	4	6	4	8	6	12	8	20	14	28	18	26	18	14	10	W
NW	4	2	4	0	4	2	4	4	6	6	12	10	20	14	22	16	8	6	SW
N (shade)	0	0	0	0	0	0	2	2	4	4	6	6	8	8	6	6	4	4	S (shade)
12-In. Concrete or Stone																			
NE	6	4	6	2	6	2	14	8	14	8	10	8	10	8	12	10	10	8	SE
E	10	6	8	6	10	6	18	10	18	12	16	10	12	10	14	10	14	10	E
SE	8	4	8	4	6	4	14	8	16	10	16	10	14	10	12	10	12	10	NE
S	6	4	4	2	4	2	4	2	10	6	14	10	16	12	14	10	10	8	N
SW	8	4	8	4	6	4	6	4	8	6	10	8	18	14	20	14	18	12	NW
W	10	6	8	6	8	6	10	6	10	6	12	8	16	10	24	14	22	14	W
NW	6	4	6	2	6	2	6	4	6	4	8	6	10	8	18	12	20	14	SW
N (shade)	0	0	0	0	0	0	0	0	2	2	4	4	6	6	8	8	6	6	S (shade)

Notes:

{Total heat transmission from solar radiation and temperature difference, Btu per hr sq ft wall area} = {Temperature differential from table} × {Heat-transmission coefficient for wall, Btu per hr sq ft deg F*}

***Select coefficients from proper tables in Chapter 4.**

1. *SOURCE.* Same as Table 6-7. A north wall has been assumed to be a wall in the shade; this is practically true. Dark colors on exterior surface of walls have been assumed to absorb 90 per cent of solar radiation and reflect 10 per cent; white colors, to absorb 50 per cent and reflect 50 per cent. This includes some allowance for dust and dirt, since clean, fresh white paint normally absorbs only 40 per cent of solar radiation.

2. *APPLICATION.* These values may be used for all normal air-conditioning estimates, usually without corrections, when the load is calculated for the hottest weather. Correction for latitude (note 3) is necessary only where extreme accuracy is required. There may be jobs where the indoor room temperature is considerably above or below 80 F, or where the outdoor design temperature is considerably above 95 F, in which case it may be desirable to make correction to the temperature differentials shown. The solar intensity on all walls other than east and west varies considerably with time of year.

3. *CORRECTIONS. Outdoor minus Room Temperature.* If the outdoor maximum design temperature minus room temperature is different from the base of 15 deg, correct as follows: When the difference is greater (or less) than 15 deg, add the excess to (or subtract the deficiency from) the above differentials.

Outdoor Daily Range Temperature. If the daily range of temperature is less than 20 deg, add 1 deg to every 2 deg lower daily range; if the daily range is greater than 20 deg, subtract 1 deg for every 2 deg higher daily range. For example, the daily range in Miami is 12 deg, or 8 deg less than 20 deg; therefore the correction is +4 deg.

Color of Exterior Surface of Wall. Use temperature differentials for light walls only where the permanence of the light wall is established by experience. For cream colors use the values for light walls. For medium colors interpolate halfway between the dark and light values. Medium colors are medium blue, medium green, bright red, light brown, unpainted wood, natural-color concrete, etc. Dark blue, red, brown, green, etc., are considered dark colors.

Latitudes Other Than 40 Deg North, and in Other Months. These table values will be approximately correct for the east or west wall in any latitude (0 deg to 50 deg north or south) during the hottest weather.

4. *INSULATED WALLS.* Use same temperature differentials used for uninsulated walls.

Table 6-7 Total Equivalent Temperature Differentials for Calculating Heat Gain through Sunlit and Shaded Roofs

DESCRIPTION OF ROOF CONSTRUCTION*	SUN TIME								
	A.M.			P.M.					
	8	10	12	2	4	6	8	10	12
Light-Construction Roofs—Exposed to Sun									
1″ wood,† or 1″ wood† + 1″ or 2″ insulation	12	38	54	62	50	26	10	4	0
Medium-Construction Roofs—Exposed to Sun									
2″ concrete, or 2″ concrete + 1″ or 2″ insulation, or 2″ wood†	6	30	48	58	50	32	14	6	2
2″ gypsum, or 2″ gypsum + 1″ insulation, or 1″ wood† or / 2″ wood† or / 2″ concrete or / 2″ gypsum } + 4″ rock wool in furred ceiling	0	20	40	52	54	42	20	10	6
4″ concrete, or 4″ concrete with 2″ insulation	0	20	38	50	52	40	22	12	6
Heavy-Construction Roofs—Exposed to Sun									
6″ concrete	4	6	24	38	46	44	32	18	12
6″ concrete + 2″ insulation	6	6	20	34	42	44	34	20	14
Roofs Covered with Water—Exposed to Sun									
Light-construction roof with 1″ water	0	4	16	22	18	14	10	2	0
Heavy-construction roof with 1″ water	−2	−2	−4	10	14	16	14	10	6
Any roof with 6″ water	−2	0	0	6	10	10	8	4	0
Roofs with Roof Sprays—Exposed to Sun									
Light construction	0	4	12	18	16	14	10	2	0
Heavy construction	−2	−2	2	8	12	14	12	10	6
Roofs in Shade									
Light construction	−4	0	6	12	14	12	8	2	0
Medium construction	−4	−2	2	8	12	12	10	6	2
Heavy construction	−2	−2	0	4	8	10	10	8	4

Table 6-7 (*Continued*)

NOTES:

$$\left\{\begin{array}{l}\text{Total heat transmission from} \\ \text{solar radiation and temperature} \\ \text{difference, Btu per hr sq ft of} \\ \text{roof area}\end{array}\right\} = \left\{\begin{array}{l}\text{Temperature differential} \\ \text{from table}\end{array}\right\} \times \left\{\begin{array}{l}\text{Heat transmission co-} \\ \text{efficient for summer,} \\ \text{Btu per hr sq ft deg F}\end{array}\right\}$$

1. *SOURCE.* Calculated by Mackey and Wright method. Estimated for July in 40 deg north latitude. For typical design day, where the maximum outdoor temperature is 95 F and minimum temperature at night is approximately 75 F (daily range of temperature, 20 F); mean 24 hr temperature 84 F for a room temperature of 80 F. All roofs have been assumed a dark color which absorbs 90 per cent of solar radiation, and reflects only 10 per cent.

2. *APPLICATION.* These values may be used for all normal air conditioning estimates, usually without correction, in latitude 0 deg to 50 deg north or south when the load is calculated for the hottest weather. Note 5 explains how to adjust the temperature differential for other room and outdoor temperatures.

3. *PEAKED ROOFS.* If the roof is peaked and the heat gain is primarily due to solar radiation, use for the area of the roof the area projected on a horizontal plane.

4. *ATTICS.* If the ceiling is insulated and if a fan is used in the attic for positive ventilation, the total temperature differential for a roof exposed to the sun may be decreased 25 per cent.

5. *CORRECTIONS. For Temperature Difference When Outdoor Maximum Design Temperature Minus Room Temperature Is Other Than 15 Deg.* If the outdoor design temperature minus room temperature is other than the base of 15 deg, correct as follows: When the difference is greater (or less) than 15 deg add the excess to (or subtract the deficiency from) the above differentials.

For Outdoor Daily Range of Temperature Other Than 20 Deg. If the daily range of temperature is less than 20 deg, add 1 deg for every 2 deg lower daily range; if the daily range is greater than 20 deg, subtract 1 deg for every 2 deg higher daily range. For example, the daily range in Miami is 12 deg, or 8 deg less than 20 deg; therefore the correction is +4 deg at all hours of the day.

For Light Colors. Credit should not be taken for light-colored roofs except where the permanence of the light color is established by experience, as in rural areas or where there is little smoke. When the exterior surface of roof exposed to the sun is a light color, such as white or aluminum (which absorb approximately 50 per cent and reflect 50 per cent of the solar radiation), add to the temperature differential for a roof in the shade 55 per cent of the difference between the roof in the sun and the roof in the shade. When the roof exposed to the sun is a medium color such as light gray, or blue, or green, or bright red, add 80 per cent of this difference.

* Includes ⅜ in. felt roofing with or without slag. May also be used for shingle roof.

† Nominal thickness of the wood.

For solving this complex problem of solar heat gain, tables have been prepared which give the representative temperature differential to use with a building structure to find the total heat transmission resulting from the combination of air-temperature difference and solar effects. The tables are based on a 15-deg (95 F − 80 F) design datum for ordinary air-temperature difference. If the air-temperature difference is other than 15 deg, a correction must be made. Tables 6-6, 6-7, and 6-8 are to be employed for walls and roofs in finding all transmission summer space heat gains where solar effects are a factor.

Example 6-2. Find the total heat gain at 4 P.M. through a roof of 6-in. concrete and 2-in. insulation when the inside temperature is 80 F and the outside design temperature is 97 F.

Solution: Read 42 deg from Table 6-7 and, from Table 6-8, $U = 0.13$. Then the uncorrected gain from solar effect is

$$q = (0.13)(42) = 5.5 \text{ Btu/h} \cdot \text{ft}^2$$

A correction should be made because the design air-temperature difference is 97 F − 80 F = 17 deg instead of the 15 deg of the table. Use note 5 of Table 6-7 and add the 2-deg excess, so that

$$q = (0.13)(42 + 2) = 5.7 \text{ Btu/h} \cdot \text{ft}^2 \qquad \textit{Ans.}$$

Table 6-8 Summer Coefficients of Heat Transmission U of Flat Roofs Covered with Built-up Roofing*
(Btu/h · ft^2 of F Temperature Difference between the Air on the Two Sides)

Type of Roof Deck (ceiling not shown)	Thickness of Roof Deck (in.)		Insulation on Top of Deck (covered with built-up roofing): No Ceiling—Underside of Roof Exposed					Furred Ceiling with Air Space, Metal Lath, and Plaster				
			No Insulation	Insulating Board † Thickness (in.)				No Insulation	Insulating Board † Thickness (in.)			
				½	1	1½	2		½	1	1½	2
Flat metal roof deck (Roofing, Insulation, Metal deck)	4-ply felt roof		0.73	0.35	0.23	0.17	0.13	0.40	0.25	0.18	0.14	0.12
	Ditto + ½-in. slag		0.54	0.30	0.20	0.16	0.13	0.34	0.22	0.16	0.13	0.11
Precast cement tile (Roofing, Insulation, Cast tile)	4-ply felt roof	1⅝	0.67	0.33	0.22	0.17	0.13	0.38	0.24	0.18	0.14	0.12
	Ditto + ½-in. slag	1⅝	0.50	0.28	0.20	0.15	0.12	0.32	0.21	0.17	0.13	0.11
Concrete (Roofing, Insulation, Concrete)	4-ply felt roof	2	0.65	0.33	0.22	0.16	0.13	0.37	0.24	0.18	0.14	0.12
		4	0.59	0.31	0.21	0.16	0.13	0.36	0.23	0.17	0.13	0.12
		6	0.54	0.30	0.20	0.16	0.13	0.33	0.22	0.17	0.13	0.11
	Ditto + ½-in. slag	2	0.49	0.28	0.20	0.15	0.12	0.31	0.21	0.16	0.13	0.11
		4	0.46	0.27	0.19	0.15	0.12	0.30	0.21	0.16	0.13	0.11
		6	0.42	0.26	0.19	0.14	0.12	0.29	0.20	0.16	0.13	0.10
Gypsum and wood fiber‡ on ½-in. gypsum board (Insulation, Roofing, Gypsum, Plaster board)	4-ply felt roof	2½	0.34	0.23	0.17	0.13	0.12	0.25	0.18	0.14	0.12	0.097
		3½	0.28	0.20	0.15	0.12	0.11	0.21	0.16	0.13	0.11	0.094
	Ditto + ½-in. slag	2½	0.29	0.20	0.16	0.13	0.11	0.22	0.16	0.13	0.11	0.093
		3½	0.25	0.18	0.14	0.12	0.10	0.19	0.15	0.13	0.10	0.090
Wood§ (Roofing, Insulation, Sheathing)	4-ply felt roof	1	0.43	0.26	0.19	0.15	0.12	0.29	0.20	0.15	0.13	0.11
		1½	0.33	0.22	0.17	0.13	0.11	0.24	0.18	0.14	0.12	0.097
		2	0.29	0.20	0.16	0.13	0.11	0.22	0.16	0.13	0.11	0.094
		3	0.22	0.16	0.13	0.11	0.09	0.17	0.13	0.12	0.10	0.085
	Ditto + ½-in. slag	1	0.35	0.23	0.17	0.14	0.11	0.25	0.18	0.14	0.12	0.10
		1½	0.29	0.20	0.15	0.12	0.10	0.21	0.17	0.13	0.11	0.093
		2	0.26	0.19	0.14	0.12	0.10	0.20	0.15	0.13	0.10	0.090
		3	0.20	0.15	0.12	0.10	0.09	0.16	0.13	0.11	0.09	0.081

* The summer coefficients have been calculated with an outdoor wind velocity of 8 mph. For summer an inside surface conductance of 1.2 has been used instead of the regular 1.65 value. In all of these roofs a 4-ply felt roof has been assumed ⅜ in. thick, thermal conductivity = 1.33. Pitch and slag have been assumed as an additional thickness of ½ in. which has been assigned a thermal conductivity of 1.0. In both cases thermal conductivity refers to one *inch* thickness.

† If corkboard insulation is used, the coefficient U may be decreased 10 per cent.

‡ 87½ per cent gypsum, 12½ per cent wood fiber. Thickness indicated includes ½ in. gypsum board. This is a poured roof.

§ Nominal thickness of wood is specified, but actual thickness was used in calculations.

A similar pattern applies for wall surfaces, for which Table 6-6 is used. The foregoing analysis of the solar and transmission load represents a method of accounting for the many variables which enter into the heat-flow pattern, and in connection with this pattern it should be realized that, during portions of the time, heat is flowing *into* the building from solar effects and thus a gradual *warming* of the masonry and structure is taking place, while at the other times heat is flowing *out of* the building and thus a gradual *cooling* of the masonry and structure is occurring. Often the full effect of direct sunlight falling on a surface does not occur until several hours subsequent to the exposure. Solar transmittance through glass causes a quick response, but on heavy, well-insulated walls the response can be much delayed. The tabular values in this chapter, when properly employed, partly compensate for the effect of the variables. However, in selecting the time at which the maximum load does occur from external effects, it is quite important to study the tables, particularly Tables 6-1, 6-2, and 6-7, to find the time at which the maximum condition does occur.

In the case of inside rooms, where heat flows through partitions, Eq. (6-1) can be employed directly, and in this equation the t_o term should, of course, be the temperature on the far side of a barrier wall.

6-4. LOAD CONTRIBUTION FROM INFILTRATION AIR, OCCUPANTS, AND MISCELLANEOUS SOURCES

HEAT AND MOISTURE INTRODUCED IN INFILTRATION AIR. Infiltration air enters by leakage through window cracks, through doors when opened, and through porous walls or other openings. Table 5-3 furnishes data on infiltration air through window cracks, and Table 6-9 gives values for determining the amount of air entering through door openings in commercial establishments.

In the case of an air-conditioning system in which there is an adequate supply of ventilating air, and in which the conditioned air is delivered into a space which has no exhaust fans, the door and infiltration loss will be greatly reduced, and may even cease, when the air supplied to the space flows outward. Expressed in other terms, whenever the fan system provides an excess pressure in the enclosure, infiltration is reduced or eliminated. Infiltration air which enters a space brings with it not only the high-temperature outside air, with its associated sensible heat as it cools to inside temperature, but also the moisture associated with the frequently humid outside air, which entails a latent heat load on the space.

Independent of infiltration air, the air-conditioning system must circulate a certain amount of fresh air required for ventilation, while it must also recirculate an amount of air sufficiently large that the total passed through the conditioning equipment and delivered into the space will absorb the sensible load from all sources and, at the same time, the supply air will increase in moisture content and absorb the latent heat load.

Table 6-9 Door Infiltration in Summer for Commercial Establishments

Revolving and Swinging Doors Opening to Outside			Average Occupancy (Patrons and Employees) on Which Values Are Based (min.)
Application	Infiltration per Person in Room (cfm)		
	72-In. Revolving Door	36-In. Swinging Door	
Bank	7.5	10.0	20
Barber shop	3.5	4.5	45
Broker's office	5.0	6.5	30
Candy and soda store	5.0	6.5	30
Cigar store	15.0	20.0	10
Department store (small)	5.0	6.5	30
Dress shop	2.0	2.5	75
Drugstore	10.0	13.0	15
Furrier	2.0	2.0	90
Lunchroom	5.0	6.5	30
Men's shop	3.5	4.5	45
Office (professional)	2.5	3.0	60
Restaurant	2.0	2.5	75
Shoe store	3.5	4.5	45

When Doors Are Left Open Continuously	
72-in. revolving door (panels open)	1200 cfm
36-in. swinging door (standing open)	800 cfm

NOTE. The values for swinging doors and for doors left open hold only where such doors are in one wall only, or where the doors in other walls are the revolving type. If swinging doors are used for access (or if doors are left open) in more than one wall, the infiltration cannot be estimated. The values for revolving doors hold regardless of number or location.

To determine the total cfm infiltration due to opening of doors, multiply the design number of occupants by the factor from the above table for the kind of establishment in question. When there is more than one door, treat them as though there were only one, except in the case of open doors.

* Reprinted, by permission, from code of Air Conditioning and Refrigerating Machinery Association.

HEAT LOAD ARISING FROM OCCUPANTS IN A GIVEN SPACE. The heat load arising from occupants in a given space can be computed with the help of values selected from Table 10-3. The heat load should be considered in terms of the type of activity of the individuals, and it should be divided into two parts—namely, that part associated with sensible cooling of the individual, and that part associated with latent cooling of the individual.

HEAT LOAD FROM MISCELLANEOUS EQUIPMENT INSTALLED IN A GIVEN SPACE. Data for computation of the heat load from miscellaneous equipment installed in a conditioned space are given in Table 6-10. This particular part of the heat load should always be considered in computing contributions to the cooling load, and one must carefully consider whether it is all sensible, or partly sensible and partly latent. Moreover, care must be exercised to ascertain whether or not this part of the heat load really appears within the space. For example, in a conditioned space a motor driving a shaft to which the load is attached in another room delivers only its electrical and mechanical losses into the conditioned space, and the shaft work (useful horsepower) is distributed elsewhere. Similarly, if a gas burner in a conditioned space is heating water or other material which is used elsewhere, all of the thermal energy combustion should not be charged to the conditioned space.

The heat equivalent of the energy delivered to air by a fan in sending the air through the ductwork usually appears in the conditioned space, and becomes part of the cooling load. If the air horsepower exerted on the air is known, this heat can be easily calculated. However, a 10% additional amount is often added to the sensible-heat load of a space to take care of contingencies, and the fan load is assumed in this amount.

Air in ductwork running through unconditioned space absorbs a certain amount of heat and entails an additional load on the conditioning system.

Table 6-10 Heat Load from Equipment

Device	Heat Dissipation During Running Time (Btuh)	
	Sensible Heat	Latent Heat
Electric lights and appliances, per installed kilowatt	3413	....
Motors, with connected load in same room,* per horsepower		
$\frac{1}{8}$–$\frac{1}{2}$ hp range	4250	....
$\frac{1}{2}$–3 hp range	3700	....
3–20 hp range	2950	....
Electric coffee urn		
3-gal	2200	1500
5-gal	3400	2300
Gas stove burner	3100	1700
Heating water	3150	3850
Domestic gas oven	8100	4000
Gas-heated coffee urn		
3-gal	2500	2500
5-gal	3900	3900
Steam-heated equipment, per square foot†		
Steam-heated surface		
Not polished	330	....
Polished	130	....
Insulated surface	80	....
Steam table	200	1000
Beauty parlor hair-driers		
Blower-type	2300	400
Helmet-type	1870	330
Restaurant, per meal served	30 [Btu]	....

* With connected load outside room, subtract 2544 Btuh per rated horsepower.
† For hooded equipment, reduce values by 50 per cent.

With long runs this heating may be considerable and should always be calculated. In Section 12-6, methods of calculation are shown.

6-5. VENTILATION AIR AND AIR COOLING

Whenever a space is to be air conditioned, ventilation air must be supplied in sufficient quantity to satisfy codes or ordinances when these are applicable; but even when such codes are not applicable, sufficient ventilation air must be supplied to control the odor level and to set up satisfactory conditions for occupants or for stored equipment. From a viewpoint of cooling-load design, the occupants of a space should be provided with not less than 5 to $7\frac{1}{2}$ cfm per person when no smoking is involved, and 25 to 40 cfm per person if smoking is a factor. Table 10-4 gives suggested values.

The ventilation load is independent of the internal space load if the ventilation air is passed through the conditioner, since under these circumstances the air is cooled before entering the space. This is in contrast to the warm infiltration air which enters directly into the space and must be removed along with the other heat loads in the space.

In analyzing the performance of cooling equipment one method is to use the so-called *by-pass factor*. When using this method, it is considered that a portion of the air moves through a cooler as though unaltered by the action of the coil, while the remainder of the air is brought completely to the coil temperature. The first part of the air is known as the by-pass air and, if considered unaltered, it can be regarded as entering directly into the conditioned space load in exactly the same way as does infiltration air. The mixture leaving is, of course, an average of the two types of air.

Manufacturers frequently describe their cooling equipment in terms of coil efficiency, which is merely another way of stating the by-pass factor. For example, to say that a coil is 75% efficient means that 25% of the air by-passes the coil, while the remaining 75% is brought to coil temperature. Thus if the by-pass analysis for cooling is used, the by-passed air would be considered part of the space load; whereas, using another viewpoint, the overall action of the conditioner is considered merely to produce supply air at its mixed temperature to serve the given space. The cooling (heat-source) load of the space is specifically called either the space load or the internal cooling load.

6-6. COOLING-LOAD CALCULATIONS

In this section, examples will be worked to illustrate detailed methods of computing the cooling load of a space. Figure 6-6 shows a representative space which will be analyzed. For the conditioned space shown in the illustration, the items entering into the cooling load are:

1. Solar-type transmission loads on the east and south walls and on the roof of the building.
2. Solar-type transmission through the glazed portions of the east and south walls.

3. Wall-transmission heat gain through the north and west partition walls and through the floor.
4. Heat gain from occupants in the space.
5. Heat gain from miscellaneous sources and equipment.
6. Heat gain from infiltration air.

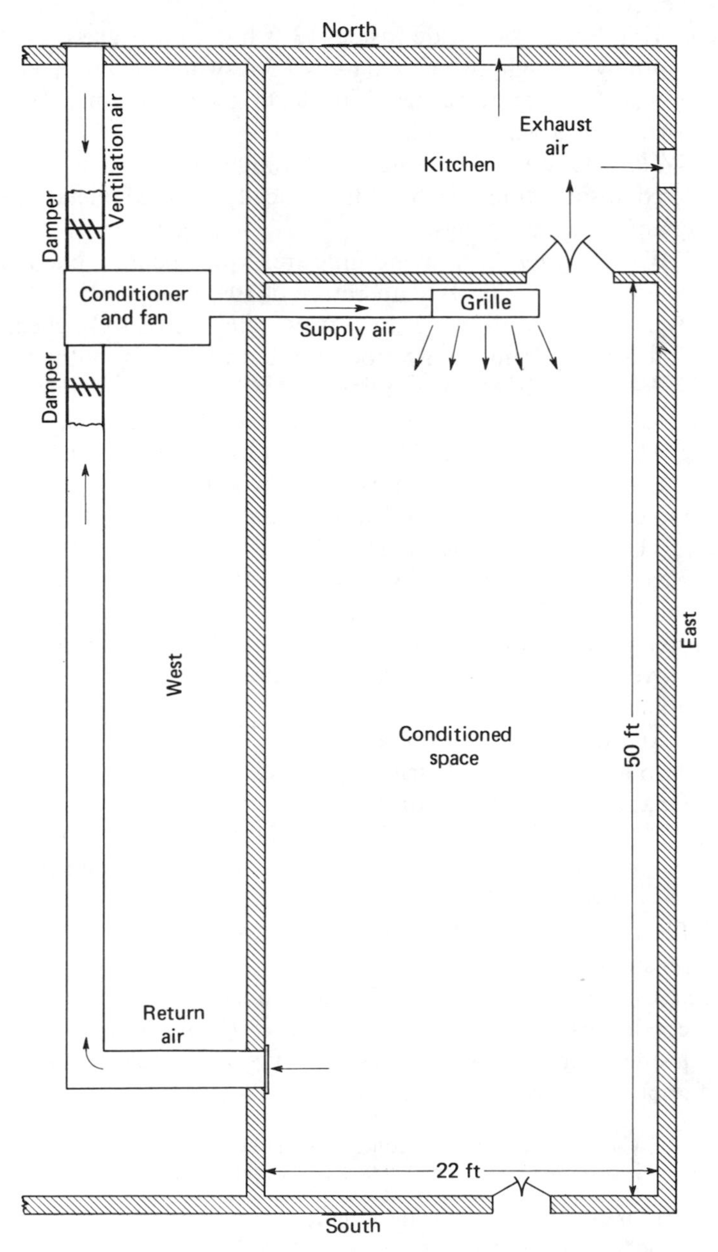

FIGURE 6-6. Representative space, to illustrate layout of a building for cooling load. Conditioner in adjacent space or basement or on roof.

The specifications for the conditioned space diagrammed in Fig. 6-6 are summarized as follows:

Specifications of space to be air conditioned in restaurant building

Location. Philadelphia, Pa., on corner; entrance and front of building face south.

South Wall. Front 22 ft wide inside, 13 ft high; plate glass, 14 ft total width by 8 ft high; two glazed 6- by 7.5-ft swinging doors; remainder of wall 12-in. brick plastered inside, thin synthetic marble outside. Awning over whole front.

East Wall. Side street; 50 ft long by 13 ft high inside; 12-in. brick, plastered inside; three 3- by 4-ft vertically pivoted metal windows equipped with awnings.

North-Wall Partition. Separates dining area from unconditioned kitchen, where a possible peak temperature of 109 F may occur; 22 ft wide by 13 ft high, with two 3- by 7-ft swinging doors to kitchen. Partition is wood frame with wood lath and plaster on both sides, and rock-wool fill between studding.

West Wall. Adjacent to unconditioned space; 12-in. brick with $\frac{3}{4}$-in. plaster on one side; 50 ft long by 13 ft high (inside dimensions).

Floor. Conventional wooden joists with wood lath and plaster below and covered with $\frac{1}{4}$-in. linoleum laid on wood flooring.

Ceiling. Under roof. Construction: 4-in. wood rafters, metal lath and plaster below, topped by 1-in. wooden roof deck, covered by roofing paper.

Occupancy. Space for total of 50 people at tables and counters; 5 employees in conditioned part of restaurant. Peak occupancy during noon hour.

Equipment. In conditioned space:

Two gas-heated coffee urns, 3-gal size.

One toaster, 2650-W rating.

(Above items hooded and vented to outside, using small exhaust fan, mounted on wall outside conditioned space)

No motors.

Electric lights: 2400 W total; never in use during peak thermal load on sunny day.

Design Conditions. From Table 5-2, for Philadelphia the outside design conditions are taken as 95 Fdb and 78 Fwb. From Table 10-1, the inside design conditions for 95 F outside are selected, for normal application, as 80 Fdb and 67 Fwb.

Example 6-3. Calculate the transmission and solar heat gains into the restaurant at peak occupancy load about 12 noon (1 : 00 P.M. daylight time).

Solution: The transmission through the walls follows the general conduction formula [(Eq. (6-1)], modified, where necessary, by solar effects:

$$Q = UA(t_o - t_i) = UA\Delta t$$

South wall. The gross area of the wall = 22 × 13 ft = 286 ft^2. The glass area = 14 × 8 ft = 112 ft^2, and the area of the door (glazed) = 6 × 7.5 ft = 45 ft^2. From Table 4-6, for the 12-in. brick wall with plaster, $U = 0.34$; and from Table 6-6, read 4 deg for the total equivalent temperature difference at 12 noon, sun time. Thus, for the net wall, the heat transmitted is

$$Q_w = (0.34)(286 - 112 - 45)(4) = 175 \text{ Btuh}$$

For the total glass area of 157 ft^2, read (for 12 noon at 40°N latitude for a south wall) the transmittance as 98 Btu/h · ft^2. Correct this by the factor $F_t = 0.87$ for plate glass, from Table 6-3, and by a factor of $F_S = 0.25$ for the awning; thus the transmittance for the glass is

$$(98)(0.87)(0.25) = 21.3 \text{ Btu/h} \cdot \text{ft}^2$$

From Table 6-2, the glass convection and radiation gain at noon is 14 for the X value, and in Table 6-4 the Y value is 19. Thus the correct convection and radiation value (not greatly reduced by the awning) is, for the plate glass,

$$1.00(X) + 0.25(Y) = 14 + 0.25(19) = 18.7 \text{ Btu/h} \cdot \text{ft}^2$$

The glass gain is therefore

$$(21.3 + 18.7)(157) = 6280 \text{ Btuh}$$

Thus the total gain for the south wall is

$$175 + 6280 = 6455 \text{ Btuh}$$

East wall. The gross area of the wall = 50 × 13 ft = 650 ft^2. The glass area = 3(3 × 4 ft) = 36 ft^2. Thus the net wall area is 614 ft^2. By Table 6-6, for the east wall (dark material) at noon, Δt = 12 deg for 12-in. brick. Thus for the net wall, the heat transmitted is

$$Q_w = (0.34)(614)(12) = 2510 \text{ Btuh}$$

For the glass, use Table 6-1 and read 19 for the east wall at noon; by Table 6-2, read 12; and by Table 6-3, the factor $F_t = 1.0$. For the awning, by Table 6-5, read $F_S = 0.25$. Thus the glass gain per square foot is

$$(19)(F_t)(F_S) + 1.0(X) + 0.0(Y) = (19)(1.0)(0.25) + (1.0)(12) = 16.7$$

The glass gain is therefore

$$(16.7)(36) = 602 \text{ Btuh}$$

Thus the total gain for the east wall is

$$2510 + 602 = 3112 \text{ Btuh}$$

North-wall partition. The gross area of the partition = 22 × 13 ft = 286 ft^2. The area of the thin panel doors = 2(3 × 7 ft) = 42 ft^2. Thus the overall coefficient of heat transfer for the partition (U_w) is 0.078 (Tables 4-8 and 4-5), and for the door (U_{door}), 0.85 (Table 4-12 footnote). For the partition, the heat transferred is therefore

$$Q_w = 0.078(286 - 42)(109 - 80) = 552 \text{ Btuh}$$

And, for the door,

$$Q_{door} = (0.85)(42)(109 - 80) = 1034 \text{ Btuh}$$

Thus the total north-wall heat gain is

$$552 + 1034 = 1586 \text{ Btuh}$$

West wall. The area of the wall = 50 × 13 ft = 650 ft², and the wall is next to a building space which is perhaps 6 deg cooler than outside, say, 89 F. For this wall, by Eq. (4-18) and Table 4-3, the thermal resistance to heat flow is

$$R = \frac{1}{U} = \frac{1}{1.65} + \frac{12}{5.0} + \frac{0.75}{5.0} + \frac{1}{1.65} = 3.76$$

Therefore the overall coefficient of heat transfer is

$$U = \frac{1}{3.76} = 0.266 \text{ (say, 0.27) Btu/h} \cdot \text{ft}^2$$

Thus the heat gain is

$$Q_w = UA(t_1 - t_2) = (0.27)(650)(89 - 80) = 1580 \text{ Btuh}$$

Floor. The area of the floor = 50 × 22 ft = 1100 ft². The overall coefficient of heat transfer, U, is 0.24 (from Table 4-10). The cellar is cooler than outside—say, 6 deg cooler. Thus the heat gain through the floor is

$$Q_f = (0.24)(1100)(89 - 80) = 2376 \text{ Btuh}$$

Ceiling and roof. The ceiling area = 50 × 22 ft = 1100 ft². Read the summer U value as 0.29 (from Table 6-8) and the Δt value at 40 deg for medium construction at 12 noon (from Table 6-7). Therefore the heat gain through the ceiling and roof is

$$Q_{\text{roof}} = (0.29)(1100)(40) = 12\,760 \text{ Btuh}$$

Thus the total transmission and solar gain through the walls and the roof, all in sensible form, is

$$6455 + 3112 + 1586 + 1580 + 2376 + 12\,760 = 27\,869 \text{ Btuh} \qquad \textit{Ans.}$$

Example 6-4. Calculate the heat gain from the occupants of the restaurant discussed in the preceding example—50 diners and 5 employees. The restaurant is maintained at 80 Fdb and 67 Fwb.

Solution: Using the values given for persons eating and for persons at moderate work, in Table 10-3, the total heat gain is found as follows:

	Sensible Heat Gain	Latent Heat Gain
50 × 250	12 500	
50 × 250		12 500
5 × 305	1 525	
5 × 545		2 725
Total heat gain	14 025	15 225

Example 6-5. Find the heat gain from the equipment and from other sources in the restaurant.

Solution: First, consider the two coffee urns. They are gas-heated and of 3-gal size. From Table 6-10, read 2500 Btuh as the sensible heat load and as the latent heat load. Thus

$$2 \times 2500 \text{ sensible} = 5000 \text{ Btuh}$$

$$2 \times 2500 \text{ latent} = 5000 \text{ Btuh}$$

Next, consider the single toaster at 2650 W (2.650 kW). From Table 6-10, read 3413 Btuh as the sensible heat load. Thus the heat gain is

$$2.650 \times 3413 = 9050 \text{ Btuh, sensible}$$

At 12 noon on a sunny day there are no motors or lights in use.

Because the cooking equipment is hooded and vented, reduce the sensible gain by 50%:

$$(5000 + 9050)(0.50) = 7025 \text{ Btuh}$$

Considering the meals served, at 30 Btu each, with 50 seats and 2 meals per seat per hour, there is an additional sensible heat gain of

$$30 \times 2(50) = 3000 \text{ Btuh}$$

Thus the total sensible heat gain is

$$7025 + 3000 = 10\,025 \text{ Btuh} \qquad \textit{Ans.}$$

The total latent heat gain, with the 50% reduction, is

$$(5000)(0.50) = 2500 \text{ Btuh} \qquad \textit{Ans.}$$

Example 6-6. For the same restaurant, find (a) the air required for ventilation, (b) the infiltration air, and (c) the cooling load for the infiltration air. (d) Sum up the total heat-gain load from all sources.

Solution: (a) For a restaurant, Table 10-4 recommends 15 cfm per person (preferred) and 12 cfm (minimum). Using 15 cfm per person, with 55 people, the outside air required for ventilation is

$$15 \text{ cfm} \times 55 = 825 \text{ cfm} \qquad \textit{Ans.}$$

The restaurant volume is $50 \times 22 \times 13$ ft $= 14\,300$ ft^3, and 825 cfm of ventilation air would indicate that the following number of positive changes of outside air per hour take place:

$$\frac{825 \text{ cfm} \times 60 \text{ min/h}}{14\,300} = 3.46$$

(b) Even though this air is being forced out, some infiltration can occur if wind is blowing outside. Assume a 10-mph wind and use Table 5-3 to determine the infiltration:

East wall. Window crack; metal windows, vertically pivoted and center-opening. Thus

$$3 \times [(3 + 4)2 + 4] = 54 \text{ ft of crack}$$

or

$$88 \text{ ft}^3\text{/h} \times 54 = 4752 \text{ ft}^3\text{/h}$$

South wall. Door-crack leakage only, since tight plate-glass show windows should show no leakage. Thus

$$3(7.5) + 2(6) = 34.5 \text{ ft of crack}$$

or

$$138 \text{ ft}^3/\text{h} \times 34.5 = 4760 \text{ ft}^3/\text{h}$$

West wall: Solid, no leakage.

North-wall Partition. The kitchen must have adequate exhauster fans and these of course exhaust some air from the conditioned space. Take crack infiltration as one-half of the total:

$$\tfrac{1}{2}(4752 + 4760) = 4756 \text{ ft}^3/\text{h}$$

An additional amount of infiltration occurs because of door opening, and can be estimated by Table 6-9. Using 2.5 cfm per transient person inside the space, the door leakage is

$$2.5 \times 50 = 125 \text{ cfm}$$

or

$$125 \times 60 = 7500 \text{ ft}^3/\text{h}$$

Thus the total infiltration is

$$4756 + 7500 = 12\,256 \text{ ft}^3/\text{h} \qquad \textit{Ans.}$$

It is improbable that this much total infiltration will occur, considering the slight positive pressure maintained by the conditioner, but it is possible—because of exfiltration into the hood outlets and into the kitchen exhaust system. In a conservative design, this total amount should be considered.

(c) The outside air is at 95 Fdb and 78 Fwb. From Plate I, or by computation, moisture = 0.01168 lb/lb; dew point = 71.6 F; volume = 14.35 ft^3/lb dry air.

The inside air is at 80 Fdb and 67Fwb. From Plate I, or by computation, the dew point = 60.1 F and 0.0112 lb/lb, and the specific volume = 13.84 ft^3/lb dry air.

The infiltration air from outside = 12 256 ft^3/h, or

$$\frac{12\,256}{14.35} = 854 \text{ lb/h}$$

The sensible heat to be removed is, by Eq. (3-34),

$$Q_s = (854)(0.244)(95 - 80) = 3140 \text{ Btu} \qquad \textit{Ans.}$$

The moisture load in the infiltration air is

$$(854)(0.0168 - 0.0112) = 4.78 \text{ lb/h}$$

The latent load, corresponding to this moisture addition can be found by using the latent heat value for steam, rounded at 1100 Btu/lb

$$Q_L = 4.78 \times 1100 = 5260 \text{ Btuh}$$

(d) Therefore the total-space cooling load is as follows, in Btu per hour:

	Sensible	Latent
Transmission and solar gain	27 869	
Human load	14 025	15 225
Miscellaneous sources	10 025	2 500
Outside infiltration air	3 140	5 260
TOTAL	55 059	22 985 *Ans.*

Note that this space cooling load (internal cooling load) is not the load which exists in the conditioner. The conditioner load must also include the additional load associated with cooling the ventilation air, and in some instances, any additional cooling required for extreme dehumidification followed by reheating.

6-7. COOLING-LOAD AIR QUANTITIES

The quantity of air circulated must be adequate to handle the cooling load as the air warms up to room temperature from its supply temperature. The lower the supply temperature the less the quantity which must be circulated, but the minimum temperature is determined by the system arrangement, the necessity of avoiding drafts and cold regions, the ceiling height, and the throw required. Summer conditioning installations are usually designed to supply air at 5 to 20 deg below room temperature. Specially designed nozzles in certain suitable locations have been used with air as low as 30 deg below room temperature. For practical purposes, a reduction of 2 deg in delivered air temperature per foot of height from floor to ceiling should not be exceeded in determining an inlet temperature of the supply air. Special grilles, nozzle arrangements, room shape, or load conditions may make other temperature differentials desirable or even necessary, but the 2-deg/ft limit should never be exceeded for preliminary calculations. Grille and nozzle recommendations from the manufacturer's data must be followed in making final designs for temperature difference, expected air throw, and air distribution.

After the space cooling load (internal cooling load) is computed by the methods shown earlier in this chapter, the air quantity can be found by making use of the method described in Section 3-8, or by the sensible heat load equation [(Eq. 3-34)] rewritten here:

$$Q_S = m_a C_p(t_i - t_S) = 0.244\, m_a(t_i - t_S) \tag{6-5}$$

where Q_S is the space (internal) cooling load, in Btu per hour; m_a is the air supplied to space, in pounds per hour; C_p is the specific heat of the moist air (approximately 0.244), in Btu per pound per degree Fahrenheit; t_i is the inside space temperature to be maintained, in degrees Fahrenheit dry bulb; t_S is the supply-air temperature entering the space, in degrees Fahrenheit dry bulb. The $\Delta\mathbf{h}/\Delta W$ slope, or the same slope, measured by the sensible-heat ratio (SHR), is extremely important in the solution of cooling-load problems.

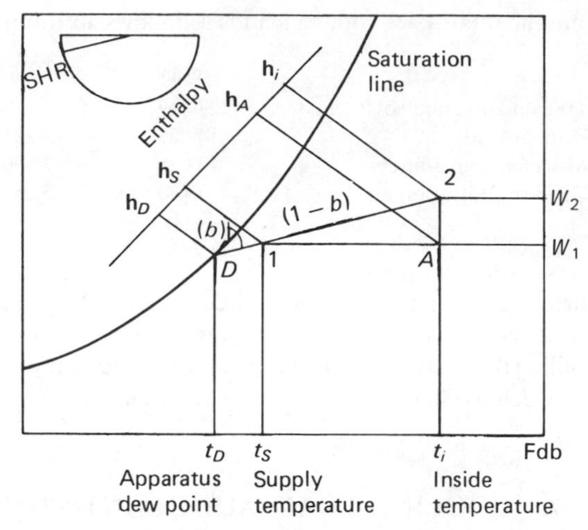

FIGURE 6-7. Skeleton psychrometric chart showing use of apparatus dew point and by-pass.

Refer to Fig. 6-7, which represents a skeleton psychrometric chart plot of a cooling and dehumidification process. On this chart a line has been drawn through 2, the inside space temperature, with the slope of the line set in terms of the SHR or $\Delta \mathbf{h}/\Delta W$ slope. It will be noticed that this line intersects the saturation line at D. This point represents the *apparatus dew-point temperature* of the conditioner coil. Conditioner coils cannot bring all of the air passing through them to the coil surface temperature. This fact makes the coil perform in a manner similar to what would be accomplished if a portion of the air were brought to coil temperature and the remainder by-passed. A dehumidifying coil thus produces unsaturated air at a higher temperature than the coil temperature. Again referring to Fig. 6-7, notice that in terms of the length of the whole line D to 2, the length D to 1 is proportional to the air by-passed (b), and the length 1 to 2 is proportional to the air not by-passed ($1 - b$). Because of the chart construction, it is also closely true that in terms of dry-bulb temperatures,

$$b \equiv \frac{t_S - t_D}{t_i - t_D} \tag{6-6}$$

And in a similar manner,

$$1 - b \equiv \frac{t_i - t_S}{t_i - t_D} \tag{6-7}$$

where b is the fraction of air by-passed, or coil by-pass factor, expressed as a decimal, and where the temperatures are dry-bulb values. Now, as before,

$$Q_S = m_a(0.244)(t_i - t_S)$$

and

$$Q_S = m_a(0.244)(t_i - t_D)(1 - b) \tag{6-8}$$

If air at average conditioner supply conditions is employed in the above equation at, say, $v = 13.5$ ft^3/lb dry air, then

$$Q_S = \frac{(\text{cfm})(60)}{13.5}(0.244)(t_i - t_D)(1 - b) \tag{6-9}$$

and therefore

$$\text{cfm} = \frac{Q_S}{(1.08)(t_i - t_D)(1 - b)} \tag{6-10}$$

where cfm is the volume flow capacity of the supply fan delivering cooled air, in cubic feet per minute; t_i is the dry-bulb temperature of conditioned space, in degrees Fahrenheit; t_D is the apparatus dew-point temperature, in degrees Fahrenheit; b is the by-pass factor of the cooling coil; and Q_S is the space sensible load, in Btu per hour.

It is frequently convenient to add the by-pass ventilation air to the sensible and latent space loads for convenience in calculation, and in this connection the equations appear as

$$Q_{vS} = (\text{cfm})(1.08)(t_o - t_i)(b) \tag{6-11}$$

$$Q_{vL} = \frac{(\text{cfm})(60)}{13.5}(W_o - W_i)(1100)(b) \tag{6-12}$$

$$Q_{vL} = 4880(\text{cfm})(W_o - W_i)(b) \tag{6-13}$$

Where Q_{vS} is the sensible heat gain from ventilation air by-passed into space, in Btu per hour; Q_{vL} is the latent heat gain from ventilation air by-passed into space, in Btu per hour; cfm is the ventilation air supplied, in cubic feet per minute; t_o, t_i are the outside temperature and space temperature, in degrees Fahrenheit; W_o, W_i is the humidity ratio of outside air and space air, in pounds per pound of dry air; and b is the by-pass factor.

Example 6-7. For the same restaurant, and its cooling load as summarized in Example 6-6, compute (a) the sensible heat ratio and (b) the enthalpy-humidity difference ratio. Assume that the minimum supply-air temperature is 64 F and find (c) the minimum supply air for the space, in pounds per hour, and (d) the humidity ratio and wet bulb of the supply air that are required in order to meet design conditions. (e) Find the amount of air delivered by the supply fan per minute. (f) When the conditioner, with its by-pass condition, cools both the ventilation and recirculated air to supply conditions, what refrigeration tonnage (200 Btu/min · ton) is absorbed by the conditioner coils?

Solution: (a) Referring to the data of Example 6-6, the sensible heat ratio is

$$\text{SHR} = \frac{55\,059}{55\,059 + 22\,985} = 0.705 \qquad \textit{Ans.}$$

(b) By Eqs. (3-49) and (3-51),

$$\frac{\Delta \mathbf{h}}{\Delta W} = \frac{55\,059 + 22\,985}{22\,985/1100} = 3730 \qquad \textit{Ans.}$$

(c) By Eq. (6-5),

$$55\,059 = m_a(0.244)(80 - 64)$$

$$m_a = 14\,000 \text{ lb/h} \qquad \textit{Ans.}$$

(d) By Eq. (3-52), with air in the room at 80 F and 67 F, with $W_2 = 0.0112$ lb/lb dry air, and with $\mathbf{h}_i = 31.5$ Btu/lb dry air,

$$22\,985 = (14\,000)(0.0112 - W_1)(1100)$$

Thus the humidity ratio is $W_1 = 0.0107$ lb/lb. *Ans.*

From Plate I, for 0.0107 lb at 64 Fdb, read the wet bulb as 59.9 F. *Ans.*

The corresponding specific volume v is 13.4 ft^3/lb, and the enthalpy is 26.0 Btu/lb dry air.

(e) The amount of supply air delivered by the fans is

$$\text{cfm} = \frac{m_a v}{60} = \frac{(14\,000)13.4}{60} = 3130 \text{ cfm} \qquad \textit{Ans.}$$

(f) Since ventilation air is supplied at 825 cfm, at 95 F and 78 F, and $v = 14.35$ ft^3/lb, then

$$m_v = \frac{(825)(60)}{14.35} = 3450 \text{ lb/h}$$

Also,

$$\mathbf{h}_o = 41.4 \text{ Btu/lb dry air}$$

and

$$W_o = 0.0168 \text{ lb/lb dry air}$$

The cooling load of the ventilation air is

$$m_v(\mathbf{h}_o - \mathbf{h}_S) = 3450(41.4 - 26.0) = 53\,100 \text{ Btuh}$$

and the cooling load of the recirculated air is

$$m_r(\mathbf{h}_i - \mathbf{h}_S) = (14\,000 - 3450)(31.5 - 26.0) = 58\,030 \text{ Btuh}$$

Thus the refrigeration absorbed by the coils is

$$\frac{53\,100 + 58\,030}{(200)(60)} = 9.2 \text{ tons} \qquad \textit{Ans.}$$

Example 6-8. Make use of the by-pass method of solution, along with apparatus dew point, to find a solution for part f of Example 6-7. Assume a by-pass factor of 0.40 as a starting point.

Solution: First, for computation purposes consider that the ventilation air which by-passes the cooling coils becomes part of the space (internal) load. Then, by Eqs. (6-11) and (6-13), and for 825 cfm,

$$Q_{vS} = (825)(1.08)(95 - 80)(0.40) = 5340 \text{ Btuh}$$

$$Q_{vL} = (4880)(825)(0.0168 - 0.0112)(0.40) = 9010 \text{ Btuh}$$

Thus the sensible heat ratio is

$$\text{SHR} = \frac{Q_s}{Q_T} = \frac{55\,059 + 5340}{78\,044 + 5340 + 9010} = 0.65$$

Use the sensible heat ratio as shown in Fig. 6-7 and following the method described in this section, find the apparatus dew point. Run the datum line through the room condition of 80 Fdb and 67 Fwb parallel to the SHR slope of 0.65 as shown in the protractor hemisphere. The intersection at the saturation line shows an apparatus dew point of 50.1 F for t_D. Using Eq. (6-6) and the by-pass factor of 0.4, given

$$b = \frac{t_S - t_o}{t_i - t_D} = \frac{t_S - 50.1}{80 - 50.1} = 0.40$$

and thus

$$t_S = 62 \text{ F}$$

If this temperature is a lower supply temperature than can be used by the occupants of the space, then a different by-pass factor must be employed or else reheat must be used. For the total conditioner load we must now add the cooling load required to bring the rest of the outside air to room conditions. In this connection, extend Eqs. (6-11) and (6-13) to cover the $(1 - b)$ fraction of the outside air:

$$\begin{aligned} Q_{oS} &= (\text{cfm})(1.08)(t_o - t_i)(1 - b) \\ &= (825)(1.08)(95 - 80)(0.60) = 8010 \text{ Btuh} \\ Q_{oL} &= (4880)(825)(0.0168 - 0.0112)(0.60) = 13\,520 \text{ Btuh} \end{aligned}$$

The conditioner load in sensible and latent forms is thus

$$\begin{aligned} Q_S &= 55\,059 + 5340 + 8010 = 68\,409 \text{ Btuh} \\ Q_L &= 22\,985 + 9010 + 13\,520 = 45\,515 \text{ Btuh} \end{aligned}$$

and the total refrigeration tonnage absorbed by the conditioner coils is therefore

$$\frac{68\,409 + 45\,515}{(200)(60)} = 9.5 \text{ tons} \qquad \textit{Ans.}$$

The designer should set operating standards to give an economically desirable pattern of operation. As it happens, the by-pass factor of 0.4 used to illustrate the foregoing problem is high; most coils operate at $b = 0.25$ or below.

6-8. SUMMARY

In this chapter a basic approach to providing space cooling was first described to give a background for development of cooling-load calculations. The items entering into cooling load were then listed. Since solar load represents the most complex component of the cooling load, this was presented in detail. The treatment involved a discussion of the earth in relation to the solar system, followed by the presentation of tabular data, and developed one method of computing the various elements constituting solar load. It should be realized that many approaches can be used and additional refinements can be carried out for computation of solar load, but additional refinement is seldom required. Tabular data for other heat sources followed and an illustrative example was given. In Chapter 17, detailed discussions of other cooling-system arrangements are presented.

PROBLEMS

6-1. Compute the instantaneous heat gain through a vertical 10- by 6-ft-high plate-glass window, facing south at latitude approximately 40°N, at noon in early August on a 95 F clear day with an inside temperature of 80 F. Modify your results to show the effect of a canvas awning tight against the building. Express all answers in Btuh and in kilowatts.

Ans. 6240 Btuh and 2182 Btuh, 1.83 kW and 0.64 kW

6-2. Rework Problem 6-1 for a 3.05- by 1.83-m window and express answers in watts and kilocalories per hour.

Ans. 1826 W and 1569 kcal/h or 638 W and 548 kcal/h

6-3. Measurements show that on August 1 at 40°N latitude at noon on a clear day the total irradiation (normal orientation) at earth level is 277 Btu/h · ft^2. That is, this much solar energy would impinge on a surface perpendicular to the rays of the sun, independent of diffuse radiation and reradiation from other surfaces. However, a horizontal surface such as a flat roof would not receive the full irradiation because the solar altitude angle is 68° and the declination angle is 22°. (a) Making use of the cos 22°, compute the irradiation (normal type) which falls on a flat roof. (b) What happens to this energy which impinges on the roof?
Ans. (a) 257 Btu/h · ft^2, (b) some reflected, some absorbed raising roof temperature, some conducted inside, remainder convected and reradiated to atmosphere and surroundings

6-4. Work Problem 6-3 to obtain the answer in watts per square meter and kilocalories per square meter.

Ans. (a) 810 W/m^2; (b) 697 kcal/m^2

6-5. Employ the data of Problem 6-3 to compute the direct irradiation (normal component only) on a south-facing vertical wall at noon on August 1. *Hint*: Give consideration to the solar-altitude angle.

Ans. 103.8 Btu/h · ft^2.

Note that some 14 Btu/h · ft^2 in addition would be contributed by diffuse radiation and reradiation from the ground and surroundings (not asked for in this problem.)

6-6. At 10 A.M. solar time at latitude 32°N at Savannah, Georgia, on July 21 the normal-oriented solar irradiation is 271 Btu/h · ft^2 on a clear day. At that time the altitude angle of the sun is 60.9° and the azimuth angle is 74.3° measured from a north-south line. Compute the solar irradiation impinging on (a) the horizontal roof surface and (b) vertical wall facing due southeast.
Ans. (a) 237 Btu/h · ft^2, (b) 115 Btu/h · ft^2. (Note that for the vertical wall an additional 18 Btuh/ft^2 from ground radiation and diffuse effects will act on this surface.)

6-7. At 10 A.M. solar time at latitude 32°N on November 21 the normal-oriented solar irradiation is 908 W/m^2. At that time and season the altitude angle is 30.8° and the azimuth angle 33.2° measured from the north-south line. Compute the direct solar irradiation impinging on a horizontal roof and compare with the answer given in Problem 6-6 for summer conditions.

6-8. At 2 P.M. solar time at latitude 40°N on March 21 the direct solar irradiation on a clear day is 297 Btu/h · ft^2, the solar altitude is 41.6°, and the solar azimuth from the north-south line is 41.9°. Two months later on May 21 the corresponding values are 277 Btu/h · ft^2, 57.5°, and 60.9°. Compute the direct solar irradiation on a flat roof (a) on March 21 and (b) on May 21.

Ans. (a) 197 Btu/h · ft^2

6-9. Find the information requested in Problem 6-8 in relation to a vertical wall facing south. To your computed answer you should add 24 Btuh/ft^2 on March 21 and 11 Btuh/ft^2 on May 21 for the effects of diffuse and gound radiation.
Ans. (a) 165 Btu/h · ft^2; (b) 72 Btu/h · ft^2 (neither answer corrected for diffuse radiation)

6-10. The following values for normal solar radiation intensity in watts per square meter are applicable on July 21 at locations 40°N latitude: 7 A.M. and 5 P.M. 655 with 24.3° alt, 97.2° az; 9 A.M. and 3 P.M. 815 with 47.2° alt, 76.7° az; 11 A.M. and 1 P.M. 862 with 66.7° alt, 37.9° az; 12 noon 870 with 70.6° alt, 0° az. Ground reradiation and atmospheric diffusion effects increase the magnitude of these values by some 10 to 15% on vertically oriented walls between the hours of 7 and 5 P.M. on clear days. Compute the intensity of direct solar radiation falling on an easterly facing wall at 9 A.M. and at 3 P.M.

Ans. 539 W/m^2 at 9 A.M., no direct radiation at 3 P.M.

6-11. Making use of data from Problem 6-10, compute the magnitude of the direct normal solar irradiation falling on a southwesterly facing vertical wall at 7 A.M. and at 5 P.M.

Partial Ans. none at 7 A.M.

6-12. At Tucson, Arizona (approximately 32°N latitude), on August 21 at 10 A.M. and 2 P.M. solar time, the direct normal solar irradiation is 275 Btu/h · ft^2, the solar altitude is 56.1°, and the solar azimuth measured from a south-north line is 75°. Compute the direct irradiation on (a) a horizontal roof and (b) a west-facing vertical wall.

6-13. A 5- by 7-ft plate-glass window is placed in the south wall of a building located near the 40°N latitude line. The outside design temperature is 93 F and the inside temperature is held at 80 F. (a) Compute the total heat gain through this window at 11 A.M. and 5 P.M., in Btu per hour. (b) What are the corresponding values of heat gain if an effective, open-side awning is installed which prevents all direct sunlight from falling on the glass?

Ans. (a) 3250 Btuh, 887 Btuh; (b) 813 Btuh, 221 Btuh

6-14. Work Problem 6-13, for 5 P.M. only, but assume that the window is in a west wall in a 30°N latitude area.

Ans. (a) 1268 Btuh; (b) 317 Btuh

6-15. Find the total instantaneous gain per square foot for single-sheet plate glass in the east wall of a building located 40°N latitude. The time is 11 A.M. Consider the outside air at 95 F and inside conditions at 80 Fdb.

Ans. 63.4 Btu/h · ft^2

6-16. On the first floor the glass area in the south wall of a certain office building at 40°N latitude consists of twenty 5- by 8-ft steel-sash windows containing single-thickness glass. At 3 P.M. in mid-August, what is the probable heat gain by solar effects through the glazed area?

Ans. 48 800 Btuh

6-17. Assume that the glazed area in Problem 6-16 is equipped with Venetian blinds of a white to cream shade and compute the probable solar transmission under the conditions indicated in the problem.

Ans. 27 300 Btuh

6-18. At 4 P.M. on a sunny afternoon in mid-August, what is the heat gain through the flat roof of a building 80 by 200 ft in size? Assume that the roof, which is approximately 6 in. thick, consists essentially of 2-in. concrete and a furred ceiling with metal lath and plaster, with no insulation. The outdoor temperature is 97 F and the indoor temperature can be considered as 80 F.

Ans. 285 000 Btuh

6-19. What is the heat transmission at 5 P.M. through the south wall of Problem 6-13 if the 140 ft^2 of nonglass area is dark red and of brick-veneer and frame construction and has a U of 0.28 $Btu/h \cdot ft^2 \cdot F$ and is not shaded in any way?

Ans. 822 Btuh

6-20. Work Problem 6-19, but assume that the wall area faces southwest instead of south.

Ans. 1252 Btuh

6-21. A survey of a theater which is to be maintained at 82 Fdb and 68 Fwb shows that at 4:00 P.M. there is a transmission and solar heat gain of 60 000 Btuh, and that on the average 500 people are then present; also, that at 8:30 P.M. there is a transmission load of 20 000 Btuh and that on the average 1000 people are then present. Determine the maximum load and the time at which it occurs. Light and equipment loads can be neglected.

Ans. 370 000 Btuh at 8:30

6-22. In the theater in Problem 6-21, how many pounds of air at 62 Fdb must be supplied to the conditioned space under maximum sensible-load conditions?

Ans. 54 300 lb/h

6-23. For a certain conditioned space, 2000 cfm of air at 62 Fdb and 59 Fwb are supplied. This consists in part of 800 cfm of outside air measured at 96 Fdb and 76 Fwb. The space is maintained at 80 Fdb and 50% relative humidity. Determine (a) the amount of air recirculated in cubic feet per minute, and (b) the tons of refrigeration.

Ans. (a) 1298 cfm; (b) 6.4 tons

6-24. Determine from the following peak-load data the total occupant and equipment load in a restaurant maintained at 83 Fdb: average number of customers, 20; waiters, 3; one 5-gal electric coffee urn; one gas-stove burner (hooded); one 48- by 18-in. steam table; and a revolving display sign carrying six 60-W lights and driven by a $\frac{1}{4}$-hp motor.

Ans. 14 358 Btuh, sensible; 15 785 Btuh, latent

6-25. A space to be conditioned has a load of 70 000 Btuh and a sensible heat ratio of 0.8. It is to be maintained at conditions not to exceed 78 F and 55% relative humidity. The direct-expansion cooling coils have an equivalent by-pass factor of 0.3. The maximum duct capacity of the system is limited to 4000 cfm. Disregard any outside air required for ventilation. (a) Make use of the SHR on the psychrometric chart (Plate I) and, by extension from room conditions, find the required temperature of the cooling coils. (b) For the by-pass factor of 0.3 and room air conditions, find the air-supply temperature required. (c) How much air, in pounds per minute and in cubic feet per minute, must pass through the system to carry the space load?

Ans. (a) 57 F; (b) 63 F; (c) 3430 cfm

6-26. Find the items called for in Problem 6-25, with similar specifications except that 1000 cfm of outside air at 85 Fdb and 75 Fwb is required for ventilation. (a) First find the mixing temperature of the outside and recirculated air (assume that the total weight of air supplied to the space is the same as in Problem 6-25; this condition, of course, sets the inlet temperature the same as before). (b) By using the inlet temperature and the mixing temperature, find the coil temperature. (c) What equivalent by-pass factor must now hold for the equipment? If this is too high, additional by-passing external to the coil or reheat can be employed.

Ans. (a) 79.9 F mixing; (b) 55.1 F dew point on coil; (c) 0.32 by-pass

6-27. A conditioner receives air at 80 Fdb and 70 Fwb, and the air leaves at 63 Fdb. The wet coils of the conditioner are at a surface temperature of 55 F. Find (a) the by-pass factor for the conditioner and (b) the amount of heat removed from each 1000 cfm supplied to the conditioner.

Ans. (a) 0.32; (b) 31 760 Btuh

6-28. A spray-type air washer with four banks of sprays cools air at 90 Fdb and 75 Fwb down to a water temperature of 60 F. The saturated air leaving the conditioner is reheated to 67 Fdb. Find (a) the weight of water removed from each pound of air entering the conditioner, (b) the heat absorbed by the spray water, and (c) the heat furnished by the reheater coils per pound of dry air supplied.

Ans. (a) 0.0042 lb; (b) 12 Btu; (c) 1.7 Btu

REFERENCES

C. O. Mackey and L. T. Wright, "Periodic Heat Flow—Homogeneous Walls or Roofs," *Trans. ASHVE*, Vol. 50 (1944), pp. 293–312.

C. O. Mackey and L. T. Wright, "Periodic Heat Flow—Composite Walls or Roofs," *Trans. ASHVE*, Vol. 52 (1946) pp. 283–96.

J. P. Stewart, "Solar Heat Gain through Walls and Roofs for Cooling Load Calculations," *Trans. ASHVE*, Vol. 54 (1948) pp. 361–88.

G. V. Parmelee, W. W. Aubele, and R. G. Huebscher, "Measurements of Solar Heat Transmission through Flat Glass," *Trans. ASHVE*, Vol. 54 (1948), pp. 165–86.

ASHRAE, Handbook of Fundamentals, 1967, chap. 28, and also 1977 issue, chap. 25.

CHAPTER

7

STEAM HEATING

7-1. STEAM IN HEATING SYSTEMS

When saturated steam is supplied to a radiator or convector the useful heating is caused primarily by condensation of the steam. The heat of condensation is of course equal to the heat of vaporization (latent heat), listed as $\mathbf{h}_{fg}$ in Tables 2-2 and 2-3. For example, at atmospheric pressure (14.7 psi) a pound of dry saturated steam, in condensing, gives up 970.3 Btu ($\mathbf{h}_{fg}$ from Table 2-3). During this process of condensation, just as in vaporization, the temperature remains constant (at 212 F when the pressure is 14.7 psi). Reference to Table 2-3 shows that when steam is allowed to condense at a lower pressure than atmospheric (partial vacuum), the $\mathbf{h}_{fg}$ values are somewhat larger than 970.3, and the saturation or condensation temperatures are lower than 212 F. In utilizing this latent heat in radiators or convectors, provision must be made for removing the condensate (water) from them. The condensate is at the same temperature as the steam until it is chilled by cool piping or radiator surface not in contact with the steam. Additional heat flow to a space, often amounting to 30 to 60 Btu/lb of steam, can occur as the liquid subcools.

Since the magnitude of $\mathbf{h}_{fg}$ does not change significantly over the pressure range employed in conventional low-pressure heating systems, pressure has little effect on the energy delivered, during condensation, by each pound of steam. However, low-pressure steam is also low-temperature steam, and since heat transfer depends primarily on the temperature difference existing between the radiator surface and the ambient air, the rate of heat flow is less with low-pressure steam than with high-pressure steam. Moreover, the specific volume of low-pressure steam is greater than that of high-pressure steam, and so a low-pressure system, in addition to requiring somewhat greater surface in the radiators, also requires larger pipe sizes to transmit the greater volume of steam.

Radiators capable of heating a building satisfactorily in zero weather would greatly overheat the building in mild weather unless the stem were turned on and off or unless the steam temperature could be varied. With subatmospheric steam systems this latter arrangement is used, and the steam pressure (and its temperature) is reduced in mild weather to suit the heat demand. Such cooler radiators supply less heat and prevent the necessity of frequent control through operation of steam valves. With many steam-heating systems, particularly single-pipe arrangements, partly closing a valve does not give satisfactory temperature modulation.

Two brief definitions will be given at this point. A *radiator* is a heating device placed in a space so that it can direct its radiation to part or all of the heated space. It delivers heat by radiation, convection, and conduction. A *convector* is a heating device arranged to deliver heat to the air largely by convection currents. Convectors are enclosed, or concealed from the heated space.

It is always desirable to express the capacity of a radiator or convector under stated conditions in terms of Btuh (Btu per hour) or Mbh (1000 Btuh). However, there still exists an old unit, called by various names, such as *equivalent direct radiation* (EDR) or *square feet of radiation*, or even *feet of radiation*. Unit EDR is defined as heat delivery at a rate of 240 Btuh. For rating of radiators and convectors, it is usually considered that the output is produced when steam is condensing at 215 F in 70 F ambient space. (In the case of hot-water heating, an EDR is often taken as 150 Btuh.)

In SI (or in the metric system), heat delivery should be expressed in kW (kilowatts) or in kJ/h, but many continue to use kcal/h (kilocalories per hour). Note that Mbh = 1000 Btuh = 0.2929 kW = 1054 kJ/h = 252 kcal/h.

7-2. STEAM FLOW IN SYSTEMS

Steam flows through pipes because of pressure differences. Heat added in a boiler causes the water to change into a vapor (steam), with a resulting increase in volume. For example, at atmospheric pressure (212 F) the volume occupied by 1 lb of saturated steam is about 1600 times as great as the volume occupied by 1 lb of water; that is, $v_g/v_f = 26.80/0.01672 = 1604$ (from Table 2-2). When 1 lb of water is converted into steam in a heating system at a pressure above atmospheric, the steam displaces and drives out the air. As heat is transferred from the pipes and radiators, the steam condenses to water and a resultant small volume. The shrinkage in volume caused a lower pressure within the system, and the flow of steam from the boiler increases because of the reduced pressure. Thus two forces influence the pressure difference necessary for steam flow: first, application of heat at the boiler, with evaporation and an accompanying increase in volume; second, heat removal in the radiators and pipes, with an accompanying shrinkage in volume and reduction in pressure as condensation occurs. If the heat removal is greater than the heat supply, the pressure of the system falls. When the heat supply at the boiler is diminished in airtight systems, transmission of heat in decreasing amount will continue while the steam pressure falls far into the vacuum region.

Flow of steam in a heating system is resisted by the skin friction of the various conduits, particularly in pipes, fittings, and valves. Condensation commences whenever saturated steam loses heat, so that in most pipes the fast-flowing steam rushes alongside a low-velocity steam of condensate. If the drainage slope of the pipe is such that the steam flow is counter to the water flow, friction is increased and therefore larger pipes must be used.

A boiler is an externally fired enclosed vessel, partly filled with water. When the water boils, the steam forming above the water in the vessel will cause an increase in pressure and compress any air in the pipes of a heating system. Since the pressures throughout the system are about equal, the water level in the down-feed pipe (drop leg) at the end of the steam supply

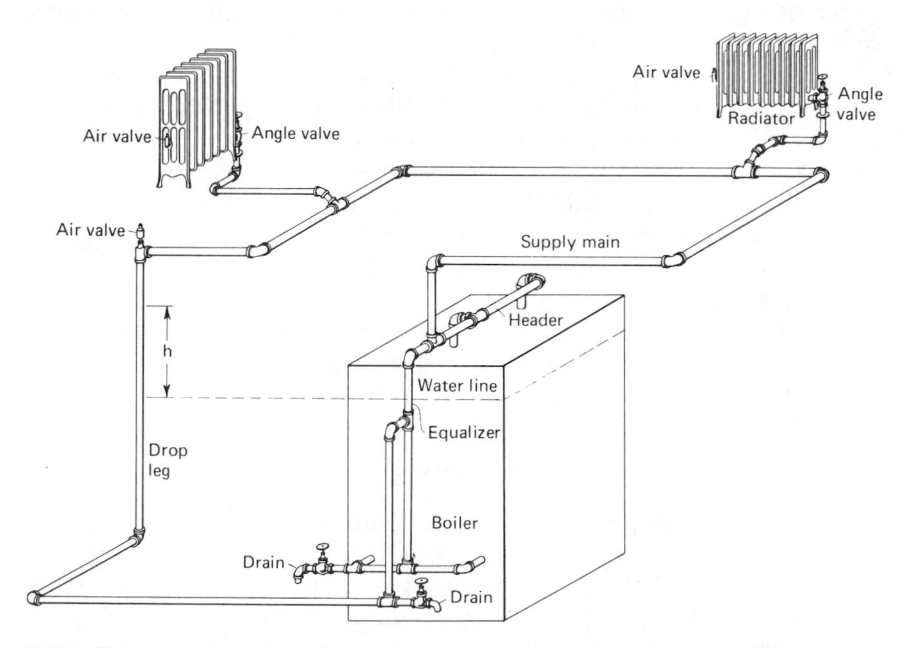

FIGURE 7-1. Layout of early design one-pipe air-vent system shown with radiators of the same early vintage in representative placement.

main will be almost level with the water level in the boiler (see Fig. 7-1). In fact, these water surfaces are exactly level with each other whenever the respective pressure out in the system and the boiler pressure are equal. When steam flows in the system the water level rises in the drop leg since the friction drop resulting from the fluid flow, in turn related to the steam condensation, reduces the pressure below that in the boiler.

When a steam coil, radiator, or convector transfers so much heat that the pressure in the drop leg is 1 psi less than boiler pressure, the water line in the drop leg will rise about 28 in. higher than the water level in the boiler. If the flow of steam is retarded, as by a valve or small-size pipe, the difference in pressure and water levels will increase. Thus in certain steam-heating systems the permissible difference between the water level in the boiler and that in the drop leg is the governing factor in choosing pipe sizes.

7-3. AIR VENTING

Water at normal temperature usually contains air, some of which separates when the water is heated. Pipes and radiators contain air when steam is first developed. To heat the surfaces the air must be removed before the steam can enter, and must be vented continually thereafter.

Acceptable air valves allow the air to pass out but close when steam starts to escape. Figure 7-2 shows a common construction. The float contains a small amount of volatile fluid which, expanding or contracting in response to temperature changes, moves the flexible bottom of the float outward, or allows it to spring inward. When the steam pressure increases above the pressure of the atmosphere, any air that is present is forced out through the escape port. After the relatively cool air escapes and the hot steam enters the valve, the confined fluid within the float chamber expands and thereby forces the flexible bottom downward. This action lifts the float and forces the valve pin into the port opening, thereby closing the valve. The float chamber will also rise if for any reason the valve becomes filled with water. Thus no water can escape through the valve. The loosely fitting siphon is used when the valve is placed on radiators, to return to the radiator any accumulation of water within the valve.

It is better to employ air valves which will retard the return of air into the system for several hours, and which thus permit steam circulation at a temperature below that of the boiling point of water at atmospheric pressures. In this connection an air valve, such as shown in Fig. 7-2, operates to discharge air and to prevent escape of water and steam, but when the steam pressure drops and the air tends to return to the system the check valve at the vent port closes to prevent such counterflow. As the steam in the system condenses, a vacuum will be formed if no air reenters the system. The check valve, having functioned first, is then supplemented by atmospheric air pressure, which, acting through the port at the bottom of the valve, presses on the vacuum diaphragm, forces the float to rise, and thus positively seal the venting port.

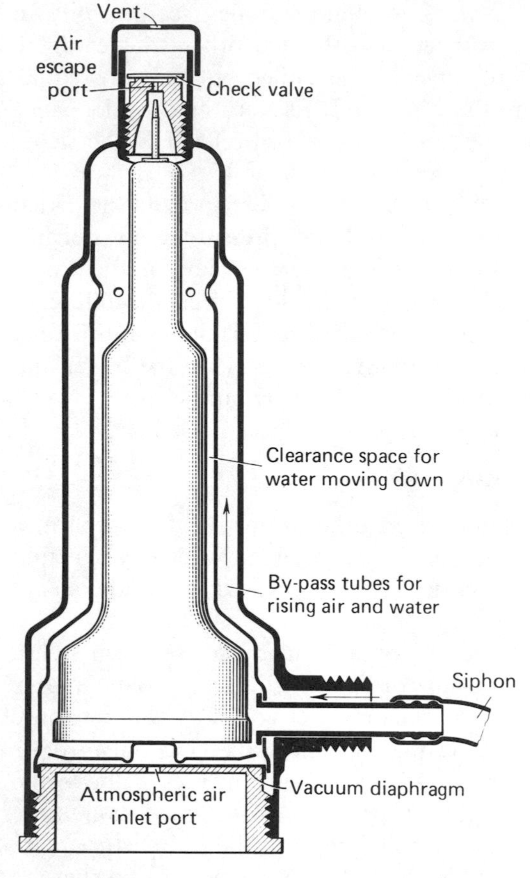

FIGURE 7-2. Vacuum-type radiator venting valve.

The advantage of a tight air valve is its ability to keep the radiator warm for some time after the fire in the boiler has been allowed to die down. This is possible because the temperature at which water boils becomes lower as the absolute pressure decreases, and steam at 160 F or an even lower temperature will circulate throughout an extensive piping and radiator system if air can be prevented from entering. The added cost for the vacuum-type valve is justified by the longer intervals between firing in mild weather, and by the better temperature control which can be obtained. Figure 7-3 shows an air-vent valve equipped with an adjustable outlet orifice. Adjustment of the orifice can control the rate at which air escapes from the radiator and consequently the rate at which the radiator warms up. In a large system, closing back on the orifice of the valves close to the boiler can prevent the radiators adjacent to the boiler from overheating before distant radiators have had a chance to warm.

The air inside the steam mains cannot all escape through the air valves on the radiators. Sometimes the radiator valves are closed, shutting off communication between the radiator air valves and the main. Therefore auxiliary air valves are required on the ends of all steam mains where they

FIGURE 7-3. Radiator-vent valve with adjustable air outlet. (Courtesy Hoffman Specialty ITT.)

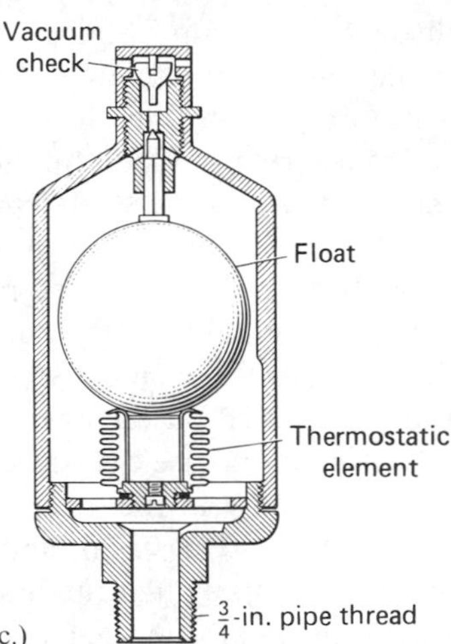

FIGURE 7-4. Air eliminator. (Courtesy Sarco Co., Inc.)

enter the drop pipes and become wet. The location most likely to trap the air accumulation is on the boiler side of the last radiator and close above the water level. However, an air vent placed too close to the water line may be closed by a backsurge of water. The best location for the air vent on a steam main, all things considered, is in the top of the main just above its final drop below the boiler water line (see Fig. 7-1 and Fig. 7-5). Figure 7-4 is applicable to hot-water systems as well as to steam systems. When water enters this type of eliminator the float rises and thereby closes the outlet, which remains closed until sufficient air collects in the chamber to drop the float, thus venting the air out at the top. If steam instead of cooled condensate enters the eliminator, the thermostatic element, which is filled with a volatile liquid, expands and forces the outlet shut, preventing escape of steam. A check valve prevents air from entering if the system is under vacuum.

7-4. TYPES OF STEAM SYSTEMS; DEFINITIONS

Numerous classifications of steam systems have been made and are in use. Many of these classifications and names, however, are merely sub-classifications of the basic types, which are summarized as follows:

One-pipe, air-vent gravity return
Two-pipe, air-vent gravity return
Vapor system, simple gravity return
Vapor system, with return trap
Mechanical return system, nonvacuum
Vacuum return system, with vacuum pump

Figure 7-1 illustrates a one-pipe, air-vent steam system. A study of the illustration will reveal the one-pipe aspect: the supply main which starts at the boiler continues as a single pipe until it finally drops down at the end of its run to constitute the vertical drop leg and, finally, the wet return, which carries the condensate back to the boiler. Steam from the supply main passes through connecting piping to the radiators. Each of the radiators has a single angle valve screwed into a low position on the radiator. The condensate which forms in the radiator moves counterflow to the direction of the steam entering the radiator, and after reentering the steam main it flows in the same direction as the steam until it enters the drop leg leading down to the wet return. Systems of this type are noisy, since water and steam can surge back and forth in the radiator. It is therefore necessary to make the connecting pipes from the radiator to the main sufficiently large to serve the double purpose of steam supply and condensate return. Notice that an air vent is required on each radiator. It is also advisable to supply one on the return main in case a closure of valves on some of the radiators makes it impossible to vent the steam main and thereby prevents satisfactory operation of the system. In this and in other systems as well, steam supply mains and return pipes often run along a basement ceiling. Where possible, mains should be pitched in such a direction as to permit condensate to flow along with the steam, and where this is done the high point of the main, exclusive of risers to other floors, is normally at the boiler. A pitch of 1 in. in 20 ft is desirable. One-pipe systems now are hardly ever installed, but many are still in use.

An *overhead main* runs horizontally, or nearly so, at an elevation higher than the radiators it serves, and is supplied by a vertical main riser. Such overhead mains are frequently placed in the attics of multistoried buildings.

Supply risers are vertical pipes that pass from story to story to convey steam to the radiators or convectors on several floors. Risers are known as either upfeed or downfeed risers, the latter being those in which the steam flows downward to the radiators or convectors from an overhead main.

Return risers are those vertical pipes that take the condensate from the radiators or coils on the several stories of a building and convey it to the return main. Return risers are always of the downfeed type.

A *return main* is a nearly horizontal line of pipe that receives the condensate from the heating system and returns it to the boiler or otherwise disposes of it. The return main is usually run in the basement, but in any event it must be below all heat-transmitting surfaces and must drain the supply mains.

A *dry-return main* is one that is run above the water line of the boiler. In some types of steam heating, a dry return conveys both water and steam, while in others which have traps at all points communicating with the steam main, a dry return carries only water and air.

A *wet-return main* is one that is run below the water line of the boiler or equivalent device and is filled with water at all times. As a rule, a wet-return main is preferable to a dry-return main, except where the main is subject to freezing temperatures. Supply mains must not be connected with other supply mains except through water seals or nonreturn traps. This type of connection is necessary because the unequal frictional resistance to steam flow, and the rate of heat dissipation in each main, produce different pressures at the respective return ends of the mains. If these ends are open through dry returns to each other, turbulence and noise will result. When the ends of supply mains in any one system are sealed separately by dropping into a wet return, the water-column level in each vertical pipe will be sufficiently different from the others to balance the pressure difference.

A *drip pipe*, or *relief*, or *bleeder*, is a pipe used to drain condensate away from the foot of supply risers or from low points, pockets, or traps in the steam main. Drip pipes usually drain into wet returns. Drip pipes in steam mains are employed at points where reduction or increase in the size of the main occurs, and where eccentric reducing fittings cannot be used. This drainage serves to prevent water hammer, by relieving the main of the water which would otherwise accumulate at such locations. The extreme end of the steam main must always be connected so as to drain the condensate into the wet-return main.

7-5. TWO-PIPE AIR-VENT SYSTEM

The two-pipe system differs from the previously described one-pipe system in that separate circuits are provided for the supply and return parts of the system. Figure 7-5 illustrates a representative two-pipe system. It will be noticed that the supply main from the boiler slopes away from the boiler and eventually connects into the return line. This connection can be made by means of a vertical downdrop water leg or, as shown, through a trap. Traps are described in detail later in this chapter, but at this point it should be mentioned that the purpose of a steam trap is to prevent the passage of steam from one part of a system to another while permitting condensate and air to pass. Thus, with the return line at lower pressure than the supply line, the trap prevents steam from passing into the return line but enables the water (condensate) and air to be removed from the supply line.

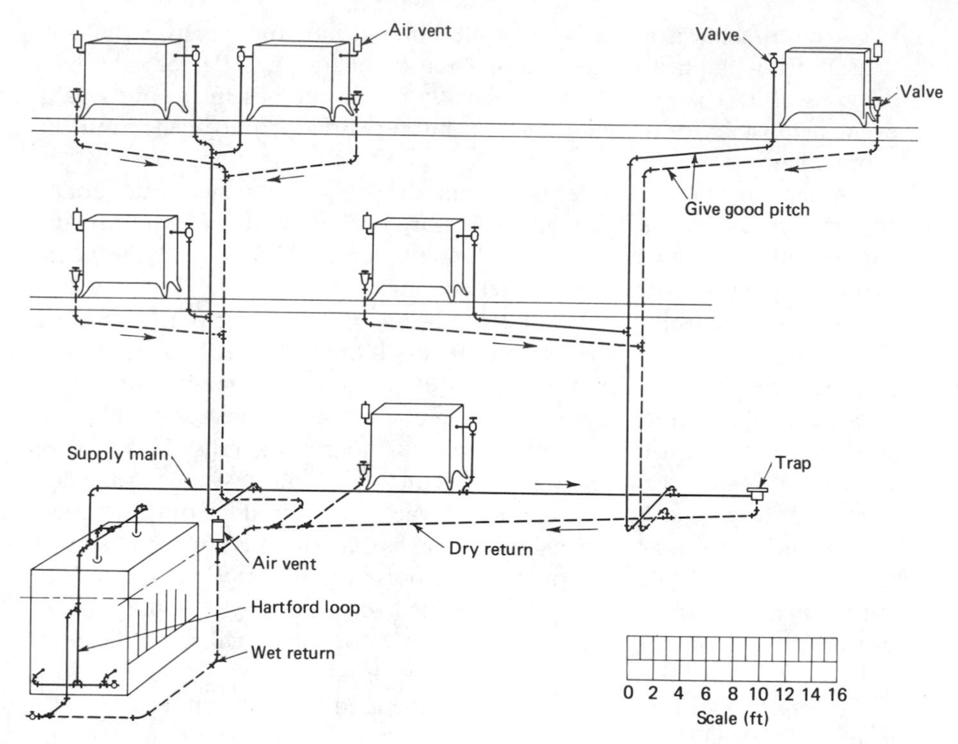

FIGURE 7-5. Two-pipe air-vent steam system.

The supply main can serve risers which supply radiators on different floors, or direct radiator connections can be made into the main. With a two-pipe system, inlet to the radiator does not have to be made at a low point. However, for condensate drainage, a low point on the heater device is connected through a valve into the return piping of the system. Valves are required at both the inlet and the outlet to the radiator, and the return lines—where they are not in the form of risers—should slope toward the boiler. In the diagram, the return is indicated as a dry return; that is, it is located at a position above the water level in the boiler. It becomes a wet return when it enters the vertical drop leg from which the condensate flows into the boiler. An air vent is supplied on each radiator and, in starting, the steam from the supply main enters the radiator and expels through the air vent the air which was previously trapped in the radiator. The return line is also supplied with an air vent and this serves to vent, from both the supply main and the return main, air which is not otherwise vented at the radiators.

7-6. VAPOR SYSTEMS

The vapor system differs from the two-pipe, air-vent system in that the air elimination is accomplished usually at a central point or points, and thus no air-vent valves are used on the radiators or convectors. Because it is possible for a system of this type to operate well in the vacuum region, it is customary to use packless valves, which have no opening to outside air around their stems. Traps are employed on each radiator or convector. As stated before, traps permit the passage of condensate and air and prevent the passage of steam. Thus, under ideal operating conditions, a supply of steam enters the radiator and as fast as this condenses it passes out through the trap to the return side. Thus the steam itself is not permitted to short-circuit through the radiator into the return side before it has given up its latent heat in condensing.

Figure 7-6 shows in diagrammatic fashion a layout of such a system. As before, it is desirable for the supply main to drain away from the boiler into the return, and for the return main to connect into the boiler feed line by means of a vertical drop leg. The extreme end of the supply main also connects through a vertical drop leg into the wet return, which feeds the boiler. Water will, of course, stand at different heights in the two vertical drop legs, depending upon the pressure difference in the two parts of the system. It is desirable in a system of this type for the lowest connected radiation to be at least 24 in. above the water level in the boiler, and where this cannot be accomplished other provisions for returning the water to the boiler must be considered. The air in the supply main which does not enter the radiators and leave then through the outlet traps can pass through a connecting trap into the return main, as shown in the illustration. Normal air found in the radiators also enters the return main and moves through

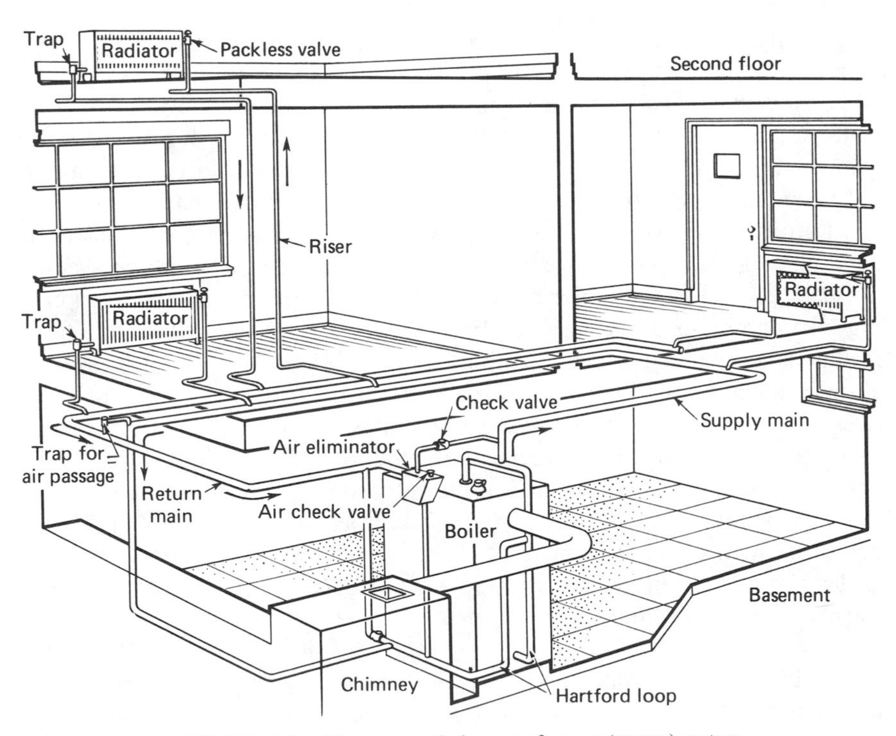

FIGURE 7-6. Diagrammatic layout of vapor (steam) system.

the dry-return part of this main to the central air eliminator of the system. Here, after some cooling, the air is eliminated to the outside, and the eliminator is so designed that when the system is under partial vacuum, outside air cannot enter through it. Because of the tightness with which vapor systems can be built, it is possible for them to operate well into the subatmospheric region. With air excluded, as long as the water-steam in the boiler is hotter than the water-steam medium in the radiators, steam can continue to flow to the radiators and maintain heating at a reduced rate even though no fuel is being burnt. It is easy to control a system of this type, because the boiler pressure can be operated over a range of positive to subatmospheric pressures and because it is possible to use a graduated type of valve at the radiator.

Vapor systems are usually satisfactory in operation because quiet circulation of steam without water hammer or air binding can be realized, and because control through variation in steam temperature is also possible. However, systems of this type usually are designed for effective operation only at low steam pressures, and consequently relatively large pipe sizes are employed. Since the condensate returns to the boiler by gravity, many systems have headroom limitations and it may be necessary to employ the return-trap arrangement described in the following section.

7-7. VAPOR SYSTEM WITH RETURN TRAP

By making use of a return-trap system, it is possible to operate the vapor system at somewhat higher steam pressures, and smaller pipe sizes are possible. However, even with a return-trap system, there are elevation limitations which must be observed in locating the return trap with respect to the return main and boiler water level. In general, slightly better control can be obtained with a return trap system than with conventional water legs.

Return traps, which are also known as alternating receivers, have, in general, a float-actuated mechanism which in one position closes a communicating port to the boiler above the water line and opens a corresponding port communicating with the dry return. In its other position, the float mechanism reverses the port openings. With the receiver open to the return system, water drains by gravity into the receiver, and as this fills the float rises. When sufficient condensate accumulates in the receiver, the float, at its high position, trips a mechanism which closes off the connection to the dry-return main and at the same time opens the return trap to the boiler. With the water in the receiver above the boiler water level, and with pressure equalized in the receiver and boiler, the water drains by gravity into the boiler. The float follows the dropping water level in the receiver until it retrips the valve mechanism to connect with the return side, and a refilling of the receiver occurs. The alternating receiver (return trap) merely makes it possible to use the vapor system more advantageously under adverse or limiting conditions; however, most new designs avoid the use of return traps.

7-8. VACUUM SYSTEM EMPLOYING VACUUM-PUMP RETURN

A vacuum steam-heating system differs from a vapor system in that a vacuum pump is used on the return side to maintain continuously a reduced pressure (usually subatmospheric) on the return side of the system. The vacuum-return pump removes the condensate and air from the return side, delivering the condensate to the boiler and the air to waste. With effective removal of the air, rapid circulation of steam is possible, smaller pipe sizes can be employed, and low-pressure steam or steam at partial vacuum can be fed to the radiators.

From a control viewpoint, the vacuum system is most versatile; however, the initial investment in the vacuum pump and the subsequent maintenance costs limit the use of this system to buildings larger than conventional residences. Vacuum systems are used in the majority of large steam-heated buildings, where the small pipe sizes are particularly desirable and where rapid well-balanced steam circulation is essential. Vacuum-system radiators and convectors have design and installation features similar to those found in vapor systems. However, the connections at the boiler return are so decidedly different that these are separately shown in Fig. 7-7, which illustrates the return arrangements of one type of vacuum system.

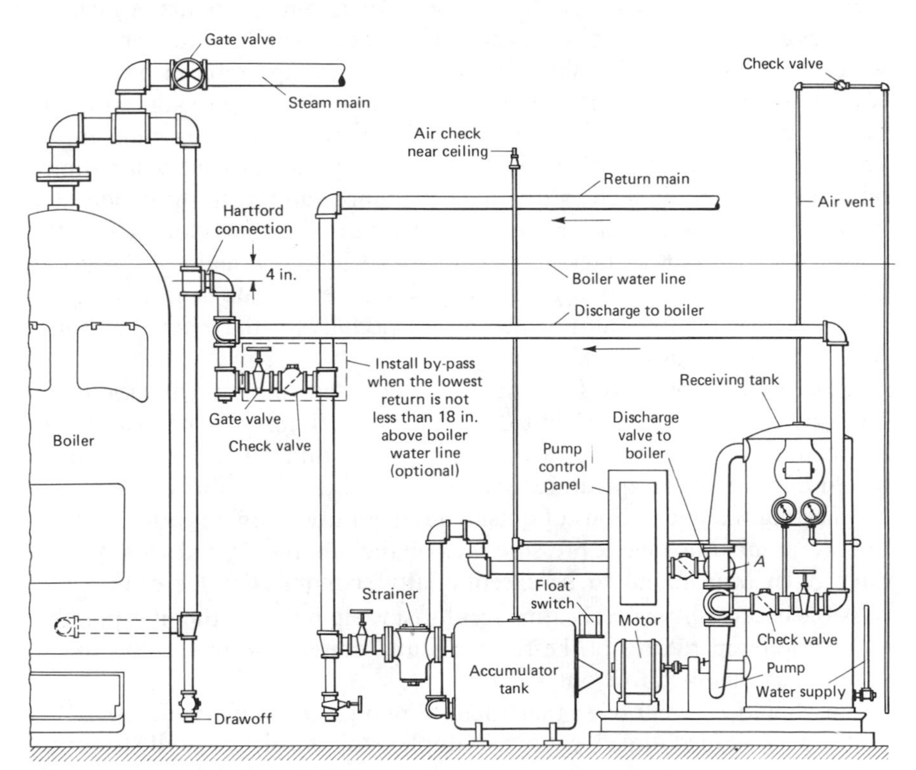

FIGURE 7-7. Vacuum-system return connections and pump. (Courtesy C. A. Dunham Company.)

Condensate and air enter the accumulator tank, which is placed at a low point to which the system can drain. If this location is below the pump suction level, lift fittings make it possible for the pump to lift water from the accumulator tank through as much as 4 ft per lift fitting. A float-operated switch on the accumulator tank starts and stops the pump motor. Since centrifugal pumps cannot effectively pump air, the pump itself is not connected to the accumulator but receives suction water from the receiving tank. The pump delivers this water at high velocity through the jets at *A*, where the kinetic effect of the water jets in a combining tube aspirates (pulls) water and air from the accumulator. In a diffuser tube the kinetic energy of the jets is sufficient to compress the air-water mixture to the receiving-tank pressure, which is slightly above atmospheric. The mixture in the receiving tank is separated, the air passing to the outside and the water remaining to recirculate through the pump. Only a portion of the water discharge from the pump is required by the aspirating jets, and the remainder is sent directly into the boiler as feed water. An automatic control maintains the proper level of water in the receiving tank by closing down on the discharge valve to the boiler feed line until the level is reestablished by additional water entering from the jet circuit.

In addition to the jet-aspirated types of vacuum pumps just described, mention should be made of the reciprocating vacuum pump, which resembles an air compressor in its general features, and of the rotor-type pump, which uses an elliptic housing. The rotor-type pump employs part of the condensate, which whirls rapidly inside an elliptic housing to compress and deliver the air. This pump may be driven by motor or by a low-differential-pressure steam turbine. Other designs employ separate air and water pumps on a common motor shaft. Most vacuum pumps can return the condensate directly to the boiler, provided the boiler pressure does not exceed the reasonable limits of a single-stage centrifugal pump. Usually if the boiler pressure is higher than 50 psig, a separate boiler-feed pump is necessary, and in any event the pump must be designed especially for the maximum boiler pressure to be encountered.

The suction pressure for a vacuum steam-heating system may be as much as 24 in. Hg—equivalent to about a 27-ft column of water—although a vacuum as high as this presents maintenance difficulties. High vacuums produce low steam temperatures and provide greater flexibility of the system for meeting the fluctuations of outside temperatures. Some heating systems operate at subatmospheric pressure even on the supply side, and such systems have often been found to be exceptionally economical. This economy is possible because little overheating results, since in mild weather it is possible to set the temperatures of the radiators and convectors in relation to the variations in temperature out of doors.

It should be mentioned that vacuum pumps are often made as double units, in a so-called duplex pattern. A duplex unit is able to handle the overloads that exist when a system is started, because then both units can run; or in the event that one pump fails, the other can continue in use and keep the system in operation. The capacity of the vacuum pump varies from some 600 000 to 24 000 000 Btuh. Representative data on the units of one manufacturer show that a 600 000 Btuh unit would handle 3.8 gpm and 1.3 cfm of air, while using a $\frac{3}{4}$-hp motor with the discharge pressure 20 psig at pump outlet. It is customary to rate pump units to maintain $5\frac{1}{2}$-in. vacuum with 160 F water. However, the usual units can operate satisfactorily at vacuums as high as 10 in. At reduced capacities or with special pumps, vacuums as high as 25 in. Hg are possible. Because pressure can be accurately controlled in the return system, and because a significant differential between inlet and outlet pressure can be maintained, it is possible to control the steam and heat input to radiators by means of variable-opening inlet valves. These restrict the flow of steam into the unit, and since steam cannot condense faster than it is supplied, the capacity of the unit is thereby limited. Some systems provide thermostatic control valves on individual radiators.

The operating pressure differential in a vacuum system is the difference between the supply pressure and the vacuum on the return side. For example, if the vacuum pump produced 6 in. Hg vacuum and if steam were supplied at 2 psig, a simple calculation would show that the operating differential would be approximately 137 in. water. Because of the significant differential pressure employed in vacuum systems, every connection between the supply main and the return main must be trapped by a thermostatic or mechanical device. One untrapped or leaking connection can defeat the operation of the entire system by passing enough steam to prevent the pump from maintaining a pressure below atmospheric. Vacuum traps are almost exclusively of the thermostatic type, with ports which open to pass water or air but which close tightly as steam, at its higher temperatures, starts to flow. Thus the return pipes of a vacuum system, when in proper condition, do not transport anything warmer than the condensate, which under vacuum is cooler than atmospheric-pressure steam. A typical thermostatic trap is show in Fig. 7-8. With such traps the metallic bellows is partly filled with a volatile liquid which expands so as to close the port when vapor warmer than the design temperature enters through the radiator or pipe drain connection. Water at steam temperature can pass through the trap, since it flows

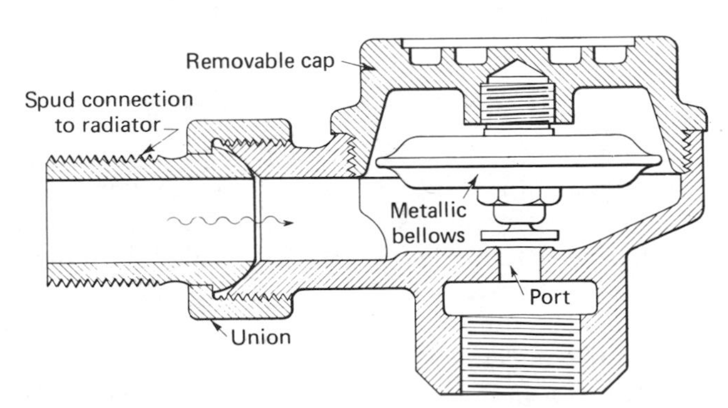

FIGURE 7-8. Thermostatic trap of radiator-type design for passing of water and air.

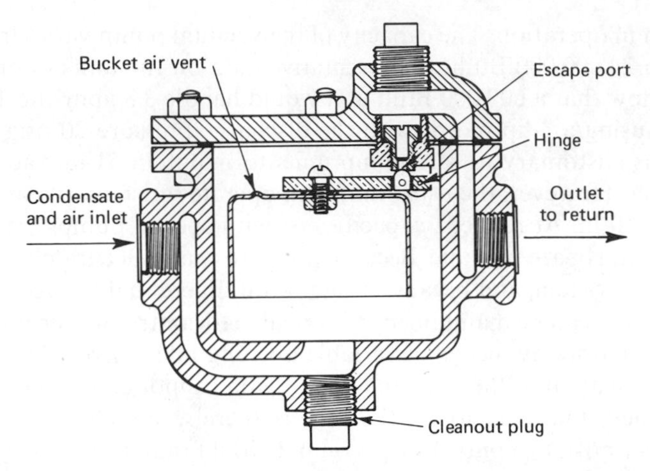

FIGURE 7-9. Inverted-bucket trap. (Courtesy of Hoffman Specialty ITT.)

along the bottom and does not come into contact with the bellows device to an extent sufficient to make it close the port.

Figure 7-9 shows an inverted-bucket trap. This type of trap is used on steam-heating systems but more particularly for a variety of industrial applications with process steam, such as industrial cookers, water heaters, and chemical-process vats. After a trap discharges, water still remains in the trap and, with steam or air in the inverted bucket, the bucket floats in the high (closed) position like an empty overturned can in a pond. As water, steam, and air enter the trap, the water level slowly rises inside the inverted bucket. Air and uncondensed steam pass to the top of the trap body through the vent hole in the bucket. When insufficient air and steam remain in the bucket to hold it up, the bucket falls and the escape port opens. The air in the top of the trap and the water outside and inside the bucket are blown out by the pressure difference between the supply and return systems until the air and vapor in the bucket again exert their buoyancy to close off the escape port. This design of trap is satisfactory for the operation of high-pressure as well as low-pressure systems, with pressure differentials of 5 to 200 psia. The units have capacities ranging from 900 to 11 000 lb/h.

7-9. STEAM-PIPE ARRANGEMENTS

When water whose temperature is higher than the temperature corresponding to the pressure is released into a pipe or other container at a reduced pressure, part of the water will flash into steam. This action occurs at the suction inlet of a pump which is lifting hot water, and at the outlet of a high-pressure steam trap. If the steam for a vacuum system comes from a high-pressure source it is necessary to provide for conservation and disposal of the high-temperature condensate from high-pressure traps. The trap outlet may be connected to a flash tank, with an interior capacity at least four times that of the trap body. From the top of the flash tank there should be a valved steam connection to the low-pressure steam main, and from the bottom of the tank a thermostatic trap connection into the vacuum return.

All vapor or temperature-modulating systems of steam heating operate on substantially the same general principle. Graduated radiator supply valves are employed, in which the area of the steam passage at the port responds to movement of the valve handle knob. To prevent air leakage into vapor and vacuum systems, as well as to contain the steam (when the pressure is above atmospheric), it is customary to employ packless valves. Such valves must be provided with a sealed diaphragm or a sealed bellows arranged so that the handle and its mechanism can operate the valve without having any passage for fluid flow into the system. Examination of Fig. 7-10 shows that by turning the handle for closure, the diaphragm is pushed down, in turn forcing the internal stem and valve washer against the valve seat. During this process the spring is compressed against its shoulder. When the handle is released to open the valve, the spring pushes up against the diaphragm and the valve stem and this pulls the valve washer off its seat. The valve can be operated at any position from closed to wide open. These valves are in sharp contrast to the packed-stem valves, which are shown in Fig. 7-12. Most of the steam-heating system layouts shown in this chapter use the Hartford loop, which is designed to reduce the possibility of water leaving the boiler by any method except evaporation. In Fig. 7-5 the loop is indicated by the arrow and legend. An equalizing pipe from the steam section of the boiler leads to that point at which the feed-water return to the boiler is connected. This point is located 2 to 4 in. below the normal boiler water level, and the downflow pipe, which is normally flooded, carries the feed water into the boiler. Under certain conditions of abnormal operation, it is possible to have a pressure much lower than boiler pressure exist in the return system and at the extreme end of the steam supply main. With such reduced pressure it would be possible for water to leave the boiler and back

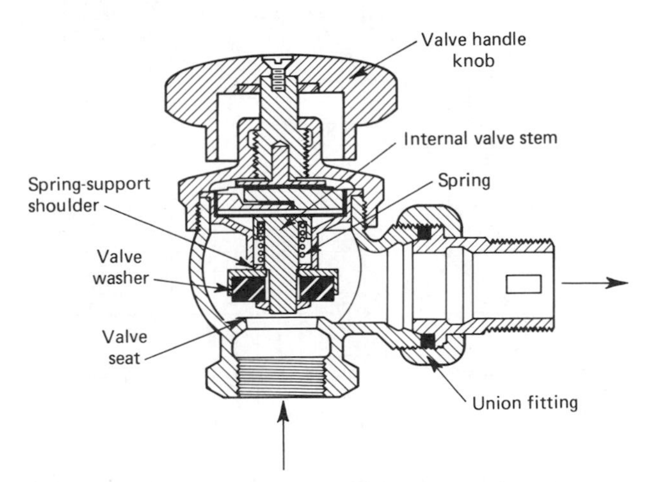

FIGURE 7-10. Cross-sectional view of packless radiator valve. (Courtesy of Hoffman Specialty ITT.)

up in the piping, and in some instances even reach the heat-transfer equipment. Such a loss of water from the boiler might endanger the combustion boiler heating surface. The Hartford loop can prevent such water loss, because when the water level drops below the return connection point, then steam from the downcomer equalizing line comes in direct contact with the return water, and the boiler pressure acting on both the return water and the boiler water stops the syphon action so that a further drop in the boiler water is avoided. A similar water-loss condition might arise if, for example, a sudden breakage occurred in the return line. Again, water would not run

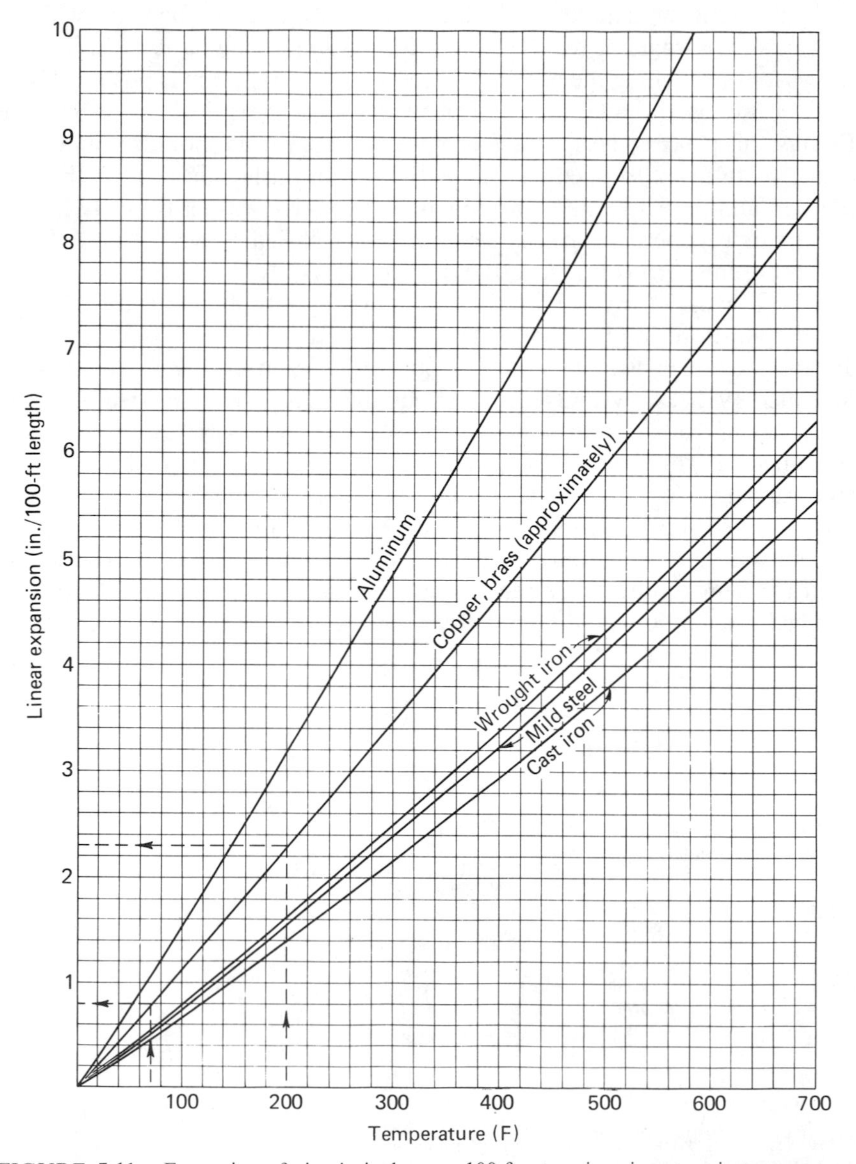

FIGURE 7-11. Expansion of pipe in inches per 100 ft, at various increases in temperature.

out of the boiler by syphon action past the point where steam acts at the return inlet. Thus it is customary to provide the Hartford loop as a safety feature for steam boilers. It has no similar use in connection with hot-water systems.

In the design and erection of steam-heating plants the problem of expansion cannot be neglected. Care must be taken to have every long steam pipe free for limited movement, and so arranged that small movements of connections or branches are not blocked by building material or by structural beams or columns. A steam main should be supported from overhead with strong hangers, but the branches to the risers must be free to move to take care of expansion. At the bottom of each vertical pipe there should be two elbows. One of these elbows directly on the riser should have more than a true 90° angle so as to facilitate drainage. This elbow and a nipple connect with an ordinary elbow placed on its side. Thus there will always be a nipple or a piece of pipe upon which some swing or spring movement is possible for expansion and contraction of the riser.

Risers, especially, require careful design and installation to provide for expansion, since the necessary drainage pitch of runouts to radiators may easily be lost by the expansion of a tall riser.

Expansion joints must be provided in the risers of multistory buildings—with guides to control, and anchors to limit, their travel. These joints are usually placed at intervals not exceeding 40 ft. Figure 7-11 indicates the variation in length of metal pipes in accordance with changes in temperature.

7-10. STEEL PIPE, COPPER TUBING, AND FITTINGS

The most common pipe metal for industrial use is steel (wrought or cast iron is also used, but less frequently). Table 7-1 gives data for standard-weight steel pipe (Schedule 40) as well as for two heavier wall types: extra-strong (Schedule 80) and double-extra-strong (for high-pressure service). For steam lines and hot-water heating lines, steel pipe is employed with threaded fittings, or in some cases it is connected by welding. Where threaded fittings are used, it is desirable to ream the ends of the pipe to eliminate burrs and to avoid the possibility of a partial closure of the internal bore. If this is not done, design computations for frictional loss are uncertain. Where steel pipe is used for supply water systems, it is desirable to use galvanized pipe, which is steel pipe covered with a corrosion-resistant zinc coating. Such coatings can prevent corrosion (rusting) for long periods of time.

Copper tubing is coming into increasing use for hot-water heating systems (and less frequently, for steam heating), and in many installations where water or refrigerants are being carried. Table 7-2 gives the characteristic data of conventional copper tubing, frequently known in the industry as *copper water tubing*. Copper water tubing is seamless, deoxidized copper tube supplied in three wall thicknesses. Type K, the heaviest, is supplied in coils 60 to 100 ft long, for the small bores of tubing, and in 12- to 20-ft lengths for the larger size tubes. It is furnished in either hard or soft temper.

Table 7-1 Standard Steel-Pipe Data

Nominal Size	External Diam (in.)	Standard-Weight Pipe (ASA Schedule 40)										Extra-Strong (ASA Schedule 80)		Double-Extra-Strong		Nominal Size
		Internal Diam (in.)	Wall Thickness (in.)	Weight per Ft Plain Ends (lb)	Threads per In.	Circumference (in.)		Transverse Area (sq in.)		Length of Pipe per Sq Ft		Wall Thickness (in.)	Weight per Ft Plain Ends (lb)	Wall Thickness (in.)	Weight per Ft Plain Ends (lb)	
						External	Internal	External	Internal	External Surface	Internal Surface					
$\frac{1}{8}$	0.405	0.269	0.068	0.244	27	1.272	0.845	0.129	0.057	9.431	14.199	0.095	0.31	......	......	$\frac{1}{8}$
$\frac{1}{4}$	0.540	0.364	0.088	0.424	18	1.696	1.144	0.229	0.104	7.073	10.493	0.119	0.54	......	......	$\frac{1}{4}$
$\frac{3}{8}$	0.675	0.493	0.091	0.567	18	2.121	1.549	0.358	0.191	5.658	7.748	0.126	0.74	......	......	$\frac{3}{8}$
$\frac{1}{2}$	0.840	0.622	0.109	0.850	14	2.639	1.954	0.554	0.304	4.547	6.141	0.147	1.09	0.294	1.71	$\frac{1}{2}$
$\frac{3}{4}$	1.050	0.824	0.113	1.130	14	3.299	2.589	0.866	0.533	3.637	4.635	0.154	1.47	0.308	2.44	$\frac{3}{4}$
1	1.315	1.049	0.133	1.678	$11\frac{1}{2}$	4.131	3.296	1.358	0.864	2.904	3.641	0.179	2.17	0.358	3.66	1
$1\frac{1}{4}$	1.660	1.380	0.140	2.272	$11\frac{1}{2}$	5.215	4.335	2.164	1.495	2.301	2.768	0.191	3.00	0.382	5.21	$1\frac{1}{4}$
$1\frac{1}{2}$	1.900	1.610	0.145	2.717	$11\frac{1}{2}$	5.969	5.058	2.835	2.036	2.010	2.372	0.200	3.63	0.400	6.41	$1\frac{1}{2}$
2	2.375	2.067	0.154	3.652	$11\frac{1}{2}$	7.461	6.494	4.430	3.355	1.608	1.847	0.218	5.02	0.436	9.03	2
$2\frac{1}{2}$	2.875	2.469	0.203	5.793	8	9.032	7.757	6.492	4.788	1.328	1.547	0.276	7.66	0.552	13.70	$2\frac{1}{2}$
3	3.500	3.068	0.216	7.575	8	10.996	9.638	9.621	7.393	1.091	1.245	0.300	10.25	0.600	18.58	3
$3\frac{1}{2}$	4.000	3.548	0.226	9.109	8	12.566	11.146	12.566	9.886	0.954	1.076	0.318	12.51	0.636	22.85	$3\frac{1}{2}$
4	4.500	4.026	0.237	10.790	8	14.137	12.648	15.904	12.730	0.848	0.948	0.337	14.98	0.674	27.54	4
5	5.563	5.047	0.258	14.617	8	17.477	15.856	24.306	20.006	0.686	0.756	0.375	20.78	0.750	38.55	5
6	6.625	6.065	0.280	18.974	8	20.813	19.054	34.472	28.891	0.576	0.629	0.432	28.57	0.864	53.16	6
8	8.625	7.981	0.322	28.554	8	27.096	25.073	58.426	50.027	0.443	0.478	0.500	43.39	0.875	72.42	8
10	10.750	10.020	0.365	40.483	8	33.772	31.479	90.763	78.855	0.355	0.381	0.593	64.3	......	......	10
12	12.750	12.000	0.375	49.562	8	40.055	37.699	127.676	113.097	0.299	0.318	0.687	88.5	......	......	12

Note. Pipe sizes 14 in. and above are designated by outside diameter, and the wall thickness is specified.

Table 7-2 Standard Dimensions, Weight, and Other Data for Copper Water Tubes

Nominal Size	Outside Diameter (in.)	Wall Thickness (in.)	Weight per Foot (lb)	Bore Area (sq in.)	Surface Area (sq ft per ft)		Safe Inside Pressure (psi)
					Outer	Inside	
Type K. Heavy Wall (Furnished Hard or Soft Temper)							
1/4	0.375	0.035	0.145	0.076	0.0981	0.0890	1120
3/8	0.500	0.049	0.269	0.127	0.1309	0.1052	1170
1/2	0.625	0.049	0.344	0.218	0.1636	0.1380	920
5/8	0.750	0.049	0.418	0.333	0.1963	0.1707	760
3/4	0.875	0.065	0.641	0.436	0.2291	0.1950	880
1	1.125	0.065	0.839	0.778	0.2945	0.2605	680
1 1/4	1.375	0.065	1.04	1.22	0.3600	0.3259	550
1 1/2	1.625	0.072	1.36	1.72	0.4254	0.3877	520
2	2.125	0.083	2.06	3.01	0.5563	0.5129	450
2 1/2	2.625	0.095	2.93	4.66	0.6872	0.6375	420
3	3.125	0.109	4.00	6.64	0.8181	0.7611	410
3 1/2	3.625	0.120	5.12	9.00	0.9490	0.8862	380
4	4.125	0.134	6.51	11.7	1.080	1.010	370
5	5.125	0.160	9.67	18.1	1.342	1.258	360
6	6.125	0.192	13.9	25.9	1.604	1.503	350
8	8.125	0.271	25.9	45.2	2.127	1.985	....
10	10.125	0.338	40.3	70.12	2.651	2.474	....
Type L. Medium Wall (Furnished Hard or Soft Temper)							
1/4	0.375	0.030	0.126	0.078	0.0981	0.0903	900
3/8	0.500	0.035	0.198	0.145	0.1309	0.1126	800
1/2	0.625	0.040	0.285	0.233	0.1636	0.1427	740
5/8	0.750	0.042	0.362	0.348	0.1963	0.1744	650
3/4	0.875	0.045	0.455	0.484	0.2291	0.2055	590
1	1.125	0.050	0.655	0.825	0.2945	0.2683	510
1 1/4	1.375	0.055	0.884	1.26	0.3600	0.3312	460
1 1/2	1.625	0.060	1.14	1.78	0.4254	0.3940	430
2	2.125	0.070	1.75	3.09	0.5563	0.5197	370
2 1/2	2.625	0.080	2.48	4.77	0.6872	0.6453	350
3	3.125	0.090	3.33	6.81	0.8181	0.7710	330
3 1/2	3.625	0.100	4.29	9.21	0.9490	0.8967	320
4	4.125	0.110	5.28	12.0	1.080	1.022	300
5	5.125	0.125	7.61	18.7	1.342	1.276	280
6	6.125	0.140	10.21	26.8	1.604	1.530	260
8	8.125	0.200	19.3	46.9	2.127	2.022	....
10	10.125	0.250	30.1	72.8	2.651	2.520	....

Table 7-2 (*Continued*)

Nominal Size	Outside Diameter (in.)	Wall Thickness (in.)	Weight per Foot (lb)	Bore Area (sq in.)	Surface Area (sq ft per ft)		Safe Inside Pressure (psi)
					Outer	Inside	
Type M. Light Wall (Furnished Hard Temper Only)							
3/8	0.500	0.025	0.145	0.159	0.1309	0.1243	....
1/2	0.625	0.028	0.204	0.254	0.1636	0.1562	....
3/4	0.875	0.032	0.328	0.517	0.2291	0.1944	....
1	1.125	0.035	0.465	0.874	0.2945	0.2852	....
1 1/4	1.375	0.042	0.682	1.31	0.3600	0.3489	....
1 1/2	1.625	0.049	0.940	1.83	0.4254	0.4124	....
2	2.125	0.058	1.46	3.17	0.5563	0.5409	....
2 1/2	2.625	0.065	2.03	4.89	0.6872	0.6698	280
3	3.125	0.072	2.68	6.98	0.8181	0.7988	260
3 1/2	3.625	0.083	3.58	9.40	0.9490	0.9268	260
4	4.125	0.095	4.66	12.2	1.080	1.055	260
5	5.125	0.109	6.66	18.9	1.342	1.312	240
6	6.125	0.122	8.92	27.2	1.604	1.571	230

Table 7-3 Standard Dimensions, Weights, and Other Data for Threaded Standard-Pipe-Size Copper Pipe and Red Brass Pipe

Nominal Size	Actual OD (in.)	Wall Thickness (in.)	Weight (lb per lin ft)		Linear Feet Containing One U.S. Gal	Safe Working Pressure (psi)	
			Copper	Red Brass		Copper	Red Brass
Regular Pipe							
1/8	0.405	0.062	0.259	0.253	310.6	533	712
1/4	0.540	0.082	0.457	0.447	175.5	845	1128
3/8	0.675	0.090	0.641	0.627	100.2	818	1090
1/2	0.840	0.107	0.955	0.934	62.9	715	952
3/4	1.050	0.114	1.30	1.27	36.2	653	868
1	1.315	0.126	1.82	1.78	21.7	511	680
1 1/4	1.660	0.146	2.69	2.63	13.1	550	732
1 1/2	1.900	0.150	3.20	3.13	9.6	505	674
2	2.375	0.156	4.22	4.12	5.8	435	580
2 1/2	2.875	0.187	6.12	5.99	3.9	364	485
3	3.500	0.219	8.75	8.56	2.6	408	544
3 1/2	4.000	0.250	11.4	11.2	2.0	450	600
4	4.500	0.250	12.9	12.7	1.5	400	534
5	5.563	0.250	16.2	15.8	0.96	324	432
6	6.625	0.250	19.4	19.0	0.65	272	362

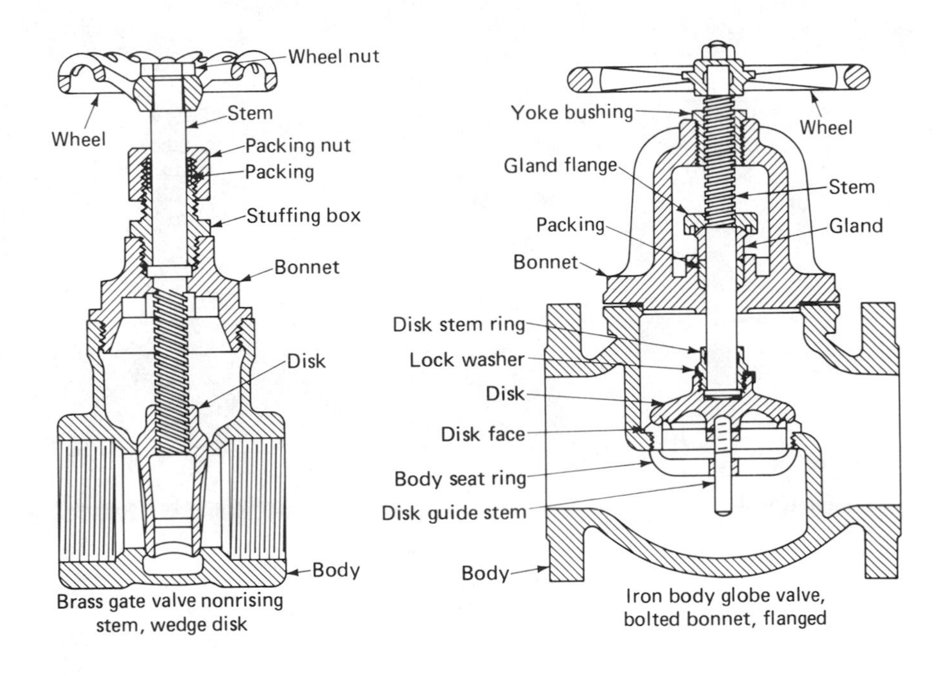

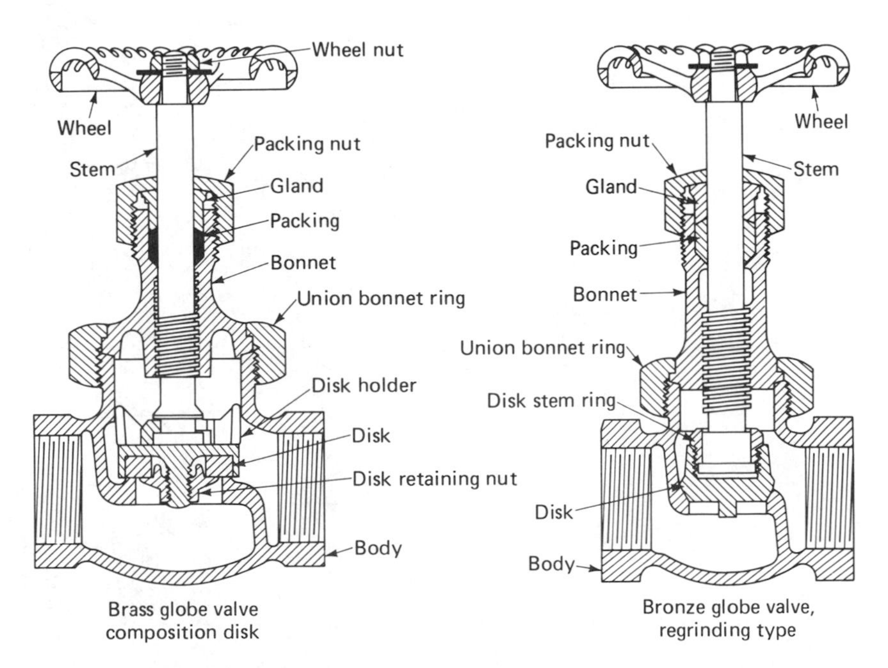

FIGURE 7-12. Packed-stem valves for steam and liquid usage.

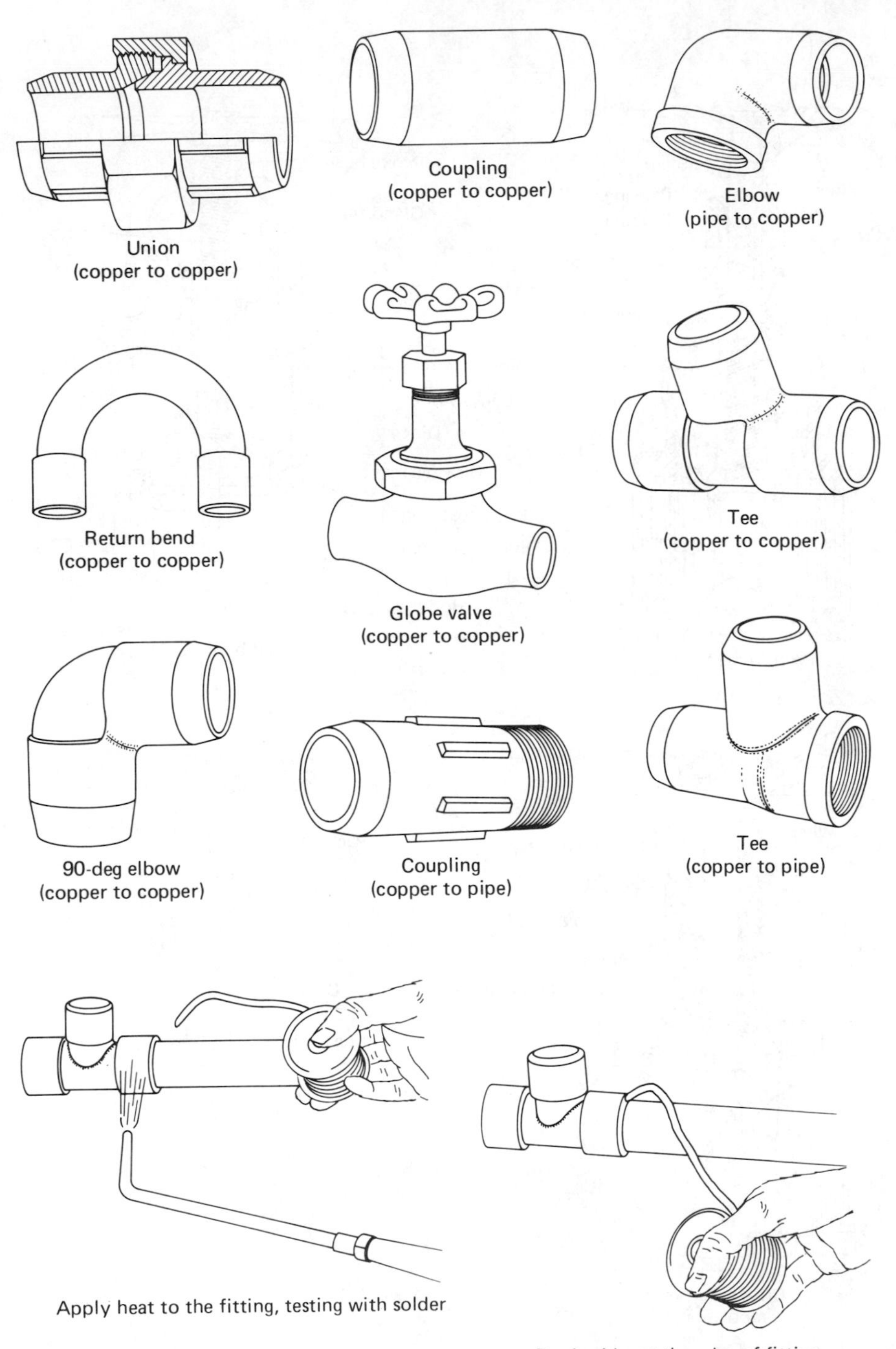

FIGURE 7-13. Representative solder-type copper fittings and diagrammatic view of method of making a soldered joint.

Type L has a lesser wall thickness. It also is supplied in 60- to 100-ft coils up to and including 1 in., and in 60-ft coils for the $1\frac{1}{4}$- and $1\frac{1}{2}$-in. sizes. Above $1\frac{1}{2}$ in. it is necessarily supplied in 12- or 20-ft lengths.

The type M tubing, which has the thinnest wall, is furnished in hard temper in 12- and 20-ft straight lengths. Whether hard or soft temper is employed depends upon the type of utilization and the appearance, which might be expected. Hard-temper tubing, which is rigid, is pleasing in appearance and is most usually used where runs of pipe are exposed. Soft-temper tubing is used for underground service or where it can be concealed in walls.

Copper water tubing, such as types K, L, and M, is not intended to be joined by threads. A most common type of connection is the solder joint, or sweated joint. In use, the pipe is cut square to size, the end mating surfaces are cleaned with steel wool, and a flux is applied to the surfaces. The pipe is inserted in the joint and the joint is then heated. Solder, usually of lead-tin or tin-antimony, melts on the hot surface and is drawn into the space between the fitting and the tubing by capillary action. After cooling and solidification a tight and permanent joint results. Where temperatures exceed the 250 F allowed for lead-tin solder, or the 312 F allowed for tin-antimony solder, it is necessary to use for the joints a brazing alloy or silver solder (Fig. 7-13). Type K and type L copper water tubes are also joined by compression fittings.

7-11. EQUIVALENT PIPE LENGTH AND STEAM FLOW

The resistance to flow of steam or other fluid through a pipe is increased by the presence of valves and fittings, and the carrying capacity is thus reduced. Investigators have determined the resistance of valves and fittings in terms of lengths of straight pipe. The resistances thus expressed are added to the measured length of the pipe, and the sum is called the equivalent pipe length. Equivalent pipe length is thus the length of straight pipe of a given size that would have the same resistance as the actual pipe with its bends, fittings, valves, and other devices.

Table 8-2, which applies for smaller pipe sizes, gives the equivalent feet of run of various pipe.

Steam flowing through pipe and fittings is retarded by friction loss between the surface of the pipe and the steam, and by turbulence losses in bends and through the restricted passages which are found in valves and other devices. The design of the system should permit uniformly controlled distribution of the steam to various outlets, with minimum noise and with provision for release of air when air is in the system. Elimination and return of condensate present an additional design problem. Furthermore, it is necessary to balance distribution in a system so that heaters remote from the steam source are not starved in relation to heaters adjacent to the steam source. This can sometimes be done by adjustment of pipe sizes, but it may also be necessary to provide orifice plates for insertion in the inlet of a radiator, to restrict or control the inflow of steam. The system and its radiators must operate satisfactorily not only under conditions of full heating load but also under conditions of partial load. Moreover, in the warming-up period the system is called upon to distribute an excessive amount of steam and to return an equivalent excess of condensate. Thus during this period the system is operating under an overload.

Because of the complexity of the problem, pressure-loss designs based purely on theory and analytical considerations are frequently unsatisfactory. It has been found better to use semiempirical formulations, or compilations of experimental test data, for laying out designs. One of the many equations used for the flow of steam in pipes is the semiempirical Babcock formula, which is

$$m = 87\sqrt{\frac{(\Delta P)D^5}{v(1 + 3.6/D)L}} \tag{7-1}$$

where m is the pounds of steam flowing per minute; ΔP is the loss of pressure, in pounds per square inch; D is the inside diameter of pipe, in inches; L is the equivalent length of pipe, in feet; and v is the specific volume of steam, cubic foot per pound, measured at average pressure.

This equation is readily used in any case where the entire load is considered at the end of the main. When the load is distributed along the main, however, the process of sizing the main is more complicated. For ordinary heating installations, tables have been developed to provide for selection of pipe sizes.

7-12. PIPE SIZES FOR STEAM-HEATING SYSTEMS

A detailed treatment of the design and pipe sizing of steam systems will not be presented here. This topic is treated completely in the ASHRAE *Handbook of Fundamentals*—1972 (chap. 16) and in other ASHRAE publications. However, to give an idea of the use of tabular pipe-sizing data, Tables 7-4, 7-5, 7-6, and 7-7 have been reproduced by permission from an earlier edition of the ASHRAE *Guide and Data Book*. Table 7-4 is satisfactory for selecting pipe sizes to use with all low-pressure steam-heating systems. In laying out a design, it is necessary to know the total pressure drop available for distributing the steam between the source and the end of the return, following which the allowable or most suitable pressure drop per 100 equivalent feet of run can be selected. The equivalent length of run of the main and return [and also the equivalent length from the steam source (boiler) to the most distant heating element] must be known, and the direction of flow of condensate—that is, whether it is with or against the direction of steam flow—must also be determined.

Table 7-4 is applicable both to one-pipe and two-pipe systems and can also be used for vapor and vacuum systems. The headings at the top of each column specifically state the conditions of use. For return piping, Table 7-5 is applicable in a similar way. It is desirable to use on the return side the same pressure drop per 100 ft that is used on the supply side of a system.

Air-vent systems of one-pipe design were formerly very common because of low initial cost. In the one-pipe system illustrated in Fig. 7-1, the single pipe from the boiler carries both the supply steam and all the condensate of the system. The combined steam main and return continues until it finally drops and becomes the wet return. A water column automatically collects in the wet return at a sufficiently high level to force the water back into the boiler and counterbalance the pressure difference between the supply and return sides of the system. In two-pipe systems, as illustrated in Figs. 7-5 and 7-6, a water column is also present to force the return water into the boiler. The height of water column needed to balance the unit pressure can be selected from the following tabulation:

$$1 \text{ oz/in.}^2 = 1.73 \text{ in. water at 70 F} = \tfrac{1}{16} \text{ psi}$$
$$= 1.78 \text{ in. water at 180 F}$$
$$= 1.80 \text{ in. water at 200 F}$$
$$1 \text{ psi} = 28.80 \text{ in. water at 200 F} = 2.4 \text{ ft}$$

Table 7-4 Steam-Pipe Capacities for Low-Pressure Systems

Pipe Size (in.)	Capacities of Steam Mains and Risers								Special Capacities for One-Pipe Systems Only		
	Direction of Condensate Flow in Pipe Line										
	With the Steam, in One-Pipe and Two-Pipe Systems						Against the Steam, Two-Pipe Only		Supply Risers, Upfeed	Radiator Valves and Vertical Connections	Radiator and Riser Runouts
	1/32-PSI or 1/2-Oz Drop	1/24-PSI or 2/3-Oz Drop	1/16-PSI or 1-Oz Drop	1/8-PSI or 2-Oz Drop	1/4-PSI or 4-Oz Drop	1/2-PSI or 8-Oz Drop	Vertical	Horizontal			
A	B	C	D	E	F	G	H*	I†	J‡	K	L†
	Capacity Expressed in Pounds per Hour										
3/4	......	......	8	......	......	......	8	......	6	......	7
1	10	12	14	20	28	40	14	9	11	7	7
1 1/4	22	25	31	43	61	87	31	19	20	16	16
1 1/2	34	39	48	67	95	135	48	27	38	23	23
2	68	79	97	137	193	273	97	49	72	42	42
2 1/2	112	130	159	225	318	449	159	99	116	......	65
3	206	237	291	411	581	822	282	175	200	......	119
3 1/2	307	355	434	614	869	1,230	387	288	286	......	186
4	435	503	614	869	1,230	1,740	511	425	380	......	278
5	806	928	1,140	1,610	2,270	3,210	1,050	788	......	......	545
6	1,320	1,520	1,870	2,640	3,730	5,280	1,800	1,400	......	......	......
8	2,750	3,170	3,880	5,490	7,770	11,000	3,750	3,000	......	......	......
10	5,010	5,790	7,090	10,000	14,200	20,000	7,000	5,700	......	......	......
12	8,040	9,290	11,400	16,100	22,700	32,200	11,500	9,500	......	......	......
16	15,100	17,400	21,200	30,300	42,400	60,500	22,000	19,000	......	......	......
	All Horizontal Mains and Downfeed Risers						Upfeed Risers	Mains and Undripped Runouts	Upfeed Risers	Radiator Connections	Runouts Not Dripped

* Do not use column H for drops of 1/24 or 1/32 psi; substitute column C or column B as required.
† Pitch of horizontal runouts to risers and radiators should be not less than 1/2 in. per ft. Where this pitch cannot be obtained, runouts over 8 ft in length should be one pipe size larger than called for in the table.
‡ Do not use column J for 1/32 psi drop except on sizes 3 in. and over; below 3 in. substitute column B.

Table 7-5 Return-Pipe Capacities for Low-Pressure Systems*
(lb/h)

Pipe Size (in.)	Capacity of Return Mains and Risers																	
	Mains																	
	$\frac{1}{32}$-PSI or $\frac{1}{2}$-Oz Drop per 100 Ft			$\frac{1}{24}$-PSI or $\frac{2}{3}$-Oz Drop per 100 Ft			$\frac{1}{16}$-PSI or 1-Oz Drop per 100 Ft			$\frac{1}{8}$-PSI or 2-Oz Drop per 100 Ft			$\frac{1}{4}$-PSI or 4-Oz Drop per 100 Ft			$\frac{1}{2}$-PSI or 8-Oz Drop per 100 Ft		
	Wet	Dry	Vac.	Wet	Dry	Vac.	Wet	Dry	Vac.	Wet	Dry	Vac.	Wet	Dry	Vac.	Wet	Dry	Vac.
M	N	O	P	Q	R	S	T	U	V	W	X	Y	Z	AA	BB	CC	DD	EE
$\frac{3}{4}$	....	....	....	....	....	42	....	....	100	....	....	142	....	....	200	....	....	283
1	125	62	....	145	71	143	175	80	175	250	103	249	350	115	350	....	....	494
$1\frac{1}{4}$	213	130	....	248	149	244	300	168	300	425	217	426	600	241	600	....	....	848
$1\frac{1}{2}$	338	206	....	393	236	388	475	265	475	675	340	674	950	378	950	....	....	1,340
2	700	470	....	810	535	815	1,000	575	1,000	1,400	740	1,420	2,000	825	2,000	....	....	2,830
$2\frac{1}{2}$	1,180	760	....	1,580	868	1,360	1,680	950	1,680	2,350	1,230	2,380	3,350	1,360	3,350	....	....	4,730
3	1,880	1,460	....	2,130	1,560	2,180	2,680	1,750	2,680	3,750	2,250	3,800	5,350	2,500	5,350	....	....	7,560
$3\frac{1}{2}$	2,750	1,970	....	3,300	2,200	3,250	4,000	2,500	4,000	5,500	3,230	5,680	8,000	3,580	8,000	....	....	11,300
4	3,880	2,930	....	4,580	3,350	4,500	5,500	3,750	5,500	7,750	4,830	7,810	11,000	5,380	11,000	....	....	15,500
5	....	....	....	....	....	7,880	....	....	9,680	....	....	13,700	....	....	19,400	....	....	27,300
6	....	....	....	....	....	12,600	....	....	15,500	....	....	22,000	....	....	31,000	....	....	43,800
	Risers																	
$\frac{3}{4}$	....	48	....	....	48	143	....	48	175	....	48	249	....	48	350	....	....	494
1	....	113	....	....	113	244	....	113	300	....	113	426	....	113	600	....	....	848
$1\frac{1}{4}$	....	248	....	....	248	388	....	248	475	....	248	674	....	248	950	....	....	1,340
$1\frac{1}{2}$	....	375	....	....	375	815	....	375	1,000	....	375	1,420	....	375	2,000	....	....	2,830
2	....	750	....	....	750	1,360	....	750	1,680	....	750	2,380	....	750	3,350	....	....	4,730
$2\frac{1}{2}$	....	....	....	....	....	2,180	....	....	2,680	....	....	3,800	....	....	5,350	....	....	7,560
3	....	....	....	....	....	3,250	....	....	4,000	....	....	5,680	....	....	8,000	....	....	11,300
$3\frac{1}{2}$	....	....	....	....	....	4,480	....	....	5,500	....	....	7,810	....	....	11,000	....	....	15,500
4	....	....	....	....	....	7,880	....	....	9,680	....	....	13,700	....	....	19,400	....	....	27,300
5	....	....	....	....	....	12,600	....	....	15,500	....	....	22,000	....	....	31,000	....	....	43,800

* Table based on pipe-size data developed through research investigations of American Society of Heating and Ventilating Engineers.

Table 7-6 Steam-Pipe Capacities for 30-psig Steam Systems (lb/h) (Steam and Condensate Flowing in Same Direction)

Pipe Size (in.)	Drop in Pressure (psi per 100-ft length)					
	$\frac{1}{8}$	$\frac{1}{4}$	$\frac{1}{2}$	$\frac{3}{4}$	1	2
$\frac{3}{4}$	15	22	31	38	45	63
1	31	46	63	77	89	125
$1\frac{1}{4}$	69	100	141	172	199	281
$1\frac{1}{2}$	107	154	219	267	309	437
2	217	313	444	543	627	886
$2\frac{1}{2}$	358	516	730	924	1,030	1,460
3	651	940	1,330	1,630	1,880	2,660
$3\frac{1}{2}$	979	1,410	2,000	2,450	2,830	4,000
4	1,390	2,000	2,830	3,460	4,000	5,660
6	4,210	6,030	8,590	10,400	12,100	17,200
8	8,750	12,600	17,900	21,900	25,300	35,100
10	16,300	23,500	33,200	40,600	46,900	66,400
12	25,600	36,900	52,300	64,000	74,000	104,500

Table 7-7 Return-Pipe Capacities for 30-psig Steam Systems (lb/h)

Pipe Size (in.)	Drop in Pressure (psi per 100-ft length)				
	$\frac{1}{8}$	$\frac{1}{4}$	$\frac{1}{2}$	$\frac{3}{4}$	1
$\frac{3}{4}$	115	170	245	308	365
1	230	340	490	615	730
$1\frac{1}{4}$	485	710	1,025	1,290	1,530
$1\frac{1}{2}$	790	1,160	1,670	2,100	2,500
2	1,580	2,360	3,400	4,300	5,050
$2\frac{1}{2}$	2,650	3,900	5,600	7,100	8,400
3	4,850	7,100	10,300	12,900	15,300
$3\frac{1}{2}$	7,200	10,600	15,300	19,200	22,800
4	10,200	15,000	21,600	27,000	32,300
6	31,000	45,500	65,500	83,000	98,000

Thus for a steam-system pressure loss of $\frac{1}{4}$ psi, the water level in the return drop pipe (with 180 F water) would be 4×1.78, or 7.12 in. higher than that in the boiler.

One procedure to follow in designing low-pressure systems is to allocate not more than one-half the initial gage pressure in computing the design pressure drop.

It is desirable for the main to pitch in the direction of flow at a rate of not less than $\frac{1}{4}$ in. in 10 ft, and preferably $\frac{1}{2}$ in. The runouts to radiators and risers should pitch toward the main at not less than $\frac{1}{2}$ in. in 10 ft. If such a pitch is not practicable, or if the runout is more than 8 ft long, pipe one size larger should be used for the runout. Further, to prevent sagging between supports, it is recommended that the main should not be smaller than nominal 2-in. pipe.

In laying out two-pipe systems, the complete piping circuit should first be sketched on the architectural plan. For vapor and return-trap systems in particular, it is desirable for the supply mains to connect the convectors and other radiation with piping circuits as short and direct as possible from the boiler. With complex circuits a reversed-return system may be desirable. In such a circuit the convectors which are first supplied with steam are connected into the far end of the return main so that, although the steam gets to these convectors first, it has a long circuit to travel before returning to the boiler. By use of a reversed-circuit arrangement in a complex system, the air travel during the warm-up period is about the same for both the close and most distant radiation. Moreover, in a reversed-return arrangement both the mains and the returns pitch in the same direction which gives a much neater appearance than is the case with a direct-return arrangement. Where possible, return mains should pitch more steeply than supply mains, 1 in. in 10 ft being desirable. On full vacuum systems, return mains need not pitch more than $\frac{1}{2}$ in. in 10 ft.

Vacuum systems are normally used in large buildings because with them it is possible to obtain both close control of temperature and well-balanced steam circulation, and also, smaller pipe sizes can be used. The differential pressure in the vacuum system consists of the suction produced by the vacuum pump on the return side, and in addition, such positive (gage) pressure as may exist on the supply side. Good design practice calls for not more than $\frac{1}{8}$-psi drop per 100 ft of equivalent run and not more than 1 psi total pressure drop in a system. Pitch of main should not be less than $\frac{1}{4}$ in. in 10 ft. Supply mains smaller than 2 in. are undesirable. Connection should not be made between the steam and return sides of a vacuum system except through steam traps, as it is necessary to prevent steam in quantity from entering the return line.

Example 7-1. A two-pipe vapor-vacuum system operates at 2 psig at maximum capacity and runs into the vacuum region on reduced capacity. The measured length of the longest run is 200 ft and the maximum heat delivery occurs when 800 lb of steam per hour are condensed off four risers. Making use of appropriate tabular data select pipe sizes for this system.

Solution: The total equivalent length is usually about double the measured length and thus 400 ft will be used. If half of the gage pressure namely 1 psi is allocated to the supply side, and the same to the return side the allowable pressure drop is 1 psi per 400 ft of run or $\frac{1}{4}$ psi per 100 ft of run. Reference to Table 7-4 (column F) shows that a $3\frac{1}{2}$-in. main would be needed.

If each riser carries 200 lb of steam per hour, column H indicates that a 2-in. riser is required.

Radiator runouts depend on the individual radiator capacity and are selected from column H.

For the return main use the same design pressure drop, namely, $\frac{1}{4}$ psi per 100 ft of equivalent run and make use of Table 7-5. Column Z here shows that a $1\frac{1}{2}$-in. wet return is satisfactory but a 2-in. pipe (column AA) is required for a dry return (running above the boiler water level). For a vacuum system with a pump a $1\frac{1}{2}$-in. pipe (column BB) is satisfactory.

When data are taken off an architectural system layout the equivalent length should then be rechecked in terms of each elbow and fitting for the pipe sizes selected. However, this is not usually necessary. In this example it should also be noted that a $\frac{1}{4}$-psi pressure loss per 100 ft is considered a high value by some designers who prefer not to exceed $\frac{1}{8}$ psi per 100 ft. However, with pipes as large as those selected, no trouble should arise provided adequate head room between the boiler and the lowest radiation is available to allow for a high water column in the drop leg of the boiler return, when a pump is not used.

7-13. HIGH-PRESSURE STEAM SYSTEMS

For heating of industrial space, high-pressure steam may be used directly in suitable radiation. However, the pressure is frequently dropped through reducing valves and the low-pressure steam is used in conventional systems. Where high-pressure steam is used directly, the load-carrying capacity of steam and condensate return pipes is given in Tables 7-6 and 7-7 for 30-psig systems. With high-pressure steam, higher pressure drops can be used to force the steam throughout the system. In the 30-psig system, for example, total pressure drops from 5 to 10 psi are customary, while in a 150-psig system, 25- to 30-psi drops are employed. High-pressure mains should pitch at least $\frac{1}{4}$ in. in 10 ft, and horizontal runouts to risers and heaters should pitch at least $\frac{1}{2}$ in./ft.

Unit pressure drops employed in design of high-pressure systems range from $\frac{1}{2}$ to 1 psi per 100 ft of equivalent run. With high-pressure steam, particularly when it exceeds 300 psig, it should be realized that the surface temperature of the pipes and radiator surface is in excess of 330 F. Where physical contact is made with surfaces of this temperature, bad burns can result. Thus the pipes must be so placed as to prevent physical contact. A common arrangement with high-pressure steam is the use of blast coils, in which air is warmed by being passed over the steam coils and the space is then warmed by the air. Unit heaters also can be employed. It is also possible to use finned-type pipe coil with suitable shielding. The same type of device is also satisfactory with low-pressure steam. In fact, it may be preferable to pipe the high-pressure steam to a reducing valve and drop the pressure in the reducing valve to the lower pressure and temperature ranges employed in conventional steam systems, Whenever the steam pressure is 100 psig or above, it is desirable to use two reducing valves in series to drop the pressure to, for example, 5 or 2 psig. In general, systems operating at 15 psig or less are called low-pressure systems, and those operating at pressures above 15 psig are called high-pressure systems.

Steam heating represents one of the earliest forms of central heating and is still used extensively, particularly in public buildings. For new installations, one-pipe systems—and in fact most air-vent systems—are outmoded. For residential heating, vapor systems are still used in new designs, but more extensive use is being made of forced-warm-air and hot-water heating systems.

PROBLEMS

7-1. Dry saturated steam at 15 psia condenses and subcools to 150 F. Making use of Tables 2-2 and 2-3, compute (a) the energy released during this process and (b) the change in volume, in cubic feet per pound, during the condensation and subcooling.

Ans. (a) 1032.8 Btu/lb; (b) 26.27 ft^3

7-2. Dry steam at 12 psia condenses and subcools to 150 F. Making use of Tables 2-2 and 2-3, compute (a) the energy released during this process and (b) the change in volume, in cubic feet per pound, during the condensation and subcooling.

Ans. (a) 1030.2 Btu/lb; (b) 32.38 ft^3

7-3. The boiler pressure in a one-pipe steam system, and at entry to the supply main, is 16 psia. As the supply main reaches its end and drops into a wet return feeding the boiler, the pressure is 14.8 psia. Compute the height of the water level in the drop leg with respect to the water level in the boiler. (See Fig. 7-1 for an illustration of a system of this type.)

Ans. 1.2 psi = 34.5 in. hot water

7-4. For a certain two-pipe system, such as the one illustrated in Fig. 7-5, the pressure in the boiler and at entry to the supply main is 16 psia. The corresponding pressure in the dry return as it turns down into the drop leg and wet return is 15 psia. Find the water-level height in the drop leg above boiler water level.

7-5. Examine Fig. 7-5 with a view to modifying the system shown there so that it becomes a vapor system. Without redrawing the figure, state exactly the minimum changes which would be required to make such a change. Assume that the mains, risers, and radiators could be used satisfactorily for either system.

7-6. Using Fig. 7-1 as a guide, diagram the system shown there and include the additions necessary to transform it into a two-pipe air-vent system. Indicate also a supply and return riser for a radiator on a higher story. Supply the valves for each end of the radiator. Notice that the line can be satisfactorily connected to the wet return at the bottom of the drop leg, below the air valve. This system reduces the noise in single-pipe systems which occurs because the steam and water flow in opposite directions from the radiator, but the two-pipe air-vent system is being superseded by more desirable two-pipe arrangements, such as the vapor and the vacuum systems.

7-7. Repeat Problem 7-6 for a two-pipe vapor system. Observe that with this system traps are used on each radiator and air valves are not required on the radiators.

7-8. A steam pipe 120 ft long is installed at 50 F. When it is filled with steam at 5.3 psig pressure, what is the increase in the length of the pipe (a) if the pipe is steel and (b) if the pipe is copper?

Ans. (a) 1.7 in.; (b) 2.7 in.

7-9. An underground steel steam line in a tunnel is 5000 ft long. It was laid in summer at an average ground temperature of 60 F, but in winter it will operate with steam at an average pressure of 105.3 psig. Compute the total change in length which must be absorbed by expansion joints or bends over summer and winter temperature ranges.

Ans. 115 in. or 9.6 ft

7-10. What size of steel pipe is required to carry steam at 6000 lb/h through a main having an equivalent length of 450 ft if the initial pressure is 100 psig and the final pressure is 95 psig?

Ans. 4 in.

7-11. A central steam plant provides steam to a remote building through an 8 in. welded steel pipe line of Schedule 40 pipe. An addition to the building is contemplated and question is raised as to whether the same line can serve both the original building and the addition. The linear run of pipe is 1200 ft and it is estimated the extension and equivalent length of fittings will add 400 ft. How many pounds of saturated steam per hour can pass through this line if a pressure drop of 10 psi is allowed for the 135 psig steam at the boiler plant?

Ans. 35 000 lb/h

7-12. Work Problem 7-11 if the allowable pressure drop is 20 psi.

Ans. 810 lb/min

7-13. Compute the pressure drop in a Schedule 40, 4-in. steel pipe of 400 ft equivalent length carrying steam at 100 lb/min if the inlet steam pressure is 60 psig.

Ans. 5.4 psi

7-14. A two-pipe vapor system serving 670 000 Btuh has a total equivalent length of 200 ft, and the total pressure drop is to be 2 oz/in.^{2}. (a) What size of main and dry return should be used? (b) To supply a radiator on the second floor delivering 18 000 Btuh, what size riser and horizontal branch should be used?

Ans. (a) 5 in., $2\frac{1}{2}$ in.; (b) $1\frac{1}{4}$ in., $1\frac{1}{4}$ in.

7-15. A vacuum system to deliver 2 160 000 Btuh has a total equivalent length of 800 ft. It is to be designed for a total pressure drop of 1 psi. What size of steam and return main should be used?

Ans. 6 in., 3 in.

7-16. For the two-pipe, air-vent, gravity-return system of Fig. 7-5, considering the extreme right circuit as having the greatest length, compute the required size of main, return, risers, and runouts for this circuit. Each unit of radiation delivers 14 400 Btuh and requires approximately 14.4 lb of steam per hour, with a total boiler output of 86.4 lb of steam per hour.

7-17. For the two-pipe, air-vent, gravity-return system of Fig. 7-5, compute the size of the risers and runouts for the radiation which in the sketch appears to be roughly over the boiler. Refer to Problem 7-16 for data regarding the system.

7-18. A central blast coil heats the air supplied to an auditorium in winter. The coil capacity at design outdoor conditions amounts to 175 000 Btuh. Steam at 5 psig is available and a vacuum return line, equipped with a pump, takes care of condensate and air venting. Find (a) the maximum steam flow to the coil, in pounds per hour, (b) size of the feed pipe from the pressure-reducing valve to the coil if its equivalent length is 200 ft, if $\frac{1}{2}$-psi pressure drop is allowed, and, (c) the return pipe size to the vacuum pump if $\frac{1}{4}$-psi drop is allowed for the 200-ft run.

Ans. (a) 175 lb/h; (b) 2 in.; (c) 1 in.

7-19. Rework Problem 7-18 under the assumption that the blast-coil capacity is 125 000 Btuh, with no other change in data.

Ans. (a) 125 lb/h; (b) 2 in.; (c) $\frac{3}{4}$ in.

CHAPTER

8

HOT-WATER (HYDRONIC) SYSTEMS

8-1. HOT-WATER (HYDRONIC) SYSTEMS

Heating systems employing hot water as a means of conveying heat to points of utilization are called hot-water heating systems. In such systems a direct-fired boiler or steam-to-water heat exchanger delivers heat into water which then flows through pipes to radiators, convectors, or other types of heat exchangers located at points of utilization. Most present-day systems employ a pump to promote the circulation of water throughout the system, but many older systems have no pump and depend upon density differences existing in hot- and cold-water circuits to promote flow through the system. Natural-circulation systems are also called gravity-circulation systems or thermal-circulation systems.

In addition to conventional hot-water heating systems employing convectors, many of the panel, or radiant-heat, systems also use hot water as the heat source. All hot-water heating systems must make provision for changes in the volume of the water contained in the system between the hot and cold states of the system. This is accomplished by means of expansion tanks of either the open or closed variety.

8-2. PRESSURE-TEMPERATURE RELATIONS

The boiling temperature of water is governed by the pressure maintained on the water. It is possible, therefore, to use water in heating systems at a temperature lower than that for saturated steam at the pressure maintained on the water. Table 2-3, which gives saturation temperatures, can be used to find limiting temperatures for hot-water heating systems operating at known pressures.

The surface of a radiator or convector will transmit heat to air at essentially the same rate from water as from steam, provided the temperature of the water is the same as that of the steam.

Because of the rapid increase in pressure which water experiences as its temperature is raised it was formerly customary to limit the top temperatures of the system to approximately 250 F (121°C). Except for most domestic systems this is no longer true and high-temperature systems in the range 300 F to 500 F (150°C to 260°C) are in common use. Reference to Table 2-2 shows that water at 500 F (260°C) has a saturation pressure of 681 psia (4694 kPa). The advantages of high-temperature distribution are particularly noteworthy in the case of isolated installations where a boiler, perhaps located at a remote point, serves a widely dispersed plant. With high-temperature water, and the resultant large energy release from the cooling of this water, it necessarily follows that smaller pipe sizes can be used in the distribution system to transfer a given amount of energy than would be required with low-temperature hot water. In fact, such high-temperature systems can employ even smaller pipe sizes than are required in high-pressure steam installations.

In hot-water heating systems the same water can be recirculated indefinitely, and thus there is minimum deposit of solids on boiler heating surfaces. Corrosion of piping is negligible, provided the oxygen-carrying fresh water for make-up is limited in amount. The temperature of the heating medium can be varied to meet weather conditions more readily than can be done with steam. For example, in mild weather the pump can circulate water to the heating surfaces at (say) 100 F to 120 F (38°C to 50°C), whereas in extremely cold weather, with greater output required from the radiation, the water can be supplied at temperatures of (say) 180 F to 240 F (82°C to 116°C).

8-3. CIRCULATION OF WATER

Any vessel which contains water and receives heat on one side will develop a circulation as the water, warmed and expanded by heat, rises, while the cooler and denser remaining water takes its place. This principle underlies the operation of all thermal-circulation (natural-circulation) hot-water heating systems. Some steam-heating systems can be changed to hot-water systems with ease; and some hot-water heating systems with slight changes, can be made to operate with steam.

The water in a thermal system circulates because of the difference between the pressure exerted by the hot-water column leaving the boiler on its way to the radiator and the pressure exerted by the relatively cool water column returning from the radiator. The unit pressure exerted by any water column is a function of water density and height.

Thus the pressure difference available to cause flow is expressed as

$$\Delta P = h(d_{fR} - d_{fD}) = h\left(\frac{1}{v_{fR}} - \frac{1}{v_{fD}}\right) \tag{8-1}$$

where ΔP is the pressure difference for water circulation, in pounds per square foot; h is the height between boiler datum and radiator datum, in feet; d_{fR}, d_{fD} are the respective densities of the return and delivery water, in pounds per cubic foot; and v_{fR}, v_{fD} are the corresponding specific volumes to be selected from Table 2-2, in cubic feet per pound.

Equation (8-1) can be transformed so that $\Delta P'$ is expressed in milinches of water; thus

$$\Delta P' = h\left(\frac{1}{v_{fR}} - \frac{1}{v_{fD}}\right) \times 0.016 \times 12\,000 = 193h\left(\frac{1}{v_{fR}} - \frac{1}{v_{fD}}\right) \tag{8-2}$$

Example 8-1. Refer to Fig. 8-1. If the average temperature of the water leaving the boiler is 180 F and the water cools in the radiator and piping to 160 F, what pressure difference is being used in overcoming frictional losses in the water system?

Solution: Use Eq. (8-1) and $h = 20$ ft. From Table 2-2, $v_{fR} = 0.01639$ ft^3/lb at 160 F, and $v_{fD} = 0.01651$ ft^3/lb at 180 F. Therefore

$$\Delta P = (20)\left(\frac{1}{0.01639} - \frac{1}{0.01651}\right) = 9.00 \text{ psf}$$

or

$$\Delta P = 9 \times 193 = 1737 \text{ milinches}$$

Ans.

Note that during operation of this boiler the 1737 milinches are entirely used in overcoming the frictional resistance to flow, and this pressure difference cannot be physically measured.

The warmer the water supplied by the boiler the greater the temperature difference from the radiator to the room, permitting less surface and better circulation. Atmospheric systems with open expansion tanks (like that in Fig. 8-1) are limited to water temperatures less than 212 F (100°C) at the top radiators, although the lower radiators can be operated hotter in a high building because of the greater water pressure at the lower levels. Closed-expansion-tank systems can be operated at temperatures appreciably above 212 F because pressures much above atmospheric can be used.

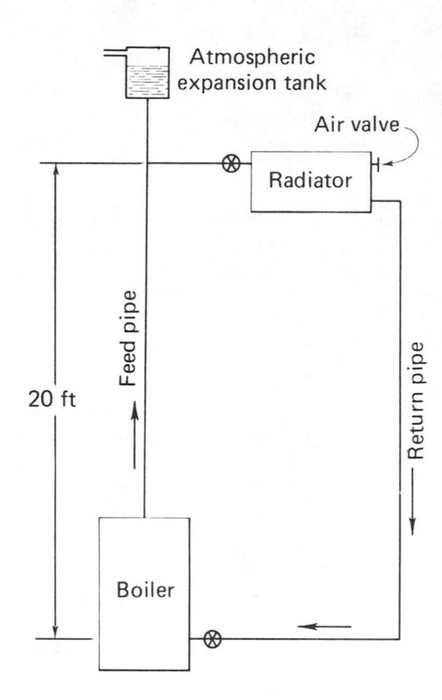

FIGURE 8-1. Elements of a thermal-circulation hot-water heating system.

8-4. HEATING-SYSTEM ARRANGEMENT

The water in a hot-water heating system may be heated either in fuel-fired boilers, which are essentially the same as boilers used with steam, or in transfer heaters, which use steam or some liquid warmer than the water as the source of heat. The water flows to the radiators and convectors, which are essentially the same as those used for steam. The supply pipes do not pitch down in the direction of flow. The flow and return pipes and branches are generally the same size, though one or the other can be reduced. A representative arrangement of accessories and control equipment for a forced-circulation-system hot-water boiler is shown in Fig. 8-2.

In the system of Fig. 8-2 the pump operates in response to a call for heat from the room thermostat. The operation of the pump is also under the control of the hot-water control *B*, which prevents the pump from starting if the water is below a set temperature and later stops the pump after operation if this low temperature is reached. A main hot-water control, *A*, usually independent of the other controls, starts and stops the oil or gas burner or the stoker, to maintain a definite water temperature in the boiler.

The expansion tank is a very essential part of a hot-water heating system, because the difference in the total volume of water between the hot and cold conditions is great. In new installations, expansion tanks located above and near the boiler are used almost exclusively. In this type of expansion tank, air trapped in the tank is compressed as the volume of water in the system increases. If the air volume of the tank is too small, and if the pressure exceeds a predetermined set value, surplus water is released through a relief valve. The pressure-relief valve of the hot-water boiler takes the place of the safety valve of the steam boiler. With a tight system and with adequate air

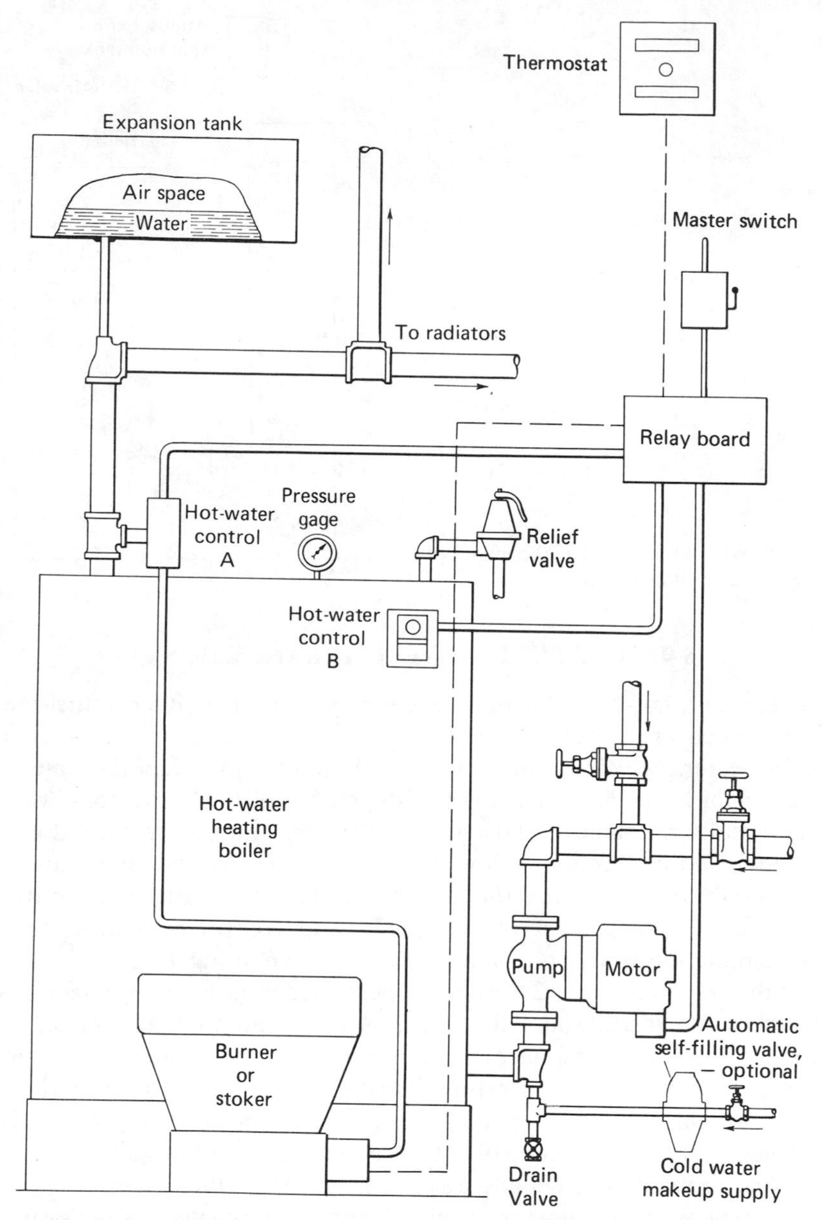

FIGURE 8-2. Representative accessories and controls for a forced-circulation hot-water boiler.

space in the expansion tank, little or no water need by wasted through the relief valve, and only occasional make-up of water by manual control is sufficient. Larger hot-water heating systems are provided with automatic filler or make-up devices.

A difficulty sometimes occurs in hot-water heating systems when air originally absorbed in the water moves with it and tends to separate at high points in radiators or other heat-transfer equipment. This is objectionable because it prevents water contact with part of the heat-transfer surface. The air has to be removed through vent valves located at high points on the units. Cold water holds more gas (air) in solution than does hot water. Consequently, if the gas is driven off from the water during heating in the boiler, it can be separated and largely delivered to the expansion tank. Figure 8-3 shows a patented design which accomplishes this end. This special fitting for outflow from the boiler connects at a high point. Air is removed around the top of the fitting and is carried to the expansion tank, while the water flowing out is taken from a lower point in the fitting remote from the point of air separation. When the fresh water entering the system is properly deaerated in the boiler at the start, little difficulty occurs from air separation in radiators.

In the closed expansion tank, the air space exists above the free water surface. As the water volume in the system increases, more water flows into the bottom of the tank and compresses the air in the upper part of the tank. As the water volume in the system decreases with lowering temperature, water flows from the bottom of the tank back into the system. It is desirable that the water-flow in and out of the tank take place as far from the air cushion as possible. It is also important that the air separated in the boiler not bubble through the water in the expansion tank and thus be reabsorbed by this relatively cool water. To prevent this, a special fitting (Fig. 8-4) has been developed which sends the air from the boiler directly into the free air space of the expansion tank. Surplus tank air can also be removed through the

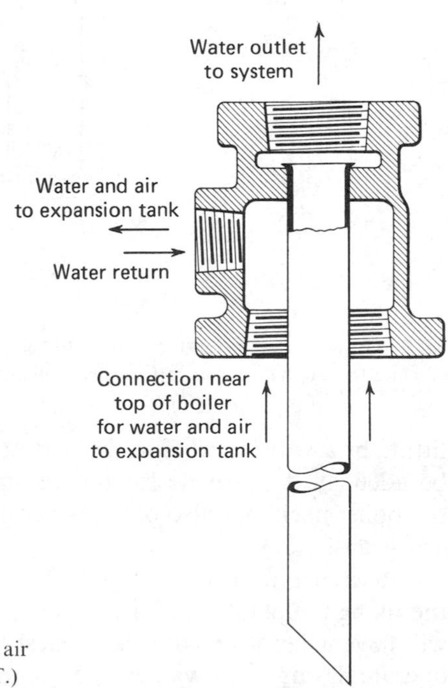

FIGURE 8-3. Boiler fitting for eliminating air at the boiler. (Courtesy Bell and Gossett, ITT.)

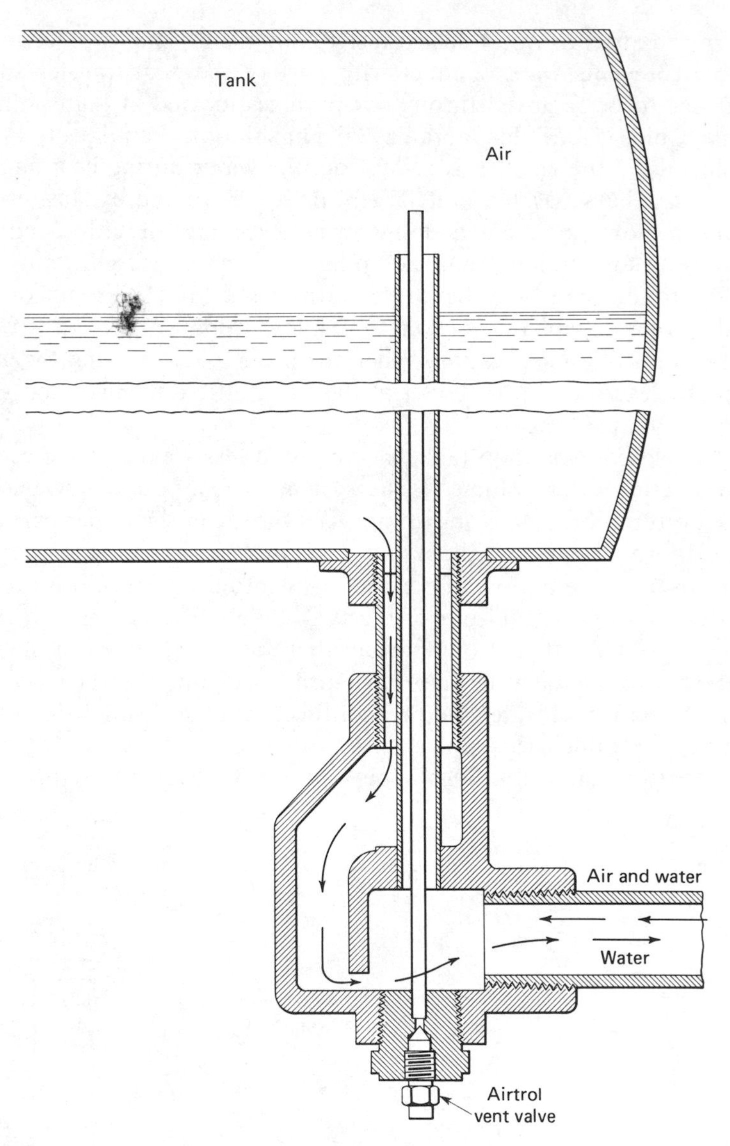

FIGURE 8-4. Expansion-tank fitting for passage of air and water into tank and for return of water to system on cooling. (Courtesy Bell and Gossett, ITT.)

fitting by a valve located in the bottom. The size of an expansion tank must be adequate to provide for the volume increase not only of the water in the boiler itself, but also of the water in all of the connecting piping and the heat-transfer devices.

Reference to Table 2-2 gives both the specific volume of cold water at the filling temperature of the system and also the specific volume the water will have when it reaches the highest temperature at which the system will operate. Using these values, the volume increase can be computed for the

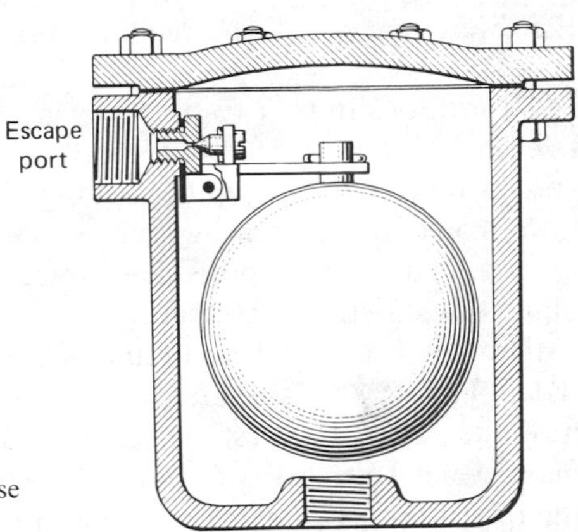

FIGURE 8-5. Automatic air valve for use on hot-water heating system line.

total weight of water in the system. This volume increase represents the minimum expansion-tank space which must be provided. Good practice calls for expansion-tank volume in excess of this amount. Considerable research has been carried out in an effort to find optimum expansion sizes. There are no universally accepted standards, but it is customary to use an 8-gal tank for systems of 60 000 Btuh or less, 15 gal for systems up to 85 000 Btuh, 18 gal for systems up to 120 000 Btuh, 24 gal for systems up to 240 000 Btuh, and 60 gal for systems up to 1 000 000 Btuh. These numbers apply to forced-circulation systems using smaller pipe sizes. For older, natural-circulation systems with large pipe sizes, it is desirable to increase the tank to one size higher.

In addition to the air which separates at the boiler and expansion tank, air can also be trapped in the lines and in the radiators. For air elimination at high points in the lines, units of the type illustrated in Fig. 8-5 can be used. These units are normally completely filled with water and the escape port remains closed. However, when air collects in sufficient quantity, the float drops, permitting the air to be eliminated, and as water flows in, causing the float to rise, the escape port closes. Air trapped in radiators must usually be removed through manually operated air-vent valves located at a high point in the radiator. In a well-adjusted system, air elimination is usually necessary only at starting, and perhaps at one other time during the heating season. Some domestic systems can function for several seasons of operation without air elimination or addition of water.

8-5. VARIETIES OF CIRCULATING SYSTEMS

Hot-water installations are now designed almost exclusively as forced-circulation systems. This choice is made because, with the pump providing a higher pressure for circulation, higher water velocities can be used, avoiding the need for the large pipe sizes required in thermal-circulation systems. Moreover, in a forced-circulation system with the pump running, some heat can be delivered to a given space even before the water temperature reaches the levels that would be required for good circulation in a gravity-type system.

The double-pipe (two-pipe) system shown in Fig. 8-6 sends water directly to each radiator essentially at boiler temperature except for any temperature loss in the connecting piping. Using a direct-return arrangement, the water from the boiler flows through the supply main to the nearest radiator, and returns as directly as possible through the return main, encountering less piping and friction loss than is met by the water going to more distant radiators. With the reversed-return system the outgoing water to the nearest radiator makes the same journey as in the direct-return system, but the return from the radiator nearest the boiler constitutes the first section of the return main, which starts away from the boiler, turning back only after picking up the return branches from the various radiators in the order of their remoteness from the boiler (see also Fig. 8-15). With this scheme the round-trip travel of the water to any radiator is the same, measured along the mains, as that to any other radiator. There may be three main pipes over part of the system, increasing the cost slightly, but the reversed-return system is almost self-equalizing so far as water flow to the different radiators is concerned.

In single-pipe systems (Fig. 8-7), the usual arrangement is for part of the water to pass successively through radiator after radiator, and as the water loses heat and drops in temperature the transmitting surface in each succeeding radiator must be increased in relation to the temperature drop if the same capacity is expected from each radiator. In small systems where the total temperature drop is small, a change in radiator size may not be required.

Some single-pipe systems use standard tee fittings and run the supply branch off the top of the main, with the return entering the side of the main. Others employ special tees (Fig. 8-8) having directed passages or deflectors for sending a definite portion of the water into the supply branch. The return runouts employ similar tees in which the passages are reversed to act as ejectors—with the water, flowing in the main, drawing a current from the return branch. In most cases one fitting on the radiator return side is sufficient to insure proper water flow for upfeed radiation. However, for

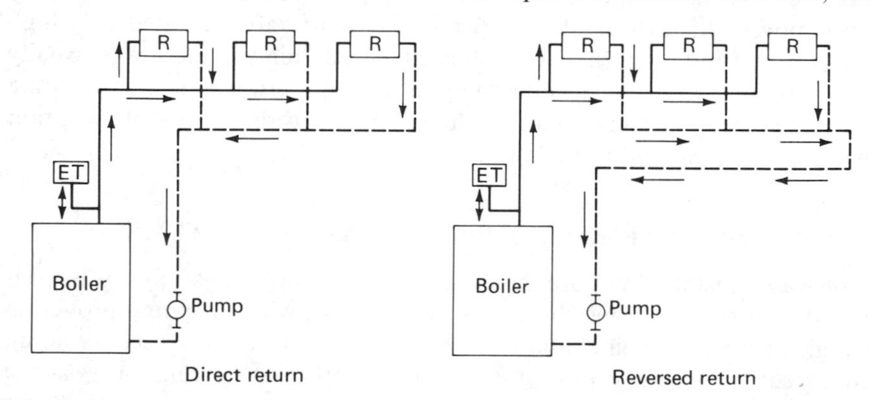

FIGURE 8-6. Diagrammatic representation of two-pipe, forced-circulation hot-water systems. R = radiator or convector; ET = expansion tank.

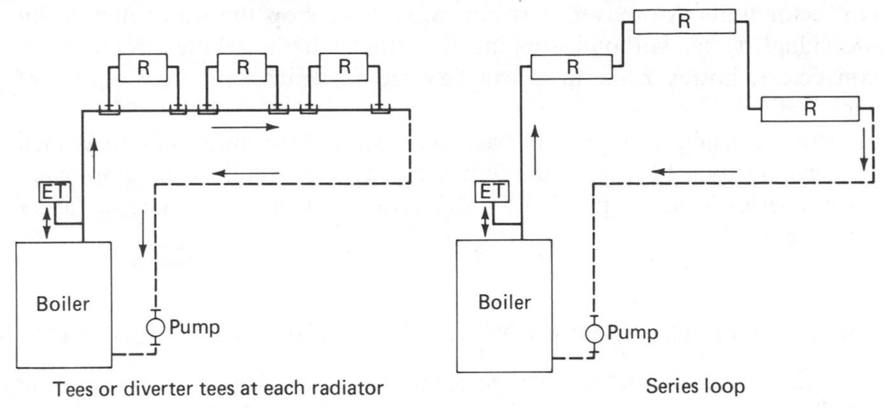

FIGURE 8-7. Diagrammatic layout of single-pipe (one-pipe) systems. R = radiator, convector, or baseboard; ET = expansion tank.

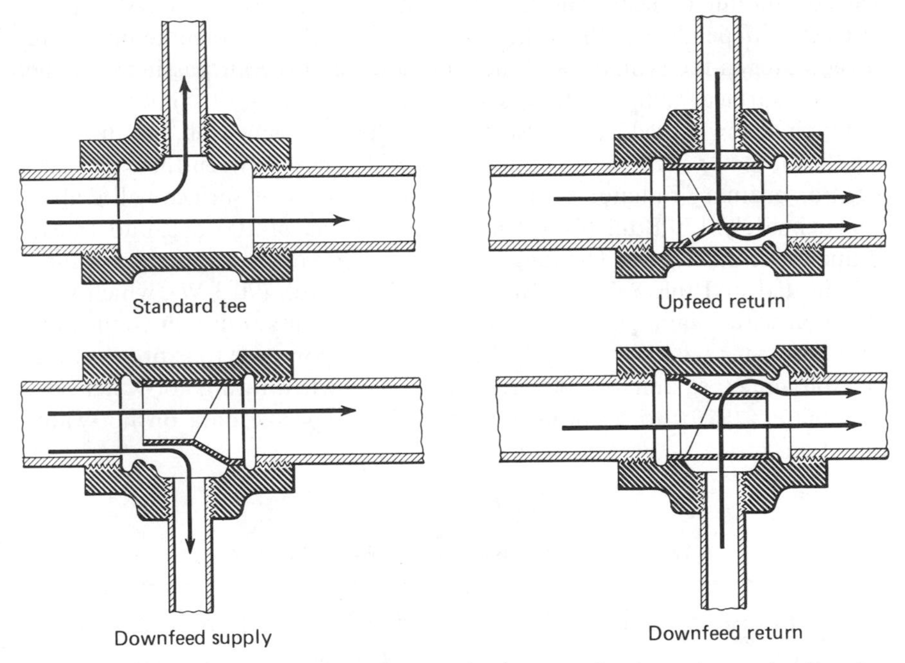

FIGURE 8-8. Piping arrangements for one-pipe hot-water heating systems using diversion tees. (Courtesy Bell and Gossett ITT.)

radiation which is awkwardly placed, or remote from the main, it may be desirable to use directional tees on both inlet and outlet to the radiation. In the case of radiators below the main, where unaided natural circulation is completely ineffective, it is necessary to supply special tees for both feed and return piping.

In the one-pipe system using a series-loop pattern (Fig. 8-7), it is obvious that the water progressively cools in passing through each heater. With the temperature lowered, the heat delivery decreases in succeeding radiator or convector units. Moreover, it is not possible to stop the water flow to an individual heater without stopping flow through the whole circuit. With convectors, however, dampers can be used to control the heat output of each individual heater. The series-loop arrangement is suitable for small apartments, using convectors or baseboard units, or for large unit areas such as a gymnasium. Flexibility in control can also be obtained by employing several series loops in parallel for different parts of a residence or larger building.

8-6. HEAT TRANSMISSION WITH HOT-WATER ELEMENTS

For each radiator the feeder-pipe sizes must be adequate, to carry the required amount of water, and the heat-transfer surface in the radiator must be sufficient to absorb from the water and dissipate to the room the needed amount of heat. The heat-transmitting capacity of a radiator or convector depends on the difference between the average temperature of the water in the radiator and the temperature of the air passing over the radiator surface. Other factors, such as the proximity of colder surfaces and high or low air velocities, also affect the heat transmission, but the most significant factor is the mean temperature difference between the water and air. For ordinary radiator surface, each square foot of surface will deliver approximately 150 Btuh (44 W) under conditions of 100 deg mtd (mean temperature difference). The range of approximate heat-transmission values is indicated in Table 8-1. The EDR unit of 240 Btuh (70.3 W), which is so common with steam heat-transfer equipment, has less utility in connection with hot-water radiation and it is therefore more common to express output in Mbh units (i.e., units of 1000 Btuh or 293 W) when rating hot-water heat transmitters. Refer to Sections 9-2 and 9-3 for specific data on hot-water radiation.

Table 8-1 Heat Emission of Hot-Water Radiators

Mean Radiator Temperature (F)	MTD between Radiator and Room Air (F)	Approximate BTU Transmitted per Hour per Sq Ft of Surface	Approximate BTUH Transmitted per Sq Ft of Surface per Degree Difference
220	150	250	1.67
215	145	240	1.65
200	130	208	1.60
190	120	188	1.57
180	110	170	1.54
170	100	150	1.50
160	90	129	1.43
150	80	110	1.39
140	70	93	1.33
130	60	76	1.27

Table 8-2 Friction Loss of Standard Pipe Fittings (in Equivalent Feet of Straight Pipe)

Fitting	Iron Pipe (in.)						Copper Tube (in.)				
	3/4	1	1 1/4	1 1/2	2	3	3/4	1	1 1/4	1 1/2	2
Elbows											
90°	1.6	2.1	2.6	3.1	4.2	6.5	1.6	2.1	2.6	3.1	4.2
45°	1.1	1.5	1.8	2.2	2.9	4.5	1.1	1.5	1.8	2.2	2.9
90° long sweep	0.8	1.0	1.3	1.6	2.1	3.0	0.8	1.0	1.3	1.6	2.1
Tees											
100% side diversion	2.8	3.8	4.7	5.6	7.5	13	1.9	2.5	3.1	3.7	5.0
50% side diversion	6.3	8.3	10.4	12.5	16.7	25	6.3	8.3	10.4	12.5	16.7
33% side diversion	14.3	18.7	23.4	28.1	37.5	56	18.7	23.5	29.4	35.2	46.9
25% side diversion	25.0	33.3	41.6	49.8	66.7	100	31.2	41.6	52.0	62.5	83.4
Valves											
Globe (full open)	18.7	25.0	33.8	36.8	50.0	66	26.6	35.4	44.2	53.0	70.8
Gate (full open)	0.8	1.0	1.3	1.6	2.1	3.0	1.1	1.5	1.8	2.2	2.9
Stopcock (full open)	1.6	2.1	2.6	3.1	4.2	6.5	1.6	2.1	2.6	3.1	4.2
Angle (full open)	3.6	4.2	5.2	6.2	8.3	12.5	4.7	6.3	7.8	9.4	12.5
Reducer coupling	0.6	0.8	1.0	1.3	1.7	2.5	0.6	0.8	1.0	1.3	1.7
Boiler or radiator	4.7	6.3	7.8	9.4	12.5	19	6.3	8.3	10.4	12.5	16.7

It has been mentioned before that the output of a hot-water radiator unit is determined by the weight of water passing per unit of time and by its temperature drop. This can be expressed by the following formula:

$$Q = 60W(1)(t_{in} - t_{out}) \tag{8-3}$$

which becomes

$$\begin{aligned} Q &= (60)(8.1)(\text{gpm})(t_{in} - t_{out}) \\ &= 490(\text{gpm})(t_{in} - t_{out}) \end{aligned} \tag{8-4}$$

where Q is the output of transmitter, in Btu per hour; $W =$ is the water flowing through, in pounds per minute; t_{in} is the temperature of water in, in degrees Fahrenheit; t_{out} is the temperature of water out, in degrees Fahrenheit; and gpm is the gallons per minute water-flow, equivalent to W lb/min. The factor 8.1 in Eq. (8-4) represents the weight per gallon of water in the range of 180 F, in contrast with the value 8.33 at 70 F. In general, it is customary to design on a 20-deg temperature drop, with 180 F entering- and 160 F leaving-water temperatures. If the water temperature drop is changed to 10 deg, then twice as much water must pass through each radiator for a given heat output. As an aid in computation, it should be mentioned that flow in pounds per hour can be transformd to flow in gallons per minute by dividing by 500 for cold water, or by 490 for hot water at about 180 F on inlet to radiators or convectors.

Compared with the standard 20-deg drop, a 30-deg drop would require only two-thirds as much water to circulate. With a direct-return piping system, the equivalent length for each radiator circuit should be computed in order to give the basis for balancing out the circuits, and in such a system it may be necessary to put restricting orifices in the shorter circuits.

In designing hydronic systems the concept of equivalent pipe length is widely used. Equivalent pipe length is the length of straight pipe of a given size that would have the same resistance as the actual pipe run with its bends, fittings, valves, and other devices. In computing equivalent length, use is often made of the *elbow equivalent*, where an elbow equivalent is equal to the resistance offered by a straight run of pipe approximately 25 times the nominal pipe diameter. Thus, for a 3-in. pipe, one elbow equivalent is $25 \times 3 = 75$ in. or 6.5 ft of linear run. Representative values of elbow equivalents show the following: elbow, 1.0; open gate valve 0.5; open globe valve 12.0; open return bend 1.0; angle radiator valve 2.0; radiator 3.0; tee, 100% side diversion 1.8; tee, 50% side diversion 4.0; boiler, 3.0; 45° elbow 0.7; reducer coupling 0.4. Table 8-2 presents the friction loss of fittings in length of straight pipe.

Example 8-2. In a certain hot-water system, from the boiler out to a given radiator and back to the boiler there are 12 elbows, 2 tees (50% side diversion), 1 angle radiator valve, 1 radiator, 1 boiler, and 90 ft of $1\frac{1}{2}$-in. iron pipe. What is the equivalent length of the system?

Solution:

		By Table 8-2
12 elbows	12 × 1 = 12	12 × 3.1 = 37.2 ft
2 tees	2 × 4 = 8	2 × 12.5 = 25.0
1 valve	1 × 2 = 2	1 × 6.2 = 6.2
1 radiator	1 × 3 = 3	1 × 9.4 = 9.4
1 boiler	1 × 3 = 3	1 × 9.4 = 9.4
TOTAL	28	87.2 ft

Therefore, by Table 8-2, the equivalent length is 90 + 87.2 = 177.2 ft. *Ans.*

By elbow equivalents, 90 ft + 28 elbow equivalents × 1.5 in. × $\frac{1}{12}$ × 25 = 177.2 ft. Solution by elbow equivalents has some convenience where the pipe size is not known at the time the layout of the circuit is started.

8-7. HOT-WATER BASEBOARD HEATING

Hot-water baseboard heating, also called perimeter heating, is a type of radiation being used in an increasing number of installations in both public and private buildings. The baseboard elements cause minimum difficulty in regard to furniture placement, and largely prevent cold-wall downdraft. Figure 8-9 shows an arrangement of baseboard radiation as it might be employed in a residence, and Fig. 9-3 shows the product of one manufacturer.

Installation practice varies, but usually the element is placed along an outside wall of a room—along part of the wall or all of it. However, it can also be placed to cover all of the walls. Ingenious arrangements are provided for extending the baseboard panels around corners. When baseboard radiation served by copper tubing is used, it is customary to connect the unit lengths of baseboard with sweated fittings; with steel and cast-iron radiation, pipe connections with special unions are employed. Apertures for valves are also provided in the baseboard panels. Dampers are provided, with at least one design, to control the air flow over the heating surface and thereby reduce the heat output. Although the heat output is largely convective, an appreciable portion of the heat is distributed by radiation. The heated air is not hot but warm, and it moves out gently from the units. This small air motion is sufficently great, however, to limit the usual variation in temperature between floor and ceiling to 3 deg or less.

In most installations it is common to design for a 10-deg temperature drop in buildings having a heat capacity of 50 000 Btuh or less, or a 20-deg temperature drop for buildings or zones having a loss greater than 50,000 Btuh. As the elements are usually in series, the water temperature progressively drops as it passes through the subsequent baseboard panels. For design with a 10-deg temperature drop, it is customary to split the circuit into two portions and design with a lower average temperature for the second part of the circuit. For a 20-deg temperature drop, it is customary to use three average temperatures, the second and third of which are respectively lower in sequence. Representative outputs for average water

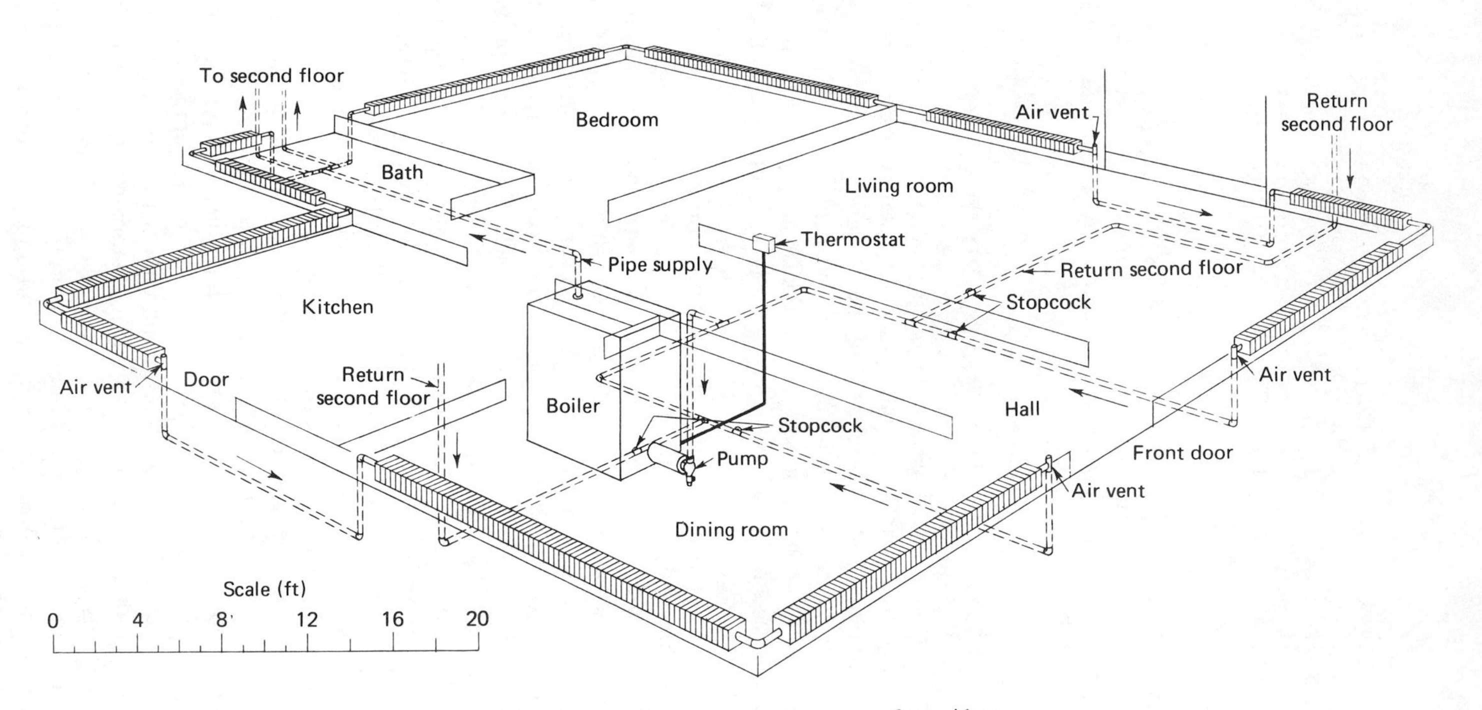

FIGURE 8-9. Layout of baseboard heating system for residence.

temperatures are shown in Tables 9-1 and 9-2 for the products of one manufacturer. Inlet temperatures about 190 F are usual.

8-8. PIPING-SYSTEM DESIGN

In the design of a hot-water heating system, it is desirable to consider the following general procedures, which are applicable to all types of layouts:

1. The system should provide means for complete and adequate drainage at convenient low points and also provide for air venting at high points in the piping system, as well as in the radiation units.
2. Straight runs in excess of 30 ft require provision for expansion.
3. Water temperature drops of 25 to 35 deg are usually assumed for gravity systems, with a 20-deg drop for forced-circulation systems. The temperature range, particularly with forced-circulation systems, can be modified where this appears to be desirable.
4. Water velocities should range from 2 to 4 ft/s where noise might be a disturbing factor. For industrial installations, and where large pipe sizes are employed, velocities up to 8 ft/s can be considered.
5. The design water supply temperatures for gravity systems normally range between 140 F and 200 F, while forced-circulation supply temperatures normally range from 170 F to 220 F. For high-temperature distribution of water, in systems also safe for high pressure, temperatures in excess of 300 F are employed.
6. For forced-circulation systems, the total design friction loss is set by the pressure (or head) characteristic of the available circulating pump.
7. It is customary to choose design friction values in the range 250 to 650 milinches/ft. Above 650 milinches/ft, high velocities occur; below 250 milinches, pipe sizes may be uneconomically large.
8. On large installations separate circulators for each run should be provided, with independent control.
9. Adjustable stopcocks should be provided, or provision for inserting flow-control orifices should be made, so that individual circuits can be balanced to give proper flow.
10. The specific design procedures for laying out a forced-circulation system involve the following:
    a. The radiation required for each room or heated space should be computed, and expressed preferably in Btu per hour (Btuh) or in equivalent thousands of Btu per hour (Mbh).
    b. A sketch on the building plans should be made to show pipe runs and the location of all radiators and of the boiler. On this sketch should also be shown the length of all pipe runs, and as computations are made the pipe sizes should be marked on the sketch.
    c. The gallons-per-minute (gpm) flow should be computed for each circuit by means of Eq. (8-4).

d. In terms of the total gallons-per-minute flow, a forced-circulation (booster) pump should be selected with the help of pump-characteristic curves (Fig. 8-10 or Fig. 8-11), picking the selection point to the right of the peak of the characteristic curve.

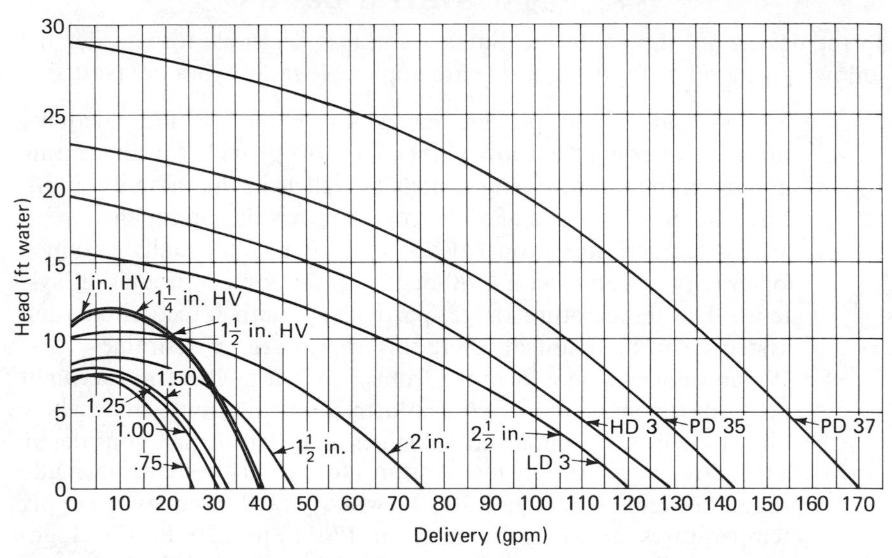

FIGURE 8-10. Representative head-capacity curves for forced-circulation (booster) pumps turning at 1750 rpm with manufacturer's designation indicated. (Courtesy Bell and Gossett, ITT.)

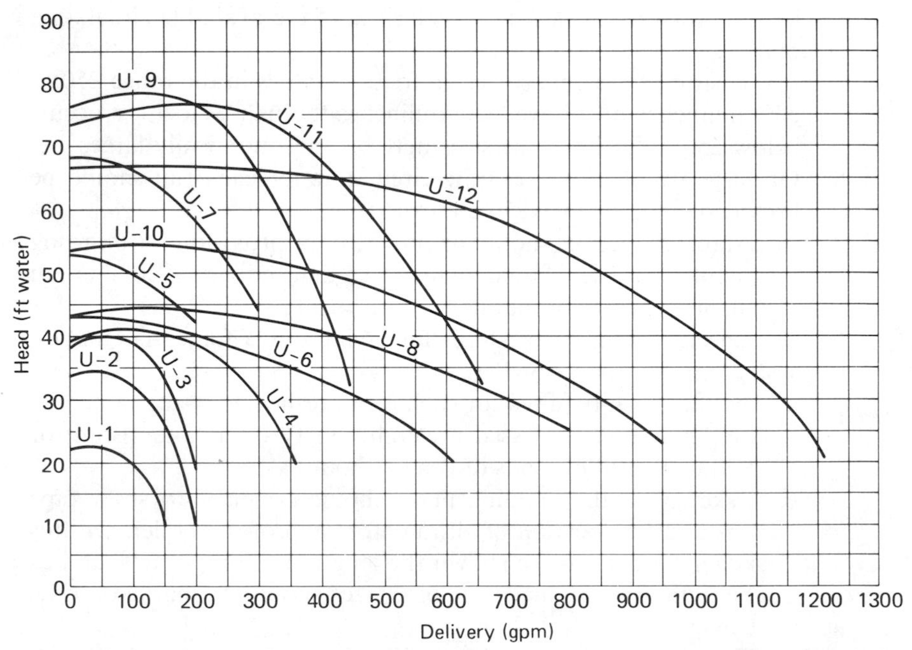

FIGURE 8-11. Representative head-capacity curves for forced-circulation pumps of large capacity turning at 1750 rpm. (Courtesy Bell and Gossett, ITT.)

e. The linear run of the longest (or highest friction loss) circuit should be observed and, for a good estimate, increased by 50 or even 100% to allow for fitting losses, in order to find the total equivalent length. (This may later be rechecked.)

f. Divide the pump head expressed in milinches (feet × 12 000) by the total equivalent length in feet to find the milinch friction loss per foot. Good design practice shows the range to be between 650 and 250 milinches/ft.

g. In terms of the milinches per foot, from the preceding, and the gallons per minute flowing, the required pipe size can be selected

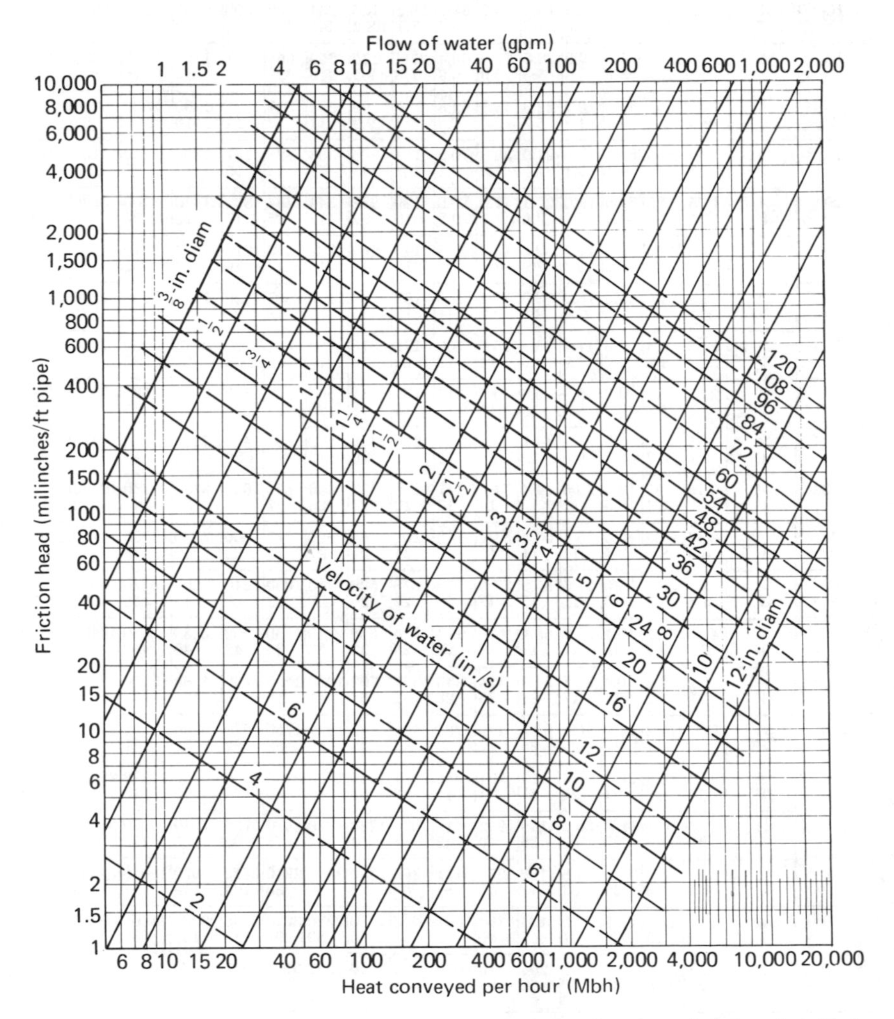

FIGURE 8-12. Friction loss for hot-water flow in steel pipe. Bottom scale is based on 20-deg temperature difference between water in supply and return lines. For other temperature differences (Δt) multiply Mbh value being considered by $20/\Delta t$ before entering chart. (Adapted, by permission, from *ASHRAE Handbook of Fundamentals, 1972*, Chap. 26.)

Table 8-3 Heat-carrying Capacity *A* (in Mbh) of Standard Black Iron (or Steel) Pipes with Temperature Drop of 20 Deg*

Milinch Friction Loss per Foot of Pipe		Nominal Pipe Size (in.)															
		3/8	1/2	3/4	1	1 1/4	1 1/2	2	2 1/2	3	3 1/2	4	5	6	8	10	12
4	*A*	0.75	1.35	2.85	5.4	11.3	17.0	33.0	53.1	95	141	197	363	596	1250	2320	3730
	V†	1.5	1.7	2.1	2.4	2.9	3.2	3.8	4.3	5.0	5.5	6.0	7.0	7.9	9.6	11	12
6	*A*	0.9	1.7	3.6	6.75	14.0	21.2	41.3	66.4	119	176	248	456	748	1570	2920	4690
	V	1.8	2.1	2.6	3.0	3.6	4.0	4.7	5.3	6.2	6.9	7.5	8.8	10	12	14	16
8	*A*	1.05	2.0	4.2	7.9	16.4	24.8	48.4	77.9	140	207	291	535	879	1850	3440	5520
	V	2.1	2.5	3.0	3.5	4.2	4.7	5.6	6.3	7.3	8.0	8.8	10	12	14	17	19
10	*A*	1.2	2.2	4.7	8.9	18.6	28.0	54.7	88.1	158	234	329	605	997	2100	3910	6270
	V	2.4	2.8	3.4	4.0	4.8	5.3	6.3	7.1	8.2	9.1	9.9	12	13	16	19	27
12	*A*	1.35	2.45	5.2	9.8	20.5	31.0	60.4	97.4	175	259	364	671	1100	2320	4330	6950
	V	2.7	3.1	3.7	4.4	5.3	5.9	6.9	7.8	9.1	10	11	13	15	18	21	24
14	*A*	1.45	2.65	5.65	10.7	22.3	33.7	65.8	106	190	282	397	731	1200	2530	4730	7590
	V	2.9	3.4	4.1	4.8	5.7	6.4	7.6	8.5	9.9	11	12	14	16	20	23	26
16	*A*	1.55	2.85	6.05	11.5	24.0	36.3	70.8	114	205	303	428	787	1300	2730	5100	8190
	V	3.1	3.6	4.4	5.1	6.2	6.9	8.1	9.7	11	12	13	15	17	21	25	28
20	*A*	1.75	3.25	6.85	13.0	27.1	41.0	80.0	129	232	344	484	892	1470	3100	5790	9300
	V	3.5	4.1	4.9	5.8	7.0	7.7	9.2	10	12	13	15	17	20	24	28	32
25	*A*	2.0	3.65	7.75	14.7	30.6	46.3	90.5	146	263	389	548	1010	1670	3510	6570	10560
	V	4.0	4.6	5.6	6.5	7.9	8.8	10	12	14	15	17	19	22	27	32	36
30	*A*	2.2	4.0	8.55	16.2	33.8	51.2	100	162	290	430	607	1120	1850	3900	7280	11710
	V	4.4	5.1	6.1	7.2	8.7	9.7	11	13	15	17	18	22	25	30	35	40
35	*A*	2.35	4.4	9.3	17.6	36.8	55.7	109	176	316	469	661	1220	2010	4250	7940	12780
	V	4.7	5.5	6.7	7.9	9.5	11	13	14	16	18	20	23	27	33	39	44
40	*A*	2.55	4.7	10.0	18.9	39.6	59.9	117	189	341	505	712	1320	2170	4580	8570	13780
	V	5.1	5.9	7.2	8.4	10	11	13	15	18	20	22	25	29	35	42	47
50	*A*	2.85	5.3	11.3	21.4	44.7	67.7	133	214	386	572	807	1490	2460	5190	9720	15650
	V	5.7	6.7	8.1	9.5	12	13	15	17	20	22	24	29	33	40	47	54
60	*A*	3.15	5.85	12.4	23.6	49.4	74.9	147	238	427	633	893	1650	2730	5760	10780	17360
	V	6.3	7.4	8.9	11	13	14	17	19	22	25	27	32	36	44	52	60
70	*A*	3.45	6.35	13.5	25.7	53.8	81.4	160	258	465	690	973	1800	2970	6280	11760	18950
	V	6.9	8.0	9.7	11	14	15	18	21	24	27	29	35	40	48	57	65
80	*A*	3.7	6.8	14.5	27.6	57.9	87.6	172	278	500	743	1050	1940	3200	6770	12690	20440
	V	7.4	8.6	10	12	15	17	20	22	26	29	32	37	43	52	62	70
100	*A*	4.15	7.7	16.4	31.1	65.4	99.0	194	314	566	840	1190	2200	3630	7680	14400	23200
	V	8.3	9.7	12	14	17	19	22	25	30	33	36	42	48	59	70	80
150	*A*	5.2	9.6	20.4	38.8	81.6	124	243	393	709	1050	1490	2760	4560	9650	18120	29220
	V	10	12	15	17	21	23	28	32	37	41	45	53	61	74	88	101
200	*A*	6.05	11.2	23.9	45.4	95.5	145	285	461	832	1240	1750	3240	5360	11350	21320	34400
	V	12	14	17	20	25	27	33	37	43	48	53	62	71	87	104	118
300	*A*	7.5	13.9	29.7	56.6	119	181	356	577	1040	1550	2190	4060	6730	14270	26830	43300
	V	15	18	21	25	31	34	41	46	54	60	66	78	90	110	131	149
400	*A*	8.75	16.2	34.7	66.2	140	212	417	676	1220	1820	2570	4780	7910	16790	31580	51000
	V	18	21	26	30	36	40	48	54	64	71	78	92	105	129	154	175
500	*A*	9.85	18.3	39.2	74.8	158	239	471	765	1380	2060	2910	5410	8970	19040	35840	57880
	V	20	23	29	33	41	45	54	62	72	80	88	104	119	147	174	199
600	*A*	10.9	20.2	43.2	82.5	174	264	521	846	1530	2280	3220	5990	9930	21100	39740	64210
	V	22	26	32	37	45	50	60	68	80	89	97	115	132	162	193	221
800	*A*	12.7	23.6	50.5	96.5	204	310	610	992	1790	2670	3780	7030	11670	24820	46780	75620
	V	25	30	37	43	52	59	70	80	94	104	114	135	155	191	228	260

* For other temperature drops the pipe capacities may be changed correspondingly. For example, with a temperature drop of 30 deg the capacities shown in this table are to be multiplied by 1.5.

† *V* = velocity, inches per second.

Table 8-4 Heat-carrying Capacity A (in Mbh) of Type L Copper Tubing with Temperature Drop of 20 Deg*

Nominal Tube Size (in.)		Milinch Friction Loss per Foot of Tube											
		720	600	480	360	300	240	180	150	120	90	75	60
3/8	A	8.9	7.8	7.0	5.9	5.4	4.7	3.9	3.6	3 1	2.7	2.3	2.1
	V†	23.6	20.8	18.6	15.7	14.4	12.5	10.4	9.6	8.2	7.2	6.1	5.6
1/2	A	16.7	15.0	13.0	11.2	10.0	8.7	7.5	6.6	5.6	5.0	4.5	3.9
	V	27.6	24.8	21.5	18.5	16.5	14.4	12.4	10.9	9.3	8.3	7.4	6.4
5/8	A	29.0	26.0	22.5	19.0	17.5	15.0	13.0	11.5	10.0	8.5	7.6	6.7
	V	32.2	28.8	25.0	21.1	19.4	16.6	14.4	12.8	11.1	9.4	8.4	7.4
3/4	A	43.5	39.0	34.5	29.0	26.5	23.0	19.6	17.5	15.0	13.0	12.0	10.5
	V	34.6	31.1	27.5	23.1	21.1	18.3	15.6	13.9	12.0	10.4	9.6	8.4
1	A	93	84	74	63	57	50	42.5	38	34	28.5	26	23
	V	43	39	34	29	27	23	20	18	16	13	12	11
1¼	A	160	145	128	107	97	85	73	65	57	48.5	44	39
	V	49	45	39	33	30	26	22	20	18	15	14	12
1½	A	260	240	206	175	160	140	118	106	93	79	71	62
	V	56	52	45	38	35	30	26	23	20	17	15	13
2	A	560	510	450	380	340	300	250	225	195	170	150	133
	V	70	64	56	47	42	37	31	28	24	21	19	17
2½	A	1100	930	820	700	630	550	470	420	370	310	280	250
	V	89	75	66	57	51	44	38	34	30	25	23	20
3	A	1650	1500	1300	1100	990	860	730	650	565	480	430	375
	V	·94	85	74	62	56	49	41	37	32	27	24	21
3½	A	2500	2250	2000	1750	1500	1320	1100	1000	860	730	660	580
	V	105	94	84	73	63	55	46	42	36	31	28	24
4	A	3600	3200	2800	2400	2150	1900	1600	1440	1250	1150	950	840
	V	116	103	90	77	69	61	51	46	40	37	31	27

* For other temperature drops the pipe capacities may be changed correspondingly. For example, with a temperature drop of 30 deg the capacities shown in this table are to be multiplied by 1.5.

† V = velocity, inches per second.

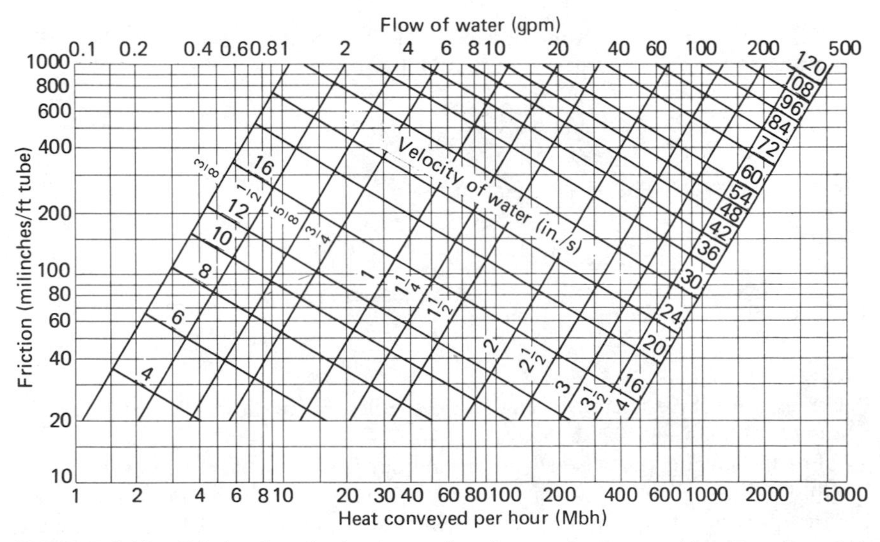

FIGURE 8-13. Friction loss for hot-water flow in type L copper tubing. Botton scale is based on 20-deg temperature difference between water in supply and return lines. For other temperature differences (Δt), multiply Mbh value being considered by $20/\Delta t$ before entering chart. (Adapted by permission from *ASHRAE Handbook of Fundamentals, 1972*, Chap. 26.)

from Fig. 8-12 or Fig. 8-13. If a standard 20-deg temperature drop is used the heat load at the bottom of the graphs can be used directly without correction or without referring to gallons per minute.

h. The radiator pipe sizes can be selected with the help of Table 8-3 or Table 8-4.

8-9. SINGLE-PIPE FORCED-CIRCULATION DESIGN PRACTICES

A representative system of this type is illustrated in Fig. 8-14. There are two circuits indicated, and in computation the longest (or highest friction) circuit must be used in sizing the pump. The radiation located below the main is not desirable and in a single-pipe system can be made to work only with the help of diversion tees.

Example 8-3. Work out the design features of the single-pipe system of Fig. 8-14. This system has seven upfeed convectors, required to deliver 9200 Btuh each, and two downfeed convectors, required to deliver 5600 Btuh each. It is assumed that the boiler supplies water at 190 F, which cools 20 deg in each radiator. With a one-pipe system such as this, the more-remote radiators will have less capacity than those radiators close to the boiler, which are supplied with the hottest water. However, this variation in capacity will not be considered sufficiently significant to upset a simple design procedure. The drawing, when scaled, shows the left circuit as having a linear run of 114 ft (rounded to the nearest foot) from boiler outlet to pump return, while the right circuit is 112 ft. The mains lie approximately 6 ft above the center line of the boiler.

Solution: The total load on the system is 75 600 Btuh, with 42 400 in the left circuit and 33 200 in the right. By Eq. (8-4), the total flow is

$$\text{gpm} = \frac{75\,600}{(490)(20)} = 7.72, \text{ or } 7.7$$

Also by Eq. (8-4), the flow in the left circuit is

$$\text{gpm} = \frac{42\,400}{(490)(20)} = 4.33, \text{ or } 4.3$$

Therefore the flow in the right circuit is

$$7.7 - 4.3 = 3.4 \text{ gpm}$$

Referring to Fig. 8-10, it can be seen that the smallest pump shown delivers 7.7 gpm at a head of 7 ft and will be used.

For the longest run, 114 ft, add an additional 50% to take care of equivalent length of fittings, giving a total of 171 ft ETL (equivalent total length) in this circuit. For the 7-ft head of the pump, or its equivalent, 84 000 milinches, the design friction loss per foot is

$$\frac{84\,000}{171} = 490 \text{ milinches}$$

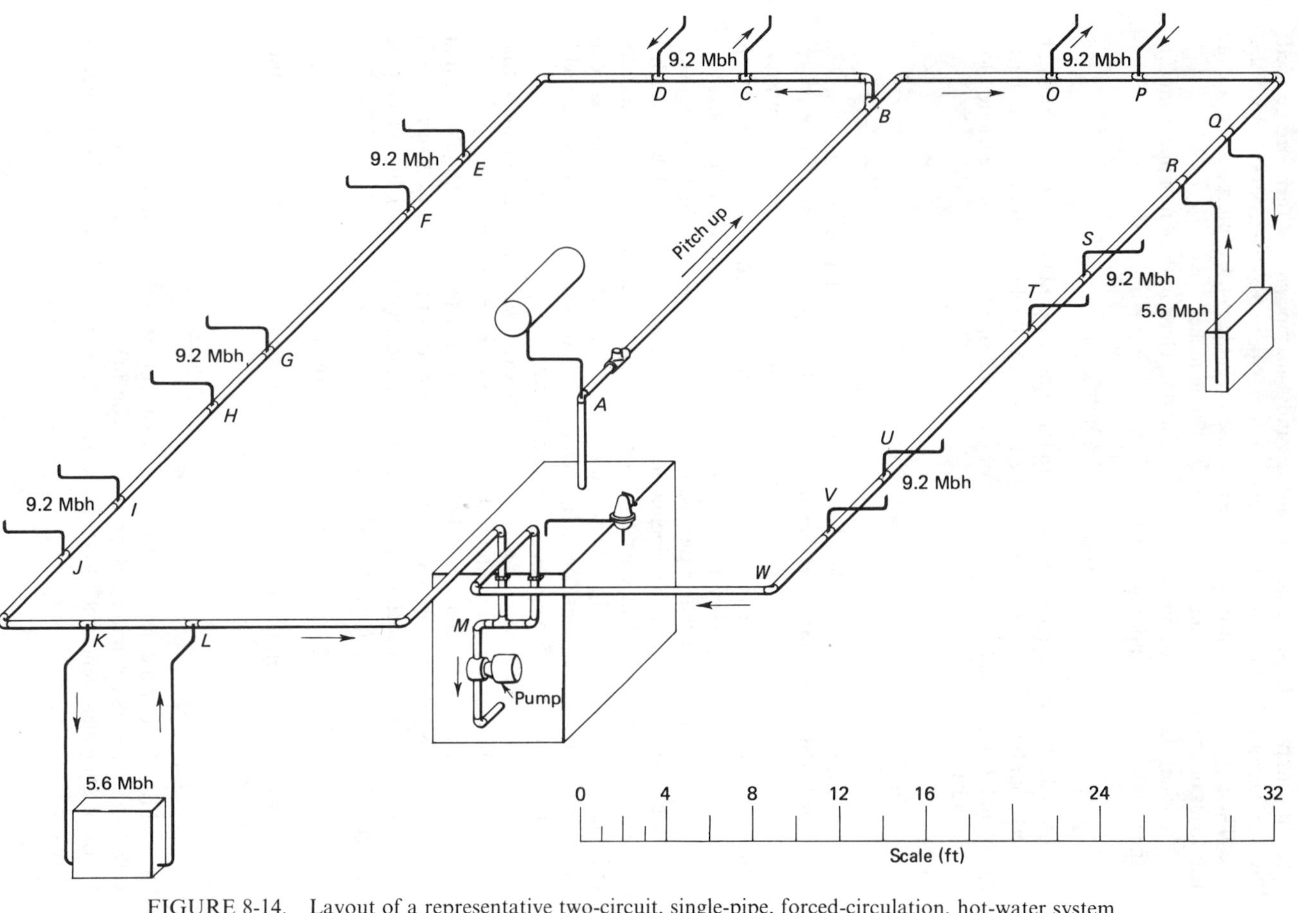

FIGURE 8-14. Layout of a representative two-circuit, single-pipe, forced-circulation, hot-water system.

Using this friction loss with Fig. 8-12, it is seen that for a flow of 7.7 gpm a 1-in. pipe is a trifle small, so a $1\frac{1}{4}$-in. pipe should be used from A to B. For the 4.3-gpm flow from B to the pump at M, $\frac{3}{4}$-in. pipe is undersize and 1-in. pipe should therefore be used; however, because the pipe run AB is oversized, it may be preferable to specify $\frac{3}{4}$-in. pipe here. The pipe sizes can also be selected by use of Table 8-4, by interpolation for the heat loads of 75 600 Btuh and 42 400 Btuh at 490 milinches.

Consider the radiator pipes attached at C and D. Here, in the 4 ft of line from C to D, the pressure drop is approximately

$$4 \times 490 = 1960 \text{ milinches}$$

This pressure differential must be sufficient to force the required flow, namely, 9200 Btuh (or 0.94 gpm by computation), through the radiator. If the linear run of this radiator circuit and its associated fittings showed, for example, an ETL of 32 ft, the friction loss for the radiator circuit would be

$$\frac{1960}{32} = 61.3 \text{ milinches/ft}$$

Choosing the pipe size from Table 8-3, at 9.2 Mbh and 61.3 milinches, it is seen that $\frac{1}{2}$ in. is too small and that a $\frac{3}{4}$-in. size is therefore required.

For the downflow radiator, two special tees arranged as in Fig. 8-8 must be employed. These cause a significant pressure drop in the line and thereby direct more flow through the radiator. They are seldom used in pairs in less than 1-in. pipe size. For the product of one manufacturer the loss in a pair of 1-in. tees at 4.3 gpm amounts to 5000 milinches, which, with the pipeline loss, could insure adequate flow through a downflow radiator. The characteristics of these tees vary greatly among manufacturers, and therefore, for precise design, the manufacturer's data should be used. For general estimation of circuit loss, consider such special tee loss to be the same as for a tee with maximum friction loss (Table 8-2), namely, minimum side flow.

The procedure for determining the pipe size for each radiation unit will follow that indicated for the radiator at CD and will therefore not be repeated here.

For the right-hand circuit, $ABOQWM$, the pressure loss for $BQWM$ should be equal to that of $BCDKLM$. The run AB, which is common to both circuits, is 25 ft long and contains an elbow, a stopcock (flow-control valve), and a wye (or tee). For $1\frac{1}{4}$-in. pipe the equivalent length of AB is, by Table 8-2,

$$25 + 2.6 + 2.6 + 10.4 = 40.6, \text{ or } 41 \text{ ft}$$

By Fig. 8-12, for 7.7 gpm flowing in $1\frac{1}{4}$-in. pipe the friction loss is 140 milinches; thus the loss in 41 ft is

$$140 \times 41 = 5740 \text{ milinches}$$

Thus, with 5740 milinches used in AB, there remains of the 84 000 milinches pump head, $84\,000 - 5750 = 78\,260$ milinches for use in circuit $BQWM$ of length $(112 - 25) = 87$ ft^2, or of equivalent length $87 + 0.5(87) = 130.5$ ft, and the design milinch value is therefore

$$\frac{78\,260}{130.5} = 600$$

Thus it can be seen, by reading at 600 milinches in Fig. 8-12, that to pass the 3.4 gpm for circuit $BQWM$ $\frac{3}{4}$-in. pipe is required. The foregoing computation was made in detail to show a method to follow, but usually this precision is not required and certainly it should not be carried out unless the left circuit, with an exact fitting computation, is also worked out. Absolute precision can never be expected in the basic design because pipe sizes are not sufficiently close together and thus the pipes are usually too large or slightly small.

When the circuits are too large the pump will not require as much head, and more water will flow until a balance is reached at a higher flow rate and lower head on the pump characteristic curve (Fig. 8-10).

The design principles outlined in this topic are equally applicable to baseboard (perimeter) radiation, which also involves a one-pipe system. Parallel circuits from the pump header should be balanced for friction losses so as to provide the required flow in each circuit. Observe also that the main flow of each circuit passes through each unit of radiation.

8-10. TWO-PIPE FORCED-CIRCULATION DESIGN PRACTICES

The design principles outlined in Section 8-8 are applicable here, with particular reference to the items listed under paragraph 10. The friction loss of each circuit is essentially the same in a reversed-return circuit, so that any one of the circuits can be used for selection of the required pump head after the total gallons per minute of the system has been found. In a direct-return system, particularly, the longest (or highest friction) circuit must be used to set the pump conditions, and more resistance must be placed in the shorter circuits either by restriction of pipe sizes or by use of an orifice or adjustable valve.

Example 8-4. For the reversed-return system shown in Fig. 8-15, with capacities and lengths as indicated, carry out the necessary design procedures to size the pipes and select a pump. Consider a 20-deg drop in water temperature, with water supplied at a design temperature of 190 F leaving the boiler. Indicate how the design would be altered with a 10-deg drop and with a 30-deg drop.

Solution: The longest circuit, *ACG* to radiator 5 to *MNOP*, has a linear run of 148 ft, allowing for a 4-ft rise to the trunk main at *A* and a 7-ft drop to the circulator at *O*. Add 50% for fittings to find an equivalent total length, giving 148 + 74 = 222 ft ETL. The longest circuit has a load of 65.6 Mbh, and the shortest circuit has a load of 56 Mbh, giving a total load of 121.6 Mbh. By Eq. (8-4), the total pump capacity is

$$\text{gpm} = \frac{121\,600}{(490)(20)} = 12.4$$

Also by Eq. (8-4), the capacity of the longest circuit is

$$\text{gpm} = \frac{65\,600}{(490)(20)} = 6.7$$

Therefore the maximum gallons per minute in the shortest circuit is

$$12.4 - 6.7 = 5.7$$

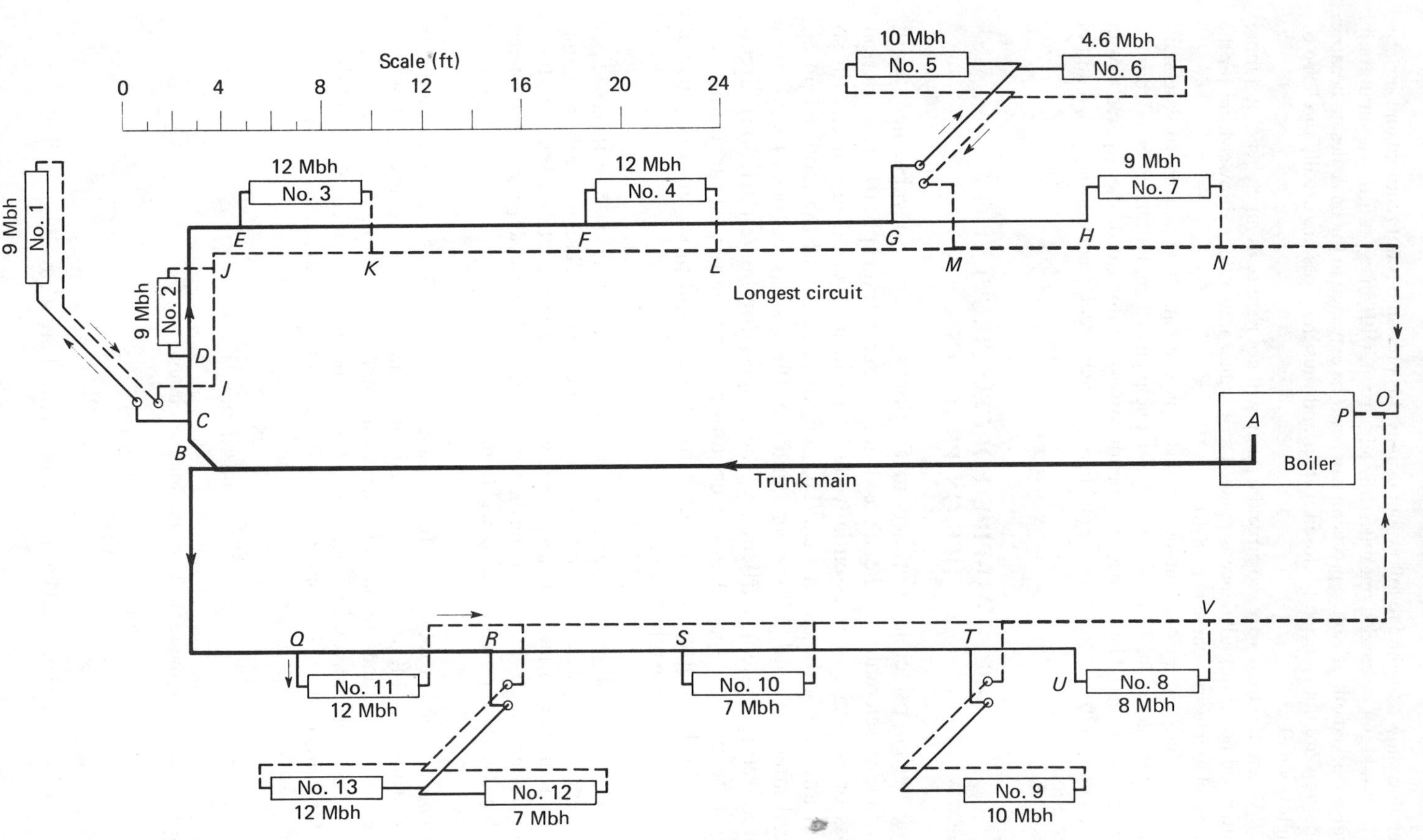

FIGURE 8-15. Layout of a representative two-pipe, reverse-return, forced-circulation, hot-water system.

By reference to Fig. 8-10, it can be seen that any one of the three smallest pumps has good head characteristics and might well be investigated. Each of these pumps is equipped with a small motor of about $\frac{1}{12}$ hp. If the highest head pump of these three is selected here, it can be seen that the head at 12.4 gpm is 7.5 ft, or $7.5 \times 12\,000 = 90\,000$ milinches.

With this pump the unit design friction head for the longest circuit is thus

$$\frac{90\,000}{222} = 405 \text{ milinches}$$

For the trunk main from *A* to the tee at *B* and *C*, Fig. 8-12 indicates that a $1\frac{1}{4}$-in. pipe size is required for 12.4 gpm, or for 121 600 Btuh. It is probably desirable to run this same size for the few feet to point *C*, although the flow there is only 6.7 gpm, corresponding to 65.6 Mbh. At *C*, 9.0 Mbh is taken off, leaving 56.6 Mbh, for which Fig. 8-12 at 405 milinches indicates that 1-in. pipe is more than adequate. Again, by Fig. 8-12,

From	At	Use
D to *E*	56.6 − 9.0 = 47.6 Mbh	1 in.
E to *F*	47.6 − 12.0 = 35.6 Mbh	$\frac{3}{4}$ in.
F to *G*	35.6 − 12.0 = 23.6 Mbh	$\frac{3}{4}$ in.
G to *H*	23.6 − 14.6 = 9.0 Mbh	$\frac{1}{2}$ in.
H to *N*	= 9.0 Mbh	$\frac{1}{2}$ in.

And, for the return pipe,

From	At	Use
C to *J*	= 9.0 Mbh	$\frac{1}{2}$ in.
D to *J*	= 9.0 Mbh	$\frac{1}{2}$ in.
J to *K*	9.0 + 9.0 = 18.0 Mbh	$\frac{3}{4}$ in.
K to *L*	18.0 + 12.0 = 30.0 Mbh	$\frac{3}{4}$ in.
L to *M*	30.0 + 12.0 = 42.0 Mbh	1 in.
M to *N*	42.0 + 14.6 = 56.6 Mbh	1 in.
N to *O*	56.6 + 9.0 = 65.6 Mbh	$1\frac{1}{4}$ in.

By the same methods, the radiator pipe size is found to be $\frac{1}{2}$ in. in every case, except that the runs from *G* and from *M*, each serving two radiators, should be $\frac{3}{4}$ in. When using steel pipe it is not desirable to use less than $\frac{1}{2}$-in. size.

In the event that this system were being sized for a 10-deg drop, the gallons-per-minute flow would have to be doubled, and the pipe would be sized thus, using 24.8 gpm for the main circuit. To use the Mbh chart at the bottom of Fig. 8-12, each value should be doubled (i.e., for the first radiator at 9.0 Mbh, the pipes must carry twice as much water as at a 20-deg drop; therefore use $2 \times 9.0 = 18$ Mbh for the chart reading). The increase in gallons per minute at 10-deg difference will probably indicate also that a different pump should be selected from Fig. 8-10. The same pump would deliver 24.8 gpm at 5.1 ft head, and the next-larger size would deliver 24.8 gpm at 7.6 ft head. For these two pumps the design milinches would then be

$$\frac{(5.1)(12\,000)}{222} = 275$$

or

$$\frac{(7.6)(12\,000)}{222} = 411$$

Using the design milinches of either of these pumps in connection with the proper gallons per minute, it will be found that larger pipes are required than for the 20-deg design.

If a 30-deg drop were to be used, the original gallons per minute would be reduced by the factor $\frac{20}{30}$. If it is desired to use the Mbh values on the chart, these are applicable after multiplying by $\frac{20}{30}$ or $\frac{2}{3}$. By similar procedure it is also possible to use Tables 8-3 and 8-4 with varying temperature drops.

For a precise check of results, each fitting in the design circuit should be summed in its proper size to get the true equivalent total length, and if the 50% allowance for fittings is greatly in error a recalculation may be required. Such a recalculation is not usually necessary in a reversed-return design, as the balance is usually close in each circuit.

Circulation pumps used as boosters in hot-water systems are usually low-head, centrifugal pumps, employing horizontal shafts and turning at conventional motor speeds. Because the water is simply being circulated in a closed circuit against resisting friction and not being lifted and delivered against external pressures, the horsepower required from the motor drive is relatively small. For example, in the representative head-capacity graphs of Fig. 8-10 the three smallest pumps have $\frac{1}{12}$-hp motors; the others, up to 2-in.-connection size, use $\frac{1}{6}$-hp motors; $2\frac{1}{2}$ in. and LD 3 use $\frac{1}{4}$ hp; HD3 is $\frac{1}{3}$ hp; PD 35 is $\frac{1}{2}$ hp; and PD 37 is $\frac{3}{4}$ hp. For the larger higher head pumps of Fig. 8-11 horsepowers range from 1 to 15. Motor speeds of 1750 rpm are most commonly employed. Booster pumps should be so made as to have long life and require a minimum of servicing.

Small centrifugal pumps of this typ have efficiencies which run from about 45 to 70%, with performance usually improving with the size of the pump. The horsepower required can be computed as

$$hp = \frac{\text{lb water/min} \times \text{ft head}}{33\,000 \times \text{pump efficiency}} \tag{8-5}$$

$$hp = \frac{(\text{gpm})(\text{weight/gal}) \times \text{ft head}}{33\,000 \times \text{pump efficiency}} \tag{8-6}$$

Motor efficiencies for fractional-horsepower motors are usually low, about 50 to 60%. Above 1 hp, efficiencies rapidly increase—reaching values of 75 to 90% with large motors.

Example 8-5. For the pump of Example 8-4, delivering 12.4 gpm against a 7.5-ft head, estimate the horsepower absorbed and the motor size which should probably be specified if the efficiency of this small pump is 52%.

Solution: By Eq. (8-6),

$$hp = \frac{(12.4)(8.1)(7.5)}{(33\,000)(0.52)} = 0.044 \qquad \textit{Ans.}$$

For this horsepower a $\frac{1}{12}$-hp motor (0.083 hp) is more than adequate but would probably represent a minimum size for such an installation.

With conventional forced-circulation systems the pump-head pressures employed run from 2 to 10 ft, with a common head pressure for design being 6 ft. Much higher heads are frequently employed in large heating systems, where the lower pipe investment cost can justify higher pump operating costs. Low heads are customary in small plants converted from thermal circulation. In a forced-circulation system, the ever-present thermal pressure head may assist the flow and reduce the pump power required, but is usually neglected in design computations.

Relatively few new systems using natural circulation are being installed at the present time because force-circulation systems use smaller pipe sizes and make possible better control and response. Nevertheless, there are some currently designed systems of this type, as well as a large number of old installations which continue to give satisfactory service. Because of the smaller pressure differences which arise from thermal effects to cause flow, greater care must be exercised in the layout and design of the system. However, the methods followed are the same as those indicated for forced circulation.

8-11. HOT-WATER PRACTICE IN LARGE BUILDINGS

The closeness of temperature control possible in spaces heated by hot water makes this medium suitable for heating large buildings. In operation the buildings are usually zoned, with each zone representing a separate hot-water distribution system. By means of combination indoor and outdoor thermostats it is possible to establish a close temperature-control system in each zone. Frequently in large systems a steam boiler supplies steam to heat

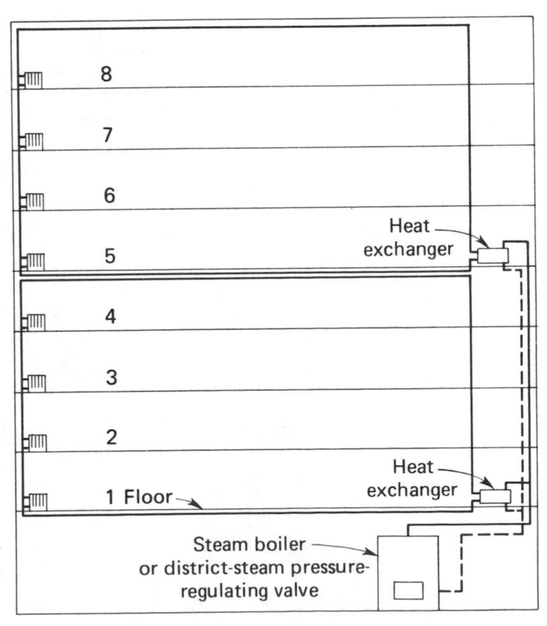

FIGURE 8-16. Arrangement for hot-water heating of multistory building.

exchangers in which the water is heated to the specified temperature called for by action of the thermostat. In multistory buildings, it is necessary to zone groups of floors in order to keep the static head on the heating elements within desirable limits. Figure 8-16 illustrates the type of arrangement which might be used in a multistory building. In large cities where district steam is available it is also possible to buy metered steam which in turn is furnished to the heat exchangers for warming the hot water used for heat distribution. The condensate is usually returned to the return mains of the steam supplier through condensate meters.

8-12. HIGH-TEMPERATURE WATER FOR HEATING AND INDUSTRY

Mention has previously been made of the use of high-temperature water as a means of distributing energy or directly heating a space. Figure 8-17 shows in diagrammatic form how a high-temperature distribution system might be arranged. The boiler in such a system is merely a heat exchanger between the burning fuel and the water, as the boiler operates at a pressure sufficiently high to keep the water from vaporizing. For example, if the water is heated to a temperature as high as 500 F, then the pressure of the system cannot be less than 681 psia wherever the water exists at this temperature. The heated water in turn is circulated to the points of utilization by a pump. Since the water gives up energy and cools throughout the system, it returns to the pump at appreciably lower temperature, perhaps at 150 F to 250 F.

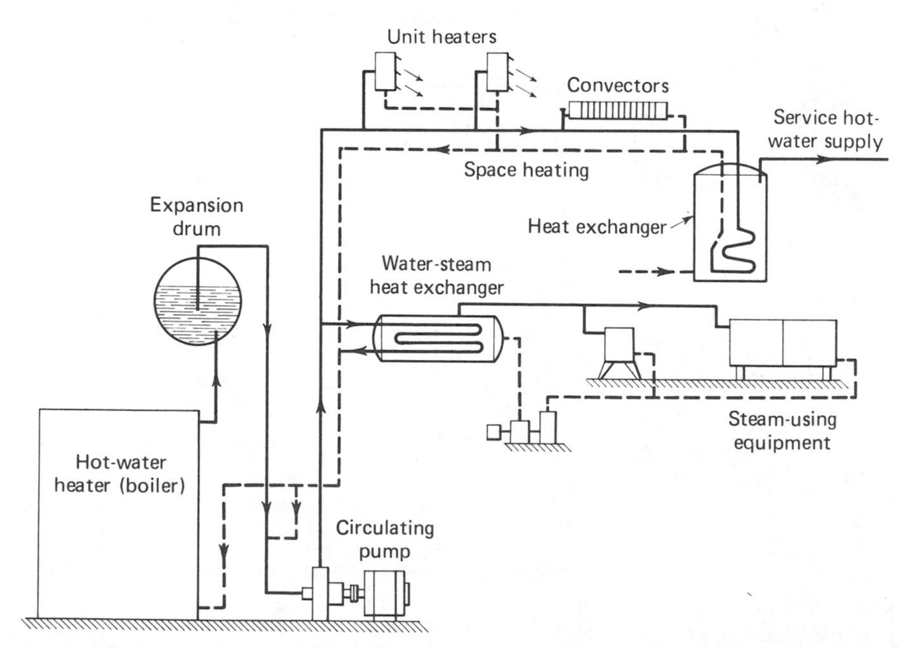

FIGURE 8-17. Heating and energy circulation system using high-temperature water.

This cooler water is delivered back into the heater-boiler, but if the designer feels it to be desirable a part of it can be recirculated into the supply water to set the temperature of the supply water at any required value.

It should be noted that a hot-water supply of this type can be used as the energy source for generating steam at lower temperature and pressure where this is desirable, as might be the case where steam is required for a constant-temperature process. In process work, temperatures can be held within very precise limits merely by adjusting the pressure of the steam.

PROBLEMS

8-1. In a certain thermal-circulating hot-water system the water leaves the boiler at 200 F and returns at 180 F. If the height between the boiler datum and the radiator datum is 10 ft, what is the pressure difference available to cause flow, expressed in milinches?
Ans. 880 milinches of water measured at 190 F

8-2. A 50 000-Btuh system is reported to have a total water content of 48 gal measured at 50 F. (a) Assuming that the system is filled at 50 F and designed for operation at a maximum temperature of 220 F, compute the increase in volume of water under the temperature extremes. (b) Determine whether the suggested expansion-tank sizes in Section 8-4 are adequate to serve this system, and if they are oversize suggest why.
Ans. (a) 2.2 gal; (b) tank must allow for air space in tank

8-3. A 100 000-Btuh system is reported to have a total water content of 82 gal measured at 50 F. (a) Assuming that the system is filled at 50 F and designed for operation at a maximum temperature of 220 F, compute the increase in the volume of the water in the system under the temperature extremes. (b) Determine whether the suggested expansion-tank sizes in Section 8-4 are adequate to serve this system, and if they are oversize suggest why.
Ans. (a) 3.8 gal; (b) tank must allow for air volume

8-4. The circuit from a boiler to a particular division of radiation, with a capacity of 40 Mbh, contains 40 lin ft of pipe, 12 elbows, 2 tees with 50% diversion to the radiator, 1 radiator valve, 1 radiator, and 1 boiler. (a) What is the total number of elbow equivalents to be added to the measured length of run? (b) Find the size of steel pipe that should be used in this radiation circuit provided the design is to be made not to exceed 350 milinches/ft and for a 20-deg design temperature drop. (c) Assume that 20 lin ft of run for this radiator circuit, along with 14 elbow equivalents, are in the main, which is $1\frac{1}{4}$-in. size, and find the precise equivalent length of run for the radiator circuit and its mains and compare this with an assumption of 50% increase in linear length or 100% increase in linear length as a means of estimating a trial equivalent length. (d) What pump head should be specified? Compute the head on the basis of actual friction loss, considering the flow in the $1\frac{1}{4}$-in. main to be twice that in the radiator run.
Ans. (a) 28; (b) 1-in. steel pipe; (c) 106 ft, use 100%; (d) 1.4 ft

8-5. Rework Problem 8-4 on the basis of copper tubing instead of steel. Assume that the largest pipe in the circuit is 1 in. Note that copper tubing has a smaller outside diameter than steel, and also a smaller bore.
Ans. (a) 31; (b) 1 in.; (c) 105 ft; (d) 2.9 ft

8-6. A particular division of radiation with four radiators in parallel has an output of 80 000 Btuh, with each radiator furnishing 20 Mbh. The total circuit from and to the boiler, and up to the longest unit of radiation served above, consists of 40 ft of $1\frac{1}{4}$-in. steel pipe, while the radiator circuit itself consists of an additional linear run of 30 ft of smaller-size pipe. The total main and radiator circuit in question consists then of 70 ft of run and includes 14 elbows, 2 tees with 50% diversion, 2 radiator valves, 1 radiator, 1 boiler, and 2 gate valves. The total boiler-radiator capacity is 120 000 Btuh. Find (a) the number of total elbow equivalents in the circuit described; (b) the size of steel pipe used for the radiator circuit if the design is not to exceed 350 milinches/ft and is based on a 20-deg design temperature drop; and (c) the precise equivalent length of the run if 15 of the elbow equivalents are in the radiator circuit and the rest are in the main circuit. (d) By what factor should the linear run of this problem have been multiplied if a trial equivalent length were required? (e) Find the pump head and the gallons-per-minute capacity which should be specified.

Ans. (a) 33; (b) $\frac{3}{4}$ in.; (c) 140 ft; (d) 2 (i.e., 100% increase); (e) 2.7 ft, 12.2 gpm

8-7. Rework Problem 8-6 but assume that copper tubing instead of steel is used. Note that the copper tubing, compared with steel, is smaller both in outer diameter and in bore, for the same nominal size.

Ans. (a) 37.4; (b) $\frac{3}{4}$ in.; (c) 140 ft; (d) 2 (i.e., 100% increase); (e) 3.7 ft, 12.2 gpm

8-8. A forced-circulation hot-water heating system is to be designed for outpoing water at 220 F and return water at 190 F. The total load is 1300 Mbh. The length of the longest circuit is 216 ft and this will be increased by 100% to obtain a trial equivalent length. The system is to be designed on the basis of 3.0 ft of water pressure drop per 100 ft of equivalent length. Find (a) the gallons per minute to be circulated, (b) the pump head required and the design milinches, and (c) the size of pipe which should be used for all parts of the circuit carrying the full flow.

Ans. (a) 90 gpm; (b) 15.7 ft, 360 milinches/ft; (c) 3 in.

8-9. Use the same conditions as in Problem 8-8, except that the supply water is at 210 F and the return water is at 200 F, and find (a) the gallons per minute which should be circulated, (b) the pump head required and the design milinches, and (c) the size of pipe required. (d) In physical size, would smaller or larger radiators be required in each of the rooms for the same heat output under the conditions of Problem 8-8 and Problem 8-9?

Ans. (a) 268 gpm; (b) 15.7 ft, 360 milinches; (c) $3\frac{1}{2}$ in., or reduce to 3 in., in parts of run; (d) same size

8-10. For the system pictured in Fig. 8-15, with capacities and lengths as indicated, complete the design with steel pipe for sizing the longest circuit when the design is based on a 30-deg drop in water temperature.

Ans. Total flow 24.8 gpm, 13.4 gpm in longest circuit; for booster pump at 7.3 ft at 24.8 gpm, use 395-milinch design basis. *A* to *BC*, 2 in.; *C* to *D*, $1\frac{1}{4}$ in.; *D* to *F*, $1\frac{1}{4}$ in.; *F* to *G*, 1 in.; *G* to *N*, $\frac{3}{4}$ in.; *N* to *O*, $1\frac{1}{4}$ in.; etc.

8-11. Work Problem 8-10 using copper tubing throughout.

8-12. For the system pictured in Fig. 8-15, with capacities and lengths as indicated, complete the design with steel pipe for sizing the longest circuit when the design is based on a 30-deg drop in water temperature.

8-13. Work Problem 8-12 using copper tubing throughout.

8-14. A forced-circulation hot-water heating system is designed for outgoing water at 200 F, with return water at 180 F. The total load is 1000 Mbh. The equivalent length of the longest circuit is thought to be 680 ft. The system is designed for use with a 15-ft-head pump. Find (a) the gallons per minute, (b) the total friction head, and (c) the size of pipe where maximum flow is required.

Ans. (a) 102 gpm; (b) 180 000 milinches; (c) $3\frac{1}{2}$ in.

8-15. For the perimeter heating system of Fig. 8-9, the dining-room–kitchen circuit has a linear run of 146 ft and serves a load of 36 Mbh. The living-room–bedroom circuit has a linear run of 122 ft and serves 30 Mbh. The upstairs loads taken from each of these circuits have 60% of the main floor load, and each has a shorter total linear run. The system operates at an average temperature of approximately 180 F and employs a 20-deg temperature drop, with finned copper radiation. Select a pump of adequate capacity to serve this building, basing its output head on the longest circuit, with its known fittings, and on a reasonable friction loss per foot.

8-16. Solve Problem 8-15 with no change except that a 10-deg drop is used.

8-17. Solve Problem 8-15 using cast-iron baseboard radiation of approximately the same output per foot when operated at 200 F average temperature and a 20-deg temperature drop.

8-18. Consideration is being given to heating a remote installation having a heat load of 1 000 000 Btuh, with the heat to be supplied from a central heating plant. (1) Hot water can be supplied in a low-pressure system at 240 F and returned to the heating plant at 140 F; or (2) high-temperature water can be supplied at 440 F and returned to the heating plant at 140 F; or (3) dry saturated steam can be supplied at 250 psia with condensate returning at approximately 200 F. (a) Making use of the water–steam table, Table 2-2, compute for each of the three foregoing systems the pounds of medium flowing per hour required to heat the remote space. (b) Making use of the specific volume of water and saturated steam at maximum temperature conditions, compute the pipe cross section which would be required to pass the hot water with the two water systems if a maximum velocity of 5 ft/s is employed, or to pass the steam if a maximum velocity of 50 ft/s is employed. From your results you will note that although steam carries the greatest energy per pound, its energy per unit of volume is smaller than that of hot water.

Ans. (a) 9960 lb/h, 3220 lb/h, 968 lb/h; (b) 1.32 in.^{2}, 0.46 in.^{2}, 1.43 sq in.^{2}

CHAPTER
9
HEAT-TRANSFER ELEMENTS, ELECTRIC HEATING, AND COMBUSTION

9-1. RADIATORS AND CONVECTORS

The heating and ventilating industry makes wide use of the terms "radiator" and "convector," although in many cases the terminology is inappropriate. Any surface warmer than its surroundings gives up some heat by radiant methods (transfer through space without the intermediary action of the medium through which the heat is passing). A person some distance away from a fire feels its warming effect although the air around him may be cold and the side of his body which does not "see" (face) the fire may be quite cool. Any surface warmer than its surroundings also gives up heat to air moving past it, by convection methods. Thus in most cases the heat is transferred by both radiation and convection.

The tendency is to use the term "convector" for any device which gives up its heat largely to air currents passing by it and in which the heating elements do not "see" the space being heated. Under this definition an ordinary steam radiator still merits its name, although the larger part of its output may be delivered by convection and not by radiation.

Originally, heaters using steam and hot water were all called radiators. If the heat-transmitting device was placed in the room which was to be heated, it was called a direct radiator. If the heater was placed in the basement or in some other room, it was called an indirect radiator, though under such a condition its radiant heat did not affect the room and the hot surface warmed the air which moved around it, and thus it was really a convector. If a heater was partly enclosed it was called a semidirect radiator. If some air from outside was made to flow past its surface to mix with the room air, the device was called a direct-indirect radiator.

The early method of rating and selling these heat-transmitting elements was to measure the surface exposed to the air and to use the square feet of area of this surface as the unit of capacity on which the price was based. This method proved highly unsatisfactory, however, because the surface area in a radiator is exceedingly difficult to measure accurately and the heat-transmitting ability of each unit of surface varies greatly.

The extensive use of fins to provide additional heat transfer surface for the air passing over the unit has made capacity based on prime surface of trivial importance, and it is necessary to use test results, such as those listed in manufacturers' catalogs, as the basis for design sizing. However, method of installation, location in a room, room temperature, and the temperature of the heating medium all offer variables to which careful consideration must be given.

Steam radiators are usually specified for test-temperature conditions of 215 F in the radiator, with 70 F air, or a temperature difference of 145 deg. Heat transfer from a radiator occurs by conduction, convection, and radiation combined, and so is not directly proportional to the temperature difference. Tests have shown that with radiators the heat transfer is proportional to the 1.3 power of the ratio of the actual temperature difference to the standard temperature difference, and to the 1.5 power with convectors. Thus a radiator using steam at 20 psia (5.3 psig) and 228 F in a room heated to 55 F would give out $[(228 - 55)/(215 - 70)]^{1.3}$, or 1.256 times as much heat as a standard radiator. A radiator operating in the subatmospheric region at 7 psia (176.85 F) would give out only $[(176.85 - 55)/(215 - 70)]^{1.3}$, or 0.797 times as much heat as one at standard conditions. When the heat transfer with 20 psia steam is compared with that for 7 psia steam it is seen that only $(0.797/1.256) \times 100 = 63.4\%$ as much heat will be transferred per hour in a given radiator operated at 7 psia (subatmospheric) as compared with one operated at 20 psia.

Figure 9-1 shows a type of free-standing radiator for steam or hot water which at one time was used extensively. Few current designs use this type of unit but so many are in use that its characteristics should be understood. This type of radiator is made of cast iron and its sections are joined to each other by malleable slip nipples at top and bottom with legs provided for each end section. The length can thus be changed to suit the heat output required. The radiator of Fig. 9-1 has 11 sections, and if it is four vertical tubes deep it will deliver some 5280 Btuh (1.55 kW) when supplied with 215 F (101.7°C)

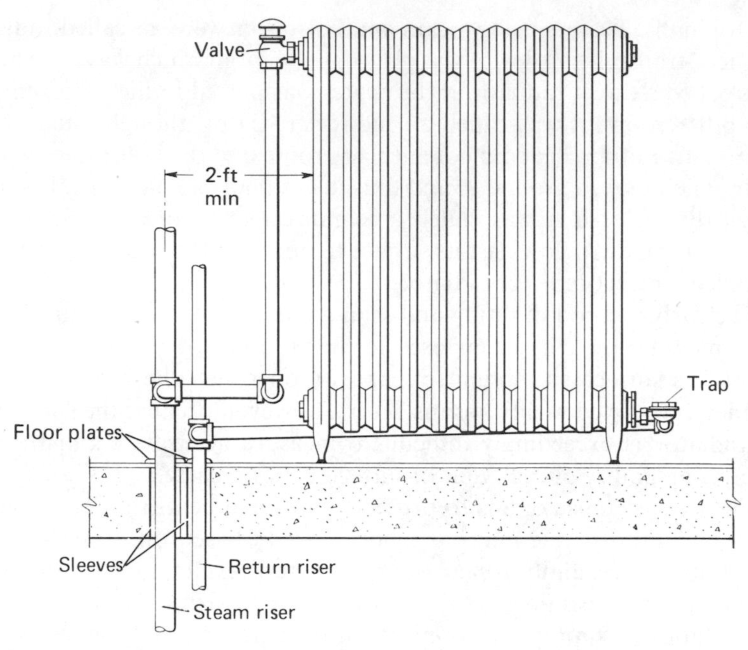

FIGURE 9-1. Representative connections for an early design steam radiator used on a vapor or vacuum system.

steam in 70 F (21.1°C) space. Its length is 19 in. (0.89 m) and its height 25 in. (0.63 m). Insulation should be provided behind each radiator or convector, either in the wall itself or in supplementary form.

9-2. CONVECTOR

A convector for use with hot water or steam is pictured in Fig. 9-2. This particular unit is a floor-mounted type, but similar units are made to be wall-hung or recessed. The front cover appears in lifted position to show the internal parts. This cover drops down and latches. When the cover is latched, the warmed air is delivered forward into the room through the grillework shown. These convectors are made in depths of 4, 6, or 8 in. and can be from 10 to 36 in. high. The length can range from 26 to 62 in., changing by 6-in. increments. The heating element at the bottom shows one end connection to which the steam or hot-water piping is joined. The actual heating element consists of $1\frac{1}{4}$-in. steel pipe with 40 or 50 steel fins per foot. The bare pipe ends attach into cast-iron headers, one of which can be seen at the far end of the element. Alternatively the element can be made of 1-in. copper tubing to which 68 aluminum fins per foot are attached, or the elements can use $1\frac{1}{4}$-in. copper tubing with 60 aluminum fins per foot. The copper tubing is also expanded into cast-iron end headers. The unit is arranged to float on a flexible steel support which allows for expansion and contraction. Not shown is a hand-operated flapper damper which is pivoted, at the back, to swing

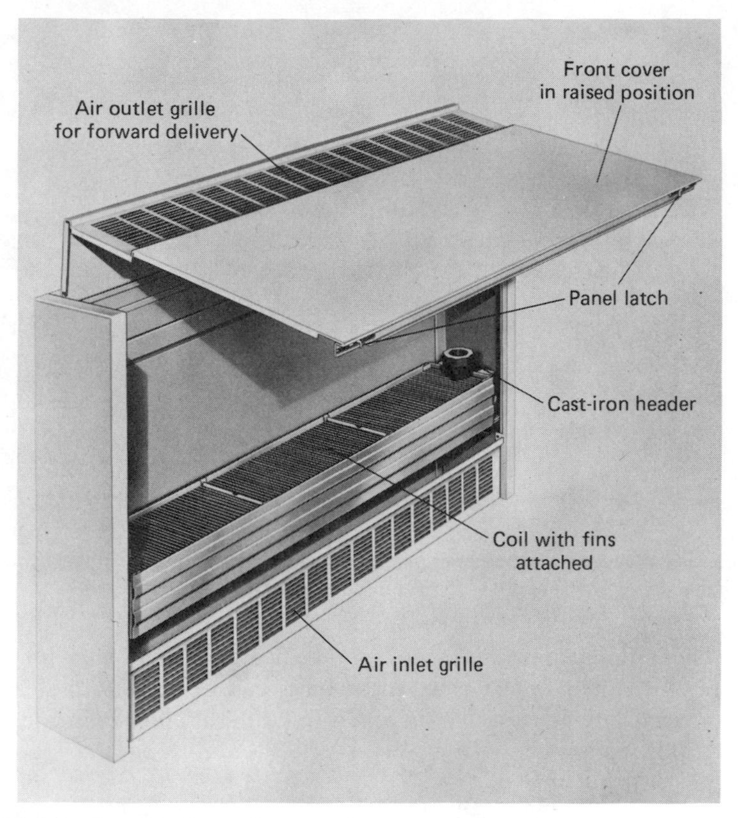

FIGURE 9-2. Front-discharge floor-mounted convector with front panel in raised position to show steam (hot-water) coil. (Courtesy of the Trane Company.)

closed and partially or fully cover the element to block the free air flow through it. This action causes a reduction in cpacity by as much as 70%.

To give some idea of the output of one of these convectors, consider a unit with one row of $1\frac{1}{4}$-in. steel pipe with $2\frac{1}{2}$ by $5\frac{1}{4}$-in. steel fins with an installed height of $22\frac{3}{4}$ in. The rating per foot of finned length with 65 F inlet air varies with average hot-water or steam temperature as follows: 220 F, 1410 Btuh; 200 F, 1150 Btuh; 180 F, 920 Btuh; 170 F, 820 Btuh; 215 F steam, 1340 Btuh. With one row of $1\frac{1}{4}$-in. copper tubing and aluminum fins, the output is greater: 220 F, 2250 Btuh; 200 F, 1840 Btuh; 180 F, 1480 Btuh; 170 F, 1310 Btuh; 215 F steam, 2140 Btuh. The capacity increases with height because of stack effect.

Figure 9-3 illustrates another arrangement in which the unit attaches to the wall with a mounting strip. The enclosures are low and can use as much length of wall space as may be required to offset the heat loss. These "wall-fin" units can be kept very low on the wall or occupy a higher position. A minimum $2\frac{1}{2}$ in. or so of space for air flow should be allowed under the coil to provide for inlet air circulation. The top of the unit can be arranged with horizontal grilles for upward air flow, with vertical grills for forward flow, or with sloping

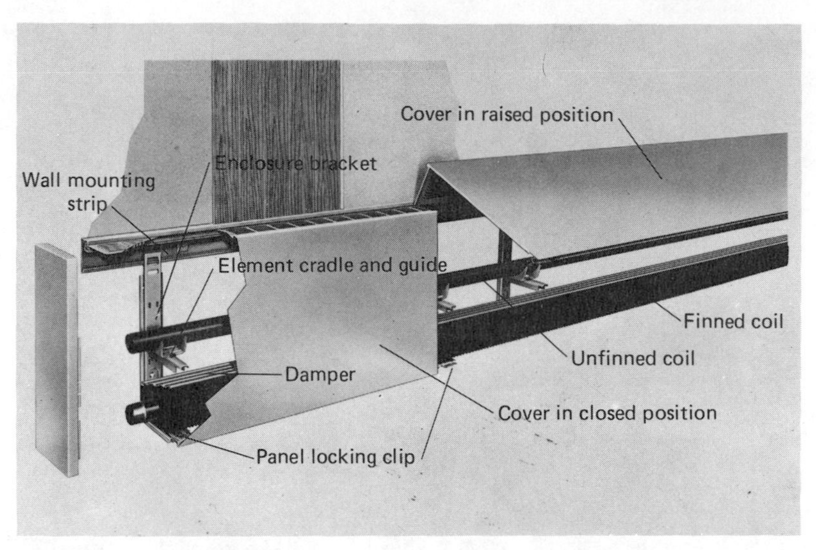

FIGURE 9-3. Diagrammatic view of Trane Wall Fin convection radiation provided with steam (hot-water) coil. (Courtesy of the Trane Company.)

grilles. The latter keep people from using the heater as a low shelf for storage of items. With their finned coils, these units can operate with steam, hot water, or electricity. In capacity, they are similar to the convectors mentioned earlier in this section.

9-3. ELECTRICALLY HEATED CONVECTOR

The price of fossil fuel has been rising at such rapid rates that the cost of heating a building electrically is competitive with heating by other methods. This is not always the case, but when a utility has excess capacity in winter and when the operational and investment costs of the heating plant are considered, electric heating may be a desirable alternative. Thus increasing usage is being made of electricity for space heating.

The convector casings of Figs. 9-2 and 9-3 can easily be adapted for electric heating. In fact no real change is involved except that a finned electric heater is substituted for the hot-water coil and wiring and controls substitute for the interconnecting piping. Precise control to meet heating demand can be provided and limit controls act to prevent fin temperatures from exceeding 260 F (142°C). A damper is not required. The finned-tube elements, similar in appearance to those for steam (hot water), are rated in watts per foot of finned length. Capacities of 120, 175, 250, 350, 500, or 700 W/ft can be provided. Wiring arrangements can be made to serve the heater elements from a single-phase or from a three-phase circuit at 208, 240 or 277 V across the element. The ordinary 110 to 120 voltage used in lighting circuits requires larger current flow (amperage) and is not usually employed with units of this type since heavier circuits for wiring and relays would be needed. Controls, very sensitive to thermostatic action, prevent overheating, and prompt response is a characteristic of electric heating. In addition, each room or space can be heated and controlled independently of other space requirements.

Space heating electrically is also accomplished in other ways. In one method, resistance heating coils are buried close to the surface in the plaster finish of a ceiling. Almost the whole ceiling area is made a heating surface so that at moderate temperature the ceiling can radiate heat into the space and offset heat losses. The response to warm-up is slower than with convector coils but is very sensitive to control under normal operation. Strip heaters in baseboard radiation are also used. Unit heaters of the type described in Section 9-5 can also be provided with electric elements. Electrically heated boilers are also available.

9-4. BASEBOARD RADIATION

We have described convectors of conventional or low height; still another possibility, known as baseboard radiation, is available both for residential and public buildings. This type of radiation has the advantage of supplying a continuous band of heat around the exposed perimeter of a room, thus effectively preventing cool downdrafts from exposed walls and windows. The baseboard elements are easy to install and they can blend into the room to create a pleasant appearance. Because the elements are close to the wall surface, they cause only minimum difficulty with regard to the placement of furniture or equipment. It should be remarked that the radiation equipment can also be recessed so as to be flush with the wall, or it can project out into the room by a few inches. The heat-transfer element consists of pipe or tubing to which metal fins are attached. The pipe can be of steel, with steel or aluminum fins, or of copper with aluminum fins. The enclosure for the convector is usually of pressed steel formed to shape.

Tables 9-1 and 9-2 give the output capacity in Btu per hour for varying lengths of heater elements in baseboard enclosures, for given operating temperatures. It should also be noted that connecting lengths of bare straight pipe, and bends at corners, contribute to the heat output of the convector, and that the output of the pipe should therefore be considered in all computations. In designing with baseboard radiation, the maximum heat output in a given room is set by the available wall space along which the elements can be installed. Thus it is possible in some cases that space limitations prevent the installation of a sufficient baseboard length to offset the heat loss from a room. Where this is the case, higher hot-water temperatures supplied to the radiation may suffice, or a conventional convector can be added. With baseboard panels, hot water is more frequently used than steam, but effective operation can also be obtained with steam.

The heat output of baseboard radiation varies greatly among manufacturers, the exact output depending on pipe and fin size, and height of the baseboard, and also on whether the lighter nonferrous or heavier cast-iron type is employed. For cast iron, the heat emission per foot of panel length

Table 9-1 Representative Capacities* of Baseboard Radiation†

Panel Length (lineal ft)	Average Water Temperature (deg F)							
	150	160	170	180	190	200	210	220
2	0.8	0.9	1.1	1.2	1.4	1.5	1.7	1.8
3	1.2	1.4	1.6	1.8	2.1	2.3	2.5	2.8
4	1.6	1.9	2.1	2.4	2.7	3.0	3.3	3.7
5	2.0	2.3	2.7	3.0	3.4	3.8	4.2	4.6
6	2.4	2.8	3.2	3.6	4.1	4.5	5.0	5.5
8	3.2	3.7	4.3	4.8	5.5	6.0	6.7	7.4
10	4.0	4.7	5.4	6.1	6.9	7.6	8.4	9.2
12	4.7	5.6	6.4	7.3	8.2	9.1	10.0	11.0
16	6.3	7.4	8.6	9.7	11.0	12.1	13.4	14.7
20	7.9	9.3	10.7	12.1	13.7	15.1	16.7	18.4
22	8.7	10.2	11.8	13.3	15.1	16.6	18.4	20.2
24	9.5	11.2	12.8	14.5	16.4	18.1	20.0	22.1
Material	Output of 1-In. Bare Pipe in Enclosure, Btuh per Lineal Foot, at Temperatures Indicated Above							
Copper	48	57	66	75	85	92	102	113
Iron	80	94	108	122	138	152	168	186

* Capacities expressed in thousands of Btu per hour (Mbh), with 65 F entering air, for copper tubing with aluminum fins.

† Table courtesy of American Radiator and Standard Sanitary Corporation (Heatrim panel).

Table 9-2 Representative Capacities* of Baseboard Radiation for Hot Water or Steam†

Panel Length (lineal ft)	Average Water Temperature (deg F)							Steam at 215 F
	150	160	170	180	190	200	220	
3	0.8	1.0	1.1	1.3	1.4	1.6	1.9	1.8
4	1.1	1.3	1.5	1.7	1.9	2.1	2.6	2.5
5	1.4	1.6	1.9	2.1	2.4	2.6	3.2	3.1
6	1.7	2.0	2.3	2.5	2.9	3.2	3.9	3.7
7	1.9	2.3	2.6	3.0	3.4	3.7	4.6	4.3
8	2.2	2.6	3.0	3.4	3.8	4.2	5.2	4.9
9	2.5	2.9	3.4	3.8	4.3	4.8	5.8	5.5
10	2.8	3.3	3.8	4.2	4.8	5.3	6.5	6.1
12	3.3	3.9	4.5	5.1	5.8	6.3	7.7	7.4
14	3.9	4.6	5.3	5.9	6.7	7.4	9.0	8.6
16	4.4	5.2	6.0	6.8	7.7	8.5	10.3	9.8
18	5.0	5.9	6.8	7.6	8.6	9.5	11.6	11.1
20	5.5	6.5	7.5	8.5	9.6	10.6	12.9	12.3
22	6.1	7.2	8.3	9.3	10.5	11.6	14.2	13.5
24	6.6	7.8	9.0	10.2	11.5	12.7	15.5	14.7

* Capacities expressed in thousands of Btu per hour (Mbh), with 65 F entering air, for cast-iron integrally-cast fin radiation and cast-iron covers.

† Table courtesy of American Radiator and Standard Sanitary Corporation (Radiantrim panel).

is usually less than for the nonferrous type because the fin area is so much less (see Table 9-2).

9-5. UNIT COOLERS AND HEATERS

When a fan is employed to force air over a heater coil surface, the output is increased approximately threefold over the corresponding heat delivery of a thermal-circulation convector of the same area, even with a high draft head. These units are often placed under a window or at least on an outside wall. Chilled water is circulated through the coil when cooling is needed, and hot water is circulated when heating is required. Only three pipe connections are required: water supply and return and a drain for condensate in summer. Electrical supply is also required for the fan motor. Recirculated air and a controlled amount of outside air for ventilation are drawn through the filter section and then pass over the coils, where the air is tempered to the desired degree. A connection is usually made through the wall in the back of the unit to provide for outside air intake, the amount of which can be adjusted by a manually operated damper. The tubes of such units are usually made of copper provided with aluminum fins.

Unit heaters are often employed in factories and warehouses, where the units are suspended from the ceiling or mounted at a high level on a wall. Figure 9-4 shows a cutaway section through a unit heater of this type. The fan mounted in back of the unit blows room air over the heat-transfer coils, and this air, in turn, is directed outward and downward by means of adjustable louvres. These units use either steam or hot water. The heat-transfer coils are made from copper tubing and have aluminum fins attached.

Figure 9-5 shows a vertical-shaft unit arranged to direct the heated air downward. In this unit room air is drawn horizontally, warms in passing through the coils, and then is directed downward into the space being heated. The top is a continuous closure except for a number of optional openings or ports which can be opened to provide cooler room air to mix with the air being heated, thus tempering it to a desired value. These units come in various capacities and in physical size they have fan-blade diameters ranging from $16\frac{3}{4}$ to 30 in.

Unit heaters are available in a wide range of capacities, varying from about 5000 Btuh for smaller units to over 1 000 000 Btuh for larger units. The capacity varies not only with the physical size of the units but also with steam pressure (temperature) or, for hot-water units, with water temperature. Many units are built with a tube-wall thickness sufficiently adequate to make the unit suitable for operation with high-pressure steam (over 100 psi). When air directly from outside is passed over the coils, the steam-condensation or water-cooling capacity of a unit is greatly increased. Motor horsepower ratings range from $\frac{1}{35}$ hp for the smallest unit to 3 hp for large units, and air is delivered at temperatures ranging from about 90 F to over 250 F. The higher temperatures are possible in the case of units remote from working space.

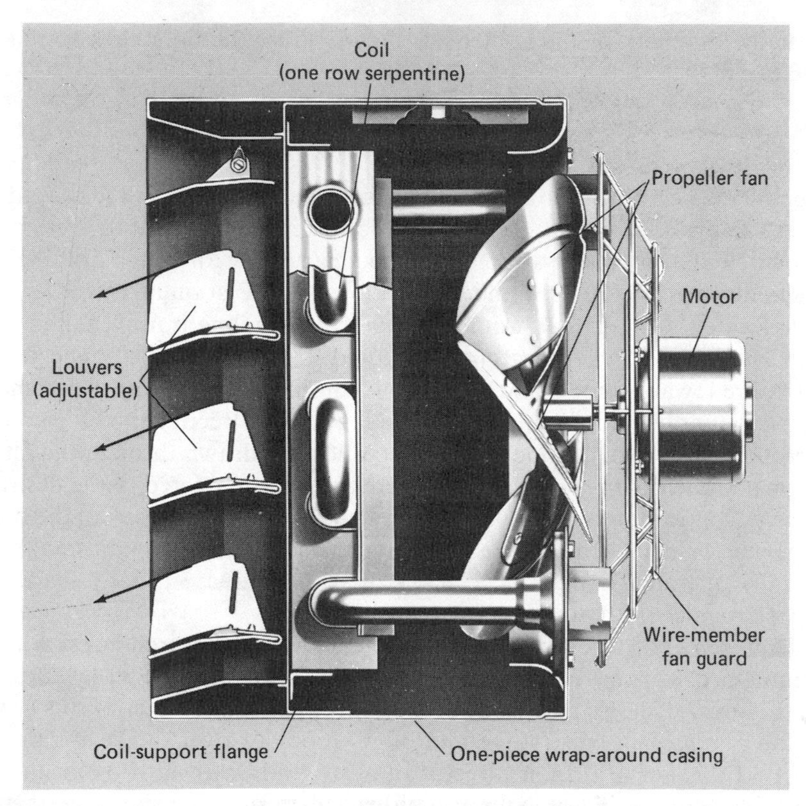

FIGURE 9-4. Cut-away section through unit heater arranged for horizontal flow. (Courtesy of the Trane Company.)

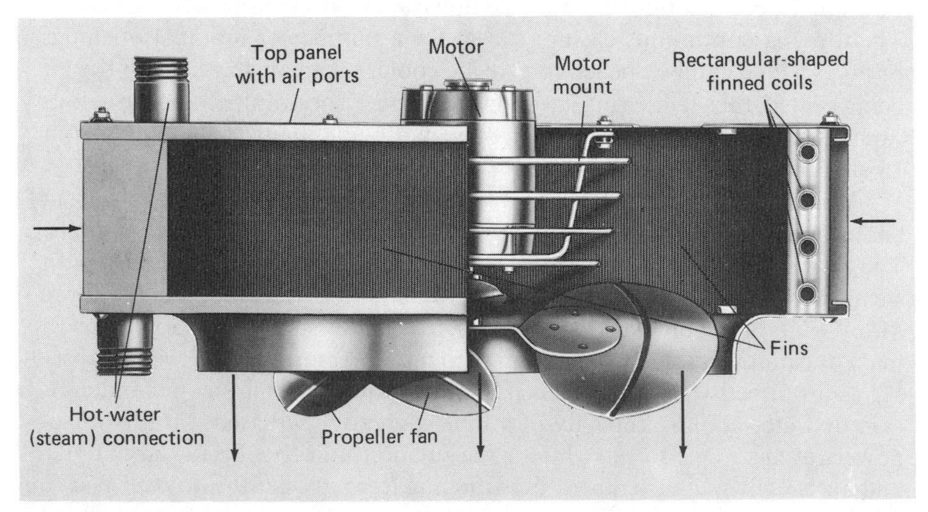

FIGURE 9-5. Cut-away view of rectangular-shaped, downflow-discharge unit heater. (Courtesy of the Trane Company.)

These units are particularly applicable in heating such industrial spaces as the single-story buildings, with or without monitor roof, used for modern factories. Such buildings usually have large glazed areas and are not as well insulated as the more massive multistory buildings. A plan and elevation of a representative factory building of this type appears in Fig. 9-6. This design shows one possible placement for the unit heaters. The placement of these units would be such as to direct warm air toward the corners and side walls so that a sheet of warmed air could move downward and across the floor, later rising upward for recirculation through the unit heaters.

Example 9-1. Consider the industrial building of Fig. 9-6 to be located in a suburb of Chicago. It is desired to maintain an inside temperature of 60 F at the breathing, or 30-in., line. Heating is accomplished by means of unit heaters suspended from the ceiling, and steam will be used at 215 F. The design temperature can be taken as −10 F, and the prevailing wind direction for the winter season can be considered as 12 mph southwest. The walls are plain brick 12 in. thick, and there are 20 steel-sash windows 6 by 9 ft on each side, with 8 windows of the same size on each end. Each end of the

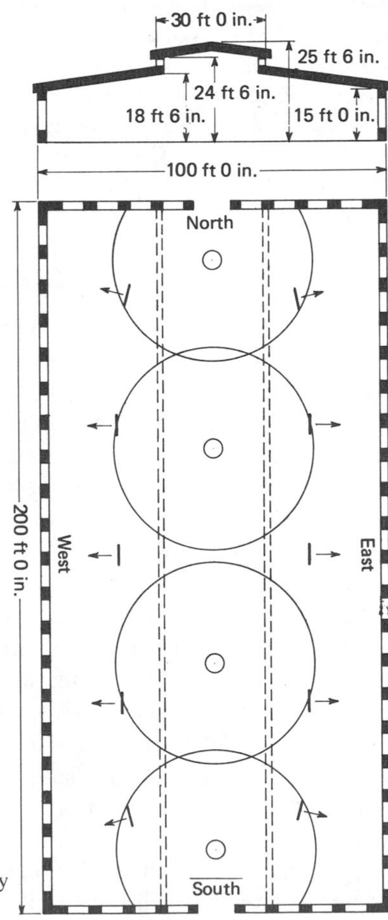

FIGURE 9-6. Plan and elevation of one-story factory building.

building has a 12- by 12-ft upward-opening steel door. The monitor section contains 40 $4\frac{1}{2}$- by $4\frac{1}{2}$-ft steel-frame windows on each side. These can be opened from below to provide ventilation. The monitor walls are of rigid asbestos shingles on solid wood sheathing mounted on studs. The roof consists of rafters on the bottom side of which is attached 1 in. of rigid insulating board. The upper surface of the roof consists of supporting sheathing covered by asphalt shingles. The floor consists of 4 in. of concrete laid on the bare ground. The design heat loss for the building was computed and found to be 1 888 600 Btuh. Plan a design pattern for the unit heaters.

Solution: This is a large building and, consequently, to give uniform coverage, it is better to use many heaters of low capacity than to use a few of large capacity. For example, 9 heaters consisting of 3 horizontal units under each roof and 3 vertical downflow units under the monitor section might serve. If each of these were of equal capacity, then each would have to deliver 1 888 600/9, or 203 200 Btuh. One manufacturer supplies such units for 60 F inlet air, 215 F steam, at 1150 fan rpm with a $\frac{1}{4}$-hp motor, to deliver 223 000 Btuh, or a 1150-rpm, $\frac{1}{2}$-hp-motor unit in vertical style, to deliver 262 000 Btuh. These could suffice. However, a better distribution, with less chance of cold spots or overheating, could be had by using 5 horizontal-discharge units on each side and 4 vertical-discharge units in the monitor section. With these 1 888 600/14, or 85 000 Btuh per unit would be required. For this arrangement, one manufacturer's catalog lists an 1800-rpm, $\frac{1}{12}$-hp-motor unit, to deliver 88 500 Btuh, for horizontal discharge, and a vertical unit at 860 rpm with a $\frac{1}{4}$-hp motor, to deliver 85 600 Btuh. These are slightly over design size, but a slight overcapacity may be desirable to allow for door openings or other contingencies. On the basis of the units indicated, a layout is shown in Fig. 9-6. Numerous other design possibilities exist, however, and could be used.

9-6. HEATING AND COOLING, SYSTEM ARRANGEMENTS

In addition to the types of heating units previously described in this chapter, extensive use is made of heat-transfer elements in central heating and cooling systems. In winter operation with these systems, recirculated and outside air is passed over heat-transfer coils which temper the air to a desired condition for delivery into the heated space. The heated air must be sufficiently warm so that in cooling it can offset the heat losses of the space. Auditoriums, lecture halls, churches, many offices, as well as other types of buildings employ this method of central heating and cooling. Steam (or sometimes hot water) is supplied to the heat-transfer surface used for tempering and warming the air.

Nonferrous heat-transfer surfaces are also used to carry chilled water or refrigerant for air chilling under summer operating conditions. The fins must be placed sufficiently far apart that any condensation of humidity from the air can drain off and be removed. Fin spacing is even more important if temperatures are sufficiently low to cause freezing on the coils, as a solid block of ice could prevent all throughflow of air if the fins are too close.

A conventional thermal-circulation radiator or convector well placed for heating a room will not be effective when cooling the room; it will cool only a layer of air close to the floor. However, if a fan is added for use when cooling, to deliver recirculated air against the heat-absorbing surfaces, the heat-transfer rate becomes rapid and the cool-air distribution may be effective and satisfactory.

9-7. HEAT TRANSFER THROUGH METAL SURFACES

Heating or cooling. Heat transfer through metal surfaces is largely limited by the film conditions which exist on the surfaces of the tubing. The transfer is affected only slightly by the conductivity of the metal itself. The film conditions depend on the fluids involved, whether these are liquid or gas, and whether vaporization or condensation may be taking place, and on velocities past or along the tubes. Convection or flow conditions in a system are classed as *free* or *forced.*

Free convection, or *natural convection*, exists when the fluid circulation is caused by changes in fluid density which occur because of temperature differences established in the fluid. *Forced convection* exists when the motion of the fluid, past the heat-transfer surfaces, is produced by some external means such as a fan or pump.

Heat transfer to or from liquids is much better than with gases, and it often happens that the capacity for heat transfer is greatly restricted by the fluid on one side of the tube while only slightly limited on the other side. An example of this is a steam radiator in a room. Here, on the side with condensing steam, the heat transfer is comparatively good, so that the temperature of the radiator metal is almost that of the steam. The heat transfer to the air moving past the outside surfaces is so poor that heat is not transferred rapidly enough to cause an appreciable temperature gradient through the metal wall. For the same surface area, changing from cast iron to a better conductor, such as copper, would not appreciably increase the heat transfer (if the differences in polish, which affects radiation slightly, were eliminated). To increase the heat transfer, extended surfaces (fins) are often attached to the side which limits the heat transfer. This is the air or gas side in the case of those devices which use either a liquid or a vapor (undergoing a change in state) as the other medium.

The temperature difference (Δt) in heat transfer is seldom constant, and therefore it is necessary to find what mean or average temperature difference actuates heat transfer. The *logarithmic mean temperature difference* (log mtd) usually gives the most reliable value of mean temperature difference to use in heat-transfer relations:

$$\text{log mtd (or } \Delta t) = \frac{\text{Gtd} - \text{Ltd}}{2.3 \log_{10}(\text{Gtd}/\text{Ltd})} \tag{9-1}$$

where log mtd is the logarithmic mean temperature difference, in either Fahrenheit or Celsius degrees; Gtd is the greater temperature difference, in either Fahrenheit or Celsius degrees; Ltd is the lesser temperature difference, in either Fahrenheit or Celsius degrees; and $\log_{10}$ is a logarithm to the base 10.

Figures 9-7 and 9-8 represent, diagrammatically, possible temperature changes occurring during counterflow or parallel flow of two fluids between which heat transfer is taking place. Counterflow is desirable, when it càn be obtained with two fluids which change in temperature, since a better temperature difference can be maintained and the leaving temperatures can be brought closer to their limiting values In case one of the fluids is a condensing or evaporating vapor, there can be no distinction between parallel flow and counterflow, as one of the fluids does not change in temperature. Many actual installations can be classed neither as parallel flow nor as counterflow.

Example 9-2. Water enters an ammonia condenser at 60 F and leaves at 70 F. The ammonia is condensing at 74 F. Find the log mtd.

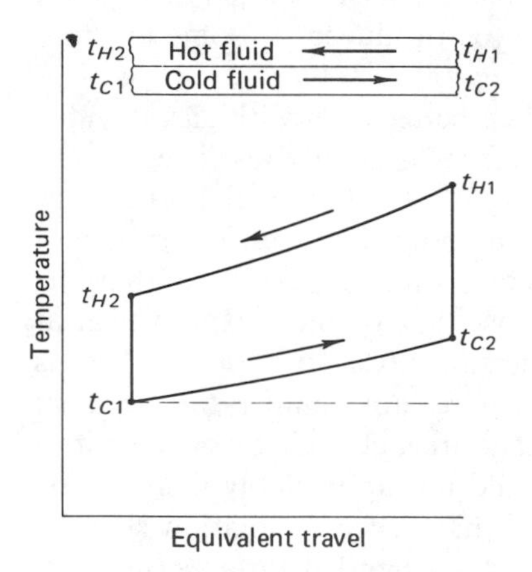

FIGURE 9-7. Counterflow heat transfer.

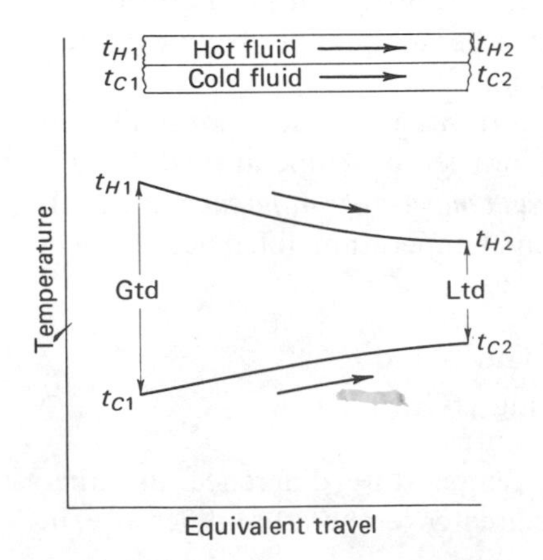

FIGURE 9-8. Parallel-flow heat transfer.

Solution: By Eq. (9-1), with Gtd = 74 − 60 = 14 deg, and Ltd = 74 − 70 = 4 deg,

$$\text{log mtd} = \frac{14 - 4}{2.3 \log_{10} 14/4} = \frac{10}{(2.3)(0.544)} = 7.98 \text{ deg} \qquad \textit{Ans.}$$

The arithmetic mean temperature difference would be

$$\frac{\text{Gtd} + \text{Ltd}}{2} = \frac{14 + 4}{2} = 9 \text{ deg}$$

and is seen to be too high. There are many cases, however, when the log mtd and the arithmetic mtd are almost identical. For values of Gtd/Ltd less than 1.5, there is little difference between the two methods of computation. In this example the terms counterflow and parallel flow had no significance, since one fluid condensed at constant temperature.

Values of heat-transfer coefficients for use in the design of units must be used with caution unless it is known that the heater (or cooler) is to be used under almost exactly the same conditions for which the coefficients were determined. In installations where convection, forced or natural, is the main factor, this rule is particularly cogent. Table 9-3 lists heat-transfer coefficients, gathered from many sources, which can be used for estimating the surface required. In selecting standard commercial equipment the manufacturers' tables for heat-transfer capacity and test coefficients should always be used if they are available. The effect of possible scaling or corrosion of the tubing after it has been placed in service should not be minimized in design, and adequate surface must always be supplied. Equations (4-14), (4-15), and (4-16) for heat transfer are applicable.

Example 9-3. A cold-storage room is to be equipped with brine-pipe coils of standard $1\frac{1}{2}$-in. pipe. Fans cause a positive air circulation in the room over the coils at about 500 ft/min. The air contacts the coils at 30 F room temperature and leaves the coils at 25 F. Brine enters the coils at 14 F and leaves at 20 F. Find (a) the mean temperature difference between the coils and the air and (b) the heat transfer per square foot of pipe surface. (c) If there are 200 lin ft of pipe, what is the probable refrigeration tonnage handled in the room?

Solution: (a) As the coils probably run parallel to the ceiling and above each other it is doubtful whether the terms counterflow and parallel flow have any significance. To make the most conservative estimate, calculate as though there is parallel flow.

Using Eq. (9-1), with Gtd = 30 − 14 = 16, and Ltd = 25 − 20 = 5,

$$\text{log mtd} = \frac{16 - 5}{2.3 \log_{10} 16/5} = \frac{11}{2.3(0.505)} = 9.47 \text{ deg} \qquad \textit{Ans.}$$

(b) From Table 9-3, select the mean value for "brine pipe to moving air" as $U = 5.5 \text{ Btu/h} \cdot \text{ft}^2 \cdot \text{F}$. *Ans.*

Table 9-3 Heat-Transfer Coefficients* for Heaters (or Coolers) and Heat Exchangers

Type of Heat Exchanger	U, Free Convection	U, Forced Convection	Remarks
Ammonia condensers			
Atmospheric			
Gas in at top	50–65		
Gas in at bottom	125–200		
Double pipe		150–250	Water velocity from 150 to 400 fpm
Shell and tube		150–300	
Submerged coil	30–40	30–40	Very ineffective
Baudelot coolers			
Cream	55	55	
Milk	70	70	
Oil	10	10	
Water	70	70	
Brine coolers			
Shell-and-tube multipass		40–120	Brine velocity 100 to 400 fpm at 0 F; U decreases below, increases above, this temperature.
Submerged refrigerant coils in brine tank	12	20	
Cooling coils			
Brine pipe			
To moving air		3.0–8.0	Air velocity 200 to 800 fpm. Increases with temperature difference.
To still air	1.5–3.0		
Direct expansion coils			
Finned air coils, sensible heat film factor (f_s)†	*300 400 500* [4.3 5.1 5.8]	*600 700 800* [6.5 7.2 7.7]	
Still air	1.0–3.0		Increases with temperature difference
Moving air		2.0–7.0	Air velocity 200 to 800 fpm
In spray of liquid		80	
Refrigerant (Freon) film surface factor (f)	[150–250]		Increases with capacity loading of tubes; also with temperature difference, metal to liquid.
Methylene chloride surface factor (f)	[100–300]		Δt metal to liquid of 5 to 25 deg
Heat exchanger			
Double pipe		80	
Radiators			
Direct steam	1.4–2.0		
Pipe coil in room, single	2.6		
Pipe coils above each other	2.6–1.6		As number increases from 1 to 8
Steam condenser			
Shell and tube type		100–600	0.5 to 6 fps water velocity
Superheat remover			
Shell and tube		25	
Tubular exchanger			
Steam to oil		75	
Oil to oil	5–10	20–50	
Water to oil		50	
Water to water	25–60	150–300	
Water heater			
Steam coil in tank (brass)	30–200		Increases with temperature difference of 10 deg to 100 deg. Less corrosion with brass and copper pipe.
Steam coil in tank (steel)	10–160		

* U in Btu per hr sq ft deg F mean temperature difference, fluid to fluid, or f in Btu per hr sq ft deg F mean temperature difference, metal to fluid.

† Italic numbers above values in second and third columns are face velocities, in feet per minute.

(c) Standard $1\frac{1}{2}$-in. pipe has an outside diameter of 1.9 in. Thus

$$\text{area per foot of pipe} = \frac{1.9 \times \pi}{12} \times 1 = 0.497 \text{ ft}^2$$

and

$$\text{area 200 lin ft} = 200 \times 0.497 = 99.4 \text{ ft}^2$$

By Eq. (4-17), the heat taken up by the brine is

$$Q = UA\Delta T = (5.5)(99.4)(9.47) = 5177 \text{ Btuh}$$

Therefore the refrigeration tonnage is

$$\frac{5177}{12\,000} = 0.431 \text{ ton} \qquad \textit{Ans.}$$

Frost (ice) on the coils will seriously decrease this capacity—by 25% or more as the thickness builds up to an inch.

The problem of selecting surface area for dehumidifiers is complicated by the fact that two simultaneous effects take place; namely, sensible heat is removed, and the latent heat from condensing moisture is also removed. The total heat transfer from air to coils with wetted surface is higher than to coils having dry surface. Condensation or dehumidification occurs when the coil temperature is lower than the dew-point temperature and sensible cooling and dehumidification can take place simultaneously. One method of treating coil performance is to use the equivalent by-pass approach. This considers that a coil performs just as if one part of the coil brought the air to coil surface temperature while the rest of the air passed through the coil (by-passed it) without being cooled or dried at all. When the cooled air and the by-passed air mix at outlet, the result is equivalent to what happens in a real coil. For dehumidifying coils it is desirable to use the manufacturer's data and recommended by-pass factors to find performance data. However, when these are lacking and design data are required, it is possible to select a sensible surface coefficient of heat-transfer factor (f_s) and use it in design even though dehumidification is taking place.

Forced-convection film factors. Forced-convection film factors can be developed by rational analysis in terms of appropriate, dimensionless number groups, for which groups the coefficients and exponents are determined by experiment.

For liquids being heated or cooled and moving in turbulent flow outside and across a single tube,

$$f = 0.385 \frac{k}{D} \left(\frac{DV\rho}{\mu}\right)^{0.56} \left(\frac{C_p \mu}{k}\right)^{0.3} \tag{9-2}$$

Multiply f by 1.3 if a staggered multiple-tube bank applies, and by 1.2 if the tubes are in line.

For liquids being heated inside tubes under conditions of turbulent flow,

$$f = 0.0225 \frac{k}{D} \left(\frac{DV\rho}{\mu}\right)^{0.8} \left(\frac{C_p\mu}{k}\right)^{0.4} \tag{9-3}$$

Use an exponent 0.3 for the last term if cooling is involved.

For gases flowing over or along tube bundles, Eq. (9-2) can also be used. For this equation, D should be taken, by Eq. (12-4), equal to 4 times the hydraulic radius, where this is found by dividing the free area between the tubes by the sum of all the tube perimeters.

The film-factor values for vapors condensing on, or liquids vaporizing from, horizontal tube bundles cannot be expressed by simple equations like those just given. Such values are high, however, and the resistance to heat transfer is small through this film. For condensing or vaporizing steam, and for ammonia, values of f run between 800 and 6000 Btu/h · ft^2 · F. For making estimates, values of f between 1000 and 3000 are suggested. For the halogenated-hydrocarbon refrigerants, vaporizing and condensing, values of f range from 150 to 350 Btu/h · ft^2 · F.

The units in Eqs. (9-2) and (9-3) may be any that are desired, provided they are dimensionally consistent. It should be observed that

$\dfrac{DV\rho}{\mu}$ is the Reynolds number (see Section 12-2)

$\dfrac{C_p\mu}{k}$ is called the Prandtl number

$\dfrac{fD}{k}$ is called the Nusselt number

All of these number groups are dimensionless. Suggested units for use in the equations are as follows: D in feet; f in Btu/h · ft^2 · F (see Section 4-2); k in Btu · ft/h · ft^2 · (in the Prandtl number, k in Btu · ft/s · ft^2 · F is suggested); V in feet per second; ρ in pounds of mass (weight) per cubic foot; μ in centipoises [centipoises × 0.000672 = lb mass (weight)/ft · s] (see Fig. 12-1; and C_p, specific heat at constant pressure, in Btu/lb mass (weight) · F (see Tables 2-1 and 4-1). Or when using SI units express f in W/m^2 · °C; D in meters; k in W/m^2 · °C; V in meters per second; ρ in kilograms per cubic meter; μ in Newton seconds per square meter; and C_p in joules per kilogram degree Celsius.

Example 9-4. Water is being cooled in a multipass shell-and-tube evaporator, with the ammonia refrigerant in the shell boiling at 56 F. Water moves through the 1-in.-O.D. steels tubes (0.05-in. wall) at a velocity of 4 ft/s. Water enters at 66 F and leaves at 60 F. Compute the film factor f on the water side and estimate the value of U, the overall coefficient of heat transfer.

Solution: For water at an average temperature of 63 F, $\mu = 1.09$ centipoises, from Fig. 12-1; $\rho = 1/0.01604 = 62.3$ lb/ft^3; $k = 4.10$ Btu · in./h · ft^2 · F = 4.10/12 Btu ·

ft/h · ft² · F (from Table 4-1); and C_p = 1.0 Btu/lb mass (weight) · F. Also, $D = (1 - 0.05 - 0.05)/12 = 0.90/12 = 0.075$ ft.

Using Eq. (9-3) for cooling conditions,

$$f = 0.0225\frac{k}{D}\left(\frac{DV\rho}{\mu}\right)^{0.8}\left(\frac{C_p\mu}{k}\right)^{0.3}$$

$$= 0.0225\left[\frac{4.1}{(12)(0.075)}\right]\left[\frac{(0.075)(4)(62.3)}{(1.09)(0.000672)}\right]^{0.8}\left[\frac{(1)(1.09)(0.000672)}{4.1/[(12)(3600)]}\right]^{0.3}$$

$$= 0.0225(4.55)(25\,520)^{0.8}(7.72)^{0.3}$$

$$= 0.0225(4.55)(3352)(1.85) = 635 \text{ Btu/h} \cdot \text{ft}^2 \cdot \text{F} \qquad \textit{Ans.}$$

The resistance R to heat flow on the tube consists of the film resistance on the water side, the negligible resistance of the metal wall itself, and the resistance of the ammonia film. Take a value of 2000 for the ammonia film, which has good heat-transfer characteristics. Then,

$$R = \frac{1}{2000} + 0 + \frac{1}{635} = 0.0005 + 0.00157 = 0.00207$$

Therefore, the overall heat-transfer coefficient (estimated) is

$$U = \frac{1}{R} = \frac{1}{0.00207} = 483 \text{ Btu/h} \cdot \text{ft}^2 \cdot \text{F} \qquad \textit{Ans.}$$

Because of deposition, tube scale, and the like, a value not exceeding 50% of the computed result might be used in a conservative design.

Example 9-5. Compute the film factor on the water side for the shell-and-tube evaporator of Example 9-4, under the conditions given, but making use of SI terminology.

Solution: Convert the given engineering units to SI form:

$D = 0.075 \times 0.3048 = 0.02286$ m
$V = 4$ ft/s $\times 0.3048 = 1.2192$ m/s
$\rho = 62.3$ lb/ft³ $\times 16.018463 = 997.95$ kg/m³
$k = 0.34167$ Btu · ft/h · ft² · F $\times 1.730742 = 0.59134$ W/m · °C
$\mu = 1.09 \times 10^{-3}$ N · s/m² (from Fig. 12-1)
$C_p = 1.0$ kcal/kg · °C $\times 4184 = 4184$ J/kg · °C

$$\text{Reynolds:} \left(\frac{DV\rho}{\mu}\right)^{0.8} = \left(\frac{(0.02286)(1.2192)(997.95)}{0.00109}\right)^{0.8} = (25520)^{0.8} = 3352$$

$$\text{Prandtl:} \left(\frac{C_p\mu}{k}\right)^{0.3} = \left(\frac{(4184)(0.00109)}{0.59134}\right]^{0.3} = (7.72)^{0.3} = 1.85$$

Substituting in Eq. (9-3),

$$f = 0.0225\left[\frac{0.59134}{0.02286}\right](3352)(1.85) = 3610 \text{ W/m}^2 \cdot {}^\circ\text{C} \qquad \textit{Ans.}$$

9-8. BOILERS

Boilers are devices in which water can absorb heat from hot products of combustion, with or without direct radiation from the burning fuel.

Steam boilers are not completely filled with water, but above the water level have a vapor space where the vapor (steam) can separate from the boiling liquid.

Hot-water boilers are not supplied with any steam-disengaging space but are completely filled with water which circulates through the boiler by thermal action (or under the acton of a pump) at a sufficiently rapid rate to prevent it from flashing into steam at the pressure under which the boiler is operating.

Boilers may be classified primarily as high-pressure or low-pressure types. Low-pressure boilers do not operate at more than 15 psig steam pressure or 30 psig water pressure. Many states and municipalities require continuous attendance of a fireman if the operating pressure exceeds these limits. Nearly all small boilers may serve hot-water heating systems as well as steam systems. Only minor changes are required for changing to either system.

Where steam is produced, the latent heat of evaporation ($\mathbf{h}_{fg}$) is available at the radiators and convectors, along with a minor amount of sensible heat. The condensate is returned to the boiler through pipes smaller than those required to bring the steam to the heat transmitters.

As there is always some steam loss from a system, the water in a steam boiler eventually becomes concentrated with salts and sediment resulting from the accumulation of ingredients which do not evaporate. Therefore the water in the boiler must be changed occasionally by *blowing down*, or discharging, part of it. The interval between blowing-down periods depends upon the amount of make-up required by the system, and on the quality of the water supply.

The unevaporated material collects upon hot surfaces of the boiler and may form a hard, stonelike deposit of scale which becomes a heat insulation on the water side of the heat-transfer surfaces. Unless these deposits are removed, the metal of the boiler will eventually be overheated, as the water cannot absorb the heat with sufficient rapidity.

Steam boilers require the constant addition of new water to replace that which is being delivered in the form of steam, and there should be a gage glass, and preferably also try cocks, to indicate the water level. A safety valve must always be provided. Automatic water-feeder and low-water cutoff seitches are highly desirable (see Fig. 9-12).

Heating boilers are of the *fire-tube type*, or *tubular type*, if, following combustion, the gases pass through tubes or comparatively small passages which are submerged in the water. They are called *water-tube boilers* if the arrangements are reversed, that is, if water is inside the tubes. Both types are effective, although when low-pressure saturated steam is used for heating, the fire-tube construction has advantages. It is easier with this type to provide adequate steam-liberating area at the water line, and the larger heat-storage capacity due to a greater volume of water in the boiler gives a more uniform rate of output.

Figure 9-9 shows a steel fire-tube boiler arranged for oil or gas firing. This horizontal boiler is a force-draft, three-pass design. The first pass consists of the single furnace tube which carries the burning combustion gases rearward from the burner while the combustion process reaches its conclusion. The still very hot gases after leaving the furnace tube enter the second-pass fire tubes, which are located in the bottom half of the boiler, and return to the front. There, restrained by the smoke-box doors, they are directed into the third-pass tubes located in the upper half of the boiler. When the gases leave the third pass they exit into a round flue stack for discharge into a chimney or into an extension stack. This is a welded all-steel unit with the tubes rolled into the front and rear headers. These boilers are characterized by adequate steam-disengaging surface, so that relatively dry steam is delivered into the steam outlet. These boilers normally use either gas or fuel oil. Heavy fuel oil, when used, may have to be preheated by means of steam from the boiler. Boilers of this particular design are made in 11 sizes ranging in capacity from 2700 to 20 700 lb/h. The unit pictured in Fig. 9-9 is a 20 700-lb/h unit (9380 kg/h).

Cast-iron sectional boilers, which are made for domestic service, are also made in larger sizes for serving public buildings, apartment houses, factories, etc. They can be arranged for firing with oil or gas, or for solid fuel with stoker or hand firing.

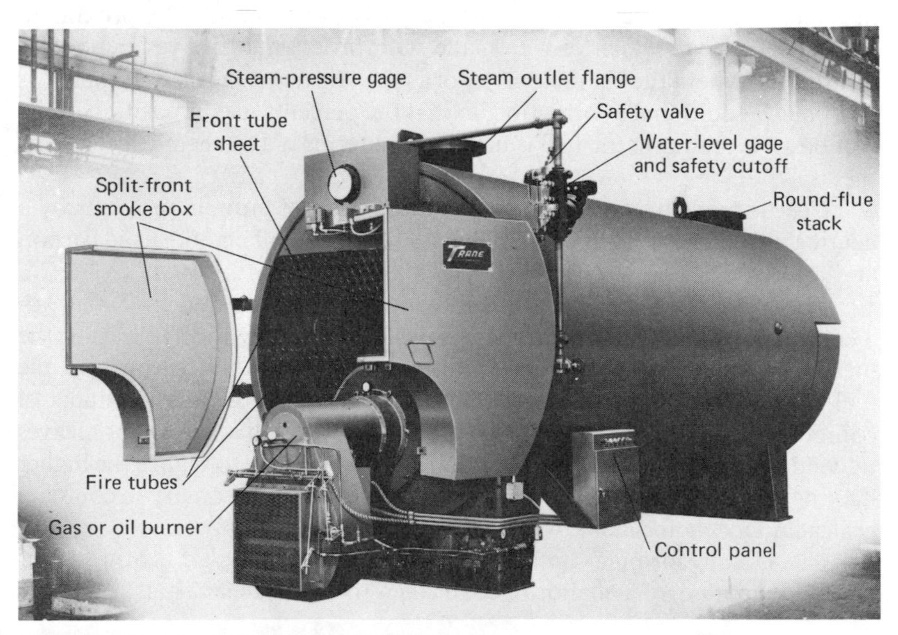

FIGURE 9-9. Steel-shell, three-pass, fire-tube boiler arranged for oil or gas firing. (Courtesy of the Trane Company.)

9-9. FUELS AND COMBUSTION

The design of a boiler or of any direct-fired heater is closely related to the combustion of the fuel. Combustion is a chemical process in which the carbon and hydrogen (and sulfur) in the fuel react with the oxygen from the air, thermal energy being released. The rate at which the fuel can be burned effectively depends largely on the intimacy with which the air is brought into contact with the fuel. This intimacy is influenced mainly by the intensity of the chimney draft, by the size of the grate, by the size of the combustion chamber, by the location of the heating surfaces, and by proper control of air and fuel feed. The combustion problem in turn is influenced by the character of the fuel—whether solid, liquid, or gas.

Solid fuels. Ordinary solid fuels are represented by the various ranks of coals, coke, and lignite. Table 9-4 shows average analyses of coals. Bituminous coals, which represent the greatest tonnage of coal mined, are high in volatile matter and have good calorific values. The volatile matter in coal is that portion of the coal which is driven off, in gaseous or vapor form, when the coal is heated. It consists largely of hydrocarbons and other gases resulting from decomposition and distillation. In carrying out a standard *proximate analysis* of a coal, a weighed sample is gently heated to 220 F to drive off the moisture, and the sample is again weighed. It is then heated for 7 min at 1700 F, out of contact with air, to drive off the volatile matter. The remaining sample consists of the fixed carbon plus the ash. The fixed carbon, which resembles coke in appearance, is burnt out when the sample is heated in air, the ash being left as a residue. The proximate analysis of a coal, which can be made with little trouble, gives important information about the characteristics of a fuel. However, the *ultimate analysis* gives much additional information, in that the percentage composition of the combustible matter in the fuel is furnished in terms of the chemical elements themselves.

Coke is made from the destructive distillation of bituminous coals by a heating process in which the volatile matter is driven off and the fixed carbon and ash remain.

The heating values of representative coals are indicated in Table 9-4. The higher heating value represents the thermal energy available from the fuel if the water vapor resulting from combustion of the hydrogen in the fuel is condensed and this latent heat is absorbed from the products of combustion. The lower heating value presupposes that this water vapor leaves uncondensed in gaseous form. Although it is not possible to realize the higher heating value of the fuel in combustion equipment, it is customary to compare boiler-furnace performance with the higher heating value.

The Dulong formula can be used to determine the higher heating value, in Btu per pound, of a coal or of coke when the ultimate analysis is known:

$$\mathrm{HHV} = 14\,540\mathrm{C} + 62\,000\left(\mathrm{H} - \frac{\mathrm{O}}{8}\right) + 4000\,\mathrm{S} \tag{9-4}$$

Table 9-4 Average Analyses of Typical U.S. Coals (%)

Number of Samples	Kind of Coal	Proximate Analysis by Weight				Ultimate Analysis by Weight						
		Volatile Matter	Fixed Carbon	Ash	Water	Carbon	Hydrogen	Sulfur	Oxygen	Nitrogen	Heating Value, As Received (Btu/lb)	
											High	Low
5	Anthracite.....	4.36	82.70	9.34	3.60	82.14	2.22	0.6	1.2	0.9	13074	12820
3	Semianthracite..	9.6	74.8	12.9	2.7	77.0	3.3	1.0	1.97	1.13	13000	12683
6	Low-volatile bituminous.	18.4	71.4	7.15	3.05	80.70	4.25	1.07	2.27	1.51	14073	13643
	High-volatile bituminous											
5	A...........	35.1	53.7	7.7	3.5	74.5	5.1	1.3	6.4	1.5	13389	12852
3	B...........	35.9	47.8	8.5	7.8	69.1	4.6	1.8	6.8	1.4	12126	11608
4	C...........	36.9	41.9	8.5	12.7	62.6	4.4	3.5	7.2	1.1	11370	10806
2	Subbituminous..	31.9	42.8	3.90	21.4	56.70	3.80	0.35	12.70	1.15	9775	9169
2	Lignite.........	28.7	30.3	6.8	34.2	42.45	2.95	0.60	12.25	0.75	7280	6614

where C, O, H, and S are the decimal fractions by weight of the carbon, hydrogen, oxygen, and sulfur in the fuel.

A serious problem to the use of coal in its various grades relates to the sulfur that is present in these solid fuels. When the coal is burned the sulfur burns to sulfur dioxide (SO_2), which is then delivered into the atmosphere. Sulfur dioxide is an irritant, toxic gas, which in large amounts can become a hazard to life, property, and vegetation. It is soluble in water, forming sulfurous acid. Thus environmental protection agencies have put restrictive measures on the burning of coal with high sulfur content. For example, a coal with 3% sulfur, by weight, in burning would release 120 lb of SO_2 into the atmosphere for each ton of coal burned. A power plant burning 250 tons of 3% sulfur coal per hour would thus release 30 000 lb of SO_2 into the air each hour. Research is actively under way to control sulfur emissions, but no easy solutions are in sight. Coal is our most abundant fuel and must be used since fuel oil and gas are becoming progressively more expensive and less available.

Fuel oil. Fuel oils are produced by distillation from crude petroleum and by cracking of heavy hydrocarbon products. They vary in character from light volatile distillates resembling kerosene to the heavy viscous fuel-oil No. 6, which resembles black molasses and is known as Bunker C oil. The fuel oils are hydrocarbons containing between 84 and 85% carbon, between 12 and 14% hydrogen, and small amounts of oxygen, nitrogen, and sulfur, with traces of moisture and solid material. Standard classifications for fuel oil place the oil in five grades. No. 1 fuel oil has a specific gravity in the neighborhood of 0.8, while No. 6 oils have specific gravities which sometimes are in excess of 0.98.

Specific gravity and degrees API of fuel oils are usually measured with suitable hydrometers. The relationship which relates degrees API to specific gravity is:

$$\text{sp gr} = 141.5/(\text{deg API} + 131.5)$$

Table 9-5 gives the API ranges of the different grades of fuel oil, and their approximate higher heating values in Btu per gallon. The heating

Table 9-5 Higher Heating Values of Fuel Oils of Various Grades

Grade No.	Representative Range (deg API)	Higher Heating Value (Btu/gal)	Representative Weight (lb/gal)
1	38–45	135 000	7.0
2	30–38	138 500	7.2
4	20–28	145 500	7.5
5	14–22	148 500	7.8
6	8–15	153 600	8.0

value may also be determined in Btu per pound through use of the Bureau of Standards formula

$$HHV = 22\,320 - (3780)(\text{sp gr at } 60\text{ F}/60\text{ F}) \qquad (9\text{-}5)$$

or by use of the Sherman-Kropf formula

$$HHV = 18\,250 + (40)(\text{deg API} - 10) \qquad (9\text{-}6)$$

Values obtained by the Sherman-Kropf formula vary somewhat from those determined by the Bureau of Standards formula.

Oils with higher specific gravity have lower heating values per pound; but, as the oils with higher specific gravity contain more pounds per gallon, the heating value on a gallon basis is higher for the higher gravity oils.

Gases. Gas is the most convenient of all fuels, and because of the network of pipelines which exists in North America and in parts of Europe it is widely available for use. However, the demand for gas is greater than the available supply and use is restricted in many areas. Gas can be manufactured from solid fuels or oil, but such gas is almost always more costly than natural gas.

Gas is an easy fuel to control and burn. It is sold either on the basis of cubic feet or in *therms*, where the therm represents the volume of gas equivalent to 100 000 Btu. Because of the wide variation in the heating value of gases per cubic foot, burners adjusted for one type of gas cannot be used with another unless readjustment of the air ratio is made. Heating values for commonly used gases in Btu per cubic foot are: carbon monoxide (CO), 322; methane (CH_4), 991; hydrogen (H_2), 315; propane (C_3H_8), 2451; coke oven gas, about 575; and natural gases range from 800 to 1125. These are higher heating values for the gases in Btu per cubic foot, measured at 62 F at 30 in. Hg and saturated with water vapor.

9-10. COMBUSTION

Combustion has been described as the application of the three *t*'s: time, temperature, and turbulence. The essentials of proper combustion should be elaborated upon, however, to include (a) proper proportioning of the air to the fuel; (b) exposure of sufficient fuel surface to permit combination with the air; (c) adequate mixing of the fuel particles with the air; (d) sufficiently high temperature to assure continuing ignition; and (e) arrangement of the combustion chamber so that sufficient time is allowed to complete burning before the gases leave. These principles apply to the combustion of any fuel, be it gaseous, liquid, or solid, but different arrangements must be made for the different fuels.

In the case of a gas, it is relatively easy to mix the gas and the air in proper proportions and to maintain ignition. Also, it is not difficult to atomize a liquid fuel into particles small enough to permit a large fuel surface to be exposed to the air. Burning of a solid fuel presents more difficulty, as the air must be brought into contact with the combustible surfaces of the fuel and must not be smothered out by the resulting ash before combustion is completed. The burning of volatile coals with a minimum of smoke presents an additional problem. These problems will be discussed further in connection with the various types of combustion equipment.

It has previously been shown how the analysis of the flue gas could be used to indicate the quantity of air used in a combustion process. Use of air greatly in excess of that theoretically required for combustion can cause appreciable loss in the operation of a boiler-furnace system. Consequently, it is desirable to adjust the air-fuel ratio to a point where good combustion is obtained and a minimum of excess air is employed. This minimum is sometimes set by other factors besides combustion, as it may be necessary to use additional air to prevent smoke or to avoid the formation of combustible products, such as carbon monoxide, in the flue gas. In addition, leaky settings permit air to enter the furnace system without contributing to the combustion process in any way.

9-11. COMBUSTION CALCULATIONS

The air theoretically required to burn a unit weight of the combustible elements in fuel can be readily calculated from the following chemical equations:

$$\begin{aligned} C + O_2 &\rightarrow CO_2 \\ 12 + 32 &= 44 \end{aligned} \qquad (9\text{-}7)$$

Thus 12 lb of carbon require 32 lb of oxygen for complete combustion, and yield 44 lb of carbon dioxide. The 12 and 16 (taken twice) are atomic weights of carbon and oxygen, respectively. For the more accurate atomic weight of carbon, 12.01, it is evident that 1 lb of C requires 32/12.01 = 2.66 lb of O_2 and yields 3.66 lb of CO_2.

As the weight ratio of oxygen to the inert gases in air is 0.2319 to 0.7681, it follows that 1 lb of C requires 2.66/0.2319 = 11.5 lb of air for complete combustion.

Similarly, other relations for combustion are:

$$\begin{aligned} 2H_2 + O_2 &\rightarrow 2H_2O \\ 4 + 32 &= 36 \end{aligned} \qquad (9\text{-}8)$$

$$1 \text{ lb of } H_2 \text{ requires } \frac{32}{4.03} = 7.94 \text{ lb of } O_2 \text{ and yields } 8.94 \text{ lb of } H_2O$$

$$1 \text{ lb of } H_2 \text{ requires } \frac{7.94}{0.2319} = 34.3 \text{ lb of air for complete combustion}$$

1 lb of S (sulfur), by like calculation, requires 4.3 lb of air for complete combustion

The hydrocarbon fuels contain no oxygen, but coal and the alcohols do. This oxygen in the fuel is already chemically combined with some of the combustible matter or can readily go into combination. Consequently, allowance must be made for this oxygen in computing the weight of air required to burn a unit weight of fuel. The allowance is usually made by considering that 1 lb of this oxygen can combine with $\frac{1}{8}$ lb of hydrogen.

Let C, H, O, and S represent the proportional parts by weight of the carbon, hydrogen, oxygen, and sulfur in a fuel. The weight of air, A, theoretically required to burn a pound of the fuel is

$$A = 11.5\mathrm{C} + 34.3\left(\mathrm{H} - \frac{\mathrm{O}}{8}\right) + 4.3\,\mathrm{S} \tag{9-9}$$

Combustion can hardly ever be carried on with this theoretical amount of air; a certain amount in excess of that theoretically required must be used to insure complete combustion and prevent carbon deposits. In combustion calculations the problem most usually presented is to find from an analysis of the waste gases what combustion conditions actually exist. Flue-gas and exhaust-gas analyses are usually given in volumetric percentages of the so-called dry flue gases CO_2, O_2, CO, and N_2. Generally, the most accurate method of making combustion calculations from flue-gas analyses is through the medium of a carbon balance.

The volume of any gas multiplied by its density gives the weight of that gas. Since the density of a gas is proportional to its molecular weight, the product of gas volume and molecular weight is also proportional to the weight of a gas. If the volume of each individual gas in a gas mixture is multiplied by its respective molecular weight, the sum of the resulting products is proportional to the weight of the gas mixture. Let CO_2, O_2, CO, and N_2 be the percentages or parts by volume of these respective gases in the dry waste gas. The proportional weights of these respective gases and of the mixture are $44CO_2$, $32O_2$, $28CO$, $28N_2$, and $44CO_2 + 32O_2 + 28CO + 28N_2$.

In carbon dioxide, CO_2, atomic weights show that $\frac{12}{44}$ of the total weight is carbon; that is, the proportional weight of the carbon in the CO_2 is $\frac{12}{44} \times 44CO_2 = 12CO_2$. Similarly, the weight of the carbon in carbon monoxide, CO, is $\frac{12}{28} \times 28CO = 12CO$.

The weight of the dry waste gases per unit weight of carbon existing in them must be

$$\frac{\text{total weight of the gases}}{\text{weight of carbon in the gases}} = \frac{44CO_2 + 32O_2 + 28CO + 28N_2}{12CO_2 + 12CO} \tag{9-10}$$

In other words, the number of pounds of dry waste gases, G_w, per pound of carbon burned, is

$$G_w = \frac{44CO_2 + 32O_2 + 28CO + 28N_2}{12CO_2 + 12CO}$$

$$= \frac{11CO_2 + 8O_2 + 7CO + 7N_2}{3(CO_2 + CO)} \tag{9-11}$$

Equation (9-11) should not be considered a formula. It is simply a statement of the weight of a certain gas mixture divided by the weight of carbon in the mixture. The equation is set up in this fractional form instead of in vertical tabular form in order that advantage can be taken of cancellation to reduce arithmetical work.

To find the number of pounds of dry waste gas per pound of fuel burned, the value of G_w obtained from Eq. (9-11) must be multiplied by C_g, where C_g is the number of pounds of carbon gasified or burnt per pound of fuel and must be determined from test data and the fuel analysis. In the case of liquid hydrocarbon fuels or gaseous fuels, essentially all of the carbon in the fuel becomes gasified or is burnt. This is not true in the case of solid fuels because some carbon nearly always appears unburned in the refuse.

The weight of air used per pound of fuel, and the combustible matter burnt per pound of fuel, appear in the waste gases either *in the dry flue gas* or *as water vapor*. Expressed in equational form, this statement becomes:

$$\text{air used} + \text{combustible burnt} = \text{dry flue gas} + \text{water vapor} \qquad \text{(9-12)}$$

or

$$\text{air used per pound of fuel} + \text{combustible burnt per pound of fuel} = G_w \times C_g + 9\text{H}$$

The combustible burnt per pound of fuel consists of all the fuel which entered into the combustion. This combustible is the whole original pound of fuel, less the ash and moisture in the fuel and that portion of the combustible matter in solid fuel which remains unburnt and appears in the refuse. The term $G_w \times C_g$ has been explained previously. For 1 lb of hydrogen burnt, 9 lb of water vapor result; therefore, 9 H gives an expression for the weight of water vapor per pound of fuel (H being the weight of hydrogen per pound of fuel). Thus the number of pounds of dry air, W_a, actually used per pound of fuel is

$$W_a = G_w \times C_g + 9\text{H} - \left(1 - \frac{\%\,\text{ash} + \%\,\text{moisture}}{100} - \frac{\text{combustible in refuse}}{\text{pound fuel}}\right) \qquad \text{(9-13)}$$

In the case of a hydrocarbon liquid fuel, this equation usually simplifies to

$$W_a = G_w \times C_g + 9\text{H} - \left(1 - \frac{\%\,\text{moisture}}{100}\right) \qquad \text{(9-14)}$$

The following method of calculating the weight of dry air used per pound of fuel can be used *if the fuel contains a negligible amount of nitrogen.* The number of pounds of nitrogen, W_N, in the waste gas per pound of fuel is

$$W_N = \frac{28.16\,N_2}{12(CO_2 + CO)}\,C_g = \frac{7.04\,N_2}{3(CO_2 + CO)}\,C_g \qquad \text{(9-15)}$$

All of this nitrogen must have come from the air (76.8% N_2, 23.2% O_2), as none was assumed in the fuel. (The value 28.16 is the molecular weight of the nitrogen–argon mixture in atmospheric air.) The number of pounds of dry air actually used per pound of fuel is

$$W_a = \frac{7.04\,N_2 \times C_g}{3(CO_2 + CO)(0.768)} \tag{9-16}$$

In the case of a fuel which contains a weight fraction of N_f parts of nitrogen, Eq. (9-16) can be modified into the following more precise form:

$$\begin{aligned} W_a &= \left[\frac{28.16\,N_2}{12(CO_2 + CO)} C_g - N_f\right] \frac{1}{0.768} \\ &= \left[\frac{2.35\,N_2}{CO_2 + CO} C_g - N_f\right] \frac{1}{0.768} \end{aligned} \tag{9-17}$$

The *percentage of excess air* used is evidently found by the relation

$$\%\text{ excess air} = \frac{\text{air actually used} - \text{air theoretically required}}{\text{air theoretically required}} \times 100 \tag{9-18}$$

The weight of any individual kind of dry waste gas per pound of fuel can be found by following the reasoning used to develop Eq. (9-11). Thus

$$\text{lb } CO_2\text{/lb fuel} = \frac{11CO_2}{3(CO_2 + CO)} C_g \tag{9-19}$$

$$\text{lb } N_2\text{/lb fuel} = \frac{7.04N_2}{3(CO_2 + CO)} C_g \tag{9-20}$$

$$\text{lb } O_2\text{/lb fuel} = \frac{8O_2}{3(CO_2 + CO)} C_g \tag{9-21}$$

$$\text{lb CO/lb fuel} = \frac{7CO}{3(CO_2 + CO)} C_g \tag{9-22}$$

In all the preceding expressions the gas items are percentages or proportional parts *by volume* and thus Orsat values can be used.

9-12. FLUE-GAS ANALYSIS AND FURNACE LOSSES

The composition of flue gases leaving a furnace can be determined by a variety of devices, of which the Orsat analyzer is perhaps the most common. In using this apparatus a sample of flue gas is aspirated from the stack and, after being measured, passes into a solution of potassium hydroxide which absorbs the carbon dioxide from the sample. The remaining gas is then passed into a second solution, usually of pyrogallic acid, which absorbs the oxygen. The carbon monoxide in the sample is finally absorbed by a solution of cuprous chloride, and the remainder of the sample can be considered to consist of nitrogen. The fraction of each gas in the original sample is found by measuring the decrease in volume after each absorption. These data furnish a volumetric analysis of the flue gas, and by using the data it is possible to analyze combustion conditions.

Example 9-6. A heavy fuel oil has the following composition in weight fractions: C, 0.855; H, 0.115; N, and other matter, 0.03. The HHV is 18 500 Btu/lb. (a) Find the volume and composition of the ideal dry products of combustion. (b) Find the air actually used if the dry-flue-gas analysis shows the following volume percentages: CO_2, 8.7; O_2, 8.5; CO, 0.1; N_2, 82.7.

Solution: (a) By Eq. (9-9) the air required per pound of fuel is

$$A = 11.5(0.855) + 34.3(0.115) = 13.77 \text{ lb}$$

As 76.8% of air by weight is N_2, there are $13.77 \times 0.768 = 10.58$ lb of N_2 in the flue gas from 1 lb of fuel.

The weight of CO_2 per pound of fuel is

$$3.66 \times C = 3.66 \times 0.855 = 3.13 \text{ lb}$$

To put the dry flue gases, consisting of 10.58 lb of N_2 and 3.13 lb of CO_2, on a volume basis, it is convenient to make use of the molal volume of 386 ft^3 at 70 F and 14.7 psi occupied by the mole weight in pounds of any gas. Thus the volume of dry gaseous product is

$$10.58 \times \frac{386}{28} + 3.13 \times \frac{386}{44} = 145.6 + 27.4 = 173.0 \text{ ft}^3 \qquad \textit{Ans.}$$

Therefore the percentage of carbon dioxide by volume is

$$CO_2 = \frac{27.4}{173.0} \times 100 = 15.82\% \qquad \textit{Ans.}$$

and if nitrogen,

$$N_2 = \frac{145.6}{173.0} \times 100 = 84.18\% \qquad \textit{Ans.}$$

Hence with ideal combustion the CO_2 cannot exceed 15.8% in the dry products.

(b) By Eq. (9-11), the weight of dry flue gas per pound of carbon is

$$G_w = \frac{(11)(8.7) + (8)(8.5) + 7(0.1 + 82.7)}{3(8.7 + 0.1)} = 28.2 \text{ lb}$$

and the weight of dry flue gas per pound of fuel is

$$28.2 \times C_g = 28.2 \times 0.855 = 24.1 \text{ lb}$$

By Eq. (9-14), the weight of air per pound of fuel is

$$W_a = 24.1 + (9)(0.115) - (1 - 0.03) = 24.2 \text{ lb} \qquad \textit{Ans.}$$

As there is little N_2 in the fuel, Eq. (9-16) can also be used. It represents a quicker method when it is applicable. In this case,

$$W_a = \frac{(7.04)(82.7)(0.855)}{3(8.7 + 0.1)(0.768)} = 24.5 \text{ lb}$$

This answer differs slightly from that secured by the previous method because there is some nitrogen which came from the fuel, and the nitrogen term has some significance. Use of Eq. (9-17) gives a more accurate answer if the nitrogen is known or assumed.

Example 9-7 An Illinois coal has 40.2% volatile matter, 39.1% fixed carbon, 8.6% ash, and 12.1% moisture. In percentages, its ultimate analysis shows C, 62.8; H, 4.6; O, 6.6; S, 4.3; and N, 1. The higher heating value is 11 480 Btu/lb and it can be shown that the maximum CO_2 by volume in the flue gas is 18%. (a) Find the HHV by Dulong's formula and compare the result with the calorimeter value. (b) The dry ashpit refuse contains 25% by weight of combustible, which is assumed to be carbon. Find the unburned carbon per pound of coal and the heat loss per pound of coal from this cause. (c) In percentages, the analysis of the dry flue gas by volume is: CO_2, 10; O_2, 8; CO, 0.1; N_2, 81.9. Compute the number of pounds of dry flue gas per pound of coal fired, the weight of moisture in the flue gas per pound of coal fired, and the weight of air used per pound of coal fired.

Solution: (a) The higher heating value is

$$\text{HHV} = (14\,540)(0.628) + (62\,000)\left(0.046 - \frac{0.066}{8}\right) + (4000)(0.043)$$

$$= 11\,587 \text{ Btu/lb} \qquad \textit{Ans.}$$

(b) The refuse contains the ash of the original coal plus unburned combustible, or 25% combustible and 75% ash matter. Thus 75% of the refuse = 8.6% of the coal, and 25% of the refuse = C_R% of the coal unburnt. Therefore the amount of combustible in the refuse per pound of coal is

$$C_R = (25)\left(\frac{8.6}{75}\right) = 2.86\%, \text{ or } 0.0286 \text{ lb}$$

If this is considered carbon with a heating value of 14 540 Btu/lb, the loss per pound of coal is

$$\text{loss} = (14\,540)(0.0286) = 416 \text{ Btu} \qquad \textit{Ans.}$$

(c) By Eq. (9-11), the weight of dry flue gas per pound of carbon is

$$G_w = \frac{(11)(10) + (8)(8) + 7(0.1 + 81.9)}{3(10 + 0.1)} = 24.7 \text{ lb}$$

and the weight of dry flue gas per pound of coal is

$$G_w C_g = 24.7(0.628 - 0.0286) = 14.8 \text{ lb} \qquad \textit{Ans.}$$

The moisture in the flue gas arises from the 0.121 lb of water in the original coal and from the combustion of the hydrogen, which equals 9 H. Thus

$$\text{moisture} = 0.121 + 9(0.046) = 0.121 + 0.414 = 0.535 \text{ lb} \qquad \textit{Ans.}$$

By Equation (9-13), the weight of air used per pound of coal is

$$W_a = 14.8 + 0.414 - (1 - 0.086 - 0.121 - 0.0286) = 14.98 \text{ lb} \qquad \textit{Ans.}$$

A nitrogen balance also gives a sufficiently close approximate solution. By Eq. (9-17), the weight of air used per pound of coal is

$$W_a = \left[\frac{(7.04)(81.9)}{3(10.1)} 0.5994 - 0.01\right]\frac{1}{0.768} = 14.86 \text{ lb}$$

It should be noted that the flue-gas analysis, as usually obtained, is on a dry basis; that is, the water vapor resulting from the combustion of hydrogen or from moisture in the fuel is condensed out and does not appear in the flue-gas analysis. Fuels which are high in hydrogen give dry-flue-gas analyses which are high in nitrogen and relatively low in carbon dioxide. In Example 9-6, where a fuel oil was treated, it was noted that the theoretical composition of the dry flue gases showed 15.8% CO_2. In contrast to this, the coal which was computed in Example 9-7 would show a maximum CO_2 of 18%, the fuel oil having the much higher hydrogen content. Natural gas, which is also high in hydrogen, would indicate a representative maximum CO_2, with no excess air, of around 12%.

Air used in excess of that required for theoretical combustion (see Table 9-6) represents a loss in operation; the excess, with its large quantities of nitrogen, is heated to the final temperature of the stack gases and passes out the chimney. In general, the sensible heat lost up the chimney with the hot flue gases represents the biggest loss in operation and for good performance should be kept to the minimum possible.

Space does not permit a complete coverage of the items entering into a boiler-furnace heat balance, but mention will be made of the most significant items. Loss due to hot dry-flue-gas constituents is represented by the following formula:

$$Q_{dfg} = W_{dfg} C_{pm}(t_g - t_a) \tag{9-23}$$

where W_{dfg} = pounds of dry flue gas formed during the burning of 1 lb of fuel; C_{pm} = the mean specific heat of the flue gases; t_g = temperature of the flue gases entering the stack; t_a = temperature of the air supplied for combustion; Q_{dfg} = Btu loss in dry flue gases per pound of fuel. For dry constituents of the flue gases in the range from (say) 70 F to 1000 F, C_{pm} ranges from 0.25 to 0.26.

Table 9-6 Air Required for Combustion

Air Usage	Anthracite per Pound	Coke per Pound	Semi-bituminous per Pound	Bituminous per Pound	Lignite per Pound	Fuel Oil per Gallon
Theoretical............	9.6 lb	11.2 lb	11.0 lb	10.7 lb	6.5 lb	115 lb
Converted to cubic feet at 70 F..............	129 cu ft	151 cu ft	148 cu ft	144 cu ft	87 cu ft	1550 cu ft
Fair practice...........	225 cu ft	260 cu ft	260 cu ft	250 cu ft	150 cu ft	2100 cu ft

A second loss occurs in connection with the water vapor formed in the flue gas from the combustion of the hydrogen in the fuel, and from any moisture which is carried by the fuel and which evaporates when the fuel burns. This water vapor leaves the stack in the form of steam and, as it is not usually condensed, carries away with it the latent heat associated with the steam. This loss can be represented by the following relationship, which is based on the formula which appears in the ASME Power Test Code for steam-generating units:

$$Q_{H_2O} = \left(\frac{M}{100} + \frac{9\,H}{100}\right)(0.46t_g + 1089 - t_a) \tag{9-24}$$

where Q_{H_2O} is the Btu loss from water vapor per pound of fuel fired; M is the percentage of water in the fuel; H is the percentage of hydrogen in the fuel; and t_g and t_a are the temperatures, in degrees Fahrenheit, of the flue gas and air supply, respectively.

Other losses from combustion of a fuel occur when combustible matter in the gaseous products is not completely burned out, as is the case with carbon monoxide (CO) or when hydrocarbons or actual solid combustible particles pass up the stack. Even though black smoke may be formed during combustion, this condition does not necessarily mean that large weights of combustible material are passing out of the stack; but poor combustion is definitely indicated. The loss from incomplete combustion of carbon monoxide, in Btu per pound of fuel fired, is

$$Q_{CO} = 10\,160C_g\left(\frac{CO}{CO_2 + CO}\right) \tag{9-25}$$

where C_g is the pounds of carbon gasified (burnt) per pound of fuel fired, and where CO and CO_2 are volumetric percentages of CO and CO_2 in the flue gases.

In addition to the previously discussed losses, which apply to all types of fuels (liquid, gaseous, and solid), there is an additional loss which occurs with solid fuels—from combustible which is lost in the refuse. With solid fuels, particles of combustible consisting largely of carbon may remain unburned in the ashy refuse. The loss from this source, in Btu per pound of fuel fired, can be computed from the relation

$$Q_{CR} = 14\,600 \times C_R \tag{9-26}$$

where C_R represents the weight fraction of solid fuel remaining unburned in the refuse.

A heat balance of a boiler or furnace is an accounting of the disposition of the energy in the fuel to all points of loss and to the useful warming of the medium being heated, such as to the water-steam (or, in a warm-air furnace, to the hot air). In a carefully conducted test, the heat balance would normally show that some 95% of the energy of the fuel can be accounted for in useful heating or as measurable losses. The remaining 5% or so represents the so-called radiation and unaccounted-for losses. Depending upon the care with which a test is conducted, or upon the insulation effectiveness of the furnace itself, radiation and unaccounted for losses may range from 3 to 12%. Large boilers, such as those connected with power plants or large heating systems, use heat-conservation devices, such as economizers (water preheaters) in the breeching, or air preheaters for the combustion air, by means of which it is possible to reduce the furnace losses to values as low as 10 to 15%. However, with small-size domestic and industrial equipment, losses in boiler-furnace operation can run from 25 to 50% of the heating value of the fuel. With residential heating equipment, the computed heat losses from the furnace are not completely dissipated, since some of the lost heat passes by indirect routes into the spaces which it is desired to warm.

Example 9-8. When the fuel oil of Example 9-6 burned, with the flue-gas analysis given, the stack temperature of the products leaving the boiler was 570 F. Air is supplied to the burner at 70 F. What loss occurs to the chimney (a) from the dry flue gas and (b) from the water vapor in the flue gas? (c) What is the loss from incomplete combustion? (d) Estimate the boiler efficiency.

Solution: (a) Reference to Example 9-6 shows that 24.1 lb of dry flue gas are generated per pound of fuel. Use this value in Eq. (9-23) and take 0.255 as a representative mean specific heat for these dry-flue-gas products when the temperature is below 800 F. Thus the loss per pound of fuel is

$$Q_{dfg} = (24.1)(0.255)(570 - 70) = 3070 \text{ Btu} \qquad \textit{Ans.}$$

Thus the percentage loss to dry flue gases, based on the HHV of the fuel, is

$$\frac{3070}{18\,500} \times 100 = 16.6\%$$

(b) By Eq. (9-24), for the dry fuel with 11.5% hydrogen, the loss from water vapor, per pound of fuel, is

$$Q_{H_2O} = \frac{(9)(11.5)}{100}[0.46(570) + 1089 - 70]$$

$$= 1315 \text{ Btu} \qquad \textit{Ans.}$$

Also, expressed as a percentage of the heating value of the fuel, the loss is

$$\frac{1315}{18\,500} \times 100 = 7.12\%$$

(c) By Eq. (9-25), the loss from incomplete combustion is, per pound of fuel,

$$Q_{CO} = 10\,160(0.855)\left(\frac{0.1}{8.7 + 0.1}\right) = 98.5 \text{ Btu} \qquad \textit{Ans.}$$

As a percentage, the loss is

$$\frac{98.5}{18\,500} \times 100 = 0.53\%$$

(d) The probable radiation and unaccounted-for losses from this unit are in the neighborhood of 3 to 7%. If 5% is taken as representative and the other losses are added, the total is

$$5 + 16.6 + 7.12 + 0.53 = 29.25\%$$

Thus 100 − 29.25 = 70.7% of the heating value of the fuel is converted into useful heating; that is, 70.7% is the probable boiler efficiency. *Ans.*

9-13. COAL COMBUSTION. STOKERS

With hand firing, coal is burned on grates; with automatic firing, it is burned on stokers which use their own fuel-bed supports (retorts, tuyères, grates). The problem associated with burning coal is that of supplying air for combustion in sufficient quantity to permit the fuel to be consumed adequately without large amounts of excess air, and yet providing enough air to prevent the occurrence of high concentrations of carbon monoxide. Smoke also raises a serious problem. In hand firing of domestic units, it is always desirable to keep a portion of the fire bed uncovered by fresh fuel so that it presents an incandescent surface which can be used to ignite the products being distilled from the fresh coal.

Stokers. A most common type of stoker, the underfeed, consists of a magazine or coal hopper, external to the boiler furnace, which delivers coal to a screw conveyor. The screw, rotating at slow speed, feeds coal to the stoker retort (or to grates), where combustion takes place. A separately operated fan forces air through the stoker air passages, called *tuyères*, and the coal is burned. The resulting ash from the combustion retort drops over the sides of the retort into a trough, from which it can be removed by hand or by a conveyor, or it can be stored for periods of time for later removal by any method desired. Such stokers are made in small sizes feeding as little as 15 to 60 lb of coal per hour, and in large sizes which can burn amounts of coal in excess of 1000 lb/h.

In bituminous stokers, mechanical coal-burning devices are not limited to the underfeed type. At the front end of the boiler, there may be a coal hopper having a very small reciprocating ram which feeds the prepared-size bituminous coal to a series of rapidly revolving distributor blades. These blades are so arranged that they throw the coal uniformly over the entire firebox area. The particles of new coal, however, are so small in proportion to the mass of coke already undergoing combustion that the volatile gases are very rapidly driven off and burn without smoke and with a minimum decrease in the temperature of the firebox at that point. Many small particles of coal are burned in suspension. With this type of spreader stoker the firebox temperature is very high, and thus it is usually necessary to water-jacket the bearings of the distributor. Air for combustion purposes is furnished by a forced-draft fan.

9-14. OIL BURNERS AND COMBUSTION

Most oil burners for domestic furnaces are of the high-pressure, mechanical-atomizing type (gun type), but burners are also made in the rotary wall-flame type, and as air-atomizing burners, and, finally, as simple vaporizing units. In the gun-type, or mechanical-atomizing, burner, illustrated in Fig. 9-10, a pump delivers No. 1 or No. 2 fuel oil, under a pressure of some 50 to 150 psi, to an atomizing nozzle, from which it leaves as a fine mist. An integrally built fan supplies the air which mixes with the oil mist. The swirling mixture of oil mist and air burns as it leaves the nozzle and passes through the combustion chamber of the furnace. Ignition of the mixture is accomplished by sparking electrodes, which receive high-tension power from a transformer. After combustion starts, the power supply to the transformer is shut off and combustion is maintained from the flame. The effectiveness of combustion in this type of burner depends on the degree of oil atomization and on the turbulence or mixing of the oil and air. Various design features are incorporated in the burner tube (gun), such as vanes or swirls for the air; and with a suitable fuel-nozzle angle and a properly designed combustion chamber, good performance can result. The design of the combustion chamber as an integral part of the burner-furnace combination is very important, as combustion is carried to completion in the chamber and proper flow patterns must exist there to complete the combustion and prevent soot formation. Combustion chambers are built from refractory bricks or furnace sections, or are frequently supplied as precast refractory units with proprietary features incorporated.

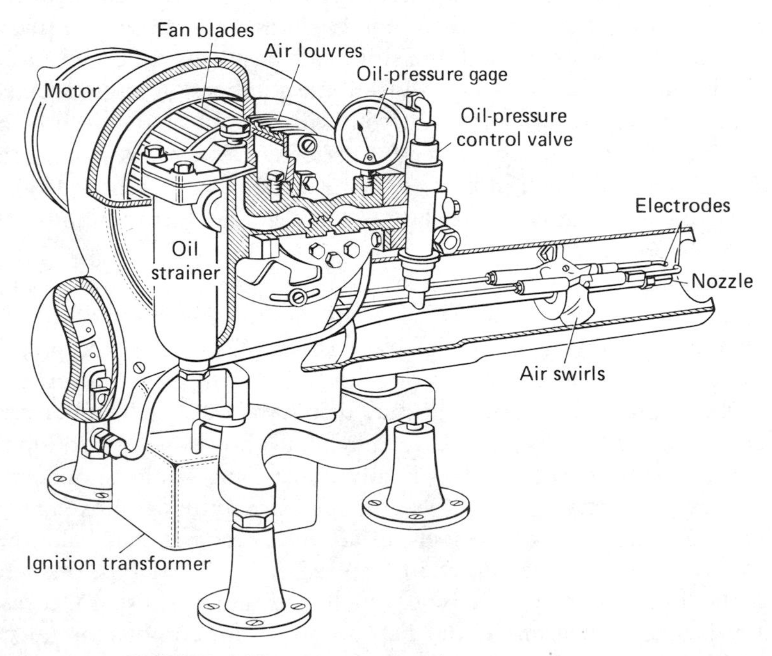

FIGURE 9-10. Pressure-atomizing gun-type oil burner.

In the wall-flame type of vaporizing burner, a centrally located rotary distributor throws the oil particles out to an outer rim, where the vaporization of the oil takes place while it mixes with the combustion air. Both the oil supply and the air can be precisely regulated, and very clean combustion can be maintained. The rotor of the oil distributor, and an attached fan, constitute the only moving parts. Ignition is usually by electrical means from a suitable electrode.

9-15. BOILERS FOR BURNING GAS

Sectional boilers are employed widely for the burning of gas. It is possible to burn gas with fair efficiency in a boiler having the large combustion chamber and wide passages which are necessary for the products of combustion for coal or oil but which are undesirable with gas. If gas is to be burned, however, it is wiser to employ a heating boiler designed especially for that fuel. Such a boiler will have a relatively small combustion chamber and will have narrow passages for the products of combustion, with many baffles. These factors will promote the turbulence which aids in rapid heat transfer, by reducing the thickness of the gaseous film which clings to the heating surfaces.

The operation of gas-fired equipment presents an element of danger, in that the flame may become extinguished and the gas continue to flow. Consequently, controls for gas-fired equipment must be extremely reliable in operation. A satisfactory type of control for gas firing is pictured in Fig. 9-11. This consists of a solenoid-operated valve which can shut off the gas supply completely when the burner is not in use, and an automatic valve which completely interrupts the flow of gas in the event that the pilot light should be extinguished. The operation of the automatic valve is responsive to the action of a thermocouple. (See Section 1-4 and Fig. 1-5.) The thermocouple is located adjacent to the pilot flame, and the warm junction of the couple is thus kept hot by the flame. The electromotive force produced by the couple causes sufficient current to flow through the magnetic holding coil to hold the valve open against the closing force of a spring. In the event that the pilot flame becomes extinguished, the current in the thermocouple circuit then ceases, or reduces to such a point that the holding coil can no longer keep the valve open, and the spring snaps the coil into the shut position, completely closing off the gas flow.

With this particular type of control, even should the pilot flame become reignited, it is necessary to reset the valve in the open position by means of the hand-operated reset button. This required action provides an additional safety feature. It is possible to design valves of this type with an automatic reset where such a device is desirable. In this particular valve, closure of the main valve also cuts off the gas supply to the pilot burner. Other control arrangements sometimes have a separate supply to the pilot direct from the main line.

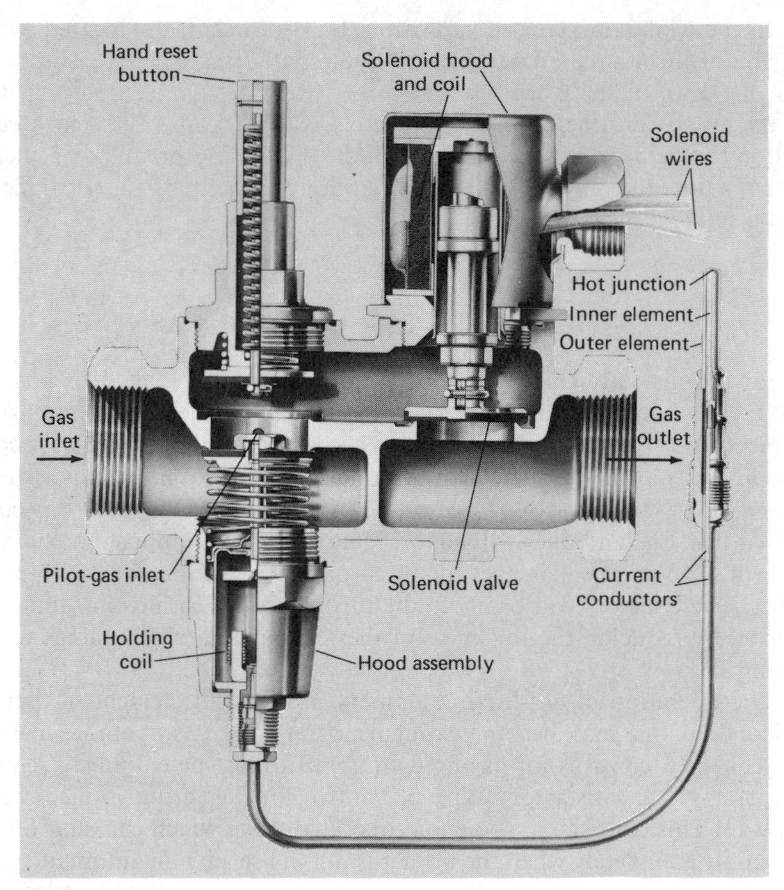

FIGURE 9-11. Combination solenoid-controlled and automatic-shutoff gas valve. (Courtesy Milwaukee Gas Specialty Company.)

9-16. CHIMNEYS

Chimneys, or *stacks*, are used to convey the gaseous products of combustion to an elevation sufficiently high above surrounding objects or buildings to keep the gases from being a nuisance. The height of a chimney directly influences the production of draft. As the column of hot gas inside the chimney is less dense than a similar column of cool air outside, the pressure inside the chimney near its base is less than that of the outside air. This pressure difference acting at any instant is known as the *draft* of the chimney and promotes flow of air through the furnace and up the chimney.

The theoretical draft depends on the difference between the density d_a of the outside air and the density d_{fg} of the hot flue gas; and, since both of these densities depend on temperature, it is possible to set up an expression for draft in terms of temperatures. Thus

$$\Delta P' = H(d_a - d_{fg}) \text{ lb/ft}^2 \tag{9-27}$$

By Eq. (2-28),

$$\Delta P' = H\left(\frac{P_B}{RT_a} - \frac{P_B}{RT_{fg}}\right)$$

which reduces to

$$\Delta P = 0.255 P_B H\left(\frac{1}{T_a} - \frac{1}{T_{fg}}\right) \text{ in. water} \tag{9-28}$$

where ΔP is the theoretical draft, in inches of water; P_B is the barometric pressure, in inches of mercury; H is the height of the stack, in feet; T_a is the air temperature, in degrees Fahrenheit absolute $(460 + t_a)$; and T_{fg} = is the average flue-gas temperature, in degrees Fahrenheit absolute $(460 + t_{fg})$.

Thus the maximum draft is directly proportional to the height of stack (H) and the atmospheric pressure (P_B), and is influenced by the air and flue-gas temperatures. This draft pressure is used in a given furnace in accelerating the air through the furnace and in overcoming frictional and impact losses in the chimney and its connections. The cross-sectional area of the chimney must be such as to permit delivering at a reasonably low velocity the volume of flue gases generated. In natural-draft chimneys, reasonable gas velocities run from less than 20 ft/s up to 40 ft/s.

The chimney diameter for the required cross-sectional area can be found from Eq. (9-29). In this rational formula, G is the flue-gas flow, in pounds per second; V is the velocity, in feet per second; P_B is the barometric pressure, in inches of mercury; T_{fg} is the average flue-gas temperature, in degrees Fahrenheit absolute; and D is the stack diameter, in inches:

$$D = 11.7\sqrt{\frac{GT_{fg}}{P_B V}} \tag{9-29}$$

Example 9-9. A fuel-oil-fired domestic unit consumes 3 gal of oil per hour. The average stack temperature is 300 F when the flue gases leave the furnace at 400 F. The barometer is at 29.9 in Hg, the outside temperature is 40 F, and the chimney height above the combustion chamber is 30 ft. (a) What theoretical draft exists and (b) what chimney size is required to carry the flue gases?

Solution: (a) Use Eq. (9-28) to find the theoretical draft. Thus

$$\Delta P = 0.255(29.9)(30)\left(\frac{1}{460 + 40} - \frac{1}{460 + 300}\right)$$

$$= 0.156 \text{ in. water}$$ *Ans.*

It is assumed that the flue gas is like air in its characteristics, and this assumption is closely true. Carbon dioxide tends to make the flue gas heavier than air, but water vapor from the combustion of hydrogen and from moisture in the fuel makes it lighter; therefore, the flue-gas density approximates that of air.

(b) Burning the fuel oil under the conditions of Example 9-7 showed that 24.2 lb of air were used per pound of fuel; or 24.2 lb of air and 1 lb of fuel formed 25.2 lb of flue gas, which we can round off to 25 lb. A gallon of domestic (No. 2 grade) fuel oil weighs (say) 7.2 lb/gal (Table 9-5). Thus for the 3 gal/h burnt, the weight of flue gas per second is

$$G = \frac{25 \times 3 \times 7.2}{3600} = 0.15 \text{ lb}$$

Using this value in Eq. (9-29) and assuming a gas velocity of 20 ft/s, we find that the required diameter of the chimney is

$$D = 11.8\sqrt{\frac{(0.15)(460 + 300)}{(29.9)(20)}} = 5.16 \text{ in.}$$

The chimney should be made a little larger to provide ample area for contingencies that might develop, such as in starting—when the burner fan forces a pulse of combustion gases into a cold connecting breeching and chimney. In this small size, enough excess capacity would be provided by going to 6 in., or even to 7 in., for the breeching-pipe diameter. The chimney is usually rectangular or square. The equivalent-diameter chart of Chapter 12 shows that a square chimney 6 by 6 in, has the same capacity as one 6.6 in. in diameter, while a chimney 7 by 7 in. has the capacity of one 7.7 in. in diameter. The 6- by 6-in. size should be a reasonable design selection. *Ans.*

Considerably less draft is required for the efficient combustion of oil or gas than is required with solid fuel, since fuel-bed resistance does not occur.

It is possible to control excessive chimney draft, or fluctuations in draft due to otherwise uncontrolled exterior conditions, by using a counter-weighted check-draft damper. With solid fuels, savings in fuel have been found where these devices have been installed. In general, no oil-burning heating boiler should be installed without such a device. In some applications the air supply for this damper control is taken directly from the boiler room, while in others, the air may be taken through a duct from out of doors. Gas stacks require an open hood or a spill vent.

9-17. BOILER SAFETY DEVICES

In the operation of boilers and other pressure vessels containing water or steam, care must be taken to see that conditions do not arise which could lead to failure of the vessel, with possible breakage or even explosion. Mention has already been made of the danger which can arise if the water level in a boiler drops blow a minimum value considered to be safe. The danger arising from low water is that when hot flue gases pass over metal surfaces not in contact with water, the surfaces may become overheated to a point where the metal is weakened and permanent damage can result. Low-water cutoff devices can be provided to stop the combustion apparatus should the water level reach a dangerously low point. In Fig. 9-9 a low-water cutoff device is shown installed on a boiler.

Water feeders are arranged to provide for a sufficiently rapid flow of water into a boiler to maintain proper water level. Figure 9-12 illustrates a combination of boiler–water feeder and low-level cutoff device, similar in

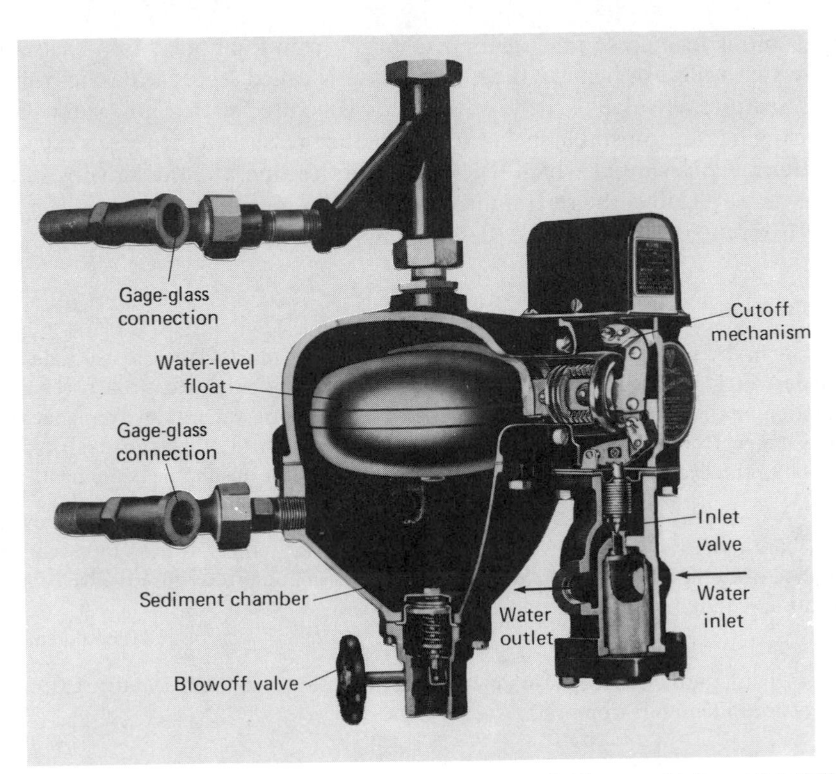

FIGURE 9-12. Boiler water feeder and feeder cutoff control. (Courtesy McDonnell & Miller ITT.)

function to the one illustrated in Fig. 9-9. The float, shown in its chamber, is responsive to variations of water level in the boiler. The top connection to the float chamber would normally be connected into the steam space of the boiler, while the bottom connection would be below ordinary normal water level. The travel of the float is conveyed to the external control system through a bellows of packless construction so that hot boiler water is neither in contact with the electrical parts of the control system nor with the feed-water supply. The water fed into the boiler is controlled by the valve shown at the lower right in the illustration. With a low water level in the boiler, the float, in its low position, keeps the feed line open, permitting water to flow into the boiler. This action continues until the level rises sufficiently to lift the float, which in turn closes off the inlet feed valve. In the upper right of the photograph the electric mechanism for cutting off the power supply to the combustion equipment is shown. This particular design is arranged to operate at a maximum steam pressure of 25 psig.

All steam boilers and water-filled pressure vessels must be provided with a pressure safety device which, when the pressure in the vessel reaches a predetermined limit, opens and permits either steam or water to leave the boiler in a quantity sufficient to bring the pressure to a safe operating condition for the system. In the case of steam boilers, the safety device is called a safety valve and is located so that steam discharges from the boiler. In the case of hot-water boilers or heaters, the safety device is called a pressure-relief valve. A pressure-relief valve of the type that would be used with a hot-water tank or heater has an adjustable spring and when the pressure in the vessels exceeds the desired pressure at which the vessel is set to operate, the valve is lifted from its seat against the restraining action of the spring, and water is eliminated from the system until the pressure falls back to a safe value.

PROBLEMS

9-1. A 16- by 14-ft corner room with a 10-ft ceiling and much glass area has a design heat loss of 14 800 Btuh. Explore the possibility of using either cast-iron baseboard radiation on the outside walls or 2 recessed convectors. Hot water at an average maximum temperature of 180 F is available. Select units from the tables in this chapter showing Btuh excess capacity or show Btuh inadequacy for unsuitable equipment.

9-2. A room has a heat loss of 7000 Btuh. It is desired to heat this room by using baseboard radiation supplied with hot water. An average temperature of 180 F can be supplied from the boiler system. Select the lineal run of nonferrous, fin-tube baseboard radiation required to serve this room.

Ans. 12 ft could suffice

9-3. Obtain the data asked for in Problem 9-2 but assume that representative cast-iron baseboard radiation is employed.

Ans. 18 ft

9-4. The manufacturer's data for a particular unit heater show that it delivers 22 600 Btuh when supplied with 60 F entering air and with 2 psig steam, using a $\frac{1}{20}$-hp motor at 1140 rpm of the fan. (a) Compute the approximate weight of steam condensed per hour in this heater when it is operating under the conditions specified. (b) When 430 cfm enter the unit, what is the approximate delivery temperature of the air? Note that C_p for air is approximately 0.245.

Ans. (a) 22.6 lb/h; (b) 108 F

9-5. For the unit heater described in Problem 9-4, the manufacturer's catalog states that when 40 F air enters the heater the heat transfer and condensate produced are greater than for normal design conditions by a factor 1.141. For the same air inflow, namely, 430 cfm measured at 60 F, (a) compute the outlet temperature of the air from the heater under 40 F inlet conditions and (b) estimate the pounds of steam condensed per hour.

Ans. (a) 94.4 F; (b) 25.8 lb/h

9-6 Cooling water in a steam condenser enters at 65 F and leaves at 80 F. (a) If the steam is condensing at 1 psia, what is the logarithmic mean temperature difference? (b) Compare this value with the arithmetic mean temperature difference.

Ans. (a) 28.7 log mtd; (b) 29.3 arithmetic mtd

9-7. A hot-water heater using a steam coil of the hairpin type (U type) is supplied with water at 60 F and delivers water at 140 F. The steam supply is at 1.3 psia. The maximum demand for water is 150 gal/h. Find the total linear feet of 1-in.-OD brass tubing required in the steam coil.

Ans. 26.3 ft (with $U = 130$)

9-8. Compute the film factor for water flowing at 0.5 ft/s through 2-in.-OD steel tubes (0.109-in. wall) in a heat exchanger. The water enters at 70 F and leaves at 90 F.

9-9. Water is heated in a single-pass shell-and-tube heat exchanger 12 ft long and of 14-in. OD. The shell contains twenty 2-in.-OD stainless-steel tubes with 0.093-in. wall thickness. Water is supplied at 70 F and is warmed to approximately 205 F. The water velocity in the twenty tubes is low at approximately 0.19 ft/s. (a) Compute the value of the film-factor (surface-transfer) coefficient on the water side of the tubes. (b) Consider the value of f for condensing steam to be 3000 Btu/h · ft^2 · F. Disregard the resistance of the clean metal itself and compute the overall thermal resistance and U of the tube. (c) Find the logarithmic mean temperature difference for the tubing with steam at 115 psia (338 F) being used for heating. (d) Find the heat transferred per hour. Note that k for water at about 140 F is 4.1 Btu · in./h · ft^2 · F.
Ans. (a) 88.4 Btu/h · ft^2 · F; (b) $R = 0.01164$, $U = 85.89$ Btu/h · ft^2 · F; (c) 192.7; (d) 1 887 000 Btuh

9-10. (a) For the 2.54-cm-OD steel tubes of Example 9-5 with a wall thickness of 0.127 cm, compute the resistance to heat flow through the tube wall, making use of the film coefficient already computed for the water side and considering that 11 360 W/m^2°C is applicable on the ammonia side of the tube. Disregard the small resistance of the metal of the tube. (b) Compute the overall coefficient of heat transfer for the tube. (c) The water cools from 18.89°C to 15.56°C and the ammonia boils at 13.33°C. Find the logarithmic mean temperature difference. (d) Find the heat transfer through each tube if it is 3 m long inside the tube sheets. Base computation on water side surface area.
Ans. (a) $R = 0.000365$; (b) $U = 2740$ W/m^2°C; (c) 3.64; (d) 2.15 kW

9-11. Water is heated in a single-pass, shell-and-tube heat exchanger 3.66 m long and 0.356 m in outside diameter. The shell contains twenty 5.08-cm-OD stainless-steel tubes with 0.236-cm wall thickness. Water is supplied at 21.11°C and warmed to 96.11°C. The water velocity in the twenty tubes because of the single-pass arrangement is low at 0.0579 m/s. (a) Compute the value of the film-factor (surface-transfer) coefficient on the water side of the tubes. Use the average water temperature for finding necessary data in this computation. The value of k for water at around 60°C is 0.649 W/m · °C. (b) Consider the film factor for condensing steam as 17040 W/m^2 · °C. Disregard the resistance of the clean metal tube but, considering the film effects on each side, find the tube resistance to heat flow and its overall U factor. (c) Find the logarithmic mean temperature difference when condensing steam at 170.0°C is used for heating. (d) Find the heat transfer rate for this unit.
Ans. (a) 498.9 W/m^2 · °C; (b) $R = 0.002063$, $U = 484.7$ W/m^2 · °C; (c) 107.0 deg; (d) 549 600 W, 549.6 kW

9-12. An Illinois bituminous coal as received has an analysis by weight, expressed in decimal fractions, as follows: 0.667 C, 0.0435 H, 0.0803 O, 0.0139 N, 0.0156 S, 0.0797 moisture, 0.100 ash. Compute the higher heating value of the coal, in Btu per pound, as received. Find the heating value per pound of dry coal. [*Hint*: Divide the as-received value by (1 − moisture).]
Ans. 11 837 and 12 862 Btu/lb

9-13. A West Virginia Pocahontas coal has a dry analysis, in percentage by weight, as follows: 85.6 C; 4.56 H; 2.85 O; 1.33 N; 0.66 S; ash, 5.0. The coal as received has 3.5% moisture. (a) Compute the analysis of this coal on an as-received basis. [*Hint*: Multiply each dry percentage by (1 − moisture).] (b) Compute the heating value per pound of dry coal and per pound of coal as received.

Ans. (a) 82.61 C, 4.40 H, etc.; (b) 15 088 and 14 560 Btu/lb

9-14. Find the amount of air theoretically required to burn the coal of Problem 9-12.

Ans. 8.89 lb/lb coal

9-15. Find the amount of air theoretically required to burn the dry coal of Problem 9-13.

Ans. 11.31 lb/lb coal

9-16. The Illinois coal of Problem 9-12, when burned in a stoker-fired furnace, showed a dry-flue-gas analysis, in percentage by volume, of 9.1 CO_2, 10.6 O_2, 0.1 CO, and 80.2 N_2. (a) Find the number of pounds of dry flue gas and the weight of air used per pound of coal, as received. (b) Find the weight of CO formed per pound of fuel burnt.

Ans. (a) 18.0, 17.6; (b) 0.017

9-17. The coal of Problem 9-13 showed a dry-waste-gas analysis, in percentage by volume, of 14 CO_2, 5.6 O_2, 0.1 CO, and 80.3 N_2. (a) Find, per pound of dry coal, the waste gases produced and the air used. (b) Find the weight of CO formed per pound of dry coal burnt.

Ans. (a) 15.4 lb dry gas, 0.41 lb water vapor, 14.8 lb air; (b) 0.014 lb

9-18. A representative No. 6 fuel oil has the following composition, in percent: C, 87.3; H, 10.8; S, 1.2; N, 0.2; O, moisture, and solids, 0.5. Its higher heating value is 18 500 Btu/lb. (a) Compute the weight of air required to burn theoretically 1 lb of this fuel; (b) considering its specific gravity as 0.985, compute the weight of air required to burn, theoretically, 1 gal of the fuel.

Ans. (a) 13.8 lb/lb; (b) 113.2 lb/gal

9-19. Assume that the No. 6 fuel oil in Problem 9-18 showed the following dry-flue-gas analysis, in percent: CO_2, 11; O_2, 7; and N_2, 82. Compute (a) the weight of air used per pound of fuel and (b) the percentage of excess air.

Ans. (a) 19.9 lb; (b) 44%

9-20. Under the combustion conditions of Problem 9-16, the waste gases left the boiler at 610 F and air entered at 70 F. Compute the loss from (a) dry waste gases, (b) water vapor in the waste gases, and (c) incomplete combustion. (d) Estimate the probable boiler efficiency if 6% of the loss can be considered as unaccounted for—and as due to combustible in refuse and to radiation.

Ans. (a) 2430; (b) 612; (c) 173; (d) 66.9%

9-21. Under the combustion conditions of Problem 9-17, the stack gases leave the boiler furnace at 550 F and air enters at 60 F. Compute, per pound of *dry* coal, (a) the loss from dry waste gases, (b) the loss from water vapor in the waste gases, and (c) the loss from incomplete combustion.

Ans. (a) 1890 Btu/lb; (b) 526 Btu/lb; (c) 142 Btu/lb

9-22. Under the combustion conditions of Problem 9-17, the stack gases leave the boiler furnace at 550 F and air enters at 60 F. Compute, per pound of coal *as received*, (a) the loss from dry waste gases, (b) the loss from water vapor in the waste gases, and (c) the loss from incomplete combustion. (d) Estimate the probable boiler efficiency if 5% loss is considered as due to radiation and unaccounted for.

9-23. A poorly adjusted oil burner uses No. 2 fuel oil with the following percentage weight composition: C, 0.85: H, 0.14: inerts, 0.01. It has a higher heating value of 19 900 Btu/lb and shows the following dry-flue-gas analysis, in percentage by volume: CO_2, 8.3; O_2, 9.4; CO, 0.3: and N_2, 82.0. Deposition of soot on the boiler heating surfaces has reduced heat transfer so much that the flue gases leave at 870 F. Air enters the burner at 70 F. Compute the loss per pound of fuel from (a) dry flue gases, (b) water vapor in flue gases, and (c) incomplete combustion. Express your answers in Btu per pound of fuel and in percentages. (d) If the radiation and unaccounted-for loss is 5%, estimate the boiler efficiency by difference calculation.

Ans. (a) 4900 Btu, 24.6%: (b) 1789 Btu, 9.0%; (c) 303 Btu, 1.5%: (d) 59.9%

9-24. An oil burner using No. 2 fuel oil of the composition and type indicated in Problem 9-23 showed a dry-flue-gas analysis, in percentage by volume, of 11.8 CO_2, 4.5 O_2, 0.2 CO, and 83.5 N_2. The flue gases enter the breeching at 600 F and air is supplied to the furnace at 70 F. Compute the loss per pound of fuel from (a) dry flue gases, (b) water vapor in the flue gases, (c) incomplete combustion. Express answers in Btu per pound of fuel and as percentages. If the radiation and unaccounted-for loss is 4%, estimate the boiler efficiency by difference calculation.

9-25. If the flue gases from the coal of Problem 9-16 have an average temperature of 540 F in the chimney when the outside temperature is 40 F, what theoretical draft is developed by the 30-ft chimney? If, under the conditions of Problem 9-16, the coal is burnt at the rate of 60 lb/h, what chimney size is required, provided the gas velocity may not exceed 25 ft/s?

9-26. Work Problem 9-25 for a 30-ft chimney if the mean temperature in the chimney is 440 F and the outside air is at 40 F, all other conditions remaining unchanged.

9-27. Fuel prices in a particular locality were:
No. 2 fuel oil (144 000 Btu), 42 cents per gallon
Bituminous coal (13 000 Btu/lb), $24 per ton
Gas per therm (100 000 Btu), 30 cents
Electricity, 3.5 cents per kilowatthour
Compute the cost of furnishing 100 000 000 Btu by each source. Assume that the efficiency of conversion is 72% for the oil, 56% for the coal, 76% for the gas, and 100% for the electricity.

9-28. Using the data of Problem 9-27, determine the cost of furnishing heat to a building for a season if the estimated requirement is 250 000 000 Btu and coal is employed.

Ans. $412

CHAPTER 10

PHYSIOLOGICAL REACTIONS TO THE ENVIRONMENT

10-1. COMFORT AIR CONDITIONS

Air conditioning as a general term implies effective control of the physical and chemical properties of air in order to produce (1) comfort air conditioning (the maintenance of the air surrounding human beings in such a way that it is in a condition most suitable for their comfort and health) or (2) industrial air conditioning (the maintenance of the air surrounding a material or product in process of manufacture or storage so as best to preserve the physical stability of the material throughout its manufacturing or storage period).

So far as air environment is concerned, the factors which affect human comfort are, in order of importance: (1) temperature, (2) humidity, (3) air motion and distribution, and (4) purity (the quality of the air in regard to odor, dusts, toxic gases, and bacteria). Unless these factors are properly controlled, comfort cannot be obtained. With complete air-conditioning systems, all four of these basic items must be considered.

Although simultaneous control of *all* these factors is required in order to produce a *fully* satisfactory environment for human comfort, many systems, as installed, do not control all of the factors and yet maintain surroundings which are pleasant and substantially conducive to comfort. This is true of many heating systems.

To understand the effect of these four factors, certain physiological and psychological responses of the human body must be taken into consideration. The objective of heating or cooling for comfort is to maintain an atmosphere of such characteristics that the people occupying the space can effectively lose enough heat to permit proper functioning of the metabolic processes in their bodies and yet not lose this heat at so rapid a rate that the body becomes chilled. The processes of combustion of food within the body produce heat in such quantity that the body temperature is normally above the temperature of the atmosphere. A complex regulating mechanism in the human body keeps the temperature of the body at or about 98.6 F. So long as an individual is capable of dissipating heat to the atmosphere at a rate equal to the rate of heat production within the body it is possible to keep the body temperature constant and no difficulty is experienced. If the body temperature rises above normal, heat prostration may result, with temporary or even permanent injury.

The body dissipates heat to the ambient air moving past its surface by ordinary methods of conduction and convection. In this process, air temperature and air motion are the essential factors causing heat transfer. The body can also lose heat by radiation to colder surroundings. Some moisture is always evaporating from the surface of the skin, but when necessary the sweat glands in the body permit large quantities of water to pass through the surface of the skin. If the air in contact with the body is not saturated, this water is taken up by evaporation into the air, with the body itself supplying an appreciable portion of the latent heat. This process of body cooling is especially effective when the air humidity is low. Heating and evaporation of moisture into the air that enters the lungs also cools the body.

The processes of heat control within the body are quite complex, and although not perfectly understood, they clearly operate in two general directions: (1) to decrease or increase internal heat production (metabolism) as the body temperature rises or falls and (2) to control the rate of heat dissipation by changing the rate of cutaneous blood circulation, and by motivating the sweat glands. When the cutaneous blood circulation is increased, more blood flows near the surface of the skin, thereby increasing the surface temperature and permitting greater heat dissipation. Similarly, under cold conditions the blood vessels near the surface of the skin contract, the blood flow at the surface is decreased, and less heat is lost. Excessive activity of the sweat glands usually comes into play following the other methods of control.

The metabolic energy in the body, if not dissipated at the same rate at which it is produced, reappears as stored energy and is evidenced by a rise in temperature of the deep body tissues. If, on the other hand, the environment is so cold that heat is lost from the body at a faster rate than that at which it is created, the deep-tissue temperature of the body slowly falls. These phenomena can be represented by the following equation:

$$S = M - E \pm R \pm C \pm \mathbf{W} \tag{10-1}$$

where S is the stored energy change in the body, evidenced by a rise or fall in deep-tissue temperature; M is the rate of metabolic heat production proportional to the total oxygen consumption of the body; E is the rate of heat loss from evaporation of body fluid; R is the rate of heat exchange with surroundings by radiation; C is the rate of heat exchange with surroundings by convection processes; and $\mathbf{W}$ is the rate at which the body is accomplishing work—that is, plus when the person is walking up stairs or minus and leaving the system when he is descending, plus when he is lifting weights or minus when he is lowering them, etc. R and C, which represent the *dry* heat-exchange effects, are minus when energy is being dissipated by the body; E, under nearly all conceivable conditions, is negative, with moisture evaporating from the surface of the body and from the respiratory system. Units for the equation must be consistent throughout, for example, watts per square meter of body surface or, similarly, Btu per square foot of body surface. However, the energy units could be referred to unit of mass of the body, or to the total mass of the body (kilograms or pounds).

The surface area of a man varies with each individual but for an average man 1.78 m tall weighing 77 kg (5 ft 10 in., 170 lb) the surface area is 1.93 m^2 (20.8 ft^2). The average woman is considered to have a representative surface area of 1.6 m^2 (17.2 ft^2).

The use of seasonable clothing minimizes variations in the load on the human system over winter and summer conditions, but it must be remembered that the same skin area which keeps the body cool in summer remains unchanged in the winter and that added clothes alone are not sufficient insulation to maintain comfort conditions for the body. Therefore homes must be heated in winter, and it is often desirable to cool them in summer to maintain reasonable comfort.

To a certain extent the human system adapts itself to extreme atmospheric conditions. This adaptation, which is termed acclimatization, is both physiological *and* psychological. In fact, people from the tropics experience some discomfort in temperate climates until several weeks have passed. Temperatures of 30°C (86 F) in winter heating are uncomfortable to most people, although in summer a temperature of 30°C (86 F) may be comfortable if the relative humidity is not high or if the air movement is pronounced. It is largely by changes in internal-combustion rates that the body adjusts itself to general atmospheric conditions. Of a similar nature, but requiring more rapid internal adjustment, are "local" changes of atmospheric conditions, such as are met when an individual goes from the hot outside in summer to a conditioned interior. If the inside temperature is in a normal comfort range, yet appreciably colder than outside, an entering occupant may feel cold until readjustment of internal-body-heat controls are made. In summer months the inside air temperatures should, to a limited extent, bear some relation to the swings in outside temperature. Table 10-1 shows a

Table 10-1 Inside Design Conditions for Summer Comfort Cooling

A. Design Conditions Related to Outside Temperature

Outside Design Dry Bulb (deg F)	Occupancy over 40 Min			Occupancy under 40 Min		
	Dry Bulb (F)	Wet Bulb (F)	Rel Hum (%)	Dry Bulb (F)	Wet Bulb (F)	Rel Hum (%)
80	75	65	60	76	66	61
	77	63	47	78	64	47
	79	61	35	80	62	36
85	76	66	61	77	67	61
	78	64	47	79	65	48
	80	62	36	81	63	36
90	77	67	61	78	69	64
	79	65	48	80	67	52
	81	63	36	82	65	40
95	78	69	64	79	70	65
	80	67	52	81	68	52
	82	65	40	83	66	41
100	79	70	65	81	71	63
	81	68	52	83	69	50
	83	66	41	85	67	38
105	80	71	65	81	72	65
	82	69	52	83	70	54
	84	67	42	85	68	41

set of recommendations which might be employed to allow for outside temperatures.

The variation of desired inside temperatures for the winter season is not critical. Recommended values are listed in Table 5-1.

10-2. COMFORT CHART

As far as the human body is concerned, air temperature, relative humidity, and air motion act together to produce the sensation of warmth or cold that is experienced. No simple method has as yet been devised which is completely adequate to evaluate the composite effect of all the variables involved in human response to the thermal environment. Formerly, extensive use was made of the concept of *effective temperature* (*ET*) as an index to express the composite effect of air temperature, relative humidity, radiation, and air motion on the human body. The numerical values of the effective temperature scale were made equal to the temperatures of calm (15 to 25 fpm) saturated air, which produces a sensation of warmth equal to that existent under the given air conditions.

Tests were conducted at the ASHRAE research laboratory, and in other physiological laboratories, to determine the effect of environmental variables on the human system. In conducting the tests, normal healthy individuals were placed in rooms (or moved from room to room) in which dry- and wet-bulb temperatures (relative humidity) and air motion could be varied, and the comfort reactions of the subjects were carefully noted.

It is obvious that if the relative humidity is low, evaporation from the surface of the skin may be so rapid as to cause undue cooling as well as some drying out of the skin surface, and even if the dry-bulb temperature is high, comfort may not prevail. On the other hand, if the relative humidity is very high, evaporation from the skin surface may practically stop, and since cooling by evaporation then ceases, comfort may be obtained with a lower dry-bulb temperature. Between such limits of humidity, varying gradations of comfort will hold. From a viewpoint of body-heat dissipation alone, the range through which relative humidity is varied does not matter, but other considerations make the most desirable relative-humidity range lie between 30 and 70%. For values much below 30% the mucous membranes and the skin surface may become uncomfortably dry. For values above the 70% (or even 60%) range, there is a tendency for a clammy or sticky sensation to develop. The effect of increased air velocity is always to increase the convection and evaporative heat dissipation from the body except at high relative humidities combined with temperatures exceeding 100 F (37.8°C).

The latest ASHRAE comfort chart appears as Fig. 10-1. It shows a number of changes over the ASHRAE 55-66 standard chart and all earlier charts. Effective temperature has been redefined with its reference base set on the 50% relative humidity line. As before, each effective-temperature line represents equivalent conditions of essentially constant physiological strain. Along the 75 F (23.9°C) line, effective temperature differs from dry-bulb temperature by hardly more than 0.5°C in the comfort region. However, the effect of relative humidity on human response is much more pronounced at higher temperatures outside of the comfort region. The cross-hatched area on the chart, which shows the comfort range of the previous standard, is applicable to individuals wearing average clothing and participating in light activity such as office work. The area bounded by the heavy-line parallelogram, shown at the center of the chart, represents a comfort range for lightly clothed individuals (summer attire). If a single comfort temperature is to be chosen it might well be selected at 76 Fdb (24.5°C) lying in the mid-humidity range. It also rests on the 76 F effective temperature line.

In general, comfort for nearly all individuals will obtain in one or both of the regions described. Frequently older poeple require slightly higher temperature by a degree or more, while active children require at least the same temperature reduction for true comfort.

There is a slight shift in the comfort range between summer and winter. Because of acclimatization and the somewhat different clothing worn in winter and summer, individuals are satisfied with slightly lower inside temperatures in winter than in summer. The difference is small, amounting to hardly more than 0.5°C. In fact, for true winter comfort, with proper but light clothing, temperatures of less than 23°C (73 F) will cause discomfort to many inactive individuals. A question then arises relative to temperatures as low as 20°C (68 F), which, because of limitation in the supply of energy, are being used in many public buildings. Here the individual to be comfortable has no choice but to wear substantially heavier clothing. Some people may be able to get by with heavier underclothing, but others will need sweaters, vests, or even coats and scarves. Substantial limitation of body heat loss is necessary for most individuals. In Europe indoor temperatures as low as 16.5°C (62 F) were often considered acceptable, but it is doubtful that many inactive individuals were thermally comfortable at such temperatures.

The comfort chart of Fig. 10-1 does not take into account variations in comfort conditions when there are wide variations in the radiant temperature of the surroundings. Basically, the chart is applicable when the walls, floor, and ceiling are essentially at the dry-bulb temperature of the space. When properly weighted, the average surface temperature of the walls, floor, and ceiling is called the mean radiant temperature (MRT). The effect of variations in MRT on comfort has been extensively investigated. In the range of 26°C (79 F) dry bulb, a rise of 1 deg in MRT produces approximately the same effect as a 1-deg rise in dry-bulb temperature. However, as temperature increases, a greater rise in mean radiant temperature is required to produce the same physiological effect. For example, at 30°C the MRT would have to move 2 deg higher, to 32°C MRT, to equal a dry-bulb change to 30.5°C. The effect of MRT on comfort is less pronounced at high temperatures than at low temperatures, but radiant effects should always be considered if the wall-surface temprature is greatly different from the room temperature. Expressed in terms of dry-bulb temperatures, a rough rule to employ is to design for a 0.3- to 1-deg increase of room temperature to offset a 1-deg drop in MRT in the conventional heating range.

Example 10-1. For the following conditions, state which can be considered in a comfort range for summer and winter occupancy: (a) 26.7°C db (80 Fdb), 15.6°C wb (60 Fwb); (b) 26.7°C db (80 Fdb), 60% relative humidity; (c) 21.1°C db (70 Fdb), 18.3°C wb (65 Fwb); (d) 21.1°C db (70 Fdb), 15.6°C wb (60 Fwb).

Summary of conclusions:

(a) Warmer than desirable but acceptable in summer to some individuals.
(b) Warmer than desirable.
(c) Too cool and relative humidity is undesirably high.
(d) Too cool even for winter except for those with heavier than normal inside clothing.

10-3. HEAT LOSS FROM THE HUMAN BODY

The heat which must be dissipated by the human body is not a constant but varies with the degree of activity, the atmospheric conditions when these are not in the comfort range, and the individual. Table 10-2 gives some of the heat-loss rates for normal adults under different conditions of activity.

Table 10-2 Total Heat Dissipation from Individuals

Type of Activity	Heat Dissipation at 60 F to 90 F Ambient (*met* units for men)	(Btuh for men)
Adults at rest		
Seated	1.0	400
Standing	1.2	480
Moderately active worker	1.5	600
Metalworker	2.0	800
Walking, 2 mph	2.0	800
Restaurant server, very busy	2.0–3.0	1000
Walking, 3 mph	2.6	1040
Walking, 4 mph; active dancing	3.8	1420
Slow run	5.0–6.0	2000
Maximum exertion	7.0–10.0	4000

Note: for women apply a factor of 0.85.

Numerous tests have shown that the metabolic rate of heat production for a fully resting man in relation to his body surface area approximates 58.2 W/m^2 or 18.4 Btu/h · ft^2. These equivalents define what is known as the body-surface *met* unit and, for a man with an average surface area of 1.93 m^2, works out to show the metabolic rate at 112 W (383 Btuh). Because full rest, even while sitting, is not possible, these values have been rounded off at a higher value of 117 W (400 Btuh) and, in turn, the latter values define the body met unit. It is convenient to express the effect of different body activities in terms of met units. This type of tabulation appears in Table 10-2, which shows met units as well as Btu per hour. For women, with on average a smaller body, the body met unit can be taken as 340 Btuh. Table 10-3 extends these data into a convenient form for making heat-gain calculations.

At low temperatures the greater part of the heat loss is by sensible-heat methods (convection, conduction, and radiation) while at higher temperatures evaporation accounts for the greater heat loss.

The cooling which the body experiences from evaporation occurs in the lungs and respiratory mucosa during the breathing process as well as almost continuously on the surface of the body. The latent heat of evaporation must always be supplied for this change of phase from water to steam and, when it is provided by the body, equivalent cooling of the body takes place. The latent heat of steam varies with the temperature of evaporation and on an overall-heat-balance basis even the energy required to raise the water an individual drinks to evaporation temperature contributes to body cooling. Although there is only a narrow working range of temperature for body functioning, it is difficult to find a single value to use for evaporative cooling. Most recently the value of 0.7 W/g of water evaporated from the body has been used. This corresponds to 1085 Btu/lb of body water evaporated. This value is usually rounded off to 1100 Btu/lb when related to air-conditioning

Table 10-3 Heat Gain from Occupants

Type of Activity	Typical Application	Total Heat Dissipation, Adult Male (Btuh)	Total Adjusted* Heat Dissipation (Btuh)	Sensible Heat (Btuh)	Latent Heat (Btuh)
Seated at rest	Theater				
	Matinee	390	330	225	105
	Evening	390	350	245	105
Seated; very light work	Offices, hotels, apartments, restaurants	450	400	245	155
Moderately active office work	Offices, hotels, apartments	475	450	250	200
Standing; light work; walking slowly	Department store, retail store	550	450	250	250
Walking; seated	Drug store	550	500	250	250
Standing; walking slowly	Bank	550	500	250	250
Sedentary work	Restaurant	590	550	275	275
Light bench work	Factory	800	750	275	475
Moderate work	Small-parts assembly	900	850	305	545
Moderate dancing	Dance hall	900	850	305	545
Walking, 3 mph; moderately heavy work	Factory	1000	1000	375	625
Bowling (participant)	Bowling alley	1500	1450	580	870
Heavy work	Factory	1500	1450	580	870

*Adjusted heat dissipation is based on the normal percentage of men, women, and children for the application listed, considering the adult female as having 85% of the value for an adult male, and a child as having 75% of the value for an adult male. With activity defined in general terms as above, these values constitute guidelines, and correction for the exact mix of a group is usually immaterial. However, when only men are involved, in the top five activities increase the sensible- and latent-heat column values by 10%.

Based on data from ASHRAE, *Handbook of Fundamentals*, chap. 22, 1972, by permission.

equipment. It should be realized that the steam produced by the occupants of a space raises the moisture content of the air; when this steam is condensed on the cooling coils of the air conditioner, it can contribute substantially to the cooling energy requirements of the unit. The so-called latent load has thus a dual nature in that it can be expressed in thermal units, Btu per hour, or in weight units, pounds of steam per hour.

Example 10-2. A certain lecture room is occupied by 100 people and is kept at a dry-bulb temperature of 74 F and a wet-bulb temperature of 62 F by the air circulated. Find (a) the relative humidity in the room, and whether the room conditions are in the comfort zone; (b) the heat given up to the air by the occupants; and (c) the portion of heat given up as latent heat by moisture evaporated into the air, and the portion given up as sensible heat.

Solution: (a) Referring to Fig. 10-1, it can be seen that 74 Fdb and 62 Fwb place the space on the cool side for comfort even at 50% relative humidity and that the room would be uncomfortable for most people in summer unless they were suitably dressed.

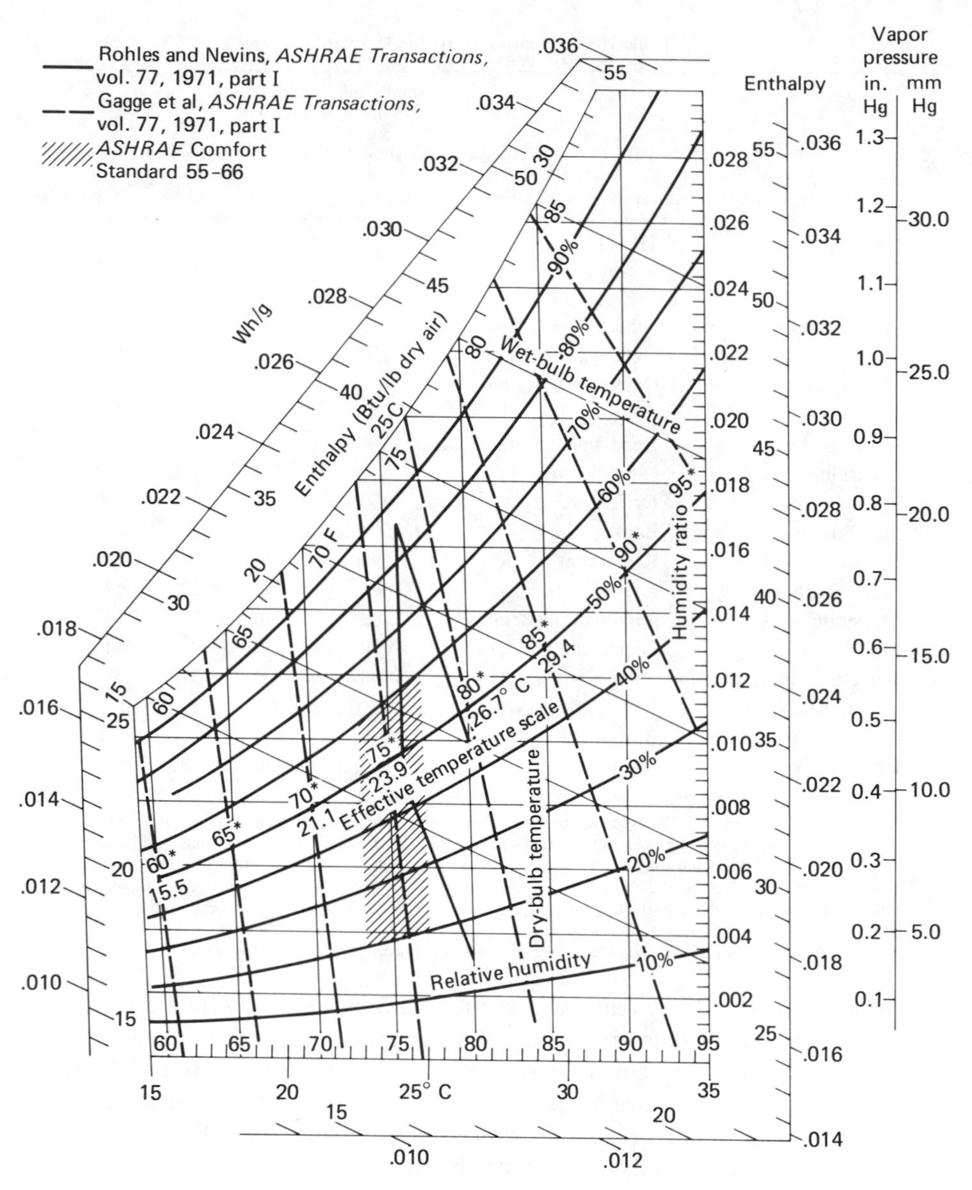

FIGURE 10-1. ASHRAE comfort chart. Based on air velocities less than 45 fpm (14 m/min) and for mean radiant temperature close to dry-bulb temperature. (Reprinted by permission from *ASHRAE Handbook of Fundamentals, 1972*, chap. 7.)

(b) From Table 10-3, read that for people seated at rest (evening), the total heat loss is 350 Btuh per person (men and women), of which 245 Btuh is sensible heat and 105 Btuh is latent heat.

$$\text{total heat delivery to space} = 100 \times 350 = 35\,000 \text{ Btuh}$$

$$\text{sensible heat delivery} = 100 \times 245 = 24\,500 \text{ Btuh}$$

(c) The latent heat delivery is

$$100 \times 105 = 10\,500 \text{ Btuh}$$

This is equivalent to

$$\frac{10\,500}{1085} = 9.67 \text{ lb/h of steam added to the room}$$

$$\text{sensible to total heat ratio} = \frac{24\,500}{35\,000} = 0.7$$

10-4. THE ENVIRONMENT UNDER STRESS CONDITIONS

Many conditions arise in industry where it is impossible to establish comfort conditions. Efforts must then be made to alleviate conditions and, in any event, to recognize the difficulties and even the changes which can arise under such conditions. For a high-temperature space the approaches that are used involve bringing in large amounts of cooler air; insulating heat sources to the greatest possible extent; keeping humidity as low as possible by venting all sources of steam; providing minimum clothing and sometimes even insulated clothing; and, from a protective viewpoint, limiting the time of exposure that is allowed for workers in the spaces. Conditioning the men to become heat-adapted to the operating conditions is also important for work in moderately hot areas where the duration period may be extended.

From the viewpoint of the individual, the time of exposure, the temperature, and the humidity in the space are important factors. It is obvious that if the space conditions are extremely hot the deep-tissue temperature of the individual rises, and exposure must be stopped before the danger point is reached. Under high temperature conditions, both the body temperature and the pulse rate rise. It is considered unsafe to let the pulse rate of an individual increase to more than 130 above the normal 75, and the rectal temperature should not be permitted to exceed 101 F when exposures exceed one hour.

High space temperatures are not unsafe if the relative humidity is low. For example, exposures of up to 0.5 h are not dangerous even at 135 F, provided the relative humidity is in the 30% or lower range. On the other hand, at about 90% humidity, temperatures appreciably above 100 F are not safe even for short periods. It is obvious, too, that the internal metabolic rate is a factor in these decisions and, for a person performing work, the time of exposure must necessarily be much shorter than for a person at rest.

No completely satisfactory indices for measuring body performance under temperature stress conditions have been developed although a number are in use. One of these is the effective temperature (ET), which has already been described. It has been extended into the hot region, but exact physiological responses are not well established (Refs. 8 and 9). Two other indices provide data, namely, the predicted Four-Hour Sweat Rate (P4SR), and the Belding-Hatch Heat Stress Index (Refs. 10 and 11).

Additional work is being carried out to obtain answers to this complex problem. Difficulty arises in carrying out experimentation with human subjects under extremely uncomfortable conditions, and it is also true that danger is involved since limiting conditions could be exceeded with temporary or permanent damage to a subject.

10-5. REQUIREMENTS FOR QUALITY AND QUANTITY OF AIR

The air in an occupied space should at all times be free from toxic, unhealthful, or disagreeable fumes; it should also be relatively free from odors and dust. To obtain these conditions, enough clean outside air must always be supplied to an occupied space to counteract or adequately dilute the sources of contamination. In industrial processes, undesirable and sometimes toxic gases may arise, but in air conditioning for human occupancy alone, the problem is largely one of supplying enough air to keep odors from arising to the point of being disagreeable.

The concentration of odors in a room depends upon numerous factors, including the dietary and hygienic customs of the occupants, the type and amount of outdoor air supplied, the room volume per occupant, and the types of odor sources. In general, where smoking is not a problem, 7.5 to 30 cfm per person will take care of all conditions. Where smoking occurs, additional outside air is necessary to counteract the effect of the smoke. It is usually considered that where smoking is taking place, not less than 15 cfm per person is desirable, and 50 cfm may be necessary. There are no absolute standards as to ventilation, although certain local codes must be satisfied and in this sense these do represent standards. Table 10-4 indicates ventilation-practice standards which have been found generally satisfactory, and these can be employed advantageously provided they do not conflict with local codes. In the case of buildings or residences where infiltration is an important factor, it may be necessary to estimate the probable air changes per hour. Table 10-5 gives data on the total air change which might be expected in certain types of building construction.

The total quantity of outside air which passes through a building or space is controlled chiefly by physical considerations concerning temperature, type of air-distribution system, and air velocities. However, the type and usage of a building, the floor area, the height of rooms, the window area, and the type of occupancy affect the operation of the air-distribution system. In most air-conditioning systems a large amount of air is recirculated over and above the amount required to satisfy minimum ventilation conditions in regard to odor and purity.

The olfactory organs in the upper nasal cavity are extremely sensitive to very slight concentrations of odoriferous matter. However, in the presence of a given odor the sensitivity of these organs gradually lessens, but it can quickly revive if the person goes to a place where the odor in question is not present and then returns. That is, breathing fresh air restores olfactory sensitivity. People on entering a place may be struck by a strong odor which the occupants do not even notice.

With recirculated air, the washing, humidifying, and dehumidifying remove a considerable proportion of body odor and may keep the outside-air requirements close to a basic minimum value of 5 cfm. For general

Table 10-4 Ventilation Standards

Application	Smoking	Air per Person (cfm)		Minimum Air per Sq Ft of Floor (cfm)
		Recommended	Minimum	
Apartment				
Average	Some	20	15	...
De luxe	Some	30	25	.33
Banking space	Occasional	10	7½	...
Barber shops	Considerable	15	10	...
Beauty parlors	Occasional	10	7½	...
Brokers' board rooms	Very heavy	50	30	...
Cocktail bars	Heavy	30	25	...
Corridors (supply or exhaust)	.....	...	...	.25
Department stores	None	7½	5	.05
Directors' room	Extreme	50	30	...
Drug stores	Considerable	10	7½	...
Factories	None	10	7½	.10
Five-and-ten-cent stores	None	7½	5	...
Funeral parlors	None	10	7½	...
Garage	.....	...	...	1.0
Hospitals				
Operating rooms *	None	...	...	2.0
Private rooms	None	30	25	.33
Wards	None	20	15	...
Hotel rooms	Heavy	30	25	.33
Kitchen				
Restaurant	.....	...	...	4.0
Residence	.....	...	...	2.0
Laboratories	Some	20	15	...
Meeting rooms	Very heavy	50	30	1.25
Office				
General	Some	15	10	...
Private	None	25	15	.25
Private	Considerable	30	25	.25
Restaurant				
Cafeteria	Considerable	12	10	...
Dining rooms	Considerable	15	12	...
Schoolrooms	None	...	...	...
Shop, retail	None	10	7½	...
Theater	None	7½	5	...
	Some	15	10	...
Toilets (exhaust)	.....	...	...	2.0

* Use outside air only, to overcome explosion hazard of anesthetics.

Table 10-5 Probable Building Air Changes per Hour by Natural Effects

Type of Building	Winter		Summer	
	Min	Max	Min	Max
Ordinary factory	$\frac{1}{2}$	2	$\frac{1}{2}$	1
Poor frame construction	$1\frac{1}{2}$	3	1	$1\frac{1}{2}$
Corrugated metal	$1\frac{1}{2}$	3	1	$1\frac{1}{2}$
Stores and shops	$\frac{1}{2}$	1	$\frac{1}{2}$	1
Commercial buildings	$\frac{1}{2}$	$1\frac{1}{2}$	$\frac{1}{2}$	1
Office building	$\frac{3}{4}$	$1\frac{1}{2}$	$\frac{1}{2}$	1
Residence				
Well-constructed	$\frac{1}{2}$	1	$\frac{1}{4}$	$\frac{1}{2}$
Poorly constructed	$1\frac{1}{2}$	2	$\frac{3}{4}$	$1\frac{1}{2}$

Note. In commercial buildings with doors leading to nonconditioned areas, add 100 cfh for each person entering or leaving. For a 36-in. door standing open, allow 48,000 cfh. Revolving doors can be calculated at 60 cu ft per person passing through, or if the door is equipped with a brake, 50 cu ft per person.

Spaces like auditoriums, clubrooms, dance halls, and theaters usually take enough outside air to offset infiltration.

application, a minimum of 10 cfm of outside air per person, mixed with perhaps 20 cfm of recirculated air, is a good general rule.

Formerly, the carbon dioxide content in an occupied space was thought to be the ultimate index of ventilation conditions, but this index has been found to be unreliable in so many cases that its use is diminishing. A normal adult at rest, in breathing, exhales air containing about 3.5 to 4.1% of carbon dioxide by volume, or about 0.5 to 0.6 ft^3/h. Combustion processes, gas burners, open fires, and tobacco fumes also contribute to the CO_2 content of a given space.

The action of CO_2 on the human system in small quantities is harmless, but in larger quantities (over 2% by volume; that is, over 200 parts in 10 000) its effect in diluting the oxygen content of the air makes breathing more rapid and causes some discomfort. With 6% CO_2, breathing is very difficult; and at 10%, a loss of consciousness usually results but is not necessarily fatal. Atmospheric air contains about 3 to 4 parts of CO_2 by volume in 10 000, and the air inside a ventilated occupied space varies from about 6 to 14 parts. There are various types of apparatus on the market available for measuring these small quantities of CO_2 in the air, among which should be mentioned the Petterson-Palmquist apparatus and the Haldane apparatus.

Example 10-3. In a given occupied room 15 cfm of outside air per person is continuously circulated. If the outside air contains 300 parts CO_2 per 1 000 000 by volume, what CO_2 contents exists in the room, assuming equilibrium has been established?

Solution: Each person contributes 0.5 to 0.6 (say, 0.55) ft^3 of CO_2 per hour to the air. During each hour, $15 \times 60 = 900$ ft^3 of air is supplied, carrying

$$900 \times \frac{300}{1\,000\,000} = 0.27 \text{ ft}^3\text{/h CO}_2$$

After equilibrium is reached, the total CO_2 associated with the air is the sum of the outside-air CO_2 plus the human CO_2, or $0.27 + 0.55 = 0.82$ ft^3/h. This amount of CO_2 is associated with the 900 ft^3 of ventilation air and therefore shows 0.82 parts of CO_2 per 900 parts of air, or

$$\frac{0.82}{900} \times \frac{1\,000\,000}{1\,000\,000} = \frac{910}{1\,000\,000} \qquad \textit{Ans.}$$

That is, 910 parts of CO_2 per million exist in the room.

Example 10-4. An auditorium seating 1500 people is to be kept at 78 Fdb and 66 Fwb when the outdoor temperature is 90 Fdb and 74 Fwb. The heat gain through the insulation, etc., amounts to 120 000 Btuh in addition to the sensible and latent heat gain from the occupants. Find (a) the outdoor air required for ventilation; (b) the sensible heat load from the occupants; (c) the latent heat load from the occupants; (d) for the total sensible heat load, the number of pounds and cubic feet of conditioned air at 65 Fdb that must be circulated; and (e) the wet-bulb temperature that is required in order to pick up the moisture load.

Solution: (a) From Table 10-4, considering the auditorium as a theater, $7\frac{1}{2}$ cfm per person is taken. This figure should be greatly increased if much smoking occurs. Some additional recirculation of the auditorium air is necessary, and its treatment in the conditioner washer will remove some odors, thus justifying the rather low figure of $7\frac{1}{2}$ cfm. Thus the amount of air supplied to the auditorium from the outside is

$$7\tfrac{1}{2} \times 1500 = 11\,250 \text{ cfm}$$

If this air is measured at inside conditions of 78 Fdb and 66 Fwb, the specific volume is 13.8 ft^3/lb (Plate I). Therefore, the outside air required is

$$11\,250 \times \frac{1}{13.8} = 8.5 \text{ lb/min} \qquad \textit{Ans.}$$

(b) Using Table 10-3, read 245 Btuh sensible heat load per person. Therefore, the sensible heat load from all occupants is

$$245 \times 1500 = 367\,500 \text{ Btuh} \qquad \textit{Ans.}$$

(c) Using Table 10.3, read 105 Btuh latent heat loss per hour per person. Thus the latent heat load from the occupants is

$$105 \times 1500 = 157\,500 \text{ Btuh} \qquad \textit{Ans.}$$

Compute the moisture load by using the latent heat factor for water evaporated from the body at 1085 Btu/lb

$$\frac{157\,500}{1085} = 145 \text{ lb/h}$$

(d) Total sensible heat load $= Q_s = 120\,000$ building $+ 367\,500$ human $= 487\,500$ Btuh. By Eq. (3-34),

$$Q_s = mC_p(t_i - t_c)$$

Substituting, we have

$$487\,500 = m(0.244)(78 - 65)$$

Therefore, for the sensible heat load, the amount of air that must be circulated is

$$m = 153\,600 \text{ lb/h} \qquad \textit{Ans.}$$

From the psychrometric chart, air at 65 F and at an estimated relative humidity of 70% shows a specific volume of 13.4 ft^3/lb of "dry" air. Therefore, in cubic feet per minute, the amount of air circulated is

$$\frac{153\,600}{60} \times 13.4 = 34\,300 \text{ cfm} \qquad \textit{Ans.}$$

(e) The amount of moisture which must be picked up per pound of "dry" air circulated is

$$\frac{145}{153\,600} = 0.00094 \text{ lb moisture/lb dry air}$$

Therefore, moisture per pound of air is:

inside air at 78 F and 66 F	0.01100 (Plate No. 1)
less moisture to be absorbed	0.00094
moisture content of entering air	0.01006 lb

From Plate I, air at 65 Fdb at 0.010 lb/lb has a 60 Fwb.

10-6. INDUSTRIAL CONTAMINANTS

Industrial activities and operation frequently produce substances which disseminate or become dispersed in the air, and, as such, these substances constitute contaminants or atmospheric impurities. Some of these substances are toxic in nature, others are relatively harmless. Some of them are potential fire hazards in certain concentrations, others are completely noncombustible.

It is, of course, most desirable to reduce the amount of contaminant at the source. If this cannot be done, solid particles such as dusts should be caught in separators and filters, or otherwise exhausted from the working area into the open atmosphere where dilution can take place to such an extent that the contaminant no longer represents a potential hazard. The process of exhausting is usually the only practicable method to follow with objectionable vapors and gases. Where it is not possible to remove completely the contaminant by exhaust ventilation, it then becomes necessary to supply dilution ventilation air in such quantities that the contaminant is reduced to a point where it is below a threshold danger level. Increasingly more attention has to be given to trapping vapors, fumes, and dusts at their source point to prevent the ever-increasing level of air pollution.

Highly toxic gases may be harmless to humans provided they appear in the atmosphere only in trace amounts. It is of importance then to know the amount of hazardous material which can exist in an atmosphere without causing significant harm or injury to those who must work or live in an atmosphere containing a hazardous material. It is clearly recognized that breathing air containing certain chemicals, in high concentration, can be lethal in a short period of time. However, as concentrations of such chemicals are reduced, the time of exposure can be increased and, if this thought is further developed, it is easy to conceive a condition where the concentration of the hazardous material may be so low as to cause no damage at all. The terms *threshold limit* and *maximum allowable concentration* imply that if the concentration of a hazardous material is kept below a certain level, activities in a contaminated space can be carried on without danger and often without annoyance to the occupants of that space.

The American Conference of Governmental Industrial Hygienists each year collects and reevaluates the data which are used in the published values of threshold limits for different contaminants in workroom air. These threshold limit values (abbreviated TLV) are the airborne concentrations of contaminants to which it is believed nearly allworkers can be exposed daily without adverse effect. Any TLV for a contaminant presumes that an individual can work a 7- to 8-h day during a 40-h work week and not suffer damage as a result of exposure to the TLV of the contaminant. In fact, appreciably higher exposures for many substances could be experienced for short periods without damage or annoyance since it is the product of contaminant concentration times length of exposure that probably represents the most important index. It should be recognized that some individuals, because of allergies or other physical characteristics, may be disturbed by concentrations appreciably lower than those indicated in the TLV tabulations.

Table 10-6 gives values for a number of common substances met in industrial and air-conditioning practice. Where the ventilation of garages

Table 10-6 Threshold Limit Values (TLV) for Selected Airborne Contaminants (Ref. 12)
(Values given in parts per million by volume and in milligrams per cubic meter at 25°C and 760 mm Hg.)

Substance	(ppm)	(mg/m³)	Substance	(ppm)	(mg/m³)
Acetone	1000	2400	Hydrogen sulfide	10	15
Benzene	25	80	Methyl alcohol	200	260
Carbon dioxide	5000	9000	Methyl chloride	100	210
Carbon monoxide*	50	55	Methylene chloride	500	1740
Carbon tetrachloride	10	65	Nitric acid	2	5
Chlorine	1	3	Nitric oxide	25	30
Chlorobenzene	75	350	Nitrogen dioxide	5	9
Chlorobromomethane	200	1050	Ozone	0.1	0.2
Dichloromonofluoromethane	1000	4200	Phenol	5	19
Dichlorotetrafluoroethane	1000	7000	Stibine	0.1	0.5
Difluorodibromomethane	100	860	Sulfur dioxide	5	13
Ethylacetate	400	1400	Tetrachloroethylene	100	670
Fluorotrichloromethane	1000	5600	Trichloroethylene	100	535
Heptane (n)	500	2000	Trifluoromonobromomethane	1000	6100
Hydrogen chloride	5	7	Turpentine	100	560
			Xylene	100	435

* Note: The Environmental Protection Agency considered 35 ppm CO to represent a maximum safe value for exposures up to 1-h.

is concerned, attention must be given primarily to keeping the carbon monoxide (CO) value at or below the threshold limit in areas of sustained occupancy. In fact, the Federal Environmental Protection Agency has called for a CO value not to exceed 35 ppm on average where 1-h occupancy is anticipated. For areas through which people move rapidly, higher concentrations are not considered unsafe.

Many dusts and particulates are also hazardous, and care must be taken that the concentration of such materials is kept within safe limits.

It is customary to express threshold limits for hazardous substances (Ref. 12) as ppm (parts per million), a volumetric evaluation, or as mg/m^3 (milligrams per cubic meter). To transform from one to the other for vapors in air it is convenient to make use of the mol volume which in SI is 24.0 liters at 20°C (68 F) and 760 mm Hg atmospheric pressure, or 22.4 at 0°C, 24.4 at 25°C (77 F). Using M as the molecular weight, at 760 mm Hg pressure and at 25°C, we can write

$$\frac{\text{mg}}{\text{liter}} \times \frac{1}{1000} \times \frac{24.4}{M} \times 1\,000\,000 = \text{ppm}$$

$$\frac{\text{mg}}{\text{liter}} = \frac{\text{ppm}}{1000} \times \frac{M}{24.4} \tag{10-2}$$

and

$$\text{ppm} = \frac{24\,400}{M} \frac{\text{mg}}{\text{liter}} \qquad \text{at } 25^\circ\text{C, } 760 \text{ mm Hg} \tag{10-3}$$

$$\% \text{ by volume} = \frac{\text{ppm}}{1\,000\,000} \times 100 \tag{10-4}$$

Example 10-5. The threshold limit for acetone in occupied space is 1000 ppm at 25°C (77 F). Convert this limit to milligrams per liter and milligrams per cubic meter.

Solution: Acetone, C_3H_6O, has a molecular weight of 58.1. By Eq. (10-2)

$$\frac{\text{mg}}{\text{liter}} = \frac{1000}{1000} \times \frac{58.1}{24.4} = 2.38 \qquad \textit{Ans.}$$

$$\frac{\text{mg}}{\text{m}^3} = 2.38 \times 1000 = 2380 \qquad \textit{Ans.}$$

also

$$\frac{\text{lb}}{\text{ft}^3} = \frac{\text{mg}}{\text{m}^3} \times \frac{1}{16 \times 10^6} = \frac{2380}{16 \times 10^6} = 0.000149$$

PROBLEMS

10-1. For the following conditions, for people at rest, indicate the probable comfort patterns for summer and winter: (a) 70 Fdb, 55 Fwb; (b) 74 Fdb, 60% relative humidity; (c) 78 Fdb, 68 Fwb. *Ans.* (a) Cold in summer, cool in winter; (b) cool in summer, comfortable in winter; (c) comfortable to warm in summer, overly warm in winter.

10-2. The air in a cinema, with an audience of 1000 people during an evening performance, is kept at a dry-bulb temperature of 72 F and at 60% relative humidity. Determine (a) whether room conditions are in the comfort zone; (b) the heat given up to the air by the occupants use Table 10-3); (c) the portion of heat given up as latent heat by moisture evaporated into the air; and (d) the portion of heat given up as sensible heat.

Ans. (a) comfortable in winter; (b) 350 000 Btuh; (c) 105 000 Btuh; (d) 245 000 Btuh

10-3. How much air at 70 F and 45% relative humidity would have to be supplied to a classroom containing 40 students if the room is to be kept at 74 F and not higher than 50% relative humidity? Neglect insulation losses.

Ans. 10 100 lb/h

10-4. Three men are doing moderate work in a room at 80 Fdb. (a) Using Table 10-3, how much heat is given off by them? (b) What portion of the total heat dissipation is by evaporation? (Use the sensible-latent ratio of Table 10-3.) (c) How much heat is given off as sensible heat? (d) Find the moisture evaporated per hour.

Ans. (a) 2700 Btuh; (b) 1730 Btuh latent; (c) 970 Btuh; (d) 1.6 lb

10-5. The threshold limit for SO_2 in air is considered to be 5 ppm by volume while for hydrogen sulfide (H_2S) the value is 15 ppm, with both determined at 25°C and 760 mm. Compute the threshold concentrations expressed in milligrams per cubic meter, atomic weight of S, 32; O, 16; H, 1.

Ans. 13, 21

10-6. How many parts of CO_2 per 10 000 by volume exist in an occupied room which is supplied with 10 cfm per person if the outside air contains 4 parts CO_2 per 10 000 by volume?

Ans. 13.2

10-7. At a certain time a residence having overall dimensions of 45 by 25 by 18 ft has $1\frac{1}{2}$ air changes per hour from infiltration. Assume that the residence is occupied by five people and compute the ventilation air from natural effects, expressed in cubic feet per minute per person.

Ans. 101 cfm

10-8. A hospital ward 130 by 22 by 10 ft occupied by a total of 30 persons (patients and staff) is to be kept at 79 Fdb and 68Fwb when the outdoor temperature is 95 F. The heat gain from outside amounts to 150 000 Btuh in addition to the heat gain from the occupants. Find (a) the minimum volume of outdoor air recommended for ventilation; (b) the sensible heat load from the occupants; and (c) the latent heat and moisture load from the occupants. (d) For the total sensible heat load, how many pounds per hour of conditioned air at 65 Fdb must be circulated? (e) What should be the wet-bulb temperature of this conditioned air in order for the air to pick up the moisture load? (f) What is the required capacity of the fan in cubic feet per minute?

10-9. A small medical center has a floor area of 80 by 80 ft and is three stories high, with a total height of 30 ft. (a) On the basis of $1\frac{1}{2}$ air changes per hour, what ventilation capacity in cubic feet per minute would apply for this building, and (b) if the maximum occupancy of this building at a given time is 80 adults, would this air flow satisfy the minimum ventilation standard? Smoking will probably be at a minimum throughout this building.

Ans. (a) 4800 cfm; (b) yes

10-10. The municipal codes of certain cities specify that certain minimum amounts of air must be supplied to buildings of varying types. Assume that for the medical center discussed in Problem 10-9 the municipal code specifies a minimum of 0.25 cfm/ft^2 of floor area. (a) For the three stories in question, compute the minimum ventilation requirements according to code. (b) Compare these with the data of the previous problem.

Ans. (a) 4800 cfm; (b) same

10-11. A certain dance hall at peak occupancy serves 150 couples. The lighting load consists of 4000 W. The space is not air-conditioned, but eight fans with $\frac{1}{4}$-hp motors are used to circulate air in the space. Compute (a) the total sensible heat load and (b) the total latent heat load. Note that 1 W = 3.41 Btuh and that 1 hp = 2545 Btuh. (The efficiency of small motors may be as low as 40%.)

10-12. In winter, the hospital ward with 30 occupants mentioned in Problem 10-8 has a heat loss of 220 000 Btuh at design conditions. It is presumed that natural ventilation effects cause $1\frac{1}{2}$ changes per hour if the ventilation system is not used. However, the ventilation-air-conditioning system is usually operated in winter as well as summer and offsets some of the infiltration air leakage. In winter 15 cfm of outside air per person, along with 20 cfm of recirculated air per person, are supplied at 88 F and enter the floor returns at 68 F. (a) How much heat is supplied by the ventilating-system air, and how much additional heat is required from steam radiation when the maximum winter design load occurs? (b) If the ventilation system is not in operation, what is the probable amount of outside air per person arising from natural ventilation effects?

Ans. (a) 196 800 Btuh from radiators; (b) 23.8 cfm

10-13. The threshold limit for ammonia NH_3 is reported as 35 mg/m^3. Compute this concentration in units of parts per million measured at 25°C and 760 mm Hg pressure.

10-14. For the ammonia of Problem 10-13, compute its threshold concentration in units of parts per million at 20°C and 70 mm Hg. Also express your answer in pounds per cubic foot.

REFERENCES

F. C. Houghten and C. P. Yaglou, "Determining Lines of Equal Comfort and Determination of the Comfort Zone," *Trans. ASHVE*, Vol. 29 (1923), pp. 163, 361.

C. P. Yaglou and W. E. Miller, "Effective Temperature with Clothing," *Trans. ASHVE*, Vol. 31 (1925), p. 89.

N. Glickman, T. Inouye, R. W. Keeton, and M. K. Fahnestock, "Physiological Examination of the Effective Temperature Index," *Trans. ASHVE*, Vol. 56 (1950), p. 51.

B. H. Jennings and B. Givoni, "Environment Reactions in the 80 F to 105 F Zone," *Trans. ASHRAE*, Vol. 65 (1959), p. 115.

W. Koch, B. H. Jennings, and C. M. Humphreys, "Environmental Study II—Sensation Responses to Temperature and Humidity under Still-Air Conditions in the Comfort Range," *Trans. ASHRAE*, Vol. 66 (1960), pp. 264–287.

R. G. Nevins, F. H. Robles, W. Springer, and A. M. Feyerherm, "A Temperature-Humidity Chart for Thermal Comfort of Seated Persons," *ASHRAE Journal*, Vol. 8 (1966), p. 61.

ASHRAE Handbook of Fundamentals, chap. 7, "Physiological Principles" (New York: ASHRAE, 1972), pp. 119–150.

C. M. Humphreys, Oscar Imalis, and Carl Gutberlet, "Physiological Response of Subjects Exposed to High Effective Temperatures and Elevated Mean Radiant Temperatures," *Trans. ASHVE*, Vol. 52 (1946), p. 153.

L. W. Eichna, W. F. Ashe, W. B. Bean, and W. B. Shelley, "The Upper Limits of Environmental Heat and Humidity Tolerated by Acclimatized Men Working in Hot Environments," *The Journal of Industrial Hygiene and Toxicology*, Vol. 27 (March, 1945), p. 59.

C. J. Wyndham, W. Bouwer, M. G. Devine, H. E. Patterson, and D. K. C. MacDonald, "Examination of Use of Heat Exchange Equations for Determining Changes in Body Temperature," *Journal of Applied Physiology*, Vol. 5 (1952), p. 299.

H. S. Belding and T. F. Hatch, "Index for Evaluating Heat Stress in Terms of Resulting Physiologic Strains," *Heating, Piping and Air Conditioning* (August, 1955), p. 129.

Air Pollution Engineering Manual, *Public Health Service Publication*, No. 999-AP-40 (Cincinnati, Ohio, 1967), pp. 872–878.

P. O. Fanger, *Thermal Comfort, Analysis and Applications in Environmental Engineering* (New York: McGraw-Hill Book Co., 1973).

B. H. Jennings and J. A. Armstrong," Ventilation Theory and Practice," *Trans. ASHRAE*, Vol. 77 (1971).

CHAPTER 11

WARM-AIR AND PANEL HEATING SYSTEMS

11-1. TYPES OF WARM AIR HEATING SYSTEMS

The general term *warm-air heating* implies that heat transfer is directly from fire-heated surfaces to the air which is used in the heated space. Any stove is a warm-air heater, and any stove becomes a furnace when it is jacketed and provided with duets so as to serve a heated space other than the one in which the furnace is placed.

The term *gravity warm-air heating*, or the synonymous term *thermal-circulation warm-air heating*, means that natural convection is the method of circulating the heated air. When air inside a duct is heated, the air expands and the weight per unit of volume becomes less than that of the surrounding cooler air. Unless the duct is horizontal, the relatively light warm air rises, and dense cool air flows in to take its place. The circulation is roughly proportional to the difference in temperature between the warm air and cool air. If a warm-air furnace is placed at the bottom of an air-duct riser, a thermally created air movement commences.

Forced-circulation warm-air heating means that mechanical energy, rather than thermal energy, does the work of moving the air which transports the heat. The power is usually furnished by a fan. Better all-around performance is possible with mechanical circulation. Such systems are also known as *mechanical warm-air systems.*

In all warm-air heating systems, the warm air which is supplied to a space, in cooling to room conditions, counteracts the heat loss from the space.

11-2. WARM AIR FURNACES

Warm-air furnaces can employ oil, gas, or hand- or stoker-fired solid fuels. With warm-air furnaces, it is difficult to transfer heat effectively from the firebox or combustion chamber to the air passing over the hot surfaces. Also, because radiant heat is not readily absorbed by an air stream, it is a necessity to provide secondary surface to contact and warm the air stream. Warm-air furnaces generally have metal casings, though brick encasements are sometimes employed.

Large warm-air furnaces, frequently installed in batteries, are in use for public buildings. These employ a fan to deliver air under pressure so that it passes upward along the combustion-chamber sides and then is deflected against secondary heat-transmitting surfaces. A battery of furnaces is usually arranged so that in mild weather some of the furnaces need not be fired.

Insofar as the heating of a building is concerned, it is essentially immaterial whether gas, oil, or solid fuel burned on a stoker, or solid fuel burned on hand grates, is employed. The availability of fuel, and the economics of the situation, determine which source of energy is to be used in a given installation.

11-3. RECIRCULATION AND YEAR ROUND OPERATION

If full recirculation for all of the furnace-warmed air throughout the heated space is employed, maximum economy will be obtained, especially in extremely cold weather. Forced-circulation warm-air systems can provide for summer operation by the addition of cooling coils so placed that chilled air can be delivered to the rooms of a building, using the same ductwork employed for heating.

In thermal-circulation systems, the vertical ducts which are built in the walls and partitions for conveying the warm air are called stacks. In cross section they can be round, but most usually they are rectangular. A stack head is the duct-fitting at the room end of a duct, and this, in turn, houses the register or grille when the location is in a wall. When the outlet is in the floor, the stack terminates in a register box. In such systems, the fitting at the bottom of a stack which connects the round leader into the vertical stack is called a boot. A leader is the duct which runs almost horizontally from the furnace to the stack.

A natural-circulation system depends for its action on a modification of chimney action whereby the pressure exerted by a vertical column of warm air is less than that exerted by a similar column of cool air. A pressure difference created in this way is sufficient to promote the required circulation of air. Theoretical analyses starting from this point and making use of the space heat loss and the required air flow, and of the friction loss in ducts, could furnish a basis for design, but they would be extremely laborious and their accuracy would be dependent on the assumptions made. Consequently, it is preferable to make use of design procedures confirmed by experience and successful designs.

The design of ducting for both natural and forced-circulation systems has been developed into simplified procedure by the National Environmental Systems Contractors and Air Conditioning Association (Refs. 1 through 6) and will not be reported here.

It should be noted that in residential structures or small buildings having design heat losses not greatly in excess of 120 000 Btuh, the temperature rise occurring in the furnace should not exceed 100 F and the static pressure, available for overcoming pressure, runs about 0.2 in. water. The furnace selected should have a bonnet capacity about 15% in excess of the design Btuh heat loss (register delivery) to take care of piping loss.

When mechanical air circulation is used, rectangular trunk-line ducts for leaders are employed almost universally, since the air velocity to each branch is independent of variations in air temperature, and since such ducts need not require a particular inclination. With trunk-line air-supply ducts there is less exterior duct surface and less heat loss than with separate, thermal-circulation supply ducts.

In general, a building is heated most satisfactorily by a forced-circulation system when the blower operates for long periods in mild weather and practically continuously in weather colder than about 40 F.

The heat input should be so controlled that in mild weather the burner operates frequently but for short periods. Frequent cycling of the source of heat, together with prolonged blower operation, insures controlled temperatures in all rooms of the building, and temperatures are maintained at nearly constant values near the floor as well as in the living zone.

With a blower under almost continuous operation, along with intermittent operation of the heat source, the temperature of the air delivered from the registers should range from about 80 F to 150 F. The registers should be located so that the air stream is never discharged directly into space normally occupied by people at rest. Deflecting-type registers control the direction of the air stream and can reduce the air velocity quickly. The use of high sidewall registers is advantageous because there is then least danger of the air stream impinging on an occupant. They are also suitable for summer cooling and do not interfere with furniture placement.

Low sidewall and baseboard registers must be more carefully located, as the problems of obstruction by furniture and impingement of air on occupants represent possible difficulties.

A ceiling register is satisfactory, provided its outlet deflects the air sideways instead of directly downward. Floor registers usually are employed only under large glass exposures.

Return Air Intakes. Return-air intakes can be located on either inside or outside walls. Where large sources of cold air exist, the intakes should be located nearby. Intake location should be near the floor level, either in the baseboard or as a floor grille, flush with the surface.

It is desirable to put return intakes in each room except bathrooms, closets, lavatories, and kitchens. The intake should be located so as to give the return air a convenient run back to the furnace, and should be so sized that excessive amounts of return air are not drawn across the floor.

11-4. WARM AIR PERIMETER HEATING

Warm-air perimeter heating is an optional approach to heating systems which use forced warm air. It can be used with almost any type of structure, either residential or industrial, but has been found to be particularly adaptable and effective for buildings without basements. It differs from conventional warm-air heating in duct arrangement and also in that the warm air is always introduced into the heated space at or near the floor along the outside walls, prefereably under windows. Moreover, the air is returned to the furnace, usually by means of high-sidewall grilles. By setting up a warm-air blanket over the cold window and wall surface, cool downdrafts are largely eliminated: the floors are characteristically warm for this reason, and also because the heat distribution system is adjacent to the floor or to part of it.

One of the simplest perimeter systems is that required for a structure without a basement, for which two representative layouts are illustrated in Figs. 11-1 and 11-2. Other variations of this are possible—as, for example, one in which the whole crawl space is enclosed and used for a plenum or pressure chamber to feed up into the outlets in the rooms. Where duct runs are made in the concrete slab around the perimeter of the building, it is

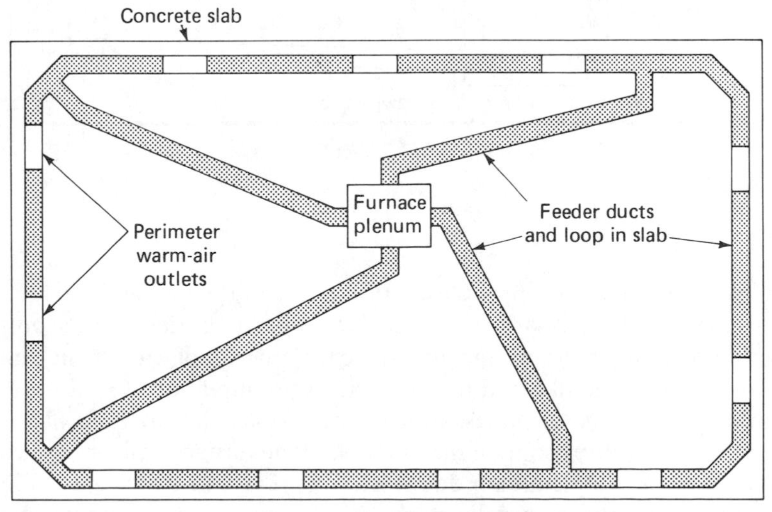

FIGURE 11-1. Perimeter loop system with feeders and loop ducts in concrete slab. (After Ref. 1.)

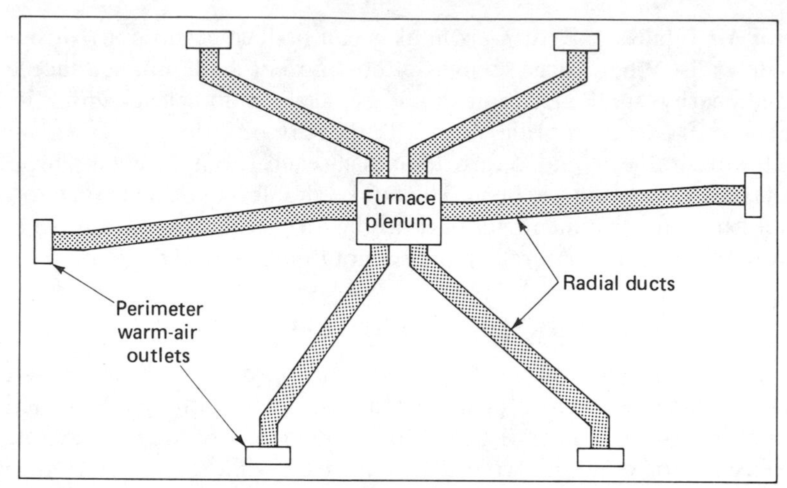

FIGURE 11-2. Perimeter radial system with feeder ducts in concrete slab or crawl space. (After Ref. 1.)

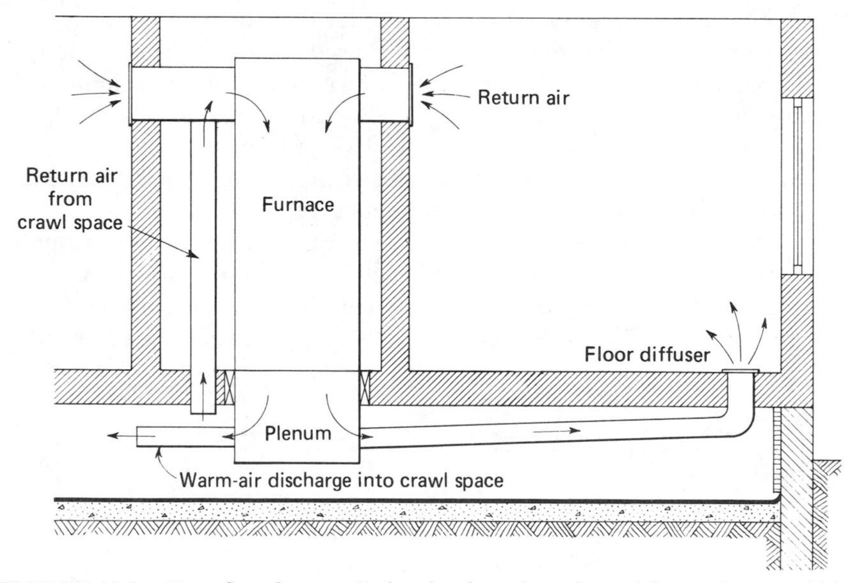

FIGURE 11-3. Downflow furnace discharging into ducts located in crawl space. (After Ref. 5.)

desirable to provide waterproof insulation between the slab and the foundation wall. Also, it is necessary to provide a waterproof barrier near the bottom of all slabs resting on the ground, to reduce the amount of water which might be drawn up into the slab and possibly cause a damp floor.

In single-story basementless homes, the furnace is usually located in a heater closet or utility room on the first floor. The furnace delivers its heated air downward into a plenum or ductwork in the floor or crawl space. Figure 11-3 shows one arrangement diagrammatically, with feed into a crawl-space plenum and ductwork. Furnaces can also be placed in attics, in which case a downcomer duct is needed to deliver the hot air into the subfloor plenum. Furnaces should be located in heater compartments or closets, with adequate insulation or cooling air space around the furnace shell. Provision must be made for combustion air grilles into the heater closet, with or without separate air ducting. The discharge-gas flue, properly encased to prevent fire hazard, is an additional requirement. In a tightly constructed building with a small air infiltration rate, a separate combustion air duct from outside is a necessity.

One simplified design system for use with residences having a crawl space or basement starts with the premise that a static pressure of 0.2 in. of static water pressure is available at the furnace and also that one or more 4-in. round ducts can be used to transmit the warm air to the room diffusers. Two arrangements that can be used in this case are the so-called individual-pipe arrangement and the extended-plenum arrangement. The plenum supply in both cases can be served either by downflow or upflow from the furnace. The air temperature rise in furnaces for this type of service ranges from 70 to 100 deg, and care must be taken to see that enough 4-in. duct outlets are provided to serve the capacity of the furnace. If under certain conditions each such outlet can serve 9000 Btuh, a 75 000-Btuh furnace would require a minimum of nine outlets. More can be used at varying capacities to serve different spaces. The furnace capacity must be sufficient to provide for the heat loss from each room, the loss from the basement or crawl space, and for outside circulation air when that is required because some or all of the delivered air cannot be returned.

In making the design for a given house, the house plan should be drawn to scale, the furnace located, and then the duct runs or plenum accurately

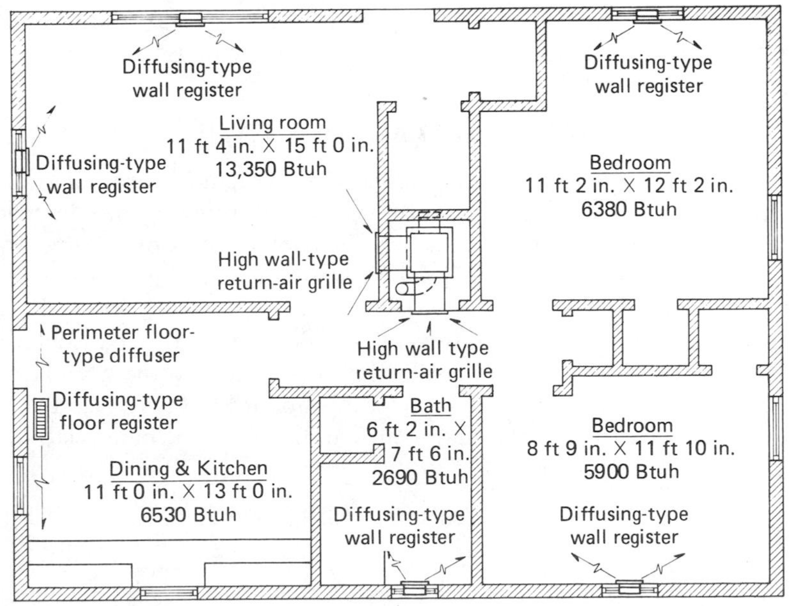

FIGURE 11-4. Floor plan of one-story building over crawl space using perimeter heating. (After Ref. 5.)

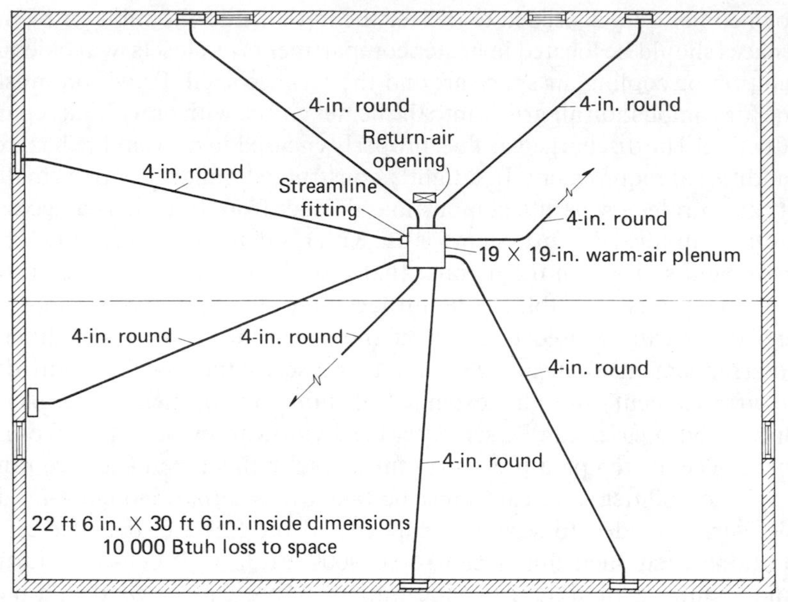

FIGURE 11-5. Crawl-space piping under one-story building using perimeter heating. (After Ref. 5.)

drawn to show the lengths to the diffusers and registers, recalling that these are to be located on the outside walls.

Figures 11-4 and 11-5 illustrate the layout of a perimeter system for a small one-story house with a crawl space beneath. This employs a down-flow furnace and two return-air runs from the house and an additional one from the crawl space. Heat loads for each part of the house are indicated.

11-5. WARM AIR QUANTITIES

In this chapter, methods have been indicated for designing warm-air systems. The heat loss of a space heated by warm air is met by bringing into the space a sufficient quantity of warm air which, in cooling to room temperature, satisfies the loss requirements. The limit setting for air temperature at the bonnet for a low-temperature air system should not exceed 200 F, but the room-air temperature at the registers must be lower than this, particularly in small rooms where the hot air could cause discomfort to the occupants. A temperature not above 150 F, and preferably about 135 F, is suggested.

The heat given up by the supply air can be expressed by a relation derived from Eq. (5-7). Thus

$$Q = (0.244)\,(d)\,(60)\,(\text{cfm})\,(t_i - 65)$$

and

$$\text{cfm} = \frac{Q}{14.7d(t_i - 65)} \tag{11-1}$$

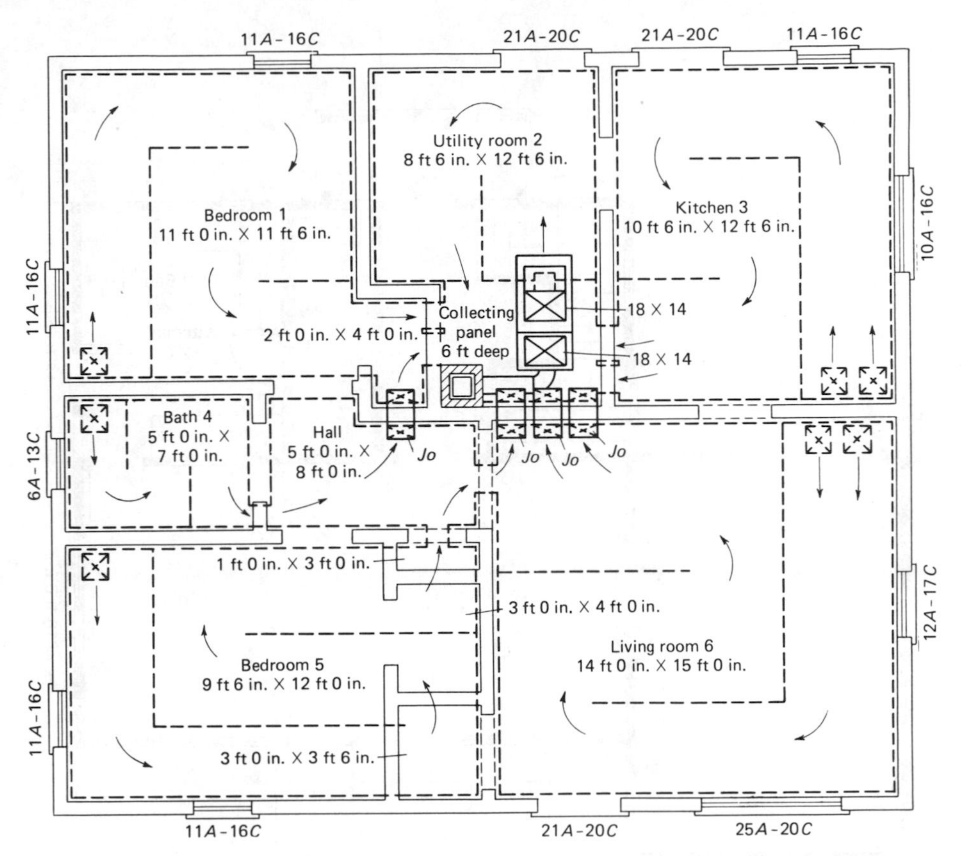

FIGURE 11-6. Layout of air-heated panel for one-story house, showing ceiling arrangement. A = full area of window (ft^2); C = running feet of window crackage; JO = jumpover. Ceiling height is 7.5 ft.

in which cfm is the air supplied to the space, in cubic feet per minute; Q is the heat loss, in Btu per hour; d is the air density at register temperature, in pounds per cubic foot; and $t_i - t_o = t_i - 65$ is the supply or register temperature minus the leaving or return-air temperature, in degrees Fahrenheit.

Example 11-1. A house has a heat loss of 50 000 Btuh. (a) With a gravity system having a 135 F register temperature and a 65 F return, find how many cubic feet per minute (measured at 65 F and 30% relative humidity) must flow through the system. (b) If the heat loss can be carried in a mechanical-circulation system at an average register temperature of 115 F and a 65 F return, find the flow at standard conditions, in cubic feet per minute. (c) How many tons of air per hour flow under the conditions in parts a and b?

Solution: (a) From the psychrometric chart (Plate I), air at 65 F and 30% relative humidity at 29.92 in. Hg has a density $d = 0.075\ \text{lb/ft}^3$. By Eq. (11-1),

$$\text{cfm} = \frac{50\,000}{(14.7)(0.075)(135 - 65)} = 648 \qquad \textit{Ans.}$$

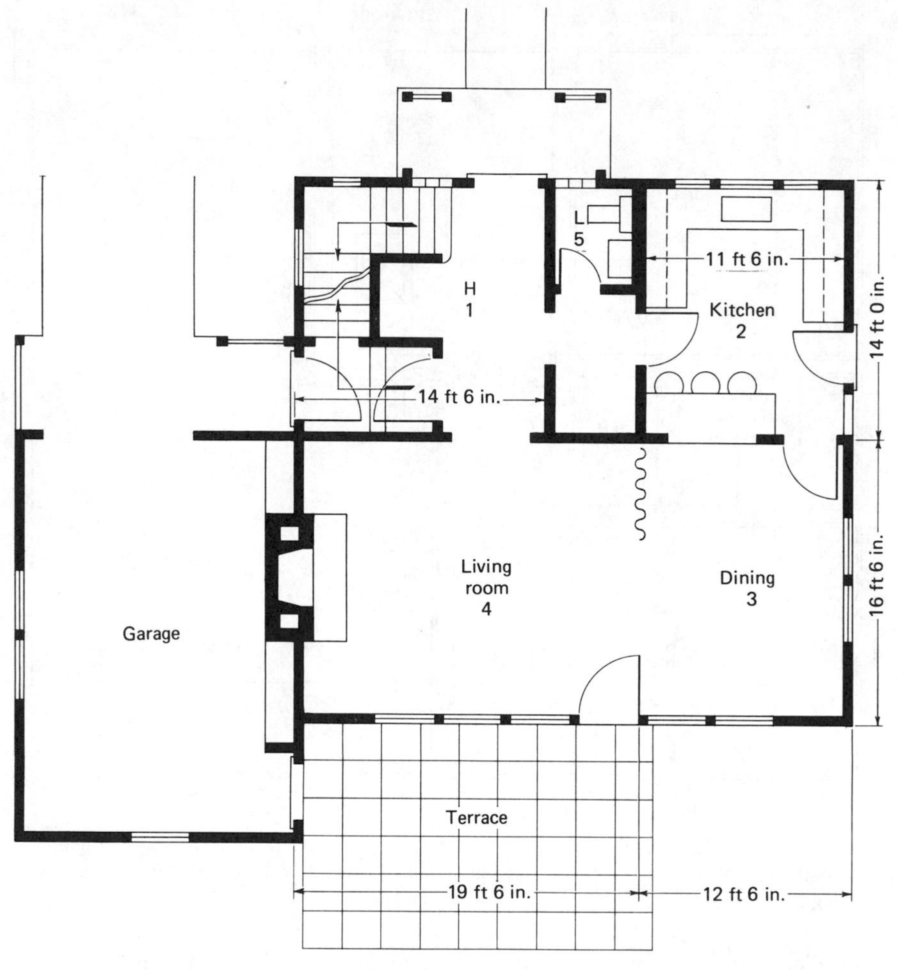

FIGURE 11-7. First-floor plan of residence for design.

Measured at 135 F, this would be closely

$$648 \times \frac{460 + 135}{460 + 65} = 734 \text{ cfm}$$

(b)
$$\text{cfm} = \frac{50\,000}{(14.7)(0.075)(115 - 65)} = 907 \qquad \textit{Ans.}$$

(c) For part a,

$$\text{tons/h} = \frac{(648)(60)(0.075)}{2000} = 1.46 \qquad \textit{Ans.}$$

For part b,

$$\text{tons/h} = \frac{(907)(60)(0.075)}{2000} = 2.08 \qquad \textit{Ans.}$$

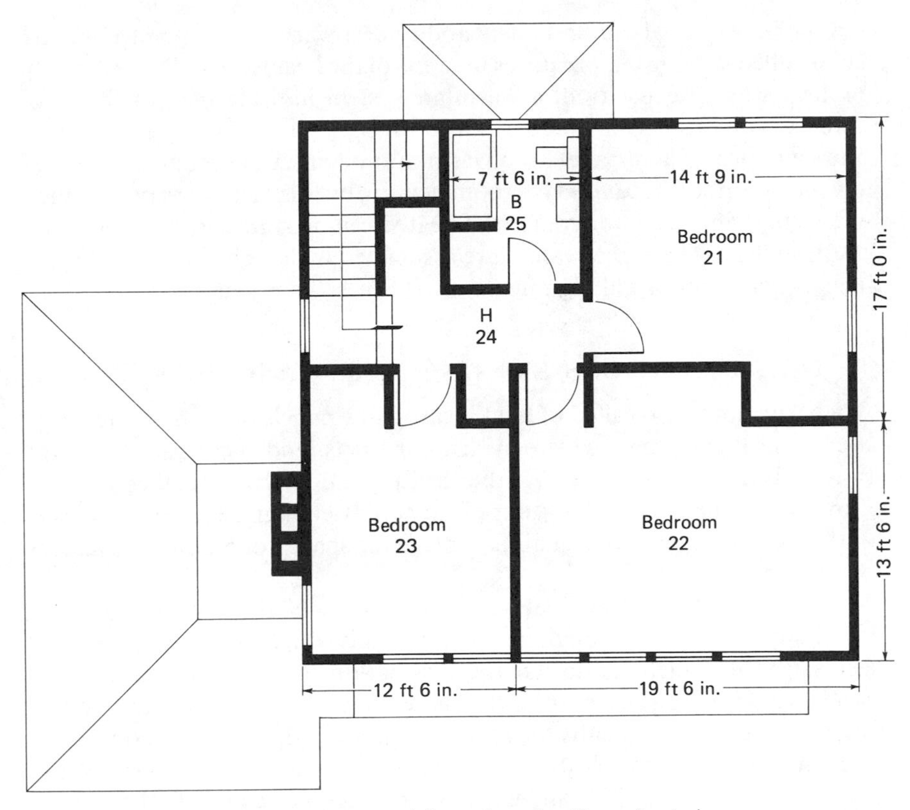

FIGURE 11-8. Second-floor plan of residence for design.

Data on Residence of Figs. 11-7 and 11-8

Walls—clapboard on $\frac{1}{2}$-in. insulating board; wood lath on studding. $U = 0.23$.

Living room—three metal casement windows; two-section, vertical axis, 37 in. total width of each, $38\frac{3}{8}$ in. high.

Other windows double-hung, wood frame, 30 in. by 48 in.; single glass.

Outside doors—front, 7 ft by 3 ft 4 in.; back and side, 7 ft by 2 ft 8 in.

First story 9 ft high; second story 8 ft 6 in.

Ceiling to unheated attic—yellow-pine flooring on joists; metal lath and plaster with $3\frac{5}{8}$-in. rock-wool fill. $U = 0.079$.

Basement warmed by furnace. The terrace side faces west.

11-6. PANELS WITH WARM AIR SYSTEMS

For certain types of installations it has been found desirable to utilize radiant heating in connection with forced-circulation warm-air systems. Most installations of this sort are arranged to operate from the ceiling of the room. The risers carry warm air through a series of passages (ducts) which are so arranged that the warm-air passages cover essentially the whole ceiling area of a room being heated.

When the system is in operation, the limit control on the furnace should be so adjusted that the air temperature does not exceed 140 F. The blower control of the furnace should start the blower in operation at 100 F and should stop the blower at about 85 F. The ability of a warm-air system to operate at controlled varying temperatures in terms of the heat load and the relatively low heat-storage capacity of a warm-air system indicate the possibility of more extended use being made of warm-air panel installations. Figure 11-6 shows a typical ceiling panel for a warm-air system. An approximate idea of how much surface is required for room heating by this method may be gained by assuming that each square foot of heated ceiling surface can supply from 50 to 100 Btuh. The higher values are associated with higher velocities in the ducts. A representative design figure of 60 Btu/h · ft^2 is suggested.

11-7. CONDITIONS EXISTING WITH PANEL HEATING

In panel heating (also called radiant heating) the panels of the building structure are heated by pipe coils or warm-air ducts, and these panels in turn transmit heat into the space. Whether heating is done by normal convection or by radiant methods, the objective is to supply enough heat to compensate for the space heat load while maintaining, in the space, conditions satisfactory for occupant comfort.

An occupant at rest dissipates approximately 400 Btuh because of metabolic body processes, although under conditions of activity the dissipation is necessarily at a higher rate. Of the dissipation loss at rest, that portion associated with moisture evaporation is relatively constant and amounts to some 100 Btuh. Consequently the remainder must be dissipated by convection to the ambient air or by radiation to cooler surfaces. In a room with an average surface temperature of 70 F and with its air temperature at 70 F, the convection loss from a human body to air is approximately 140 Btuh and the radiation loss is 160 Btuh. The net energy transferred by radiation always flows from the warmer surfaces to the cooler surfaces; and for an individual wearing conventional indoor winter clothing the average surface temperature of the body, including clothed and exposed surface, is in the neighborhood of 85 F. Thus when the average temperature of enclosure (room) surfaces is 70 F, a definite radiation loss from the body to the room surface takes place. It is naturally desirable to know how the radiation and convection heat losses vary as the air temperature and average surface temperature change from 70 F. It is difficult to predict the amount of such losses accurately for a change in temperature of more than a few degrees; but in a narrow range (± 8 deg) above and below 70 F, it has been found that each 1-deg rise in the average enclosure temperature decreases the radiant heat loss by very nearly the same amount as a 1-deg decrease in air temperature increases the convection heat loss. This statement can be formulated as follows:

$$\frac{t_a + t_e}{2} = 70 \tag{11-2}$$

where t_a is the air temperature, in degrees Fahrenheit; t_e is the average surface temperature of the enclosure, in degrees Fahrenheit; and 70 is the operative temperature for office and residence conditions, in degrees Fahrenheit. Thus if $t_a = 68$ F and $t_e = 72$ F for one case, whereas $t_a = 73$ F and $t_e = 67$ F for another case, the conditions in both cases are equivalent for body heat loss or body comfort.

For a store with continuous activity, an operative temperature of 66 should be substituted for the 70 in Eq. 11-2; for a manufacturing plant with production activity, 64 should be used for the operative temperature.

11-8. PANEL LOCATION

Panels are placed in the floor or in the ceiling but rarely in the side walls. Although panels are thought of as radiant heat devices, a large portion of the energy supplied from the heating source to the panel is actually delivered into the enclosure by convection methods. It is customary to consider the fraction of the heat transferred by convection as approximately 30% from a ceiling panel, 40% from a wall panel, and 48% from a floor panel. Because of the large convection loss from floor panels and because for occupancy requirements the floor should not have a surface warmer than 85 F in walking spaces, interest is being directed to greater use of ceiling panel surface. Panels in this location can be operated to 115 F without discomfort if the average mean radiant temperature is not too high—say, 77 to 78 F or less.

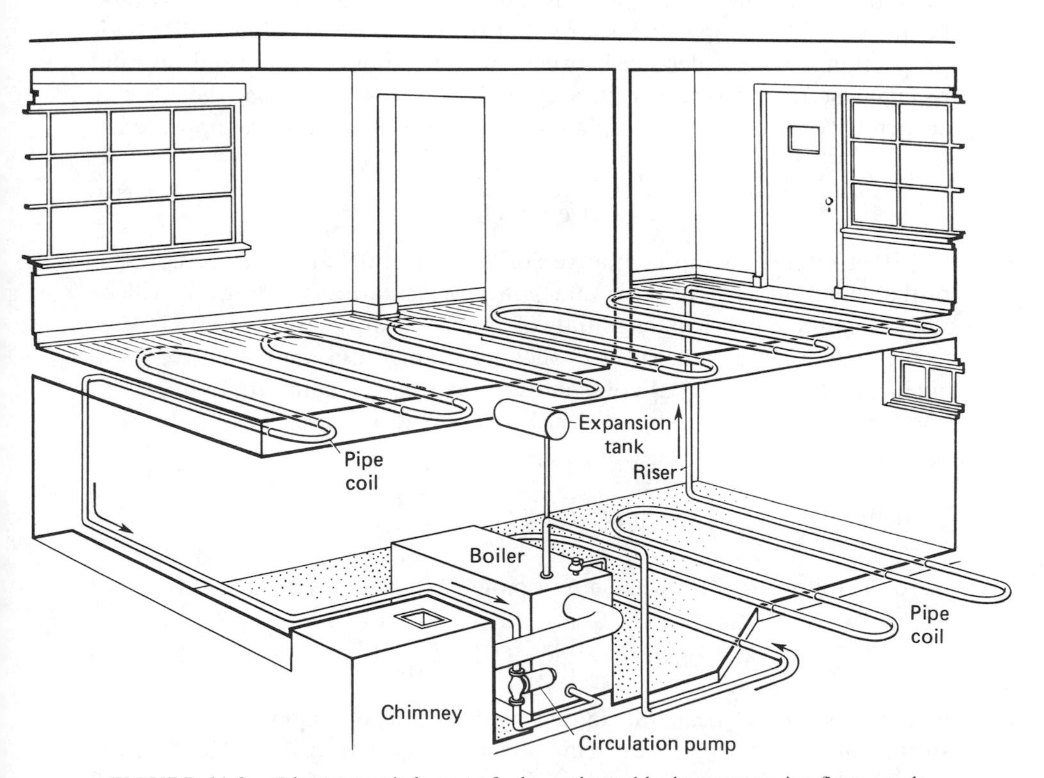

FIGURE 11-9. Diagrammatic layout of a house heated by hot water using floor panels.

Figure 11-9 is a diagrammatic layout of a hot-water panel system, with the floors of the building structure holding embedded pipe coils. This diagram illustrates the general principles of heat circulation and is not intended to be used as a detailed installation guide. Figure 11-6 shows a representative layout for a warm-air ceiling-panel system for a one-story house. In this arrangement the warm air is delivered into the attic through a centrally located trunk riser, from which ducts lead over to the seven inlets that supply air to the passage spaces between the true room ceiling and a second continuous closure directly under the joists. The warm air flows in this inner space, following guided passageways as indicated in the illustration, and finally returns to the furnace through the six return ducts shown. The passage space between the ceiling and the gypsum (rock) lath under the joists should be $3\frac{1}{4}$ in. deep, and insulation of not less than 2 in. of rock wool or its equivalent must be provided over the passage. In fact, insulation should always be provided on the back side of panel surfaces in order that the uncontrolled effect of the panel heat will not pass into spaces above or below that which the panels serve.

11-9. PANEL OUTPUT

To find the heat output of a panel by radiation effects alone, reference should be made to Eq. (4-34), which shows that the heat transferred by radiation is proportional to the difference between the fourth powers of the emitting and receiving sources. Moreover, in Section 4-10 it was stated that the effective absorptivity factor for parallel planes and large enclosed surfaces appeared as

$$e = \frac{1}{(1/e_s) + (1/e_r) - 1} \tag{11-3}$$

If, in Eq. (11-3), representative emissivities of 0.9 are used to represent probable surface absorptivity values in representative buildings, it will be found that the numerical magnitude of Eq. (11-3) works out to be 0.82. Also, in the case of large plane surfaces such as are met in a representative room enclosure, the geometrical configuration factor F_a is equal to 1, and when $e = 0.82$ and $F_a = 1$ are inserted in Eq. (4-34), it is found that it reduces to the form

$$q'_r = 0.142A\left[\left(\frac{T_s}{100}\right)^4 - \left(\frac{T_r}{100}\right)^4\right] \tag{11-4}$$

or to the following form when a unit panel area (1 ft^2) is considered:

$$q_r = 0.142\left[\left(\frac{T_s}{100}\right)^4 - \left(\frac{T_r}{100}\right)^4\right] \tag{11-5}$$

where q_r is the net heat transmitted by radiation per square foot of panel surface, in Btuh, when T_s is the panel surface temperature, in degrees Fahrenheit absolute ($t_s + 460$), and when T_r is the temperature of absorbing wall

Table 11-1 Heat Delivery by Radiation from Panels (Btu/h · ft²)

Temperature of Panel Surface	Unheated Mean Radiant Temperature of Room (UMRT)								
	50 F	55 F	60 F	65 F	70 F	75 F	80 F	85 F	90 F
60 F	8	4	...	...	...	...	...	...	...
70 F	16	12	8	5	...	...	...	...	...
80 F	26	22	17	13	9	5	...	...	...
85 F	30	26	22	18	14	10	5	...	...
90 F	35	31	27	22	18	14	10	5	...
100 F	44	40	36	32	28	24	19	15	10
105 F	49	45	41	37	32	28	24	20	15
110 F	53	50	46	42	37	33	28	24	20
115 F	60	55	52	47	43	40	35	31	16
120 F	65	62	57	53	50	45	40	37	32

surfaces, usually considered as the wall *unheated mean radiant temperature* (UMRT), in degrees Fahrenheit absolute.

This equation is not rigorously precise because of the assumptions involved. However, under most conditions the radiant output of a room panel computed by this formula is correct to within 8 to 12%. Table 11-1 gives radiant heat delivery from a panel, expressed in terms of panel surface temperature and average temperature of surroundings and facing surfaces, the so-called unheated mean radiant temperature (UMRT).

Heat transfer by convection per unit area from a flat surface to air moving by natural methods takes the general form

$$q_c = f(t_s - t_a)^n \tag{11-6}$$

where q_c is the heat transferred by convection, in Btu/h · ft² of panel surface; t_s is the surface temperature of the panel, in degrees Fahrenheit; t_a is the representative bulk temperature of air adjacent to the panel, in degrees Fahrenheit; n is an exponent whose value, although related primarily to plate location, is also affected by temperature difference (for heat flow upward from horizontal surfaces, $n = 1.12$; for heat flow downward, as for example from ceiling panels, $n = 1.25$); and f is the coefficient of surface conductance, in Btu/h · ft² · F temperature difference between surface and air (the value of f varies with panel location; it can be taken as 0.81 for upflow panels and 0.22 for downflow panels). For floor panels, Eq. (11-6) becomes

$$q_c = 0.81(t_s - t_a)^{1.12} \tag{11-7}$$

and for ceiling panels,

$$q_c = 0.22(t_s - t_a)^{1.25} \tag{11-8}$$

Table 11-2 gives values computed from Eqs. (11-7) and (11-8) for a range of temperature differences.

Table 11-2 Heat Delivery by Convection from Panels to Air ($Btu/h \cdot ft^2$)

Temperature Difference between Panel Surface and Air (deg F)	Panel Location		
	Ceiling	Wall	Floor
5	1.8	3.0	5.0
10	4.0	6.8	11.0
15	6.8	11.2	17.5
20	9.2	16.0	23.0
25	12.5	21.2	31.0
30	16.0	26.7	40.0
35	18.9	32.6	52.0
40	22.4	38.4	...
45	25.7	44.6	...
50	29.0	50.4	...
55	33.0	56.7	...
60	36.6	62.4	...

Example 11-2. For a ceiling panel with a surface temperature of 100 F, find the heat delivery by radiation and by convection into a room having a UMRT of 60 F and an air temperature of 76 F under the ceiling.

Solution: By Eq. (11-5), the heat addition by radiation from this panel is

$$q_r = 0.142\left[\left(\frac{460 + 100}{100}\right)^4 - \left(\frac{460 + 60}{100}\right)^4\right]$$

$$= 35.9 \text{ Btu/h} \cdot \text{ft}^2$$

By Table 11-1, the corresponding value is 36.

By Eq. (11-8), for the heat added by convection,

$$q_c = 0.22(100 - 76)^{1.25} = 11.7 \text{ Btu/h} \cdot \text{ft}^2$$

By Table 11-2, the corresponding value is found to be 11.8.

In a room or enclosure the surface temperature is related to the overall wall resistance R_w, to the temperatures on both sides of the wall, and to the inside film factor R_f. For most cases the film factor of the inside wall is considered to be 1.65 and the corresponding resistance is 0.606, or approximately 0.6. By making use of resistance concepts, the inside surface temperature can easily be found (see Section 4-4). Designate the outside design temperature as t_o and the inside air temperature as t_a; then, as indicated previously in Eq. (4-27)

$$\frac{\Delta t_f}{R_f} = \frac{t_a - t_o}{R_w}$$

and the temperature drop through the air film Δt_f is

$$\Delta t_f = (t_a - t_o)\left(\frac{R_f}{R_w}\right) \tag{11-9}$$

or

$$\Delta t_{\text{film}} = (t_a - t_o)(R_{\text{film}})(U_{\text{wall}}) \tag{11-10}$$

And when $t_a = 70$ F, and $R_f = 0.6$,

$$\Delta t_f = (70 - t_o)\frac{0.6}{R_w} = \frac{42 - 0.6t_o}{R_w} \tag{11-11}$$

Then the wall surface temperature t_w is

$$t_w = t_a - \Delta t_f \tag{11-12}$$

and, for $t_a = 70$ F and $R_f = 0.6$,

$$t_w = 70 - (70 - t_o)\frac{0.6}{R_w}$$

$$= 70 + \frac{0.6t_o - 42}{R_w} \tag{11-13}$$

Example 11-3. An exposed side of a room is 16 by 8 ft and has two 3- by 5-ft single-glazed windows. The wall has an overall coefficient of heat transfer U, 0.25 Btu/h · ft² · F, considering film effects; and for the glass, $U = 1.13$. When it is 0 F outside and at 70 F design inside air temperature, what is (a) the surface temperature of the wall and of the glass and (b) the average weighted surface temperature of the exposed wall?

Solution: (a) The resistance of the wall is

$$R_w = \frac{1}{U} = \frac{1}{0.25} = 4.0\,\frac{\text{h} \cdot \text{ft}^2 \cdot \text{F}}{\text{Btu}}$$

and the resistance of the glass is

$$R_g = \frac{1}{1.13} = 0.885$$

For the wall, by Eq. (11-13),

$$t_w = 70 + \frac{(0.6)(0) - 42}{4.0} = 59.5 \text{ F} \qquad \textit{Ans.}$$

and for the glass,

$$t_w = 70 + \frac{(0.6)(0) - 42}{0.885} = 22.6 \text{ F} \qquad \textit{Ans.}$$

(b) Of the 16 by 8 ft = 128 ft² of surface, there are 2(3 × 5) = 30 ft² of glass, and thus the remaining 98 ft² is wall surface.

We then make use of the areas and their respective temperatures to find

$$t_{\text{avg}} = \frac{(98)(59.5) + (30)(22.6)}{128} = 50.8 \text{ F} \qquad \textit{Ans.}$$

11-10. PANEL HEATING CALCULATIONS

1. In the layout of a panel heating system it is first necessary to compute the design heat loss from the room or space. In computing the heat load, one uses the same method that is used in computing the heat load for a convection heating system. Allowance for air change should also be made, with at least one air change per hour being considered. Although it is possible to operate radiant systems at air temperatures somewhat below 70 F, from a viewpoint of design 70 F is employed except for unusual applications.
2. After the approximate location and extent of the heating-panel area have been selected from a plan of the space, the unheated mean radiant temperature (UMRT) of the room surface should be found. The UMRT represents an average temperature for the unheated surfaces of the enclosure, with each surface weighted according to its area and temperature. Whenever the surfaces of inside walls, floors, and ceilings are not exposed to unheated space on their far side, the surface temperature is usually considered to be 70 F, although this figure should be modified for the case of adjacent spaces definitely operated at lower design temperatures. The method of finding the surface temperature of an outside (exposed) wall was discussed in Section 11-9.
3. Two approaches can be made to finding the panel surface. One is to select the maximum or desired surface temperature which might be used for the panel in question. This temperature is, of course, related to whether the panel is in the floor, the ceiling, or a side wall. Then, in terms of this surface temperature and the UMRT, Table 11-1 is used to find the radiant heat output; and, employing the respective surface and air temperatures, Table 11-2 is used to find the convection heat output. The sum of these two items represents the total heat output per square foot of panel surface. This unit figure can now be used in connection with the total room heat loss to find the required panel area. If this area is larger than the space available in the ceiling or floor, it is then necessary either to raise the operating temperature of the panel surface or, if this is not possible, to supply supplementary heat from another source or from an additional panel surface in another location.

The other approach is to make use of tabular data based on confirmed field and test data and listed here in Table 11-3. The manufacturer's literature also contains a variety of data of this type for use in design.

4. The coil layout and spacing should be designed to produce the desired surface temperature and heat output. The necessary flow through the system is then computed.

Figure 11-10 shows a few of the arrangements which are employed in using copper-tubing or steel-pipe coils for panel heating. For floor and structural slabs, $\frac{3}{4}$-, 1-, or $1\frac{1}{4}$-in. sizes, of steel or wrought-iron pipe, and $\frac{3}{4}$-in.

Table 11-3 Approximate Heat Delivery of Panels Serving Spaces Designed for Normal Comfort Occupancy (Btuh per Pane) with Tubes at Various Center to Center Distances)

Panel Area* (sq ft)	Location and Surface Temperature			
	Ceiling at 100 F	Ceiling at 110 F	Ceiling at 120 F	Floor at 85 F
	Tube Spacing and Approximate Average Water Temperature			
	$\frac{3}{8}$-In. Tube 4 In. C to C at 121 F 6 In C to C at 133 F	$\frac{3}{8}$-In. Tube 4 In. C to C at 137 F 6 In. C to C at 153 F	$\frac{3}{8}$-In. Tube 4 In. C to C at 155 F	$\frac{1}{2}$- and $\frac{3}{4}$-In. Tube 9 In. C to C at 120 F 12 In. C to C at 130 F
100	5000	6500	8000	4000
200	10000	13000	16000	8000
500	25000	32500	40000	20000

*** For other panel areas, prorate the base values listed.**

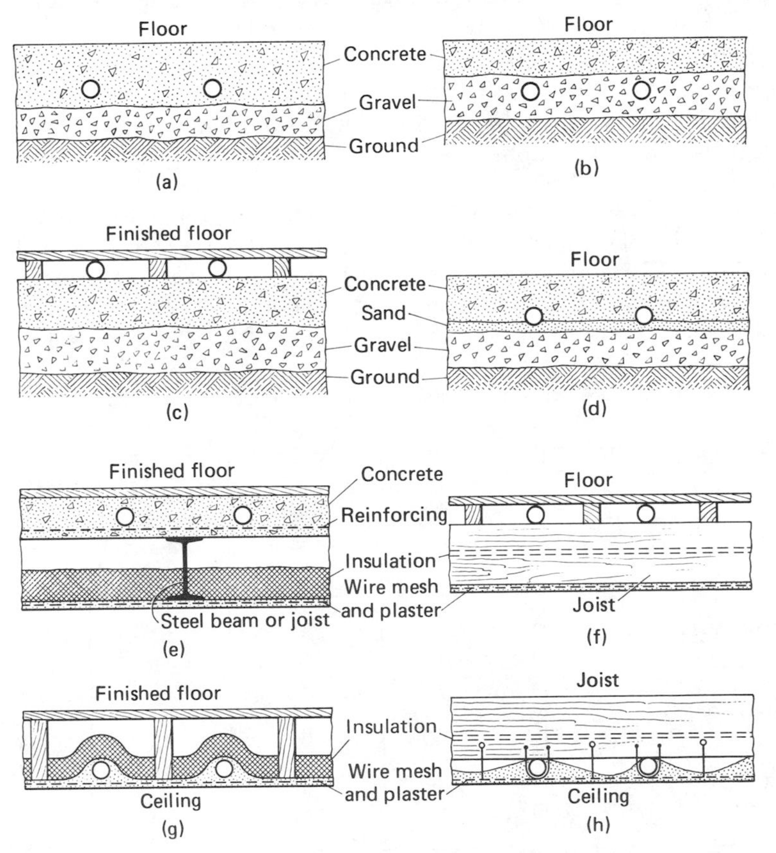

FIGURE 11-10. Typical pipe-coil arrangements of radiant-heating coils installed in floor or ceiling.

or 1-in. sizes of copper water tubing, are most common. For ceilings, $\frac{3}{8}$-in. or $\frac{1}{2}$-in. copper tubing, or $\frac{1}{2}$-in. steel (or wrought-iron) pipe sizes are most common. The permissible top water temperatures used are determined, to a large degree, by the necessity of limiting the surface temperature below values which could cause discomfort to occupants and, to a lesser degree, by the necessity of preventing damage to the plaster, concrete, or other surface employed. During the initial drying-out (stabilizing) period, care must be exercised to keep temperatures below 90 F for at least 48 h after the plaster or concrete has set.

Surface temperatures for *floor panels* should hardly ever exceed 85 F, with 80 F being preferable for many installations. Water temperatures in floor panels reach 120 F for conventional design, and go to 130 F for an acceptable maximum. For *ceiling panels* the height of the room and the room arrangement dictate the surface temperatures acceptable. Up to 120 F is considered possible; but for residential application, temperatures of 110 F or below are usually preferable. Top water temperatures of from 155 F to 165 F are used with ceiling panels. *Wall panels* are governed by considerations similar to those which hold for ceiling panels if the wall panels are built high in the side wall; if low, they are in a category similar to floor panels in regard to temperature effects on occupants. Top water temperatures employed range from 150 F to 160 F. Table 11-4 gives approximate output values for panels using copper pipe when operated with water temperatures in the range indicated in this paragraph.

Example 11-4. Compute the unheated mean readiant temperature (UMRT) for the combined living-dining room of the residence in Figs. 11-7 and 11-8. The overall U for the walls is 0.23, and the windows are single glass. The outside design temperature is −10 F, and the assumed air temperature is to be taken as 70 F. Take all inside walls as being at 70 F, but find the temperature of each of the other surfaces before starting the

Table 11-4 Tube Spacing and Panel Output for Copper-Water-Tube Panels at High Ratings

Panel Position	Nominal Tube Size (in.)	Tube Center-to-Center Spacing	Panel Output (Btu per hr sq ft)	
			Nominal Design	Maximum Output
Ceiling	$\frac{3}{8}$	$4\frac{1}{2}$	65	75
	$\frac{3}{8}$	6	50	60
	$\frac{3}{8}$	9	35	40
Wall	$\frac{3}{8}$	$4\frac{1}{2}$	65	75
	$\frac{3}{8}$	6	50	60
	$\frac{3}{8}$	9	35	40
Floor	$\frac{1}{2}$ or $\frac{3}{4}$	9	40	50
	$\frac{3}{4}$ or 1*	12	40	50

* Because of the low surface temperature for floor panels (80 F to 85 F) which can be tolerated, the coil output is not limited by tube size.

computation. Exclude the ceiling in this calculation, as it is desired to add a panel surface there. The basement is heated.

Solution: The dining room has a gross outside wall area of

$$12.5 \times 9 + 16.5 \times 9 = 261 \text{ ft}^2$$

and a glass area of

$$(2.5 \times 4)(4) = 40 \text{ ft}^2$$

There are 221 ft^2 of net outside wall area in the dining room.

The living room has a gross outside wall area of

$$(19.5 + 16.5)(9) = 324 \text{ ft}^2$$

and a glass and door area of

$$\frac{37 \times 38.375}{144}(3) + (2.67)(7) = 48.3 \text{ ft}^2$$

The net outside wall area of the living room is 275.7 ft^2 (note that the room has two exposed sides, since the garage is largely at a lower level than the room). Also,

$$\text{floor area} = 16.5(12.5 + 19.5) = 528 \text{ ft}^2$$
$$\text{inside wall area} = 9(12.5 + 19.5) = 288 \text{ ft}^2$$

The inside surface temperature of the outside wall can be found by Eqs. (11-10) and (11-12). Recall that the inside film factor is $f_i = 1.65$ and $R_{\text{film}} = 1/1.65 = 0.606$. Then, (11-12)

$$\Delta t_{\text{film}} = [70 - (-10)](0.606)(0.23) = 11.5 \text{ deg}$$

and the wall surface temperature is therefore

$$70 - 11.5 = 58.5 \text{ F}$$

For the glass surface temperature,

$$\Delta t_{\text{film}} = [70 - (-10)](0.606)(1.13) = 54.8 \text{ deg}$$

and the glass temperature is therefore

$$70 - 54.8 = 15.2 \text{ F}$$

A summary of the results is shown in the accompanying tabulation. Thus the unheated mean radiant temperature (UMRT), exclusive of the ceiling and its panels, is 62.4 F.

Surface Location	Net Area (ft^2)	Surface Temperature (F)	Area Times Surface Temperature
Outside area of dining room	221	58.5	12 930
Outside area of living room	275.7	58.5	16 130
Glass area of dining room	40	15.2	608
Glass area of living room	48.3	15.2	734
Floor area	528	70	36 960
Inside wall area	288	70	20 160
TOTAL	1401.0	[62.4]	87 522

Example 11-5. For the combined living-dining room of the residence illustrated in Figs. 11-7 and 11-8 and considered in Example 11-4, prepare a design using (a) ceiling-panel surface and (b) coils imbedded in the floor slab.

Solution: The heat loss from this space, computed by conventional methods and allowing for one air change per hour, was found to be 23 920 Btuh with 70 F inside air temperature and −10 F outside air temperature.

(a) Assume that the ceiling panel surface temperature will be held at 100 F for design conditions. Since the UMRT from Example 11-4 is 62.4 F, the radiant heat output, by Table 11-1, is 34 Btu/h · ft². Table 11-2, for 100 F − 70 F, or 30 deg difference, shows the convection output to be 16.0. Thus there is a total output of 34 + 16 = 50 Btu/h · ft² of heated ceiling panel. The panel area needed can now be found as

$$(A_p)(50) = 23\,920$$

and

$$A_p = 478 \text{ ft}^2$$

This panel area requires the use of a large portion of the ceiling area (528 ft² available), but the requirement is not excessive. The mean radiant temperature can now be computed with the help of the tabulated data which is already available in Example 11-4 and to which the ceiling data can now be added:

$$\begin{array}{rcrcr} 1401 & \times & 62.4 & = & 87\,522 \\ 478 & \times & 100 & = & 47\,800 \\ 50 & \times & 70 & = & 3\,500 \\ \hline 1929 & \times & [\ 72.0] & = & 138\,822 \end{array}$$

Thus the mean radiant temperature of the heated room under design heat loading is 72 F. This is a satisfactory temperature and it might be possible and even desirable to reduce the air temperature to slightly below 70 F.

(b) For the floor panel, if 80 F is selected as the surface temperature for a UMRT of 62.4 F, Table 11-1 shows a radiant output of 15.5 Btu/h · ft² and, for a 10-deg floor-to-air temperature difference, Table 11-2 shows a convection heat delivery of 11.0. The total is thus 26.5 Btuh/ft², which is too small to carry the design heat loss. If 85 F is tried, the corresponding figures are 20.0 and 17.5, or 37.5 Btu/h · ft² of floor area. For this,

$$(A_f)(37.5) = 23\,920$$

and thus

$$A_f = 637 \text{ ft}^2$$

Thus more panel surface is required than the available floor area of 528 ft². A floor panel, then, will not suffice unless supplementary heat from another source is provided—or unless the floor temperature is raised above 85 F, which is not a desirable solution.

As the ceiling panel design is satisfactory, reference to Table 11-4 shows that the design conditions can be met using $\frac{3}{8}$-in. copper water tubing spaced on 6-in. centers. A similar result could also be reached using $\frac{1}{2}$-in. nominal-size steel pipe on 8-in. centers, with possibly a slightly higher water temperature employed.

11-11. WATER REQUIREMENTS FOR PANELS

In previous sections, methods were developed for finding the required panel area to offset the heat loss of a room and to maintain desired mean radiant and air temperatures in the space. It is obviously necessary to circulate adequate amounts of water at proper mean temperatures, in order to hold the panel surfaces at desired temperatures. The amount of water and its allowed temperature drop must be sufficient not only to offset the heat losses of a space but also to offset the heat loss through the far side of the panel—as, for example, from a floor panel into the ground, or from a ceiling panel into a room above or into an unheated attic. The latter type of loss is called the reversed heat loss. To provide for reversed heat loss, up to 10% additional input energy to the panel should be allowed in the case of ceiling panels, and with floor panels up to 25% should be allowed.

For a given room design, when the heat loss is increased by an adequate factor to compensate for reversed loss, water calculations can then be made. Call the heat loss from the panel, with its associated reversed loss, Q_p. The water supplied to the panel in cooling must exactly balance this loss. Thus we can write

$$\begin{aligned} Q_p &= 60WC_p(t_{\text{in}} - t_{\text{out}}) = 60W(1)(t_{\text{in}} - t_{\text{out}}) \\ &= (60)(\text{gpm})(8.1)(t_{\text{in}} - t_{\text{out}}) \\ &= 490(\text{gpm})(t_{\text{in}} - t_{\text{out}}) \end{aligned} \tag{11-14}$$

where Q_p is the total required input from water, including reversed heat loss, in Btu per hour; W is the water flow, in pounds per minute; gpm is the flow equivalent to W, in gallons per minute; t_{in} is the temperature of water in, in degrees Fahrenheit; and t_{out} is the temperature of water out, in degrees Fahrenheit.

With the required pump flow to the coil determined, it is next necessary to find whether one or more than one circuit will be required for the panel in question. Two factors enter into this consideration: (1) the frictional loss of the water in flowing through its circuit and (2) whether it is possible to maintain the desired temperature difference and mean water temperature

Table 11-5 Approximate Length of Tubing per Square Foot of Panel Surface for Various Center-to-Center Spacing of Tubes

Spacing (in.)	Length (ft)
4	3.0
$4\frac{1}{2}$	2.7
6	2.0
8	1.4
9	1.3
12	1.0

in the flow circuit. Table 11-5 furnishes the approximate length of tubing required to serve a given panel area. For example, if a panel requiring 500 ft^2 of surface had tubes spaced on 6-in. centers, the factor 2.0 from Table 11-5 would show that $2 \times 500 = 1000$ lin ft of tubing are required. Unquestionably, such a length is too long to put into one circuit—particularly if the tubing is small, such as $\frac{3}{8}$ or $\frac{1}{2}$ in. In general, circuit lengths should not exceed 200 ft from header to header unless tubing $\frac{3}{4}$ in. or larger is employed. For ceiling panels, it is generally considered that with $\frac{3}{8}$-in. tubing, circuits should not greatly exceed 120 to 140 ft with standard pumps, or 160 to 200 ft with high-head pumps. For floor panels with larger size tubing, the corresponding figures might be 200 ft for standard pumps and up to 350 ft for high-head circulating pumps.

Coils are most usually sinuous in character, looping back and forth. However, sometimes headers on each side of a panel are used with tubing running across the space. It is always advisable to insulate the reverse side of a coil, not only to control the reversed heat loss but also to avoid the possibility of undesired and uncontrolled heat leakage into adjacent spaces. In laying out the coils, care must always be taken to see that adequate vent points are provided so that air caught in the system can be vented and not interfere with the free flow of water in the various circuits.

11-12. PANEL HEATING CONTROL

Radiant heating can be controlled by ordinary thermostatic controls, but under certain conditions the space may be overheated or underheated. This condition is particularly prevalent, with floor coils embedded in heavy masonry slabs, during periods of rapid fluctuation in outside temperature. It should be realized that, as the outside temperature falls, a radiant panel can provide only for the additional heat load required by the space by means of an increase in surface temperature. The end result is to bring the room temperature up to the desired point but at the expense of having a panel temperature which is so high as to cause overheating and discomfort. With a thermally heavy structure, even after the thermostat has cut off the heat supply, the energy stored in the structure will maintain the space temperature for a delay period and prevent the heat supply from coming on sufficiently soon, producing a period of low-temperature discomfort. Consequently, with a simple thermostat and a heavy masonry-panel structure, "hunting" will result. On the other hand, with a light panel system (small thermal storage), a simple indoor thermostat with some manual adjustment available can be perfectly satisfactory.

Because of thermal lag it is necessary, in a fully automatic control system, that provision for change in the temperature of the panel be anticipated before the space conditions change, if the space conditions are to be held essentially constant. As the space internal load does not greatly vary, a control responsive to change in outdoor temperature might be used to actuate a change in panel temperature. One commercial control system employs an outdoor bulb which in turn sets the control-point limits of a heating-medium bulb. Thus the panel water temperature is set at predetermined values and the heating medium is controlled in response to outside temperature rather than the temperature of the space itself. Actually, a room thermostat may be superposed on a system like that just described, to take care of zone problems or to make desired modifications.

11-13. SNOW MELTING

Active interest exists for using panels to melt snow from roadways, pavements, runways, and the like. Most snow-melting systems of this type consist of ferrous or nonferrous coils embedded in the concrete. Instead of water, an antifreeze mixture, frequently ethylene glycol, is employed in the snow-melting circuit. The antifreeze mixture in turn is warmed through a steam or hot-water heat exchanger. The problem with these devices is to maintain the surface at a sufficiently high temperature, at least above 32 F (0°C), so that snow falling on the surface will melt, later to run off or evaporate.

11-14. PANEL MATERIALS AND PRACTICES

In this chapter we have developed methods for use in designing radiant (panel) heating systems. However, it should be mentioned that throughout the industry there is a divergency of opinion as to the best overall design methods for all conditions. This is true not only in design but also in the selection and installation of controls. Many successful installations using the design methods indicated in this chapter are in use.

Both copper tubing and steel (or wrought-iron) piping are used for panels. Corrosion is not a serious problem, since the water in the system is continuously recirculated, and when, during the process of recirculation, it has once been deoxygenated, the basis for corrosion disappears. The coefficient of linear expansion of steel or iron (0.0000122 m/m · °C) is close to that of concrete or plaster. The expansion coefficient for copper is somewhat higher, but for the temperature ranges employed the cracking of plaster because of unequal expansion is rare for any of the metals unless a very sudden heating (temperature rise) is imposed on the panel. Surface coverings over panel surfaces do not greatly limit output unless the coverings are of high insulating capacity. Rugs over panels do reduce the output somewhat. The radiation emissivity of woods, rugs, plaster, painted surfaces, etc.—in fact, of almost everything except polished metal—lies in a range of 0.85 to 0.90, and the type of surface is not too significant from a radiation viewpoint.

Initial costs of panel heating systems are usually somewhat higher than costs of most other good heating systems, but in many cases the difference is not great. A study made of typical residences showed hydronic panel-system costs ranging from 6.8 to 11.0% of the cost of the residence. In contrast, conventional systems run from about 5 to 10%.

The electrically heated panel is finding increasing usage as a satisfactory heating method. Here the wiring circuits are usually plastered close to the ceiling surface. The surface temperature, held at some 113 F (45°C) for maximum load conditions, does not produce cracking nor interfere with surface finish. Simple thermostatic controls in each room can be adjusted to provide a high degree of comfort for the occupants of each individual space.

PROBLEMS

11-1. In a gravity warm-air heating system a room on the second floor of a certain residence has a computed heat loss of 8000 Btuh. The boot to its riser is 12 ft from the bonnet of the furnace. Tabular design data show that a 10-in. round leader, a 14- by $3\frac{1}{4}$-in. stack, and a 12- by 8-in. sidewall register could deliver sufficient air at 135 F with a 65 F return to heat the space. Compute the necessary flow into the room in cfm, measured at 135 F.

11-2. Tabular data for the room on the first floor, corresponding to the second-floor room of Problem 11-1, also show a heat loss of 8000 Btuh. Tabular design data show that a 12-in. round leader and a 13- by 11-in. side wall register are required. Explain why larger sized connections are needed for the first floor as compared to the second for the same air flow and heating load.

11-3. In the residence of Figs. 11-7 and 11-8 the computed heat losses in Btuh for certain design conditions in the different rooms are as follows, by room number: (1) 7200; (2) 6400; (3) 7710; (4) 9400; (5) 1548; (21) 7725; (22) 9373; (23) 6631; (24) 5196; (25) 2195. One of the flues at the left of the living room is the furnace chimney. Place the furnace in the basement near the easterly chimney of the fireplace and using this as a starting point, graph the position a warm-air feeder trunk duct might take. Also sketch in the feeder and riser ducts from the trunk for heating the various rooms on both the first and second stories. Locate the outlets under or adjacent to windows on outside walls, and more than one may be required for large rooms. Also sketch in a trunk return duct with connections to all major rooms. One central return should suffice for the hall area.

11-4. Consider that the residence of Figs. 11-7 and 11-8 is to be served by an extended plenum (large trunk duct), with a small-pipe warm-air perimeter system. The furnace is located under the living room close to the innermost chimney, which rises alongside the fireplace. The plenum runs outward under the living room and dining room. From the plenum the small pipes for the different rooms of each floor extend to the outer walls. (a) Make a rough sketch of the house to scale, locate the furnace, the plenum, and the individual 4-in. individual supply runs for the rooms on each floor. Returns should be located on the inner walls but not necessarily one to each room.

11-5. Compute, in cubic feet per minute, the amount of air supplied for heating each first story space of the residence of Figs. 11-7 and 11-8. Use the heat-load data given in Problem 11-3. Consider the air supply to be at 140 F and the return air to be at 65 F.

11-6. For a residence with a heat loss of 94 600 Btuh, compute how many cubic feet per minute of standard air and how many tons of air per hour must be circulated if a gravity system with 140 F air temperature is used.

Ans. 1140 cfm, 2.6 tons/h

11-7. Solve Problem 11-6 but assume that the residence uses a forced-circulation system and that the air-supply temperature is 120 F.

Ans. 1560 cfm, 3.5 tons/h

11-8. Making use of basic equations, compute the heat output per hour of a 10- by 10-ft ceiling panel having a surface temperature of 105 F in a room with a UMRT of 61 F and an air temperature of 68 F. Compare your answer with Tables 11-1 and 11-2.

Ans. 6010 Btuh

11-9. If the panel surface of Problem 11-8 is extended along the top side wall instead of on the ceiling, what is the heat output in Btu per hour? Assume that the panel-surface and room-temperature conditions are the same as in Problem 11-8. Use Table 11-1 and Table 11-2.

Ans. 7500 Btuh

11-10. Rework Problem 11-8 by any method, assuming the following conditions: a ceiling-panel surface temperature of 95 F, a UMRT of 60 F, and an air temperature of 68 F.

Ans. 4540 Btuh

11-11. Rework Problem 11-9 for a panel surface temperature of 95 F, a UMRT of 60 F, and an air temperature of 68 F.

Ans. 5490 Btuh

11-12. Compute the unheated mean radiant temperature for the combined living-dining room of the residence shown in Figs. 11-7 and 11-8, with no change except that the outside design temperature is +5 F. Refer to Example 11-4 for the area and for other data on the room.

11-13. Consider the room described in Problem 5-1 to use panel heat in its ceiling, and find the UMRT for the conditions indicated. In a radiant-heated room the temperature gradient from floor to ceiling is small and no correction for vertical temperature gradient need be made. Except for the exposed walls, all other surfaces are adjacent to heated spaces.

Ans. 61.7 F

11-14. A recreaction lounge in a resort hotel is 60 ft long, 30 ft deep, and 10 ft high. Along one of the long sides is a glass area amounting to 300 ft^2. The space is heated by panel coils in the ceiling, which are designed to operate at a surface temperature of 100 F. Surface temperatures are as follows: the glass, 50 F; the exposed wall, 60 F; the end walls, 64 F; and the back wall, the floor, and the unheated ceiling, 70 F. The design heat loss, including infiltration, amounts to 19 000 Btuh. (a) Compute the unheated mean radiant temperature, considering that half the ceiling is not provided with panel surface. (b) What area of panel surface is needed to heat the space if a 73 F air temperature is maintained?

Ans. (a) 67.2 F; (b) 430 ft^2

11-15. (a) For 430 ft^2 of panel area in Problem 11-14, at an average temperature of 100 F, refer to Tables 11-3 and 11-4 and select appropriate tubing. (b) Find the gpm required to serve the panel with a 20-deg drop in water temperature. (c) Find an appropriate number of circuits through the panel and a probable pumping head.

Ans. (a) 1290 ft of $\frac{3}{8}$-in. tubing on 4-in. centers; (b) (c) 1.94 gpm in 7 circuits each 185 ft long or 0.28 gpm per circuit, 3.4 ft head loss

11-16. The room of Problem 11-13, under the design conditions indicated and under certain infiltration conditions, has a total loss of 6150 Btuh. (a) Find the panel area which would be required at surface temperatures of 115 F and 110 F and (b) estimate the lowest surface temperature which would suffice with the 80 ft^2 of ceiling area available. If this temperature is above 110 F, supplementary heat should be used.

Ans. (a) Available area inadequate (81.4 ft^2); (b) 115 F

11-17. Refer to Example 11-4 and find the room UMRT for a location where −5 F is the outside design temperature.

11-18. Carry through the ceiling panel design on the basis of the data in Problem 11-17, with a recomputation showing the heat load to be 22 400 Btuh. Find (a) the panel area, (b) the length of the $\frac{3}{8}$-in. copper tubing on 6-in. centers which is to be used, and (c) a desirable number of circuits to use.

11-19. (a) For Example 11-4, compute the gpm of water required in the coil if it enters at 145 F and cools to 121 F. (b) Assume that this flow is handled in 20 parallel circuits each 48 ft long using $\frac{1}{2}$-in. copper tubing, and find the friction drop. Refer to Chapter 8 for friction-loss charts.

REFERENCES

Manual 4. *Warm-Air Perimeter Heating*, 8th ed., 1964.

Manual 5. *Gravity Code and Manual for the Design and Installation of Gravity Warm-Air Heating Systems*, 5th ed., 1954.

Manual 7 *Code and Manual for the Design and Installation of Warm-Air Winter Air Conditioning Systems*, 1953.

Manual 7A *Code and Manual for Ceiling Panel Systems*, 1950.

Manual 10 *Small-Pipe Warm-Air Perimeter Heating*, 1953.

Manual 9. *Code and Manual for the Design and Installation of Warm-Air Winter Air Conditioning Systems for Large Structures*, 7th ed., 1960.

References 1 to 6 are publications of the National Warm Air Heating and Air Conditioning Association also known as the National Environmental Systems Contractors and Air Conditioning Association.

ASHRAE, *Handbook-Systems 1973*, Chapter 12, "Forced Air Systems;" Chapter 8, "Panel Heating."

CHAPTER 12

FLUID FLOW, DUCT DESIGN, AND AIR-DISTRIBUTION METHODS

12-1. BASIC THEORY

In normal flow of a fluid (liquid or gas) in a restraining channel or duct, a drop of pressure occurs. The magnitude of this pressure drop depends on various factors: fluid velocity; diameter or shape of duct section, and condition of its surface; viscosity; density; temperature and pressure of the fluid; heat transfer to or from the fluid; and type of flow, viscous or turbulent. These seemingly numerous variables can be correlated into simple relationships.

When a fluid moves in a pipe or duct, there is always a thin film of the fluid which clings to the side of the pipe and does not move appreciably. In *laminar* (*viscous*) *flow*, each particle of the fluid moves parallel to the motion of the other particles. No crosscurrents occur, and the velocity of the fluid particles increases as their distance from the walls of the conduit increases. The maximum velocity occurs at the center of the conduit, and the average velocity over the entire cross section is equal to one-half the maximum. In this laminar flow the pressure drop, after equilibrium of flow is established, is all used in the shearing or sliding of the various layers of fluid against each other. The magnitude of the pressure drop for viscous flow can be calculated by the Poiseuille relationship:

$$\Delta P = \frac{32\mu L V}{g_c D^2} \tag{12-1}$$

The significance of the terms in this equation will be discussed later, but here it should be mentioned that the pressure drop (ΔP, pascals) is directly proportional to the viscosity (μ, newton seconds per square meter), to the equivalent length of pipe (L, meters), to the average velocity, (V, meters per second) and inversely proportional to (g_c = unity) and to the square of the equivalent diameter (D, meters).

When the fluid flow in a pipe increases above a certain critical velocity, laminar flow as just described can no longer continue and the flow becomes turbulent. In *turbulent flow* there are numerous eddies and cross currents in the stream and Eq. 12-1 does not apply. The average velocity over the cross section for turbulent flow is usually about 0.8 of the maximum. Turbulent flow is the type of flow most commonly met in engineering practice.

The pressure drop in turbulent flow is closely approximated by the Darcy equation

$$\Delta P = f\left(\frac{L}{D}\right)\rho\left(\frac{V^2}{2g_c}\right) \tag{12-2}$$

or

$$\Delta P = f\,\frac{L\rho V^2}{8g_c m} \tag{12-3}$$

where ΔP is the pressure drop in pascals; f is the friction factor to be obtained from Fig. 12-2; L is the equivalent length of pipe in meters; ρ is the density of fluid flowing, in kilograms per cubic meter; V is the velocity in meters per second; D is the diameter of pipe or conduit, in meters; g_c is unity for SI units; m is the hydraulic radius in meters. The hydraulic radius (m), is the cross-sectional area of the conduit, through which flow occurs, divided by the perimeter of the cross section in contact with the flowing fluid (gas or liquid). For a circular duct of radius (r)

$$m = \frac{\pi r^2}{2\pi r} = \frac{r}{2} = \frac{D}{4} \tag{12-4}$$

The magnitude of the friction factor (f) depends on many variables, the most important being roughness of the contact flow surface, along with the viscosity and other characteristics of the moving fluid. Fortunately by making use of the Reynolds number, it is possible to correlate data from pressure-loss tests of diverse fluids to find appropriate values of the friction factor for many other fluids. In the technical literature Eqs. 12-2 and 12-3 may appear with a factor 4 on the right-hand side and when this is the case tabular or chart values of f are consequently one-fourth as great.

Before discussing the Reynolds number and its relation to friction factor (f) certain aspects of viscosity will be presented. The *dynamic (absolute) viscosity* or the *coefficient of viscosity* of a fluid is defined as the tangential force required to move a plane fluid surface of unit area, at unit velocity, relative to a parallel plane surface of the fluid at unit distance away when the intervening space is filled with the fluid. Use the Greek letter μ (Mu) to designate dynamic viscosity and develop appropriate units in terms of this definition

$$\mu = \frac{\text{force}}{\dfrac{(\text{area} \times \text{velocity})}{\text{unit distance}}} = \frac{(\text{newton (N)}}{\dfrac{\text{m}^2 \cdot \text{m/s}}{\text{m}}} = \frac{\text{N} \cdot \text{s}}{\text{m}^2} = \frac{\text{kg}}{\text{m} \cdot \text{s}} \qquad (12\text{-}5)$$

This SI unit of viscosity is unnamed but a metric unit called the Poise is still in use. Note the following relations between SI, metric, and foot-pound-second units

$$\frac{\text{N} \cdot \text{s}}{\text{m}^2} = 10 \text{ Poise} = \frac{\text{kg}}{\text{m} \cdot \text{s}} \qquad (12\text{-}6)$$

$$\frac{\text{lb}}{\text{ft} \cdot \text{s}} \times 1.488 = \frac{\text{N} \cdot \text{s}}{\text{m}^2} \qquad \frac{\text{N} \cdot \text{s}}{\text{m}^2} \times 0.672 = \frac{\text{lb}}{\text{ft} \cdot \text{s}}$$

$$\frac{\text{lb}}{\text{ft} \cdot \text{s}} \times 1488 = \text{centipoise} \qquad \text{centipoise} \times 0.000672 = \frac{\text{lb}}{\text{ft} \cdot \text{s}}$$

$$\frac{\text{lbf} \cdot \text{s}}{\text{ft}^2} \times 47.88 = \frac{\text{N} \cdot \text{s}}{\text{m}^2} \qquad \frac{\text{N} \cdot \text{s}}{\text{m}^2} \times 0.0209 = \frac{\text{lbf} \cdot \text{s}}{\text{ft}^2} \qquad (12\text{-}7)$$

$$\frac{\text{lbf} \cdot \text{s}}{\text{ft}^2} \times 47\,880 = \text{centipoise} \qquad \text{centipoise} \times 0.0000209 = \frac{\text{lbf} \cdot \text{s}}{\text{ft}^2}$$

The viscosities of fluids vary over wide limits. For liquids, viscosity decreases substantially with rise in temperature; while for gases and vapors viscosity increases both with increase of temperature and with rise of pressure. Fig. 12-1 gives viscosity values for a number of fluids. Pure water at 20°C has a viscosity of 1.008×10^{-3} N · s/m² = 1.008 centipoises.

Kinematic viscosity (ν) is defined as the dynamic viscosity divided by the fluid density: $\nu = \mu/\rho$. Its SI units are square meters per second. Kinematic viscosity can be used in computing Reynolds numbers when this approach is convenient. Many viscosimeters (instruments for measuring viscosity) determine kinematic viscosity by a simple direct reading. For example, the widely used Saybolt viscosimeter relates the seconds (t), required for a sample of the fluid to flow through an orifice, under controlled conditions, to the kinematic viscosity in accordance with this equation:

$$\nu = 2.201 \times 10^{-7} t - \frac{1.802 \times 10^{-4}}{t} \qquad (12\text{-}8)$$

where ν is expressed in square meters per second.

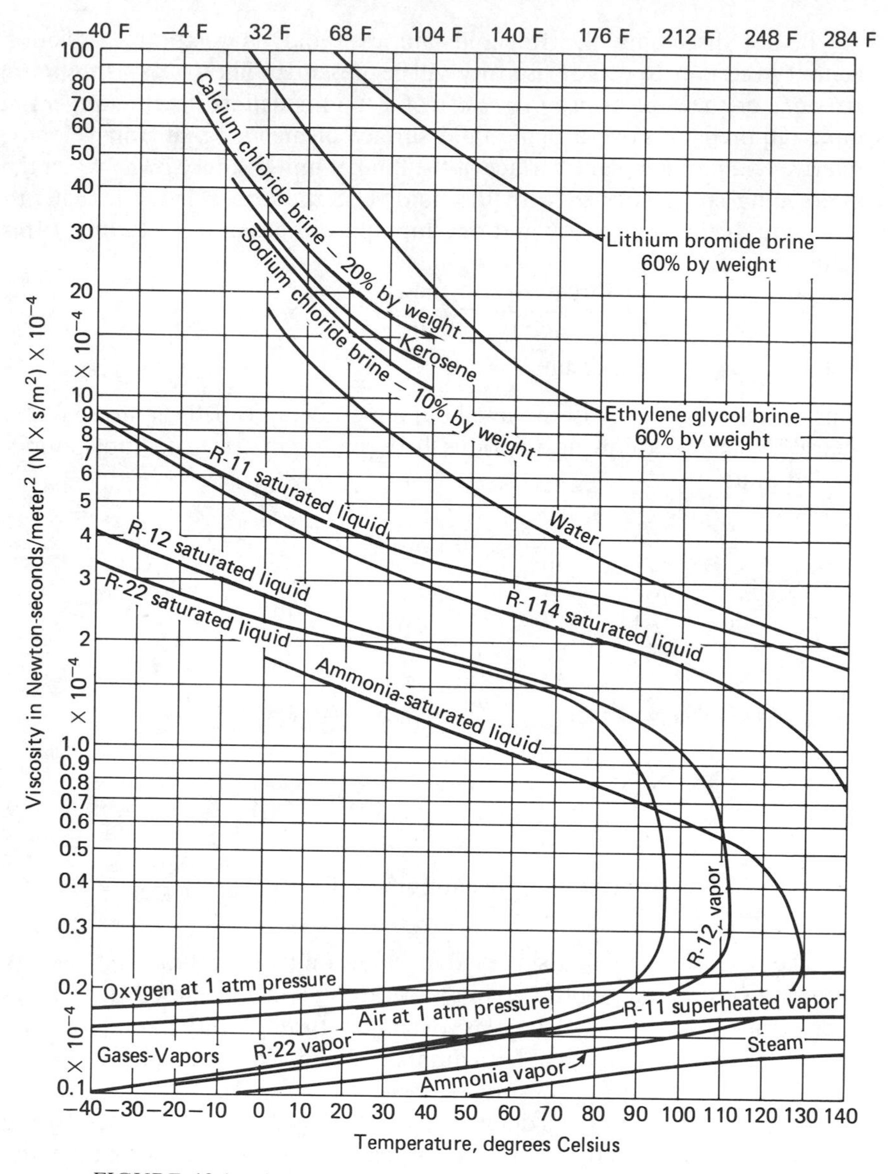

FIGURE 12-1. Absolute viscosities of selected refrigerants and other fluids.

12-2. REYNOLDS NUMBER

The Reynolds number (Re) which is dimensionless, is usually written with the following symbols.

$$Re = \frac{VD\rho}{\mu} = \frac{VD}{\nu} \tag{12-9}$$

where V is the velocity in meters per second for SI; or feet per second in foot-pound-second units; D is the representative dimension such as diameter in case of a circulat duct, in consistent units, meters or feet; ρ is the density in consistent units of kilograms per cubic meter, or lb per cubic foot; μ is the dynamic viscosity in consistent units of kilograms per meter second or pounds per foot second; ν is the kinematic viscosity in consistent units, square meters per second or square feet per second.

To show that Re is dimensionless in any consistent set of units substitute appropriate units for each term. In SI units

$$Re = \frac{VD\rho}{\mu} = \frac{(\text{m/s})(\text{m})(\text{kg/m}^3)}{\text{N} \cdot \text{s/m}^2 \equiv \text{kg/m} \cdot \text{s}} = 1$$

In foot-pound-second units

$$Re = \frac{VD\rho}{\mu} = \frac{(\text{ft/s})(\text{ft})(\text{lb/ft}^3)}{\text{lb/ft} \cdot \text{s}} = 1$$

In a mixed system it may be convenient to employ

$$Re = \frac{VD\rho}{\mu} = 1.488 \frac{(\text{ft/s})(\text{ft})(\text{lb/ft}^3)}{\text{N} \cdot \text{s/m}^2} \tag{12-10}$$

12-3. FRICTION FACTORS

One use of the Reynolds number R is to have it serve as a base on which to show variations in the friction factor f. Figure 12-2 covers an extended range of Reynolds numbers in the turbulent-flow region. In this region, surface roughness of the pipe or conduit has a significant influence on the values of the friction factor. The effect of surface roughness is best represented by the ratio of the surface-variation height to the diameter (or depth) of the conduit. The ordinate of Fig. 12-2 is thus relative roughness (ε/D). Values of ε and selected ε/D values, for various types of pipes or conduits, are given in Table 12-1. When these are used with the pipe-diameter scale at the top of Fig. 12-2, a family of lines for different kinds of pipes can be represented on the chart. It should be observed that the values of Fig. 12-2 are independent of the fluid and apply equally well to water or air, or to other gases when these are evaluated in terms of the Reynolds number. Reynolds numbers ranging between 2000 and 2500 are in a transition zone. On the low side laminar (viscous) flow occurs, while on the high side turbulent flow takes place. A value of 2300 is sometimes used as the most representative critical value.

In considering Table 12-1 it should further be noted that the roughness of a conduit is related not only to the material from which it is made but also depends on the kind of workmanship used in the fabrication of the conduit and on the number and size of burrs and the amount of incrustation which may be present.

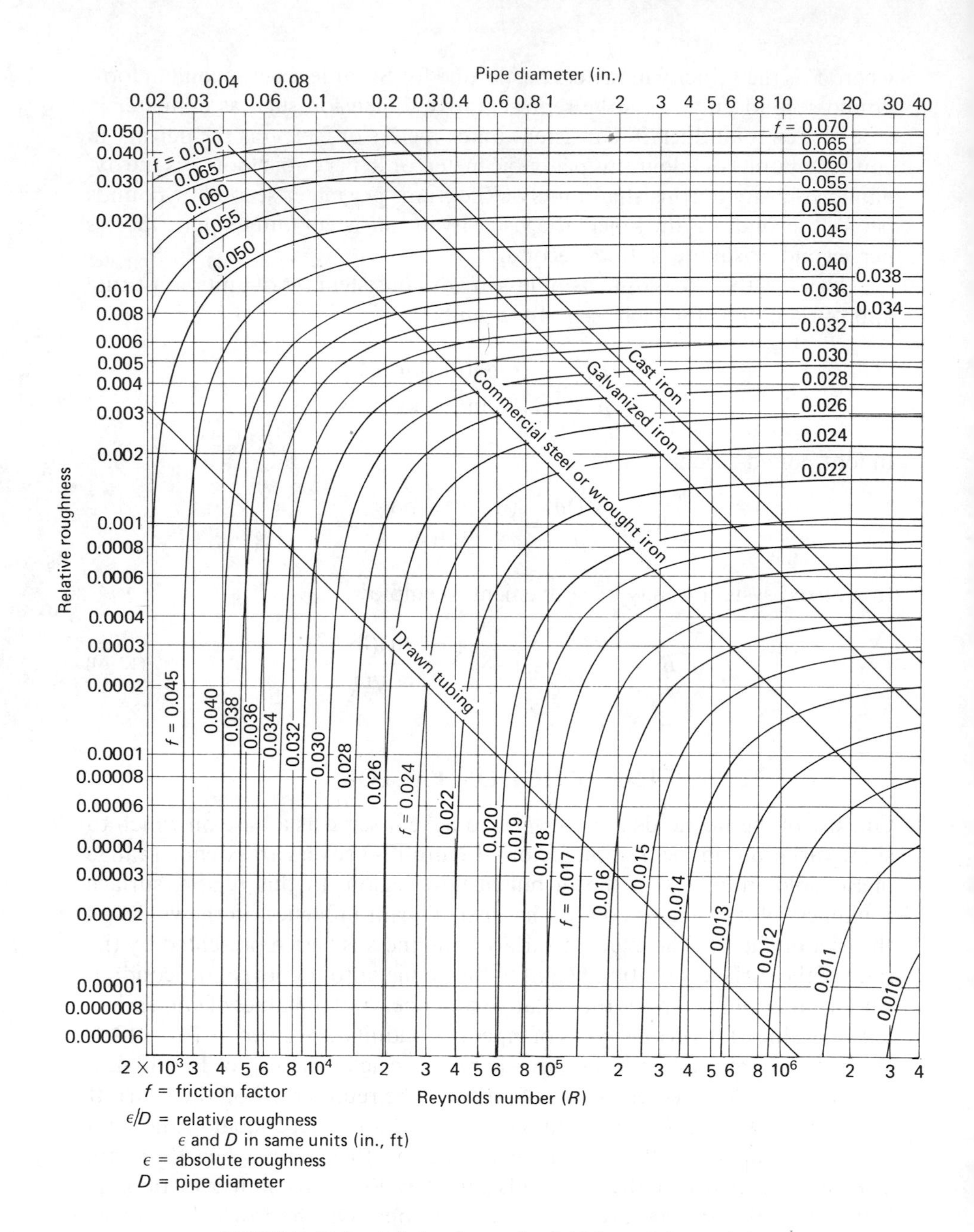

FIGURE 12-2. Friction factors for fluid flow. (After Ref. 4.)

Example 12-1. Water at 93.3°C (200 F) is flowing at 0.1524 m/s through 4-in. standard steel pipe (ID = 4.03 in.) of relatively smooth surface. Find (a) the Reynolds number, (b) whether the flow is laminar or turbulent, and (c) the frictional loss per 30.5 m of equivalent pipe length, (d) Rework parts a, b, and c directly in foot-pound-second units.

Table 12-1 Representative Values of Surface Roughness (ε) and Relative Roughness (ε/D)

Material	ε (mm)	ε/D for Conduits				
		D = 50 mm	150 mm	250 mm	500 mm	750 mm
Cast iron,						
asphalt dipped	0.12	0.0024	0.0008	0.0005	0.0002	0.0002
as cast	0.26	0.0052	0.0017	0.0010	0.0005	0.0003
Concrete,	0.31 to 3.1	—	—	—	—	—
average	0.91	0.018	0.006	0.0036	0.0018	0.0012
Drawn tubing of lead						
or brass; glass	0.0015	0.0003	0.00001	0.00001	—	—
Galvanized steel duct	0.15	0.003	0.001	0.0006	0.0003	0.0002
Pipe, clean steel	0.05	0.001	0.0003	0.0002	0.0001	0.00007
Pipe, galvanized	0.15	0.003	0.001	0.0006	0.0003	0.0002
Riveted steel,	0.91 to 9.1	—	—	—	—	—
average	3.0	0.060	0.020	0.012	0.006	0.004
Stainless steel, or						
aluminum duct	0.10	0.002	0.007	0.0004	0.0002	0.0001
Wood, planed panels						
or staves	0.18 to 0.91	—	—	—	—	—

Solution: (a)

$$D = \frac{4.03}{12} \times 0.3048 = 0.1024 \text{ m}$$

$\mu = 0.296 \times 10^{-3} = 2.96 \times 10^{-4}$ N · s/m² (from Fig. 12-1). Density of water at 93.3°C (200 F) can be found from Table 2-2 as

$$\frac{1}{0.01663} = 60.1 \text{ lb/ft}^3$$

$$60.1 \text{ lb/ft}^3 \times 16.0185 = 962.7 \text{ kg/m}^3$$

$$R = \frac{DV\rho}{\mu} = \frac{(0.1024)(0.1524)(962.7)}{2.96 \times 10^{-4}} = 50\,760 \qquad \textit{Ans.}$$

(b) As R at 50 760 is far above 2300, turbulent flow is taking place. *Ans.*

(c) In using Fig. 12-2 to find f, read the pipe diameter of 4.03 at the top of the chart, drop down to the line for commercial steel, and then move horizontally over until the Reynolds number 5.08×10^4 is reached; here read $f = 0.0225$, say, 0.023. To find f without the use of a line on the chart, such as the one shown for commercial steel, it would be necessary to find an ε/D value for the material in question. In this case,

$$\frac{\varepsilon}{D} = 0.00048$$

Find this value on the left axis of Fig. 12-2, run horizontally over to the Reynolds number of 5.08×10^4, and read $f = 0.0225$.

Using Eq. (12-2), the pressure drop in 30.5 m of equivalent pipe length is

$$\Delta P = 0.023\left(\frac{30.5}{0.1024}\right)(962.7)\frac{(0.1524)^2}{2} = 76.6 \text{ Pa} \qquad \textit{Ans.}$$

(d) For the Reynolds number, $D = 4.03/12 = 0.336$ ft, $V = 0.1534 \times 3.28 = 0.50$ ft/s, $\rho = 60.1$ lb/ft^3, and $\mu = 2.96 \times 10^{-4}$ N · s/m^2. Substituting in Eq. (12-10),

$$R = 1.488\frac{(0.366)(0.50)(60.1)}{2.96 \times 10^{-4}} = 50\,760 \qquad \textit{Ans.}$$

Flow is turbulent and, as before, $f = 0.023$. Using Eq. (12-2),

$$\Delta P = 0.023\left(\frac{30.5 \times 3.28}{0.336}\right)(60.1)\frac{(0.50)^2}{(2)(32.17)} = 1.6 \text{ lbf/ft}^2$$

$$\Delta P = \frac{1.6}{144} = 0.0111 \text{ psi} \qquad \textit{Ans.}$$

For the rather unusual cases of viscous flow, the friction factor f is practically independent of the surface because of the clinging undisturbed film and therefore does not appear in Eq. (12-1). However, if it is desired to calculate pressure drop in viscous flow by using Eq. (12-2) or (12-3), in which equations friction factors appear, a working value of f can be found by equating Eq. (12-1) to Eq. (12-2):

$$\frac{32\mu LV}{gD^2} = (f)\left(\frac{L}{D}\right)\frac{\rho V^2}{2g}$$

Or, for laminar flow only,

$$f = \frac{64\mu}{DV\rho} = \frac{64}{R} \tag{12-11}$$

Thus the friction factor f for laminar (streamline) flow, which can be used in Eq. (12-2) is found by dividing 64 by the Reynolds number. In turbulent flow, f must always be found from experimental data (using any suitable test fluid).

For conduits of noncircular cross section, friction factors may vary greatly from those for round sections. For example, a shallow rectangular duct offers much more resistance to fluid flow than does a square or round duct when each has the same cross-sectional area. In general, the largest area for the least perimeter of duct gives the least frictional drop. The *hydraulic radius* (m), which is defined as the cross-sectional area divided by the fluid-contact perimeter, is useful in making comparisons of odd-shaped sections.

For a circular section of diameter D the hydraulic radius by Eq. 12-4 is

$$r = 2m \tag{12-12}$$

and

$$D = 4m \tag{12-13}$$

For a rectangular section of sides H and W, in consistent units of feet or inches,

$$m = \frac{H \times W}{2(H + W)} \tag{12-14}$$

For a square conduit of side a feet or inches, the hydraulic radius is

$$m = \frac{a^2}{4a} = \frac{a}{4} \tag{12-15}$$

and thus

$$a = 4m \tag{12-16}$$

In the case of a round duct (conduit), it is seen that $D = 4m$. Thus if any duct is not too flattened, a Reynolds number can be calculated, using as D in Eq. (12-9) or (12-10) the calculated value of $4m$ for that duct. From this evaluation a friction factor can be found for use in Eq. (12-2) or (12-3) with the odd-shaped duct.

The preceding discussion gives a logical basis for calculating frictional losses from fluid flowing in straight runs of piping. Losses through fittings, bends, and elbows must be obtained from experimental data and are usually stated as equivalent in loss to a certain length (in diameters) of straight run of pipe. For a given type of piping or duct it is convenient to construct tables or diagrams to facilitate computations. Some of these may be found at several places in this book. In some cases it has been found that the turbulent basic-loss equations are not exactly proportional to the square of the velocity but may vary by a power somewhat less than 2 (that is, 1.8 and 1.9), but for most cases use of the square of the velocity gives the best results in making computations.

12-4. AIR DUCT FLOW AND LOSSES

The energy equation [Eq. (2-8)] is applicable to air that is flowing in a duct; and by eliminating from the equation irrelevant terms such as potential energy (Z) and heat transferred ($Q_{(1-2)}$), there remains the expression

$$P_1 v_1 + \frac{V_1^2}{2g} = P_2 v_2 + \frac{V_2^2}{2g} + 778(u_2 - u_1) = P_2 v_2 + \frac{V_2^2}{2g} + \text{lost head} \tag{12-17}$$

This equation states that at a given point the mechanical energy terms, namely, the flow work ($P_1 v_1$) plus the kinetic energy ($V_1^2/2g$), must be equal to the flow work and kinetic energy at any other point plus any dissipation of mechanical energy to internal energy (lost head) which has meanwhile occurred. In the case of air, use is made of the term *static head*, or *static pressure*, for Pv, and *velocity head*, or *velocity pressure*, for $V^2/2g$; and the sum of static pressure and velocity pressure is the *total pressure*, or *total dynamic pressure*, or *head* (h_T). The energy units of this equation are foot-pounds per pound of air or can be expressed simply as feet of the fluid (air) considered. In the case of fans the static pressure is usually expressed in gage pressure (excess of pressure above that of the atmosphere) instead of in absolute pressure.

The static and velocity pressures of air, instead of being expressed in "feet of air" units, are more commonly changed to "inches of water" units. The transformation can be made by the use of the fundamental relation [Eq. (1-23)],

$$h_a d_a = \frac{h}{12} d_w$$

where h_a represents feet of air; d_a represents density of air, in pounds per cubic foot; h represents inches of water (gage); and d_w represents the density of 68 F water, 62.3 lb/ft^3. Thus, employing appropriate values at 68 F, it is seen that

$$h_a = \frac{h}{12}\frac{62.3}{d_a} = 5.19\frac{h}{d_a} \tag{12-18}$$

For the velocity head, h_v, in inches of water,

$$\frac{V^2}{2g} = 5.19\frac{h_v}{d_a}$$

or

$$h_v = \frac{V^2 d_a}{(5.19)2g}$$

With V_m as fpm (feet per minute),

$$h_v = \frac{V_m^2 d_a}{(3600)(5.19)(2g)} = \frac{V_m^2 d_a}{1\,202\,000} = \left(\frac{V_m}{1096.5}\right)^2 d_a \tag{12-19}$$

$$V_m = 1096.5\sqrt{\frac{h_v}{d_a}} \tag{12-20}$$

Standard air is air having a density of 0.075 lb/ft^3 at 29.92 in. Hg. This is essentially equivalent to dry air at 70 F or to air at 30% relative humidity and 65 F (actually, $d_a = 0.07494$); and

$$V_m = 4005\sqrt{h_v}$$

or

$$h_v = \left(\frac{V_m}{4005}\right)^2 \tag{12-21}$$

Call h_s the static head of the air in inches of water-gage pressure; then the total head becomes

$$h_T = h_s + h_v = h_s + \left(\frac{V_m}{1096.5}\right)^2 \times d_a = h_s + \left(\frac{V_m}{4005}\right)^2_{\text{std air}} \tag{12-22}$$

The total head h_T would remain undiminished and h_s and h_v would be mutually interchangeable from one form to the other, if it were not for frictional loss, which resists flow, and for impact losses due to sudden enlargements, contractions, and bends in the duct. These items dissipate the total head or mechanical energy, which eventually reappears as increased internal energy of the air.

Figure 12-3 has been constructed to show in graphic form the interrelationships between total, static, and velocity pressures (heads) along with their relationships to frictional loss in an illustrative duct system. The duct system is at the top of the figure while below it appears a graph of the pressure (head) variations that take place as the air or other fluid moves along the passageway. Two important generalizations are applicable: (1) The total pressure (head), because of friction and turbulence, always diminishes in the direction of flow; (2) the velocity pressure (head) is determined solely by the stream velocity, which in turn is set by the total flow (cubic feet per minute) and by the cross-sectional area (for fluids of constant density during the process). Static and not total pressure is usually measured in a duct system and thus only by the addition of velocity pressure is the total pressure found. It is, of course, necessary to know or measure the quantity of air flowing before velocity pressure can be computed (see Chapter 13).

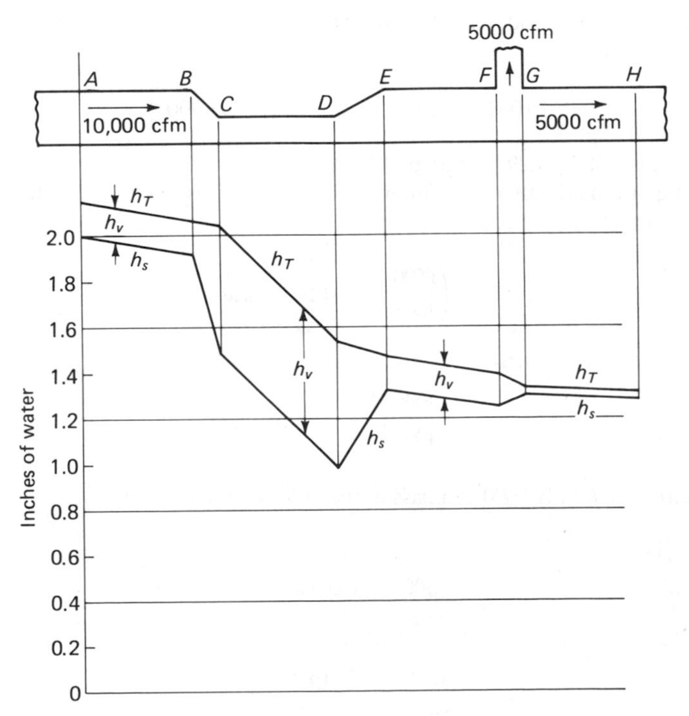

FIGURE 12-3. Diagram of the representative duct system of Example 12-2, with a plot of total, static, and velocity heads to show their interrelationships.

Example 12-2. The duct run shown in Fig. 12-3 carries 10 000 cfm of standard-density air (0.075 lb/ft^3). The run from A to B is square in cross section and of such size that the velocity is 1500 fpm. In order to pass under an obstruction at point C, the duct is reduced in depth by one-half and the air velocity increases to 3000 fpm. From D to E the duct is returned to its original size and continues to F where 5000 cfm is delivered to a side branch. The remaining air passes on with no change in duct size. (*a*) Compute the dimensions of the duct in its different sections. (*b*) Compute the velocity head in each part of the duct. (*c*) Assume that 50% of the maximum velocity-pressure regain is realized at G and compute its magnitude. (*d*) The frictional pressure losses in the duct parts are as follows, in inches of water: AB, 0.08; BC, 0.02; CD, 0.50; DE, 0.07; EF, 0.08; and GH, 0.023. Make use of these data and prior computations to find the static and total pressure at each point and plot their magnitudes to scale when the static pressure at A has a measured value of 2.0 in. water.

Solution: (a) For a medium of constant density flowing in a duct, at average velocity of V fpm,

$$Q = AV$$

Express Q in cfm (cubic feet per minute) and area A in square feet; then for the runs AB and EF,

$$10\,000 = (A)(1500)$$
$$A = 6.67 \text{ ft}^2$$

and the resulting square sides are 2.58 by 2.58 ft or 31 by 31 in. *Ans.*

For the run CD,

$$10\,000 = (A)(3000) = (2.58)(1.29)(3000)$$

The size is thus 2.58 by 1.29 ft or 31 by 15.5 in. *Ans.*

(b) Equation (12-20), or its simplified form (12-21), is applicable to find velocity head. For AB and EF,

$$h_v = \left(\frac{1500}{4005}\right)^2 = 0.14 \text{ in. water} \qquad \textit{Ans.}$$

For CD,

$$h_v = \left(\frac{3000}{4005}\right)^2 = 0.56 \text{ in. water} \qquad \textit{Ans.}$$

For the section GH, only 5000 cfm pass in the full-size duct, 6.67 ft^2,

$$Q = AV$$
$$5000 = (6.67)(V)$$

and thus

$$V = 750 \text{ fpm}$$

$$h_v = \left(\frac{750}{4005}\right)^2 = 0.035 \text{ in. water} \qquad \textit{Ans.}$$

(c) Except for turbulence and friction, all of the surplus velocity head in EF could be realized as a static-pressure regain at G following the takeoff at FG.

$$\text{ideal } \Delta h \text{ at } FG = 0.14 - 0.035 = 0.105$$

If 50% is actually realized, $\Delta h = 0.052$ in. water and the same amount is lost. *Ans.*

(d) The total pressure at A is

(A) $h_T = h_s + h_v = 2.00 + 0.14 = 2.14$ in. water

(A) $h_s = 2.0$ in. water as measured

(B) $h_{TB} = h_{TA} - h_f = 2.14 - 0.08 = 2.06$ in. water
$h_{sB} = h_{sA} - h_f = 2.00 - 0.08 = 1.92$ in. water

(C) $h_{TC} = h_{TB} - h_f = 2.06 - 0.02 = 2.04$ in. water
$h_{sC} = h_{sB} - h_f - \Delta h_v = 1.92 - (0.56 - 0.14) = 1.48$ in. water

(D) $h_{TD} = h_{TC} - h_f = 2.04 - 0.50 = 1.54$ in. water
$h_{sD} = h_{sC} - h_f = 1.48 - 0.50 = 0.98$ in. water

(E) $h_{TE} = h_{TD} - h_f = 1.54 - 0.07 = 1.47$ in. water
$h_{sE} = h_{sD} - h_f + \Delta h_v = 0.98 - 0.07 + (0.56 - 0.14) = 1.33$ in. water

(F) $h_{TF} = h_{TE} - h_f = 1.47 - 0.08 = 1.39$ in. water
$h_{sF} = h_{sE} - h_f = 1.33 - 0.08 = 1.25$ in. water

(G) $h_{TG} = h_{TF} - h_f = 1.39 - 0.052 = 1.338$ in. water
$h_{sG} = h_{sF} + \Delta h_v - h_f = 1.25 + 0.105 - 0.052 = 1.303$ in. water

(H) $h_{TH} = h_{TG} - h_f = 1.338 - 0.023 = 1.315$ in. water
$h_{sH} = h_{sG} - h_f = 1.303 - 0.023 = 1.280$ in. water

The value of total pressure at H must also correspond as

$$h_{TH} = h_{sH} + h_v = 1.280 + 0.035 = 1.315 \text{ in. water.}$$

For frictional loss in air ducts, Eq. (12-2) or (12-2a) applies. In pounds per square foot,

$$\Delta P = (f)\frac{L\rho V^2}{8gm}$$

And in feet of air,

$$\Delta h_a = \frac{\Delta P}{\rho} = f\left(\frac{L}{m}\right)\frac{V^2}{8g}$$

When for a round duct the pressure loss is expressed in inches of water and the constants given in Eq. (12-19) are used,

$$\Delta h = f\left(\frac{L}{D}\right)\left(\frac{V_m}{1096.5}\right)^2 d_a. \tag{12-23}$$

The results of tests on friction loss for air flowing in galvanized ducts have been assembled by the ASHRAE and are plotted in Fig. 12-4. This chart is based on standard air but can also be used, with little error, for air in the range of 50 F to 90 F. Moreover, a correction is not required for humidity variations or for small deviations in barometric pressure (not exceeding ± 0.5 in. Hg). It is desirable to apply a correction factor to the chart value for

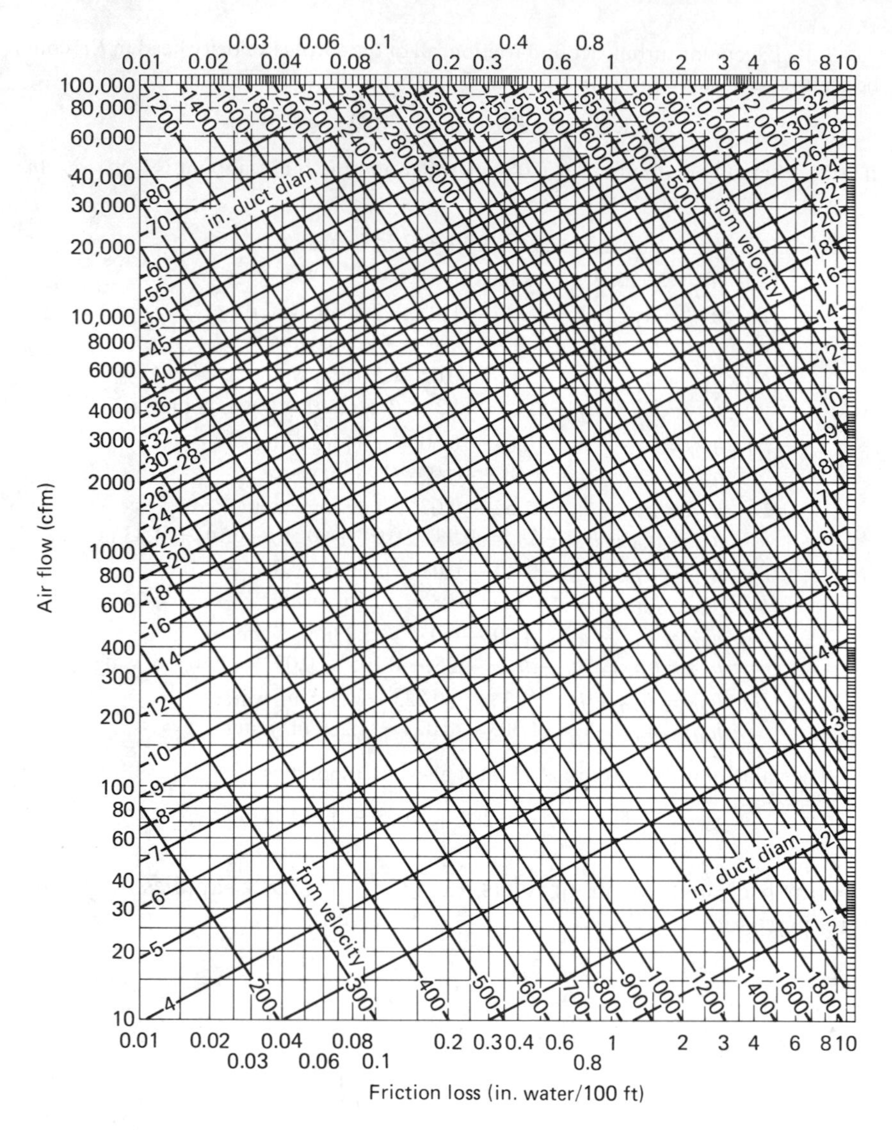

FIGURE 12-4. Friction loss in air ducts. (Reproduced, by permission, from *Heating Ventilating Air Conditioning Guide* 1949.)

unusual construction or rough ductwork. Figure 12-5 gives representative correction factors. In addition, the final design should include an overall safety factor of (say) 10% to cover possible construction contingencies.

Where the air in the duct is at a density significantly different from that for which the chart was constructed, the pressure loss should be multiplied by a factor as follows:

$$\Delta h_c = (\Delta h)\left(\frac{d_a \text{ at actual conditions}}{0.075}\right) \text{ in in. of water} \tag{12-24}$$

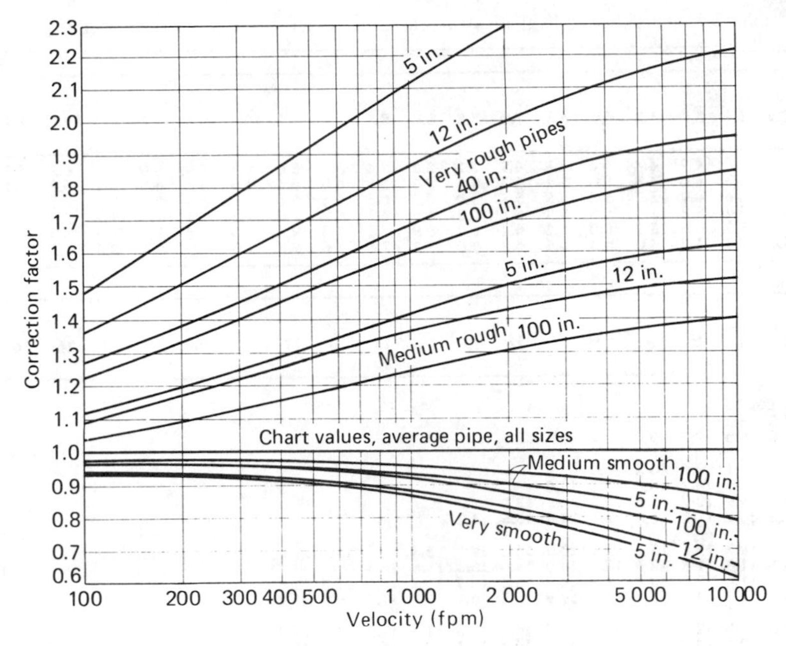

FIGURE 12-5. Friction correction factors for pipe and duct roughness. (After Ref. 6.)

When the temperature is the important variable, the pressure loss should be multiplied by a temperature correction factor, giving

$$\Delta h_c = (\Delta h)\left(\frac{460 + 70}{460 + t}\right) \text{ in in. of water} \tag{12-25}$$

In the last two equations, Δh_c is the corrected value of pressure loss.

A greater frictional loss occurs in a rectangular duct than in a circular duct of the same cross-sectional area. If the frictional loss per foot of length in a circular duct is to be made equal to that in a rectangular duct, fundamental expressions for loss for the two types of ducts are equated; and with suitable friction factors being used, there evolves

$$D = 1.3\left[\frac{(HW)^5}{(H + W)^2}\right]^{1/8} = 1.3\,\frac{(HW)^{0.625}}{(H + W)^{0.25}} \tag{12-26}$$

where H, W are the two sides of a rectangular duct, respectively, in feet or inches; D is the diameter of a circular duct having the same frictional loss as a rectangular duct of the same length delivering the same quantity of air, in feet or inches; and H/W is the aspect ratio of a rectangular duct. In case of bends, H is the duct dimension perpendicular to the plane of the bend and W is the duct dimension in the plane of the bend.

Equation (12-26) was developed by Huebscher (Ref. 3), who further concluded that for most practical purposes a rectangular duct with an aspect ratio not exceeding 8 to 1 will have the same static friction pressure loss for

Table 12-2 Circular Equivalents of Rectangular Ducts for Equal Friction and Capacity

Side Rectangular Duct	4.0	4.5	5.0	5.5	6.0	6.5	7.0	7.5	8.0	8.5	9.0	9.5	10.0	10.5	11.0	11.5	12.0	12.5	13.0	13.5
3.0	3.8	4.0	4.2	4.4	4.6	4.8	4.9	5.1	5.2	5.4	5.5	5.6	5.7	5.9	6.0	6.1	6.2	6.3	6.4	6.5
3.5	4.1	4.3	4.6	4.8	5.0	5.2	5.3	5.5	5.7	5.8	6.0	6.1	6.3	6.4	6.5	6.7	6.8	6.9	7.0	7.1
4.0	4.4	4.6	4.9	5.1	5.3	5.5	5.7	5.9	6.1	6.3	6.4	6.6	6.8	6.9	7.1	7.2	7.3	7.5	7.6	7.7
4.5	4.6	4.9	5.2	5.4	5.6	5.9	6.1	6.3	6.5	6.7	6.9	7.0	7.2	7.4	7.5	7.7	7.8	8.0	8.1	8.2
5.0	4.9	5.2	5.5	5.7	6.0	6.2	6.4	6.7	6.9	7.1	7.3	7.4	7.6	7.8	8.0	8.1	8.3	8.4	8.6	8.7
5.5	5.1	5.4	5.7	6.0	6.3	6.5	6.8	7.0	7.2	7.4	7.6	7.8	8.0	8.2	8.4	8.6	8.7	8.8	9.0	9.2

Side Rectangular Duct	6	7	8	9	10	11	12	13	14	15	16	17	18	19	20	22	24	26	28	30
6	6.6																			
7	7.1	7.7																		
8	7.5	8.2	8.8																	
9	8.0	8.6	9.3	9.9																
10	8.4	9.1	9.8	10.4	10.9															
11	8.8	9.5	10.2	10.8	11.4	12.0														
12	9.1	9.9	10.7	11.3	11.9	12.5	13.1													
13	9.5	10.3	11.1	11.8	12.4	13.0	13.6	14.2												
14	9.8	10.7	11.5	12.2	12.9	13.5	14.2	14.7	15.3											
15	10.1	11.0	11.8	12.6	13.3	14.0	14.6	15.3	15.8	16.4										
16	10.4	11.4	12.2	13.0	13.7	14.4	15.1	15.7	16.3	16.9	17.5									
17	10.7	11.7	12.5	13.4	14.1	14.9	15.5	16.1	16.8	17.4	18.0	18.6								
18	11.0	11.9	12.9	13.7	14.5	15.3	16.0	16.6	17.3	17.9	18.5	19.1	19.7							
19	11.2	12.2	13.2	14.1	14.9	15.6	16.4	17.1	17.8	18.4	19.0	19.6	20.2	20.8						
20	11.5	12.5	13.5	14.4	15.2	15.9	16.8	17.5	18.2	18.8	19.5	20.1	20.7	21.3	21.9					
22	12.0	13.1	14.1	15.0	15.9	16.7	17.6	18.3	19.1	19.7	20.4	21.0	21.7	22.3	22.9	24.1				
24	12.4	13.6	14.6	15.6	16.6	17.5	18.3	19.1	19.8	20.6	21.3	21.9	22.6	23.2	23.9	25.1	26.2			
26	12.8	14.1	15.2	16.2	17.2	18.1	19.0	19.8	20.6	21.4	22.1	22.8	23.5	24.1	24.8	26.1	27.2	28.4		
28	13.2	14.5	15.6	16.7	17.7	18.7	19.6	20.5	21.3	22.1	22.9	23.6	24.4	25.0	25.7	27.1	28.2	29.5	30.6	
30	13.6	14.9	16.1	17.2	18.3	19.3	20.2	21.1	22.0	22.9	23.7	24.4	25.2	25.9	26.7	28.0	29.3	30.5	31.6	32.8
32	14.0	15.3	16.5	17.7	18.8	19.8	20.8	21.8	22.7	23.6	24.4	25.2	26.0	26.7	27.5	28.9	30.1	31.4	32.6	33.8
34	14.4	15.7	17.0	18.2	19.3	20.4	21.4	22.4	23.3	24.2	25.1	25.9	26.7	27.5	28.3	29.7	31.0	32.3	33.6	34.8
36	14.7	16.1	17.4	18.6	19.8	20.9	21.9	23.0	23.9	24.8	25.8	26.6	27.4	28.3	29.0	30.5	32.0	33.0	34.6	35.8
38	15.0	16.4	17.8	19.0	20.3	21.4	22.5	23.5	24.5	25.4	26.4	27.3	28.1	29.0	29.8	31.4	32.8	34.2	35.5	36.7
40	15.3	16.8	18.2	19.4	20.7	21.9	23.0	24.0	25.1	26.0	27.0	27.9	28.8	29.7	30.5	32.1	33.6	35.1	36.4	37.6
42	15.6	17.1	18.5	19.8	21.1	22.3	23.4	24.5	25.6	26.6	27.6	28.5	29.4	30.4	31.2	32.8	34.4	35.9	37.3	38.6
44	15.9	17.5	18.9	20.2	21.5	22.7	23.9	25.0	26.1	27.2	28.2	29.1	30.0	31.0	31.9	33.5	35.2	36.7	38.1	39.5
46	16.2	17.8	19.2	20.6	21.9	23.2	24.3	25.5	26.7	27.7	28.7	29.7	30.6	31.6	32.5	34.2	35.9	37.4	38.9	40.3
48	16.5	18.1	19.6	20.9	22.3	23.6	24.8	26.0	27.2	28.2	29.2	30.2	31.2	32.2	33.1	34.9	36.6	38.2	39.7	41.2
50	16.8	18.4	19.9	21.3	22.7	24.0	25.2	26.4	27.6	28.7	29.8	30.8	31.8	32.8	33.7	35.5	37.3	38.9	40.4	42.0
52	17.0	18.7	20.2	21.6	23.1	24.4	25.6	26.8	28.1	29.2	30.3	31.4	32.4	33.4	34.3	36.2	38.0	39.6	41.2	42.8
54	17.3	19.0	20.5	22.0	23.4	24.8	26.1	27.3	28.5	29.7	30.8	31.9	32.9	33.9	34.9	36.8	38.7	40.3	42.0	43.6
56	17.6	19.3	20.9	22.4	23.8	25.2	26.5	27.7	28.9	30.1	31.2	32.4	33.4	34.5	35.5	37.4	39.3	41.0	42.7	44.3
58	17.8	19.5	21.1	22.7	24.2	25.5	26.9	28.2	29.3	30.5	31.7	32.9	33.9	35.0	36.0	38.0	39.8	41.7	43.4	45.0
60	18.1	19.8	21.4	23.0	24.5	25.8	27.3	28.7	29.8	31.0	32.2	33.4	34.5	35.5	36.5	38.6	40.4	42.3	44.0	45.8
62	18.3	20.1	21.7	23.3	24.8	26.2	27.6	29.0	30.2	31.4	32.6	33.8	35.0	36.0	37.1	39.2	41.0	42.9	44.7	46.5
64	18.6	20.3	22.0	23.6	25.2	26.5	27.9	29.3	30.6	31.8	33.1	34.2	35.5	36.5	37.6	39.7	41.6	43.5	45.4	47.2
66	18.8	20.6	22.3	23.9	25.5	26.9	28.3	29.7	31.0	32.2	33.5	34.7	35.9	37.0	38.1	40.2	42.2	44.1	46.0	47.8
68	19.0	20.8	22.5	24.2	25.8	27.3	28.7	30.1	31.4	32.6	33.9	35.1	36.3	37.5	38.6	40.7	42.8	44.7	46.6	48.4
70	19.2	21.1	22.8	24.5	26.1	27.6	29.1	30.4	31.8	33.1	34.3	35.6	36.8	37.9	39.1	41.3	43.3	45.3	47.2	49.0
72															3916	41.8	43.8	45.9	47.8	49.7
74															40.0	42.3	44.4	46.4	48.4	50.3
76															40.5	42.8	44.9	47.0	49.0	50.8
78															40.9	43.3	45.5	47.5	49.5	51.5
80															41.3	43.8	46.0	48.0	50.1	52.0
82															41.8	44.2	46.4	48.6	50.6	52.6
84															42.2	44.6	46.9	49.2	51.1	53.2
86															42.6	45.0	47.4	49.6	51.6	53.7
88															43.0	45.4	47.9	50.1	52.2	54.3
90															43.4	45.9	48.3	50.6	52.8	54.8
92															43.8	46.3	48.7	51.1	53.4	55.4
94															44.2	46.7	49.1	51.6	53.9	55.9
96															44.6	47.2	49.5	52.0	54.4	56.3

* **From R. G. Huebscher, *Trans. ASHVE*, Vol. 54 (1948), pp. 112–13. Reprinted by permission.**

14.0	14.5	15.0	15.5	16
6.6	6.7	6.8	6.9	7.0
7.2	7.3	7.4	7.5	7.6
7.8	7.9	8.1	8.2	8.3
8.4	8.5	8.6	8.7	8.9
8.9	9.0	9.1	9.3	9.4
9.4	9.5	9.6	9.8	9.8

32	34	36	38	40	42	44	46	48	50	52	56	60	64	68	72	76	80	84	88	Side Rectangular Duct
																				6
																				7
																				8
																				9
																				10
																				11
																				12
																				13
																				14
																				15
																				16
																				17
																				18
																				19
																				20
																				22
																				24
																				26
																				28
																				30
35.0																				32
36.0	37.2																			34
37.0	38.2	39.4																		36
38.0	39.2	40.4	41.6																	38
39.0	40.2	41.4	42.6	43.8																40
39.9	41.1	42.4	43.6	44.8	45.9															42
40.8	42.0	43.4	44.6	45.8	46.9	48.1														44
41.7	43.0	44.3	45.6	46.8	47.9	49.1	50.3													46
42.6	43.9	45.2	46.5	47.8	48.9	50.2	51.3	52.6												48
43.5	44.8	46.1	47.4	48.8	49.8	51.2	52.3	53.6	54.7											50
44.3	45.7	47.1	48.3	49.7	50.8	52.2	53.3	54.6	55.8	56.9										52
45.0	46.5	48.0	49.2	50.6	51.8	53.2	54.3	55.6	56.8	57.9										54
45.8	47.3	48.8	50.1	51.5	52.7	54.1	55.3	56.5	57.8	58.9	61.3									56
46.6	48.1	49.6	51.0	52.4	53.7	55.0	56.2	57.5	58.8	60.0	62.3									58
47.3	48.9	50.4	51.8	53.3	54.6	55.9	57.1	58.5	59.8	61.0	63.3	65.7								60
48.0	49.7	51.2	52.6	54.2	55.5	56.8	58.0	59.4	60.7	62.0	64.3	66.7								62
48.7	50.4	52.0	53.4	55.0	56.4	57.7	59.0	60.3	61.6	62.9	65.3	67.7	70.0							64
49.5	51.1	52.8	54.2	55.8	57.2	58.6	59.9	61.2	62.5	63.9	66.3	68.7	71.1							66
50.2	51.8	53.5	55.0	56.6	58.0	59.5	60.8	62.1	63.4	64.8	67.3	69.7	72.1	74.4						68
50.9	52.5	54.2	55.8	57.3	58.8	60.3	61.7	63.0	64.3	65.7	68.3	70.7	73.1	75.4						70
51.5	53.2	54.9	56.5	58.0	59.6	61.1	62.6	63.9	65.2	66.6	69.2	71.7	74.1	76.4	78.8					72
52.1	53.9	55.6	57.2	58.8	60.4	61.9	63.3	64.8	66.1	67.5	70.1	72.7	75.1	77.4	79.9					74
52.7	54.6	56.3	57.9	59.5	61.2	62.7	64.1	65.6	67.0	68.4	71.0	73.6	76.1	78.4	80.9	83.2				76
53.3	55.2	57.0	58.6	60.3	62.0	63.4	64.9	66.4	67.9	69.3	71.8	74.5	77.1	79.4	81.8	84.2				78
53.9	55.8	57.6	59.3	61.0	62.7	64.1	65.7	67.2	68.7	70.1	72.7	75.4	78.1	80.4	82.8	85.2	87.5			80
54.5	56.4	58.2	60.0	61.7	63.4	64.9	66.5	68.0	69.5	71.0	73.6	76.3	79.0	81.4	83.8	86.2	88.6			82
55.1	57.0	58.9	60.7	62.4	64.1	65.7	67.3	68.8	70.3	71.8	74.5	77.2	79.9	82.4	84.8	87.2	89.6	91.9		84
55.7	57.6	59.5	61.3	63.0	64.8	66.4	68.0	69.5	71.1	72.6	75.4	78.1	80.8	83.3	85.8	88.2	90.6	92.9		86
56.3	58.2	60.1	62.0	63.7	65.4	67.0	68.7	70.3	71.8	73.4	76.3	79.0	81.6	84.2	86.8	89.2	91.6	93.9	96.3	88
56.9	58.8	60.7	62.6	64.4	66.0	67.8	69.4	71.1	72.6	74.2	77.1	79.9	82.5	85.1	87.8	90.2	92.6	94.9	97.3	90
57.4	59.4	61.3	63.2	65.0	66.8	68.5	70.1	71.8	73.3	74.9	77.8	80.8	83.4	86.0	88.7	91.2	93.6	95.9	98.3	92
57.9	60.0	61.9	63.8	65.6	67.5	69.2	70.8	72.5	74.1	75.6	78.6	81.7	84.3	86.9	89.6	92.1	94.6	96.9	99.3	94
58.4	60.5	62.4	64.4	66.2	68.2	69.8	71.5	73.2	74.8	76.3	79.4	82.6	85.2	87.8	90.5	93.0	95.6	97.9	100.3	96

a certain length and mean velocity of flow as a circular duct of the same hydraulic radius. By using Eq. (12-26). Table 12-2 was constructed to give values of circular equivalents for rectangular and square ducts.

Example 12-3. (a) Find the friction loss in 60 ft of 30-in.-diam duct carrying 10 000 cfm of 70 F air, by the use of the friction-loss chart in Fig. 12-4. (b) Assume that the duct is made of poor construction in the field and estimate the friction loss. (c) Use Table 12-2 to find the corresponding size of rectangular duct not more than 16 in. wide to carry this air. (d) Assume that the air is at 150 F instead of 70 F and compute the pressure loss.

Solution: (a) On Fig. 12-4, locate the 10 000-cfm line on the left axis and follow this horizontally to its intersection with the diagonal line for the 30-in. duct diameter.

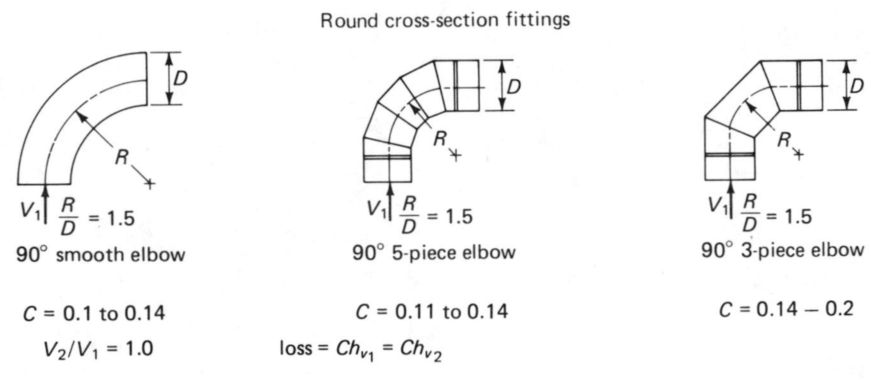

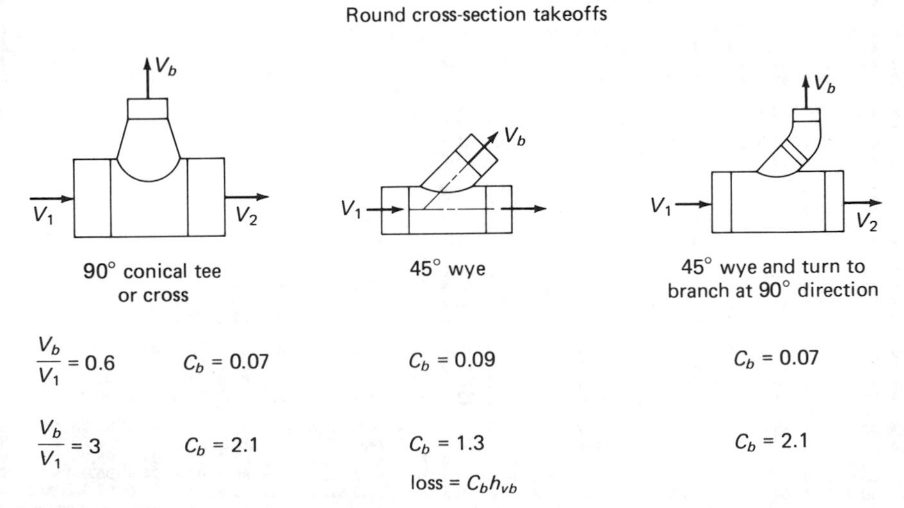

FIGURE 12-6. Factors for determining static pressure loss in elbows and takeoffs of round cross-section ducts expressed in terms of velocity heads.

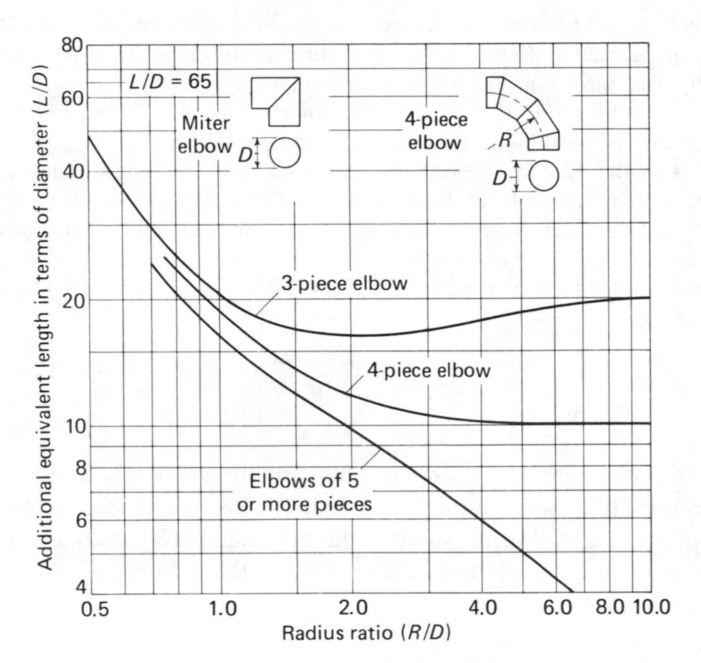

FIGURE 12-7. Loss in additional elbow equivalents for 90° elbows of round cross section. (Reprinted, by permission, from *Heating Ventilating Air Conditioning Guide 1955*, chap. 32.)

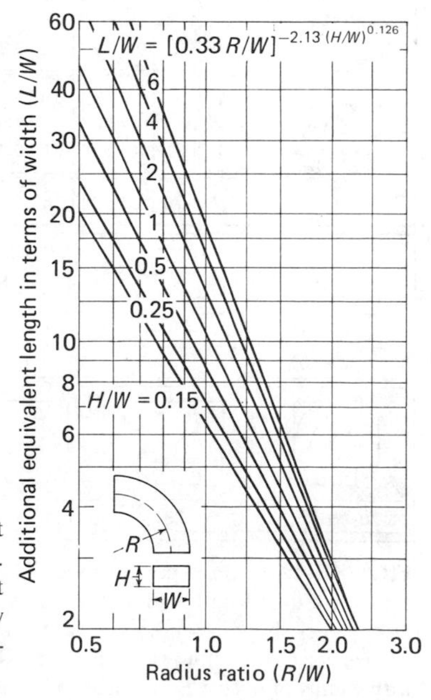

FIGURE 12-8. Loss in additional equivalent widths for 90° elbows of rectangular cross section. To find additional equivalent length L, multiply duct width W in feet by L/W value shown. (Reprinted, by permission, from *Heating Ventilating Air Conditioning Guide 1955*, chap. 32.)

At this point it will be noted that the velocity is above 2000 fpm, or at (say) 2060 fpm; the friction loss read at the top or bottom of the chart is 0.16 in. water per 100 ft of run. For a 60-ft run, the loss is $0.16 \times \frac{60}{100} = 0.096$ in. water. *Ans.*

(b) Read from Fig. 12-5, at 2060 fpm for medium rough pipe, a factor of 1.35. Therefore the corrected loss $= 0.096 \times 1.35 = 0.130$ in. water. *Ans.*

(c) In Table 12-2 locate, in the column for a duct of 16-in. side, a circular-duct diameter of 30 in. Interpolation between 30.3 and 29.8 shows that a 30-in. duct corresponds to a 51-in. side. Hence employ a 16- by 51-in. rectangular duct for the same friction loss. *Ans.*

(d) To correct for temperature, use Eq. (12-25). Thus the pressure loss, in inches of water, is

$$\Delta h_c = (0.130)\left(\frac{460 + 70}{460 + 150}\right) = 0.113 \qquad \textit{Ans.}$$

Data on representative pressure loss in elbows, bends, and fittings are given in Figs. 12-7, 12-8, and 12-9, in which the loss is expressed in equivalent length of duct (L) and is measured in duct diameters (D) or duct widths (W). This method of expression is independent of the velocity of the air or gas

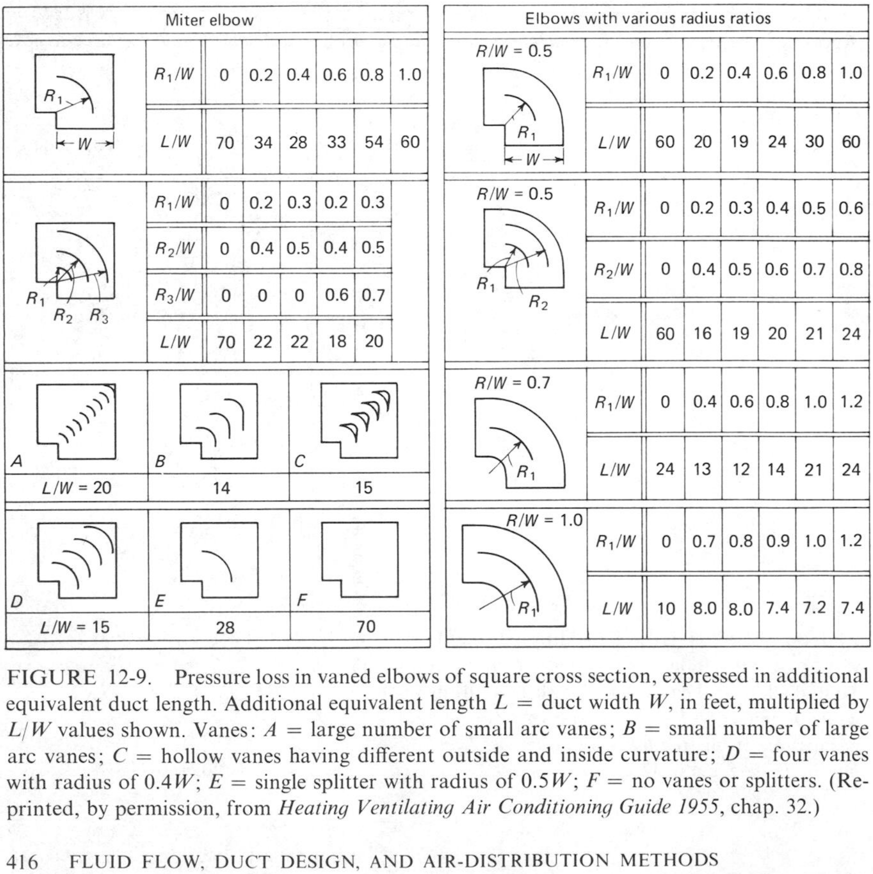

Miter elbow

R_1/W	0	0.2	0.4	0.6	0.8	1.0
L/W	70	34	28	33	54	60

R_1/W	0	0.2	0.3	0.2	0.3
R_2/W	0	0.4	0.5	0.4	0.5
R_3/W	0	0	0	0.6	0.7
L/W	70	22	22	18	20

A	B	C
$L/W = 20$	14	15

D	E	F
$L/W = 15$	28	70

Elbows with various radius ratios

$R/W = 0.5$

R_1/W	0	0.2	0.4	0.6	0.8	1.0
L/W	60	20	19	24	30	60

$R/W = 0.5$

R_1/W	0	0.2	0.3	0.4	0.5	0.6
R_2/W	0	0.4	0.5	0.6	0.7	0.8
L/W	60	16	19	20	21	24

$R/W = 0.7$

R_1/W	0	0.4	0.6	0.8	1.0	1.2
L/W	24	13	12	14	21	24

$R/W = 1.0$

R_1/W	0	0.7	0.8	0.9	1.0	1.2
L/W	10	8.0	8.0	7.4	7.2	7.4

FIGURE 12-9. Pressure loss in vaned elbows of square cross section, expressed in additional equivalent duct length. Additional equivalent length L = duct width W, in feet, multiplied by L/W values shown. Vanes: A = large number of small arc vanes; B = small number of large arc vanes; C = hollow vanes having different outside and inside curvature; D = four vanes with radius of $0.4W$; E = single splitter with radius of $0.5W$; F = no vanes or splitters. (Reprinted, by permission, from *Heating Ventilating Air Conditioning Guide 1955*, chap. 32.)

flowing and gives the amount of straight run which will give the same loss as is occasioned by the elbow or elbows in question. Figures 12-7, 12-8, and 12-9. which should be used in designing duct systems, represent design values which include some allowance for actual construction. The square duct with built-in vanes, Fig. 12-9, shows how much the loss figure can vary for different arrangements. Losses expressed in velocity heads are given in Table 12-3 and also in Fig. 12-6. Figure 12-11 is a chart for converting given velocities to velocity heads.

Example 12-4. In Fig. 12-10 is shown a rectangular duct having constant area throughout. The elbow at BC with width (W) in the plane of the bend and with depth (H) has an aspect ratio H/W of 2.5 and a radius ratio, R/W, of 1.0. The elbow at BC has an aspect ratio, H/W, of 0.4 and a radius ratio, R/W, of 0.5. (a) Make use of values from Fig. 12-8 and compute the additional length of duct run, expressed in duct widths, for both elbows. (b) Compute the corresponding lengths in feet if the duct is 20 by 50 in. at each elbow. (c) An equal-area transition exists in this duct at DE. Compute the loss in velocity heads and in inches of water for standard air when the duct velocity is 1000 fpm. (d) Carry out the transition computation for DE in SI (metric) units when the velocity is given at 304.8 m/min and the air density is 1.201 kg/m^3.

Solution: (a) Refer to Fig. 12-8 and for $H/W = 2.5$ at $R/W = 1.0$, read $L = 13.5W$; for $H/W = 0.4$ at $R/W = 0.5$, read $L = 30\ W$. *Ans.*

(b) For elbow BC,

$$L = 13.5W = (13.5)\left(\frac{20}{12}\right) = 22.5\ \text{ft} \qquad \textit{Ans.}$$

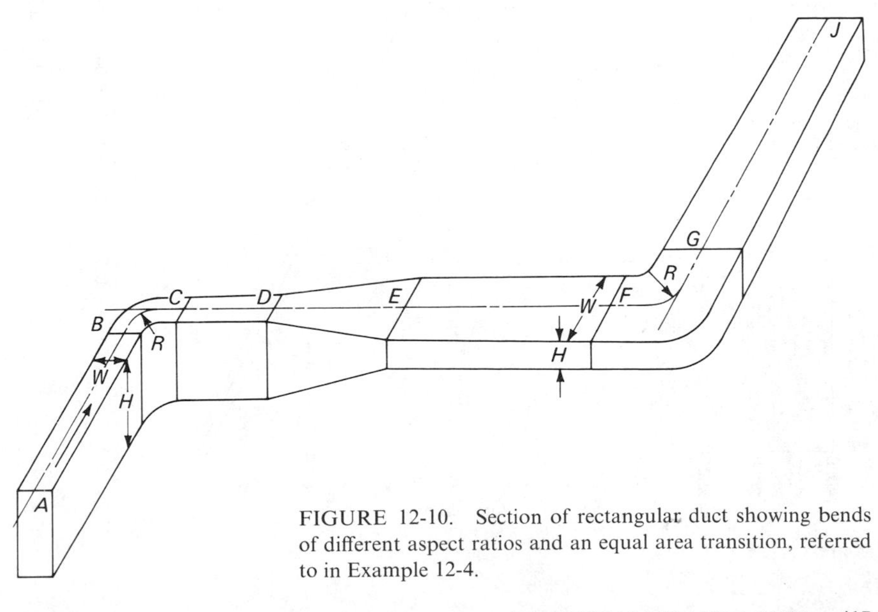

FIGURE 12-10. Section of rectangular duct showing bends of different aspect ratios and an equal area transition, referred to in Example 12-4.

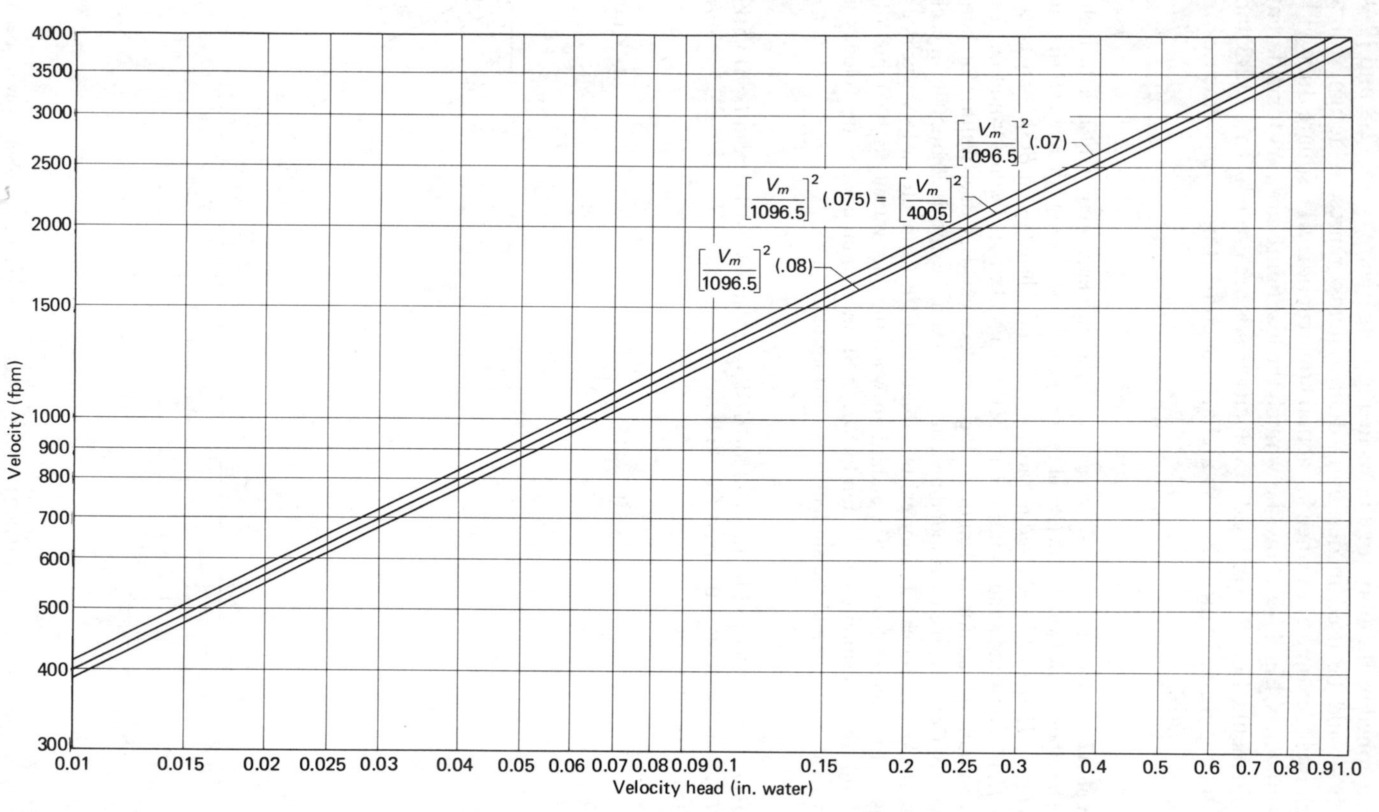

FIGURE 12-11. Velocity heads for gases of various densities.

Table 12-3 Pressure Loss in Duct Systems under Conditions of Area Change (Loss coefficient factor C_1 applies to upstream velocity head, h_{v1}; factor C_2 applies to downstream velocity head, h_{v2}.)

Contraction, Abrupt with Sharp Edge Inlet: Loss $= C_2 h_{v2}$

$$\frac{V_1}{V_2} = 0.2,\ C_2 = 0.32;\ \frac{V_1}{V_2} = 0.4,\ C_2 = 0.25;\ \frac{V_1}{V_2} = 0.6,\ C_2 = 0.16$$

Contraction, Gradual: Loss $= C(h_{v2} - h_{v1})$

30° slope, $C = 1.02$; 45° slope, $C = 1.04$; 60° slope, $C = 1.07$

Expansion, Abrupt: Loss $= C_1 h_{\ 1}$, and for

$$\frac{V_2}{V_1} = 0.1,\ C_1 = 0.81;\ \frac{V_2}{V_1} = 0.2,\ C_1 = 0.64;\ \frac{V_2}{V_1} = 0.4,\ C_1 = 0.36$$

$$\frac{V_2}{V_1} = 0.5,\ C_1 = 0.25;\ \frac{V_2}{V_1} = 0.6,\ C_1 = 0.16;\ \frac{V_2}{V_1} = 0.8,\ C_1 = 0.04$$

Expansion, Gradual, with Regain: Loss $= C(h_{v1} - h_{v2})$, and for

$$\frac{V_2}{V_1} - 0.2\text{: at } 10^\circ,\ C = 0.74;\text{ at } 20^\circ,\ C = 0.62;\text{ at } 30^\circ,\ C = 0.52$$

$$\frac{V_2}{V_1} = 0.4;\text{ at } 10^\circ,\ C = 0.83;\text{ at } 20^\circ,\ C = 0.74;\text{ at } 30^\circ,\ C = 0.68$$

$$\frac{V_2}{V_1} = 0.6;\text{ at } 10^\circ,\ C = 0.87;\text{ at } 20^\circ,\ C = 0.82;\text{ at } 30^\circ,\ C = 0.79$$

Equal-Area Transition for Rectangular Duct (Wide Side Becoming Narrow, Narrow Side Becoming Wide, 15° or Less Slope): Loss $= C_1 h_{v1}$

$C_1 = C_2 = 0.15$

Fan Entrance Louvers: Loss $= C_1 h_{v1} = 0.5\ h_{v1}$

Grille, Net Area Equal to Duct Area: Loss $= C_1 h_{v1} = 1.25\ h_{v1}$

Obstruction of Dimension W Running Across Duct of Height D: Loss $= C_1 h_{v1}$

Rectangular obstruction of width W

$$\frac{W}{D} = 0.10,\ C_1 = 0.7;\ \frac{W}{D} = 0.25,\ C_1 = 1.4;\ \frac{W}{D} = 0.5,\ C_1 = 4.0$$

Round obstruction (pipe) of diameter W

$$\frac{W}{D} = 0.10,\ C_1 = 0.20;\ \frac{W}{D} = 0.25,\ C_1 = 0.55;\ \frac{W}{D} = 0.5,\ C_1 = 2.0$$

Streamlined obstruction of width W

$$\frac{W}{D} = 0.10,\ C_1 = 0.07;\ \frac{W}{D} = 0.25,\ C_1 = 0.23;\ \frac{W}{D} = 0.5,\ C_1 = 0.9$$

Orifice, Square-Edge Inlet: Loss $= C_2 h_{v2}$ (use orifice efflux velocity), for

$$\frac{A_o}{A_1} = 0.2,\ C_2 = 2.44\text{: } \frac{A_o}{A_1} = 0.4,\ C_2 = 2.66;\ \frac{A_o}{A_1} = 0.8,\ C_2 = 1.54$$

Note. These loss coefficients are fully applicable only under incompressible-flow conditions, such as exist in ducts under small pressure variations.

For elbow FG,

$$L = 30W = (30)\left(\frac{50}{12}\right) = 125 \text{ ft} \qquad \textit{Ans.}$$

The comparative greater length for the bend FG (greater friction loss) points to the desirability of adding vanes or having a larger turning radius for a flat bend such as this one.

(c) Refer to Table 12-3 and select the coefficient for a gradual-slope, equal-area transition as 0.15 velocity head. Loss is thus $0.15 \times h_v$. The numerical value of velocity head for this case can be found by use of Eq. (12-19) or (12-21).

$$h_v = \left(\frac{1000}{1096.5}\right)^2 (0.075) = \left(\frac{1000}{4005}\right)^2 = 0.0624 \text{ in. water}$$

It can also be read directly from Fig. 12-6 as 0.062 in. water.

$$\text{loss} = (0.15)(0.0624) = 0.0094 \text{ in. water} \qquad \textit{Ans.}$$

(d) To develop a working equation for velocity head in SI (metric) units it is merely necessary to express the velocity head in *meters of air* and then convert to *centimeters (or millimeters) of water*, using SI units throughout. With V expressed in meters per second,

$$h_a = \frac{V^2}{2g_c} = \frac{V^2}{(2)(9.80665)} \text{ m of air}$$

Using consistent units, h_a and h_w in meters and d_a and d_w in kilograms per cubic meter.

$$h_a d_a = h_w d_w$$

$$h_a = h_w\left(\frac{d_w}{d_a}\right) = \frac{V^2}{(2)(9.80665)} = \frac{V^2}{19.613}$$

$$h_w = \frac{V^2(d_a)}{19.613 d_w}$$

Let us now change V in meters per second to V_m in meters per minute, and h_w in meters to h_v in centimeters of water, and note that for water $d_w = 998 \text{ kg/m}^3$ and $d_a = 1.201 \text{ kg/m}^3$ for standard air.

$$h_v = \frac{(V_m)^2}{(60)^2}\frac{(d_a)(100)}{(19.613)(d_w)} = \frac{V_m^2 d_a}{704\,660} = \left(\frac{V_m}{839.4}\right)^2 d_a \text{ cm water}$$

or, following the same pattern,

$$h_v = \left(\frac{V_m}{265.5}\right)^2 d_a \text{ mm water}$$

The 0.15 velocity head loss thus becomes

$$\text{loss} = 0.15h_v = (0.15)\left(\frac{V_m}{265.5}\right)^2 1.201 = 0.237 \text{ mm water}$$

$$0.237 \times 9.789 = 2.32 \text{ Pa} \qquad \textit{Ans.}$$

Except for very long runs of ducting, the largest pressure losses are associated with the duct fittings, elbows, bends, transitions, branch connections, dampers, and other obstructions that are required in connection with duct layout and construction.

12-5. DUCT DESIGN PROCEDURE

In any mechanical-circulation heating, cooling, or ventilating system the fan or fans must have adequate capacity to deliver the air quantity required at a static pressure equal to or slightly greater than the total resistance offered by the duct system. The sizes of the ducts are set by the maximum air velocities which can be used without causing undue noise or without causing excessive friction loss. Large ducts will reduce frictional losses, but the space and investment requirements offset the power saving at the fan. An economic balance must be made in designing the installation. In general, however, the layout should be made as direct as possible, sharp bends should be avoided, and the duct cross section, if rectangular, should not be too flattened. For a rectangular duct, a ratio of long to short sides of up to 6 to 1 is good practice, and the ratio should never exceed 10 to 1.

Table 12-4 Recommended and Maximum Duct Velocities for Low-Velocity Systems

	Recommended Velocities (fpm)			Maximum Velocities (fpm)		
Designation	Residences	Schools, Theaters, Public Buildings	Industrial Buildings	Residences	Schools, Theaters, Public Buildings	Industrial Buildings
Outside air intakes[a]	500	500	500	800	900	1200
Filters[a]	250	300	350	300	350	350
Heating coils[a]	450	500	600	500	600	700
Cooling coils[a]	450	500	600	450	500	600
Air washers	500	500	500	500	500	500
Suction connections	700	800	1000	900	1000	1400
Fan outlets	1000–1600	1300–2000	1600–2400	1700	1500–2200	1700–2800
Main ducts	800–900	1000–1300	1200–1800	800–1000	1100–1600	1300–2200
Branch ducts	600	600–900	800–1000	700–1000	800–1300	1000–1800
Branch risers	500	600–700	800	650–800	800–1200	1000–1600

[a] These velocities are for total face area, not the net free area.
Adapted by permission from *ASHRAE Handbook of Fundamentals*, 1972, chap. 25.

Table 12-5 Typical Frictional Losses for Duct System Equipment

Item	Possible Range of Loss* (in. water)
Air intake or fan entry	0.005 to 0.1
Air heaters or coolers, one row to several rows	0.1 to 0.35
Air washer	0.2 to 0.35
Air filters	0.2 to 0.4
Duct system (calculated for worst run)	0.04 to 0.4
Miscellaneous, screens, grilles, etc.	0.1 to 0.2
Nozzle-type outlets	0.1
Less any velocity head regain	0.01 up
Total static pressure loss for system (fan)	1.0 to 1.6 usual

*Selected from representative manufacturer's data, or calculated.

The possible resistances that a fan must overcome in delivering its air are tabulated in Table 12-5. These items are only representative; they differ for each system.

In the duct design the following procedure may be followed:

1. Lay out the most convenient system of placing the various ducts to obtain adequate distribution and to facilitate construction.
2. From the heating or cooling load, calculate the air requirements cubic feet per minute (cfm) at each duct outlet, zone, or division of the building.
3. Determine the sizes of these outlet branches, using a proper velocity or pressure drop to deliver the required quantity.
4. Calculate the size of each duct by one of the following methods:
   a. *Assumed-velocity method*. The velocity in each of the various sections of the duct is assumed, in accordance with good practice, and the separate losses of each part of a definite system or circuit are added to find the total pressure loss. A modification of this approach, known as the *velocity-reduction method*, follows the general procedure indicated, but the assumed velocities are reduced progressively in the duct sections. The highest velocity is taken at the fan outlet, and velocities are lowered in the main after various branches are taken off. In general, the assumed-velocity method without refinement should be used only for relatively simple duct-system layouts. The control of flow in the various branches will be largely dependent on dampering.
   b. *Constant-pressure-drop method* (*equal-friction method*). The duct is proportioned so that the frictional loss per foot of length is constant. It is then possible for the resistances of the branches to be made essentially equal unless they are of greatly different lengths. When this method is used, it is customary to establish

the constant pressure drop on the basis of the desirable velocity in the duct main beyond the fan. The branches must be dampered for control.

   c. *Balanced-pressure-loss method.* This method employs pertinent procedures of method a or b, but every branch is designed to have the same pressure loss from the fan in order that minimum dependence on dampering is required. Expressed in another way, the static pressure required for the flow in any branch from its point of attachment is made equal to the static pressure of the main system at the juncture point. Theoretically, dampers for adjustment of the design flow in any branch would not be required in such a system, but it is always desirable to supply them.
   d. *Static-regain method.* This method endeavors to meet the objectives in method c but employs static regain as an accessory device. The use of static regain will be discussed later.
5. Determine, from the calculations described under item 4 the circuit offering the greatest frictional resistance. While the circuit thus selected is frequently the longest circuit, this is not necessarily true. The maximum resistance determines the static pressure the fan must deliver to supply the air through the ducts. Note that the fan has to supply more air than is indicated in item 2, in order to make up for leakage in the duct system (often about 10%) and in order to allow for heat transfer to or from the air in the duct while passing through unconditioned spaces.

In the case of factory-built conditioners, including a fan, the available static pressure of the conditioner fan determines the total allowable resistance in the duct system. In cases such as this it may be necessary to redesign the duct system for a smaller or greater resistance, to accommodate it to the characteristics of the fan supplied.

Pressure loss from air flowing in a duct occurs from frictional resistance to flow, from the shock losses of abrupt area changes, and from the shock and turbulence associated with change in direction. Pressure decreases also when the velocity of the air stream is increased. This change naturally occurs when the cross-sectional area for flow decreases (that is, in any converging section). Conversely, in a diverging or increasing section a velocity decrease can result in a pressure rise. Unfortunately, this reversed process of pressure rise when velocity decreases (known as diffusion) is a more difficult transformation. The air is necessarily flowing in the direction of increasing pressure, and turbulence loss due to separation of the stream from the walls of the passage can occur unless the divergence of the section is gradual (the included angle is less than 20°). With a very smooth passage and gradual taper, 85 to 75% of the ideal pressure rise might occur, but with conventional duct arrangements 60 and 50% are more representative limits. Pressure rise of this kind is called *regain.* Regain occurs also in ducts of constant cross-sectional area, as in a trunk duct beyond an outlet. For example, if a constant-diameter duct carrying 4000 cfm at 1600 fpm delivers 1500 cfm to a branch, and if the remaining 2500 cfm continue on, the velocity beyond the outlet can be only 1000 fpm, since $\frac{4000}{2500} = \frac{1600}{1000}$.

Regain Equations. By using representative coefficients for typical duct construction, the following equations for standard air can give the expected increase or decrease in pressure.

Static pressure regain (SPR), in inches of water, resulting from a lower final velocity (V_f fpm), with V_s fpm the velocity at the start of the transformation, is

$$\text{SPR} = 0.5\left[\left(\frac{V_s}{4005}\right)^2 - \left(\frac{V_f}{4005}\right)^2\right] \tag{12-27}$$

or

$$\text{SPR} = 0.5\left[\frac{(V_s - V_f)(V_s + V_f)}{16\,040\,000}\right] \tag{12-28}$$

A static pressure loss (SPL) occurs when the velocity increases; and, if a formula of the form of Eq. (12-27) or (12-28) is used, the answer is minus, indicating a negative static pressure regain. The coefficient must also be modified. The relation to be used is

$$\text{SPL} = 1.05\left[\left(\frac{V_s}{4005}\right)^2 - \left(\frac{V_f}{4005}\right)^2\right] \tag{12-29}$$

or

$$\text{SPL} = 1.05\left[\frac{(V_s - V_f)(V_s + V_f)}{16\,040\,000}\right] \tag{12-30}$$

In the static-regain method of design, mentioned earlier in this section, the velocity in the main duct is reduced after each branch or takeoff, and thus the static-pressure recovery from velocity reduction offsets or at least reduces the pressure loss in a following section. Because of this, essentially the same static pressure is available for all outlets, and remote branches are not at a disadvantage in terms of pressure required for distribution.

Example 12-5. Design a typical year-round air-conditioning duct system supplying several offices, as shown in Fig. 12-12. The maximum air capacity required in summer or winter is shown. The building limitations are such that the depth of the trunk ducts cannot exceed 16 in. and the vertical flues cannot be deeper than 10 in.

Solution [by the Constant Pressure Drop (Equal Friction) Method]: The total air to be delivered, neglecting duct leakage and heat gain or loss in the ducts, is indicated for the fan as 4000 + 3600 = 7600 cfm.

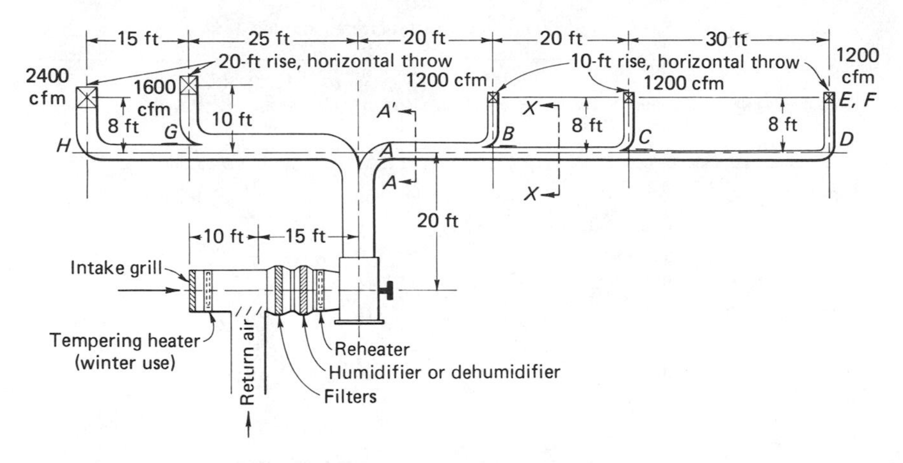

FIGURE 12-12. Duct system.

The inlet duct is designed for maximum capacity at 7600 cfm, with a velocity of 1000 fpm selected from Table 12-4. The velocity in the trunk duct at the fan outlet is taken as 1200 fpm (Table 12-4), as noise should be minimized. With Q in cfm and V in fpm,

$$Q = AV = \frac{HW}{144} V$$

With a maximum depth, H is 16 in. for the selected conditions, and

$$7600 = \left(\frac{16 \times W}{144}\right)(1200) \qquad \text{or} \qquad W = 57 \text{ in.}$$

The trunk-duct dimensions are 57 by 16 in. For a duct of this size, Table 12-2 shows the corresponding round size to be 31.45 (say, 31.5) in. The friction chart in Fig. 12-4 shows, for 7600 cfm in a 31.5-in. round duct, a loss of 0.076 in water per 100 ft of run.

The right branch of the trunk duct at AA' carries 3600 cfm; and, for 0.076 in. water loss, Fig. 12-4 indicates an equivalent round size of 23.4 in. Similarly, at XX, where the required capacity is 2400 cfm, read 20.3 in.; and for run CDE at 1200 cfm, read 15.2-in. diam. Table 12-2 shows a 32- by 16-in. rectangular duct for the 23.4-in. round size at AA'; a 22- by 16-in. rectangular duct for the 20.3-in. round size at XX; a 12- by 16-in. rectangular duct for the 15.2-in. round size for CD and DE; and 20- by 10-in. rectangular size for EF. The rectangular sizes have been rounded off to the nearest inch (half-inch for small ducts).

The velocity in the riser EF is

$$V = \frac{144Q}{HW} = \frac{(144)(1200)}{(20)(10)} = 864 \text{ fpm}$$

As this is below the maximum recommended in Table 12-4, it will be considered acceptable.

Tabulation of Data and Results for Example 12–5

Section or Elbow	Capacity (cfm)	Velocity (fpm)	Duct Diameter (in.)	Rectangular Size (in.)	Depth H Perpendicular to Plane of Bend	Width W in Plane of Bend	H/W	L/W	L (ft)
Fan	7600	1200	31.5	16 × 57	16	57	0.28	—	—
AA'	3600	1010	23.4	16 × 32	16	32	—	—	—
A	—	—	—	—	—	—	0.5[a]	4.0	10.7
XX	2400	980	20.3	16 × 22	16	22	0.73	—	—
C to E	1200	900	15.2	16 × 12	16	12	—	—	—
D	—	—	—	—	—	—	1.33*	4.7	4.7
E to F	1200	864	15.2	20 × 10	20	10	—	—	—
E	—	—	—	—	—	—	2†	25.0	21.0
F	—	—	—	—	—	—	2†	25.0	21.0
Total of additional equivalent length for elbows									57.4

* R/W taken as 1.5.
† R/W taken as 0.75.

These data are expressed in tabular form in the accompanying table, for the right-hand trunk and longest branch, with Fig. 12-8 being used to find the L/W values for the appropriate elbows.

For the left branch, employ the same methods. For a 0.076-in. loss per 100 ft, the trunk to the right of G at 4000 cfm shows 26.6 in. round; and the trunk GH at 2400 cfm shows 20.3 in. round. The corresponding rectangular sizes are 39 by 16 in. before G, 22 by 16 in. from G to H, and 38 by 10 in. in the riser above H.

Calculation of Frictional Loss. The frictional loss is obviously greatest in the extreme right run, but if there should be any doubt the extreme left run and other runs should also be checked.

The right run from the fan out has a length of

$$
\begin{aligned}
&20 + 20 + 20 + 30 + 8 + 10 = 108 \text{ ft} \\
&\text{From the table, equivalent} \\
&\quad \text{additional length for elbows} = 57.4 \\
&\text{Total equivalent length} \qquad = 165.4 \text{ ft}
\end{aligned}
$$

A minimum factor of safety of 10% should be added to this length. As the loss is 0.076 in. water/100 ft, the total duct and elbow friction loss is:

$$165.4 \times 1.10 \times \frac{0.076}{100} = 0.138 \text{ in. water}$$

Some overall static regain occurs in this run, and it can be found by Eq. (12-28). The starting and final velocities in the fan run and in the riser EF being used, we find that

$$\text{SPR} = 0.5\left[\frac{(1200 - 864)(1200 + 864)}{16\,040\,000}\right] = 0.022 \text{ in. water.}$$

The net static pressure loss in the duct, bends, etc., is thus

$$0.138 - 0.022 = 0.116 \text{ in. water}$$

In addition to this net static pressure loss in the duct system, the fan is required to overcome the losses in the accessory equipment and in the return system. Computed or representative values for these losses will now be tabulated. The first item is the static pressure loss in moving the air through the intake grille and louvers, or 0.50 velocity head (from Table 12-3).

$(1200/4005)^2 \times 0.5 = 0.090 \times 0.5$	0.045
Tempering heater (mfr.'s data)	0.10
Filters (mfr.'s data)	0.25
Dehumidifier (mfr.'s data)	0.22
Reheater (mfr.'s data)	0.10
Outlet grille (mfr.'s data)	0.057
Net static pressure loss, computed before	0.116
Total static pressure	0.888 in. water

The fan selected must have a static pressure not less than 0.888 in. water when delivering 7600 cfm. The total pressure of the fan, which takes into consideration the fact that the air is inducted and accelerated to the fan outlet velocity, is obviously greater than 0.888 in. water. Fan selection methods are considered later in the chapter. In the event that the return system is also handled by this fan, pressure to cover the return-system loss should also be added.

For the other branches and the left run, similar methods of computation can be employed.

The preceding example was worked by the constant-pressure-drop method; that is, the frictional loss per foot of equivalent duct length was assumed constant. This method is generally to be preferred for solving problems in duct design, because less experience in selecting velocities is required. The branches closest to the fan have the highest static pressure, and to prevent them from getting more than their share of air they must have adequate damper control. This general method applies equally well to exhaust systems, and the procedure to be followed is the same. If it is desired to avoid much use of dampers, earlier branches can be designed for higher velocities and therefore, higher losses. This possibility is often limited by the maximum allowable velocity dictated by considerations of noise generation and duct vibration or bulging.

The previously mentioned assumed-velocity method may be used satisfactorily in simple systems where the duct losses form a relatively small part of the total loss. Even with the constant pressure drop, to start the problem a trial velocity may be assumed, as in Example 12-5. A rational friction drop could have been assumed instead of a velocity, if this procedure had been desired.

Many variations in design procedure can be made in accordance with the basic rules previously outlined. A frequently used design involves delivering the air into a plenum (pressure) chamber and then employing a number of separate small ducts leading to the point of use. The plenum pressure must be maintained high enough to deliver the required quantity of air through the duct of greatest resistance.

Fundamentally the balanced-pressure-loss method should be most satisfactory, and the extra labor of adjusting the whole design to bring about equal total friction loss may not be difficult. However, for a short duct directly off a plenum chamber or in the main duct close to the outlet of a fan, the static pressure is close to the maximum of the system. It may thus be impossible, merely by increasing the velocity within reasonable limits, to set up the required pressure loss in this short duct; and dampers must be employed to utilize the available pressure with or without a restricted opening, such as an orifice at the supply point to supplement the dampers. In this method the basic problem is simply to design the system so that the total pressure drop from the fan to any grille is the same, and a dependence on dampers only for minor final adjustments is the desired end.

Example 12-6. For the duct system of Fig. 12-12, design the first branch on the right side to have the same total friction loss as the extreme right (longest) branch.

Solution: As determined in Example 12-5, the static pressure on discharge from the fan, is 0.888 in. water, less the pressure utilized for bringing in and carrying the air through the various devices preceding it, or $0.888 - (0.045 + 0.10 + 0.25 + 0.22 + 0.10) = 0.888 - 0.715$ in. water. At leaving the fan, the total static pressure is thus 0.173 in. water, and the velocity head is 0.09 in. for 1200 fpm (read from Fig. 12-11).

For the main duct system from the fan to point B, there is 40 ft of run plus 10.7 ft of equivalent length for elbow A; and, as this was designed for 0.076 in. of loss per 100 ft, the pressure loss is $0.076 \times 50.7/100 = 0.037$ in. water. Thus the static pressure at B is $0.173 - 0.037 = 0.037 = 0.136$ in. water. The grille absorbs 0.057 in., leaving $0.136 - 0.057 = 0.079$ in. water; thus, disregarding velocity heads for the moment, 0.079 in. water must be absorbed in the run of duct from B to the grille outlet—which consists of 18 ft and three elbows and in which the entry into the branch is considered as an elbow.

Consider first that the elbows in the circuit from B to its outlet grille are the same in size and type as that of branch DEF. Then the equivalent length would be $18 + 4.7 + 21 + 21 = 64.7$ ft. With the same design friction loss of 0.076, the friction would be $0.076 \times 64.7/100 = 0.049$ in. water, which is less than the 0.079 in. water required.

The inlet to this branch is essentially an elbow, and if so taken the circuit at B to its outlet resembles that of branch DEF, and its elbow data can be used as a first trial for equivalent length, giving $4.7 + 21 + 21 + 18$ ft of run $= 64.7$ ft. Thus

$$\frac{64.7}{100} \times F = 0.079$$

and $F = 0.122$ (say, 0.12) in. water/100 ft is the trial friction loss factor to be used with Fig. 12-4 for resizing the branch starting at B. On this basis, to carry the 1200 cfm a round size 14.5 in. in diameter is indicated. For this the corresponding rectangular sizes are 16 by 11 in. in the horizontal run and 10 by 18 in. in the vertical run.

The equivalent additional lengths for the elbows, on the basis of the new sizes, and using elbow data similar to that in the table of Example 12-5, are

$$18 \times \tfrac{10}{12} + 18 \times \tfrac{10}{12} + 4.7 \times \tfrac{11}{12} = 34.3 \text{ ft}$$

and $18 + 34.3 = 52.3$ ft equivalent length is sufficiently close to the assumed 64.7 ft as not to require a significant change in the indicated duct sizes of 16 by 11 in. and 10 by 18 in. for the design modified to absorb more friction. For any branch a similar procedure can be followed, namely, that of estimating a total equivalent length and finding the needed friction loss for 100 ft to absorb the static head available. Another procedure where leeway is available is to select a higher velocity in the branch and carry out trial computations on the basis of the assumed velocity.

In the present case, the velocity is somewhat high from a viewpoint of noise and good practice, namely,

$$V = \frac{1200 \times 144}{18 \times 10} = 960 \text{ fpm}$$

However, if it is felt that this is definitely too high, little can be done to absorb the static pressure except through the use of dampers which in themselves are noisy, or through a redesign of the main duct system. The use of sound-deadening material in the riser may also be considered.

A reasonable grille belocity based on effective face area is 600 fpm. By Fig. 12-11, velocities of 600 fpm and 960 fpm show velocity pressures of 0.023 and 0.056 in. water, respectively. The leaving velocity pressure of 0.023 in. water is dissipated directly into the room space, and thus, of the original velocity pressure of 0.056 in. water in the riser, 0.033 in. water has not been accounted for. However, if the factor 0.5 in Eq. (12-27) is considered, it appears that only 0.033×0.5, or 0.016 in. water, is not accounted for. By using a nondirect takeoff approaching a full right angle, the velocity head transmitted into the branch can be reduced to minor proportions.

Even with this method, which equalizes the pressure losses in each and every circuit, it is still necessary to supply dampers to make adjustments for minor inaccuracies in design and for variations in construction, as well as to permit modification of operating conditions, if desired, at some later time.

Table 12-6 Recommended Maximum Pressure Losses or Gains in Branch Outlet Runs*
(with Pressures at First and Last Outlet)

Item	Average Outlet Pressure (in. of water)									
	.025	.05	.075	.10	.15	.20	.25	.30	.40	.50
Maximum allowable pressure loss in outlet run (in.) †	.01	.02	.03	.04	.06	.08	.10	.12	.16	.20
Pressure at first outlet (in.)	.03	.06	.09	.12	.18	.24	.30	.36	.48	.60
Pressure at last outlet (in.)	.02	.04	.06	.08	.12	.16	.20	.24	.32	.40

* Reprinted by permission of the Carrier Corporation.
† Consideration of the outlet selected and the individual job requirement will often indicate a selection of a loss less than the maximum value.

The previous discussion of this article are broadly applicable but are most suitable in relation to conventional low-pressure, low-velocity duct systems. Systems are classified as follows:

1. *Low pressure*, employing velocities of 2000 fpm or less and static pressures under 2 in. water pressure.
2. *Medium pressure*, employing velocities in excess of 2000 fpm and static pressures up to 6 in. water pressure or slightly more.
3. *High pressure*, employing velocities greater than 2000 fpm and static pressures in the range of 6 to 10 in. water pressure.

Ducting for medium- and high-pressure systems instead of being constructed in rectangular form is most usually made in round or oval form. Round ducts, which can readily operate under high static pressures without distortion or damage, with appropriate wall thickness can provide effective structural rigidity. Round high-velocity ducts most frequently are manufactured from spiral-seam tubes, although grooved or welded longitudinal-seam tubing is widely used as well. Connections are made in various ways, with slip joints, by gasketed flange connections, by overlapping sleeves, or even by butt welding of the pipe ends. Refer to Table 12-7 and note that for duct sizes up to 2 ft diam, U.S. Gage 22 will suffice; for 2 to 4 ft, Gage 20 is desirable; for sizes above 4 ft, Gage 18 should be employed. Spiral lock-seam ducting permits the use of the next thinner gage compared to plain-wall ducting because of the rigidity provided by the lock seam.

Table 12-4 shows the velocities which are employed in low-velocity (low-pressure) systems. A very superficial examination of these will show that for the large volumes of air that are necessary for multistory and complex buildings the size of ducting required becomes almost prohibitive. To install such ducting the ceiling height and resulting space between floors

Table 12-7 Gage Number, Thickness, and Weight of Sheet Metal Used in Ducts

	Galvanized-Sheet Steel Galvanized-Sheet Gage		Rolled-Sheet Steel U.S. Standard Gage		Aluminum Alloy B & S Gage	
Gage Number	Thickness (in.)	Weight (lb/ft^2)	Thickness (in.)	Weight (lb/ft^2)	Thickness (in.)	Weight (lb/ft^2)
26	0.0217	0.906	0.0179	0.75	—	—
24	0.0276	1.156	0.0239	1.0	0.020	0.285
22	0.0336	1.406	0.0299	1.25	0.025	0.356
20	0.0396	1.656	0.0359	1.5	0.032	0.456
18	0.0516	2.156	0.0478	2.0	0.040	0.570
16	0.0635	2.656	0.0598	2.5	0.050	0.713
14	0.0785	3.281	0.0747	3.125	0.063	0.898
12	0.1084	4.531	0.1046	4.375	0.071	0.914
10	0.1382	5.781	0.1345	5.625	—	—

must be increased significantly and large areas of otherwise usable floor space must be allocated for the vertical supply and return ducting. If, for example, a supply riser can operate at 6000 fpm instead of 2000 fpm its cross section can be reduced to one-third of its original size. The saving in material cost and building space can often offset the higher fan horsepower needed and the greater care which must be exercised for noise control in a high-pressure (high-velocity) system. Round duct of the character shown in Fig. 12-6 is usually employed in high-velocity systems. Velocities running to 3000 fpm are used in systems of 6000 cfm or less, with the velocity increasing to 4000 fpm for systems of 10 000 to 15 000 cfm, reaching as high as 6000 fpm for systems over 40 000 cfm. The terms high-pressure and high-velocity are used almost interchangeably in describing these systems.

12-6. STATIC-REGAIN METHOD OF DESIGN

The static pressure regain as the velocity is reduced in a duct system can be employed in many variations. One of the most satisfactory modifications consists of reducing the velocity after each branch by an amount sufficient to compensate for the frictional pressure loss in the succeeding section. It must be realized that in any duct system the total pressure always drops in the direction of flow. Static pressure regain merely relates to interrelationships that exist between velocity pressure and static pressure at different locations in the duct system. Only when the velocity decreases can static pressure regain occur. Ideally, pressure regain is independent of the manner in which the velocity is reduced, but the actual regain is highly dependent on the physical configuration. Where the velocity reduction takes place by virtue of branch-duct takeoffs from a main of constant cross section, realization of regain is usually taken as only 50% of the ideal. Attention is here called to the duct system of Fig. 12-3 which illustrates these comments. Note that static pressure regain occurs at *DE* because of an increase in the size of ducting and also in smaller amount at *FG* because of a takeoff from a main of uniform cross section.

Nomographs can be prepared to make easier the task of balancing regain to offset static pressure loss in succeeding sections of a duct system. None will be used here because, by simple trial and adjustment, the same result can be obtained and the nomographs tend to obscure the principles involved.

A simple example will be worked in two ways to illustrate the procedure in using static-regain methods.

Example 12-7. A duct system for a school must satisfy the conditions indicated in the diagrammatic sketch in Fig. 12-13. Size the duct system, following a reasonable design and using a constant-pressure-drop method. The maximum depth of duct cannot exceed 10 in. on one side, and the pressure at the outlet grilles must not be less than 0.05 in. water.

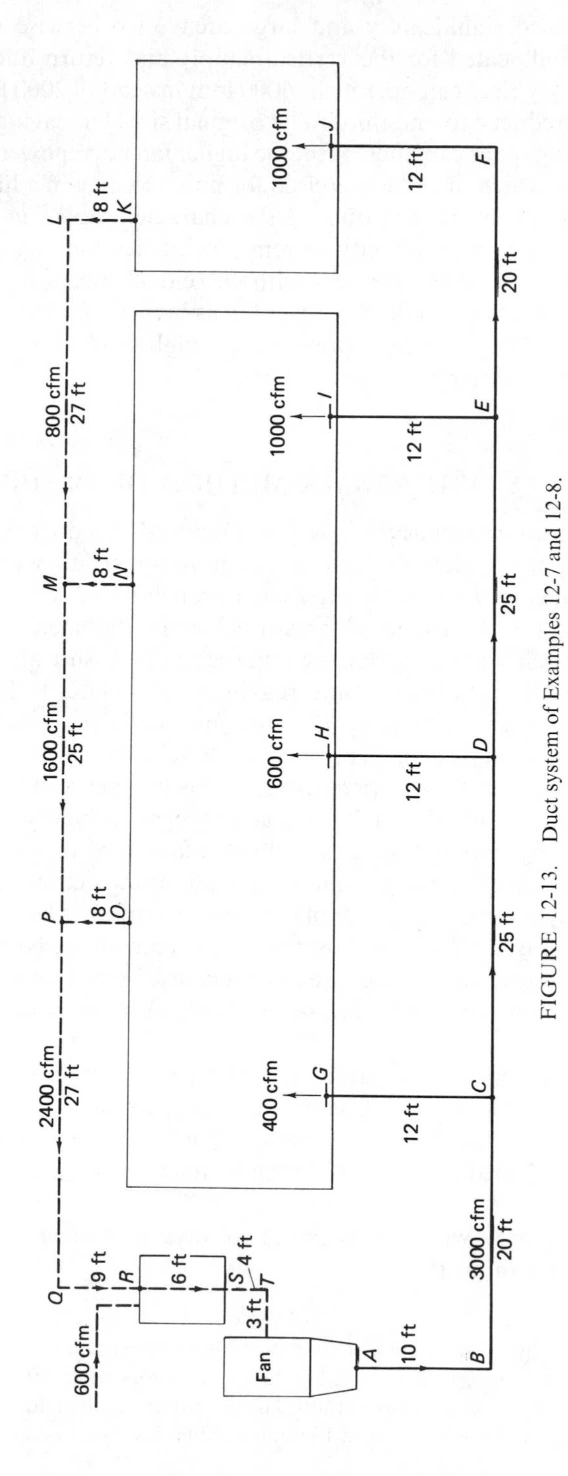

FIGURE 12-13. Duct system of Examples 12-7 and 12-8.

Solution: From Table 12-4 select a 1300-fpm velocity in the 3000-cfm section *ABC*. Find the duct size from the relation

$$A = \frac{W \times H}{144} = \frac{Q}{V}$$

Thus

$$\frac{W \times 10}{144} = \frac{3000}{1300}$$

and

$$W = 33.2 \text{ in.}$$

Round off to 33 by 10 in.

From Table 12-2, a 33- by 10-in. duct corresponds to a 19.05-in. round duct. Figure 12-4 shows that 3000 cfm in a 19.05-in. duct produces a loss of 0.16 in. water/100 ft. Consider the fan to deliver into a duct-run just under a basement ceiling.

Run *ABC* has one bend at *B*, which can be assumed to be a hard bend (i.e., on the 33-in. width, in contrast to one on the 10-in. depth). By Fig. 12-8, for a 1.5 radius ratio, R/W, and for $H/W = 10/33 = 0.3$,

$$L = 3.6W = 3.6 \times \tfrac{33}{12} = 10 \text{ ft}$$

$$\text{total EL of run } ABC = 10 + 20 + 10 = 40 \text{ ft}$$

$$\text{loss in } ABC = 0.16 \times \tfrac{40}{100} = 0.064 \text{ in. water}$$

For flow in *CD* at 0.16 in. water/100 ft and 2600 cfm, Fig. 12-4 shows an 18-in. round duct; and from Table 12-2 the corresponding rectangular size is 29 by 10 in. Therefore,

$$\text{velocity in } CD = \frac{Q}{A} = \frac{2600 \times 144}{29 \times 10} = 1292 \text{ fpm}$$

$$\text{loss for } CD = 0.16 \times \tfrac{25}{100} = 0.04 \text{ in. water}$$

For flow in *DE* at 0.16 in. water/100 ft and 2000 cfm, Fig. 12-4 shows a 16.4-in. round duct; and from Table 12-2 the rectangular size is 23.4 by 10 in. Round off to 23 by 10 in. Therefore,

$$\text{velocity in } DE = \frac{Q}{A} = \frac{2000 \times 144}{23 \times 10} = 1252 \text{ fpm}$$

$$\text{loss for } DE = 0.16 \times \tfrac{25}{100} = 0.04 \text{ in. water}$$

For flow in *EF* at 0.16 in. water/100 ft and 1000 cfm, the required round size is 12.5 in. and the corresponding rectangular size is 13.2 by 10 in. Round off to 13 by 10 in. Therefore,

$$\text{velocity in } EF = \frac{1000 \times 144}{13 \times 10} = 1107 \text{ fpm}$$

$$\text{loss for } EF = 0.16 \times \tfrac{20}{100} = 0.032 \text{ in. water}$$

The flow in branch FJ is 1000 cfm, and the velocity should be lowered from the 1107 fpm in EF. From Table 12-4 select 800 fpm; then the duct size is 18 by 10 in., since

$$\frac{W \times 10}{144} = \frac{1000}{800}$$

and

$$W = 18 \text{ in.}$$

From Table 12-2, this size corresponds to a 14.5-in. round duct; and from Fig. 12-4 the loss rate is 0.075 in. water/100 feet. Consider the bend at F to be hard (on the 18-in. width) and assume that a second bend at J just before the grille is easy (on the 10-in. dimension). From Fig. 12-8, for F at $R/W = 1.5$ and $H/W = 10/18$, read $L = 4.0W = 6$ ft; and for J leading into the outlet grille, use $R/W = 1.0$, read $H/W = 18/10$, and compute L as 12 ft. Therefore the total equivalent length from bend F to bend J is $6 + 12 + 12 = 30$ ft, and

$$\text{loss in } FJ = 0.075 \times \tfrac{30}{100} = 0.0225 \text{ in. water}$$

Even though a regain-type design is not considered here, there is necessarily some regain throughout the whole system; however, for this problem, regain will be computed and used only for the branch run into FJ, for which there is a large change in velocity. By Eq. (12-28),

$$\text{SPR} = 0.5\left[\frac{(1107 - 800)(1107 + 800)}{16\,040\,000}\right] = 0.018 \text{ in. water}$$

The results are shown in the accompanying tabulation. It can be seen that the total pressure loss from A to J is 0.181 in. water. If 0.050 in water is allowed at the outlet grille, it is obvious that the static pressure at A (the fan outlet) must be 0.181 + 0.05, or 0.231 in. water. Notice that the respective pressures at branch outlets C, D, and E are higher than that at F, and thus

Combined Tabulation of Results for Examples 12-7 and 12-8

Section	Flow (cfm)	Velocity Assumed or by Constant Pressure Drop	Duct Size $W \times H$ (in.)	Diameter of Equivalent Round (in.)	Equivalent Length (ft)	Pressure Loss (in. water)	Static Pressure (in. water)
A	—	—	—	—	—	—	0.231 (A)
ABC	3000	1300	33 × 10	19.05	40	0.064	0.167 (C)
CD	2600	1292	29 × 10	18.0	25	0.040	0.127 (D)
DE	2000	1252	23 × 10	16.4	25	0.040	0.087 (E)
EF	1000	1107	13 × 10	12.5	20	0.032	0.055 (F)
FJ	1000	800	18 × 10	14.5	30	0.023	0.050 (J)
						−0.018	
A–J	—	—	—	—	—	0.181	0.050 (J)

decided dampering is required. Computations for these branches have not been included in the solution, nor has the return circuit been considered.

Example 12-8. Develop a design for the supply-main duct of Fig. 12-13 that will employ high-velocity medium-pressure components and use round spiral ducting. Employ static-pressure regain wherever possible. The outlet pressure for branches should average about 0.2 in. water. The basic velocity initially will approximate 2500 fpm. A five-piece elbow is employed at B.

Solution: *Duct diameter for ABC.*

$Q = AV$

$$3000 = \frac{\pi}{4} \times \frac{D^2}{144} \times 2500$$

$$D = 14.8 \text{ in. (round off to 15 in., an obtainable size)}$$

The velocity for this diameter is

$$V = \frac{3000}{(\pi/4)(15^2/144)} = 2445 \text{ fpm}$$

$$h_{vABC} = \left(\frac{2445}{4005}\right)^2 = 0.373 \text{ in. water}$$

by Eq. (12-21) or Fig. 12-11. From Fig. 12-6, for a five-part elbow, $C = 0.11$.

$$\text{loss in elbow } B = Ch_{vB} = 0.11 \times 0.373 = 0.041 \text{ in. water}$$
$$\text{loss in } ABC \text{ (from Fig. 12-4)} = \tfrac{30}{100} \times 0.52 = 0.156 \text{ in. water}$$
$$\text{total loss } ABC = 0.197 \text{ in. water}$$

Refer to Table 12-6 for the advisory suggestion that if 0.2 in. water average pressure is required for points C, D, E, and F, it would be reasonable to set point C at 0.24 in. water and have point F at 0.16 in. water. Taking this suggestion, the static fan pressure at A must be 0.24 + loss in ABC = 0.24 + 0.197 = 0.437 in. water.

Section CD. Note that the 400 cfm taken off at C is so small that the static regain cannot offset much friction drop in CD. However, it is inadvisable to increase the size of duct so it will be kept uniform to D. For (3000 − 400) = 2600 cfm in a 15-in. duct,

$$V = \frac{2600}{(\pi/4)(15^2/144)} = 2119 \text{ ft/min}$$

$$h_{vCD} = \left(\frac{2119}{4005}\right)^2 = 0.280 \text{ in. water}$$

static pressure regain = $0.5(h_{vBC} - h_{vCD}) = 0.5(0.373 - 0.280) = 0.047$ in.
loss in run CD (from Fig. 12-4) = $\tfrac{25}{100} \times 0.39 = 0.098$ in. water
net loss in CD considering regain = 0.098 − 0.047 = 0.051 in. water
static pressure at $D = h_{SC}$ − loss = 0.240 − 0.051 = 0.189 in. water

This is close enough to the 0.2 in. water static pressure to be acceptable.

Section DE. Let us first explore the static regain if the diameter is not changed. For 3000 − 400 − 600 = 2000 cfm in a 15-in. duct, $V = 1630$ fpm and from Fig. 12-11, $h_{vDE} = 0.166$ in. water.

Results for Example 12-8, High Velocity and Static Regain Used

Section	Flow (cfm)	Velocity Used (fpm)	Diameter of Duct (in.)	Pressure Loss (in. water)	Static-Pressure Regain (in. water)	Net Pressure Loss (in. water)	Static Pressure (in. water)
A	—	2445	15.0	—	—	—	0.437 (*A*)
ABC	3000	2445	15.0	0.197	0.000	0.197	0.240 (*C*)
CD	2600	2119	15.0	0.098	0.047	0.051	0.189 (*D*)
DE	2000	1630	15.0	0.065	0.057	0.008	0.181 (*E*)
EF	1000	1273	12.0	0.040	0.033	0.007	0.174 (*F*)
FJ	1000	1273	12.0	0.024	0.000	0.024	0.150 (*J*)
J	—	1273	12.0	—	—	—	0.150 (*J*)

possible regain in $DE = 0.5(0.280 - 0.166) = 0.057$ in. water
loss (using Fig. 12-4) $= \frac{25}{100}(0.26) = 0.065$ in. water

Here the static pressure regain almost offsets the friction loss $(0.065 - 0.057)$ and the static pressure at E is $h_{SD} - 0.008 = 0.189 - 0.008 = 0.181$ in. water. A change in duct size hardly appears worthwhile.

Section EF. Here we can expect a substantial static-pressure regain if the duct size remains unaltered; thus a lower trial diameter appears indicated, and 12 in. will be chosen.

$$Q = AV \qquad 1000 = \frac{\pi}{4}(12)^2 \frac{V}{144} \qquad V = 1273 \text{ fpm}, \ h_{vEF} = 0.100$$

$$\text{regain} = 0.5\,(0.166 - 0.100) = 0.033 \text{ in. water}$$

$$\text{loss (using Fig. 12-4)} = \tfrac{20}{100}(0.20) = 0.04 \text{ in. water}$$

$$\text{net loss in } EF = 0.040 - 0.033 = 0.007 \text{ in. water}$$

static pressure at $F = h_{sE} - \text{loss} = 0.181 - 0.007 = 0.174$ in. water. The 12-in. diameter appears as a good choice and will not be changed. The average static pressure at C, D, E, and F is 0.196 in. water, which is close to the 0.2 value sought. If there were need of raising its value it would only be necessary to operate the fan at a slightly higher static pressure.

Section FJ. There appears no need to change the size of this run. The friction loss, using Fig. 12-4, is

$$\tfrac{12}{100} \times 0.20 = 0.024 \text{ in. water}$$

Summary: A designer could justifiably have made different choices of variables and arrived at other acceptable solutions in both Examples 12-7 and 12-8, but no great variation in results would be expected. When the tabulations for both problems are compared it can be seen that the supply pressures for the branches are much more uniform in the high velocity, static-regain solution. The higher pressures involved in the high velocity system require more fan horsepower, but the ducts of this system are smaller and thus require less material and less building space.

Return Ducts. The computations in the design of the return duct system are most usually based on the constant-pressure-drop (equal-friction) method. The net amount of air to be returned must be known, and the total pressure drop of the return system must not exceed the allowable negative suction pressure of the fan. The return system must be laid out as carefully as the supply system. Dampers should be supplied in the branches to permit adjustment of flow. If the system as originally designed has greater loss than that permissible for the fan, it must be resized (increased) to reduce the loss.

Heat Loss or Gain to Ducts. When ductwork, carrying cooled air, passes through unconditioned (warm) spaces, its temperature is raised by the heat which is transmitted into the duct. The resulting warming of the air is a complex phenomenon, and its rate depends on the ratio of surface perimeter to cross-sectional area; the length of duct; the temperature difference; the air velocity as it affects surface convection coefficients; the composite radiation and convection effects on both sides of the duct; the type of duct surface; and the effectiveness and kind of insulation. In the case of polished aluminum or galvanized iron, the low surface emissivity to radiation may make the bare duct more effective in retarding heat transfer than a duct covering of relatively poor insulation effectiveness.

Figures 12-14 and 12-15 have been prepared for ducts with aspect ratios not greater than 2 to 1. For ducts flatter than this, the following multiplying correction factors should be applied: 1.10 for a 3:1 ratio of sides; 1.18 for 4:1; 1.26 for 5:1; 1.35 for 6:1; 1.47 for 7:1; 1.50 for 8:1. Furred ducts are covered with $\frac{3}{4}$ in. of metal lath and gypsum plaster.

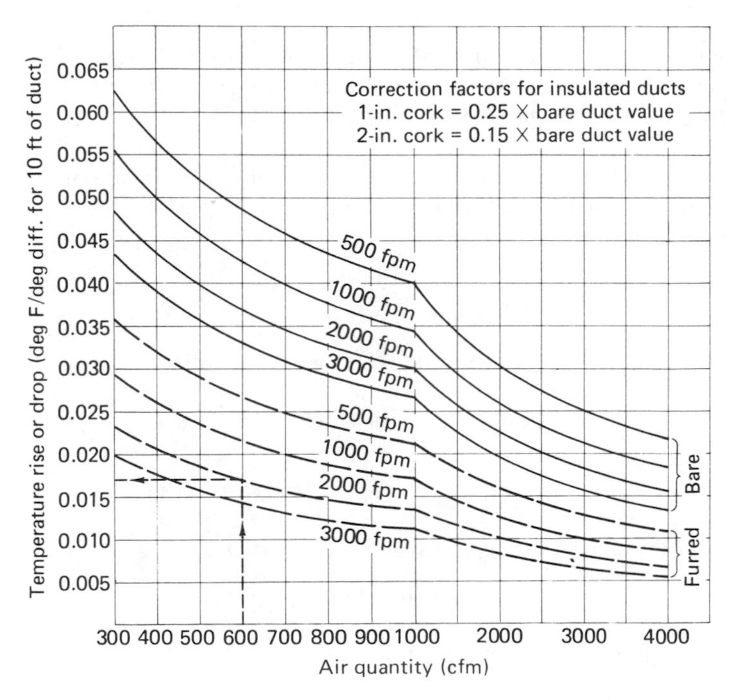

FIGURE 12-14. Temperature rise or drop in ducts (300 to 4000 cfm). (Courtesy Carrier Corporation.)

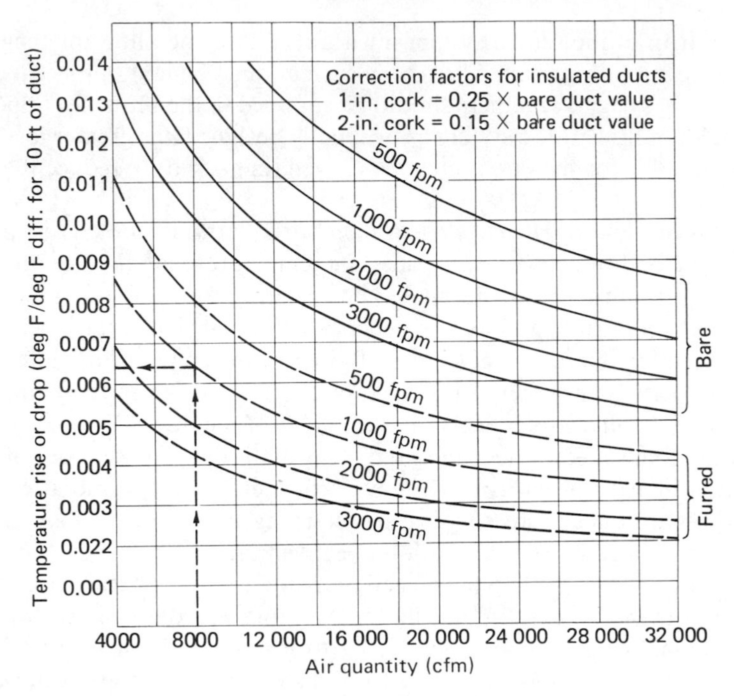

FIGURE 12-15. Temperature rise or drop in ducts (4000 to 32 000 cfm). (Courtesy Carrier Corporation.)

Example 12-9. A 12- by 36-in. duct 80 ft long carries 3000 cfm through a space at 85 F. Chilled air enters the duct at 64 F. Find the temperature rise of the air leaving (a) if the duct is bare and (b) if the duct has a 1-in cork covering.

Solution: (a) The temperature difference is based on conditions at start of the duct passage absorbing (or delivering) heat. In this case $\Delta t = 85 - 64 = 21$ deg. The velocity can be found from the relation

$$V = \frac{\text{cfm}}{\text{ft}^2 \text{ flow area}} = \frac{3000 \times 144}{12 \times 36} = 1000 \text{ fpm}$$

In Fig. 12-14, start at 3000 cfm, run up to the intersection with the 1000-fpm bare-duct line, and, from this intersection, run over to the left axis. Read 0.0212 F as the temperature rise per degree Fahrenheit difference for each 10 ft of duct run. The temperature rise is thus

$$0.0212 \times \Delta t \times L = 0.0212 \times 21 \times \tfrac{80}{10} = 3.6 \text{ F}$$

Apply an aspect-ratio correction of 1.10 for 3 to 1 sides. Thus

$$3.6 \times 1.10 = 3.96 \text{ (say, 4) F} \qquad \textit{Ans.}$$

(b) For the insulated duct apply the factor of 0.25 from Fig. 12-14. Thus the temperature rise is

$$4 \times 0.25 = 1.0 \text{ F} \qquad \textit{Ans.}$$

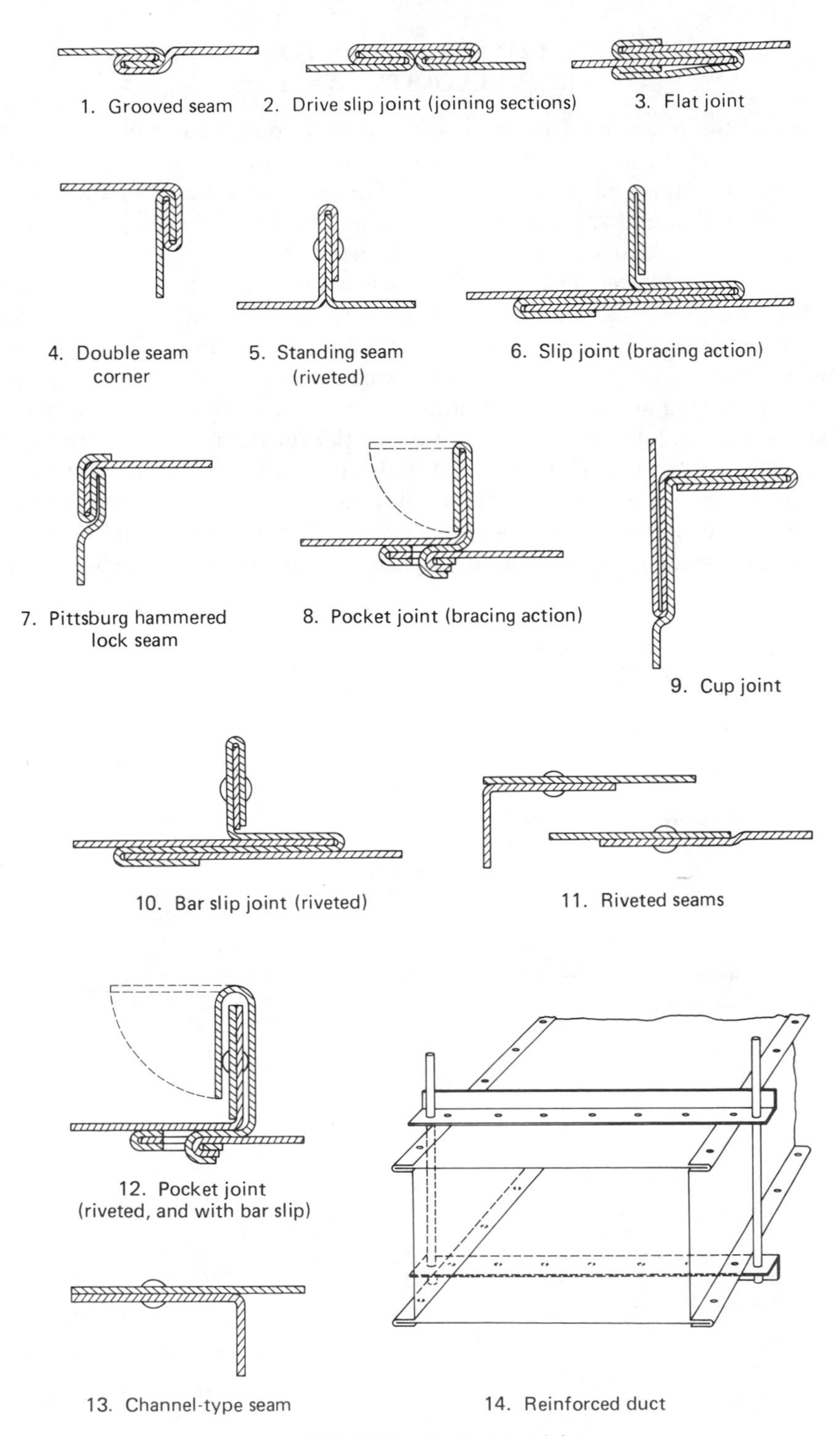

FIGURE 12-16. Typical duct joints.

12-7. DUCT CONSTRUCTION FOR LOW AND HIGH VELOCITY SYSTEMS

Some of the seams used in making low-velocity ducting are illustrated in Fig. 12-16. The type of joint used is influenced somewhat by the construction facilities available and the gage of metal. The metal must be heavy enough to resist vibration and sagging between braces, and therefore the thickness is increased as the diameter or width of the duct increases.

Galvanized sheet steel is most extensively used for ducting, but rolled carbon-steel sheeting, aluminium, stainless steel, and even copper sheeting can be employed. Large-size rectangular ducts (over 30 in.) must always be braced with transverse exterior tees (or angles), at intervals close enough to prevent sagging and possible vibration. Aluminium is light in weight and is not difficult to fabricate; and because it is thicker than steel, for the same weight, it tends to have greater rigidity. It does have a higher coefficient of thermal expansion, which must be allowed for where there are extreme temperature changes. Thickness designations of aluminum should be given in decimal fractions of inches, although gage numbers are frequently used.

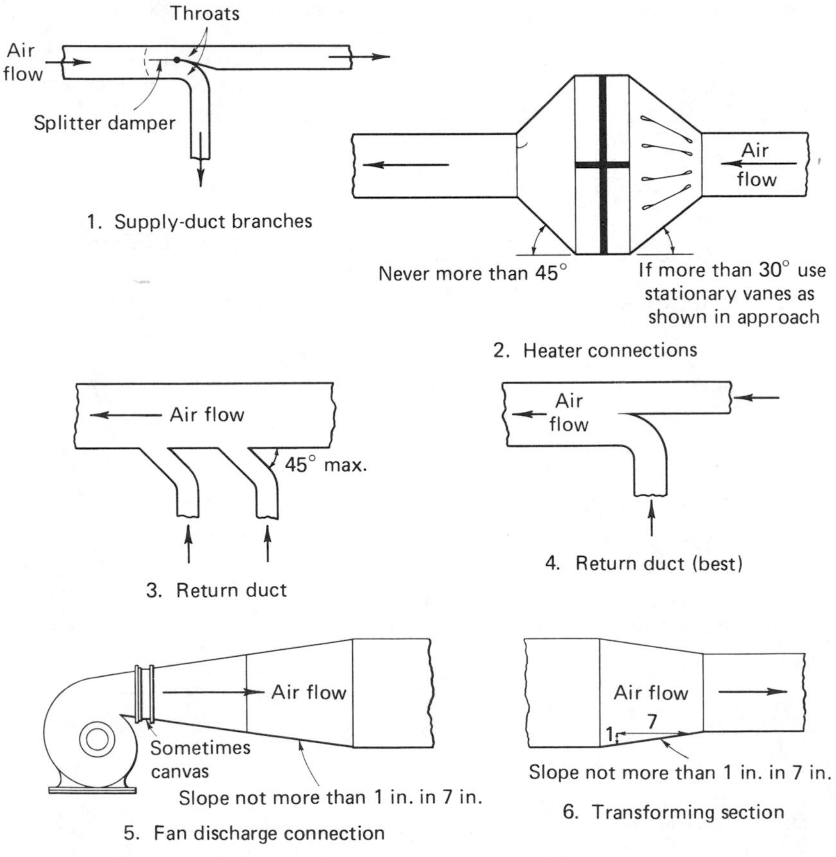

FIGURE 12-17. Equipment in ducts and junction arrangements.

It is unfortunate that the gage system used for aluminum differs from the system used for steel. Table 12-7 shows these gage numbers and corresponding sheet thicknesses.

It should be recognized that conventional ductwork is not leakproof and that all of the air is not delivered by the fan in the space intended. It is customary to allow 10% more air flow from the fan to compensate for leakage, if the ductwork runs outside of the treated space.

In the installation of equipment, such as heaters and filters in air ducts, abrupt changes in size should be avoided. Figure 12-17 shows some arrangements which have been found satisfactory. Abrupt changes in direction and other resistance-creating conditions cause noise and reduce volume. Interior vanes should therefore be used at elbows, and obstructions should be streamlined. Some ducts must have sound-absorbing material applied on the inside. Insulation against heat transfer is usually applied on the outside, except for the reflecting types used in ducts carrying high-temperature air.

12-8. AIR DELIVERY

Air must be delivered into the heated or conditioned space and distributed to the desired points at temperatures and velocities which are not objectionable to the occupants. Temperature differentials in the occupied zone of a room should not vary more than 2 deg, although in the cooling season slightly greater differentials may be permissible. An air movement of about 25 fpm around the bodies of the people is desirable, but this velocity is usually exceeded when an appreciable volume of air has to be handled. About 50 fpm is a maximum for comfort with people seated, and a slightly higher maximum can be tolerated where people are moving. Air directed into the faces of occupants is to be preferred to air directed toward their backs or sides. Downflow with regard to occupants is preferable to upflow.

The *throw*, or *carry*, of an air stream is the distance, perpendicular to the face of the outlet, through which a directed air stream travels on leaving the outlet. When the stream velocity is reduced to a value of about 75 fpm, the directed energy, or throw, has largely been expended. Air will drop or rise as it leaves the outlet, the direction depending on temperature (density) differences between the entering and room air and the velocity of the entering stream. Supply air that is cooler than the air in the room falls, and warmer supply air rises. Every moving air stream mixes with and carries along some of the room air, and the momentum of the original air stream is decreased in accelerating this room air. This process of mixing, or *induction* causes the air stream to spread out as it progresses forward. The induction of room air increases as the perimeter of the air cross section leaving the grille is increased. This induction is thus a maximum with a flat rectangular-shaped outlet, and a minimum with a round cross-sectioned outlet of the same area.

In the case of an open outlet without vanes, the spread of the air leaving the outlet gives an included angle between the sides of the air stream of about 12 to 15 deg. The spread and induction of an air stream can be visualized by referring to Fig. 12-18. Straight vanes do not change this angle to any extent, and about 13 deg can be taken as average. Converging vanes in the outlet make the air stream converge at first, but the spread which then follows is greater than that produced with straight vanes. Diverging vanes greatly increase the spread and are used to decrease the throw when this is desired. Where outlets without straightening vanes are used on the side of a duct, the air stream leaving the outlet will not flow out perpendicular to the duct but will make an angle of less than 90 deg with the direction of motion of the air in the duct. This divergence is caused by the velocity of the air in the duct tending to make the air carry forward as it leaves the grille. With high duct velocities a considerable deflection of the air stream from normal may occur.

The volume of air and the allowable velocity from the standpoint of noise determine to a great extent the number of outlets or grilles to be used, but their location must be carefully considered. Figure 12-18 shows a very desirable arrangement in which a return is located on the same wall as the supply. A return on the far wall tends to create a stagnant space beneath the supply outlet. Figure 12-19 shows a *pan-type outlet* used on a ceiling. This appliance gives a rather uniform discharge around its perimeter and can be used either with the return located as shown or with a return duct built up vertically through and inside the outlet. For cooling, and even for heating, inlets at high levels are, in general, desirable. For the heating returns, the low levels are preferable. But in many cases the same system serves either function in different seasons. Satisfactory arrangements which can be used for either heating or cooling are shown in Fig. 12-20. The air stream should be kept from impinging on walls or ceilings, to reduce formation of dirty spots and discoloration from deposited dirt.

Conventional distribution systems for theaters or auditoriums are shown in Figs. 12-21 and 12-22. The ejector system of air delivery at high velocity with a long throw can be used to advantage if the ceiling is relatively free from

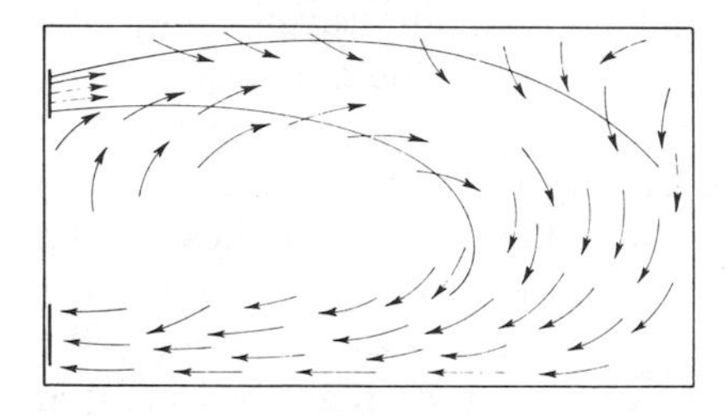

FIGURE 12-18. Desirable supply and return grille locations.

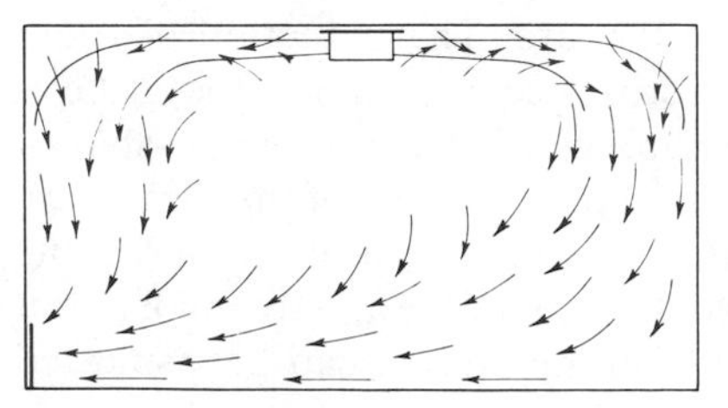

FIGURE 12-19. Ceiling outlet, conditioned air.

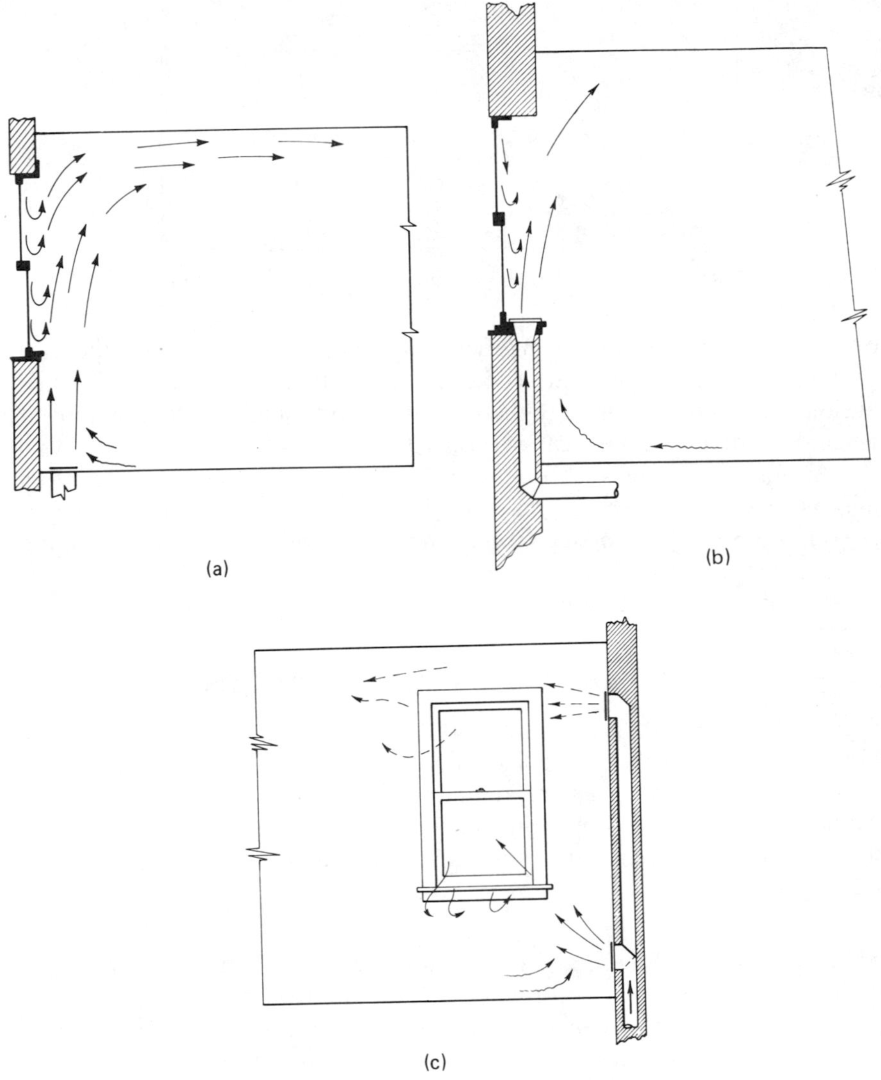

FIGURE 12-20. Supply air locations: (a) and (b), winter or summer; (c) top outlet for summer, bottom outlet for winter.

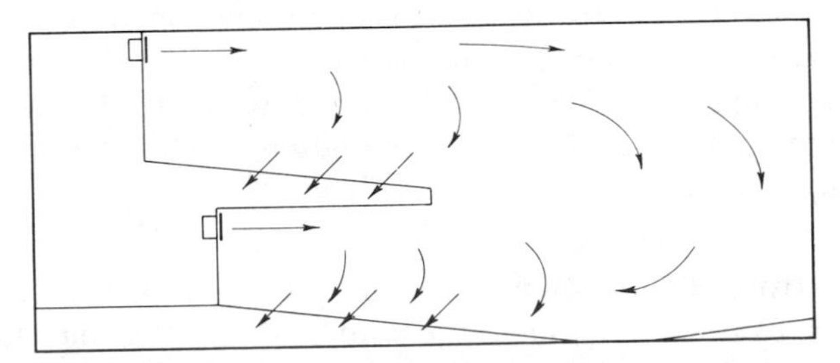

FIGURE 12-21. Forward-throw distribution system for theater with plain ceiling.

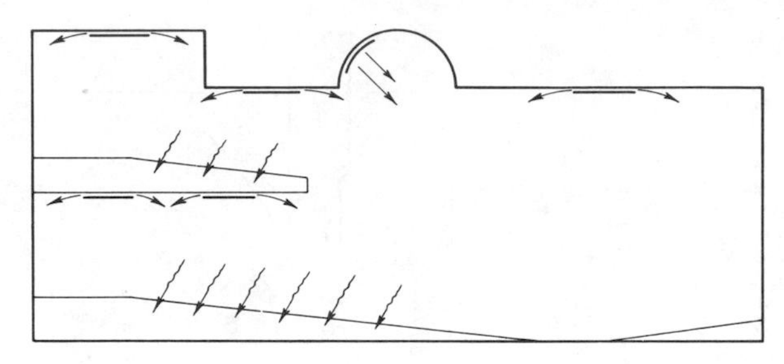

FIGURE 12-22. Overhead distribution system for any type of theater.

obstructions. Returns should, in general, be under the seats, or along the rear, or along the sides at the rear. The overhead system with downward distribution at very low velocity gives better control and usually must be employed when the ceilings are not clear of obstructions.

Air for ventilating washrooms, kitchens, and other places where odors may arise should never be recirculated; it must be continually vented to the outside. Supply air for such rooms should generally enter from adjoining rooms, such as corridors, anterooms, and dining rooms, so as to insure absence of odors in the latter rooms.

12-9. AIR DISTRIBUTION METHODS

Figure 12-23 shows the air flow patterns which develop in one type of diffuser. The passages are designed so that the delivered air aspirates (inducts) room air to mix with the leaving air and temper it closer to room conditions. Additional induction and mixing also occurs after the air stream moves into the occupied space. Such units are also built in square, half-round, and rectangular form.

Air Diffuser Box. Figure 12-24 shows a square-shaped diffuser in position under its environmental control box. When used on high-velocity (high-pressure) systems, boxes of this type can be *system-powered*. This means that with static pressures of the system ranging from 0.75 to some 4.0 in. water, the space thermostat initiates movement of the control-box dampers to change the volume of air flowing to adjust for varying room-load conditions. This arrangement avoids the necessity of having to install a separate pneumatic or electric control system for operating the control box. This of box is most useful for meeting cooling-load requirements for interior spaces both summer and winter. For external spaces in winter the box could provide tempered ventilation air while direct radiation would be needed to serve the heating load.

Dual Duct Distribution. Many other arrangements of control boxes are in use to serve special functions of cooling, ventilating, and heating. One of the most versatile and widely used boxes is the dual-duct mixing box. In

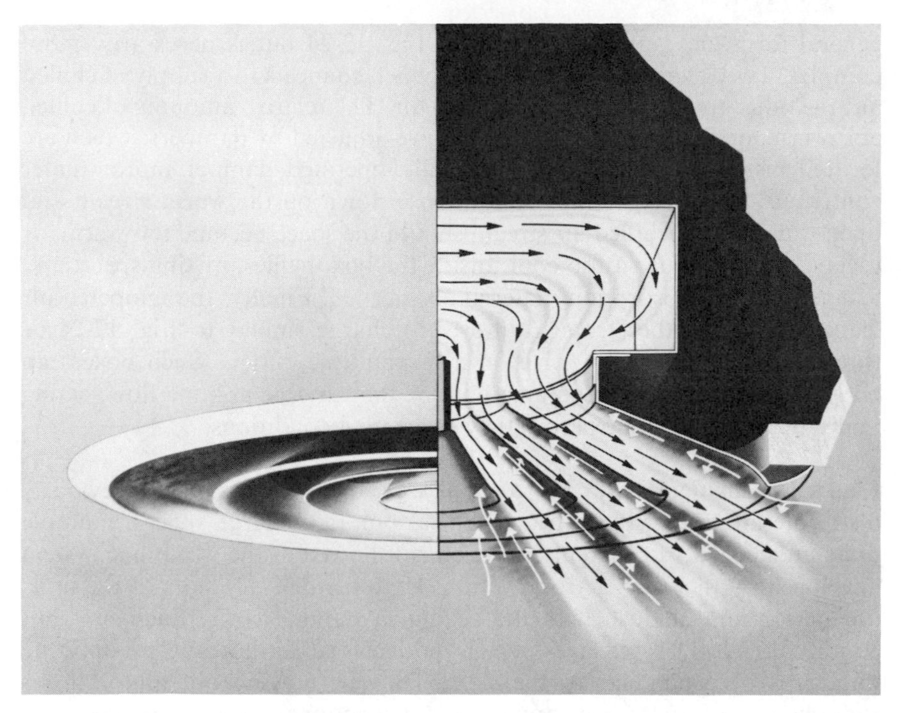

FIGURE 12-23. Anemostat air diffuser in sectional view showing principle of aspiration (induction) of some space air. (Courtesy Anemostat Products Division/Dynamics Corporation of America.)

FIGURE 12-24. System-powered control box with square air diffuser. (Courtesy Anemostat Products Division/Dynamics Corporation of America.)

general form and outline it resembles Fig. 12-24 but is necessarily more complex. Two inlets are needed, one of which connects to a supply of chilled air, the other to a supply of warm (hot) air. The relative amounts of chilled and warm air flowing through the box are adjusted by dampers, which are located inside the box. A pneumatically operated damper motor under control of the space thermostat will close down on the warm stream and open wider on the chilled-air stream should the space become too warm, or reverse the procedure if too cool. Inside the box, baffles, mixing space, and extensive sound absorption material are needed. Finally, the tempered air leaves the box, either from a diffuser discharge similar to Fig. 12-24 or through an end discharge, or even from multiple outlets. Such boxes can be arranged for constant volume flow or for variable volume flow, acting under thermostatic control to satisfy space load conditions.

Linear Distribution Unit. For a small area, the units described previously distribute air primarily from one central point; for large spaces, multiple units are needed. A different approach is followed in the Moduline system developed by the Carrier Corporation. Here the delivery slots of the units are placed almost flush with the ceiling in parallel rows. Each unit has continuous slotlike passages which project the air stream in opposite directions perpendicular to the slots. The air, flowing out almost horizontally, has adequate opportunity to induct room air and mix further with it before dropping down into the occupied space to provide required cooling. Such a unit is pictured in Fig. 12-25.

These units and their controls are placed above a false ceiling. As usually constructed, the ceiling contains, in addition to the units, lighting fixtures and filler (or acoustic) panels to give a pleasing streamlined appearance overhead. The Moduline units are operated by system pressure. Dampers are not required because, under thermostatic action, system pressure is directed into bellows-like devices which expand sufficiently to restrict the air flow down to the two side slots and provide partial or even complete closure, should this be required. When more air flow is needed the bellows pressure is reduced, allowing the bellows to contract and provide a larger opening for air flow. Each unit has its own plenum for directing air from the high-velocity (high-pressure) system. Because the control system is so simple yet effective, in a large building it is easy to set up numerous modules, each one of which can be operated and controlled by its own thermostat.

Perforated Ceilings. In many types of public buildings, use is made of ceiling panels consisting of metal sheets with many uniformly distributed perforations, the space being supplied with air from a plenum chamber above the sheets. Valves for delivering air to the plenum must be arranged to provide even distribution over the conditioned space. When the valves are properly adjusted, the downward movement of air is so slow that the occupants do not notice a draft. It is sometimes possible to deliver large quantities of air through such panels without annoyance to occupants, since

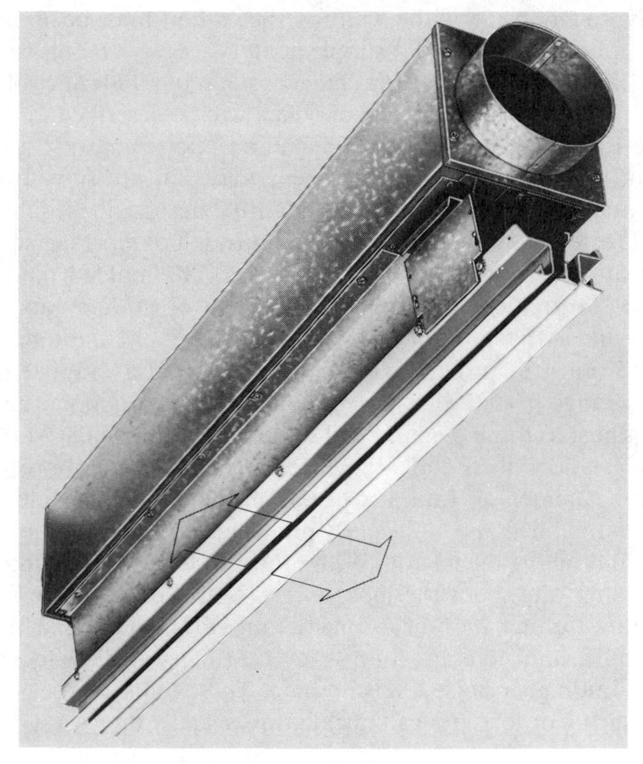

FIGURE 12-25. Moduline air diffuser unit with direction of air flow from slots indicated by arrows. (Courtesy Carrier Corporation.)

the diffusing effect of the streams of air causes rapid mixing with the room air.

Conduit Systems and Induction Units. Most air-conditioning systems provide not only for supplying conditioned air to a space but also for returning a portion of this air to the conditioner for recirculation. With central systems this involves the use of both supply ducts and return ducts, and these ducts require the use of valuable space in the building—perhaps for false ceilings, exposed ductwork overhead, or vertical areas which carry ductwork and services from floor to floor. To reduce the demands on available space, which become particularly critical where older buildings are being modernized, extensive use is being made of high-velocity systems, also known as conduit systems. Static pressures of from 1.0 to 6 in. water pressure are common. The air in turn is delivered through special diffusers such, as already described in this section, or through air-terminal induction units.

Air Terminal Induction Units. Where the air supplied to a space does not exceed ventilation needs, air return is not necessary, but for air circulation in amounts seriously above this the economies involved must be analyzed thoroughly. In any event, with no return, the air flow must be minimized and this end can be accomplished if supplementary *wet heating* or *wet cooling* is piped into the area. Manufacturers have brought out a number of induction designs to meet this objective, one of which will be described in detail.

Figure 12-26 is a schematic drawing of an "Economaster" air terminal. Primary air from the high-velocity system is used at pressures from 0.35 to 1.85 in. water, and to minimize the quantity needed it is delivered at a relatively low temperature. The air passes through dampers, a sound baffle, and a reheat coil before reaching the nozzles. The molded plastic nozzles are designed for inducting an optimum amount of room air to temper the primary air and at the same time provide for good air circulation in the room. Warm water must be provided for the reheat coil at average temperatures which may range from 100 F to some 200 F. Water temperatures can be altered to adjust to room demand and seasonal requirements. At the central-conditioner station, air is chilled and dehumidified to an appropriate temperature for summer operation, whereas in winter the outside air would need to be humidified and sufficiently preheated to avoid making an overly heavy demand on the reheat coil. With other settings adjusted into a proper operating range, the room thermostat can control capacity by varying the hot-water flow through the reheat coil in a range of 0.5 to 2 gpm. Up to 225 cfm per unit can be supplied at quiet noise level. Primary air flow to 375 cfm can be reached, but higher noise levels develop. These units can be arranged for ceiling mounting or for side-wall installation, usually under a window.

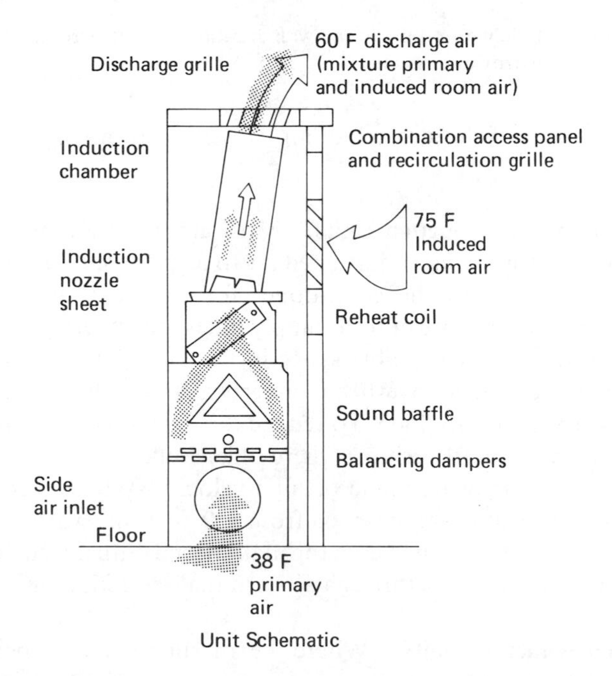

FIGURE 12-26. Schematic diagram showing operation of an "Economaster" induction-type air terminal. (Courtesy Carrier Corporation.)

Other induction units most frequently are designed to have the induced air pass only over a cooling or heating coil with the mixed stream then entering the space. The flow resistance of the coil reduces the ratio of the inducted air to the primary air and thus the temperature of the supply air must be reasonably close to room temperature to keep the mixed air temperature from being too cold for comfort. This means that a larger flow of supply air is required to offset the space load than is required with a unit of the type illustrated in Fig. 12-26. Moreover, conventional induction units must have both chilled and hot water supplied to them if these units are to be completely flexible for year-round operation. This makes it necessary to install four pipes, two for cold and two for hot water. In addition a drain for condensate is usually needed.

Cooling Load Estimation. Before a final design for any building is completed, a careful analysis of the probable cooling and heating load must be made. Following this, a decision has to be made as to the type of system which will be employed for distribution of cooling and heating (see Chapter 17). The details of equipment selection and sizing of pipe and duct runs can then be carried out. It is sometimes desirable to make estimates of what cooling capacity is needed before making such detailed calculations. In this

Table 12-8 Representative Cooling-Load and Air-Flow Values for Building Types and Usage

	Cooling in Relation to Area		Air-Flow Quantities Needed in an Air-Flow Cooling System					
			North Exposure		East, South, or West Exposure		Inside Space	
Type of Space	$\left(\frac{ft^2}{ton}\right)$	$\left(\frac{kW}{ft^2}\right)$	$\left(\frac{cfm}{ft^2}\right)$	$\left(\frac{m^3}{[min]m^2}\right)$	$\left(\frac{cfm}{ft^2}\right)$	$\left(\frac{m^3}{[min]m^2}\right)$	$\left(\frac{cfm}{ft^2}\right)$	$\left(\frac{m^3}{[min]m^2}\right)$
Apartments	400	0.0088	0.8	0.24	1.2	0.37	0.9	0.27
Department stores	250	0.0141	—	—	—	—	1.4	0.43
Drug stores	140	0.0250	1.4	0.43	2.3	0.70	1.0	0.30
Factories, light production	150	0.0235	—	—	—	—	3.0	0.91
Hospital, patient rooms	220	0.0160	0.5	0.15	0.6	0.18	—	—
Motels, hotels	300	0.0117	1.2	0.37	1.4	0.43	—	—
Offices	280	0.0126	0.5	0.15	0.5	0.15	1.1	0.33
Residences	500	0.0070	0.8	0.24	1.2	0.37	0.9	0.27
Restaurants	120	0.0293	1.4	0.43	2.0	0.61	1.0	0.30
Schools, colleges	190	0.0185	1.3	0.40	1.6	0.50	1.2	0.37
Theaters, auditoriums	250	0.0141	—	—	—	—	2.0	0.61

connection, Table 12-8 can be of use. It should be noted that the values in this table are merely representative or average values, and installations can be found operating with values up to 50% above or below those given in the table.

PROBLEMS

12-1. Compare the viscosity of water at 200 F expressed in SI units of $N \cdot s/m^2$ and in $lb/ft \cdot s$, with that of saturated steam at the same temperature.

Ans. Water 2.95×10^{-4}, 1.98×10^{-4}, steam 0.01185×10^{-3}, $0.00863 \times 10^{-3} = 8.63 \times 10^{-6}$

12-2. Air at standard atmospheric pressure and 70 F is flowing in a 2-ft ID (inside diameter) round duct at 3600 fpm. (a) Find the viscosity of the air in SI units and in units of $lb/ft \cdot s$. (b) Compute the Reynolds number for this air if its density is $0.075\ lb/ft^3$. (c) Read the friction factor from Fig. 12-2 and find the frictional loss in a 100-ft run of duct.

Ans. (a) $0.0178 \times 10^{-3}\ N \cdot s/m^2$, 1.196×10^{-5}; (b) 752 000; (c) 0.70 in. water

12-3. Air at standard atmospheric pressure and 21.1°C (70 F) is flowing in a 0.61-m-ID duct at a velocity of 18.3 m/s. (a) Find the viscosity of the air in SI units. (b) Compute the Reynolds number for this air if its density is $1.20\ kg/m^3$. (c) Read the friction factor from Fig. 12-2 and find the frictional loss in 30.5 m of duct run.

Ans. (a) 0.0178×10^{-3}; (b) 753 000; (c) $173.8\ N/m^2 = 1.78$ cm water

12-4. Derive an expression similar to Eq. (12-21) for standard air of density $1.2\ kg/m^3$ when velocity is expressed in meters per minute and the velocity head has units of centimeters of water gage. Note that the density of water at room conditions is $998\ kg/m^3$.

Ans. $h_v = (V_m/765)^2$ cm

12-5. What is the friction loss in a 120-ft length of duct 45 in. in diameter and carrying 15 000 cfm of air? The air is at 70 F and 50% relative humidity and under a pressure of 14.7 psi.

Ans. 0.05 in. water

12-6. After computation of the duct system of Example 12-5 in Section 12-5, it was decided to increase the size of the last room on the right branch duct so that 2400 cfm of air were required. (a) Calculate the new duct dimensions and (b) find the fan requirements.

Ans. (a) At fan—16 by 66 in., *AA*—16 by 38 in., *XX*—16 by 31 in. 16 by 22 in., vertical—10 by 39 in.; (b) 8800 cfm at 0.88 in. water

12-7. A duct carries 6000 cfm, and at 60 ft from the fan a branch is taken off which carries 4000 cfm through three elbows and 30 ft of run for delivery through a grille. The remaining 2000 cfm continue in a 60-ft run in which there are two elbows and one grille. Size these runs to have the same pressure drop, and state the required fan capacity if a static pressure drop of 0.5 in. water is used on the suction side of the fan, exclusive of the velocity head required to accelerate the air-to-fan outlet velocity. The velocity in the duct on outlet from the fan is 1400 fpm. The maximum duct depth is 12 in. What is the total pressure the fan must deliver?

12-8. Refer to Example 12-7 in Section 12-6 of this chapter and redesign the supply side of the system under the assumption that each outlet requires 0.8 of the quantity of air indicated in Fig. 12-13. Work by a method essentially independent of regain methods.

12-9. Work Problem 12-8 by regain methods.

12-10. Standard air is flowing in a 20- by 40-in. duct at a velocity of 1500 fpm at a static pressure of 1 in. water. After 200 ft of duct run, a branch receives 4000 cfm while the main duct continues with the same dimensions at reduced flow. Find (a) the drop in pressure in the first 200 ft of run; (b) the velocity in the duct after 4000 cfm are taken off; (c) the probable static regain as a result of velocity decrease; and (d) the probable static pressure 400 ft down the duct from the original reference point.

Ans. (a) 0.21 in.; (b) 779 fpm; (c) 0.051 in. SPR;
(d) $1.00 - 0.21 - 0.061 + 0.051 = 0.78$ in. water

12-11. The return system of Fig. 12-13 feeds into a low-wall-type grille inlet with a static-pressure loss of 0.02 in. water. Each branch has a bend after the grille, drops 8 ft vertically, and makes a hard bend (on greater width) into the trunk return duct. The trunk duct makes an easy bend at Q and runs for 9 ft before mixing with outside air at entry to the filters and coils. Leaving the last coil, the horizontal duct is transformed to round and, with a radius ratio of 2 (Fig. 12-7), bends into the fan inlet cone with a 10% slope. From S to T, in addition to the bend, there are 7 ft of straight run. At design flow through the heaters and filters there is a total pressure drop of 0.6 in. for RS. Design the return duct system so that the static pressure loss at the fan suction for the return system does not exceed 0.80 in. water. The static-pressure depression in accelerating the air to fan inlet velocity has been allowed for, and thus the 0.80 in. of water pressure drop need not include this item. Fix one dimension at a 16-in. maximum for the rectangular duct, and vary the other as required. Size each branch, the trunk main, and the fan inlet to satisfy design specifications.

12-12. Work Problem 12-11 on the assumption that only half as much air is returned for circulation—namely, 400 cfm in each branch, with 1800 cfm supplied from outside. All other specifications are unchanged.

12-13. A 20- by 40-in. duct 110 ft long carries 8330 cfm of chilled air at 52 F through a space at 72 F. What is the temperature of the air on outlet (a) if the duct is bare galvanized metal and (b) if the duct is covered with 1-in. cork?

Ans. (a) 54.6 F; (b) 52.6 F

12-14. A warm-air duct, carrying air at 140 F, runs for 300 ft in a factory where the air temperature is 80 F. Its size is 30 by 60 in. and the velocity of the air is 2000 ft/min. Find the final air temperature at exit from the duct when the duct is (a) bare galvanized iron and (b) furred and covered with $\frac{3}{4}$-in. metal lath and plaster. (c) Estimate the heat saving, in Btu per hour, resulting from covering the duct.

Ans. (a) 127.6 F; (b) 134.8 F; (c) 170 600 Btuh

12-15. A fan supplies 6000 cfm to a duct system. The system consists of a duct run of 300 ft, at the end of which there is a 90° bend through which 4000 cfm flow into a space. At 150 ft from the fan, a branch turns off the main duct at right angles and travels for 100 ft. This branch then turns at right angles again and delivers 2000 cfm of air. The duct has a maximum depth of 20 in., which is used throughout the design. Size the main and branch so that each of these has the same pressure drop from fan to outlet. Assume that the air velocity in the main duct, shortly after the air leaves the fan, is 1000 ft/min, and then design the main run for the same friction drop throughout. Assume that each bend has a radius ratio of 1.5, and that the outlet grilles on each branch have equal resistances.

Ans. Main run: 20 by 36 in. and 20 by 26 in.; branch: 20 by 14 in.

12-16. In order to construct ductwork in a narrow passage through a heavy concrete wall, it was found necessary to reduce a 30- by 60-in. duct carrying 20 000 cfm of standard air to a size of 20 by 50 in. The static pressure in the duct, just before the duct merged into the gradual sheet-metal transformation, was 1.5 in. water. What is the velocity and probable static pressure of the air in the duct when it has reached the far side of the wall?

Ans. 2880 fpm; 1.5 − 0.375 = 1.125 in. water

12-17. Refer to Fig. 12-13 and redesign the supply main for this system on the basis that because of a change in building plans the last branch will need to carry 2000 cfm. This means that a 4000-cfm fan will be required. The system will employ high-velocity, medium-pressure components, use round spiral ducting, and employ static-pressure regain wherever feasible. The basic velocity is initially 3000 fpm. A three-piece elbow is employed at *B*. The average outlet pressure at each branch should be planned as 0.25 in. water gage.

Ans. Size *AB*—16 in. and one solution would employ 16-in. duct throughout.

12-18. An air-terminal induction unit is supplied with 170 cfm of primary air at 40 F, no reheat is used, and the mixture of primary and inducted air is directed upward at a temperature of 61 F. (a) For the conditions indicated, make a simple heat balance to find the amount of room air inducted at 75 F. (Note that the primary air in warming to 61 F cools the induced air to the same temperature.) For the total air flow from the unit, compute the amount of sensible cooling provided by the unit. (c) Compute the same result in terms of primary air. Note that 0.018 Btu/ft^3 · F = 1.08 Btuh/cfm · F can be used with no essential inaccuracy since the specific heat of the air changes little for the small temperature and volume changes involved.

Ans. (a) 255 cfm; (b) and (c) 6426 Btuh

12-19. Rework Problem 12-18 for the same unit when it is supplied with 170 cfm of primary air at 50 F and delivers air at 64.4 F for a 76 F room temperature.

Ans. (a) 211 cfm, (b) and (c) 4773 Btuh

12-20. What pressure drop will occur when 90 lb/min of dry saturated refrigerant-12 vapor flow in a steel pipe of 3-in. ID at a pressure of 38.57 psia? Table 15-5, Fig. 12-1, and Fig. 12-2 can provide necessary data.

Ans. 0.72 psi

12-21. Rework Problem 12-20 on the basis of using 3-in. ID copper tubing in place of steel.

Ans. 0.49 psi

REFERENCES

M. C. Stuart, C. F. Warner, and W. C. Roberts, "Effect of Vanes in Reducing Pressure Loss in Elbows in 7-Inch Square Ventilating Duct," *Trans. ASHVE*, Vol. 48 (1942), pp. 409–24.

M. C. Stuart, C. F. Warner, and W. C. Roberts, "Pressure Loss Caused by Elbows in 8-Inch Round Ventilating Duct," *Trans. ASHVE*, Vol. 48 (1942), pp. 335–50.

R. G. Huebscher, "Frictional Equivalents for Round, Square and Rectangular Ducts," *Trans. ASHVE*, Vol. 54 (1948), pp. 111–17.

A. B. Stickney, "Friction in Pipes," *Refrigerating Engineering*, Vol. 53 (1947), pp. 129–31.

S. F. Gilman, "Pressure Losses of Divided-Flow Fittings," *Trans. ASHAE*, Vol. 61 (1955), pp. 281–96.

C. M. Ashley, S. F. Gilman, and R. A. Church, "Branch Fitting Performance at High Velocity," *Trans. ASHVE*, Vol. 62 (1956), pp. 279–94.

ASHRAE Handbook of Fundamentals 1972, Chapter 25, "Air Duct Design Methods," pp. 465–92.

ASHRAE Guide and Data Book, Equipment 1972, Chapter 1 and 2, "Duct Construction and Air Diffusing Equipment," pp. 1–16.

CHAPTER

13

FANS AND AIR DISTRIBUTION

13-1. FAN TYPES

Fans, which are almost universally used for the circulation of air or other gases through low-pressure systems, are made in the four general types or patterns illustrated in Fig. 13-1.

The *centrifugal fan* is widely used, since it is the most versatile type and can efficiently move large or small quantities of air over an extended range of pressures. It consists of a rotor or wheel mounted in a scroll type of housing. The wheel is turned either by direct drive or, more frequently, by motor drive employing pulleys and belts. The fan wheel, one of which can be seen in Fig. 13-2, is supplied with forward-curved, backward-curved, or radial (straight) blades. By varying the design shape of the blades, the characteristics of the fan can be changed over wide limits. The housing of the fan is usually constructed of sheet metal, although cast metal is sometimes used for smaller sizes. Where corrosive gases are handled by a fan, protective coatings of rubber, lead, asphaltum, or paint are often supplied to the housing and rotor. Details on centrifugal-fan performance will be considered later in this chapter.

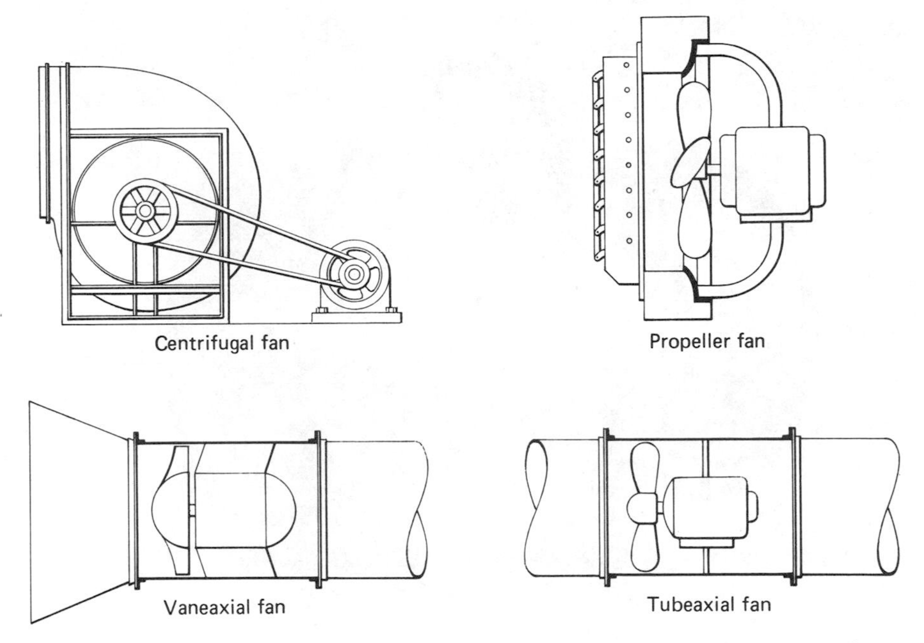

FIGURE 13-1. Classification of fan types. Each type can have either belt drive or direct connection. (Courtesy National Association of Fan Manufacturers.)

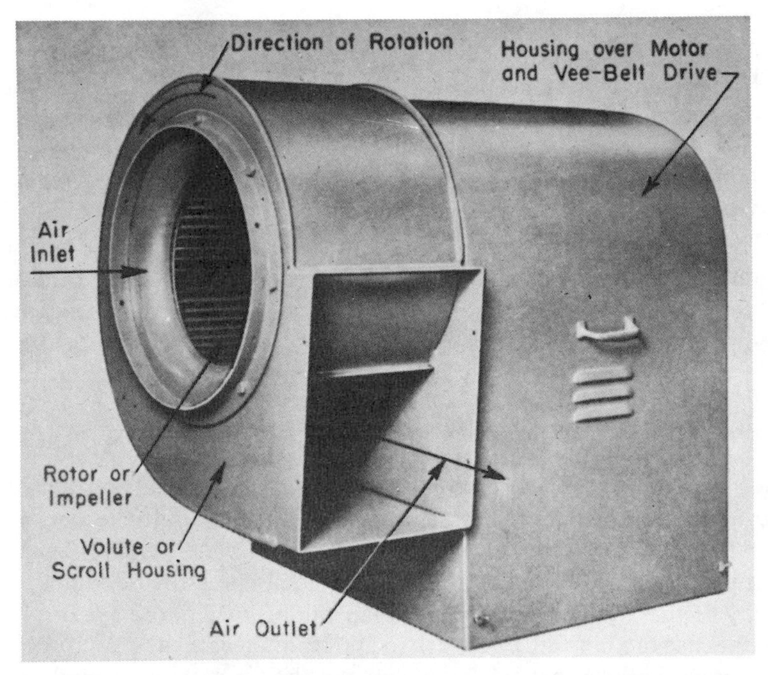

FIGURE 13-2. Centrifugal fan and motor unit for ventilation service. (Courtesy Sturtevant Division, Westinghouse Electric Corporation.)

FIGURE 13-3. Axial-flow fan. Straight-line flow provided by conversion vanes located behind four-bladed wheel. (Courtesy American Standard, Inc.)

The *vaneaxial fan* produces an axial flow of the gases through the wheel and blades. The wheel and its blading are located in a cylindrical housing and the fan is provided with air-guide vanes before or after the wheel.

The *tubeaxial fan* is similar to the vaneaxial fan and can move air over a wide range of volumes and through medium pressure ranges.

An axial fan is shown in Fig. 13-3.

The *propeller fan* can move large quantities of air but is unable to produce a significant pressure increase on the air being moved. Its main field of usefulness is in producing air motion or in moving air into or from a space against trivial pressure differences.

13-2. STATIC AND TOTAL AIR HORSEPOWER

In Section 12-4 the terms *static pressure*, *velocity pressure*, and the sum of these, *total pressure*, were discussed in connection with air. In the case of a fan, the static pressure, usually expressed in inches of water, is the pressure increase produced by the fan on the air. The velocity pressure is found from the velocity of the air stream leaving the fan outlet, by Eq. (12-19) or (12-21); or, in the case of a fan with ductwork to and from the fan, the velocity pressure is concerned with the net velocity change in the air passing through the fan. If a fan delivers directly into a free open space (not a plenum chamber), the discharge static pressure is considered zero, and all the energy of the air stream leaving the fan is in kinetic (velocity) form.

The useful work that a fan does consists in raising the air speed up to the discharge velocity and increasing the static pressure. This total pressure of the fan is the sum of static and velocity pressure changes, and thus the horsepower imparted to the air in static and velocity form is

$$\text{total air hp} = \frac{[(h_T/12)(62.3)](\text{cfm})}{33\,000} = \frac{(h_T)(\text{cfm})}{6350} \tag{13-1}$$

It then follows that

$$\text{total fan efficiency} = \frac{\text{total air hp}}{\text{shaft hp input to fan}} = \frac{(h_T)(\text{cfm})}{(6350)(\text{hp input})} \tag{13-2}$$

where cfm is the air delivered, in cubic feet per minute; h_T is the total pressure, in inches of water; and hp is the horsepower. If the velocity energy of the air leaving the fan is not utilized, or if it is desired to disregard that energy, the static pressure (h_s in. water) should be used instead of the total pressure. Then

$$\text{static air hp} = \frac{[(h_s/12)(62.3)](\text{cfm})}{33\,000} = \frac{(h_s)(\text{cfm})}{6350} \tag{13-3}$$

and it follows that

$$\text{static fan efficiency} = \frac{(h_s)(\text{cfm})}{(6350)(\text{hp input})} \tag{13-4}$$

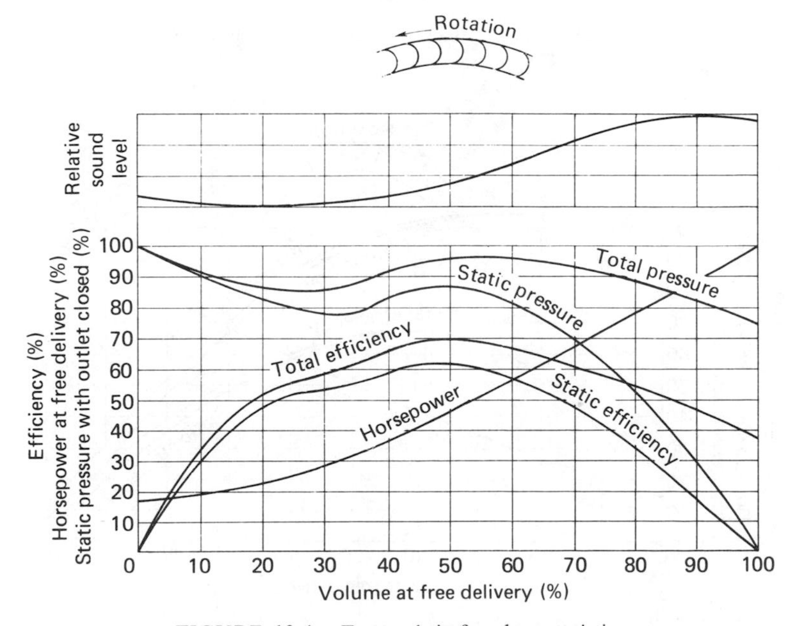

FIGURE 13-4. Forward-tip fan characteristics.

Typical performance curves for certain types of fans are shown in Figs. 13-4, 13-5, and 13-6. In these figures the abscissas represent the range of air (gas) flow capacity, expressed as a percentage of the amount delivered when the fan is discharging freely into open space and not against the resistance of a duct system. The horsepower for this condition is usually taken as a basis for comparison, and the requirement for any other discharge can be read on

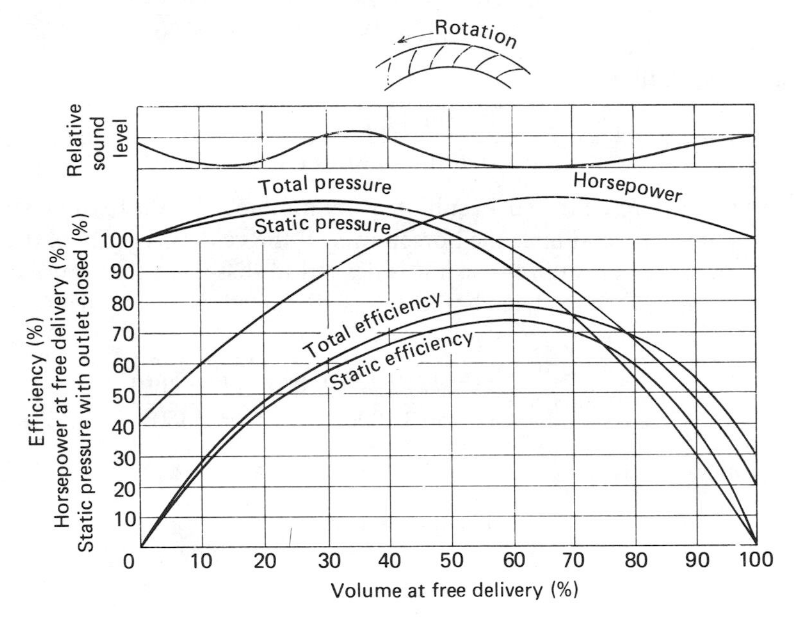

FIGURE 13-5. Backward-tip fan characteristics.

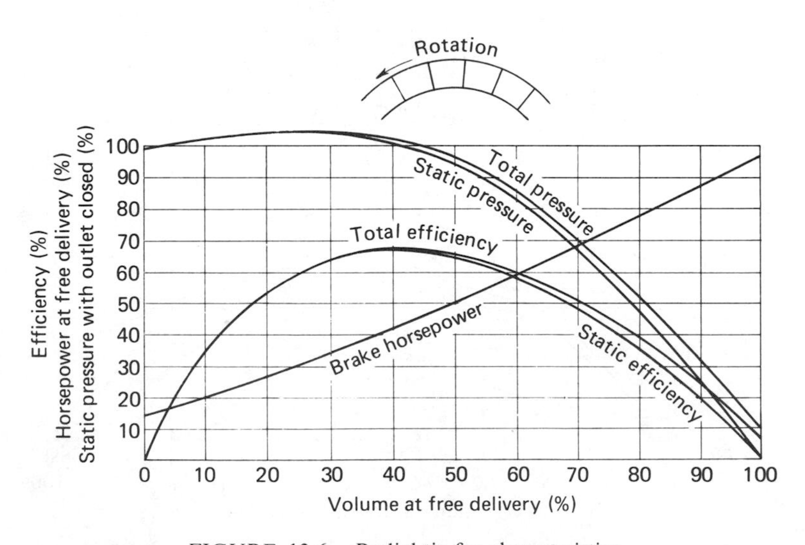

FIGURE 13-6. Radial-tip fan characteristics.

the horsepower curve as a percentage of free-delivery horsepower. The static- and total-pressure curves are also expressed as relative percentages, usually in terms of the static pressure with the outlet closed. The efficiency curves are not relative but express the probable actual efficiencies for a fan of each type. Each set of curves is drawn for a constant speed (rpm) of the fan in question.

13-3. CENTRIFUGAL FANS

Centrifugal fans develop pressure largely by converting a portion of the kinetic energy imparted to the air by the impeller into a rise in pressure. In addition to the pressure rise created in this manner, a smaller pressure increase is developed in the centrifugal field created in the rotating rotor blades. However, when the blades are shallow in radial depth, the contribution to pressure rise from the centrifugal field is small, as is also the pressure rise resulting from relative velocity change in the rotor blading. The air, on leaving the impeller of the fan, enters the scroll (volute) section, which is designed to bring about a decrease in the velocity of the air. This is accomplished by fabricating the scroll as a diffusing passage (a passage with increasing cross section in the direction of flow).

Figure 13-7 shows representative impeller-exit velocity diagrams for the three conventional types of fans. The air (gas) moves outward through the rotor-blade passage with a relative velocity w_2, while at the same time the rotor itself is turning and imparting to the air a tangential velocity indicated as u_t. The resultant velocity that the air possesses as it moves into the scroll (volute) is the vector sum of w_2 and u_t. In the case of backward-curved impeller blading the direction of w_2 is such that it may decrease the magnitude

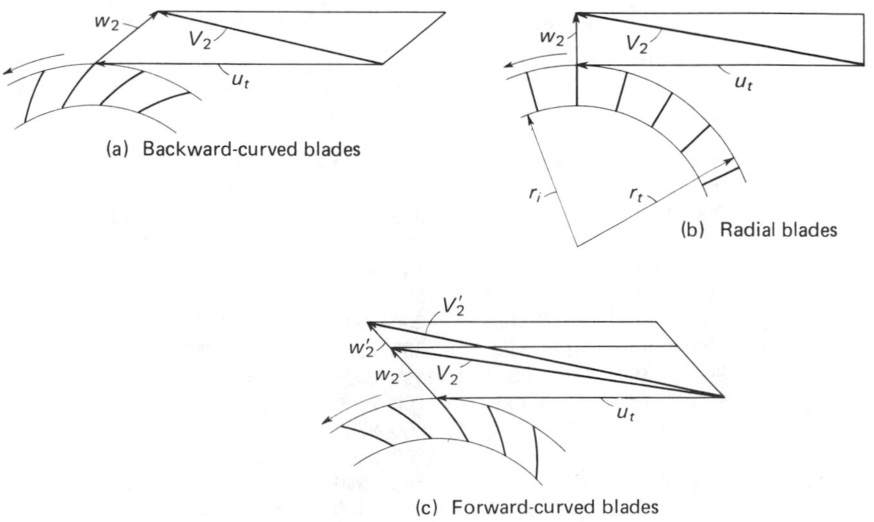

FIGURE 13-7. Representative velocity diagrams for varying types of centrifugal-fan blades. u_t = velocity of blade tip (fps) = $2\pi(r_t/12)$ (rpm/60); w_2 = relative velocity of air (gas) stream leaving blade passage (fps); V_2 = absolute velocity of air (gas) stream at entry to scroll (fps).

of V_2 compared with that of u_t. In the case of the radial (straight) blades, V_2 is increased but usually so slightly that it is not appreciably greater than u_t. On the other hand, forward-curved impeller blading produces a significant increase in the resultant velocity V_2 entering the scroll (volute). Since it is the magnitude of the velocity V_2 entering the volute, combined with effectiveness of conversion, that largely produces the pressure rise, it is obvious that forward-curved blading produces the maximum pressure rise for any fixed tip speed of an impeller. In contrast to this, backward-curved blading produces the lowest pressure rise for a given impeller tip speed.

The velocity w_2 of the air flowing through the impeller blade passage is directly related to the quantity of air being delivered by the fan. Consequently, as delivery increases under conditions of decreased system pressure the magnitude of w_2 increases. This is illustrated in Fig. 13-7c by the vectors w'_2 and V'_2. The energy absorbed in a fan is a function of the quantity of flow and the pressure against which the air is delivered. In the case of a forward-tipped fan, the air flow quantity increases at a rapid rate as external (system) pressure is reduced, with the result that the horsepower required to drive the fan increases with lessened resistance on the system. Consequently, when a driving motor is selected for a fan with forward-curved blades delivering into a fixed air system, if the flow capacity of the system is underestimated and more air flows in the system for a given inlet pressure, there is an ever-present tendency to overload the motor. This can be seen in the horsepower curve of Fig. 13-4. Fans with backward-curved blades have a horsepower characteristic which reaches a peak (Fig. 13-5) and then falls off as the

Table 13-1 Good Operating Velocities and Tip Speeds for Ventilating Fans*

Static Pressure (in. of water)	Fans with Forward-Curved Blades		Fans with Backward-Tipped and Double-Curved Blades		Tubeaxial and Vaneaxial Fans
	Outlet Velocity (fpm)	Tip Speed (fpm)	Outlet Velocity (fpm)	Tip Speed (fpm)	Wheel Velocity† (fpm)
1/4	1000–1100	1520–1700	800–1100	2600–3100	1100–1500
3/8	1000–1100	1760–1900	800–1150	3000–3500	1250–1700
1/2	1000–1200	1970–2150	900–1300	3400–4000	1400–1900
5/8	1200–1400	2225–2450	1000–1500	3800–4500	1500–2100
3/4	1300–1500	2480–2700	1100–1650	4200–5000	1650–2350
7/8	1400–1700	2660–2910	1200–1750	4500–5300	1800–2500
1	1500–1800	2820–3120	1200–1900	4800–5750	1900–2700
1 1/4	1600–1900	3162–3450	1300–2100	5300–6350	2150–3000
1 1/2	1800–2100	3480–3810	1400–2300	5750–6950	2350–3300
1 3/4	1900–2200	3760–4205	1500–2500	6200–7550	2500–3550
2	2000–2400	4000–4500	1600–2700	6650–8050	2700–3800
2 1/4	2200–2600	4250–4740	1700–2800	7050–8550	
2 1/2	2300–2600	4475–4970	1800–2950	7450–9000	
3	2500–2800	4900–5365	2000–3200	8200–9850	

* Reprinted, by permission, from *Heating Ventilating Air Conditioning Guide 1955*, Chapter 33.
† Wheel velocity is the axial mean air velocity through the inside diameter of the housing cylinder at the point of wheel location.

flow increases, while radial-bladed fans have intermediate characteristics between the two.

Table 13-1 indicates tip speeds and operating velocities of fans used in ventilating systems. It is of interest to use these figures to obtain some idea of the pressure rise created in the scroll of a fan. To do this simply, the approximate assumption is made that the air (gas) being compressed behaves as a relatively incompressible fluid. The pressure rise can be expressed, in pounds per square foot, as

$$\Delta p = K_p \frac{\rho}{2g}(V_2^2 - V_3^2) \tag{13-5}$$

where ρ is the density of the medium being compressed, in pounds per cubic foot; g is the gravitational constant, 32.2; V_2 is the velocity of the air stream leaving the impeller, in feet per second; V_3 is the reduced velocity the air stream possesses on exit from the fan, in feet per second; and K_p is the performance factor relating the conversion of velocity into pressure rise. For estimates, take K_p as ranging from 0.7 to 0.8.

Example 13-1. A fan with an impeller 21 in. in diameter is designed to turn at 600 rpm. Compute the probable pressure increase this fan might develop in its volute when passing standard air ($\rho = 0.075$ lb/ft^3) if the blades are forward-curved in such manner that, at a desired operating point and flow, $V_2 = 1.3u_t$. Consider K_p, the pressure-realization factor, to be 0.75, and take the outlet velocity from the fan as 1800 fpm, or 30 fps.

Solution: The blade tip speed can be found as

$$u_t = 2\pi \frac{r_t}{12}\left(\frac{\text{rpm}}{60}\right) = \frac{\pi(2r_t)(\text{rpm})}{720}$$

$$= \frac{\pi(21)(600)}{720} = 54.9 \text{ fps, or } 3290 \text{ fpm}$$

The absolute air velocity is thus

$$V_2 = 1.3u_t = (1.3)(54.9) = 71.4 \text{ fps}$$

By Eq. (13-5), the pressure rise is

$$\Delta p = 0.75\left(\frac{0.075}{64.4}\right)(71.4^2 - 30^2) = 3.67 \text{ psf}$$

or

$$\Delta p = (3.67)(0.192) = 0.704 \text{ in. water}$$

where 0.192 is the conversion factor for changing pounds per square foot to inches of water.

A certain manufacturer's catalog shows that a fan similar in size and operating in this range produces 1.25 in. water, so that $1.25 - 0.704 = 0.55$ in. water pressure is produced in the rotor of the fan by the centrifugal field and the diffusion effect.

Table 13-2 Comparative Characteristics of Centrifugal-Fan Types

Item	Forward-curved Blades	Radial (Straight) Blades	Backward-curved Blades
Efficiency	Medium	Medium	High
Stability of operation	Poor	Good	Good
Space required	Small	Medium	Medium
Tip speed for given pressure rise	Low	Medium	High
Resistance to abrasion	Poor	Good	Medium
Noise characteristic	Poor	Fair	Good

Table 13-2 lists relative characteristics of centrifugal fans in a general way but, in addition to the three types listed, there are a number of intermediate types by means of which designers have been able to produce characteristics to serve particular purposes. For example, one design uses backward-curved blade at the tip (outlet) edge of the blade, with forward curving at the inside (inlet) edge. Such a fan has a continuously rising pressure characteristic, but the horsepower reaches a peak and then falls off—and thus the overloading characteristic is eliminated.

The noise emitted by a fan is of greatest importance in certain types of applications. For example, in the air supply of a lecture hall a noisy fan could not be tolerated. On the other hand, exhaust fans for industrial plants, or fans used as blowers under boilers, are not critical in regard to noise characteristics. Fan noise is frequently associated with the high tip speeds which are required to produce high pressures. For a given pressure, noise is related primarily to two quantities: the tip speed of the impeller and the air velocity leaving the wheel. Expressed in another way, noise is roughly proportional to the pressure developed, whether this is produced by a low-speed forward-curved blade or by a higher speed backward-curved blade. For the same pressure rise, the noise in the two types will be of almost the same magnitude. In general, however, backward-curved fan blading is considered to be less sensitive to noise than is forward-curved fan blading. The graphs at the top of Figs. 13-4 and 13-5 give an indication of sound-level intensities for forward- and backward-curved fan blading. It will be noted that the forward-tip fan blading rapidly increases the intensity of the noise level as the air flow increases.

The pressure produced by the fan is naturally limited by the maximum allowable tip speed and by the design of the blading. For the production of high pressure and where noise production is not critical, rotors with straight blades, deep in a radial direction, are frequently used. These give the benefit of the more efficient type of pressure build-up which occurs in the impeller, and the absolute velocity of the medium entering the scroll is also aided by the outward-flow component (Fig. 13-7). If specially designed high-speed fans do not supply adequate pressure, one or more fans must be used in series. Little or no difficulty need develop if fans are operated in series. However, difficulty may be encountered when fans operate in parallel, as under these circumstances each fan will deliver air into a system at fixed pressure in accordance with the characteristic static-pressure curve of the fan—and if instabilities exist in the fans, there can be some surging back and forth between the fans, particularly so if the system capacity fluctuates. Fans with forward-curved blades, in particular, cannot be operated near their point of maximum efficiency because of inherent instabilities which exist in such fans if the rate of flow through the fan reduces to values of less than about 40% of wide-open delivery.

13-4. AXIAL FLOW FANS

Axial-flow fans, while incapable of developing high pressures, are nevertheless capable of forcing air through ductwork and low-resistance equipment which may be placed in such ducts, such as light filters and also heating or cooling coils. The fans are particularly suitable for handling large volumes of air at relatively low pressures. They are low in first cost and possess good efficiency, and since they are directly in the duct system, with through air-flow, they eliminate the 90-deg change of direction which is a characteristic of the centrifugal fan.

Axial-flow fans, in general, have a large hub and blades of airfoil shape. The blades are stubby and are not close together. Figure 13-3 shows a representative fan of this type. The inlet guide vanes, when employed in the fan, align and direct the air into the fan blades. The blades are visible in the foreground of the picture. The fan blades impart energy to the air and deliver it into the exhaust diffuser section of the fan.

To reduce losses, the bearing and pulley are streamlined, as shown, and a streamlined covering can also be supplied for the vee belts from the motor. In some arrangements, the fan is direct-driven and the motor itself is also mounted in the duct.

The fan blades are made in many forms, but the most effective have airfoil sections. Angle change and twist are given the blade at various positions outward from the hub to the tip. The simplest fans, however, may have little more than modified flat blades.

Axial fan units have now been developed to a high degree of effectiveness. They show good efficiencies, and they can operate at high static pressures if such operation is necessary. The fan can be so designed that the horsepower characteristic is quite flat and nonoverloading. A diffusing-section outlet cone is customarily provided on the discharge side of the fan. The swirl imparted to the air by the fan blades can be eliminated by proper guide vanes on the inlet side and, in some designs, on the outlet side as well.

13-5. FAN SYSTEM CHARACTERISTIC CURVES

The importance of selecting a fan to suit accurately the characteristics of the duct system cannot be overestimated. Figure 13-8 shows the *system characteristic curves* of two different duct systems, X and Y. Both of these curves are

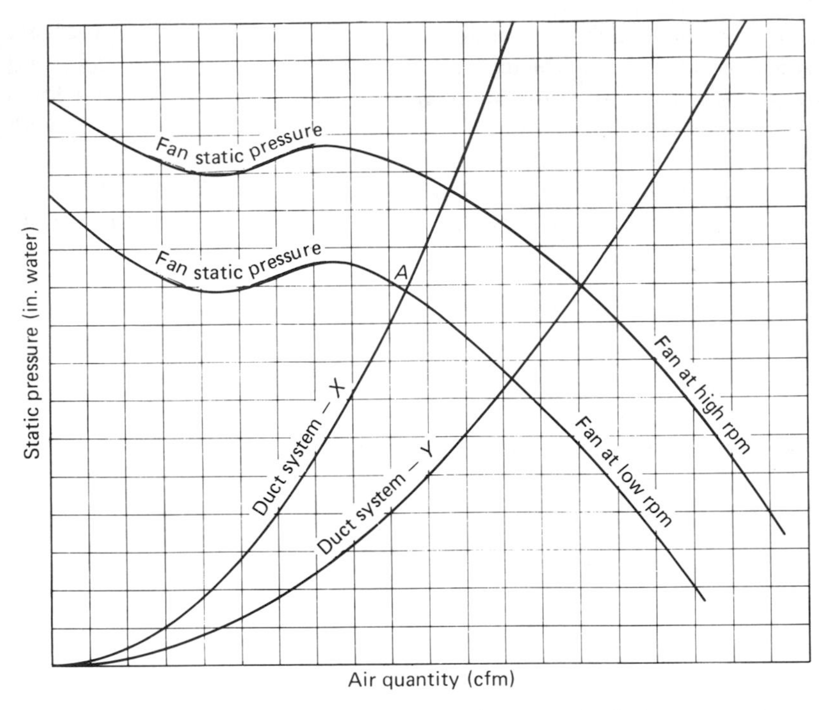

FIGURE 13-8. Characteristic resistance-quantity curves for duct systems and fans.

of parabolic shape and can be represented by equations of the form

$$\text{static pressure} \propto (\text{cfm})^2 \text{ or } (\text{cfm})^n$$

That is, the static pressure required to send air through a system is proportional to the square of the quantity (cfm) delivered. The exponent n in this equation is for systems other than those having smooth ducts. Its value is more or less than 2. For example, in sending air through a grate and furnace for forced draft, n may have a value of about 1.3.

The characteristic pressure curves for a given fan when it operates at a low speed and at a high speed are also shown in Fig. 13-8. These curves are for a fan having forward-curved blades; and, if the X-system characteristics are met by the fan for a given quantity of air at A, the choice would be satisfactory. However, if the fan were fixed to run at the higher speed shown, it would not deliver the quantity of air A without wasteful dampering because the fan pressures for quantity A become higher, as shown. For maximum efficiency, a fan having forward-curved blades should usually be selected so that the system characteristic crosses to the right of the hump. A fan having backward-curved blades is not so critical as to point of operation, but it should be selected to operate near its point of maximum efficiency. If the intersection point is selected near the steep part of the pressure curve, dampering to throttle the quantity delivered will cause a considerable pressure rise and will create high velocities and resultant noise at the dampers or registers.

Notice that changing the fan speed will not change the relative point of intersection of the fan and system-characteristic curves. This relation can only be changed by selecting a different size of fan.

13-6. FAN LAWS

When a given fan is used with a given system the following fan laws hold:

1. The capacity (cfm) is directly proportional to the fan speed.
2. The pressure (static, velocity, or total) is proportional to the square of the fan speed.
3. The horsepower required is proportional to the cube of the fan speed.
4. At constant speed and capacity, the pressure and the horsepower are proportional to the density of the air.
5. At constant pressure, the speed, capacity, and horsepower are inversely proportional to the square root of the density.
6. At constant weight delivered, the capacity, speed, and pressure are inversely proportional to the density, and the horsepower is inversely proportional to the square of the density.

Example 13-2. A fan described in a manufacturer's table is rated to deliver 17 500 cfm at a static pressure (gage) of 1 in. water when running at 256 rpm and requiring 4.54 hp. If the fan speed is changed to 300 rpm, what are the capacity, the static pressure, and the horsepower required?

Solution: By fan laws 1, 2, and 3,

$$\text{capacity} = 17\,500 \times (\tfrac{300}{256}) = 20\,500 \text{ cfm} \qquad \textit{Ans.}$$

$$\text{static pressure} = 1.0 \times (\tfrac{300}{256})^2 = 1.37 \text{ in. water} \qquad \textit{Ans.}$$

$$\text{hp} = 4.54 \times (\tfrac{300}{256})^3 = 7.31 \qquad \textit{Ans.}$$

Example 13-3. If in Example 13-2, in addition to changing the speed, the air handled were at 150 F instead of standard 70 F, what capacity, static pressure, and horsepower would be required?

Solution: The density of standard dry air at 70 F and 29.92 in. Hg is 0.075 lb/ft^3, and therefore at 150 F the density is

$$0.075 \times \frac{460 + 70}{460 + 150} \times \frac{29.92}{29.92} = 0.0651 \text{ lb/ft}^3$$

The density at any temperature and any barometer is obtained by multiplying by absolute temperature and pressure ratios. The capacity, on a volume basis, is unchanged, and thus

$$\text{capacity} = 20\,500 \text{ cfm at } 150 \text{ F} \qquad \textit{Ans.}$$

By fan law 4,

$$\text{static pressure} = (1.37)\frac{0.0651}{0.075} = 1.19 \text{ in. water}$$

and

$$\text{hp} = (7.31)\frac{0.0651}{0.075} = 6.35 \qquad \textit{Ans.}$$

The following information is required to select properly a fan for a given installation: (a) The number of cubic feet of air per minute to be moved; (b) the static pressure that must be developed to move the air through the system; (c) the type of motive power available; (d) whether the fans are to operate singly or in parallel on any one duct; (e) the degree of noise permissible; and (f) the nature of the load, such as variable quantities or pressures of air.

When the requirements of the system are known, the main points to be considered in selecting a fan are (a) efficiency, (b) reliability of operation, (c) size and weight, (d) speed, (e) noise, and (f) cost.

To assist in choosing apparatus to the best advantage, manufacturers of fans supply tables or curves that show the following factors for each size of fan, over a wide range of static pressure:

1. The volume of standard air (0.075 lb/ft^3) handled, in cubic feet per minute.
2. The air velocity at the outlet.
3. The fan speed, in revolutions per minute.
4. The brake horsepower.
5. The peripheral speed, or speed of the blade tip, in feet per minute (if not listed, it can be computed from data shown).
6. The static pressure, in inches of water.

In the tables of fan capacities, the most efficient point of operation of the fan is sometimes indicated by printing the values in italic or boldface type.

Two especially important factors in selecting fans for ventilating systems are the efficiency, which affects the cost of operation, and the noise. First cost and space occupied are secondary considerations. The fans should be selected to operate at maximum efficiency without noise. Noise may be caused not only by the fan but also by other conditions—for example, by too high a velocity of air in the ductwork, and by improper construction of ducts and airways, as well as of foundations, housings, floors, and walls. Where noise is chargeable directly to the fan, it may be caused by improper selection of fan type or by too high speed for the size. The tip speed required for a specified capacity and pressure varies with the type of blade, and a tip speed that is excessive for forward-curved blades is not necessarily so for the type curved backward or slightly backward (see Table 13-1). A fan that is operated at a point considerably beyond maximum efficiency is usually noisy.

Table 13-3 Pressure-Capacity Table for Two American Blower Corporation Type-ACH Fans

	Outlet Velocity	$\frac{1}{2}$-in. SP		$\frac{5}{8}$-in. SP		$\frac{3}{4}$-in. SP		1-in. SP		$1\frac{1}{4}$-in. SP		$1\frac{1}{2}$-in. SP	
(cfm)	(fpm)	(rpm)	(bhp)	(rpm)	(bhp)	(rpm)	(bhp)	(rpm)	(bhp)	(rpm)	(bhp)	(rpm)	(bhp)
851	1200	848	0.13	933	0.16	1018	0.19	—	—	—	—	—	—
922	1300	866	0.15	945	0.18	1019	0.21	—	—	—	—	—	—
993	1400	884	0.17	957	0.20	1030	0.23	1175	0.30	—	—	—	—
1064	1500	901	0.19	973	0.22	1039	0.26	1182	0.32	—	—	—	—
1134	1600	926	0.22	997	0.24	1057	0.29	1190	0.35	1320	0.43	—	—
1205	1700	954	0.25	1020	0.27	1078	0.31	1200	0.38	1325	0.46	1436	0.55
1276	1800	983	0.28	1044	0.31	1100	0.34	1210	0.42	1330	0.50	1440	0.59
1347	1900	1011	0.31	1068	0.35	1126	0.38	1230	0.46	1341	0.54	1447	0.63
1418	2000	1039	0.35	1092	0.39	1152	0.42	1250	0.50	1352	0.59	1458	0.66
1489	2100	1068	0.39	1115	0.43	1178	0.47	1275	0.54	1370	0.62	1470	0.72
1560	2200	1096	0.44	1147	0.47	1204	0.51	1300	0.59	1390	0.67	1482	0.77
1631	2300	1124	0.48	1179	0.52	1230	0.56	1325	0.64	1420	0.73	1500	0.83
1702	2400	1152	0.53	1210	0.58	1256	0.62	1350	0.70	1448	0.78	1525	0.88
5136	1200	372	0.75	407	0.91	444	1.06	512	1.37	—	—	—	—
5564	1300	375	0.83	412	1.00	448	1.19	513	1.51	566	1.86	617	2.37
5992	1400	380	0.94	417	1.12	452	1.30	517	1.69	572	2.02	619	2.41
6420	1500	385	1.05	421	1.23	458	1.46	521	1.84	575	2.21	622	2.68
6848	1600	390	1.16	426	1.35	462	1.60	526	1.98	578	2.39	626	2.80
7276	1700	395	1.30	433	1.51	466	1.74	528	2.31	580	2.59	629	3.04
7704	1800	401	1.40	438	1.64	469	1.87	530	2.47	583	2.82	633	3.29
8132	1900	409	1.56	444	1.81	475	2.04	535	2.60	590	3.02	637	3.59
8560	2000	416	1.72	449	1.98	480	2.22	540	2.72	596	3.21	641	3.90
8988	2100	423	1.98	457	2.18	488	2.43	546	2.98	601	3.53	645	4.02
9416	2200	429	2.24	465	2.35	493	2.59	550	3.12	604	3.74	648	4.28
9844	2300	435	2.31	472	2.55	500	2.78	554	3.47	610	4.02	654	4.55
10272	2400	452	2.59	477	2.76	507	2.82	562	3.65	614	4.27	658	4.90

Note. The upper data are for the model 109H fan, which has a 9-in. wheel diameter and an outlet of 0.709 ft^2. The lower data are for the model 121H fan, which has a 21-in. wheel diameter and an outlet of 4.280 ft^2.

* Courtesy American Blower Corporation.

Table 13-3 is a capacity table for two typical fans with rotors of the squirrel-cage type and indicates the kind of data customarily supplied. The manufacturer's model 109H is a small-size fan, and model 121H is a large fan; both are types common in air-conditioning work. The fan must be selected to supply the needed cfm at the proper static pressure, and the outlet velocity must also be in the range required or suitable corrections must be made.

Example 13-4. A fan in a certain duct system must deliver 1200 cfm at a static pressure (SP) of 0.76 in. water. The duct velocity is 1500 fpm. Ascertain if one of the fans in Table 13-3 is satisfactory.

Solution: Fan model 109H, running at 1078 rpm, delivers 1205 cfm at $\frac{3}{4}$-in. SP and requires 0.31 bhp. This should be satisfactory, particularly as the 1700-fpm outlet velocity can build up some pressure regain when air slows down in the duct with the larger cross section. By Eq. (12-28), the static regain is

$$\text{SPR} = 0.5\left[\frac{(1700 - 1500)(1700 + 1500)}{16\,040\,000}\right] = 0.02 \text{ in. water}$$

Hence the static pressure in the duct is 0.75 + 0.02 = 0.77 in. water. If necessary, a slight change of the speed through pulley adjustment could bring this fan into the required range.

The blade-tip speed can be found by computation of the peripheral velocity from the wheel diameter and the fan speed. Thus

$$\text{tip speed} = \frac{\pi(D_t)(\text{rpm})}{12} = \frac{(3.14)(9)(1078)}{12} = 2540 \text{ fpm}$$

13-7. MEASUREMENT OF AIR FLOW

The quantity of air flowing in ducts can be measured by installing orifices or nozzles which develop a measurable pressure difference. Where such a construction is not feasible, a traverse across the air stream may be made with some form of impact (pitot) tube. In Fig. 13-9 is shown a typical pitot tube

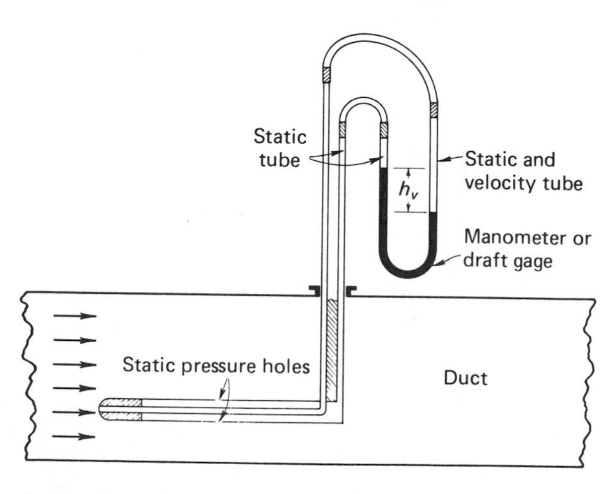

FIGURE 13-9.
Pitot tube in duct.

inserted in a duct to measure the flow of air or other fluid. A fluid in a duct exerts its static pressure in all directions, and if the fluid is in motion it also exerts a velocity pressure because of the kinetic energy of the stream. In Section 12-4 it was shown that the pressure equivalent to a given velocity could be represented for low and moderate speeds by expressions involving the velocity squared:

$$h = \frac{V^2}{2g} \tag{13-6}$$

A typical pitot tube, shown diagrammatically in Fig. 13-9, and in detail in Fig. 13-10, consists of two concentric tubes independently sealed from each other. The outer tube has small holes drilled in its side, through which the static pressure is transmitted to the manometer. The outlet of the inner tube faces the air current and consequently has two pressure effects impressed on it—that developing from the velocity of the stream, plus the static pressure existing in the duct. If the two tubes are connected to opposite legs of a manometer the static pressures which appear in each tube will equalize and the manometer will indicate only the pressure equivalent to velocity (h_v). The sum of the static pressure and the pressure equivalent to velocity is known as the *impact pressure*, or *total pressure*.

There are many sources of inaccuracy in using pitot tubes. The tube itself should be built in accordance with a well-tested, successful design (see Fig. 13-10), as with such a tube 100% transformation of velocity to pressure will be obtained.

If excessive cross currents or eddy currents exist in a duct, the static-pressure holes may not be at right angles to the main stream flow, and impact pressures may be indicated in the static tube. If the pressure is pulsating, inaccuracies will ensue. The velocity of the air current varies between different points in the cross section of the duct, so a traverse to obtain an average

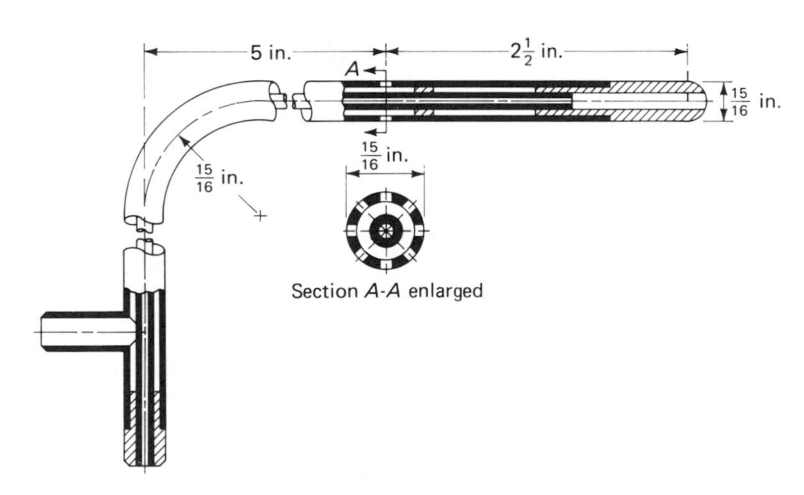

FIGURE 13-10. Acceptable design of pitot tube.

velocity should always be made. This is done by dividing the cross section into a series of imaginary areas of equal size and finding the velocity pressure (h_v) in the effective center of each such division. The average of the readings thus measured will give the value from which to calculate the average velocity in the duct.

In the case of a circular duct it is customary to divide the cross section into one central area and four concentric rings each of equal area. By placing the pitot tube at the radius, corresponding to the mean area of each concentric area, the velocity pressure for that area can be found. The average velocity found from these readings, multiplied by the total area of the duct, will give the flow. The locations of these points in a duct of radius r are $0.316r$, $0.548r$, $0.707r$, $0.837r$, and $0.949r$. It is more accurate to take these readings on each side of the center line of the duct, making ten readings in all and giving rise to the descriptive name *ten-point method* for making such a traverse. For more-precise work it is desirable to make two traverses at right angles to each other, giving twenty readings.

The manometer pressure in air ducts is usually measured in inches of water. The equation $V = \sqrt{2gh}$, in which h is expressed in equivalent feet for the fluid flowing, must therefore be put in terms of inches of water. By Eq. (12-17) it is possible to develop the relationship

$$V = \sqrt{2g\frac{h_v d_w}{12d_a}} = 2.31\sqrt{h_v\frac{d_w}{d_a}} \tag{13-7}$$

where V is the fluid velocity, in feet per second; g is 32.2; h_v is the velocity pressure from the indicating manometer, in inches; d_w is the density of the measuring fluid, usually water, in pounds per cubic foot; and d_a is the density of the air or fluid flowing in the duct, in pounds per cubic foot.

For water at 68 F, $d_w = 62.3\ \text{lb/ft}^3$. With 68 F water as the measuring fluid, and with velocity expressed in feet per minute, V_m appears, from Eq. (13-7), as

$$V_m = 1096.5\sqrt{\frac{h_v}{d_a}} \tag{13-8}$$

Example 13-5. A duct 4 ft in diameter carrying air at 72 F and 55% relative humidity, with the barometer at 29.9 in. Hg at very low static pressure, is traversed by a pitot tube across a diameter, using the ten-point method with five equal areas. The readings, in inches of water of the velocity pressure from one side of the duct to the other, are respectively 0.210, 0.216, 0.220, 0.219, 0.220; and 0.220, 0.218, 0.219, 0.220, 0.216. Find the air flow, in cubic feet per minute and in pounds per minute.

Solution: The velocity at each point can be calculated and these results then averaged, but it is quicker, and equally accurate, to average the square roots of the velocity pressures. Thus

$$\sqrt{h_v} = \frac{\sqrt{0.210} + \sqrt{0.216} + \sqrt{0.220} + \sqrt{0.219} + \sqrt{0.220} + \cdots + \sqrt{0.216}}{10} = 0.4668$$

The density of air at the conditions indicated can be calculated as shown in Section 2-5. Finding the weight of 1 ft^3 of dry air by $PV = mRT$ and adding the weight of steam in 1 ft^3, it is seen that $d_a = 0.0736$ lb/ft^3 and $d_w = 62.3$ lb/ft^3. By Eq. (13-7),

$$V = 2.31\sqrt{h_v \frac{d_w}{d_a}} = 2.31(0.4668)\sqrt{\frac{62.3}{0.0736}} = 31.37 \text{ ft/s}$$

The quantity flowing is

$$q = (\text{ft}^2)_{\text{duct area}} \times \frac{\text{ft}}{\text{s}} \times 60 = \text{cfm}$$

$$= \frac{\pi}{4}4^2 \times 31.37 \times 60 = 23\,650 \text{ cfm} \qquad \textit{Ans.}$$

And

$$\text{lb/min} = \text{cfm} \times \text{air density} = 23\,650 \times 0.0736 = 1740.6 \qquad \textit{Ans.}$$

Anemometer. An anemometer, illustrated in Fig. 13-11, can be used for finding the air velocity in ducts or at outlet grilles. An anemometer contains a miniature wind wheel which revolves on bearings having slight friction and thus turns a pointer in front of a dial calibrated to indicate air travel in linear feet. It must be used in connection with a timepiece, preferably a stop watch. An anemometer is a useful tool for comparative readings, but it is not adapted for very high velocities or extreme accuracy. An anemometer should be calibrated frequently, since the condition of the bearings affects the indications and since the instrument is easily damaged by rough handling.

For measuring the air volume through supply grilles the surface of the grille should be apportioned, by eye or measurement, into equal areas about

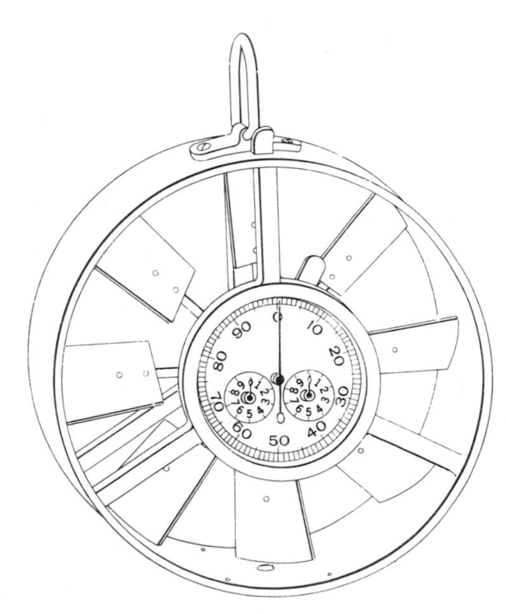

FIGURE 13-11. Anemometer.

6-in. square. A 4-in. anemometer should be used and it should be held very close to (or even against) the grille, with the dial facing the operator. The average readings taken at each of the area divisions should be used. In each square, generally not less than 0.5-min, and preferably 1-min, readings should be taken. Then

$$\text{cfm} = CV\left(\frac{A + a}{2}\right) \qquad (13\text{-}9)$$

where V is the average velocity from corrected anemometer readings, in feet per minute; A is the gross area of the grille, in square feet; a is the net free area of the grille, in square feet; and C is a flow factor determined by experiment (0.97 for velocities of 150 to 600 fpm, and 1.0 for velocities over 600 fpm).

For measuring the air volume through exhaust grilles the same subdividing of the area should be done, and the same average velocities should be obtained, but the dial of the instrument should face the grille. Then

$$\text{cfm} = YVA \qquad (13\text{-}10)$$

where V is the average velocity, in feet per minute; A is the gross area of the grille, in square feet; and Y is a flow factor determined by experiment (0.8).

Anemometers are almost universally used in measuring the air flow through grilles and registers in ordinary ventilating work, where velocities usually are comparatively low (not over 800 fpm).

Kata Thermometer. The Kata thermometer is essentially an alcohol thermometer developed for determining very low air velocities. The bulb is heated in water until the alcohol expands and rises to a reservoir above the graduated tube. The time required for the liquid to cool by 5 F is observed by use of a stop watch, and this time is a measure of the air movement.

Velometer. The velometer consists of a delicately poised vane within a substantial housing. The vane actuates a pointer calibrated to read directly in feet per minute the speed of the air flow, as with a pitot tube, without the necessity for timing. The velometer may be placed directly in the air stream, or it may be connected through a flexible tube to special jets which permit taking accurate velocity readings in locations inaccessible for an anemometer or pitot tube. The ordinary accuracy is within 3%.

Electrical Anemometers. Electrical anemometers are useful in measuring air currents. They operate on the principle of the variation in electrical resistance of a hot wire with temperature. The rate of cooling of this wire is responsive to the velocity of the air passing around the wire.

Orifices for Gas Flow. Orifices for measuring the flow of air, steam, and other gases are in extensive use for industrial flow-measurement. The orifice is usually a carefully fabricated opening made in a metal plate, with the plate designed for mounting between two mating flanges in a pipe or duct. Most orifices are made with a sharp (square) edge facing upstream to receive the flow, and the trailing orifice edge is chamfered away. If properly manufactured and used, the orifices can give reliable flow data. Coefficients for orifices are most frequently given for flange taps or for radius taps. For flange taps, the orifice pressure taps are located close to the orifice plate and the connections reach into the stream through the sides of the flanges. For radius taps, the upstream pressure tap is located one pipe diameter (two pipe radii) upstream, and the downstream tap is one pipe radius downstream from the orifice. Values of the flow coefficient K are given in Table 13-4, where this coefficient as given includes the velocity-of-approach correction favor. For conditions where the density of the gas does not appreciably change in passing through the orifice, the following equations apply:

$$Q = KA_o\sqrt{2gh_g} \tag{13-11}$$

$$Q = 8.02KA_o\sqrt{\frac{h_w d_w}{12d_g}} \tag{13-12}$$

where Q is the gas flow, in cubic feet per second; A_o is the orifice area, in square feet; K is the flow coefficient, including velocity of approach; h_g is the differential pressure across orifice, measured in feet of gas flowing; h_w is the differential pressure across the orifice, measured in inches of the measuring medium (usually water or mercury); d_w is the density of the measuring fluid, in pounds per cubic foot (for water, 62.3 lb/ft^3; for mercury, 62.3 $\times$ 13.6 lb/ft^3); d_g is the density of the gas under measurement, in pounds per

Table 13-4 Flow Coefficients K for Square-edged Orifice Plates, with Velocity of Approach Included

Reynolds Number	Diameter Ratio of Orifice to Inside of Approach Duct D_o/D_1						
	0.30	0.40	0.50	0.55	0.60	0.65	0.70
Flange Taps							
5×10^4	0.61	0.62	0.64	0.66	0.67	0.71	0.75
10^5	0.60	0.62	0.63	0.65	0.66	0.68	0.72
5×10^5	0.60	0.61	0.63	0.63	0.65	0.67	0.69
10^6	0.60	0.61	0.62	0.63	0.65	0.67	0.69
10^7	0.60	0.61	0.62	0.63	0.65	0.67	0.68
Radius Taps							
10^5	0.60	0.62	0.63	0.65	0.67	0.68	0.72
5×10^5	0.60	0.61	0.62	0.64	0.65	0.67	0.71
10^6	0.60	0.61	0.62	0.64	0.65	0.67	0.70
10^7	0.60	0.61	0.62	0.63	0.65	0.67	0.70

cubic foot; and D_o is the orifice diameter, in inches. When a water manometer is used and the orifice diameter (D_o) is in inches,

$$Q = K \frac{\pi}{4} \frac{D_o^2}{144} (8.02) \sqrt{\frac{h_w(62.3)}{12 d_g}}$$

which becomes

$$Q = 0.0996 K D_o^2 \sqrt{h_w / d_g} \qquad \text{ft}^3/\text{s} \tag{13-13}$$

Example 13-6. The air flow into a certain high-velocity duct system is measured on discharge from the outlet fan by a square-edge orifice 8.0 in. in diameter, located in a pipe of closely 12-in. inside diameter. The pressure drop across the orifice is 10 in. water and flange taps are used. What is the flow, in cubic feet per minute, entering the duct system if the air density is 0.078 lb/ft³? Also express the flow in pounds per minute.

Solution: Use Eq. (13-13) and read $K = 0.68$ from Table 13-4 for a diameter ratio of $8.0/12.0 = 0.667$ at a representative Reynolds number of 5×10^5. Then

$$Q = (0.0996)(0.68)(8.0)^2 \sqrt{10/0.078}$$
$$= 49.2 \text{ ft}^3/\text{s}$$
$$\text{flow} = (49.2)(60) = 2952 \text{ cfm} \qquad \textit{Ans.}$$
$$\text{flow} = (2952)(d_g) = (2952)(0.078) = 230 \text{ lb/min} \qquad \textit{Ans.}$$

The preceding equations, which have been developed in the foot-pound-second system, can readily be arranged for use in SI terminology. For conditions where the density of the fluid does not appreciably change in passing through an orifice, the familiar relation $V = \sqrt{2 g_c h_g}$ is applicable to gases as well as to liquids. When this relation is combined with $Q = AV$, there results

$$Q = AV = K A_o \sqrt{2 g_c h_g} \tag{13-11}$$

where in SI, Q is cubic meters per second of gas flow; K is the flow coefficient, including velocity of approach; A_o is the orifice area, in square meters; g_c is 9.8067; h_g is the differential pressure across the orifice, in meters of the gas under measurement; h_w is the same differential pressure, usually measured by a column of water, or by a column of mercury, expressed in meters; d_w is the density of the measuring fluid, in kilograms per cubic meter (water at 20°C (68 F) is 998 kg/m³, mercury 998 × 13.6 kg/m³); d_g is the density of the gas under measurement, in kilograms per cubic meter; and D_o is the orifice diameter, in meters.

Making appropriate substitutions in Eq. (13-11), we find

$$Q = K \frac{\pi}{4} D_o^2 \sqrt{(2)(9.8067)(h_w)(998)/d_g}$$

$$Q = 109.9 K D_o^2 \sqrt{h_w / d_g} \tag{13-14}$$

Example 13-7. The air flowing in a high-velocity duct is measured by use of a square-edged orifice, 0.2032 m in diameter bolted between flanges in a round duct of 0.3048-m inside diameter. Flange taps across the orifice on one occasion measured the pressure differential at 0.254 m water. Compute for this occasion the air flow, in cubic meters per second and in kilograms per second. if the density of the air was 1.249 kg/m³.

Solution: Use Eq. (13-14). Select $K = 0.68$ from Table 13-4 for a diameter ratio of $0.2032/0.3048 = 0.667$ at an assumed Reynolds number in the 10^5 range. Then

$$Q = 109.9(0.68)(0.2032)^2\sqrt{\frac{0.254}{1.249}}$$

$$= 1.392 \text{ m}^3\text{/s} \qquad \textit{Ans.}$$

$$\text{mass flow rate} = (1.392)(1.249) = 1.74 \text{ kg/s}. \qquad \textit{Ans.}$$

Rounded-entry orifices, or flow nozzles, are also used in measurement of gas flow. The flow coefficients K for these also vary with the diameter ratio and the Reynolds number but are so close to unity over a broad range of flow conditions that for approximate computation $K = 1.0$ can be used.

PROBLEMS

13-1. A fan delivers 3000 cfm under a static pressure of 1.0 in. water. The air is at 68 F (20°C) and has a specific volume of 13.33 ft³/lb. (a) Compute the static air horsepower without making use of the specific volume of the air. (b) Compute static air horsepower by means of a mass flow computation. (c) Find the required shaft horsepower if the static fan efficiency is 60% under these delivery conditions.

Ans. (a) and (b) 0.472 hp; (c) 0.786 hp

13-2. A fan delivers 1.42 m³/s of air under a static pressure of 2.54 cm water. The air at 20°C (68 F) has a density of 1.20 kg/m³. (a) Find the factor to convert centimeters of water pressure to pascals when the density of water at 20°C is 998 kg/m³. (b) Compute the static air in kilowatts, making use of only volume flow and static pressure. (c) Compute the necessary shaft power in kilowatts when the static fan efficiency is 60%.

Ans. (a) cm water × 97.9 = N/m²; (b) 0.356 kW; (c) 0.59 kW

13-3. Air flows in a duct at 80 F (26.7°C) under a static pressure of 3 in. water gage with the barometer at 29.4 in. Hg. The average velocity in the duct is 1500 fpm. Find the total pressure (head) in the duct.

Ans. 3.0 + 0.136 = 3.136 in. water

13-4. Air flows in a duct at 26.7°C (80 F) under a static pressure of 7.62 cm water. The barometer is 0.747 m Hg. The average air velocity in the duct is 457.2 m/min. (a) Convert the static pressure to pascals. (b) Find the density of the air flowing. (c) Compute the velocity pressure. (Note Example 12-4, part d.) (d) Find the total pressure head in the duct.

Ans. (a) 746 Pa; (b) 1.16 kg/m³; (c) h_v = 0.344 cm water; (d) 7.96 cm, 779 Pa

13-5. (a) If air leaves the fan of Problem 13-1 at a velocity of 1050 fpm what is the total pressure at the fan outlet when the static pressure is 1.0 in. water as indicated? The air density is 0.075 lb/ft^3. (b) What is the total air horsepower and the total fan efficiency based on the fan shaft input recorded for Problem 13-1?

Ans. (a) 1.0 + 0.0687; (b) 0.505 hp, 64.2%

13-6. (a) The fan of Problem 13-2 delivers its air at a static pressure of 2.54 cm water at a velocity of 5.33 m/s. Compute the velocity head pressure when the air has a density of 1.20 kg/m^3. (b) Find the total fan head at the delivery outlet. (c) Compute the total air in kilowatts under delivery conditions of 1.42 m^3/s. (d) As the shaft power of 0.59 kW, found in Problem 13-2, is still applicable, what is the fan efficiency based on total delivery conditions?

Ans. (a) 0.174 cm water; (b) 2.714; (c) 0.377 kW, 64%

13-7. In a certain duct, for a flow of 6000 cfm a static pressure of 0.86 in. is needed. (a) If the duct velocity at the fan is 1400 fpm, will one of the fans for which data are given in Table 13-3 be satisfactory? If so, at how many revolutions per minute will it run?

Ans. (a) Yes; (b) closely 483

13-8. By use of the ten-point method, pitot-tube readings across a duct 2 ft in diameter and carrying standard air (0.075 lb/ft^3) were 0.15, 0.15, 0.14, 0.15, 0.14, 0.15, 0.14, 0.16, 0.15, and 0.14 in. water. Compute (a) the average velocity and (b) the air flow, in cubic feet per minute.

13-9. Air at 80 F and 60% relative humidity is flowing through a duct of 2-ft diameter. The barometric pressure is 29.92 in. Hg. A pitot tube inserted in the duct indicates a velocity pressure of 0.28 in. water. Assuming that the value for this one point can be considered a representative value, (a) compute the air velocity and (b) find the flow through the duct in cubic feet per minute and in pounds per minute.

13-10. Solve Problem 13-9 on the assumption that dry carbon dioxide is flowing through the duct and that the same readings are observed.

13-11. Standard air of density 0.075 lb/ft^3 is flowing through a duct and a ten-point traverse is made of the duct, which is 6 ft in diameter. The following readings, in inches of water, are observed: 0.080, 0.095, 0.086, 0.084, 0.087, 0.083, 0.090, 0.088, 0.086, and 0.084. (a) Compute the flow through the duct in cubic feet per minute and in pounds per minute. (b) Compute the flow through the third annulus, for which the readings are 0.086 and 0.088. (c) Assuming that the average of these two values can be considered to represent the flow through the whole duct, multiply this value by 5, and compare the answer with the answer obtained by the longer computation.

13-12. Standard air of density 0.075 lb/ft^3 is flowing through an 8- by 12-in. duct. The duct cross section is split into four imaginary rectangles, and a pitot reading is taken at the center of each rectangle, yielding the following values in inches of water: 0.162, 0.168, 0.164, and 0.166. Find (a) the average velocity in feet per minute and (b) the flow in cubic feet per minute.

13-13. A certain fan has an impeller (rotor) 21 in. in diameter, turns at 580 rpm, delivers 7276 cfm at a static pressure of 1.25 in. water, and requires 2.59 hp. The outlet velocity is 1700 ft/min. Assume that the fan speed is increased to 610 rpm and estimate for this new speed (a) the probable capacity, (b) the horsepower, and (c) the static pressure. Explain why your answers differ from the data given in Table 13-3 for this fan. A study of Fig. 13-8 should suggest an answer.

Ans. (a) 7650 cfm; (b) 3.0 hp; (c) 1.38 in. water

13-14. The fan in problem 13-13 delivers 7276 cfm at a static pressure of 1.25 in. water, and the velocity in the outlet is 1700 fpm. If 2.59 brake horsepower are supplied by the motor, what is (a) the static air horsepower, (b) the total air horsepower, and (c) the fan efficiency, based on total pressure?

Ans. (a) 1.43 hp; (b) 1.64 hp; (c) 63.3%

13-15. A 4-in. anemometer is used in making a 12-reading traverse over the face of a 12- by 20-in. rectangular exhaust grille. The average of the anemometer readings, with each reading taken for 1 min, is 1265, and the resultant units are necessarily in feet per minute. Assume that the factor Y for a grille of this type can be considered 0.8. Compute the probable exit flow from the grille.

Ans. 1686 cfm

13-16. A square-edged orifice of 15-in. diameter is mounted in a round duct which is 25 in. in diameter and which carries a gas of density 0.065 lb/ft^3. The pressure across the orifice is 8 in. water, measured at radius-tap positions. Compute the gas flow in cubic feet per second and in pounds per minute.

13-17. A square-edged orifice of 0.381-m diameter is mounted in a round duct which is 0.635 m in diameter and carries a gas having a density of 1.041 kg/m^3. The pressure differential across the orifice is 0.2032 m water, measured at radius-tap positions. Compute the gas flow in cubic meters per second and in kilograms per second.

Ans. 4.581 m^3/s, 4.769 kg/s

13-18. Air is flowing in a duct under a static pressure of 0.508 m water and the barometer is at 0.748 m Hg. A square-edged orifice with flange taps shows a differential pressure of 0.2388 m water. The diameter of the orifice is 0.2032 m and that of the duct 0.4064 m. (a) Compute the density of the air, considering it to be dry and at 15.6°C (60 F). (b) Find the air flow in cubic meters per second and in kilograms per second.

Ans. (a) 1.264; (b) 1.243, 1.571 kg/s

13-19. Air flowing in a duct is under a static pressure of 20 in. water, and the barometer is at 29.45 in. Hg. A square-edged orifice with flange taps shows a differential pressure of 9.4 in. water. The diameter of the orifice is 8 in., and that of the duct is 16 in. (a) Compute the density of the air, considering it to be dry and at a temperature of 60 F. (b) Find the air flow in cubic feet per minute and in pounds per minute.

13-20. Dry saturated steam at 50 psia is flowing in a 4-in. steam line (4.03-in. ID). A 3-in. orifice with flange taps shows a differential pressure of 5 in. Hg. Compute the steam flow in the line in pounds per hour. Refer to Table 2-3 for necessary data on the steam.

CHAPTER

14

PRINCIPLES OF REFRIGERATION

14-1. REFRIGERATION WITH ICE

Refrigeration may be considered the development, in a given space, of a temperature lower than that which exists in some other, or adjacent, space.

The melting of ice or snow was one of the earliest methods *of* refrigeration and is still employed. Ice (or snow) melts at 32 F;[1] so when ice is placed in a given space warmer than 32 F, heat flows into the ice and the space is cooled or refrigerated. The capacity of the ice to absorb heat arises from the fact that when ice changes state from solid to liquid, the latent heat of fusion (143.3 Btu/lb) must be supplied from the surroundings.

In the simple case of ice melting, enough heat-transfer surface must be provided to permit a desired rate of heat flow into the ice from air or water moving past it, or from metal surfaces in contact with it.

Another medium of refrigeration is solid carbon dioxide (dry ice). At atmospheric pressure carbon dioxide (CO_2) cannot exist in a liquid state and consequently, when solid CO_2 absorbs heat, it sublimes or goes directly

[1] Under super pressures, ice does not melt at 32 F; for example, at 14200 psi, ice melts at 16 F.

from the solid into the vapor state. At 14.7 psia, sublimation occurs at about -109.3 F, and the heat of sublimation is approximately 246 Btu/lb at this temperature. Thus dry ice is well suited for low-temperature refrigeration. It also is used widely for the refrigeration of small packages, where the absence of any resulting liquid adds to its convenience.

Water ice was formerly employed, to a limited extent, as a cooling medium in small-size air-conditioning installations (less than 10 to 20 tons of refrigeration). In most such installations water, cooled by spraying and trickling over the ice, served the conditioning equipment. Ice systems performed well under fluctuating loads, since the rate of ice melting could be made to follow closely the cooling-load variations, and the efficiency was not greatly reduced under light loads. Although the cost of ice was relatively high and presented some inconvenience in handling, the low investment cost, compared with that for mechanical refrigeration equipment, justified the use of ice for some cooling installations. Ice and ice-salt mixtures are still used on refrigerated railway cars to maintain the quality of perishable foods in transit.

14-2. MECHANICAL VAPOR REFRIGERATION

Just as a solid, in changing its state into liquid (or gaseous) form, absorbs heat from its surroundings or other sources, so a liquid, in vaporizing, must absorb heat. For example, ammonia, at atmospheric pressure, boils at -28 F and possesses a latent heat of about 589.3 Btu/lb. Ammonia, if placed in surroundings warmer than -28 F, would cool the surrounding space as it evaporated. If the pressure on the ammonia were increased to 30.4 psia, it would boil to 0 F and thus cooling could be accomplished, but in a higher temperature range. Mechanical refrigeration makes possible the control of pressure and temperature of the boiling refrigerant and also makes possible the use of the same refrigerant over and over again with little or no loss of the refrigerant. Theoretically, almost any stable, noncorrosive liquid can operate as a refrigerant if its pressure-temperature relations are suitable for the conditions desired. However, certain features, discussed in detail in the next chapter, determine the desirability of a given refrigerant. The temperature-pressure relationship, which is an important characteristic, can be quickly observed for many refrigerants in Fig. 15-1.

The elements of a compression refrigeration system are shown in Fig. 14-1. In the evaporator the liquid refrigerant, in vaporizing, absorbs heat from brine (or from water, or directly from the air of the space to be cooled). The low-pressure (l-p) refrigerant vapor from the evaporator is inducted into the compressor, which raises the vapor in pressure and temperature for delivery to the condenser.The refrigerant must be compressed sufficiently to have a saturation temperature higher than the temperature of the cooling medium employed so that heat can be dissipated in the condenser. After heat removal and condensation in the condenser, the liquid refrigerant may pass to a receiver, or storage tank. The high-pressure (h-p) liquid refrigerant

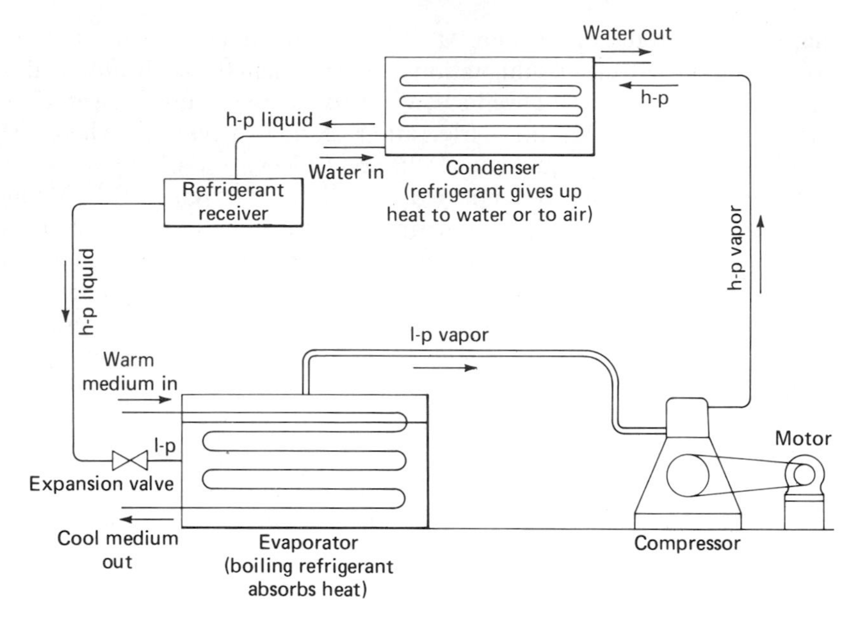

FIGURE 14-1. Elements of a mechanical (compression) refrigerating system with water-cooled condenser.

next passes through the expansion valve, where the refrigerant throttles (drops) to the evaporator pressure of the system. In passing through the expansion valve the liquid refrigerant cools itself at the expense of evaporating a portion of the liquid. In a refrigeration system the low pressure in the evaporator is determined by the temperature which it is desired to maintain in the cooled space. The high pressure in the condenser is determined ultimately by the temperature of the available cooling medium, that is, the circulating water or the atmosphere (air temperature). The process is one in which the refrigerant absorbs heat at a low temperature and then, by the action of mechanical work, the refrigerant is raised to a sufficiently high temperature to allow rejection of this heat. Note that the mechanical work or energy supplied to the compressor is the means used to raise the temperature of the system. Compressors can be powered by any convenient means: electric motors, steam engines, internal combustion engines, etc.

14-3. THE IDEAL (CARNOT) CRITERIA OF REFRIGERATION

In thermodynamic analyses, wide use is made of the Carnot criterion (Carnot cycle) as a standard against which the performance of a prime mover (turbine or engine) can be compared. The reasoning underlying the Carnot criterion is also applicable to an idealized refrigeration system, constituting, as it does, a reversed heat engine or heat pump. The criterion, in its conventional form, presupposes two constant temperature levels of heat exchange and two reversible-adiabatic processes, none of which is completely possible of realization. Nevertheless, the criterion will be presented here because of its importance as a measure of maximum performance.

The efficiency of a Carnot heat engine is expressed as

$$\text{efficiency} = \frac{Q_C - Q_R}{Q_C} = \frac{T_C - T_R}{T_C} \tag{14-1}$$

where T_C is the high temperature of the system; Q_C is the heat interchange at T_C; T_R is the low temperature of the system; and Q_R is the heat interchange at T_R. The values Q_C and Q_R can be expressed in any consistent energy units, and T_C and T_R are expressed in absolute degrees, that is, degrees Fahrenheit + 460 = degrees Rankine; or, alternatively, for SI in Kelvin = degrees Celsius +273.16.

The cycle can advantageously be shown on the temperature-entropy plane. This plane is convenient for representation, since in this plane, for reversible processes, areas can show magnitudes of heat interchange. Thus in Fig. 14-2, for a heat engine, the heat added (Q_C) from 2 to 3 is represented by the area *a*23*b*, and the path of reversible-adiabatic expansion, with temperature dropping from T_C to T_R, is shown by 3 to 4 (no area). The heat rejected (Q_R) is represented by the area 4*ba*1. The path 1 to 2 represents reversible-adiabatic compression from temperature T_R to T_C. In a power cycle the work produced (W) is equal to the difference between the heat added and the heat rejected, or

$$W = Q_C - Q_R \tag{14-2}$$

Consequently the work area is 1234.

For refrigeration the process is reversed, with heat added at the low temperature T_R, in amount Q_R, and represented by the area 14*ba*. The temperature of the cycle medium is raised by compression, following the path

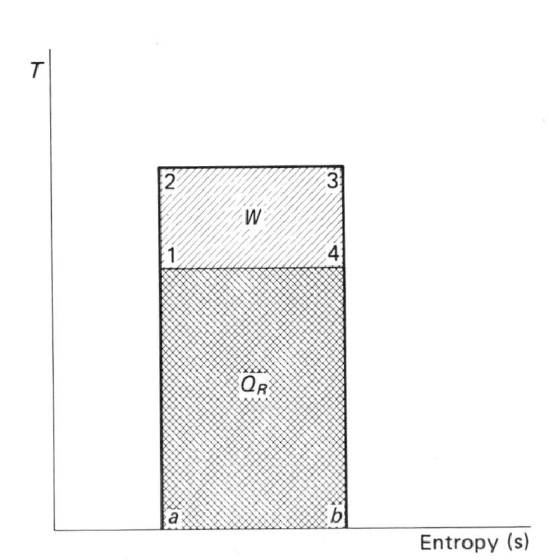

FIGURE 14-2. Ideal Carnot cycle for power or refrigeration.

4 to 3. Heat is rejected at the high temperature T_C, in amount Q_C, and represented by the area 32*ab*. The work. which must be provided from an external source, is $-(Q_R - Q_C)$ and is represented by the area 3214.

For the refrigeration (heat pump) type of operation, conventional efficiency has little significance and it is desirable to introduce the term coefficient of performance (CP). The coefficient of performance for any refrigeration system, either ideal or actual, is expressed as the refrigeration created divided by the work required to produce it. Thus

$$CP = \frac{Q_R}{W} \tag{14-3}$$

For the Carnot (ideal) basis, Q_R/W appears as

$$CP = \frac{Q_R}{Q_C - Q_R} \tag{14-4}$$

$$= \frac{T_R}{T_C - T_R} \tag{14-5}$$

Any consistent energy or power units for Q_C, Q_R, and W can be employed with Btu, Btu per pound, and Btu per hour, in engineering units, and in joules or joules per kilogram, or kilowatthours or kilowatthours per kilogram, and with the kilowatt as the usual power unit.

Example 14-1. An ideal (Carnot) refrigeration system operates between temperature limits of −20 F and 80 F (−28.89°C and 26.67°C). Find (a) the ideal coefficient of performance and (b) the horsepower required from an external source to absorb 12 000 Btuh at the low temperature. (c) Rework part a using SI units. (d) Rework part a when 3.51 kW needs to be absorbed at the low temperature.

Solution: (a) By Eq. (14-5),

$$CP = \frac{-20 + 460}{(80 + 460) - (-20 + 460)} = \frac{440}{540 - 440} = 4.4 \qquad \textit{Ans.}$$

That is, 4.4 times as much refrigeration is produced (heat is absorbed) as is required as work.

(b) A horsepower is 33 000 ft lbf/min. Thus

$$\begin{aligned} 33\,000 \text{ ft lbf/min} &\equiv 33\,000/778 \text{ or } 42.4 \text{ Btu/min} \\ &\equiv 2545 \text{ Btuh} \end{aligned}$$

With a CP of 1, the horsepower required would be 12 000/2545. However, since the ideal CP is 4.4,

$$\text{hp} = \frac{12\,000}{(\text{CP})(2545)} = \frac{12\,000}{(4.4)(2545)} = 1.07 \qquad \textit{Ans.}$$

(c) By Eq. (14-5),

$$CP = \frac{-28.89 + 273.16}{(26.67 + 273.16) - (-28.89 + 273.16)} = 4.4 \qquad \textit{Ans.}$$

(d) With a CP of 4.4, the power requirement by Eq. (14-3) is

$$W = \frac{Q_R}{\text{CP}} = \frac{3.51}{4.4} = 0.798 \text{ kW} \qquad \textit{Ans.}$$

$$\text{hp} = \frac{0.798}{0.746} = 1.07 \qquad \textit{Ans.}$$

14-4. REFRIGERANT DIAGRAMS, TEMPERATURE ENTROPY

It has been mentioned that a refrigerant can be any stable, noncorrosive substance having suitable liquid-vapor phase characteristics to serve satisfactorily in the evaporator-condenser temperature range required. The mediums actually used must satisfy a number of other criteria, such as availability, cost, low toxicity, pressure range, behavior with lubricants, and the like. These criteria are discussed in detail in Chapter 15, whereas at this point the basic thermodynamics of the refrigerating cycle as related to any refrigerant will be primarily considered. However, in this chapter reference to tabular properties of refrigerants given in Chapter 15 will be made, with frequent discussions related to the refrigerants ammonia and dichlorodifluoromethane. The latter is also known as refrigerant-12.

The properties of a refrigerant or of any thermodynamic substance can be represented to advantage by plotting property values on coordinate diagrams. In common use are the temperature-entropy plane (*Ts*), the pressure-enthalpy plane (*p***h**), and, of less frequent use, the pressure-volume plane and the enthalpy-entropy plane. In refrigeration work the pressure-enthalpy diagram is by far the most useful type of plot, although the temperature-entropy plot, on which areas can represent the heat interchange, is also useful.

Figure 14-3 is a temperature-entropy plot of carbon dioxide. This substance was chosen for discussion because it illustrates a medium used, to a limited extent, as a refrigerant in all three phases (solid, liquid, and vapor). The critical point for CO_2, indicated at P_0 on the diagram, has a value of 87.8 F, with a corresponding saturation pressure of 1072 psia. At temperatures higher than its critical temperature it is impossible to liquefy a gas, no matter what pressure is exerted on it. Below the critical point a definite pressure-temperature relationship applies; that is, for each pressure there is a corresponding temperature at which vaporization (or condensation) occurs. For example, liquid CO_2 at a pressure of 852.5 psia (point *A* on Fig. 14-3) will boil (vaporize) at 70 F, and such a liquid when heated would change completely from liquid to vapor, as shown at point *B*. If further heat is added to the dry saturated vapor, with pressure remaining constant at *B*, the temperature of the vapor will rise as it moves toward point *C*, and we speak of this vapor as being superheated. The region to the right of the vapor line is known as the superheated (or dry-gas) region. The region between the liquid and vapor lines consists of saturated mixtures of liquid and vapor.

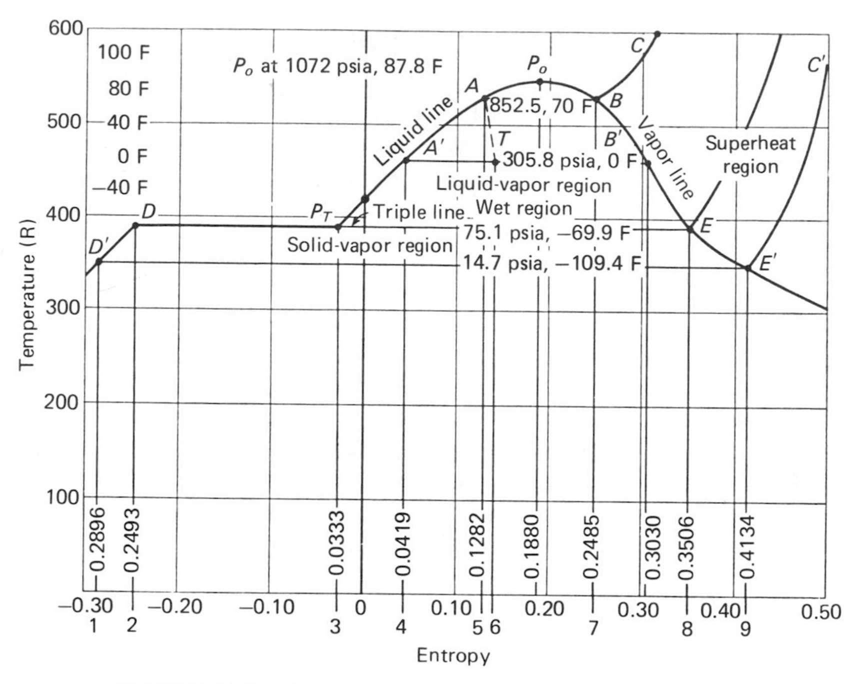

FIGURE 14-3. Temperature-entropy diagram for carbon dioxide.

It is possible for a liquid, such as the saturated liquid at A, to be cooled with the imposed pressure remaining constant, as for example, to the temperature A', for which the saturation pressure is lower than for point A. Such a liquid existing at a higher pressure than the saturation pressure corresponding to its temperature is known as *compressed liquid* or as *subcooled liquid*. The entropy of compressed liquid is trivially lower than the entropy of saturated liquid at the same temperature. However, the difference cannot be shown on a Ts diagram of this type; thus the liquid line represents saturated and compressed-liquid conditions. In most cases the properties of a liquid, particularly enthalpy and specific volume, are found in tables under the temperature listing and are independent of pressure. If a saturated (or compressed) liquid is cooled sufficiently, it will ultimately reach a point at which solidification or fusion takes place. The particular temperature and pressure at which saturated liquid, solid, and vapor are in equilibrium is known as the *triple point*, marked P_T in Fig. 14-3. If heat is removed from liquid at the triple point, solidification starts and continues from the liquid-solid mixture until all of the liquid has fused. This point is represented by D on the diagram, and the line D to P_T is thus a solid-liquid line. The triple point conditions for CO_2 are at -69.9 F and 75.1 psia. An interesting consequence of this fact is that liquid CO_2 cannot exist at atmospheric pressure (14.7 psia), since this is lower than 75 psia. Solid CO_2 (dry ice) is very familiar at atmospheric conditions, but an open container of liquid CO_2, no matter what its temperature, cannot exist. In contrast to CO_2, water (H_2O) has a triple point of 32 F at 0.0885 psia, and thus we are quite familiar with both water and ice at atmospheric pressure. Melting of solid CO_2 to liquid can take place only at pressures equal or higher than 75.1 psia.

The line DD' in Fig. 14-3 represents loci of solid CO_2 saturation conditions, because, as is true for the liquid phase, there are similar solid phase saturation-temperature-pressure relationships. For each pressure there is a corresponding temperature at which sublimation takes place and the solid becomes vapor, as at E or E'. The vapor at E is unique, in that it is at the triple point and can be in equilibrium either with solid or liquid. It should also be mentioned that the triple point on the Ts diagram is the line DE.

Water ice can also sublime and go directly from the solid to the vapor (steam) phase. For example, wet articles hung outside to dry, in 0 F temperature weather, first freeze and then slowly dry as the water sublimes. Snow and ice on the ground, under subfreezing conditions, also slowly disappear by sublimation without ever reaching the liquid phase.

The thermodynamic properties of all the commonly used refrigerants have been computed and usually can be found in tabular form. The datum of reference for the tabular values is usually taken at -40 F and the corresponding saturation pressure. At this reference condition, enthalpy and entropy are usually arbitrarily given values of zero. This represents a satisfactory procedure, since our interest rests in changes in these properties and not in absolute values of them.

Areas on the Ts diagram, when properly interpreted, can represent heat energy added or withdrawn during a process. For such areas to be to scale, it is necessary to draw the diagram using temperature expressed in absolute degrees. The entropy, however, can be from an arbitrary datum, but of course in consistent units (usually Btu/lb · deg R if the area and result are desired in Btu). For example, in Fig. 14-3 the area under $A'A$—namely, $A'A54$—represents the heat addition required to warm saturated liquid CO_2 from conditions at A' to those at A; correspondingly, the area $AB75$ represents the heat addition required to change liquid to vapor or the enthalpy of vaporization at 852.5 psia, and the area $D'E'91$ represents the enthalpy of sublimation at 14.7 psia. As most refrigeration is concerned only with the refrigerant in the liquid and vapor phases, diagrams are not usually drawn for conditions below triple point, nor are they extended up to the critical point (P_o).

14-5. REFRIGERANT DIAGRAMS, PRESSURE ENTHALPY

A skeleton pressure-enthalpy (p**h**) diagram for dichlorodifluoromethane (refrigerant-12, popularly known as Freon-12 or genetron-12) appears in Fig. 14-4. As in the Ts diagram, the key lines are those for saturated liquid and saturated vapor. If these lines were extended higher than shown in the figure, they would meet at the critical point (P_o), which for this refrigerant happens to be at 596.9 psia and 233.6 F. The region enclosed between the saturated liquid and vapor lines is the wet-vapor region. To the right of the

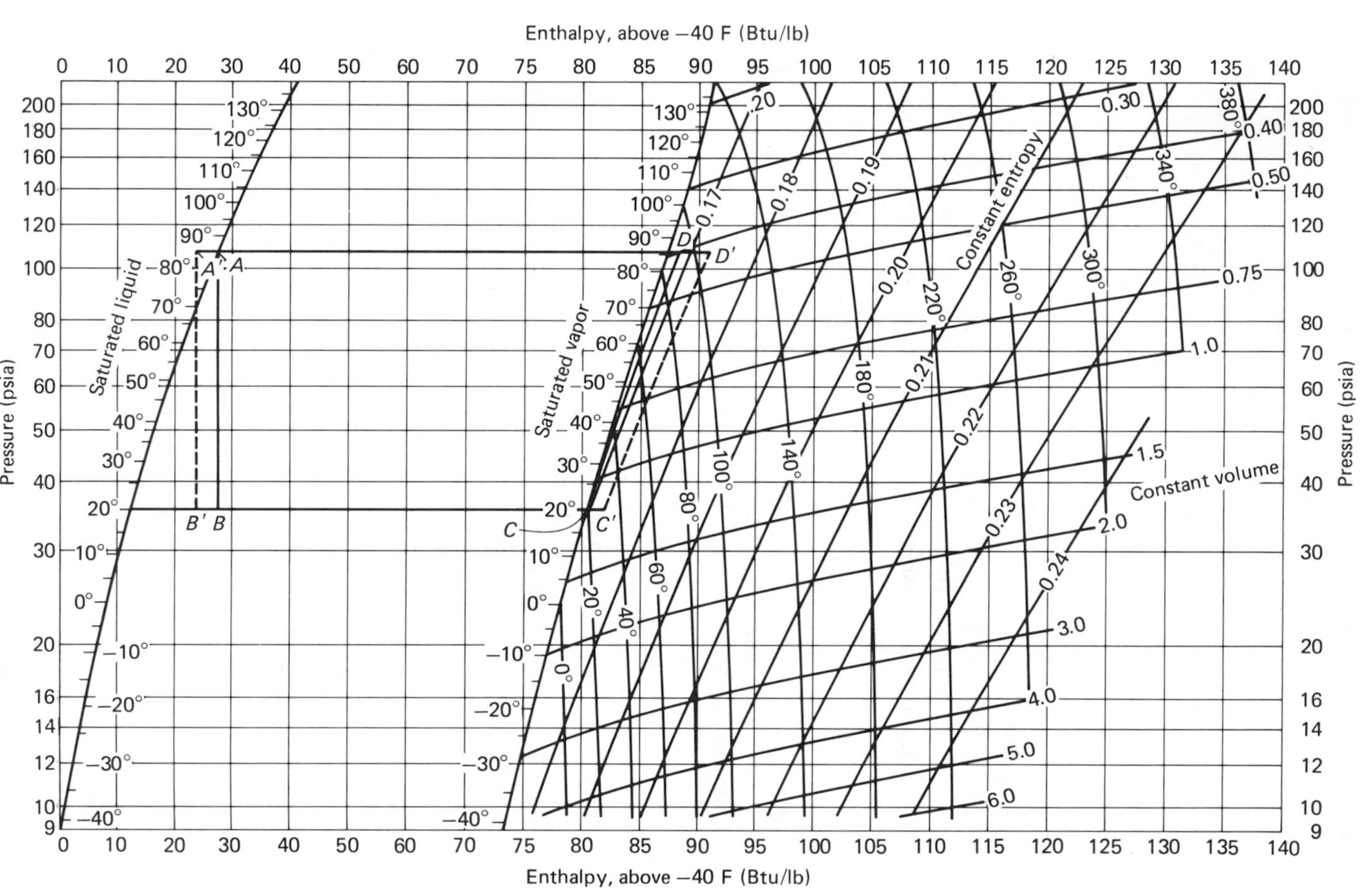

FIGURE 14-4. Refrigerant cycle superposed on pressure-enthalpy diagram for dichlorodifluoromethane (refrigerant-12).

vapor line is the superheated-vapor region. To the left of the liquid (saturated) line lies the compressed-liquid region. The conditions of compressed liquid can advantageously be shown on this type diagram at A and A'. If the liquid temperature is constant, no significant change in the enthalpy of compressed liquid occurs as the pressure on the liquid is varied. There is, however, a slight increase in compressed-liquid enthalpy with pressure increase, and although this is significant in power-plant operations (as feed-pump work for the extensive pressure ranges), in refrigeration computations the trivial enthalpy change can be disregarded.

The pressure-enthalpy diagram is important in refrigeration computations because on this diagram, enthalpy changes can be read with ease, and such changes can represent heat added or withdrawn, or work received (in adiabatic compression). Moreover, the refrigeration cycle largely employs two pressures, namely, the pressure in the evaporator associated with the useful refrigeration and that in the condenser associated with the heat-dissipation phase of the cycle. Both of these pressures are readily shown on the $p\mathbf{h}$ diagram.

14-6. REFRIGERATION CYCLE, EXPANSION AND EVAPORATION

Expansion Valve (Throttling). Liquid refrigerant, after it leaves the condenser (or the receiver), enters the expansion valve. This valve serves the purposes of controlling the refrigerant flow and of dropping the refrigerant, both in temperature and pressure, from condenser to evaporator conditions. A flow process which takes place adiabatically without work production is a throttling process, and in such it is easy to show, from the steady-flow energy equation [Eq. (2-9)], after eliminating irrelevant terms, that

$$\frac{V_1^2}{2gJ} + \mathbf{h}_1 = \frac{V_2^2}{2gJ} + \mathbf{h}_2 \tag{14-6}$$

Because the kinetic energy at entry to and exit from an expansion valve seldom differs greatly, it is customary to disregard these terms and write the expansion valve equation as

$$\mathbf{h}_1 = \mathbf{h}_2 \tag{14-7}$$

This equation states that the enthalpy remains constant in the throttling process of an expansion valve.

A certain amount of the liquid, however, always flashes into gas, because a pound of liquid at the low pressure and temperature has less enthalpy than a pound at higher temperature. Thus some vaporization must take place. The process may also be considered in this light: the bringing of warm liquid to cold evaporator temperature occurs at the expense of evaporating a portion of the liquid, with the loss of some refrigerating effect. This process results in the production of flash vapor which can accomplish no useful cooling in the evaporator. Thus the expansion-valve equation is

$$\mathbf{h}_{f1} = \mathbf{h}_{fR} + x\mathbf{h}_{fgR} \tag{14-8}$$

where $\mathbf{h}_{f1}$ is the enthalpy of liquid refrigerant at the temperature at which it enters the expansion valve, in Btu per pound; $\mathbf{h}_{fR}$ is the enthalpy of liquid at evaporator pressure, in Btu per pound; $\mathbf{h}_{fgR}$ is the latent heat of refrigerant at evaporator pressure, in Btu per pound; and x is the quality, expressed as a decimal, of the refrigerant after passing through the expansion valve, also the pound (weight) of flash gas formed per pound of refrigerant. The expansion valve (throttling process) is represented on the $p\mathbf{h}$ chart (Fig. 14-4) by the straight line AB (or by $A'B'$ if the liquid is originally subcooled).

On the Ts chart, a line of constant enthalpy swings down and to the right, and the area representing liquid enthalpy at the beginning of the process must equal the area for wet vapor at the end of the process. Using the Ts chart of Fig. 14-3 for illustration, the liquid condition is shown at A, before throttling, and the wet-vapor condition is indicated at T, after throttling. Realizing that the entropy datum is not significant, let us select $A'4$ as a convenient datum; then for the throttling process we can state that the area $A'A54$, before throttling, must equal the area $A'T64$, after throttling.

Example 14-2. Saturated liquid dichlorodifluoromethane (refrigerant-12) at a temperature of 86 F at 108.04 psia enters an expansion valve and expands into an evaporator in which the refrigerant is boiling at 20 F. Find (a) the enthalpy of the refrigerant at entry to and exit from the expansion valve and (b) the weight of flash gas formed per pound of refrigerant entering the valve.

Solution: (a) In Table 15-5 the enthalpy of liquid refrigerant-12 at 86 F is shown to be $\mathbf{h}_f = 27.769$ Btu/lb. The enthalpy at exit from the valve is likewise 27.769 Btu/lb. *Ans.*

(b) At 20 F evaporator temperature, from Table 15-5, the saturation pressure is 35.736 psia, $\mathbf{h}_{fR} = 12.863$, and $\mathbf{h}_{fgR} = 66.522$. By Eq. (14-8)

$$27.769 = 12.863 + x(66.522)$$

$x = 0.224$ lb of flash gas formed per pound of refrigerant entering the expansion valve *Ans.*

Example 14-3. If the liquid refrigerant of Example 14-2 is subcooled to 70 F before entering the expansion valve, find the items asked for in Example 14-2.

Solution: Table 15-5 shows that liquid refrigerant-12 has an enthalpy at 70 F of 24.050 Btu/lb. This is true for 70 F compressed liquid at 108.04 psia just as it is true for saturated 70 F liquid at 84.888 psia; that is, the enthalpy of liquid is a function of temperature. Thus for part a the enthalpy is 24.050 Btu/lb before entry and at exit from the expansion valve. *Ans.*

(b) At 20 F,

$$24.050 = 12.863 + x(66.522)$$

$x = 0.168$ lb of flash gas formed per pound of refrigerant entering the expansion valve *Ans.*

Note that subcooled (compressed) liquid produces less flash gas than liquid at the higher temperature corresponding to saturation.

Evaporator. In the evaporator the liquid from the expansion valve changes into vapor as it absorbs heat from the space being cooled. The heat absorbed appears as increased enthalpy of the refrigerant. The vapor leaving the evaporator may be dry-saturated as at C in Fig. 14-4, super-heated as at C', or slightly wet. Thus

$$Q_R = \mathbf{h}_R - \mathbf{h}_{f1} \tag{14-9}$$

where Q_R is the absorbed per pound of refrigerant in evaporator; $\mathbf{h}_R$ is the enthalpy of vapor leaving the evaporator, in Btu per pound; and $\mathbf{h}_{f1}$ is the enthalpy of liquid refrigerant at the temperature supplied to the expansion valve, in Btu per pound.

Refrigerant flow is often expressed in pounds per minute per ton of refrigeration. As indicated before, the ton of refrigeration is a rate of heat exchange (absorption); by definition, 200 Btu/min or 12 000 Btuh. Thus

$$w_R = \frac{200}{Q_R} = \frac{200}{\mathbf{h}_R - \mathbf{h}_{f1}} \tag{14-10}$$

where w_R represents pounds of refrigerant circulated per minute per ton. The values Q_R, $\mathbf{h}_R$, and $\mathbf{h}_{f1}$ are as they appear in Eq. (14-9).

Example 14-4. Refrigerant-12 is vaporizing in an evaporator at 20 F. Assuming that dry-saturated vapor leaves the evaporator, and that liquid is supplied to the expansion valve at 86 F, find the rate of refrigerant flow required per ton of refrigeration and also per 10 000 Btuh.

Solution: Reference to Table 15-5 shows that dry-saturated vapor ($\mathbf{h}_g$) at 20 F and 35.736 psia has an enthalpy of 79.385 Btu/lb. By Eq. (14-9),

$$Q_R = \mathbf{h}_R - \mathbf{h}_{f1} = 79.385 - 27.769 = 51.616 \text{ Btu/lb}$$

By Eq. (14-10),

$$w_R = \frac{200}{51.616} = 3.87 \text{ lb/min} \cdot \text{ton} \qquad \textit{Ans.}$$

For 10 000 Btuh,

$$\text{flow} = \frac{10\,000}{(60)(51.616)} = 3.23 \text{ lb/min} \qquad \textit{Ans.}$$

14-7. REFRIGERATION CYCLE, COMPRESSION

To maintain a given evaporator pressure the compressor must remove the vapor as fast as it is formed. If the evaporator load is small, that is, if the temperature difference between the medium cooled and the evaporator is slight, little refrigerant can evaporate and the compressor suction causes a reduction in the evaporator pressure. This pressure decrease will continue until the temperature difference between the refrigerated space and the colder evaporator becomes just sufficient to generate enough vapor to supply the effective piston displacement of the compressor. If, on the other hand, the temperature difference is great (excessive load), vapor will be generated rapidly at relatively high evaporator temperature and the compressor may be overloaded. The heat-transfer load characteristics of the evaporator must be such as not to overload the compressor.

More work is required to compress a pound of refrigerant from a low pressure (temperature) to a given condenser pressure (temperature) than from a higher pressure to the same condenser pressure. Yet, in the actual machine, the greater weight of refrigerant handled at a higher suction pressure (because of lower specific volume and compressor-clearance behavior) usually causes a greater power requirement on the driving motor as the evaporator pressure increases.

Compression of a vapor, under adiabatic conditions, requires the least work when performed *isentropically* (at constant entropy). Isentropic compression cannot be attained in actual equipment, but it forms a basis for computing ideal work under different operating conditions. From this basis the actual work can be estimated. The entropy of a given vapor is always listed in complete tables of properties of the refrigerant and is also plotted on pressure-enthalpy charts.

The theoretical (or isentropic) work of compression, using the symbols of Fig. 14-4, is found by

$$W_T = (\mathbf{h}_D - \mathbf{h}_C)_s \qquad \text{or} \qquad (\mathbf{h}_{D'} - \mathbf{h}_{C'})_s \qquad \text{Btu/lb}$$

or in general symbolism

$$W_T = (\mathbf{h}_D - \mathbf{h}_R)_s \qquad \text{Btu/lb of refrigerant} \tag{14-11}$$

where $\mathbf{h}_D$ is the enthalpy at discharge pressure and conditions and $\mathbf{h}_R$ is the enthalpy of the vapor entering the compressor, with entropy s assumed constant during compression to $\mathbf{h}_D$.

The theoretical horsepower per ton of refrigeration is

$$\text{hp} = \frac{\text{lb refrigerant}}{\text{min} \times \text{ton}} \times W_T \times \frac{1}{42.4} \tag{14-12}$$

To find the actual work or power required by a compressor from the theoretical work of compression, a factor must be employed. For rough approximations it may be considered that the isentropic work, increased by 30 to 50%, approximates the shaft work that is required to carry out the compression. This factor must account for friction in the packing glands of the compressor, in the bearings, and from piston rings, heat exchanges, irreversibilities in the compressor, and the like. The factor is higher with small, inefficient compressors, and lower with larger-size, well-designed units. Centrifugal compressors, because of higher internal loss and turbulence, might be expected to lie at the higher end of the range of factor values, but, since these machines are usually large and well-designed, the factor values for centrifugal compressors usually lie at about 30%.

Example 14-5. A refrigeration compressor takes refrigerant-12, dry-saturated vapor at 20 F (at 35.736 psia), and compresses and delivers it to a condenser operating at a condensing temperature of 86 F. Compute (a) the ideal work of compression, in Btu per pound, (b) the theoretical horsepower required per ton of refrigeration if this compressor serves an evaporator operating under the conditions of Example 14-4, and (c) the probable actual input shaft horsepower required for a 1-ton-capacity reciprocating compressor.

Solution: (a) Refer to the ***ph*** chart for refrigerant-12 vapor (Fig. 15-4) and find the isentropic work of compression such as would be indicated by the line *CD* of Fig. 14-4. For the given compressor inlet condition, namely, dry-saturated vapor at 20 F, the enthalpy $\mathbf{h}_R$ or $\mathbf{h}_C$ on the diagram is 79.4 Btu/lb. Follow a path parallel to the lines of constant entropy until the pressure corresponding to 86 F condensation (namely, 108.04 psia) is reached at which point $\mathbf{h}_D = 87.9$ Btu/lb. Then by Eq. (14-11),

$$W_T = (\mathbf{h}_D - \mathbf{h}_R)_s = 87.9 - 79.4 = 8.5 \text{ Btu/lb} \qquad \textit{Ans.}$$

(b) From Example 14-4, the refrigerant flow per ton is 3.87 lb/min, and the theoretical power is

$$w_R W_T = (3.87)(8.5) = 32.9 \text{ Btu/min}$$

and

$$\text{theoretical hp/ton} = \frac{32.9}{42.4} = 0.78 \qquad \textit{Ans.}$$

where $2545/60 = 42.4$ Btu/min $\equiv$ 1 hp.

(c) If we take a representative shaft work factor of 1.4, the probable shaft horsepower is

$$\text{compressor shaft hp} = (1.4)(0.78) = 1.09 \qquad \textit{Ans.}$$

The capacity of a reciprocating compressor for handling refrigerant depends on the piston displacement per minute and on the volumetric efficiency. Volumetric efficiency (charge efficiency) of a compressor is defined as the amount of vapor handled, measured in cubic feet per minute at suction pressure and temperature, divided by the piston displacement per minute. This efficiency is always less than unity because of (1) superheating of the entering vapor as it enters the warm cylinder; (2) throttling or friction loss through ports and valves, which reduces cylinder pressure; and (3) reexpansion of compressed vapor from the clearance volume, which reduces the effective volume for fresh vapor charge at suction pressure. Values of volumetric efficiency will thus vary with the type of compressor; with the operating conditions, as affected by the condition of the suction vapor and by the pressure ratio of compression; and with the refrigerant. Volumetric efficiencies η_v of reciprocating compressors range from about 76 to 90%. A representative value of 86% is suggested for general usage with current compressor designs. If liquid enters the compressor along with vapor, a fictitious volumetric efficiency in excess of 100% may arise. A further discussion of volumetric efficiency is presented at the end of this chapter.

The effective piston displacement (PD) of a single-acting compressor can be expressed, in cubic feet per minute, as

$$\mathrm{PD} = C \frac{\pi d^2 S N}{4 \times 1728} (\eta_v) \qquad (14\text{-}13)$$

where C is the number of cylinders in the compressor; d is the compressor bore, in inches; S is the compressor stroke, in inches; N is revolutions per minute; η_v is the volumetric efficiency, expressed as a decimal; and PD is the effective piston displacement, for induction of vapor at low-pressure (suction) conditions, in cubic feet per minute.

14-8. REFRIGERATION CYCLE, CONDENSATION

The condenser, whether water-cooled or air-cooled, must remove heat from the refrigerant to change the superheated gas leaving the compressor into saturated or subcooled liquid. Again referring to Fig. 14-4, the condenser must remove heat ideally to change the gas with enthalpy $\mathbf{h}_D$ (or $\mathbf{h}_{D'}$) to liquid with enthalpy $\mathbf{h}_{fA}$ (or if subcooled to $\mathbf{h}_{fA'}$). That is,

$$Q_C = \mathbf{h}_D - \mathbf{h}_A \qquad \text{Btu/lb} \qquad (14\text{-}14)$$

or for other operating conditions

$$Q_C = \mathbf{h}_{D'} - \mathbf{h}_{A'} \qquad \text{Btu/lb} \qquad (14\text{-}15)$$

Expressed another way, the heat removed in the condenser is equal to the heat absorbed at low temperature in the evaporator plus the heat equivalent of the work performed on the refrigerant in the compressor or, symbolically,

$$Q_C = Q_R + W \qquad (14\text{-}16)$$

where Q_C is the heat removed from refrigerant in condenser, in Btu per pound; Q_R is the refrigerating effect (heat absorbed) in the evaporator, in Btu per pound of refrigerant; and W is the work energy added to the refrigerant, in Btu per pound.

Example 14-6. A two-cylinder, 3- by 3-in. refrigerant-12 compressor running at 1140 rpm operates on a system with condensation occurring at 86 F and evaporation at 20 F. Assume that dry-saturated vapor enters the compressor and the condensed refrigerant is not subcooled. Compute (a) the piston displacement; (b) the compressor capacity, in pounds of refrigerant per minute, if the volumetric efficiency is 82%; (c) the tons of refrigeration produced; (d) the idealized condenser heat load, in Btu per pound of refrigerant and (e) in Btu per hour; and (f) the actual coefficient of performance (CP).

Solution: (a) Use Eq. (14-13) or equivalent reasoning to find the true piston displacement D.

$$D = 2\frac{(\pi)(3^2)(3)(1140)}{(4)(1728)} = 27.95 \text{ cfm} \qquad \textit{Ans.}$$

The true piston displacement is independent of volumetric efficiency ($\eta_v = 1.0$).

(b) PD = (27.95)(0.82) = 22.9 cfm effective piston displacement

Because the vapor supplied to the compressor is dry and saturated and at 20 F, its specific volume can be read directly from Table 15-5 as $v_g = 1.0988$ ft^3/lb,

$$w = \frac{\text{PD}}{v_g} = \frac{(\text{ft}^3) \times (\text{lb})}{(\text{min}) \times (\text{ft}^3)} = \frac{22.9}{1.0988} = 20.8 \text{ lb/min refrigerant flow rate}$$

Ans.

(c) From Example 14-4, the refrigeration effect per pound of refrigerant is 51.616 Btu/lb. Thus

$$\text{refrigeration produced} = (51.616)20.8 = 1074 \text{ Btu/min}$$

$$= \frac{1074}{200} = 5.37 \text{ tons} \qquad \textit{Ans.}$$

(d) For the given operating conditions, it was shown in Example 14-5 that the isentropic work per pound of refrigerant was 8.5 and the actual work was taken 40% higher, or

$$W = (8.5)(1.4) = 11.9 \text{ Btu/lb}$$

From part c the refrigeration effect is $Q_R = 51.616$ Btu/lb and by Eq. (14-16),

$$Q_C = 51.616 + 11.9 = 63.5 \text{ Btu/lb} \qquad \textit{Ans.}$$

(e)
$$Q = (63.5)(20.8)(60) = 79\,200 \text{ Btuh} \qquad \textit{Ans.}$$

(f) By Eq. (14-3),

$$\text{CP} = \frac{51.616}{11.9} = 4.3 \qquad \textit{Ans.}$$

14-9. COMPRESSOR WORK BY THE GAS EQUATION

The work required for compression of a refrigerant should always be found from tabulations of properties (or charts) of the refrigerant when these are available, and for the commonly used refrigerants complete tables have been prepared. However, for gases and vapors used as refrigerants and when complete property tabulations are not available, use can be made of relationships developed from perfect-gas laws. The general form of the expression for isentropic work of compression, as derived in standard thermodynamics textbooks, appears as

$$W_T = \frac{144}{778}\frac{k}{k-1} p_R v_R\left[\left(\frac{p_D}{p_R}\right)^{(k-1)/k} - 1\right] \quad \text{Btu/lb} \tag{14-17}$$

$$W_T = \frac{k}{(778)(k-1)} R T_R\left[\left(\frac{p_D}{p_R}\right)^{(k-1)/k} - 1\right] \quad \text{Btu/lb} \tag{14-18}$$

$$\text{hp} = \frac{144}{33\,000}\frac{k}{k-1} p_1 V_1\left[\left(\frac{p_2}{p_1}\right)^{(k-1)/k} - 1\right] \tag{14-19}$$

where $k = C_p/C_v$, is the ratio of specific heats of the refrigerant (Table 2-1); p_R or p_1 is the pressure at the inlet to the compressor, in pounds per square inch absolute; v_R or v_1 is the specific volume at compressor inlet, in cubic feet per pound; T_R or T_1 is the temperature at the compressor inlet, in degrees Rankine; R is the gas constant, $1545.3/M$, in foot-pounds per pound degree Rankine (see Section 2-5); and V_R or V_1 is the volume, in cubic feet per minute at inlet conditions, entering the compressor.

Moreover, in terms of temperature it is closely true that

$$\frac{T_2}{T_1} = \left(\frac{p_2}{p_1}\right)^{(k-1)/k} = \left(\frac{v_1}{v_2}\right)^{k-1} \tag{14-20}$$

where the subscripts 1 and 2 refer to conditions before and after compression with temperature and pressure in same units as previous equations. Thus when a gas is compressed it may increase greatly in temperature, provided the compression is carried out so quickly that the gas has little chance to cool. In fact, gas compression in reciprocating compressors is closely adiabatic, although not truly isentropic. Moreover, when a gas expands in an expander engine or gas turbine and produces work, the gas temperature decreases and Eq. (14-20) (when inverted) is equally applicable to rapid (adiabatic) expansion. The work and power Eqs. (14-17) through (14-19) are also applicable to the expansion process if the inlet terms and ratios are appropriately modified.

14-10. AIR CYCLE REFRIGERATION

Before the development of the halogenated-hydrocarbon refrigerants, with their low toxicity or even nontoxic characteristics, air-cycle systems were often used where absolute assurance was required against possible escape of toxic refrigerant. In comparison to mechanical vapor-liquid systems, air systems were bulky and heavy; the horsepower per ton was many times greater than with vapor systems; and the cooling medium, instead of acting at a fixed temperature, warmed over a temperature range. Most of the early air systems were closed, so-called dense-air systems. A closed system made possible a plant of smaller physical size, because the air was compressed at all stages of the cycle, and humidity, as a problem, was largely eliminated.

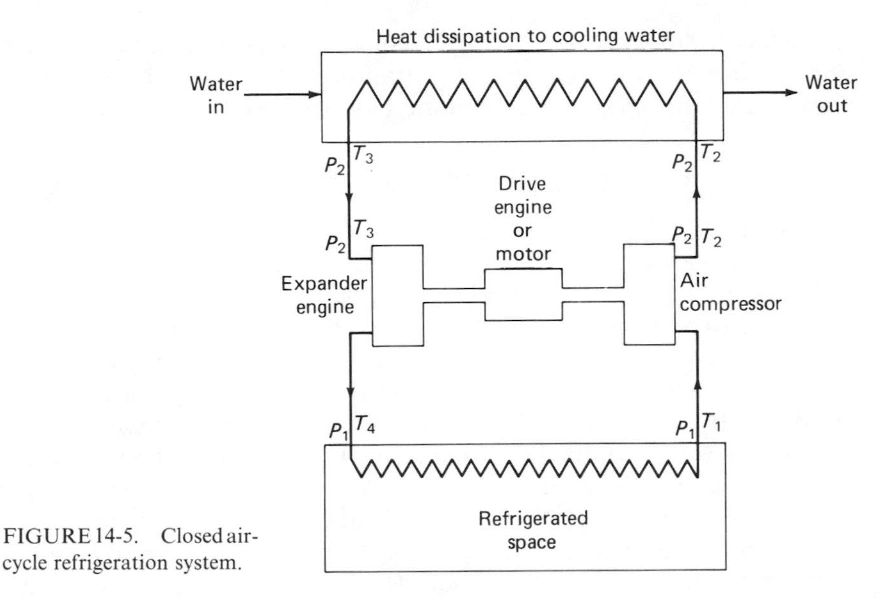

FIGURE 14-5. Closed air-cycle refrigeration system.

A diagrammatic layout of a closed system is shown in Fig. 14-5. In this system, air at temperature T_1, several degrees lower than the space temperature, and at pressure p_1 enters the compressor. During compression, the air increases in pressure to p_2 and in temperature to T_2. The compressed air then enters the water-cooled heat exchanger, which corresponds to the condenser of a vapor system, and is cooled to temperature T_3, which is a few degrees warmer than the inlet water temperature. The cooled compressed air then enters the expander engine, where it drops to the low pressure of the system, at the same time decreasing to temperature T_4. The useful refrigeration produced is

$$Q_R = wC_p(T_1 - T_4) \qquad \text{Btu/min} \tag{14-21}$$

The heat dissipation is

$$Q_C = wC_p(T_2 - T_3) \qquad \text{Btu/min} \tag{14-22}$$

The isentropic work input to the air compressor is

$$W_C = wC_p(T_2 - T_1) = wC_p T_1\left[\left(\frac{p_2}{p_1}\right)^{(k-1)/k} - 1\right] \tag{14-23}$$

The isentropic work delivered by the expander engine is

$$W_E = wC_p(T_3 - T_4) = wC_p T_3\left[1 - \left(\frac{p_4}{p_3}\right)^{(k-1)/k}\right] \tag{14-24}$$

The net isentropic work required from the motor, in Btu per minute, is

$$W_N = wC_p[T_2 - T_1 - (T_3 - T_4)] \tag{14-25}$$

$$= wC_p\left\{T_1\left[\left(\frac{p_2}{p_1}\right)^{(k-1)/k} - 1\right] - T_3\left[1 - \left(\frac{p_4}{p_3}\right)^{(k-1)/k}\right]\right\} \tag{24-26}$$

In an idealized system, in which there is no pressure loss, the compression pressure ratio p_2/p_1 equals the expansion ratio p_3/p_4. Letting r represent each of these ratios, the expression for net work simplifies to

$$W_N = wC_p\left\{T_1\left[r^{(k-1)/k} - 1\right] - T_3\left[1 - \frac{1}{r^{(k-1)/k}}\right]\right\} \tag{14-27}$$

In the preceding expressions w is air flow in the system, in pounds per minute; C_p is the specific heat of air, approximately 0.24; p_1, p_2, p_3, and p_4 are pressures, in any consistent units but generally in pounds per square inch absolute; T_1, T_2, T_3, and T_4 are temperatures in degrees Rankine (except where temperature difference only is involved, and degrees Fahrenheit may be used); and W_C, W_E, and W_N are values of work, in Btu per minute.

Figure 14-6 illustrates the diagrammatic layout of an air-cycle refrigeration system as employed for air conditioning the cockpit and cabin space of an airplane. The system illustrated is primarily applicable to turbojet and turbopropellor airplanes, where it is possible to bleed (draw) a supply of compressed air from the compressor. This compressed air would be at a pressure of perhaps 60 to 80 psia and at a temperature of 300 F to 400 F. Thus it is necessary to cool the compressed air, and, as shown, both a precooler and a heat exchanger, which employ ram air from the atmosphere, are used for this purpose. The fan employed for drawing air from the atmosphere through the heat exchanger is driven by the expander turbine that produces chilled air. The fan constitutes the load for the expander turbine. The temperature of the supply air entering the turbine depends on outside

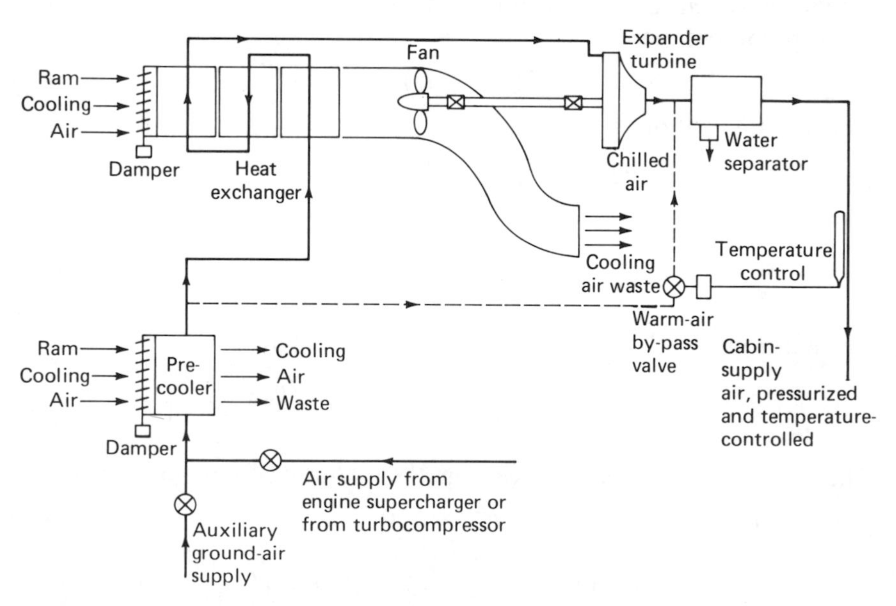

FIGURE 14-6. Diagrammatic layout of an air-cycle refrigeration system for an airplane.

air temperature, and with 60 F outside air, at sea level, might be at a temperature of about 90 F. At high-altitude temperatures, which range from 20 F to −80 F, the compressed air would be cooled appreciably more.

It has been mentioned that the performance of air refrigeration is poor, in that some 4 to 5 hp are required per ton on refrigeration. In contrast, a representative 1 hp/ton is required for vapor refrigeration in the air-conditioning range. However, the light weight of the expander turbine and its fan running at some 30 000 to 40 000 rpm does give a weight advantage to the air system.

For comparison, in Fig. 14-7, an open-air system, such as might be used on an airplane, and a vapor system are shown together on a *Ts* diagram. The air cycle indicates compression taking place isentropically from ambient air pressure (p_a) and ambient air temperature (T_1) to p_2 and T_2. (In the actual compressor, because of internal irreversibilities, the final temperature is higher than T_2 at a value T'_2.) The hot air at the high pressure of the system cools in the heat exchanger to the temperature T_3, and this value is shown as corresponding to the condensing temperature (T_c) of the vapor system. In the expander engine, the air drops to the pressure corresponding to the variable cabin pressure p_x, ideally reaching temperature T_4 (actually T'_4). The cool air warms to final temperature T_5, producing useful refrigeration, which is measured by the area $T_4 jlT_5$ for ideal conditions (or actually $T'_4 klT_5$). The heat dissipation load is the area $T'_2 T_3 jn$.

The conventional vapor-liquid system operates between a condensing temperature T_c and an evaporator temperature T_r. Note that the useful

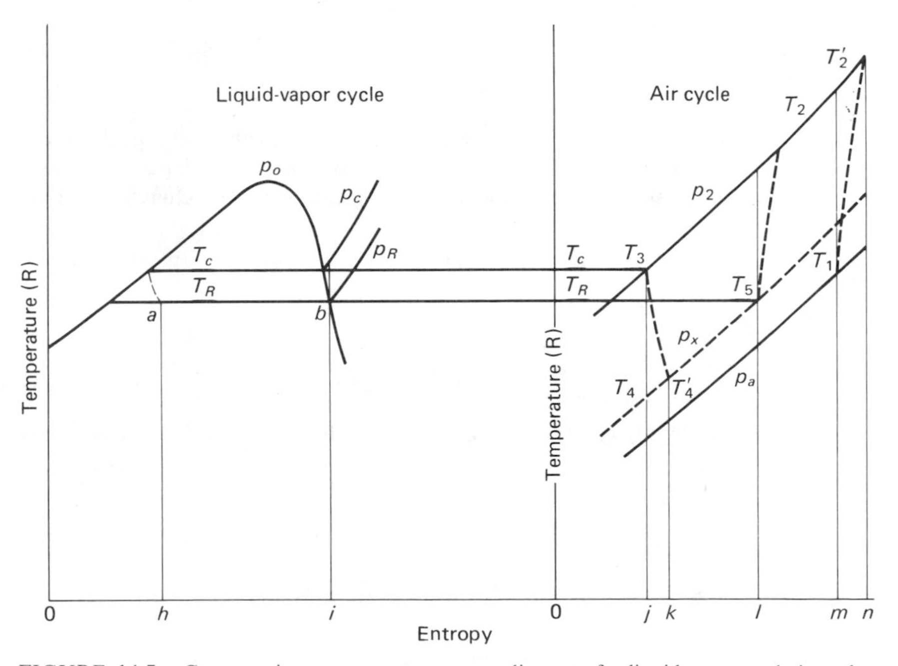

FIGURE 14-7. Comparative temperature-entropy diagrams for liquid-vapor and air cycles.

refrigeration per pound of refrigerant is the area *ahib*, and note also that the refrigeration is all produced at a fixed temperature T_r. This area should be contrasted with the much smaller ideal air-cycle area $T_4 j l T_5$, which represents useful refrigeration per pound of air. Notice also that the air-cycle refrigeration has the disadvantage of being produced at a variable temperature. The turbines and centrifugal compressors in the small sizes, used in the range of 10 to 40 hp, are not highly efficient. A general value for efficiency of 65% relative to isentropic conditions can be used in preliminary design layout.

Example 14-7. An air-cycle refrigeration system for an airplane is supplied with compressed air bled off from the main compressor at 66 psia and 400 F. This air, after passing through four heat-exchanger elements, arrives at the expander turbine at 60 psia and 90 F. The turbine delivers its output air at 12 psia and on test shows an isentropic performance efficiency of 65%. The flow rate of air to the turbine under maximum conditions is 60 lb/min. Compute (a) the minimum possible exit temperature from the air turbine under the conditions indicated, (b) the probable exit temperature attained, and (c) the horsepower delivered to a fan coupled to the turbine. (d) The cold air is tempered by mixing with recirculated or other warm air and is then supplied to the cabin-cockpit areas, where it warms to 78 F before exhausting from the plane. Compute the total tons of refrigeration equivalent to warming the chilled air from its low temperature to 78 F.

Solution: (a) By Eq. (14-20):

$$\frac{T_4}{T_3} = \frac{T_4}{460 + 90} = \left(\frac{12}{60}\right)^{(1.4-1)/1.4} = \left(\frac{1}{5}\right)^{0.286} = \frac{1}{1.584}$$

$$T_4 = 347 \text{ R}$$

$$t_4 = 460 - 347 = -113 \text{ F} \qquad \textit{Ans.}$$

(b) This low temperature is not reached because of inefficiencies in the small, high-speed turbine, and the presence of water vapor or humidity in the air also alters the performance. Considering merely the turbine efficiency here, the actual temperature drop in the turbine is

$$\Delta t = 0.65(460 + 90 - 347) = 132 \text{ deg}$$

The probable exit temperature is

$$T'_4 = 550 - 132 = 418 \text{ R}$$

$$t'_4 = 460 - 418 = -42 \text{ F} \qquad \textit{Ans.}$$

(c) Use Eq. (14-24) with $C_P = 0.24$ for air. The theoretical work of the expander engine is

$$W_E = (60)(0.24)(550 - 347) = 1626 \text{ Btu/min}$$

Actual work of expander engine is

$$W'_E = (60)(0.24)[0.65(550 - 347)] = 1058 \text{ Btu/min}$$

$$\text{expander hp} = \frac{1058}{42.4} = 24.9 \qquad \textit{Ans.}$$

(d) Refrigeration produced is

$$Q_R = wC_p(T_1 - T_4')$$
$$= (60)(0.24)[78 - (-42)] = 1728 \text{ Btu/min}$$
$$\text{tons} = \frac{1728}{200} = 8.6 \qquad \textit{Ans.}$$

The horsepower required to compress the air was provided by the main compressor drive and was not computed in this example. However, a simple computation will show that at 60 lb/min air compressed requires a substantial amount of power, far in excess of that produced by the expander engine.

14-11. VOLUMETRIC EFFICIENCY OF RECIPROCATING COMPRESSORS

In a reciprocating compressor, when the piston reaches the end of its stroke (dead-center position), a portion of gas always remains in and is not discharged from the cylinder. The space in the cylinder which this gas occupies is known as the clearance volume, and the relative magnitude of this volume varies greatly with compressor design and valve arrangement. *Clearance* is usually expressed as a percentage of the stroke-displacement volume (i.e., the volume swept through by the piston in one stroke). Representative clearance values lie in the range of 8 to 2% with 6 to 4% being a common range for most refrigerant compressors now being built. Figure 14-8 is a pressure-volume diagram of events in a compressor cylinder, with V_P representing the stroke-displacement volume, and V_{C1} representing the clearance volume. The percentage clearance (m) is therefore

$$m = \frac{V_{C1}}{V_P} 100 \qquad (14\text{-}28)$$

In internal combustion engines, clearance is expressed in another way as the volume ratio of compression (R); namely, the total gas volume in the cylinder at the start of a stroke divided by the gas-space volume present with the piston on dead center at the end of its stroke. Thus in Fig. 14-8

$$R = \frac{V_P + V_{C1}}{V_{C1}} \qquad (14\text{-}29)$$

A relationship between m and R can easily be found.

Some clearance is necessary and desirable to provide cushioning and prevent piston slap and also to provide passage space up to and adjacent to the valve seats. However, clearance is objectionable because the capacity of two similar compressors each having the same piston displacement will be smaller for the machine having the greater clearance. In fact, some large compressors are provided with variable clearance pockets in the cylinder head, so

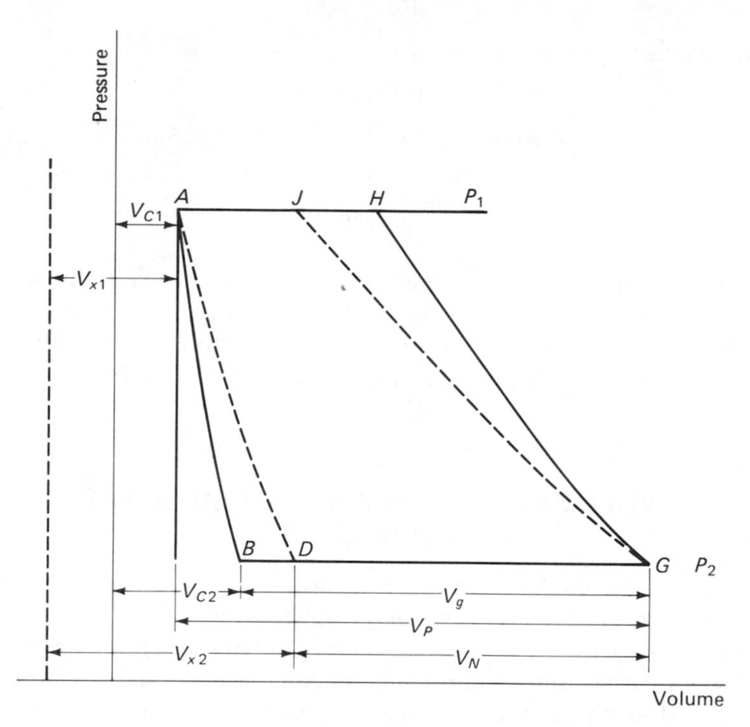

FIGURE 14-8. Pressure-volume diagram of events in a compressor cylinder showing the effect of clearance.

arranged that a valve can be turned to open a pocket to increase the clearance space and thereby reduce the capacity of constant-speed compressors.

The volume (V_{C1}) of gas in the clearance space of a compressor is at discharge pressure p_1, and as the piston moves out on its suction stroke, this gas reexpands, finally reaching volume V_{C2} when the cylinder pressure has dropped to the system suction pressure p_2. The valves on the compressor operate by differential pressure and therefore cannot open until the pressure in the cylinder is less than the pressure in the suction line from the evaporator. Thus a fresh charge of gas cannot enter until the piston has swept out slightly more than the volume (V_{C2} – V_{C1}). The effective piston-stroke displacement remaining for new charge is thus only V_g instead of the full displacement V_P, which would be available if no clearance existed.

A gas expanding quickly does so almost adiabatically, and the reexpansion stroke is closely approximated by the equation

$$p_1 V_{C1}^k = p_2 V_{C2}^k \tag{14-30}$$

where k is the ratio of specific heats, or the polytropic coefficient. Refer to Fig. 14-8; by algebra, note that

$$V_{c2} = V_{C1}\left(\frac{p_1}{p_2}\right)^{1/k} \tag{14-31}$$

and

$$V_g = V_P - (V_{C2} - V_{C1}) = V_P - V_{C1}\left[\left(\frac{p_1}{p_2}\right)^{1/k} - 1\right] \tag{14-32}$$

The volumetric efficiency based on clearance alone (η_{vC}) is thus

$$\eta_{cC} = \frac{V_g}{V_P} = \frac{V_P - V_{C1}[(p_1/p_2)^{1/k} - 1]}{V_P} \tag{14-33}$$

But by Eq. (14-28),

$$V_{C1} = m\frac{V_P}{100}$$

therefore

$$\eta_{cC} = 1 - \frac{m}{100}\left[\left(\frac{p_1}{p_2}\right)^{1/k} - 1\right] \tag{14-34}$$

or

$$\eta_{vC} = 1 + \frac{m}{100} - \frac{m}{100}\left(\frac{p_1}{p_2}\right)^{1/k} \tag{14-35}$$

where η_{vC} is the volumetric efficiency related to clearance alone, expressed as a decimal fraction; m is the clearance, expressed as a percentage and based on piston displacement; p_1 is the compressor discharge pressure, usually in pounds per square inch absolute; and p_2 is the compressor suction pressure, usually in pounds per square inch absolute.

Consider what might happen if a cylinder with stroke-displacement volume V_P had its clearance volume increased to a value V_{x1} (formerly V_{C1}). This might be accomplished by installing a new cylinder head or by using a clearance pocket. With the greater volume of gas in the clearance, the re-expansion line would take the position AD instead of AB, and the new charge would only be inducted during the stroke in amount V_N. Thus capacity would be decreased closely in the ratio of $1 - V_N/V_g$.

Clearance is only one of the factors affecting volumetric efficiency. In Section 14-7 it was mentioned that superheating of the gas to a higher temperature as it entered the cylinder, and throttling or friction loss to gas flow through ports and valves, thereby reducing the pressure, also contributed to lowering volumetric efficiency. Thus mainly three items act to produce volumetric efficiencies of less than unit value. If these two final factors are incorporated, a relation for probable actual volumetric efficiency can be written as

$$\eta_v = (\eta_{vC}) \times \frac{p_{\text{cyl}}}{p_e} \times \frac{T_e}{T_{\text{cyl}}}$$

$$= \left[1 + \frac{m}{100} - \frac{m}{100}\left(\frac{p_1}{p_2}\right)^{1/k}\right] \times \frac{p_{\text{cyl}}}{p_e} \times \frac{T_e}{T_{\text{cyl}}} \tag{14-36}$$

where p_{cyl} is the pressure of the vapor in the cylinder at the start of compression, in pounds per square inch absolute, essentially equal to p_2; p_e is the pressure of the vapor in the evaporator suction pipe at the compressor inlet, in pounds per square inch absolute; T_e is the temperature of the vapor in the evaporator suction pipe at entry to the compressor, in degrees Rankine; and T_{cyl} is the temperature of the vapor in the cylinder at the start of compression, in degrees Rankine.

Example 14-8. A $2\frac{1}{2}$-in.-bore, $1\frac{1}{4}$-in.-stroke compressor runs at 800 rpm. The compressor has a clearance of 4.2%. Refrigerant-12 is used, the discharge pressure is 108 psia at the compressor, and the evaporator operates at 20 F and 35.7 psia. It is thought that the gas in the cylinder warms to 40 F, and there is a pressure drop of 1.5 psi on entry to the cylinder. The value of k for the reexpansion line is closely 1.14. Compute (a) the volumetric efficiency based on clearance volume (η_{vC}) and (b) the probable actual volumetric efficiency.

Solution: (a) The true pressures in the cylinder control reexpansion and must be used in Eq. (14-35). Thus

$$\eta_{vC} = 1 + \frac{4.2}{100} - \frac{4.2}{100}\left(\frac{108}{34.2}\right)^{1/1.14}$$

$$= 1.042 - 0.042(3.16)^{1/1.14} = 1.042 - 0.042(2.74) = 0.927 \qquad \textit{Ans.}$$

(b) By Eq. (14-36),

$$\eta_v = 0.927 \times \frac{34.2}{35.7} \times \frac{460 + 20}{460 + 40} = 0.852 \qquad \textit{Ans.}$$

The effect of volumetric efficiencies less than 100% is naturally to lower compressor capacity. This also contributes indirectly to an increase in the power consumed by the compressor per pound of refrigerant compressed. This additional power is required because at reduced capacity the compressor has to make more strokes for a given output, and the total frictional power is therefore increased per unit of refrigerant handled. Moreover, the lowering in volumetric efficiency contributed by pressure loss (throttling) adds to the power consumed. The reexpansion process produces power and is not inherently a power-absorbing process.

Mention has been made of the fact that, if the isentropic work of compression is multiplied by a factor ranging from 1.3 to 1.5, a close measure of the input shaft work required for compression can be obtained. The basis on which this factor is obtained involves, among many items, turbulence of gas flow into and from the compressor, heat-transfer effects, and mechanical efficiency, with the latter in turn partly affected by volumetric efficiency. In general, about 20% added to the isentropic work of compression gives a reasonable estimation figure for the indicated work of the compressor pistons. Mechanical-efficiency values for compressors vary over wide limits, but for commercial machines it should be higher than 80%, say,

85% for a representative value. Using these values, if the isentropic factor of 1.20 has a representative mechanical efficiency of 0.85 superposed, the overall factor becomes 1.41, which can be seen to lie midway in the range of indicated factors for shaft work input. To arrive at electrical input to the driver, note that motor efficiency varies from over 90% for large-horsepower motors to some 85% for motors of about 5 hp, and falls to very low values, say, 50%, for fractional-horsepower motors.

Note. In this chapter the various operations involved in the creation of useful refrigeration have been discussed in detail. It was felt advisable that the numerical examples should be presented only in engineering units since the presentation here of SI units in addition might confuse rather than help the understanding of these operations. After the processes are better understood and developed it becomes a simple matter to illustrate how easily SI computations are carried out for each of them. In the next chapter this is done and attention is called to Example 15-2.

PROBLEMS

14-1. What is the coefficient of performance for an idealized (Carnot) refrigeration system operating between temperature limits of 5 F (−15°C) and 86 F (30°C)? What horsepower is needed to produce 1 ton of refrigeration (12 000 Btuh) under these idealized conditions?

Ans. 5.74, 0.82 hp

14-2. A ton of refrigeration represents heat interchange occurring at 3.514 kW. Rework Problem 14-1 in SI units and find in addition the kilowatt power requirement per ton of refrigeration.

Ans. 5.74, 0.611 kW, 0.82 hp

14-3. A certain building in winter has a heating loss of 150 000 Btuh when it is 0 F (−17.78°C) outside and 68 F (20°C) is needed inside. Consider that it can be heated by a reversed refrigeration system. Assume that such a system is an idealized Carnot system for which it is desired to find the ideal CP, the ideal kilowatt power requirement, and the ideal horsepower needed. It should be realized that for any system the absorbing unit will have to operate at not less than 10 F lower than outside temperature to abstract heat from outside air (say, −10 F) and at least a 10 deg differential is needed on the heater elements to force heat into the living area (say, 78 F). Apply these temperature differentials and compute the desired results.

Ans. 5.12 CP, 7.18 kW, 9.62 hp

14-4. A refrigerant-12 system (Table 15-5) is operating between 20 F and 100 F. If the liquid is subcooled to 80 F, find the pounds of flash vapor formed in the expansion valve per pound of refrigerant.

Ans. 0.203

14-5. Compare the refrigerating effect obtained from ammonia operating between 0 F evaporating temperature and 100 F condensing temperature, with (a) no subcooling and dry compression from saturated vapor and (b) subcooling to 75 F and dry compression. Refer to Table 15-3.

Ans. (a) 456.6; (b) 485.6 Btu/lb

14-6. Ammonia is used for a refrigeration system between 26.92 and 169.2 psia. Find (a) the pounds of refrigerant circulated per minute per ton, (b) the isentropic work of compression, and (c) the coefficient of performance (CP). Dry-saturated vapor enters the compressor.

Ans. (a) 0.424; (b) 116; (c) 4.0

14-7. Find the bore and stroke of a two-cylinder, 200-rpm, single-acting refrigerant-12 compressor to handle 8 tons of refrigeration when operating between 0 F and 90 F condensing, with liquid subcooling to 80 F. Make bore and stroke equal.

Ans. 7 × 7 at 0.82 vol eff

14-8. A 100-ton refrigerant-12 (Table 15-5, Fig. 15-4) compressor runs with a discharge pressure of 108.0 psia and a suction pressure of 26.5 psia. If the liquid is subcooled to 70 F before expansion and if the vapor entering the compressor is superheated by 15 deg, find (a) the quality leaving the expansion valve, (b) the heat absorbed in the evaporator per pound of refrigerant circulated. (c) the weight of refrigerant per minute per ton of refrigeration, (d) the theoretical piston displacement per minute per ton, (e) the theoretical horsepower, (f) the probable actual horsepower, and (g) the probable actual piston displacement per minute per ton.

Ans. (a) 0.21; (b) 56.4; (c) 3.5; (d) 5.4 cfm/ton; (e) 102; (f) 143; (g) 6.1

14-9. If a four-cylinder, 250-rpm, 75-ton ammonia compressor is operating between 90 F condensation and 30 F evaporator conditions without subcooling and with dry compression, find (a) the quality of ammonia leaving expansion valve, (b) the heat absorbed in the evaporator per pound of ammonia circulated, (c) the weight of ammonia per minute per ton of refrigeration, (d) the piston displacement, (e) the size of the compressor, if it is single-acting and the bore-stroke ratio = 1, and (f) the theoretical horsepower.

Ans. (a) 0.124; (b) 477.0; (c) 0.42; (d) 152.3, 173; (e) 7.25 × 7.25; (f) 52.5

14-10. (a) A cylinder of refrigerant-12, partly filled with liquid, lie in a room at a temperature of 90 F. What pressure exists in the cylinder? (b) What pressure exists in similar cylinders of ammonia, sulfur dioxide, and trichloromonofluoromethane located in the same room? See tables in Chapter 15.

Ans. (a) 114.3 psia

14-11. Using data for monochlorodifluoromethane (refrigerant-22), construct a ***ph*** chart to approximate scale on a sheet of graph paper. Select enough points to indicate the liquid line and the saturated-vapor line, and show two constant-temperature lines in the superheat region. Use Table 15-6 and superheat data from Plate V.

14-12. Repeat Problem 11 for refrigerant-12, excluding data in the superheat region.

14-13. A certain new refrigerant is known to have a $C_p/C_v = k$ ratio of 1.16, and its specific volume at 10 F is computed to be 1.13 ft^3/lb at a pressure of 48 psia. (a) Compute the ideal horsepower required to compress 5 lb/min of this refrigerant to 144 psia. (b) Compute the temperature of the refrigerant on discharge from the compressor.

Ans. (a) 1.4; (b) 547 R

14-14. Repeat Problem 14-13 for air if $k = 1.4$ and the pressures are the same. Use Eq. (2-24) to find the specific volume of air.

Ans. (a) 5.4; (b) 644 R

14-15. In the expander turbine of an air-cycle machine, air at 90 F and 72 psia expands to 12 psia. (a) Find the temperature of the air after ideal expansion. (b) Compute the actual temperature after expansion if the adiabatic turbine efficiency is 70%. (c) Compute the actual turbine horsepower produced for 2 lb/s air flow. (d) Find the useful refrigeration possible if the cool turbine exhaust air warms to 70 F.

Ans. (a) 329.5 R; (b) 395.7 R; (c) 104; (d) 3870 Btu/min

14-16. A reciprocating compressor has a clearance volume of 5%. Compute the volumetric efficiency based on clearance alone for a compressor with compression ratio of (a) 2.5 and (b) 4.0. Assume that the refrigerant has a k value of 1.16.

Ans. (a) 94%; (b) 87%

14-17. Repeat Problem 14-16 for a refrigerant with a k value of 1.31.

Ans. (a) 94.4%; (b) 90.6%

14-18. If the refrigerant of Problem 14-17 warms from −10 F to 5 F in entering the compressor and the pressure drops from 23.7 to 23.1 psia after entry to the cylinder, estimate the actual volumetric efficiency.

Ans. (a) 89.6%; (b) 85.4%

CHAPTER

15

REFRIGERANTS AND THEIR CONTROL

15-1. REFRIGERANT CHARACTERISTICS

An ideal refrigerant as such does not exist, and even if an almost ideal chemical could be found, as a refrigerant it could never cover the complete spectrum of ranges man would need to have it serve. Thus any refrigerant selected will be a compromise, but for any particular service it should possess as many as practicable of the following qualities:

1. *Condensing pressures that are not excessive* so that extra-heavy construction will be unnecessary.
2. *Low boiling temperatures* at atmospheric pressure, so that the system does not require vacuum operation with attendant possibility of leakage of damp air into the system.
3. *High critical temperature.* It is impossible to liquefy (condense) a vapor at a temperature higher than the critical temperature, no

matter how high the pressure is raised. With air-cooled equipment this fact makes it desirable to have critical temperatures higher than 130 F. With the exception of CO_2 and ethane, which have critical temperatures of 87.8 and 89.8 F, respectively, all the common refrigerants have critical temperatures higher than 200 F.

4. *High latent heat of vaporization.* The higher the latent heat, the less the weight of refrigerant which must be circulated per minute per unit of capacity.
5. *Low specific heat of liquid* is a desirable quality, since the expansion valve throttles the liquid, and the liquid refrigerant must be cooled at the expense of partial evaporation.
6. *Low specific volume of vapor* is essential with reciprocating machinery, but may not be an important item with centrifugal machines.
7. *Absence of corrosive action on metals used* is necessary.
8. *Chemical stability* of the compound is essential.
9. The refrigerant should be *nonflammable* and *nonexplosive.*
10. That the refrigerant be *nontoxic* to lungs, eyes, and general health is an important item in air-conditioning installations.
11. *Ease of locating leaks* by odor or suitable indicator is an important consideration.
12. *Availability, low cost,* and *ease of handling* are obviously desirable features.
13. *Action of the refrigerant on lubricants* must be such as not to ruin their lubricating value.
14. *Satisfactory heat transfer* and *viscosity* coefficients are required.
15. The *freezing temperature of the liquid* should be appreciably below any temperature at which the evaporator might operate.
16. For the ratios of the compression to be used, *low, compressor-discharge temperatures* are desirable to prevent possible breakdown or deterioration of refrigerant and lubricant in the system.

Thermodynamically, on the ideal cycle the performance of all refrigerants is exactly the same between the same temperature limits, but in actual cases the irreversible action through the expansion valve and the relative amount of compression in the superheat region cause deviations from the ideal.

Figure 15-1 is a plot of the pressure-temperature relations of a large number of refrigerants. Freezing-point, triple-point, and critical-point conditions are also marked on this chart. Complete and thermodynamically consistent tabulations of properties are now available for almost all of the refrigerants which are used to any extent. Properties of the most extensively used refrigerants are given in this text, but by far the most extensive and complete coverage of all refrigerant properties will be found in the *ASHRAE Handbook of Fundamentals* (Ref. 1).

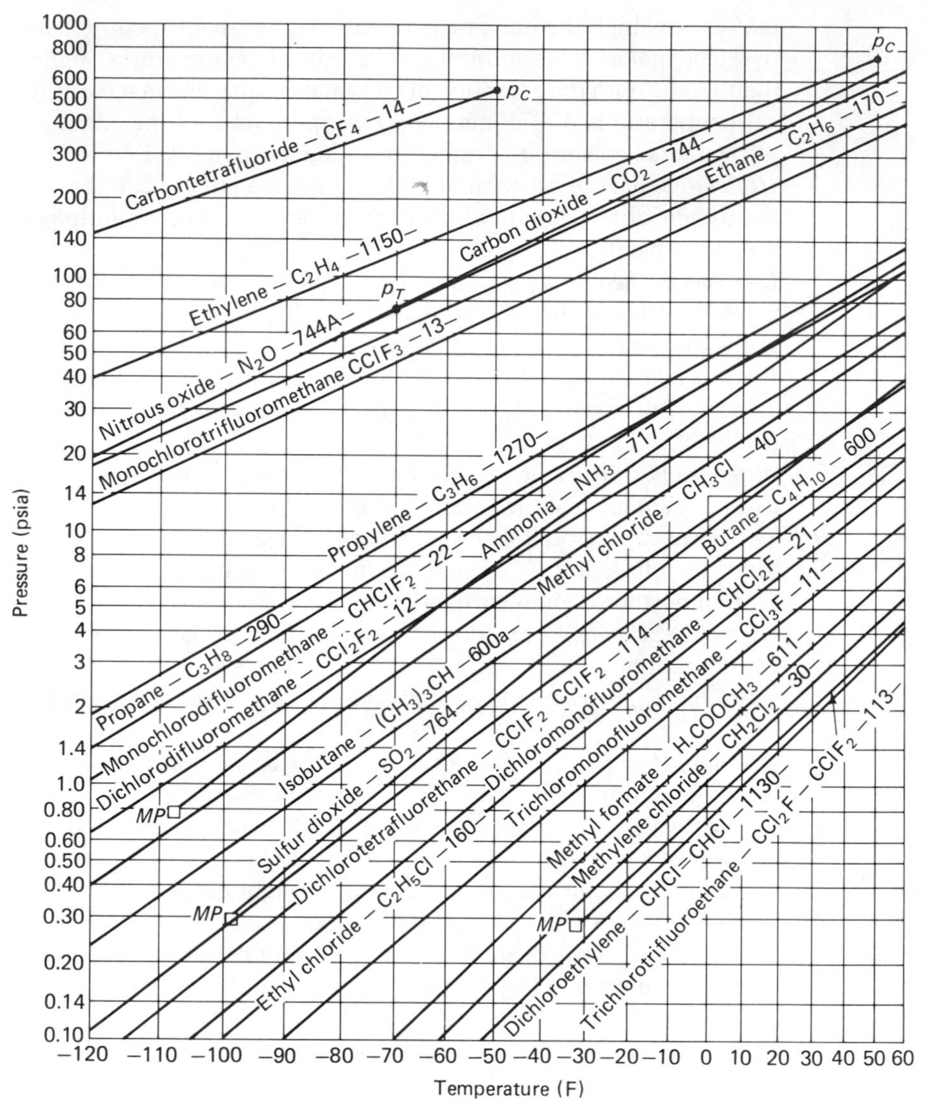

FIGURE 15-1. Saturation pressure-temperature relationships for common refrigerants listed by name and number.

15-2. THE HALOCARBON (HALOGENATED HYDROCARBON) REFRIGERANTS

Toward the end of the 1920 decade a team of engineers and scientists working with Dr. Thomas Midgley, Jr., successfully developed a new family of refrigerants having exceptionally desirable characteristics. The most outstanding feature of the new group was its extremely low toxicity, and this characteristic contributed to early acceptance of one of these refrigerants for widespread use in air-conditioning installations.

Refrigerants of this group are substitution refrigerants in that halogen atoms, mainly chlorine and fluorine, are substituted in a hydrocarbon structure for hydrogen atoms. The hydrocarbon methane, with a formula CH_4, has been most widely featured in this pattern. For example, assume that two chlorine atoms and two fluorine atoms are used to replace the hydrogen atoms in methane. The resulting formula becomes CCl_2F_2. This can properly be called dichloro-difluoro-methane, where the prefix *di* (or *bi*) represents 2 and similarly, *mono* 1, *tri* 3, *tetra* 4, etc. Another refrigerant is $CHClF_2$, which can be named monochlorodifluoromethane; or since the prefix *mono* is redundant in this case, the spelling can be simplified to chlorodifluoromethane.

Such names, although logical, are long and difficult to remember and so proprietary names were largely used instead, with dichlorodifluoromethane being called Freon-12, genetron-12, isotron-12, etc., and monochlorodifluoromethane being called Freon-22, genetron-22, frigen-22, etc. A larger number of proprietary names coming into use began to confuse the issue further, and ASHRAE decided to adopt a standard system of designation and call these refrigerants merely by number, with CCl_2F_2 being designated as refrigerant-12, $CHClF_2$ as 22, CHF_3 as 23, CH_2F_2 as 32, etc. Tables 15-1 and 15-2 list a large number of refrigerants tabulated in accordance with the ASHRAE standard system.

The numbers employed in the standard for the halocarbon refrigerants follow a definite pattern originally developed in connection with the Freon refrigerants. The halogen substitutions are based on two hydrocarbons, methane, CH_4 (for which two-digit numbers are used to indicate the refrigerants), and ethane, C_2H_6 (for which three-digit numbers are used to indicate the different refrigerants). For the methane derivatives, the first number is always one greater than the number of hydrogen atoms appearing in the refrigerant molecule, whereas the second digit gives exactly the number of fluorine atoms appearing in the refrigerant molecule. For example, in refrigerant-12, the number 1, being one greater than the number of hydrogen atoms, indicates no hydrogen to be present, while the 2 indicates that two fluorine atoms exist in the molecule so that the formula is CCl_2F_2. It is also possible to name this refrigerant in terms of its constituents, as dichlorodifluoromethane. For the 110-series, the first digit merely indicates that ethane is the basic hydrocarbon. The second 1 indicates no hydrogen atoms in the formula. The third digit indicates, as before, the number of fluorine atoms appearing in the molecule. The respective boiling points at atmospheric pressure of each of the refrigerants are also given in Tables 15-1 and 15-2. It will be noticed that the boiling point decreases with the increasing number of fluorine atoms in any particular grouping. No specific formulation can be made as to toxicity, but of all the refrigerants listed, refrigerant-12 (Freon-12, genetron-12) is least toxic. The other refrigerants in the table (except for the straight hydrocarbons which essentially fit the halocarbon numbering pattern) are provided with distinctive numbers. The distinctive first digit 7 in a three-figure number applies to the inorganic refrigerants.

Table 15-1 American Society of Heating, Refrigerating, and Air-Conditioning Engineers (ASHRAE) Refrigerant Numbering System

ASHRAE Standard Designation	Chemical Name	Chemical Formula	Molecular Weight	Boiling Point (F at 14.7 psi)
Halocarbon and Hydrocarbon Compounds				
10	Carbontetrachloride	CCl_4	153.8	170.2
11	Trichloromonofluoromethane	CCl_3F	137.4	74.8
12	Dichlorodifluoromethane	CCl_2F_2	120.9	−21.6
13	Monochlorotrifluoromethane	$CClF_3$	104.5	−114.6
13B1	Monobromotrifluoromethane	$CBrF_3$	148.9	−72.0
14	Carbontetrafluoride	CF_4	88.0	−198.4
20	Chloroform	$CHCl_3$	119.4	142
21	Dichloromonofluoromethane	$CHCl_2F$	102.9	48.1
22	Monochlorodifluoromethane	$CHClF_2$	86.5	−41.4
23	Trifluoromethane	CHF_3	70.0	−119.9
30	Methylene chloride	CH_2Cl_2	84.9	105.2
31	Monochloromonofluoromethane	CH_2ClF	68.5	48.0
32	Methylene fluoride	CH_2F_2	52.0	−61.4
40	Methyl chloride	CH_3Cl	50.5	−10.8
41	Methyl fluoride	CH_3F	34.0	−109
50	Methane	CH_4	16.0	−259
110	Hexachloroethane	CCl_3CCl_3	236.8	365
111	Pentachloromonofluoroethane	CCl_3CCl_2F	220.3	279
112	Tetrachlorodifluoroethane	CCl_2FCCl_2F	203.8	199.0
112a	Tetrachlorodifluoroethane	CCl_3CClF_2	203.8	195.8
113	Trichlorotrifluoroethane	CCl_2FCClF_2	187.4	117.6
113a	Trichlorotrifluoroethane	CCl_3CF_3	187.4	114.2
114	Dichlorotetrafluoroethane	$CClF_2CClF_2$	170.9	38.4
114a	Dichlorotetrafluoroethane	CCl_2FCF_3	170.9	38.5
114B2	Dibromotetrafluoroethane	$CBrF_2CBrF_2$	259.9	117.5
115	Monochloropentafluoroethane	$CClF_2CF_3$	154.5	−37.7
116	Hexafluoroethane	CF_3CF_3	138.0	−108.8
120	Pentachloroethane	$CHCl_2CCl_3$	202.3	324
123	Dichlorotrifluoroethane	$CHCl_2CF_3$	153	83.7
124	Monochlorotetrafluoroethane	$CHClFCF_3$	136.5	10.4
124a	Monochlorotetrafluoroethane	CHF_2CClF_2	136.5	14
125	Pentafluoroethane	CHF_2CF_3	120	−55
133a	Monochlorotrifluoroethane	CH_2ClCF_3	118.5	43.0
140a	Trichloroethane	CH_3CCl_3	133.4	165
142b	Monochlorodifluoroethane	CH_3CClF_2	100.5	12.2
143a	Trifluoroethane	CH_3CF_3	84	−53.5
150a	Dichloroethane	CH_3CHCl_2	98.9	140
152a	Difluoroethane	CH_3CHF_2	66	−12.4
160	Ethyl chloride	CH_3CH_2Cl	64.5	54.0
170	Ethane	CH_3CH_3	30	−127.5
218	Octafluoropropane	$CF_3CF_2CF_3$	188	−36.4
290	Propane	$CH_3CH_2CH_3$	44	−44.2
600	Butane	$CH_3CH_2CH_2CH_3$	58.1	31.3
600a	Isobutane	$CH(CH_3)_3$	58.1	14
Inorganic Compounds				
717	Ammonia	NH_3	17	−28.0
718	Water	H_2O	18	212
727	Air		29	−318
744	Carbon dioxide	CO_2	44	−109(subl.)
744A	Nitrous oxide	N_2O	44	−127
764	Sulfur dioxide	SO_2	64	14.0

Table 15-2 Azeotrope Refrigerants

ASHRAE Standard Designation	Name	Chemical Formula	Composite Molecular Weight	Boiling Point at 14.7 psia
500	R-12, R-152a azeotrope	CCl_2F_2	106.55	−28.3 F
	73.8% and 26.2% by weight	CH_3CHF_2		−33.5°C
502	R-12, R-115 azeotrope	CCl_2F_2	138.10	−49.8 F
	48.8% and 51.2% by weight	$CClF_2CF_3$		−45.4°C
503	R-23, R-13 azeotrope	CHF_3	90.66	−127.6 F
	40.1% and 59.9% by weight	$CClF_3$		−88.7°C
504	R-32, R-115 azeotrope	CH_2F_2	105.09	−71.0 F
	48.2% and 51.8% by weight	$CClF_2CF_3$		−57.2°C
505	R-12, R-31 azeotrope	CCl_2F_2	94.56	21.3 F
	78.0% and 22.0% by weight	CH_2ClF		−5.9°C

15-3. AMMONIA (NH_3), AND NONHALOGEN REFRIGERANTS

Ammonia is the most extensively used refrigerant, particularly in industrial and commercial refrigeration. When water-free, it is known as *anhydrous ammonia*, and when mixed with water (as used in absorption refrigeration systems) it is known as *aqua ammonia*. By reference to Tables 15-3 and 15-4 and Figs. 15-2 and 15-3 it can be seen that the operating pressure range is moderate, subatmospheric only for temperatures below −28 F and usually not exceeding 200 psia in the condenser. Its latent heat is exceptionally high, over 500 Btu/lb. No lubrication difficulties exist with ammonia, provided a proper mineral oil is selected. Water mixed with ammonia will not freeze at expansion valves; freezing can happen with most other refrigerants. The thermodynamic performance in a refrigerating cycle is high. Ammonia also has a high critical temperature, 271.4 F at 1657 psia.

Ammonia is noncorrosive to iron and steel materials, but rapidly corrodes copper and copper or zinc alloys (brass, bronze), so care should be exercised that these metals are not used in contact with ammonia.

Ammonia gas is irritating particularly to eyes and mucous membranes. Small amounts of ammonia in air appear to be more annoying than harmful, but in quantities approaching 0.5% by volume in air, serious effects may result if exposure is prolonged beyond a few minutes. Ammonia is rarely used with direct-expansion evaporators in the air ducts for comfort air conditioning and its use in this manner is usually prohibited by law.

Ammonia burns with difficulty, but can form explosive mixtures with air between mixture ratios of 16 to 25% by volume. Such mixtures do not ignite easily, however. No harmful decomposition products are formed.

Carbon dioxide (CO_2), refrigerant-744, although formerly used in some air-conditioning installations, has now been almost completely supplanted by lower pressure refrigerants. Carbon dioxide is noncorrosive and inert,

Table 15-3 Temperature Table for Saturated Ammonia

Temp F t	Pressure psia p	Sp Vol of Liquid cu ft/lb v_f	Sp Vol of Vapor cu ft/lb v_g	Density of Vapor lb/cu ft $\frac{1}{v_g}$	Enthalpy Btu/lb above −40 F			Entropy		Temp F t
					Liquid h_f	Vaporization h_{fg}	Vapor h_g	Liquid s_f	Vapor s_g	
−40	10.41	0.02322	24.86	0.04022	0.0	597.6	597.6	0.0000	1.4242	−40
−30	13.90	.02345	18.97	.05271	10.7	590.7	601.4	.0250	1.4001	−30
−20	18.30	.02369	14.68	.06813	21.4	583.6	605.0	.0497	1.3774	−20
−10	23.74	0.02393	11.50	0.08695	32.1	576.4	608.5	0.0738	1.3558	−10
− 9	24.35		11.23	.08904	33.2	575.6	608.8	.0762	.3537	− 9
− 8	24.97		10.97	.09117	34.3	574.9	609.2	.0786	.3516	− 8
− 7	25.61		10.71	.09334	35.4	574.1	609.5	.0809	.3495	− 7
− 6	26.26		10.47	.09555	36.4	573.4	609.8	.0833	.3474	− 6
− 5	26.92	0.02406	10.23	0.09780	37.5	572.6	610.1	0.0857	1.3454	− 5
− 4	27.59		9.991	.1001	38.6	571.9	610.5	.0880	.3433	− 4
− 3	28.28		9.763	.1024	39.7	571.1	610.8	.0904	.3413	− 3
− 2	28.98		9.541	.1048	40.7	570.4	611.1	.0928	.3393	− 2
− 1	29.69		9.326	.1072	41.8	569.6	611.4	.0951	.3372	− 1
0	30.42	0.02419	9.116	0.1097	42.9	568.9	611.8	0.0975	1.3352	0
2	31.92		8.714	.1148	45.1	567.3	612.4	.1022	.3312	2
4	33.47		8.333	.1200	47.2	565.8	613.0	.1069	.3273	4
5	34.27	0.02432	8.150	0.1227	48.3	565.0	613.3	0.1092	1.3253	5
6	35.09		7.971	.1254	49.4	564.2	613.6	.1115	.3234	6
8	36.77		7.629	.1311	51.6	562.7	614.3	.1162	.3195	8
10	38.51	0.02446	7.304	0.1369	53.8	561.1	614.9	0.1208	1.3157	10
11	39.40		7.148	.1399	54.9	560.3	615.2	.1231	.3137	11
12	40.31		6.996	.1429	56.0	559.5	615.5	.1254	.3118	12
13	41.24		6.847	.1460	57.1	558.7	615.8	.1277	.3099	13
14	42.18		6.703	.1492	58.2	557.9	616.1	.1300	.3081	14
15	43.14	0.02460	6.562	0.1524	59.2	557.1	616.3	0.1323	1.3062	15
16	44.12		6.425	.1556	60.3	556.3	616.6	.1346	.3043	16
17	45.12		6.291	.1590	61.4	555.5	616.9	.1369	.3025	17
18	46.13		6.161	.1623	62.5	554.7	617.2	.1392	.3006	18
19	47.16		6.034	.1657	63.6	553.9	617.5	.1415	.2988	19
20	48.21	0.02474	5.910	0.1692	64.7	553.1	617.8	0.1437	1.2969	20
21	49.28		5.789	.1728	65.8	552.2	618.0	.1460	.2951	21
22	50.36		5.671	.1763	66.9	551.4	618.3	.1483	.2933	22
23	51.47		5.556	.1800	68.0	550.6	618.6	.1505	.2915	23
24	52.59		5.443	.1837	69.1	549.8	618.9	.1528	.2897	24
25	53.73	0.02488	5.334	0.1875	70.2	548.9	619.1	0.1551	1.2879	25
26	54.90		5.227	.1913	71.3	548.1	619.4	.1573	.2861	26
27	56.08		5.123	.1952	72.4	547.3	619.7	.1596	.2843	27
28	57.28		5.021	.1992	73.5	546.4	619.9	.1618	.2825	28
29	58.50		4.922	.2032	74.6	545.6	620.2	.1641	.2808	29
30	59.74	0.02503	4.825	0.2073	75.7	544.8	620.5	0.1663	1.2790	30
31	61.00		4.730	.2114	76.8	543.9	620.7	.1686	.2773	31
32	62.29		4.637	.2156	77.9	543.1	621.0	.1708	.2755	32
33	63.59		4.547	.2199	79.0	542.2	621.2	.1730	.2738	33
34	64.91		4.459	.2243	80.1	541.4	621.5	.1753	.2721	34
35	66.26	0.02518	4.373	0.2287	81.2	540.5	621.7	0.1775	1.2704	35
36	67.63		4.289	.2332	82.3	539.7	622.0	.1797	.2686	36
37	69.02		4.207	.2377	83.4	538.8	622.2	.1819	.2669	37
38	70.43		4.126	.2423	84.6	537.9	622.5	.1841	.2652	38
39	71.87		4.048	.2470	85.7	537.0	622.7	.1863	.2635	39
40	73.32	0.02533	3.971	0.2518	86.8	536.2	623.0	0.1885	1.2618	40
41	74.80		3.897	.2566	87.9	535.3	623.2	.1908	.2602	41
42	76.31		3.823	.2616	89.0	534.4	623.4	.1930	.2585	42
43	77.83		3.752	.2665	90.1	533.6	623.7	.1952	.2568	43
44	79.38		3.682	.2716	91.2	532.7	623.9	.1974	.2552	44
45	80.96	0.02548	3.614	0.2767	92.3	531.8	624.1	0.1996	1.2535	45
46	82.55		3.547	.2819	93.5	530.9	624.4	.2018	.2519	46
47	84.18		3.481	.2872	94.6	530.0	624.6	.2040	.2502	47
48	85.82		3.418	.2926	95.7	529.1	624.8	.2062	.2486	48
49	87.49		3.355	.2981	96.8	528.2	625.0	.2083	.2469	49
50	89.19	0.02564	3.294	0.3036	97.9	527.3	625.2	0.2105	1.2453	50

Table 15-3 (*Continued*)

Temp F t	Pressure psia p	Sp Vol of Liquid cu ft/lb v_f	Sp Vol of Vapor cu ft/lb v_g	Density of Vapor lb/cu ft $\frac{1}{v_g}$	Enthalpy Btu/lb above −40 F			Entropy		Temp F t
					Liquid h_f	Vaporization h_{fg}	Vapor h_g	Liquid s_f	Vapor s_g	
51	90.91		3.234	.3092	99.1	526.4	625.5	.2127	.2437	51
52	92.66		3.176	.3149	100.2	525.5	625.7	.2149	.2421	52
53	94.43		3.119	.3207	101.3	524.6	625.9	.2171	.2405	53
54	96.23		3.063	.3265	102.4	523.7	626.1	.2192	.2389	54
55	98.06	0.02581	3.008	0.3325	103.5	522.8	626.3	0.2214	1.2373	55
56	99.91		2.954	.3385	104.7	521.8	626.5	.2236	.2357	56
57	101.8		2.902	.3446	105.8	520.9	626.7	.2257	.2341	57
58	103.7		2.851	.3508	106.9	520.0	626.9	.2279	.2325	58
59	105.6		2.800	.3571	108.1	519.0	627.1	.2301	.2310	59
60	107.6	0.02597	2.751	0.3635	109.2	518.1	627.3	0.2322	1.2294	60
61	109.6		2.703	.3700	110.3	517.2	627.5	.2344	.2278	61
62	111.6		2.656	.3765	111.5	516.2	627.7	.2365	.2262	62
63	113.6		2.610	.3832	112.6	515.3	627.9	.2387	.2247	63
64	115.7		2.565	.3899	113.7	514.3	628.0	.2408	.2231	64
65	117.8	0.02614	2.520	0.3968	114.8	513.4	628.2	0.2430	1.2216	65
66	120.0		2.477	.4037	116.0	512.4	628.4	.2451	.2201	66
67	122.1		2.435	.4108	117.1	511.5	628.6	.2473	.2186	67
68	124.3		2.393	.4179	118.3	510.5	628.8	.2494	.2170	68
69	126.5		2.352	.4251	119.4	509.5	628.9	.2515	.2155	69
70	128.8	0.02632	2.312	0.4325	120.5	508.6	629.1	0.2537	1.2140	70
71	131.1		2.273	.4399	121.7	507.6	629.3	.2558	.2125	71
72	133.4		2.235	.4474	122.8	506.6	629.4	.2579	.2110	72
73	135.7		2.197	.4551	124.0	505.6	629.6	.2601	.2095	73
74	138.1		2.161	.4628	125.1	504.7	629.8	.2622	.2080	74
75	140.5	0.02650	2.125	0.4707	126.2	503.7	629.9	0.2643	1.2065	75
76	143.0		2.089	.4786	127.4	502.7	630.1	.2664	.2050	76
77	145.4		2.055	.4867	128.5	501.7	630.2	.2685	.2035	77
78	147.9		2.021	.4949	129.7	500.7	630.4	.2706	.2020	78
79	150.5		1.988	.5031	130.8	499.7	630.5	.2728	.2006	79
80	153.0	0.02668	1.955	0.5115	132.0	498.7	630.7	0.2749	1.1991	80
81	155.6		1.923	.5200	133.1	497.7	630.8	.2769	.1976	81
82	158.3		1.892	.5287	134.3	496.7	631.0	.2791	.1962	82
83	161.0		1.861	.5374	135.4	495.7	631.1	.2812	.1947	83
84	163.7		1.831	.5462	136.6	494.7	631.3	.2833	.1933	84
85	166.4	0.02687	1.801	0.5552	137.8	493.6	631.4	0.2854	1.1918	85
86	169.2		1.772	.5643	138.9	492.6	631.5	.2875	.1904	86
87	172.0		1.744	.5735	140.1	491.6	631.7	.2895	.1889	87
88	174.8		1.716	.5828	141.2	490.6	631.8	.2917	.1875	88
89	177.7		1.688	.5923	142.4	489.5	631.9	.2937	.1860	89
90	180.6	0.02707	1.661	0.6019	143.5	488.5	632.0	0.2958	1.1846	90
91	183.6		1.635	.6116	144.7	487.4	632.1	.2979	.1832	91
92	186.6		1.609	.6214	145.8	486.4	632.2	.3000	.1818	92
93	189.6		1.584	.6314	147.0	485.3	632.3	.3021	.1804	93
94	192.7		1.559	.6415	148.2	484.3	632.5	.3041	.1789	94
95	195.8	0.02727	1.534	0.6517	149.4	483.2	632.6	0.3062	1.1775	95
96	198.9		1.510	.6620	150.5	482.1	632.6	.3083	.1761	96
97	202.1		1.487	.6725	151.7	481.1	632.8	.3104	.1747	97
98	205.3		1.464	.6832	152.9	480.0	632.9	.3125	.1733	98
99	208.6		1.441	.6939	154.0	478.9	632.9	.3145	.1719	99
100	211.9	0.02747	1.419	0.7048	155.2	477.8	633.0	0.3166	1.1705	100
102	218.6		1.375	.7270	157.6	475.6	633.2	.3207	.1677	102
104	225.4		1.334	.7498	159.9	473.5	633.4	.3248	.1649	104
105	228.9	0.02769	1.313	0.7615	161.1	472.3	633.4	0.3269	1.1635	105
106	232.5		1.293	.7732	162.3	471.2	633.5	.3289	.1621	106
108	239.7		1.254	.7972	164.6	469.0	633.6	.3330	.1593	108
110	247.0	0.02790	1.217	0.8219	167.0	466.7	633.7	0.3372	1.1566	110
115	266.2	.02813	1.128	.8862	173.0	460.9	633.9	.3474	1.1497	115
120	286.4	.02836	1.047	.9549	179.0	455.0	634.0	.3576	1.1427	120
125	307.8	.02860	0.973	1.028	185.1	448.9	634.0	.3679	1.1358	125

Table 15-4 Properties of Superheated Ammonia Vapor

PRESSURE, IN PSIA (SATURATION TEMPERATURE IN ITALICS)

Temp F	Sp Vol Vapor Cu Ft/Lb v	Enthalpy Btu/Lb h	Entropy s
	34 *4.66* F		
Sat	*8.211*	*613.2*	*1.3260*
10	8.328	616.4	1.3328
20	8.542	622.3	.3452
30	8.753	628.0	.3570
40	8.960	633.6	.3684
50	9.166	639.2	1.3793
60	9.369	644.7	.3900
70	9.570	650.1	.4004
80	9.770	655.5	.4105
90	9.969	660.9	.4204
100	10.17	666.3	1.4301
110	10.36	671.6	.4396
120	10.56	677.0	.4489
130	10.75	682.3	.4581
140	10.95	687.7	.4671
150	11.14	693.0	1.4759
160	11.33	698.4	.4846
170	11.53	703.8	.4932
180	11.72	709.2	.5017
190	11.91	714.5	.5101
200	12.10	720.0	1.5183
220	12.48	730.8	.5346
240	12.86	741.7	.5504
260	13.24	752.7	.5659
280	13.62	763.8	.5811

Temp F	Sp Vol Vapor Cu Ft/Lb v	Enthalpy Btu/Lb h	Entropy s
	48 *19.80* F		
Sat	*5.934*	*617.7*	*1.2973*
20	5.937	617.8	1.2976
30	6.096	624.0	.3103
40	6.251	630.0	.3225
50	6.404	635.9	1.3341
60	6.554	641.6	.3453
70	6.702	647.3	.3561
80	6.848	652.9	.3666
90	6.993	658.5	.3768
100	7.137	664.0	1.3868
110	7.280	669.5	.3965
120	7.421	675.0	.4061
130	7.562	680.5	.4154
140	7.702	685.9	.4246
150	7.842	691.4	1.4336
160	7.981	696.8	.4425
170	8.119	702.3	.4512
180	8.257	707.7	.4598
190	8.395	713.2	.4683
200	8.532	718.7	1.4766
220	8.805	729.6	.4930
240	9.077	740.6	.5090
260	9.348	751.7	.5246
280	9.619	762.9	.5399
300	9.888	774.1	1.5548

Temp F	Sp Vol Vapor Cu Ft/Lb v	Enthalpy Btu/Lb h	Entropy s
	60 *30.21* F		
Sat	*4.805*	*620.5*	*1.2787*
40	4.933	626.8	1.2913
50	5.060	632.9	1.3035
60	5.184	639.0	.3152
70	5.307	644.9	.3265
80	5.428	650.7	.3373
90	5.547	656.4	.3479
100	5.665	662.1	1.3581
110	5.781	667.7	.3681
120	5.897	673.3	.3778
130	6.012	678.9	.3873
140	6.126	684.4	.3966
150	6.239	689.9	1.4058
160	6.352	695.5	.4148
170	6.464	701.0	.4236
180	6.576	706.5	.4323
190	6.687	712.0	.4409
200	6.798	717.5	1.4493
210	6.909	723.1	.4576
220	7.019	728.6	.4658
230	7.129	734.1	.4739
240	7.238	739.7	.4819
260	7.457	750.9	1.4976
280	7.675	762.1	.5130
300	7.892	773.3	.5281

Table 15-4 (*Continued*)
Properties of Superheated Ammonia Vapor

Temp F	120 (66.02 F)		
Sat	*2.476*	*628.4*	*1.2201*
70	2.505	631.3	1.2255
80	2.576	638.3	.2386
90	2.645	645.0	.2510
100	2.712	651.6	1.2628
110	2.778	658.0	.2741
120	2.842	664.2	.2850
130	2.905	670.4	.2956
140	2.967	676.5	.3058
150	3.029	682.5	1.3157
160	3.089	688.4	.3254
170	3.149	694.3	.3348
180	3.209	700.2	.3441
190	3.268	706.0	.3531
200	3.326	711.8	1.3620
210	3.385	717.6	.3707
220	3.442	723.4	.3793
230	3.500	729.2	.3877
240	3.557	734.9	.3960
250	3.614	740.7	1.4042
260	3.671	746.5	.4123
270	3.727	752.2	.4202
280	3.783	758.0	.4281
290	3.839	763.8	.4359
300	3.895	769.6	1.4435

Temp F	140 (74.79 F)		
Sat	*2.132*	*629.9*	*1.2068*
80	2.166	633.8	1.2140
90	2.228	640.9	.2272
100	2.288	647.8	1.2396
110	2.347	654.5	.2515
120	2.404	661.1	.2628
130	2.460	667.4	.2738
140	2.515	673.7	.2843
150	2.569	679.9	1.2945
160	2.622	686.0	.3045
170	2.675	692.0	.3141
180	2.727	698.0	.3236
190	2.779	704.0	.3328
200	2.830	709.9	1.3418
210	2.880	715.8	.3507
220	2.931	721.6	.3594
230	2.981	727.5	.3679
240	3.030	733.3	.3763
250	3.080	739.2	1.3846
260	3.129	745.0	.3928
270	3.179	750.8	.4008
280	3.227	756.7	.4088
290	3.275	762.5	.4166
300	3.323	768.3	1.4243
320	3.420	780.0	.4395

Temp F	170 (86.29 F)		
Sat	*1.764*	*631.6*	*1.1900*
90	1.784	634.4	1.1952
100	1.837	641.9	1.2087
110	1.889	649.1	.2215
120	1.939	656.1	.2336
130	1.988	662.8	.2452
140	2.035	669.4	.2563
150	2.081	675.9	1.2669
160	2.127	682.3	.2773
170	2.172	688.5	.2873
180	2.216	694.7	.2971
190	2.260	700.8	.3066
200	2.303	706.9	1.3159
210	2.346	713.0	.3249
220	2.389	719.0	.3338
230	2.431	724.9	.3426
240	2.473	730.9	.3512
250	2.514	736.8	1.3596
260	2.555	742.8	.3679
270	2.596	748.7	.3761
280	2.637	754.6	.3841
290	2.678	760.5	.3921
300	2.718	766.4	1.3999
320	2.798	778.3	.4153
340	2.878	790.1	.4303

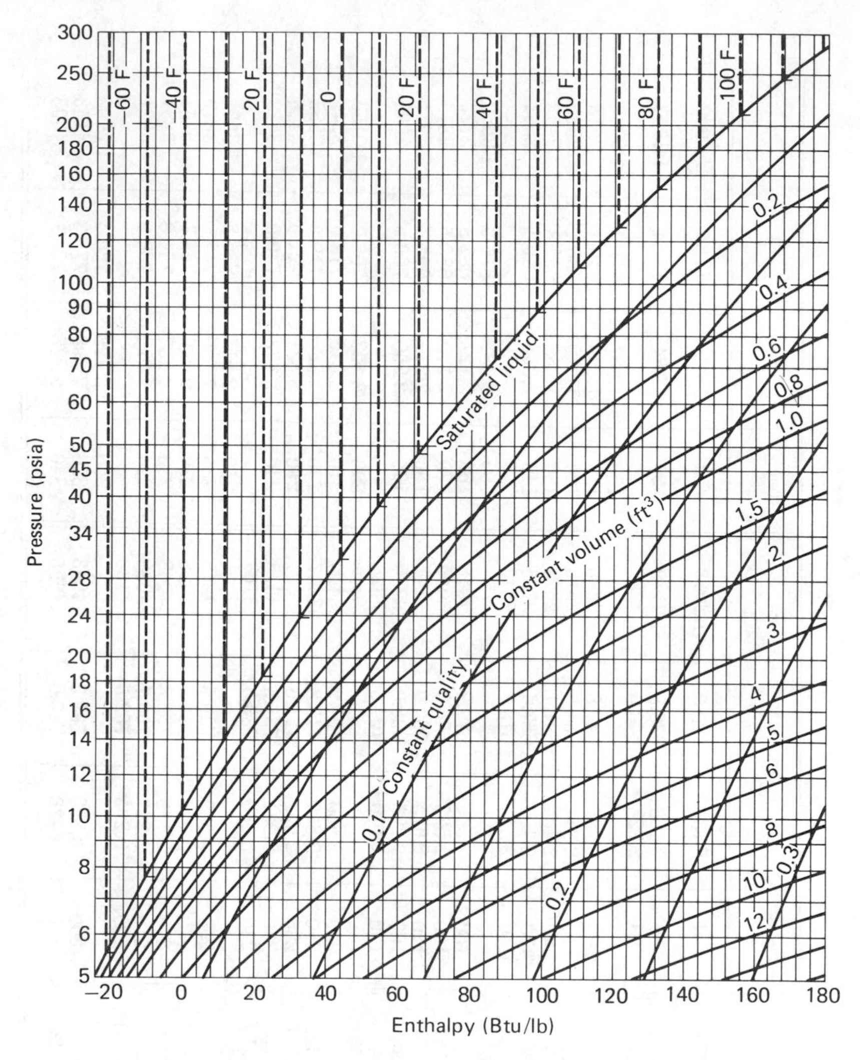

FIGURE 15-2. Pressure-enthalpy chart for liquid ammonia.

as well as relatively nontoxic, odorless, and nonirritating. However, in high concentrations, over 6% by volume, much discomfort is experienced and loss of consciousness and ultimately death can result if the person exposed is not moved into fresh air. Carbon dioxide is a very high-pressure refrigerant (Fig. 14-3 and Table 15-7). Thus all equipment must be made extra heavy. Lubrication is simple and CO_2 is very suitable for operation at temperatures down to about −60 F. Solidification occurs at −69.88 F.

Sulfur dioxide (SO_2), refrigerant-764, which has a very irritating acrid vapor, was used formerly in domestic refrigerators. It is a moderately low-pressure refrigerant (Table 15-7) and must be kept dry in use because it forms an acid in the presence of water.

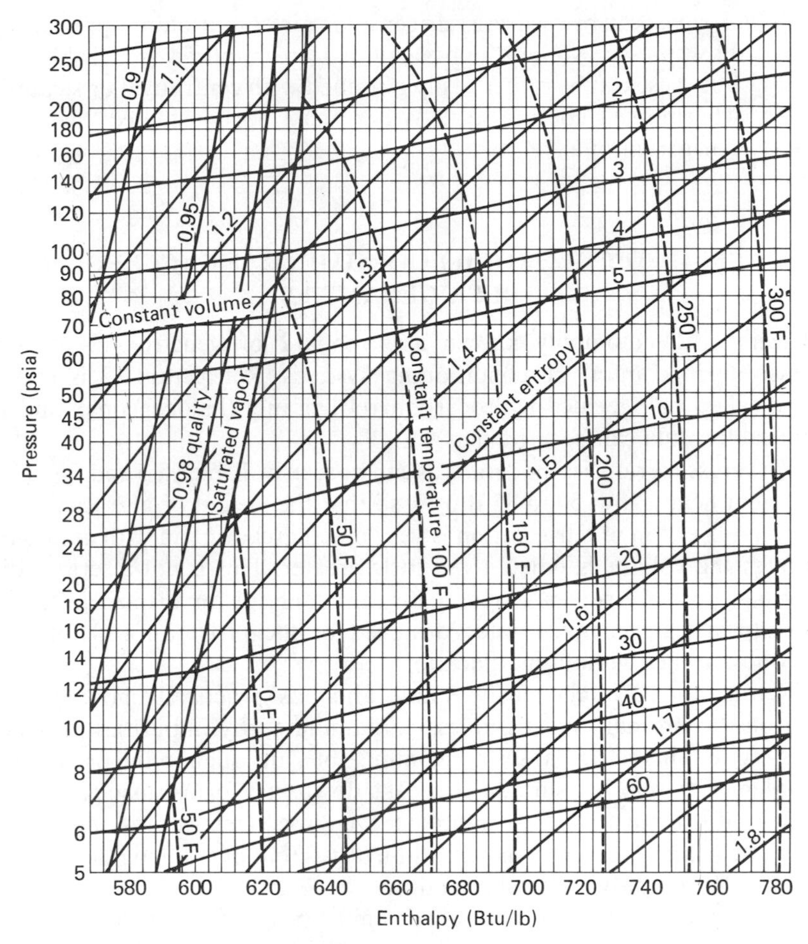

FIGURE 15-3. Pressure-enthalpy chart for ammonia, vapor region.

Hydrocarbon refrigerants are all very inflammable and explosive. They are in general not very toxic. They are somewhat soluble in lubricating oil. Different hydrocarbons can be selected to work in desired pressure and temperature ranges. Among those used are:

Butane (C_4H_{10}), refrigerant-600, has a moderately low pressure and exerts a pressure of 13.1 psia at 5 F and 59.5 psia at 86 F.

Propane (C_3H_8), refrigerant-290, is in an intermediate pressure range, 42.1 psia at 5 F and 185.3 psia at 86 F.

Ethane (C_2H_6), refrigerant-170, lies in a high-pressure range, about 240 psia at 5 F and 676 psia at 86 F.

Ethylene (C_2H_4), refrigerant-1150, has been used to some extent for very low-temperature work as in gas liquefaction and separation. It boils at −154.7 F at 14.7 psia and has a critical temperature of 49.3 F at 743 psia.

Water-steam can be used as a refrigerant in air-conditioning applications. Low pressures (very high vacuums) are required (see Table 2-2), but these can be rather easily obtained with centrifugal or steam-jet (thermo-jet) compressors. Water may be considered the safest refrigerant of all.

15-4. DICHLORODIFLUOROMETHANE (CCl_2F_2), REFRIGERANT 12

Refrigerant-12, commonly known as Freon-12 or genetron-12, is used extensively in comfort air-conditioning systems. Refrigerant-12 is one of the so-called halogenated-hydrocarbon refrigerants (halocarbons), and its chemical formula can be written as CCl_2F_2.

The pressure range is moderate, as can be seen in Table 15-5 and Fig. 15-4. The latent heat is low, 50 to 85 Btu/lb, so that the weight of "12" circulated per minute per ton of refrigeration is very much greater than with ammonia, although the volume handled is but little greater. This is not a serious disadvantage but necessitates designing for the proper piston displacement and valve capacity. In general, a system designed for one refrigerant will not work well with another without modification.

Refrigerant-12 is chemically stable and has practically no corrosive effect on the ordinary metals unless contaminated by impurities of which water is one. This refrigerant is noncombustible, although in the presence of open flames or very hot surfaces it breaks down and forms toxic gases. The vapor itself is almost perfectly nontoxic, even in concentrations above 20% by volume in air; apparently the only difficulty is that of reducing the amount of oxygen. It has only a very slight odor.

Pipe joints for all of the halocarbons must be carefully made to prevent leakage of gas. Copper tubing with sweated joints is customary in small installations. Mineral oil of selected grade and free of water is used. Some difficulty may occur in the lubrication systems, since oil and refrigerant are mutually soluble in each other. Care must be exercised to see that too much oil from the compressor is not carried over into the evaporator, endangering compressor lubrication as well as reducing heat transfer in the evaporator. Rubber gasket material is inadvisable with most of the halogenated-hydrocarbon refrigerants, but various synthetic gasket compositions (neoprene) are satisfactory in many cases.

Water must be completely removed from refrigerant systems, with the exception of those using ammonia, and the systems must be kept continuously dry to prevent hydrolysis, as well as freezing and clogging at the expansion valves. Various types of dehydrators, using such materials as silica gel, activated alumina, drierite (calcium sulfate), and so on, can be attached in systems either temporarily or permanently for removing water. In many cases, before starting, a high vacuum is held on the warm system to evaporate (dry out) any moisture in the piping.

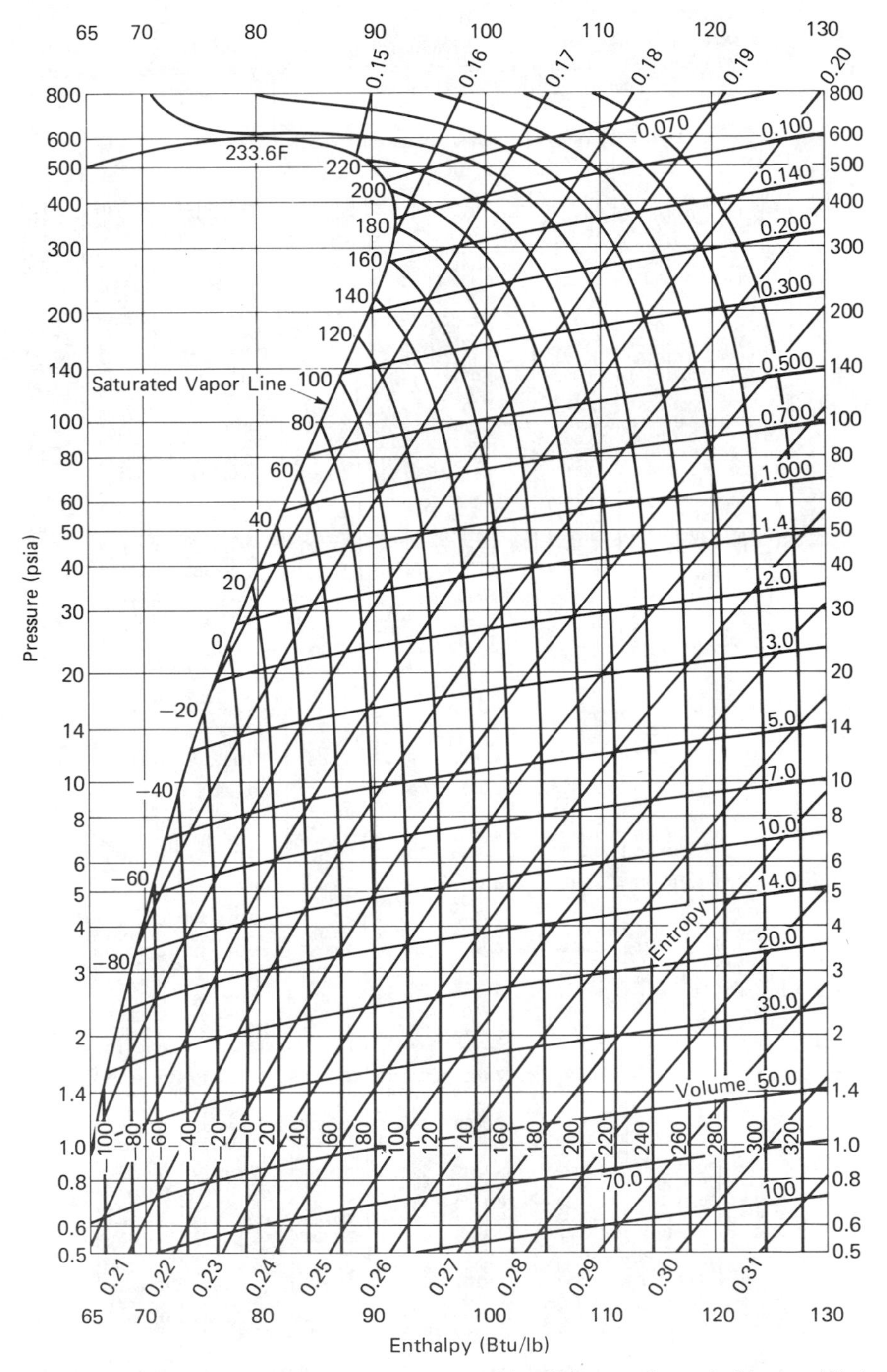

FIGURE 15-4. Pressure-enthalpy chart for dichlorodifluoromethane (refrigerant-12) in vapor region.

Table 15-5 Saturation Properties of Dichlorodifluoromethane (Refrigerant-12)*

Temp F	Pressure psia	Volume cu ft/lb Liquid v_f	Volume cu ft/lb Vapor v_g	Enthalpy Btu/lb Liquid h_f	Enthalpy Btu/lb Latent h_{fg}	Enthalpy Btu/lb Vapor h_g	Entropy Btu/(lb)(R) Liquid s_f	Entropy Btu/(lb)(R) Vapor s_g	Temp F
−140	0.2562	0.0096579	110.46	−20.652	82.548	61.896	0.056123	0.20208	−140
−130	0.4122	0.0097359	70.730	−18.609	81.577	62.968	−0.049830	0.19760	−130
−120	0.6419	0.0098163	46.741	−16.565	80.617	64.052	−0.043723	0.19359	−120
−110	0.9703	0.0098992	31.777	−14.518	79.663	65.145	−0.037786	0.19002	−110
−100	1.4280	0.0099847	22.164	−12.466	78.714	66.248	−0.032005	0.18683	−100
−90	2.0509	0.010073	15.821	−10.409	77.764	67.355	−0.026367	0.18398	−90
−80	2.8807	0.010164	11.533	−8.3451	76.812	68.467	−0.020862	0.18143	−80
−70	3.9651	0.010259	8.5687	−6.2730	75.853	69.580	−0.015481	0.17916	−70
−60	5.3575	0.010357	6.4774	−4.1919	74.885	70.693	−0.010214	0.17714	−60
−50	7.1168	0.010459	4.9742	−2.1011	73.906	71.805	−0.005056	0.17533	−50
−45	8.1540	0.010511	4.3828	−1.0519	73.411	72.359	−0.002516	0.17451	−45
−40	9.3076	0.010564	3.8750	0	72.913	72.913	0	0.17373	−40
−39	9.5530	0.010575	3.7823	0.2107	72.812	73.023	0.000500	0.17357	−39
−38	9.8035	0.010586	3.6922	0.4215	72.712	73.134	0.001000	0.17343	−38
−37	10.059	0.010596	3.6047	0.6324	72.611	73.243	0.001498	0.17328	−37
−36	10.320	0.010607	3.5198	0.8434	72.511	73.354	0.001995	0.17313	−36
−35	10.586	0.010618	3.4373	1.0546	72.409	73.464	0.002492	0.17299	−35
−34	10.858	0.010629	3.3571	1.2659	72.309	73.575	0.002988	0.17285	−34
−33	11.135	0.010640	3.2792	1.4772	72.208	73.685	0.003482	0.17271	−33
−32	11.417	0.010651	3.2035	1.6887	72.106	73.795	0.003976	0.17257	−32
−31	11.706	0.010662	3.1300	1.9003	72.004	73.904	0.004469	0.17243	−31
−30	11.999	0.010674	3.0585	2.1120	71.903	74.015	0.004961	0.17229	−30
−29	12.299	0.010685	2.9890	2.3239	71.801	74.125	0.005452	0.17216	−29
−28	12.604	0.010696	2.9214	2.5358	71.698	74.234	0.005942	0.17203	−28
−27	12.916	0.010707	2.8556	2.7479	71.596	74.344	0.006431	0.17189	−27
−26	13.233	0.010719	2.7917	2.9601	71.494	74.454	0.006919	0.17177	−26
−25	13.556	0.010730	2.7295	3.1724	71.391	74.563	0.007407	0.17164	−25
−24	13.886	0.010741	2.6691	3.3848	71.288	74.673	0.007894	0.17151	−24
−23	14.222	0.010753	2.6102	3.5973	71.185	74.782	0.008379	0.17139	−23
−22	14.564	0.010764	2.5529	3.8100	71.081	74.891	0.008864	0.17126	−22
−21	14.912	0.010776	2.4972	4.0228	70.978	75.001	0.009348	0.17114	−21
−20	15.267	0.010788	2.4429	4.2357	70.874	75.110	0.009831	0.17102	−20
−19	15.628	0.010799	2.3901	4.4487	70.770	75.219	0.010314	0.17090	−19
−18	15.996	0.010811	2.3387	4.6618	70.666	75.328	0.010795	0.17078	−18
−17	16.371	0.010823	2.2886	4.8751	70.561	75.436	0.011276	0.17066	−17
−16	16.753	0.010834	2.2399	5.0885	70.456	75.545	0.011755	0.17055	−16
−15	17.141	0.010846	2.1924	5.3020	70.352	75.654	0.012234	0.17043	−15
−14	17.536	0.010858	2.1461	5.5157	70.246	75.762	0.012712	0.17032	−14
−13	17.939	0.010870	2.1011	5.7295	70.141	75.871	0.013190	0.17021	−13
−12	18.348	0.010882	2.0572	5.9434	70.036	75.979	0.013666	0.17010	−12
−11	18.765	0.010894	2.0144	6.1574	69.930	76.087	0.014142	0.16999	−11
−10	19.189	0.010906	1.9727	6.3716	69.824	76.196	0.014617	0.16989	−10
−9	19.621	0.010919	1.9320	6.5859	69.718	76.304	0.015091	0.16978	−9
−8	20.059	0.010931	1.8924	6.8003	69.611	76.411	0.015564	0.16967	−8
−7	20.506	0.010943	1.8538	7.0149	69.505	76.520	0.016037	0.16957	−7
−6	20.960	0.010955	1.8161	7.2296	69.397	76.627	0.016508	0.16947	−6
−5	21.422	0.010968	1.7794	7.4444	69.291	76.735	0.016979	0.16937	−5
−4	21.891	0.010980	1.7436	7.6594	69.183	76.842	0.017449	0.16927	−4
−3	22.369	0.010993	1.7086	7.8745	69.075	76.950	0.017919	0.16917	−3
−2	22.854	0.011005	1.6745	8.0898	68.967	77.057	0.018388	0.16907	−2
−1	23.348	0.011018	1.6413	8.3052	68.859	77.164	0.018855	0.16897	−1
0	23.849	0.011030	1.6089	8.5207	68.750	77.271	0.019323	0.16888	0
1	24.359	0.011043	1.5772	8.7364	68.642	77.378	0.019789	0.16878	1
2	24.878	0.011056	1.5463	8.9522	68.533	77.485	0.020255	0.16869	2
3	25.404	0.011069	1.5161	9.1682	68.424	77.592	0.020719	0.16860	3
4	25.939	0.011082	1.4867	9.3843	68.314	77.698	0.021184	0.16851	4
5	26.483	0.011094	1.4580	9.6005	68.204	77.805	0.021647	0.16842	5
6	27.036	0.011107	1.4299	9.8169	68.094	77.911	0.022110	0.16833	6
7	27.597	0.011121	1.4025	10.033	67.984	78.017	0.022572	0.16824	7
8	28.167	0.011134	1.3758	10.250	67.873	78.123	0.023033	0.16815	8
9	28.747	0.011147	1.3496	10.467	67.762	78.229	0.023494	0.16807	9

Table 15-5 (*Continued*)

Temp F	Pressure	Volume cu ft/lb		Enthalpy Btu/lb			Entropy Btu/(lb)(R)		Temp F
	psia	Liquid v_f	Vapor v_g	Liquid h_f	Latent h_{fg}	Vapor h_g	Liquid s	Vapor s_g	
10	29.335	0.011160	1.3241	10.684	67.651	78.335	0.023954	0.16798	10
11	29.932	0.011173	1.2992	10.901	67.539	78.440	0.024413	0.16790	11
12	30.539	0.011187	1.2748	11.118	67.428	78.546	0.024871	0.16782	12
13	31.155	0.011200	1.2510	11.336	67.315	78.651	0.025329	0.16774	13
14	31.780	0.011214	1.2778	11.554	67.203	78.757	0.025786	0.16765	14
15	32.415	0.011227	1.2050	11.771	67.090	78.861	0.026243	0.16758	15
16	33.060	0.011241	1.1828	11.989	66.977	78.966	0.026699	0.16750	16
17	33.714	0.011254	1.1611	12.207	66.864	79.071	0.027154	0.16742	17
18	34.378	0.011268	1.1399	12.426	66.750	79.176	0.027608	0.16734	18
19	35.052	0.011282	1.1191	12.644	66.636	79.280	0.028062	0.16727	19
20	35.736	0.011296	1.0988	12.863	66.522	79.385	0.028515	0.16719	20
21	36.430	0.011310	1.0790	13.081	66.407	79.488	0.028968	0.16712	21
22	37.135	0.011324	1.0596	13.300	66.293	79.593	0.029420	0.16704	22
23	37.849	0.011388	1.0406	13.520	66.177	79.697	0.029871	0.16697	23
24	38.574	0.011352	1.0220	13.739	66.061	79.800	0.030322	0.16690	24
25	39.310	0.011366	1.0039	13.958	65.946	79.904	0.030772	0.16683	25
26	40.056	0.011380	0.98612	14.178	65.829	80.007	0.031221	0.16676	26
27	40.813	0.011395	0.96874	14.398	65.713	80.111	0.031670	0.16669	27
28	41.580	0.011409	0.95173	14.618	65.596	80.214	0.032118	0.16662	28
29	42.359	0.011424	0.93509	14.838	65.478	80.316	0.032566	0.16655	29
30	43.148	0.011438	0.91880	15.058	65.361	80.419	0.033013	0.16648	30
31	43.948	0.011453	0.90286	15.279	65.243	80.522	0.033460	0.16642	31
32	44.760	0.011468	0.88725	15.500	65.124	80.624	0.033905	0.16635	32
33	45.583	0.011482	0.87197	15.720	65.006	80.726	0.034351	0.16629	33
34	46.417	0.011497	0.85702	15.942	64.886	80.828	0.034796	0.16622	34
35	47.263	0.011512	0.84237	16.163	64.767	80.930	0.035240	0.16616	35
36	48.120	0.011527	0.82803	16.384	64.647	81.031	0.035683	0.16610	36
37	48.989	0.011542	0.81399	16.606	64.527	81.133	0.036126	0.16604	37
38	49.870	0.011557	0.80023	16.828	64.406	81.234	0.036569	0.16598	38
39	50.763	0.011573	0.78676	17.050	64.285	81.335	0.037011	0.16592	39
40	51.667	0.011588	0.77357	17.273	64.163	81.436	0.037453	0.16586	40
41	52.584	0.011603	0.76064	17.495	64.042	81.537	0.037893	0.16580	41
42	53.513	0.011619	0.74798	17.718	63.919	81.637	0.038334	0.16574	42
43	54.454	0.011635	0.73557	17.941	63.796	81.737	0.038774	0.16568	43
44	55.407	0.011650	0.72341	18.164	63.673	81.837	0.039213	0.16562	44
45	56.373	0.011666	0.71149	18.387	63.550	81.937	0.039652	0.16557	45
46	57.352	0.011682	0.69982	18.611	63.426	82.037	0.040091	0.16551	46
47	58.343	0.011698	0.68837	18.835	63.301	82.136	0.040529	0.16546	47
48	59.347	0.011714	0.67715	19.059	63.177	82.236	0.040966	0.16540	48
49	60.364	0.011730	0.66616	19.283	63.051	82.334	0.041403	0.16535	49
50	61.394	0.011746	0.65537	19.507	62.926	82.433	0.041839	0.16530	50
51	62.437	0.011762	0.64480	19.732	62.800	82.532	0.042276	0.16524	51
52	63.494	0.011779	0.63444	19.957	62.673	82.630	0.042711	0.16519	52
53	64.563	0.011795	0.62428	20.182	62.546	82.728	0.043146	0.16514	53
54	65.646	0.011811	0.61431	20.408	62.418	82.826	0.043581	0.16509	54
55	66.743	0.011828	0.60453	20.634	62.290	82.924	0.044015	0.16504	55
56	67.853	0.011845	0.59495	20.859	62.162	83.021	0.044449	0.16499	56
57	68.977	0.011862	0.58554	21.086	62.033	83.119	0.044883	0.16494	57
58	70.115	0.011879	0.57632	21.312	61.903	83.215	0.045316	0.16489	58
59	71.267	0.011896	0.56727	21.539	61.773	83.312	0.045748	0.16484	59
60	72.433	0.011913	0.55839	21.766	61.643	83.409	0.046180	0.16479	60
61	73.613	0.011930	0.54967	21.993	61.512	83.505	0.046612	0.16474	61
62	74.807	0.011947	0.54112	22.221	61.380	83.601	0.047044	0.16470	62
63	76.016	0.011965	0.53273	22.448	61.248	83.696	0,047475	0.16465	63
64	77.239	0.011982	0.52450	22.676	61.116	83.792	0.047905	0.16460	64
65	78.477	0.012000	0.51642	22.905	60.982	83.887	0.048336	0.16456	65
66	79.729	0.012017	0.50848	23.133	60.849	83.982	0.048765	0.16451	66
67	80.996	0.012035	0.50070	23.362	60.715	84.077	0.049195	0.16447	67
68	82.279	0.012053	0.49305	23.591	60.580	84.171	0.049624	0.16442	68
69	83.576	0.012071	0.48555	23.821	60.445	84.266	0.050053	0.16438	69

Table 15-5 (*Continued*)

Temp F	Pressure	Volume cu ft/lb		Enthalpy Btu/lb			Entropy Btu/(lb)(R)		Temp F
	psia	Liquid v_f	Vapor v_g	Liquid h_f	Latent h_{fg}	Vapor h_g	Liquid s_f	Vapor s_g	
70	84.888	0.012089	0.47818	24.050	60.309	84.359	0.050482	0.16434	70
71	86.216	0.012108	0.47094	24.281	60.172	84.453	0.050910	0.16429	71
72	87.559	0.012126	0.46383	24.511	60.035	84.546	0.051338	0.16425	72
73	88.918	0.012145	0.45686	24.741	59.898	84.639	0.051766	0.16421	73
74	90.292	0.012163	0.45000	24.973	59.759	84.732	0.052193	0.16417	74
75	91.682	0.012182	0.44327	25.204	59.621	84.825	0.052620	0.16412	75
76	93.087	0.012201	0.43666	25.435	59.481	84.916	0.053047	0.16408	76
77	94.509	0.012220	0.43016	25.667	59.341	85.008	0.053473	0.16404	77
78	95.946	0.012239	0.42378	25.899	59.201	85.100	0.053900	0.16400	78
79	97.400	0.012258	0.41751	26.132	59.059	85.191	0.054326	0.16396	79
80	98.870	0.012277	0.41135	26.365	58.917	85.282	0.054751	0.16392	80
81	100.36	0.012297	0.40530	26.598	58.775	85.373	0.055177	0.16388	81
82	101.86	0.012316	0.39935	26.832	58.631	85.463	0.055602	0.16384	82
83	103.38	0.012336	0.39351	27.065	58.488	85.553	0.056027	0.16380	83
84	104.92	0.012356	0.38776	27.300	58.343	85.643	0.056452	0.16376	84
85	106.47	0.012376	0.38212	27.534	58.198	85.732	0.056877	0.16372	85
86	108.04	0.012396	0.37657	27.769	58.052	85.821	0.057301	0.16368	86
87	109.63	0.012416	0.37111	28.005	57.905	85.910	0.057725	0.16364	87
88	111.23	0.012437	0.36575	28.241	57.757	85.998	0.058149	0.16360	88
89	112.85	0.012457	0.36047	28.477	57.609	86.086	0.058573	0.16357	89
90	114.49	0.012478	0.35529	28.713	57.461	86.174	0.058997	0.16353	90
91	116.15	0.012499	0.35019	28.950	57.311	86.261	0.059420	0.16349	91
92	117.82	0.012520	0.34518	29.187	57.161	86.348	0.059844	0.16345	92
93	119.51	0.012541	0.34025	29.425	57.009	86.434	0.060267	0.16341	93
94	121.22	0.012562	0.33540	29.663	56.858	86.521	0.060690	0.16338	94
95	122.95	0.012583	0.33063	29.901	56.705	86.606	0.061113	0.16334	95
96	124.70	0.012605	0.32594	30.140	56.551	86.691	0.061536	0.16330	96
97	126.46	0.012627	0.32133	30.380	56.397	86.777	0.061959	0.16326	97
98	128.24	0.012649	0.31679	30.619	56.242	86.861	0.062381	0.16323	98
99	130.04	0.012671	0.31233	30.859	56.086	86.945	0.062804	0.16319	99
100	131.86	0.012693	0.30794	31.100	55.929	87.029	0.063227	0.16315	100
101	133.70	0.012715	0.30362	31.341	55.772	87.113	0.063649	0.16312	101
102	135.56	0.012738	0.29937	31.583	55.613	87.196	0.064072	0.16308	102
103	137.44	0.012760	0.29518	31.824	55.454	87.278	0.064494	0.16304	103
104	139.33	0.012783	0.29106	32.067	55.293	87.360	0.064916	0.16301	104
105	141.25	0.012806	0.28701	32.310	55.132	87.442	0.065339	0.16297	105
106	143.18	0.102829	0.28303	32.553	54.970	87.523	0.065761	0.16293	106
107	145.13	0.012853	0.27910	32.797	54.807	87.604	0.066184	0.16290	107
108	147.11	0.012876	0.27524	33.041	54.643	87.684	0.066606	0.16286	108
109	149.10	0.012900	0.27143	33.286	54.478	87.764	0.067028	0.16282	109
110	151.11	0.012924	0.26769	33.531	54.313	87.844	0.067451	0.16279	110
112	155.19	0.012972	0.26037	34.023	53.978	88.001	0.068296	0.16271	112
114	159.36	0.013022	0.25328	34.517	53.639	88.156	0.069141	0.16264	114
116	163.61	0.013072	0.24641	35.014	53.296	88.310	0.069987	0.16256	116
118	167.94	0.013123	0.23974	35.512	52.949	88.461	0.070833	0.16249	118
120	172.35	0.013174	0.23326	36.013	52.597	88.610	0.071680	0.16241	120
122	176.85	0.013227	0.22698	36.516	52.241	88.757	0.072528	0.16234	122
124	181.43	0.013280	0.22089	37.021	51.881	88.902	0.073376	0.16226	124
125	183.76	0.013308	0.21791	37.275	51.698	88.973	0.073800	0.16222	125
130	195.71	0.013447	0.20364	38.553	50.768	89.321	0.075927	0.16202	130
135	208.22	0.013593	0.19036	39.848	49.805	89.653	0.078061	0.16181	135
140	221.32	0.013746	0.17799	41.162	48.805	89.967	0.080205	0.16159	140
145	235.00	0.013907	0.16644	42.495	47.766	90.261	0.082361	0.16135	145
150	249.31	0.014078	0.15564	43.850	46.684	90.534	0.084531	0.16110	150
170	313.00	0.014871	0.11873	49.529	41.830	91.359	0.093418	0.15985	170
190	387.98	0.015942	0.089418	55.769	35.792	91.561	0.10284	0.15793	190
210	475.52	0.017601	0.064843	62.959	27.599	90.558	0.11332	0.15453	210
230	577.03	0.021854	0.039435	72.893	12.229	85.122	0.12739	0.14512	230
233.6 (Critical)	596.9	0.02870	0.02870	78.86	0	78.86	0.1359	0.1359	233.6 (Critical)

15-5. CHLORODIFLUOROMETHANE ($CClHF_2$), REFRIGERANT 22

Refrigerant-22, also known as Freon-22 and as genetron-22 is in extensive use for reciprocating compressors. Like refrigerant-12, it is chemically stable, almost odorless and nonirritating, and exhibits no permanent deleterious effects in concentrations up to 18% by volume for exposures of less than 2-h duration.

A study of the thermodynamic properties of refrigerant-22 in Table 15-6 and Plate V indicates that 22 is particularly suitable for use in the low-temperature field (−40 to −100 F) because its pressure is higher than that of ammonia, refrigerant-12, and, in fact, of most refrigerants with the exception of CO_2 and some of the hydrocarbons. The specific volume is moderate at low pressures, and, with the relatively good latent heat which this refrigerant possesses, it can be seen that reasonable piston displacements per ton are possible even at low-suction temperatures. This refrigerant is also being used extensively in the moderate-temperature (air-conditioning) range because its low specific volume makes it possible, with a compressor of given physical size (piston displacement), to serve a greater tonnage load than is possible with refrigerant-12. It is practically noninflammable and nonexplosive. The specific heat of the vapor C_p, in Btu per pound degree Fahrenheit, is 0.138 at −30 F, 0.143 at −10 F, 0.152 at 40 F, and 0.160 at 100 F. The value of C_p/C_v is 1.16 at 86 F. It has a critical temperature of 205 F at 722 psia.

15-6. MISCELLANEOUS HALOCARBON REFRIGERANTS

Trichloromonofluoromethane (CCl_3F), refrigerant-11, also known as Carrene No. 2, is a so-called vacuum refrigerant; that is, refrigeration temperatures in the evaporator can be obtained only at subatmospheric pressures (Table 15-7).

Refrigerant-11 is practically odorless and relatively nontoxic, even in high concentrations, provided the exposure is not continued over very long periods. It is nonexplosive and practically noninflammable, but in the presence of flames and very hot surfaces it forms toxic decomposition products. There is no corrosive action on the common metals of construction, but rubber gaskets are attacked.

The specific volume of a vacuum refrigerant is usually so great that the large volume of gas can be handled effectively only by high-speed centrifugal compressors. For very large vapor volumes the physical dimensions required for reciprocating compressors become prohibitive. With the centrifugal compressors ordinarily used, the range of pressure is small, usually less than 30 psi.

Methyl chloride (CH_3Cl), refrigerant-40, now has limited use in small commercial (storage-compartment) units. The pressure range for methyl chloride is moderate, 21.1 psia to 94.7 psia at 5 F to 86 F. Latent heat of evaporation in a temperature-working range of −10 F to +20 F varies from 184 to 177 Btu/lb.

Table 15-6 Saturation Properties of Chlorodifluoromethane (Refrigerant-22)

Temp. (F)	Pressure (psia)	Sp. Vol. (ft^3/lb) Liquid v_f	Sp. Vol. (ft^3/lb) Vapor v_g	Enthalpy (Btu/lb) Liquid h_f	Enthalpy (Btu/lb) Latent h_{fg}	Enthalpy (Btu/lb) Vapor h_g	Entropy (Btu/lb · R) Liquid s_f	Entropy (Btu/lb · R) Vapor s_g	Temp. (F)
−150	0.27163	0.010180	141.23	−25.974	113.495	87.521	−0.07147	0.29501	−150
−140	0.44692	0.010271	88.532	−23.725	112.405	88.681	−0.06432	0.28729	−140
−130	0.71060	0.010365	57.356	−21.463	111.311	89.848	−0.05736	0.28027	−130
−120	1.0954	0.010462	38.280	−19.185	110.205	91.020	−0.05055	0.27388	−120
−110	1.6417	0.010561	26.242	−16.886	109.082	92.196	−0.04389	0.26805	−110
−100	2.3983	0.010664	18.433	−14.564	107.935	93.371	−0.03734	0.26274	−100
−90	3.4229	0.010771	13.235	−12.216	106.759	94.544	−0.03091	0.25787	−90
−80	4,7822	0.010881	9.6949	−9.838	105.548	95.710	−0.02457	0.25342	−80
−70	6.5522	0.010995	7.2318	−7.429	104.297	96.868	−0.01832	0.24932	−70
−60	8.8180	0.011113	5.4844	−4.987	103.001	98.014	−0.01214	0.24556	−60
−50	11.674	0.011235	4.2224	−2.511	101.656	99.144	−0.00604	0.24209	−50
−48	12.324	0.011261	4.0140	−2.012	101.381	99.369	−0.00483	0.24143	−48
−46	13.004	0.011286	3.8179	−1.511	101.103	99.592	−0.00361	0.24078	−46
−44	13.712	0.011311	3.6334	−1.009	100.832	99.814	−0.00241	0.24014	−44
−42	14.451	0.011337	3.4596	−0.505	100.541	100.036	−0.00120	0.23951	−42
−40	15.222	0.011363	3.2957	0.000	100.257	100.257	0.00000	0.23888	−40
−39	15.619	0.011376	3.2173	0.253	100.114	100.367	0.00060	0.23858	−39
−38	16.024	0.011389	3.1412	0.506	99.971	100.477	0.00120	0.23877	−38
−37	16.437	0.011402	3.0673	0.760	99.826	100.587	0.00180	0.23797	−37
−36	16.859	0.011415	2.9954	1.014	99.682	100.696	0.00240	0.23767	−36
−35	17.290	0.011428	2.9256	1.269	99.536	100.805	0.00300	0.23737	−35
−34	17.728	0.011442	2.8578	1.524	99.391	100.914	0.00359	0.23707	−34
−33	18.176	0.011455	2.7919	1.779	99.244	101.023	0.00419	0.23678	−33
−32	18.633	0.011469	2.7278	2.035	99.097	101.132	0.00479	0.23649	−32
−31	19.098	0.011482	2.6655	2.291	98.949	101.240	0.00538	0.23620	−31
−30	19.573	0.011495	2.6049	2.547	98.801	101.348	0.00598	0.23591	−30
−29	20.056	0.011509	2.5460	2.804	98.652	101.456	0.00657	0.23563	−29
−28	20.549	0.011523	2.4887	3.061	98.503	101.564	0.00716	0.23534	−28
−27	21.052	0.011536	2.4329	3.318	98.353	101.671	0.00776	0.23506	−27
−26	21.564	0.011550	2.3787	3.576	98.202	101.778	0.00835	0.23478	−26
−25	22.086	0.011564	2.3260	3.834	98.051	101.885	0.00894	0.23451	−25
−24	22.617	0.011578	2.2746	4.093	97.899	101.992	0.00953	0.23423	−24
−23	23.159	0.011592	2.2246	4.352	97.746	102.098	0.01013	0.23396	−23
−22	23.711	0.011606	2.1760	4.611	97.593	102.204	0.01072	0.23369	−22
−21	24.272	0.011620	2.1287	4.871	97.439	102.310	0.01131	0.23342	−21
−20	24.845	0.011634	2.0826	5.131	97.285	102.415	0.01189	0.23315	−20
−19	25.427	0.011648	2.0377	5.391	97.129	102.521	0.01248	0.23289	−19
−18	26.020	0.011662	1.9940	5.652	96.974	102.626	0.01307	0.23262	−18
−17	26.624	0.011677	1.9514	5.913	96.817	102.730	0.01366	0.23236	−17
−16	27.239	0.011691	1.9099	6.175	96.660	102.835	0.01425	0.23210	−16
−15	27.865	0.011705	1.8695	6.436	96.502	102.939	0.01483	0.23184	−15
−14	28.501	0.011720	1.8302	6.699	96.344	103.043	0.01542	0.23159	−14
−13	29.149	0.011734	1.7918	6.961	96.185	103.146	0.01600	0.23133	−13
−12	29.809	0.011749	1.7544	7.224	96.025	103.250	0.01659	0.23108	−12
−11	30.480	0.011764	1.7180	7.488	95.865	103.352	0.01717	0.23083	−11

Table 15-6 (*Continued*)

Temp. (F)	Pressure (psia)	Sp. Vol. (ft³/lb) Liquid v_f	Sp. Vol. Vapor v_g	Enthalpy (Btu/lb) Liquid h_f	Enthalpy Latent h_{fg}	Enthalpy Vapor h_g	Entropy (Btu/lb · R) Liquid s_f	Entropy Vapor s_g	Temp. (F)
−10	31.162	0.011778	1.6825	7.751	95.704	103.455	0.01776	0.23058	−10
−9	31.856	0.011793	1.6479	8.015	95.542	103.558	0.01834	0.23033	−9
−8	32.563	0.011808	1.6141	8.280	95.380	103.660	0.01892	0.23008	−8
−7	33.281	0.011823	1.5812	8.545	95.217	103.762	0.01950	0.22984	−7
−6	34.011	0.011838	1.5491	8.810	95.053	103.863	0.02009	0.22960	−6
−5	34.754	0.011853	1.5177	9.075	94.889	103.964	0.02067	0.22936	−5
−4	35.509	0.011868	1.4872	9.341	94.724	104.065	0.02125	0.22912	−4
−3	36.277	0.011884	1.4574	9.608	94.558	104.166	0.02183	0.22888	−3
−2	37.057	0.011899	1.4283	9.874	94.391	104.266	0.02241	0.22864	−2
−1	37.850	0.011914	1.4000	10.142	94.224	104.366	0.02299	0.22841	−1
0	38.657	0.011930	1.3723	10.409	94.056	104.465	0.02357	0.22817	0
1	39.476	0.011945	1.3453	10.677	93.888	104.565	0.02414	0.22794	1
2	40.309	0.011961	1.3189	10.945	93.718	104.663	0.02472	0.22771	2
3	41.155	0.011976	1.2931	11.214	93.548	104.762	0.02530	0.22748	3
4	42.014	0.011992	1.2680	11.483	93.378	104.860	0.02587	0.22725	4
5	42.888	0.012008	1.2434	11.752	93.206	104.958	0.02645	0.22703	5
6	43.775	0.012024	1.2195	12.022	93.034	105.056	0.02703	0.22680	6
7	44.676	0.012040	1.1961	12.292	92.861	105.153	0.02760	0.22658	7
8	34.591	0.012056	1.1732	12.562	92.688	105.250	0.02818	0.22636	8
9	46.521	0.012072	1.1509	12.833	92.513	105.346	0.02875	0.22614	9
10	47.464	0.012088	1.1290	13.104	92.338	105.442	0.02932	0.22592	10
11	48.423	0.012105	1.1077	13.376	92.162	105.538	0.02990	0.22570	11
12	49.396	0.012121	1.0869	13.648	91.986	105.633	0.03047	0.22548	12
13	50.384	0.012138	1.0665	13.920	91.808	105.728	0.03104	0.22527	13
14	51.387	0.012154	1.0466	14.193	91.630	105.823	0.03161	0.22505	14
15	52.405	0.012171	1.0272	14.466	91.451	105.917	0.03218	0.22484	15
16	53.438	0.012188	1.0082	14.739	91.272	106.011	0.03275	0.22463	16
17	54.487	0.012204	0.98961	15.013	91.091	106.105	0.03332	0.22442	17
18	55.551	0.012221	0.97144	15.288	90.910	106.198	0.03389	0.22421	18
19	56.631	0.012238	0.95368	15.562	90.728	106.290	0.03446	0.22400	19
20	57.727	0.012255	0.93631	15.837	90.545	106.383	0.03503	0.22379	20
21	58.839	0.012273	0.91932	16.113	90.362	106.475	0.03560	0.22358	21
22	59.967	0.012290	0.90270	16.389	90.178	106.566	0.03617	0.22338	22
23	61.111	0.012307	0.88645	16.665	89.993	106.657	0.03674	0.22318	23
24	62.272	0.012325	0.87055	16.942	89.807	106.748	0.03730	0.22297	24
25	63.450	0.012342	0.85500	17.219	89.620	106.839	0.03787	0.22277	25
26	64.644	0.012360	0.83978	17.496	89.433	106.928	0.03844	0.22257	26
27	65.855	0.012378	0.82488	17.774	89.244	107.018	0.03900	0.22237	27
28	67.083	0.012395	0.81031	18.052	89.055	107.107	0.03958	0.22217	28
29	68.328	0.012413	0.79604	18.330	88.865	107.196	0.04013	0.22198	29
30	69.591	0.012431	0.78208	18.609	88.674	107.284	0.04070	0.22178	30
31	70.871	0.012450	0.76842	18.889	88.483	107.372	0.04126	0.22158	31
32	72.169	0.012468	0.75503	19.169	88.290	107.459	0.04182	0.22139	32
33	73.485	0.012486	0.74194	19.449	88.097	107.546	0.04239	0.22119	33
34	74.818	0.012505	0.72911	19.729	87.903	107.632	0.04295	0.22100	34

Table 15-6 (*Continued*)

Temp. (F)	Pressure (psia)	Sp. Vol. (ft³/lb) Liquid v_f	Sp. Vol. Vapor v_g	Enthalpy (Btu/lb) Liquid $\mathbf{h}_f$	Enthalpy Latent $\mathbf{h}_{fg}$	Enthalpy Vapor $\mathbf{h}_g$	Entropy (Btu/lb · R) Liquid s_f	Entropy Vapor s_g	Temp. (F)
35	76.170	0.012523	0.71655	20.010	87.708	107.719	0.04351	0.22081	35
36	77.540	0.012542	0.70425	20.292	87.512	107.804	0.04407	0.22062	36
37	78.929	0.012561	0.69221	20.574	87.316	107.889	0.04464	0.22043	37
38	80.336	0.012579	0.68041	20.856	87.118	107.974	0.04520	0.22024	38
39	81.761	0.012598	0.66885	21.138	86.920	108.058	0.04576	0.22005	39
40	83.206	0.012618	0.65753	21.422	86.720	108.142	0.04632	0.21986	40
41	84.670	0.012637	0.64643	21.705	86.520	108.225	0.04688	0.21968	41
42	86.153	0.012656	0.63557	21.989	86.319	108.308	0.04744	0.21949	42
43	87.655	0.012676	0.62492	22.273	86.117	108.390	0.04800	0.21931	43
44	89.177	0.012695	0.61448	22.558	85.914	108.472	0.04855	0.21912	44
45	90.719	0.012715	0.60425	22.843	85.710	108.553	0.04911	0.21894	45
46	92.280	0.012735	0.59442	23.129	85.506	108.634	0.04967	0.21876	46
47	93.861	0.012755	0.58440	23.415	85.300	108.715	0.05023	0.21858	47
48	95.463	0.012775	0.57476	23.701	85.094	108.795	0.05079	0.21839	48
49	97.085	0.012795	0.56532	23.988	84.886	108.874	0.05134	0.21821	49
50	98.727	0.012815	0.55606	24.275	84.678	108.953	0.05190	0.21803	50
51	100.39	0.012836	0.54698	24.563	84.468	109.031	0.05245	0.21785	51
52	102.07	0.012856	0.53808	24.851	84.258	109.109	0.05301	0.21768	52
53	103.78	0.012877	0.52934	25.139	84.047	109.186	0.05357	0.21750	53
54	105.50	0.012989	0.52078	25.429	83.834	109.263	0.05412	0.21732	54
55	107.25	0.012919	0.51238	25.718	83.621	109.339	0.05468	0.21714	55
56	109.02	0.012940	0.50414	26.008	83.407	109.415	0.05523	0.21697	56
57	110.81	0.012961	0.49606	26.298	83.191	109.490	0.05579	0.21679	57
58	112.62	0.012982	0.48813	26.589	82.975	109.564	0.05634	0.21662	58
59	114.46	0.013004	0.48035	26.880	82.758	109.638	0.05689	0.21644	59
60	116.31	0.013025	0.47272	27.172	83.540	109.712	0.05745	0.21627	60
61	118.19	0.013047	0.46523	27.464	82.320	109.785	0.05800	0.21610	61
62	120.09	0.013069	0.45788	27.757	82.100	109.857	0.05855	0.21592	62
63	122.01	0.013091	0.45066	28.050	81.878	109.929	0.05910	0.21575	63
64	123.96	0.013114	0.44358	28.344	81.656	110.000	0.05966	0.21558	64
65	125.93	0.013136	0.43663	28.638	81.432	110.070	0.06021	0.21541	65
66	127.92	0.013159	0.42981	28.932	81.208	110.140	0.06076	0.21524	66
67	129.94	0.013181	0.42311	29.228	80.982	110.209	0.06131	0.21507	67
68	131.97	0.013204	0.41653	29.523	80.755	110.278	0.06186	0.21490	68
69	134.04	0.013227	0.41007	29.819	80.527	110.346	0.06241	0.21473	69
70	136.12	0.013251	0.40373	30.116	80.298	110.414	0.06296	0.21456	70
71	138.23	0.013274	0.39751	30.413	80.068	110.480	0.06351	0.21439	71
72	140.37	0.013297	0.39139	30.710	79.836	110.547	0.06406	0.21422	72
73	142.52	0.013321	0.38539	31.008	79.604	110.612	0.06461	0.21405	73
74	144.71	0.013345	0.37949	31.307	79.370	110.677	0.06516	0.21388	74
75	146.91	0.013369	0.37369	31.606	79.135	110.741	0.06571	0.21372	75
76	149.15	0.013393	0.36800	31.906	78.899	110.805	0.06626	0.21355	76
77	151.40	0.013418	0.36241	32.206	78.662	110.868	0.06681	0.21338	77
78	153.69	0.013442	0.35691	32.506	78.423	110.930	0.06736	0.21321	78
79	155.99	0.013467	0.35151	32.808	78.184	110.991	0.06791	0.21305	79

Table 15-6 (*Continued*)

Temp. (F)	Pressure (psia)	Sp. Vol. (ft^3/lb)		Enthalpy (Btu/lb)			Entropy (Btu/lb · R)		Temp. (F)
		Liquid v_f	Vapor v_g	Liquid $\mathbf{h}_f$	Latent $\mathbf{h}_{fg}$	Vapor $\mathbf{h}_g$	Liquid s_f	Vapor s_g	
80	158.33	0.013492	0.34621	33.109	77.943	111.052	0.06846	0.21288	80
81	160.68	0.013518	0.34099	33.412	77.701	111.112	0.06901	0.21271	81
82	163.07	0.013543	0.33587	33.714	77.457	111.171	0.06956	0.21255	82
83	165.48	0.013569	0.33083	34.018	77.212	111.230	0.07011	0.21238	83
84	167.92	0.013594	0.32588	34.322	76.966	111.288	0.07065	0.21222	84
85	170.38	0.013620	0.32101	34.626	76.719	111.345	0.07120	0.21205	85
86	172.87	0.013647	0.21623	34.931	76.470	111.401	0.07175	0.21188	86
87	175.38	0.013673	0.31153	35.237	76.220	111.457	0.07230	0.21172	87
88	177.93	0.013700	0.30690	35.543	75.968	111.512	0.07285	0.21155	88
89	180.50	0.013727	0.30236	35.850	75.716	111.566	0.07339	0.21139	89
90	183.09	0.013754	0.29789	36.158	75.461	111.619	0.07394	0.21122	90
91	185.72	0.013781	0.29349	36.466	75.206	111.671	0.07449	0.21106	91
92	188.37	0.013809	0.28917	36.774	74.949	111.723	0.07504	0.21089	92
93	191.05	0.013836	0.28491	37.084	74.690	111.774	0.07559	0.21072	93
94	193.76	0.013864	0.28073	37.394	74.430	111.824	0.07613	0.21056	94
95	196.50	0.013893	0.27662	37.704	74.168	111.873	0.07668	0.21039	95
96	199.26	0.013921	0.27257	38.016	73.905	111.921	0.07723	0.21023	96
97	202.05	0.013950	0.26859	38.328	73.641	111.968	0.07778	0.21006	97
98	204.87	0.013979	0.26467	38.640	73.375	112.015	0.07832	0.20989	98
99	207.72	0.014008	0.26081	38.953	73.107	112.060	0.07887	0.20973	99
100	210.60	0.014038	0.25702	39.267	72.838	112.105	0.07942	0.20956	100
101	213.51	0.014068	0.25329	39.582	72.567	112.149	0.07997	0.20939	101
102	216.45	0.014098	0.24962	39.897	72.294	112.192	0.08052	0.20923	102
103	219.42	0.014128	0.24600	40.213	72.020	112.233	0.08107	0.20906	103
104	222.42	0.014159	0.24244	40.530	71.744	112.274	0.08161	0.20889	104
105	225.45	0.014190	0.23894	40.847	71.467	112.314	0.08216	0.20872	105
106	228.50	0.014221	0.23549	41.166	71.187	112.353	0.08271	0.20855	106
107	231.59	0.014253	0.23209	41.485	70.906	112.391	0.08326	0.20838	107
108	234.71	0.014285	0.22875	41.804	70.623	112.427	0.08381	0.20821	108
109	237.86	0.014317	0.22546	42.125	70.338	112.463	0.08436	0.20804	109
110	241.04	0.014350	0.22222	42.446	70.052	112.498	0.08491	0.20787	110
111	244.25	0.014382	0.21903	42.768	69.763	112.531	0.08546	0.20770	111
112	247.50	0.014416	0.21589	43.091	69.473	112.564	0.08601	0.20753	112
113	250.77	0.014449	0.21279	43.415	69.180	112.595	0.08656	0.20736	113
114	254.08	0.014483	0.20974	43.739	68.886	112.626	0.08711	0.20718	114
115	257.42	0.014517	0.20674	44.065	68.590	112.655	0.08766	0.20701	115
120	274.60	0.014694	0.19238	45.705	67.077	112.782	0.09042	0.20613	120
130	311.50	0.015080	0.16661	49.059	63.877	112.936	0.09598	0.20431	130
140	351.94	0.015518	0.14418	52.528	60.403	112.931	0.10163	0.20235	140
150	396.19	0.016025	0.12448	56.143	56.585	112.728	0.10739	0.20020	150
160	444.53	0.016627	0.10701	59.948	52.316	112.263	0.11334	0.19776	160
170	497.26	0.017367	0.091279	64.019	47.419	111.438	0.11959	0.19490	170
180	554.78	0.018332	0.076790	68.498	41.570	110.068	0.12635	0.19133	180
200	686.36	0.022436	0.047438	80.862	21.990	102.853	0.14460	0.17794	200
204.81	721.91	0.030525	0.030525	91.329	0.000	91.329	0.16016	0.16016	204.81

Table 15-7 Saturated Liquid and Vapor Tables for Various Refrigerants

Temp F t	Pressure psia p	Sp Vol of Liquid cu ft/lb v_f	Sp Vol of Vapor cu ft/lb v_g	Density of Vapor lb/cu ft $\frac{1}{v_g}$	Enthalpy Btu/Lb above −40 F			Entropy		Temp F t
					Liquid h_f	Vaporization h_{fg}	Vapor h_g	Liquid s_f	Vapor s_g	
Carbon Dioxide, CO_2 (Refrigerant-744)										
−40	145.8	0.01437	0.6113	1.64	0.00	137.8	137.8	1.0000	1.3285	−40
−30	177.8	0.01466	0.5029	1.99	4.5	133.7	138.2	1.0107	1.3218	−30
−20	214.9	0.01498	0.4168	2.40	9.1	129.4	138.5	1.0212	1.3154	−20
−10	257.3	0.01532	0.3472	2.88	13.9	124.8	138.7	1.0314	1.3091	−10
0	305.5	0.01570	0.2904	3.44	18.8	120.1	138.9	1.0418	1.3029	0
4	326.5	0.01588	0.2707	3.69	20.8	118.0	138.8	1.0460	1.3006	4
6	337.4	0.01596	0.2614	3.83	21.8	116.9	138.7	1.0481	1.2994	6
10	360.2	0.01614	0.2437	4.10	24.0	114.7	138.7	1.0536	1.2980	10
20	421.8	0.01663	0.2049	4.88	29.4	108.9	138.3	1.0648	1.2919	20
30	490.8	0.01719	0.1722	5.81	35.4	102.4	137.8	1.0768	1.2859	30
40	567.8	0.01787	0.1444	6.93	41.7	95.0	136.7	1.0884	1.2786	40
50	653.6	0.01868	0.1205	8.30	48.4	86.6	135.0	1.1010	1.2709	50
60	748.6	0.01970	0.0994	10.06	55.5	76.6	132.1	1.1145	1.2618	60
70	853.4	0.02112	0.08040	12.44	63.7	63.8	127.5	1.1292	1.2497	70
80	968.7	0.02370	0.06064	16.49	73.9	44.8	118.7	1.1486	1.2314	80
86	1043.0	0.02686	0.04789	20.88	83.3	27.1	110.4	1.1646	1.2143	86
87.8	1066.2	0.03454	0.03454	28.95	97.0	0.0	97.0	1.1890	1.1890	87.8
Trichloromonofluoromethane (Refrigerant-11)										
−80	0.157	0.00961	189	0.0053	−7.89	90.68	82.79	−0.0197	0.1995	−80
−60	0.356	0.00974	87.5	0.0114	−3.94	89.06	85.12	−0.0096	0.2037	−60
−40	0.739	0.00988	44.2	0.0226	0.00	87.48	87.48	0.0000	0.2085	−40
−20	1.420	0.01002	24.06	0.0415	3.94	85.93	89.87	0.0091	0.2046	−20
−10	1.920	0.01010	18.17	0.0550	5.91	85.16	91.07	0.0136	0.2030	−10
0	2.555	0.01018	13.94	0.0718	7.89	84.38	92.27	0.0179	0.2015	0
5	2.931	0.01022	12.27	0.0815	8.88	84.00	92.88	0.0201	0.2009	5
10	3.352	0.01026	10.83	0.0923	9.88	83.60	93.48	0.0222	0.2003	10
20	4.342	0.01034	8.519	0.1174	11.87	82.82	94.69	0.0264	0.1991	20
30	5.557	0.01042	6.776	0.1476	13.88	82.03	95.91	0.0306	0.1981	30
40	7.032	0.01051	5.447	0.1836	15.89	81.22	97.11	0.0346	0.1972	40
50	8.804	0.01060	4.421	0.2262	17.92	80.40	98.32	0.0386	0.1964	50
60	10.90	0.01069	3.626	0.2758	19.96	79.57	99.53	0.0426	0.1958	60
70	13.40	0.01079	2.993	0.3342	22.02	78.71	100.73	0.0465	0.1951	70
80	16.31	0.01088	2.492	0.4012	24.09	77.84	101.93	0.0504	0.1947	80
86	18.28	0.01094	2.242	0.4461	25.34	77.31	102.65	0.0527	0.1944	86
90	19.69	0.01098	2.091	0.4783	26.18	76.95	103.12	0.0542	0.1942	90
100	23.60	0.01109	1.765	0.5666	28.27	76.03	104.30	0.0580	0.1938	100
110	28.09	0.01119	1.499	0.6671	30.40	75.08	105.47	0.0617	0.1935	110
120	33.20	0.01130	1.281	0.7808	32.53	74.10	106.63	0.0654	0.1933	120
Sulfur Dioxide, SO_2 (Refrigerant-764)										
−40	3.136	0.01044	22.42	0.04460	0.00	178.61	178.61	0.00000	0.42562	−40
−20	5.883	0.01063	12.42	0.08052	5.98	175.09	181.07	0.01366	0.41192	−20
−10	7.863	0.01072	9.44	0.10593	9.16	172.97	182.13	0.02075	0.40544	−10
0	10.35	0.01082	7.280	0.13736	12.44	170.63	183.07	0.02795	0.39917	0
5	11.81	0.01087	6.421	0.15574	14.11	169.38	183.49	0.03155	0.39609	5
10	13.42	0.01092	5.682	0.17599	15.80	168.07	183.87	0.03519	0.39306	10
20	17.18	0.01102	4.487	0.22287	19.20	165.32	184.52	0.04241	0.38707	20
30	21.70	0.01114	3.581	0.27925	22.64	162.38	185.02	0.04956	0.38119	30
40	27.10	0.01126	2.887	0.34638	26.12	159.25	185.37	0.05668	0.37541	40
50	33.45	0.01138	2.348	0.42589	29.61	155.95	185.56	0.06370	0.36969	50
60	40.93	0.01150	1.926	0.51921	33.10	152.49	185.59	0.07060	0.36405	60
70	49.62	0.01163	1.590	0.62893	36.58	148.88	185.46	0.07736	0.35846	70
80	59.68	0.01176	1.321	0.75700	40.05	145.12	185.17	0.08399	0.35291	80
86	66.45	0.01184	1.185	0.84388	42.12	142.80	184.92	0.08783	0.34954	86
90	71.25	0.01190	1.104	0.90580	43.50	141.22	184.72	0.09038	0.34731	90
100	84.52	0.01204	0.9262	1.07968	46.90	137.20	184.10	0.09657	0.34173	100
110	99.76	0.01219	0.7804	1.28139	50.26	133.05	183.31	0.10254	0.33611	110
120	120.93	0.01236	0.6598	1.51561	53.58	128.78	182.36	0.10829	0.33046	120

Methyl chloride is moderately inflammable and can be explosive between limits of 8.1 and 17.2% by volume in air. As a halocarbon it forms toxic decomposition products in the presence of open flames and very hot surfaces.

Methyl choride is a sweet-smelling vapor that is not seriously offensive and so should have odorous agents added to it to give active warning of its presence since it is a mild anesthetic and is toxic in concentrations above 2% by volume on continuing exposure.

Dichloromonofluoromethane ($CHCl_2F$), refrigerant-21, also called Freon-21, is a vacuum-type refrigerant. It exerts 5.24 psia pressure at 5 F and 31.23 psia at 86 F. It is somewhat less toxic than methyl chloride and is practically noninflammable and nonexplosive.

Dichlorotetrafluoroethane ($C_2Cl_2F_4$), refrigerant-114a and -114, is a low-pressure refrigerant of the halogenated-hydrocarbon class. The particular isomer of dichlorotetrafluoroethane, called 114a, has a molecular arrangement CCl_2FCF_3, and for this the boiling point at 14.7 psia is 37.6 F; at 5 and 86 F the respective saturation pressures are 7.03 psia and 22.7 psia. The specific volume of saturated vapor at 5 F is 4.04 ft^3/lb, and the ratio of specific heats (k) is 1.01. Refrigerant-114a is characterized by low toxicity, approaching that of refrigerant-12. Moreover, for a vacuum refrigerant, the required compressor displacement per ton is very moderate, being only 18.8 cfm at 5 F evaporator temperature.

Methylene chloride (CH_2Cl_2), refrigerant-30, also known as Carrene No. 1 or dichloromethane, is another vacuum refrigerant (Fig. 15-1) suitable for centrifugal compressors. It has a mild odor similar to that of chloroform. It is nonexplosive and relatively noninflammable, although, like the other halogenated hydrocarbons, it breaks down in the presence of flames and very hot surfaces to form toxic products. It is toxic in concentrations of over 5% by volume.

Chloropentafluoroethane or *monochloropentafluoroethane* ($CClF_2CF_3$), refrigerant-115, can reach a temperature as low as −37.7 F (—38.7°C) before dropping into subatmospheric pressures. At 5 F (−15°C) its saturation pressure is 38.6 psia and at 86 F (30°C) it is 150.5 psia. This refrigerant is important as being a component of two refrigerant azeotropes, namely, refrigerant-502 and refrigerant-504 (Table 15-2 and Section 15-7).

15-7. AZEOTROPE HALOCARBON REFRIGERANTS

An azeotrope is a mixture of two or more chemicals which maintain the same ratio of constituent chemicals in both the liquid and vapor phase. For example, the constituents of an azeotropic mixture cannot be separated by distillation. A number of the halocarbon azeotropes, with the weight concentrations of the constituent refrigerants listed, appear in Table 15-2. The azeotropic condition represents almost an abnormality because when two miscible but different liquids are mixed it is most usual on boiling for a higher concentration of the more volatile constituent to appear in the vapor phase than exists in the liquid phase. The true azeotrope acts as though it is a distinct and new substance not subject to separation, possessing its own pressure-temperature relationship independent of that possessed by either of its constituents. The refrigerant azeotropes essentially meet these qualifications; in fact, they exhibit a higher saturation pressure at any temperature than either constituent component and essentially behave as unique substances over their evaporation, compression, and condensation ranges. Numbers for azeotrope refrigerants are all assigned in the 500s for their ASHRAE designations.

The question naturally arises as to why an azeotrope mixture is used in preference to the underlying constituents. Some reasons may be apparent but performance experience has to show the others. In this connection let us consider refrigerant-502 and refrigerant-503 which have gained substantial usage.

Refrigerant-502 is a mixture of 48.8% refrigerant-22 and 51.2% refrigerant-115 by weight. It is now widely used for moderately low-temperature applications such as are encountered in frozen-food cases and display cabinets. A $p\mathbf{h}$ chart of this refrigerant appears as Plate VI. Its saturation pressure at −40 F (−40°C) is 18.8 psia and its specific volume 2.05 ft^3/lb. This pressure is higher and the specific volume is lower than are the corresponding values for refrigerant-22, one of its constituents (15.22 psia and 3.296 ft^3/lb). For the refrigerant-115 constituent, the specific volume for refrigerant-502 is slightly larger but the pressure-compression ratio is favorable for the refrigerant-502. It also happens that the discharge temperature following compression is lower for refrigerant-502 than for refrigerant-22. The stability of refrigerant-502 and its interaction with lubricating oil appears less serious than is the case with refrigerant-22.

The azeotrope refrigerant-503 is composed of 40.1% refrigerant-23 and 59.9% refrigerant-13 by weight. At 14.7 psia it boils at −127.6 F (−88.7°C) while at −40 F (−40°C) its saturation pressure is 122.6 psia and the specific volume for saturated vapor is 0.3474 ft^3/lb. This volume is smaller than exists for either of its constituent refrigerants at the same saturation temperature, and consequently greater refrigeration capacity is possible for the same compressor displacement with refrigerant-503 than with refrigerant-23 or refrigerant-13. For low-temperature application, refrigerant-503 is now supplanting refrigeration-13 as the preferred refrigerant.

The first azeotrope employed was refrigerant-500, also known as Carrene No. 7. Its azeotropic composition is 73.8% refrigerant-12 and 26.2% refrigerant-152a by weight. It possesses saturation pressures of 31.1 psia at 5 F (−15°C) and 127.6 psia at 86 F (30.0°C).

15-8. TOXICITY OF REFRIGERANTS

All gaseous substances, with the possible exception of air, are to a certain extent toxic. There are various degrees or levels of toxicity, since some substances produce toxic effects and danger to life merely because they exclude or reduce the amount of oxygen necessary for living processes, while others are truly poisonous. Carbon dioxide, for example, is sometimes considered harmless (since it is exhaled every time a person breathes) in amounts ranging from 3.5 to 4% by volume. Nevertheless, if a person enters an atmosphere containing 6% CO_2, breathing becomes difficult, and with 10% CO_2 in air, loss of consciousness can quickly result. The danger in this case does not result from the fact that oxygen is excluded, but from the fact that the regulating mechanism in the respiratory passages becomes disturbed and may cease to function. However, refrigerant-12 (Freon-12), which is remarkably nontoxic, can be breathed safely in concentrations up to 20% by volume without seriously upsetting normal life processes. At the toxic end of the scale are gases, such as sulfur dioxide, which are so irritating to the mucosa that, even in concentrations of less than 1% by volume, they cause surface damage. This condition is also true with ammonia if its concentration exceeds the amount which can readily be diluted by surface moisture. Still other gases, when breathed, are absorbed into the system and can cause damage to internal organs. A case in point is the manner in which protracted exposures to carbon tetrachloride cause kidney damage.

Toxicity is related (1) to the nature of the material, (2) to the relative amount of the chemical in the air, and (3) to the length of time during which

Table 15-8 Relative Toxicity and Safety of Refrigerants

Refrigerant	Toxicity Number of National Fire Underwriters	LETHIALITY Kills or Seriously Injures: Duration of Exposure (h)	LETHIALITY Kills or Seriously Injures: Percentage by Volume	Flammable or Explosive (% by vol.)	US ASA B-9.1 Code Group
Ammonia (R 717)	2	0.5	0.5–0.6	16–25	2
Butane (R 600)	5	0.2	37.5	1.6–6.5	3
Carbon dioxide (R 744)	5	0.5–1	5–7*	not	—
Ethane (R 170)	5b	2	37	3.3–10.6	3
Methane (R 50)	5b	2	>20	4.9–15	3
Methyl Choride (R 40)	4	2	2–2.5	8.1–17.2	2
Propane (R 290)	5b	2	>20	2.3–7.3	3
Refrigerant-11	5	2	10	not	1
Refrigerant-12	6	2	28.5–30.4	not	1
Refirgerant-13	6	2	>20	not	1
Refrigerant-13B1	6	2	>20	not	1
Refrigerant-14	6	2	>20	not	1
Refrigerant-21	4–5	0.5	10	not	1
Refrigerant-22	5a			not	1
Refrigerant-30	4a			not	1
Refrigerant-113	4–5			not	1
Refrigerant-114A	6	2	20.1	not	1
Refrigerant-500	5a			not	1
Refrigerant-502	5a			not	1
Sulfur dioxide (764)	1	0.08	0.7	not	2

*Not toxic, but heavy concentrations upset the regulatory mechanisms of breathing and cause suffocation.

exposure (breathing of the vapor) takes place. For many years The National Board of Fire Underwriters has coordinated a program to measure the relative toxicity of refrigerants. The board has set up a classification scheme with numbers from 1 to 6, the refrigerants in classification 1 are extremely toxic or dangerous whereas those in classification 6 are essentially non-toxic. Each number has definite specifications; for example, 4 applies to refrigerant vapors which at concentrations of 2 to $2\frac{1}{2}\%$ by volume under durations of exposure approximating 2 h are lethal or cause serious injury. Vapors in group 5 are less toxic than those in group 4 but more toxic than those in group 6, while 5a and 5b represent low toxicity levels comparable to 6 but about which there is uncertainty in the light of available data. Table 15-8 lists the toxicity and combustibility hazards of many refrigerants.

In addition to possible danger from the refrigerant itself, a second type of hazard arises in case of fire, because certain refrigerants can burn or are explosive, and other refrigerants, in the presence of an open fire or of a hot incandescent surface, break down and can form poisonous by-products. The halogenated hydrocarbons are particularly dangerous; they readily form toxic combustion products in the presence of flames because of the chlorine and fluorine they contain. The B-9.1 Safety Code of ASHRAE and the American Standards Association classifies the hazards of refrigerants unter three numbers:

1. Flammable vapor-air mixtures, only dangerous within narrow limits of concentration.
2. The refrigerant vapor is under sufficiently high pressure that it tends to extinguish flame.
3. Because of its offensive odor or because of performance failure, the loss of refrigerant is usually known before a combustible concentration occurs.

15-9. REFRIGERANT LEAK DETECTION AND GAS PURGERS

In spite of the care with which refrigerant systems are fabricated, there is always some possibility that leakage may occur. Leakage of refrigerant is expensive and can also be objectionable, particular in the case of refrigerants with unpleasant odor and those toxic in nature. Various methods have been developed to determine points of leakage. Ammonia leaks can often be found by smell or by burning sulfur candles or wicks, which generate a dense cloud of white smoke in the presence of ammonia vapor.

Leaking halogenated hydrocarbon refrigerants were formerly extensively detected by means of the *halide torch*. In this torch an alcohol flame is arranged to burn in the presence of copper and, when air with traces of a halogen gas passes through the flame, the flame takes on a decidely greenish color. The torch unit is provided with a search tube or hose which can be moved into the region of each suspected leak point. The combustion process aspirates air through the hose and into the flame which can then be observed for color change.

The *electronic leak detector* for halogen gases makes use of a specially designed platinum diode for which the positive ion emission greatly increases when it contacts refrigerant molecules containing chlorine, fluorine, or bromine. These instruments are very sensitive and can detect gas leakages of a very minute nature. However, the calibrated scale of the instrument is responsive as well to higher concentrations of gas, thereby indicating leaks of greater magnitude. The instruments are portable and direct-reading, operating from either batteries or a direct power source. As is true with the halide torch, the search tube must be brought adjacent to each suspected leak point.

In high concentrations of refrigerant gases, except possibly refrigerant-11, -12 or, -13, an appropriate gas mask should always be worn.

The reverse problem also exists, since it is possible for extraneous noncondensable gases to intermix with the refrigerant inside a refrigeration system. In the case of vacuum systems, whenever there are leaks in the piping, air, which contains moisture, enters the system to cause trouple. Noncondensable gases also appear in any closed system whenever there is breakdown of the refrigerant or lubricant as a result of high compression temperatures or from chemical reaction. Chemical reaction can take place when unwarranted moisture in a system promotes hydrolysis and subsequent gas formation. Air is also frequently present because of improper removal when the system is started. Noncondensable gases most frequently collect in the condenser, and their presence is indicated whenever the condenser pressure is found to be far out of agreement with the pressure which the temperature of the cooling water or circulating air shows should be indicated. When this is the case, the noncondensable gases must be purged out, usually with some loss of refrigerant. However, excessive condenser pressures are not always caused by noncondensable gases, since dirty tubes, which interfere with effective cooling, cause high pressures, and an overloaded condenser, fed with more refrigerant than that for which it was designed, can also show high pressures.

A purger of proprietary design for exhausting noncondensable gases with minimum loss of refrigerant is illustrated in Fig. 15-5. This uses refrigerant from the main system, fed into a coil at *A* and withdrawn at *B*, to cool the gas-laden vapor in chamber *C*. This cooling condenses out refrigerant which drains back to the receiver, while the chamber *C* holds the gases at high pressure. These gases can be discharged to waste. In operation, refrigerant feeding through the coil in chamber *C* chills this chamber, and liquid refrigerant is drawn into the chamber through *E*. After chamber *C* is filled with liquid, valve *F* is opened slightly and gas and refrigerant vapor pass into the space above the liquid refrigerant. If noncondensable gas is present, it collects in chamber *C* above the liquid, the level of which falls as the gas flows in. If no gas is present the refrigerant vapors merely condense in space *C* and the liquid level remains constant. When *C* does fill with gas, the waste valve *D* is cracked and the gas is discharged to waste. At the same time, the liquid level should be observed through the gage glass to prevent loss of refrigerant.

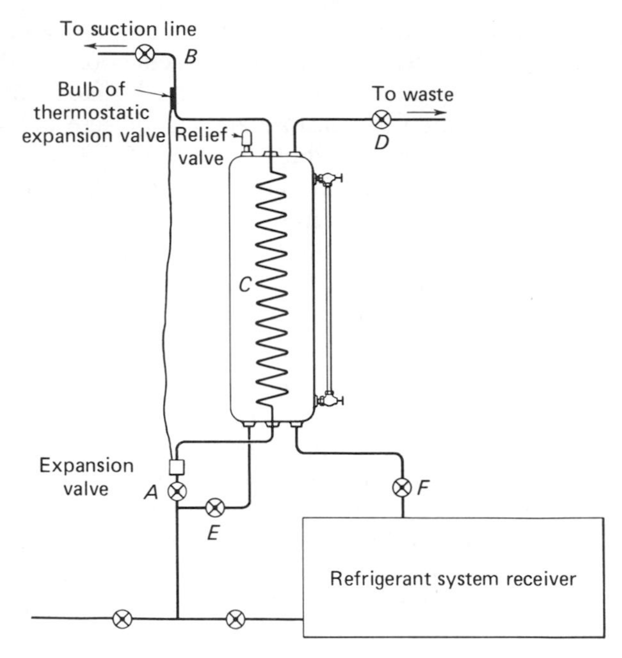

FIGURE 15-5. Noncondensable gas purger for refrigeration system.

This level will rise as the gas is forced out. When no more gas collects in *C*, the refrigerant feed at *A* is closed. Chamber *C* then warms and the refrigerant drains back through *E*. All valves on the purger can then be closed completely, isolating it from the system. This type of purger must always be placed higher than the receiver or condenser.

Refrigerant Driers. The presence of water and water vapor in a refrigerant system is so noxious that steps must always be taken to remove or isolate this moisture. Water vapor can enter a system whenever it is opened for repairs or adjustment. It is always present on metal surfaces during installation, particularly on equipment that was not properly dehydrated during manufacture, as well as in tube and pipe runs. Traces of water also are present in the refrigerant charge. Systems that operate at subatmospheric pressure are subject to water vapor entry from the air, even though only trace amounts leak into the system. To remove water and water vapor, whatever its source, it is customary to install a *drier*. This is usually located in the liquid line before the expansion valve to reduce the possibility of a freeze-up there. Driers are manufactured in differing design arrangements but all of them contain a desiccant material which traps out and holds the moisture being carried with the liquid refrigerant.

The most common desiccant materials used in driers are activated alumina, silica gel, zeolite, and anhydrous calcium sulfate. A number of propriatory-combination driers make it possible for the drier also to serve as a filter to entrap particles and absorb acidic material should this develop in the system.

15-10. EXPANSION VALVES

Every mechanical vapor refrigeration system operates in two distinct pressure regimes. In the low-pressure (l-p) part of the system, the useful cooling or refrigeration takes place as the refrigerant vaporizes in the evaporator. The pressure there, for any refrigerant, is set by the temperature which the system must maintain to meet its cooling objectives. The high pressure (h-p) part of the system, involving the compressor discharge piping and the condenser, has its temperature set at a value slightly higher than the temperature of the available cooling water or ambient air. For any given refrigerant, the resulting condensing temperature sets the high pressure of the system. At the outlet of the evaporator it is the compressor that raises the vapor from the low- to the high-pressure regime.

At the inlet or evaporator feed point the expansion valve must perform the function of dropping the liquid refrigerant to evaporator pressure. In this process some flash gas forms while the remainder of the liquid reaches evaporator temperature. It is essential that the expansion valve pass sufficient liquid to provide the cooling requirements of the evaporator, in terms of the liquid needed for evaporation. At the same time the flow must be limited to keep the low side from being over supplied or flooded with liquid. In properly selecting an expansion device (valve), consideration must be given to the relative storage capacity for refrigerant on both the high- and low-pressure sides of the system.

Manual expansion valves are specially built valves, with needle- or cone-pointed stems, which can be opened or closed by hand in amount to feed adequate refrigerant. This arrangement is satisfactory only for systems in which the cooling load is relatively constant.

Capillary tubes are often used with domestic refrigerators and small air-conditioning units in place of a more complex expansion valve. These consist of very small bore (capillary) tubes of appropriate length installed between the high- and low-pressure parts of the system. Such a tube offers a restriction to flow but can pass much greater amounts of liquid than of gas. The tube is sized to allow liquid flow up to the cooling capacity of the unit. On intermittent operation during shutdown, when all of the liquid has passed from the high to the low side, gas will continue to flow but at such a limited rate that the high and low pressures equalize at a very slow rate. Capillary tubes range in bore from 0.66 mm (0.026 in.) to 2.3 mm and are installed in lengths from some 50 to 350 cm.

Automatic high-side valves are built along the lines of a float-operated steam trap and deliver all the liquid coming from the condenser to the evaporator. The refrigerant charge in a system with high-side float valve must be such that the liquid can largely be stored in the evaporator without danger of sending liquid slugs over to the compressor.

Automatic low-side float valves operate to maintain a definite level in an evaporator of the flooded type. The valve chamber is equalized to the evaporator with pipes on top and bottom and is placed at about the liquid level to be held in the evaporator. As refrigerant in the evaporator vaporizes, the float level drops to permit liquid to enter the evaporator until the proper level is regained.

Thermal-expansion valves are of several types, in all of which a thermostatic bulb, clamped to the side of the suction pipe or actually mounted inside this pipe, reacts to the temperature of the suction gas leaving the evaporator, if the rate of refrigerant flow to the evaporator is inadequate, it will be indicated by a leaving refrigerant temperature much higher than the saturation refrigerant temperature in the evaporator. This relatively high temperature reacts on the fluid in the thermostatic bulb to increase its pressure. This increased pressure is transmitted back to a bellows, or diaphragm chamber, and then operates against a spring resistance and push rod to open the refrigerant needle valve a greater amount. These controls can be adjusted after installation on a given job or may be set for a definite temperature before installation. From 3 to 20 deg of superheat may be necessary to give desired rates of flow. The thermostatic fluid used in the bulb is frequently the same refrigerant used in the system (thus ammonia liquid is loaded into the expansion-valve bulb of an ammonia system). Figure 15-6 is a section through such a valve.

Automatic diaphragm expansion valves (constant-pressure type) have spring-loaded diaphragms which are acted on by the evaporator pressure (Fig. 15-7). As this pressure lowers, indicating insufficient refrigerant in the evaporator, the pressure on the diaphragm cannot prevent the spring on the far side from moving the diaphragm and this motion in turn is used to increase the reducing valve opening; increased refrigerant flow is thus permitted. This

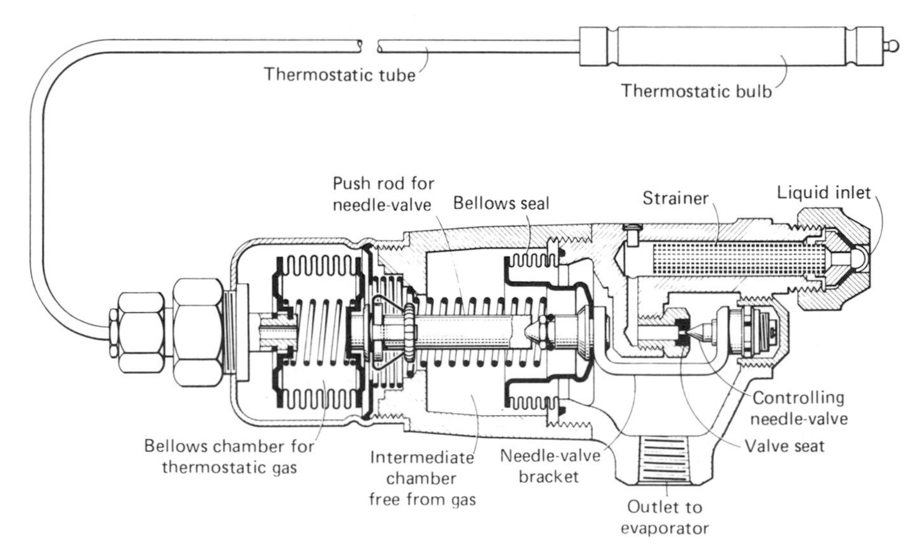

FIGURE 15-6. A thermostatic expansion valve.

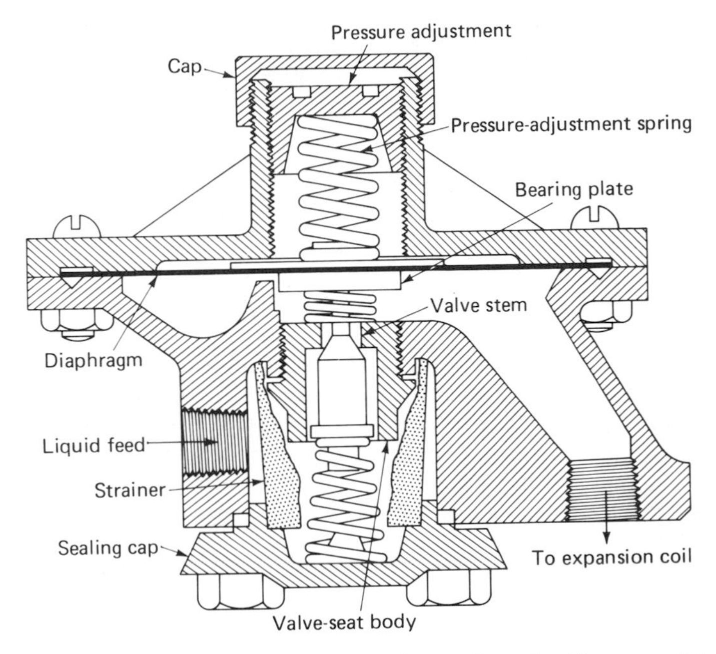

FIGURE 15-7. Constant-pressure type of expansion valve (diagrammatic).

type of valve operates to keep the suction pressure essentially constant and is called a constant-back-pressure valve.

Automatic electrically operated valves connected from thermostatically controlled relay circuits are also used as expansion valves. In these a solenoid arrangement usually holds the valve open against a spring which closes the valve when the electric circuit is broken.

Example 15-1. A 20-ton (70.3-kW) refrigeration system using refrigerant-22 operates at an evaporating temperature of 20 F (−6.67°C) and condensation takes place at 95 F (35°C). Assume that essentially dry-saturated vapor enters the compressor. (a) Find the average refrigerant flow in pounds per minute that a constant-pressure type expansion valve must pass when the system is operating at its normal capacity. (b) Compute the compressor size for a two-cylinder hermetic compressor serving this unit if it runs at 1800 rpm and has a volumetric efficiency of 86%.

Solution: (a) Table 15-6 shows that for saturated vapor leaving the evaporator $\mathbf{h}_R = 106.383$ Btu/lb and for feed liquid at 95 F, $\mathbf{h}_f = 37.704$ Btu/lb. Refrigeration effect [Eq. (14-9)] is

$$Q_R = 106.383 - 37.704 = 68.679 \text{ Btu/lb}$$

For 20 tons, refrigerant flow per minute is

$$w_R = \frac{(20)(200)}{68.679} = 58.24 \text{ lb/min} \qquad \textit{Ans.}$$

(b) Specific volume of saturated refrigerant at 20 F from Table 15-6 is 0.9363 ft^3/lb. The theoretical piston displacement is thus

$$PD = 58.24 \times 0.9363 = 54.53 \text{ cfm}$$

Make use of Eq. (14-3), assuming a bore-stroke ratio of unity $d \equiv S$.

$$54.53 = C\frac{\pi d^2 SN}{4 \times 1728}\eta_v = (2)\frac{\pi d^3(1800)}{4 \times 1728}0.86$$

$d = 3.39$ in. (say, $3\frac{25}{64}$ in.) and $S = 3\frac{25}{64}$ in. *Ans.*

$d = S = 86.1$ mm *Ans.*

Example 15-2. A two-cylinder hermetic refrigeration system operating at 1800 rpm using refrigerant-12 produces 70.3 kW of cooling when operating at an evaporating temperature of −6.67°C (20 F) and a condensing temperature of 35°C (95 F). Essentially dry-saturated vapor enters the compressor and the liquid is not subcooled. (a) Find the average refrigerant flow in kilograms per minute through the constant-pressure type expansion valve at rated cooling capacity. (b) Using SI (metric) units, compute the compressor size to produce this capacity if the volumetric efficiency is 86%. (c) Compare the compressor size of this unit with that of Example 15-1, realizing that the unit of Example 15-1 operates under exactly the same conditions but employs refrigerant-22.

Solution: Use Table 15-5 and for the respective evaporator and condenser temperatures read $\mathbf{h}_R = 79.385$ and $\mathbf{h}_f = 29.901$ Btu/lb.

$$79.385 \times 2.326 = 184.65$$

and

$$29.901 \times 2.326 = 69.55 \text{ kJ/kg}$$

$$Q_R = 184.65 - 69.55 = 115.10 \text{ kJ/kg}$$

The refrigerating capacity of 70.3 kW = 70.3 kJ/s.

$$w'_R = 70.3 \text{ kJ/s} \div 115.1 \text{ kJ/kg} = 0.6108 \text{ kg/s}$$

$w_R = 0.6108 \times 60 = 36.65$ kg/min *Ans.*

(b) The specific volume of saturated refrigerant-12 at 20 F (−6.67°C) is 1.0988 ft^3/lb or $1.0988 \times 62\,428 = 68\,596$ cm^3/kg. Make use of Eq. (14-3) employing SI (metric) units and a bore-stroke ratio of unity, $d \equiv S$. The theoretical piston displacement (PD) for 36.65 kg/min becomes

$$PD = 68\,596 \text{ cm}^3/\text{kg} \times 36.65 \text{ kg/min} = 2\,514\,000 \text{ cm}^3/\text{min}$$

$$2\,514\,000 = C\frac{\pi}{4}d^2 SN\eta_v = (2)\left(\frac{\pi}{4}\right)d^3(1800)(0.86)$$

$d = 10.11$ cm $= 101.1$ mm $\quad S = 101.1$ mm *Ans.*

Observe that for the same operating conditions and capacity (see Example 15-1), larger piston displacement and increased compressor size are required with refrigerant-12 as compared with refrigerant-22. Horsepower requirements for the two refrigerants are essentially equivalent.

15-11. SUCTION LINE HEAT EXCHANGERS

The liquid refrigerant leaving the condenser is at high temperature relative to the evaporator. Subcooling of this liquid improves performance and reduces the amount of flash gas formed as the liquid passes through the expansion valve. In the suction-line heat exchanger, the liquid refrigerant on its way to the expansion valve passes in counterflow direction over the heat-transfer surface chilled by the cold gas from the evaporator on its way to the compressor. This makes possible liquid sub-cooling from merely warming the gas entering the compressor. This warming increases the specific volume of the gas entering the compressor and thereby slightly reduces its capacity. In addition the compressed gas will leave the compressor more highly superheated than would otherwise be the case. These disadvantages are small compared to the benefits of the sub-cooling, and in addition the heat exchanger can vaporize any inadvertent liquid carryover from the evaporator and prevent its causing damage to the compressor.

The heat exchangers are made in different construction patterns. In most, the vapor line is made of a straight central copper tube of substantial cross section, to reduce pressure drop, and often is provided with parallel internal fins to aid heat transfer from the gas to the metal wall. A small-diameter copper coil for the liquid line is wrapped, in close contact, around the tube and usually is brazed to it as well. End connections are provided for both the tube and the coil, and all are mounted in a suitable casing, brazed or soldered to make an integral unit.

Example 15-3. A refrigeration system using refrigerant-22 operates at an evaporator temperature of −20 F (−28.9°C), 24.845 psia, and a condensation temperature of 120 F (48.9°C), 274.6 psia. A suction-line heat exchanger is used in which the temperature of the suction gas is raised to 65 F (18.3°C). (a) Compute the required refrigerant flow per ton of refrigeration when the heat exchanger is not in operation, the liquid is not subcooled, and dry-saturated vapor leaves the evaporator. (b) With the heat exchanger in use for the conditions indicated, find the temperature reached by the subcooled refrigerant. (c) Compute the refrigerant flow per ton of refrigeration with the heat exchanger in operation.

Solution: (a) From Table 15-6, read $\mathbf{h}_g$ at −20 F as 102.415 Btu/lb and $\mathbf{h}_f = 45.705$ Btu/lb at 120 F.

$$Q_R = 102.415 - 45.705 = 56.710$$

$$w_R = \frac{200}{56.710} = 3.42 \text{ lb/min} \cdot \text{ton of refrigeration} \qquad \textit{Ans.}$$

(b) To find the enthalpy increase in the vapor, read the enthalpy at 65 F and 24.845 psia from Plate V as 115.5 Btu/lb and note above that saturated vapor leaving the evaporator is at 102.4 Btu/lb.

$$\Delta\mathbf{h} = 115.5 - 102.4 = 13.1 \text{ Btu/lb}$$

With trivial radiation loss or gain in the exchanger, the liquid in turn decreases by the same $\Delta\mathbf{h}$, as it cools; thus

$$45.705 - \Delta\mathbf{h} = 45.7 - 13.1 = 32.6 \text{ Btu/lb}$$

From Table 15-6 note that this enthalpy corresponds to liquid at 78.3 F. This is the temperature reached by the subcooled liquid prior to entering the expansion valve.

Ans.

(c) The useful cooling in the evaporator is

$$\Delta\mathbf{h} = 102.4 - 32.6 = 69.8 \text{ Btu/lb} = Q_R$$

$$w_R = \frac{200}{69.8} = 2.87 \text{ lb/min} \cdot \text{ton of refrigeration} \qquad \textit{Ans.}$$

To find w_R it is not necessary to find the enthalpy of the subcooled liquid since the same result appears when we consider that the warming effect of the vapor in the heat exchanger is the equivalent of subcooling the liquid and

$$Q_R = \mathbf{h}_{v65\text{F}} - \mathbf{h}_{f120\text{F}} = 115.5 - 45.7 = 69.8 \text{ Btu/lb}$$

Thus unless the liquid temperature is actually needed for some other purpose, there is no need to calculate it to find w_R.

PROBLEMS

15-1. A refrigeration system using refrigerant-22, chlorodifluoromethane condenses refrigerant at 100 F (37.8°C) with no subcooling. Evaporation takes place at 0 F (−17.8°C) and essentially dry-saturated refrigerant enters the compressor. (a) Find the refrigerant flow in pounds per minute through the expansion valve per ton of refrigeration produced. (b) Find the cylinder dimensions required for a 4-cylinder reciprocating compressor directly connected to a 1750-rpm motor if the required capacity is 100 tons and the volumetric efficiency is 86%.

Ans. (a) 3.07; (b) $d = S = 5.36$ in., or select $5\frac{3}{8} \times 5\frac{3}{8}$ in.

15-2. Rework Problem 15-1 with no change except that the azeotrope, refrigerant-502, is employed. For this azeotrope $\mathbf{h}_f$ at 100 F (230.9 psia) is 37.563 Btu/lb, the $\mathbf{h}$ and v of dry-saturated vapor at 0 F (−17.8°C) and 45.78 psia are 77.690 Btu/lb and 0.8814 ft^3/lb. *Ans.* (a) 4.98; (b) $d = S = 5.431$ in., or select $5\frac{7}{16} \times 5\frac{7}{16}$ in.

15-3. Make use of the $\mathbf{ph}$ chart for refrigerant-22, and using appropriate refrigerant flow, compute for Problem 15-1 (a) the isentropic horsepower needed to produce 100 tons of refrigeration under the operating conditions specified. (b) Find the probable shaft horsepower needed for the driving motor. For this large unit, a value of 1.41 is suggested.

Ans. (a) 135 hp; (b) 190 hp

15-4. A 100-ton refrigeration system operates with a 0 F (−17.8°C) evaporator and with condensation at 100 F (37.8°C) and employs refrigerant-502. For these same operating conditions it was shown in Problem 15-2 that 4.98 × 60 lb/ton · h of refrigerant flow was needed and the enthalpy of the dry-saturated vapor entering the compressor is 77.7 Btu/lb at 0 F and 45.78 psia. (a) Compute the isentropic horsepower required for the compression phase. (b) Find the probable shaft horsepower of the driving motor if a factor of 1.41 is considered applicable. (c) Find the kilowattage drawn from the power line if the motor is 85% efficient.

Ans. (a) 148 hp; (b) 208 hp; (c) 183 kW

15-5. A four-cylinder packaged unit using refrigerant-12 rotates at 1750 rpm. On a particular test it produced 32 000 Btuh of cooling and drew 2.92 kW with an evaporator temperature of 46 F. (a) Compute the refrigerant flow in pounds per minute, assuming no liquid subcooling and that the total evaporator capacity is realized in useful cooling. (b) For this refrigerant flow and with 105 F (40.6°C) condensation, compute the isentropic horsepower and isentropic kilowattage required. (c) Estimate the probable shaft horsepower and shaft kilowattage required. (d) Assuming a motor efficiency of 80%, compare the kilowatt input with the test value.

Ans. (a) 10.73; (b) hp = 1.82, kW = 1.36; (c) 2.73, 2.04; (d) kW = 2.55 (good check considering no allowance made for condenser fan power or heat gains through piping)

15-6. A four-cylinder packaged unit using refrigerant-22 rotates at 1750 rpm. It is rated to produce 32 000 Btuh of cooling when operating at an evaporator temperature of 46 F (7.78°C) and with condensation at 105 F (40.6°C) and no subcooling. (a) For the conditions indicated, compute the minimum refrigerant flow needed in pounds per minute. (b) Using data from Plate V compute the isentropic horsepower and isentropic kilowattage needed to produce this refrigeration. (c) Use a factor of 1.5 and estimate the shaft horsepower and shaft kilowattage needed. (d) If the efficiency of this small motor is 80%, find the kilowattage drawn from the power lines.

15-7. Consider the unit of Problem 15-6 operating to produce 32 000 Btuh of cooling with the same evaporating and condensing temperatures but with a suction-line heat exchanger added. In the heat exchanger the suction gas is warmed to 70 F and the liquid is subcooled from 105 F to a lower temperature. Compute the items called for in parts (a), (b), and (c) of Problem 15-6 under liquid subcooling conditions.

Ans. (a) 7.39 lb/min; (b) 1.76 hp, 1.31 kW; (c) 2.64 hp, 1.97 kW

15-8. Refrigerant-502 is used in a system with evaporation at −20 F (−28.9°C), 30.01 psia, and condensation at 120 F (48.9°C), 297.4 psia. A heat exchanger subcools the liquid from the condenser at the expense of warming the evaporator gas so that it enters the compressor at 65 F (18.3°C) with an enthalpy of 89.03 Btu/lb. Enthalpy of liquid after condensation at 120 F is 43.765 Btu/lb. The specific volume of the vapor at entry to the compressor is 1.6222 ft^3/lb. Make use of additional data from Plate VI. (a) Compute the net refrigeration effect per pound of refrigerant circulating. (b) Compute the refrigerant circulation per ton in pounds per minute and as theoretical piston displacement in cfm. (c) Compute the isentropic horsepower per ton.

Ans. (a) 45.265 Btu/lb; (b) 4.418, 7.167 cfm; (c) 2.01

15-9. A 200-ton refrigeration unit, which operates with ammonia, provides the cooling for a factory building. The evaporator operates at 41 F (5°C) to chill water, used in the cooling units of the plant. The condenser operates at 115 F (46.1°C) and there is minimal or no subcooling of the liquid. (a) Find the refrigerant flow in pounds per minute for the conditions indicated. (b) Compute the ideal (isentropic) horsepower for 200-ton capacity. (c) Estimate the probable shaft horsepower and shaft kilowattage required. (d) Find the probable line power demand.

15-10. A refrigeration unit operating with ammonia provides chilled water for use in a factory building. The evaporator operates at 5°C (41 F) and ammonia condenses at 46.1°C (115 F). There is minimal to zero subcooling of the liquid refrigerant. The unit is required to produce 350 kW of chilled water cooling. For the conditions indicated, (a) find the refrigerant flow in kilograms per hour; (b) compute the ideal (isentropic) kilowattage; and (c) estimate the probable shaft kilowattage needed. (d) Estimate the kilowattage drawn from the power lines for these operating conditions. *Suggestion*: select values from Table 15-3 and Fig. 15-3 and convert to SI (metric) units.

Ans. (a) 1200; (b) 59.6; (c) 84 shaft kW; (d) 99 kW

15-11. A certain refrigeration system using dichlorodifluoromethane, refrigerant-12, condenses refrigerant at 100 F (131.7 psia), and the evaporator refrigerant temperature is 36 F (48.12 psia). The liquid at the expansion value is not subcooled, and dry-saturated vapor enters the compressor. Find (a) the enthalpy of the liquid at the expansion-valve inlet; (b) the useful refrigeration in the evaporator, in Btu per pound; (c) the isentropic work of compression, in Btu per pound; (d) the probable shaft horsepower required by this compressor per ton of refrigeration; and (e) the probable condenser heat load per ton of refrigeration.

Ans. (a) 31.1 Btu/lb; (b) 49.9 Btu/lb; (c) 7.9 Btu/lb; (d) 1.05 shaft kW; (e) 14 700 Btuh

15-12. What size of four-cylinder, 1140-rpm, single-acting compressor is required in order to develop 12 tons of refrigeration when the compressor is operating under the conditions of Problem 15-11? Consider the bore-stroke ratio as unity, and the volumetric efficiency as 87%. Round off the answer to the nearest larger sixteenth inch.

Ans. $2\frac{13}{16}$ by $2\frac{13}{16}$ in.

15-13. The bulb of a thermostatic expansion valve is filled with refrigerant-12 and controls an evaporator which is to be held at 20 F in the coils. The expansion valve is actuated and causes refrigerant flow when 6 deg of superheat exist in the return line at the point of attachment of the bulb. Under these conditions, find what pressures exist in the bellows chamber for thermostatic gas (Table 15-5) and near the valve outlet to the evaporator and thus the pressure difference causing the valve to operate.

Ans. $\Delta p = 4.3$ psi

15-14. Work Problem 15-13 with no change except that refrigerant-22 is used in the system and appropriate tabular values for this refrigerant must be used.

Ans. $\Delta p = 6.9$ psi.

REFERENCES

ASHRAE Handbook of Fundamentals, 1972, Chapter 14, "Refrigerants," and Chapter 31, "Refrigerant Tables and Charts." Available from ASHRAE, 345 East 47th St., New York, N.Y.

U.S. Bureau of Standards, *Tables of Thermodynamic Properties of Ammonia*, Circular No. 142 (1945).

"Freon" Products Div. E. I. du Pont de Nemours & Co., *Thermodynamic Properties of Freon-12, Dichlorodifluoromethane*. (Wilmington, Del.: 1956).

"Freon Products Div., E. I. du Pont de Nemours and Co., *Thermodynamic Properties of Freon-22 (Chlorodifluoromethane)* (Wilmington, Del.: 1964).

C. H. Meyers and M. S. Van Dusen, *The Vapor Pressure of Liquid and Solid Carbon Dioxide*, Bureau of Standards Research Paper No. 538, (March, 1933).

E. I. du Pont de Nemours & Co., *Thermodynamic Properties of Trichloromonofluoromethane "Freon-11" At Low Pressures* (Wilmington, Del.: 1942).

"Freon" Products Div. E. I. du Pont de Nemours & Co., *Thermodynamic Properties of "Freon" 502* (Wilmington, Del.: 1969).

National Board of Fire Underwriters' Laboratories, *The Comparative Life, Fire and Explosion Hazards of Common Refrigerants* (1933).

Underwriters Laboratories, *Reports MH2375, MH2630, MH3072.* Available from Underwriters Laboratories, 207 East Ohio St., Chicago, Il. 60611.

United States of America Standards Institute, *Standard Safety Code for Mechanical Refrigeration B9.1-1964.* Available from ASI, 10 E. 40th St., New York, N.Y. 10016.

CHAPTER

16

REFRIGERATION EQUIPMENT AND ITS ARRANGEMENT

16-1. EVAPORATORS

Refrigeration evaporators are classified in regard to the way they are used—direct-expansion or indirect-expansion. A direct-expansion evaporator is one in which the boiling refrigerant, in the evaporator coils, cools the air, or substance being refrigerated, by direct contact. In an indirect-expansion evaporator, water, brine, or some other medium is cooled by the refrigerant, and this medium is pumped or delivered to take up the heat load from the air or product. While both systems have advantages, distribution of refrigeration from a central point by means of chilled water (or by means of brine for low-temperature refrigeration) is most common in nonresidential installations.

Water is well adapted for indirect expansion in comfort air-conditioning systems because temperatures below 32 F (0°C) are not needed. Brines, made up of solutions of calcium chloride ($CaCl_2$), in proper concentrations to prevent freezing, are widely used in industrial systems but aqueous solutions of the glycols and alcohols are also employed. Of the glycols, ethylene and propylene are in most extensive use. Both salt and glycol brines must be provided with inhibitors to prevent internal corrosion of the metal vessels and piping.

Shell and tube heat exchangers are constructed by welding ends (tube sheets) to a length of large-diameter pipe (or a welded shell) and tightly rolling or welding tubes into the previously drilled and reamed tube sheets. The upper part of Fig. 16-1 shows a condenser of shell-and-tube design. The left-end cut-away view shows the tube sheet into which the tubes are rolled. The rolling process mechanically swages the tube wall to lock it in place and make a leak-proof junction with the sheet. Outside of the tube sheet is shown the external header. Heavy machine screws (or studs and nuts) hold the header tightly against a gasket to prevent water leakage. As can be seen, this exchanger is designed for two-pass operation; the water, which enters at the bottom left, flows through tubes in the lower half of the shell to rise at the far end and finally flow back through the tubes in the upper half of the shell. With two-pass flow, water inlet and outlet are at the same end of the exchanger, whereas with single-pass flow inlet and outlet are at opposite ends. By appropriate design of the external headers to direct the water flow, more water passes can be planned if this is desired. Heat transfer on the water side is improved as water velocity is increased, which is one of the reasons why multipass designs are employed. In this condenser the refrigerant vapor

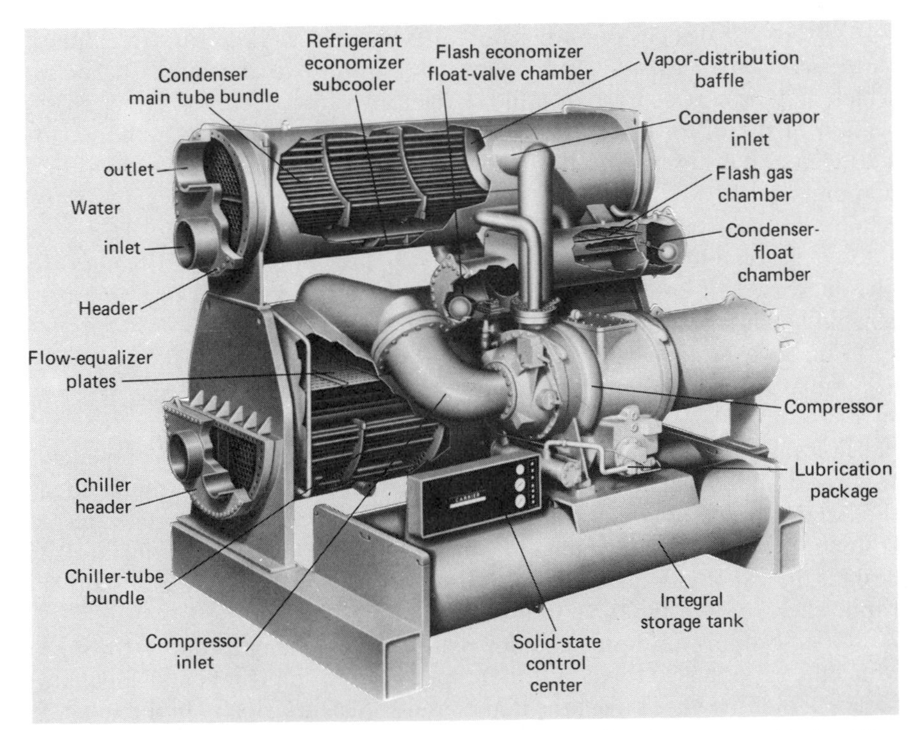

FIGURE 16-1. Cut-away view of a Carrier hermetic, centrifugal-compressor, liquid chiller unit. (Courtesy Carrier Corporation.)

is shown entering the shell at the upper right. The vapor then passes over the outside of the tubes where it gives up heat and condenses. To improve the heat-transfer capacity when using halocarbon refrigerants, it is customary to provide the tubes with low fins to increase heat-transfer area on the refrigerant side. The ends of the tubes are bare of fins where they are rolled into the tube sheets. The outside diameter of the fins does not exceed the tube end diameter so that tubes can be inserted through the tube sheets during construction and replacement. Shells are made of steel with the tubes often made of copper to prevent rust-fouling on the water side.

Evaporator. Directly below the condenser in Fig. 16-1 can be seen the shell-and-tube evaporator. This is fabricated in the same manner as the condenser, described above. This is a two-pass evaporator employing a vertical baffle in the external header. This evaporator and many like it are operated *flooded*, that is, with liquid refrigerant surrounding the tubes. Vapor generated at the tube surfaces causes rapid ebullition in the liquid as vapor rises to the surface, and this action promotes rapid heat transfer. The liquid level must be kept at the proper height in the evaporator; otherwise, liquid carryover may take place into the compressor. Above the tubes of the evaporator is shown a flow equalizer plate. This acts as a liquid-carryover eliminator and takes care of the surges which could occur under certain boiling (evaporating) conditions.

With other designs of evaporators it is sometimes necessary to employ *surge drums* or *accumulators* to permit separation of the liquid particles which may be carried over with the refrigerant vapor. Here the larger cross section of the drum in contrast to connecting piping allows the liquid to settle out and drain back to the evaporator while the vapor passes on to the compressor.

Finned evaporators are frequently used for direct air cooling. The outside of the pipe for such evaporators is supplied with extended flanges to give increased air-contact surface and may be arranged with one or more hairpin bends so as to lengthen the individual circuits.

Finned, as well as bare-pipe, evaporators may be operated flooded with accumulators, may be partly flooded with no accumulator, or may be dry. In dry evaporators the refrigerant liquid supply is controlled at the expansion valve in such amount that, in passing through the coil, it is completely vaporized and is usually somewhat superheated. In contrast to the flooded shell-and-tube design, where the refrigerant outside the tubes has a large volume in which to change phase from liquid to vapor, in these finned-coil designs the vapor can make room for itself only by pushing liquid and other vapor ahead of it in the tube until all of the liquid is vaporized.

Good evaporator design calls for relatively short lengths of coil through which the vapor has to travel, adequate space to permit prompt and adequate separation of the vapor from the liquid, and sufficient velocity in the tube to sweep the vapor from the tube surface as it forms. Low-temperature work also calls for limitation of high static head (deep liquid depth) in the evaporator, as this causes a higher pressure in the lower part of the evaporator and, consequently, a higher boiling temperature.

16-2. RECIPROCATING COMPRESSORS

Both reciprocating- and centrifugal-type compressors are in use. Centrifugal compressors can use low-pressure (high-specific-volume) refrigerants as well as conventional refrigerants and usually are of large capacity, 75 tons of refrigeration or over. Reciprocating machines are more common and more widely distributed and run in capacity from a fraction of a ton to more than 100 tons per unit. These machines are built as vertical or horizontal units, as radial-cylinder units, or in V, W, Y, or X cylinder arrangements. Most units are single-acting, with double-acting units (refrigerant compressed by both sides of pistons) usually of the horizontal type. Compressor piston speeds for any type seldom exceed 700 fpm and usually are much lower.

Figure 16-2 shows a two-cylinder, single-acting refrigerant compressor of a type widely used with ammonia before the halocarbons reached the extensive use they now enjoy. The suction and discharge valves in this compressor are spring-loaded poppet valves which are operated in response to the different pressures on the two sides of the valve. The suction valves are located in the head (top) of the piston, and the crankcase and underside of the piston are open to the suction vapor pressure. When a piston moves down, the discharge valve will be shut and the pressure above the piston will be reduced below suction-vapor (crankcase) pressure. When this reduction occurs, the suction valves are lifted from their seats by the pressure difference and vapor passes into the cylinder above the piston. When the piston is on its upstroke, the vapor being compressed forces the suction valves tightly against their seats. The pressure increases as the piston rises until it exceeds condenser pressure, when the discharge valves in the head of the cylinder will lift from their seats and the compressed vapor will pass to the condenser. The internal head of the cylinder is held by heavy springs so that, if a slug of liquid should get above the piston, the head will yield enough to prevent serious damage. This permits the close clearances between the pistons and the cylinder head, which are necessary for efficient operation.

The shaft packing on this compressor (Fig. 16-2) is metallic or semimetallic and is held in place by the adjustable packing gland. An opening for inspection, called a *lantern*, is used with this packing and furnishes a means of lubricating and oil discharge. All packed glands leak to some extent and require constant attention.

Figure 16-3 shows one type of revolving shaft seal widely used to eliminate the packed joint on modern-type compressors. In this, a spring holds a revolving, hardened-steel seal ring tightly against a similar stationary surface fastened to the compressor frame. This surface is lubricated and the refrigerant or air will not leak past the rubbing surfaces. While the sylphon bellows illustrated in Fig. 16-3 turns with the shaft, this flexible element in other makes

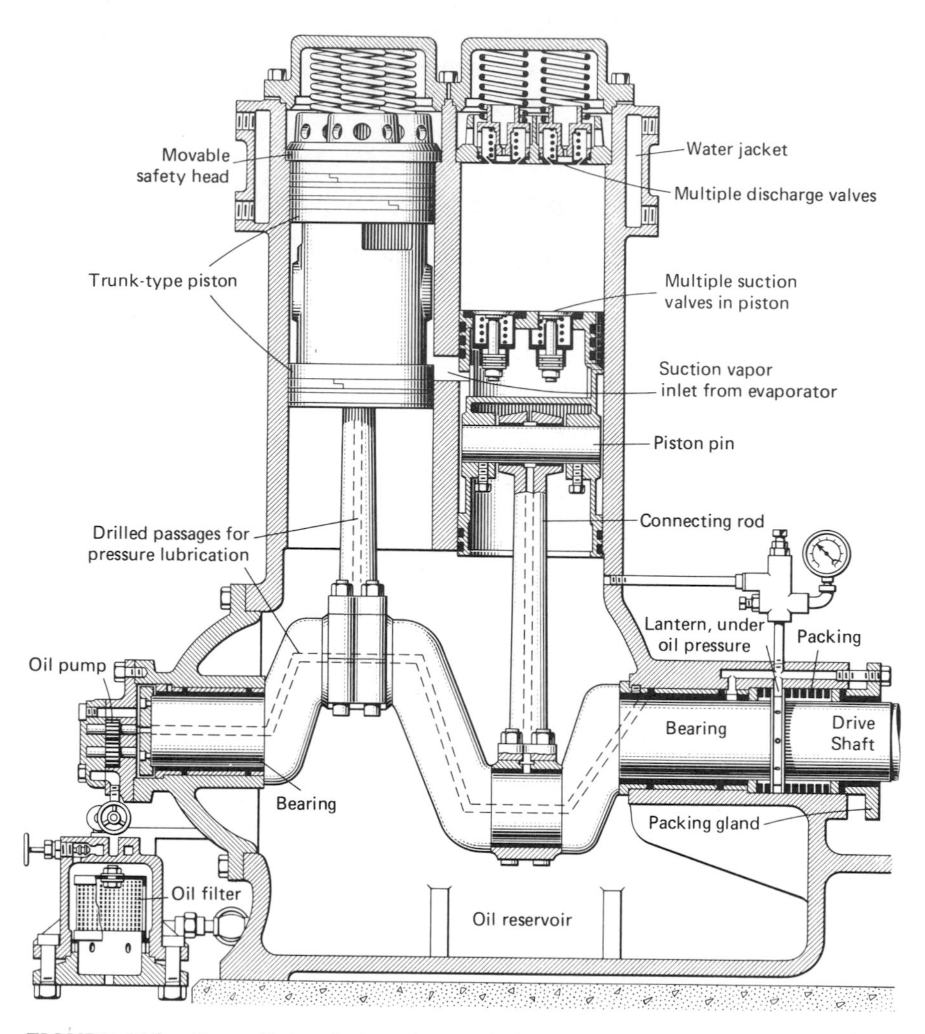

FIGURE 16-2. Two-cylinder, single-acting, vertical ammonia compressor of historic design. Many used in early years of refrigeration.

of seal may be stationary. Most small compressors have the motor and compressor hermetically sealed within a refrigerant-containing vessel and require no shaft seal. With such units the compressor and motor are placed in the container which is welded shut to enclose the unit for its life of operation. Larger size units, which are built as semihermetics, also require no shaft seal. These employ gasketed cover plates, pulled to leakproof tightness by bolts. By removing the cover plates the motor and compressor can be disassembled, removed, and repaired to provide indefinite life.

The refrigerant valves of recprocating compressors nearly always operate under the action of differential pressure. They are usually held in closed position during part of the piston stroke with the help of light springs. Valve construction takes several forms. Spring-loaded *poppet valves* are shown

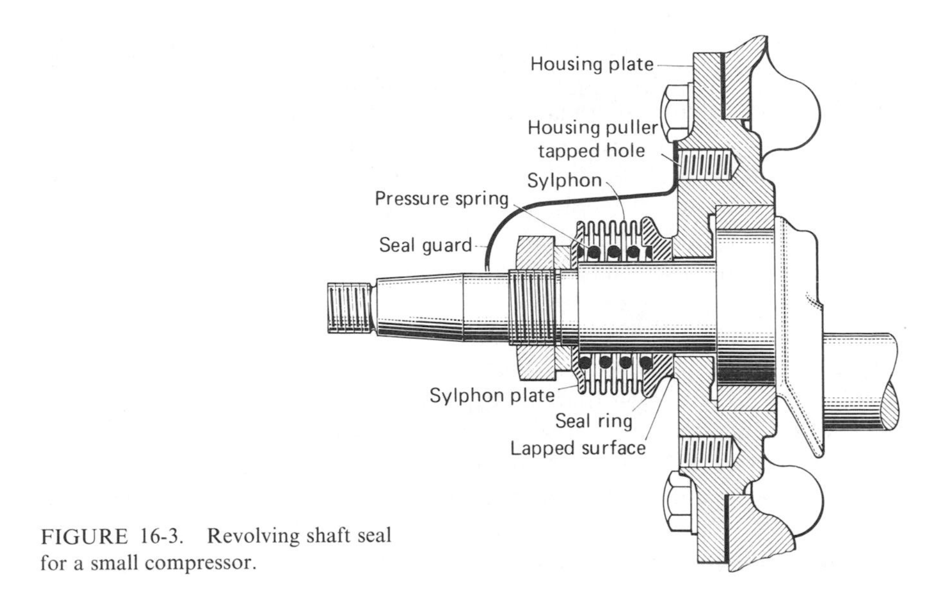

FIGURE 16-3. Revolving shaft seal for a small compressor.

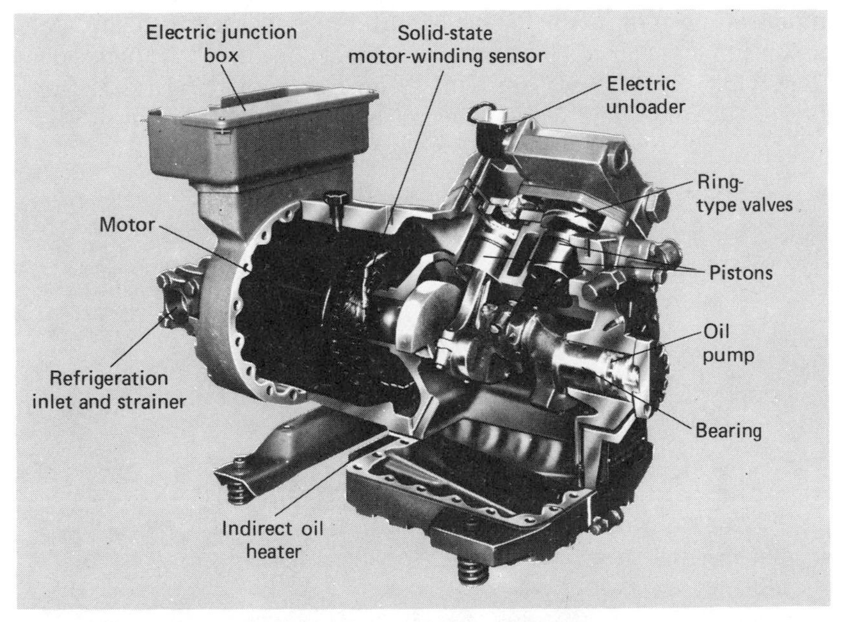

FIGURE 16-4. Trane industrial-duty, semihermetic, Model-M compressor. (Courtesy Trane Corporation.)

in Fig. 16-2, and spring-loaded *ring valves* are used in the compressor of Fig. 16-4. Those valves which employ thin ribbons of steel held in a protective cage to deflect open near their center are called *feather valves*. *Flapper valves* of thin steel, rigidly fixed at one end, flex to open and close under differential pressure while the spring action of the steel returns the valve to a neutral closed position. The valve opening can be circular in cross section or rectangular, or in the form of an annulus, as may be required.

Figure 16-4 illustrates a Trane Company semihermetic compressor of a type built in 10 to 30 tons of refrigeration capacity (35 to 105 kW) and designed for use with refrigerant-22 or refrigerant-12. This design is built as a three- or four-cylinder unit arranged in V form or as a six-cylinder unit in W form. Here the two trunk pistons on the right-hand side of the V can be seen in their respective cylinders. The unit rotates at 1750 rpm and has a $2\frac{11}{16}$-in. bore and a $2\frac{1}{4}$-in. stroke. Ring-type valves are used on both suction and compression. A fast-start, forced-feed lubrication system provides oil to all bearings. When less than full capacity is needed from the compressor, electrically operated unloaders can inactivate the ouput refrigerant flow from one or both of the cylinders in each bank. This reduces both the capacity and the power requirement. The three- and six-cylinder units can operate at 100, $66\frac{2}{3}$, and $33\frac{1}{3}$% of capacity, the four-cylinder at 100, 50, and 25% of capacity. The motor is kept cool by the refrigerant flow passing over and around it.

Figure 16-5 shows a Carrier Corporation semihermetic reciprocating compressor positioned on its shell-and-tube condenser with most of its component parts named. Such an arrangement constitutes the high side of a system. Units of this style are built to operate with refrigerant-22 or refrigerant-502 and come with six, eight, or twelve cylinders in a V or W

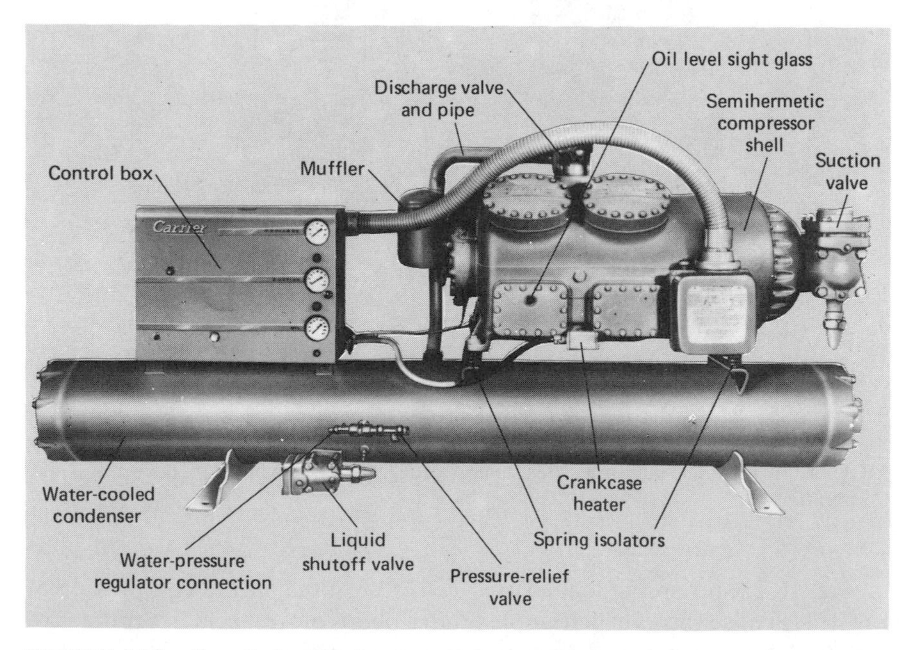

FIGURE 16-5. Six-cylinder Carrier Corporation semihermetic reciprocating compressor supported on base of water-cooled condenser. (Courtesy Carrier Corporation.)

arrangement. The pistons, which have a $3\frac{1}{4}$-in. bore and a $2\frac{3}{4}$-in. stroke, run at motor-operating speed, approximately 1725 rpm.

In operation, the compressed gas moves through the discharge valve downward into the water-cooled condenser. The condenser can be arranged for three-pass or six-pass operation. Since the water velocity increases when using a greater number of passes, the heat transfer is improved but there is naturally an increase in water-pressure loss through the tubes. The capacity ranges from 60 to 80 tons (218 to 290 kW) for the six-cylinder unit, when operating in the air-conditioning range, and proportionally doubles for the twelve-cylinder unit. The tonnage capacity is of course reduced at lower evaporator temperatures. Each unit is provided with capacity control steps and can operate as low as 33.3% of full-load capacity by use of suction-pressure controlled unloader valves on selected cylinders. The condenser length of the six-cylinder unit is approximately 8 ft.

Compressor Operation. Compressors are nearly always equipped with hand-operated shutoff valves placed close to the compressor, one in the suction line and one in the discharge line. Very large machines, when these are not equipped with internal cylinder-valve unloaders, usually have a valve in a pipe line connecting the suction and discharge lines; the valve can be operated manually when the need arises. For example, this by-pass valve may have to be opened at starting to reduce starting load. Or instead of this approach, when the compressor is started against the high suction pressure of a warm evaporator, it may be desirable to prevent overloading by throttling the suction valve, although this sometimes causes crankcase oil to froth badly. Nearly all compressors have a safety valve which is set above normal discharge pressure and can discharge into the suction side of the system in case excessive pressures are developed in the cylinder.

Constant-speed operation makes regulation of output difficult under varying load conditions. Intermittent operation can be used with small compressors, but frequent starting and stopping of big machines is undesirable. Clearance pockets are built into some compressors by which, through opening a valve, the volume of the cylinder clearance can be increased. This operation in turn reduces the capacity by reducing the amount of fresh vapor drawn in per stroke.

The capacity of a refrigeration machine in relation to the load on a system can be varied in the following ways:

1. By-passing gas from the high-pressure to low-pressure sides of the compressor.
2. By-passing internally, in the compressor, by holding open a suction valve or valves.
3. Throttling the amount of suction gas entering the compressor.
4. Using variable speed motors.

Bypassing of the internal variety is in most extensive use at the present time. When the suction valve of a cylinder is held open so that the charge of gas merely surges back and forth in the cylinder, that cylinder is inoperative as far as gas delivery is concerned. Such unloaders can be operated by a solenoid mechanism built into the compressor or act under the action of suction-pressure reduction. The mechanism was mentioned in connection with two of the compressors described in this section.

An illustrative control system which has been successfully used operates in the following manner. Consider a direct air-cooling system or one using a water chiller. In either system, thermostats would be used. The first thermostat is set to operate a cylinder unloader when the air (water-chiller) temperature reaches a certain minimum. The thermostat causes unloading of one or more compressor cylinders. If the temperature of the air (chilled-water) continues to drop, the second thermostat actuates the controller unloader to set a second cylinder (or pair of cylinders) out of operation. the second thermostat may have a final step possible, such that, if the temperature reaches an absolute minimum, the main power switch is opened to stop the unit. In reverse sense, as the air (chilled-water) temperature rises, the thermostats act to put the by-passed cylinders back into operation. Provision can also be made so that when the compressor is started again, a time-delay mechanism holds some suction-valve controls open and thus permits the unit to start under light load.

An additional control which must be provided on a machine of this type is the high-low pressure controller, which acts to stop the compressor in the event of either excessively high pressure or excessively low pressure in the refrigeration system. There is also usually a protective device on the lubrication system such that if the pressure of the lubricant is too low, the compressor is automatically stopped.

16-3. CENTRIFUGAL COMPRESSORS

Centrifugal compressors are limited in the pressure ratio through which they can operate and this puts them at a slight disadvantage compared to positive-displacement compressors, but their ability to handle large volumes of gases more than offsets this slight disadvantage. Units smaller than 75 tons are seldom seen, but larger units running up to 10 000 tons (35 000 kW) of cooling can be purchased. They are flexible in that they can utilize either the low-pressure (vacuum) refrigerants or operate with conventional higher pressure refrigerants.

Figure 16-1 in cut-away section shows a centrifugal compressor unit using a hermetic-motor-compressor design. This type of unit with its framing pattern is built for cooling capacities of 1100 to 1600 tons (3800 to 5600 kW) using either refrigerant-12 or refrigerant-500. The needed motor power is some 800 to 1300 kW, with such units operating for air-conditioning application at leaving chilled water temperatures of 38 F (2.2°C) to 48 F (8.9°C). In operation, refrigerant, vaporized in the evaporator as it absorbs heat from the water being cooled in the tubes, passes through flow-equalizing (liquid-eliminator) plates and enters the two-stage centrifugal compressor. Compressed vapor then enters the condenser where a distributor baffle directs the vapor to pass over the outside of the water-cooled tubes, condense on their surface, and drop down into the lower part of the condenser for subcooling. The resulting high-pressure liquid goes through a two-stage expansion; first a float valve (similar in design to a float steam trap) drops the pressure down to that of the suction pressure of the second stage, and then the remaining liquid is expanded through a second float feeding the liquid down to evaporator pressure. The economizer action of this two-step expansion saves power and is well justified, since the flash gas from the first expansion has to be raised only from the intermediate pressure and not from evaporator pressure. At the bottom of the unit and constituting part of the supporting base is an integral storage tank for the refrigerant. This storage chamber carries the refrigerant charge when a new machine is shipped to a site for installation as a package unit and it can also store the charge when a machine is shut down for repairs.

At capacities above some 1600 tons (5600 kW), physical size becomes so prohibitive that it is difficult to build hermetic units, and consequently open units are required. This involves the possibility of leakage at the shaft seals, and also separate shipment of components and field assembly are usually required. Centrifugal machines run at high speeds and consequently gear up from the usual 3600-rpm motor speed to some 7200 rpm or more on the compressor shaft. A velocity limit is set by the maximum tip speed at which the impeller tip can safely operate.

Figure 16-6 is a cut-away section of a centrifugal compressor. At the left note that vapor from the evaporator first passes through inlet guide vanes, which can be moved by an internal mechanism, to permit optimum

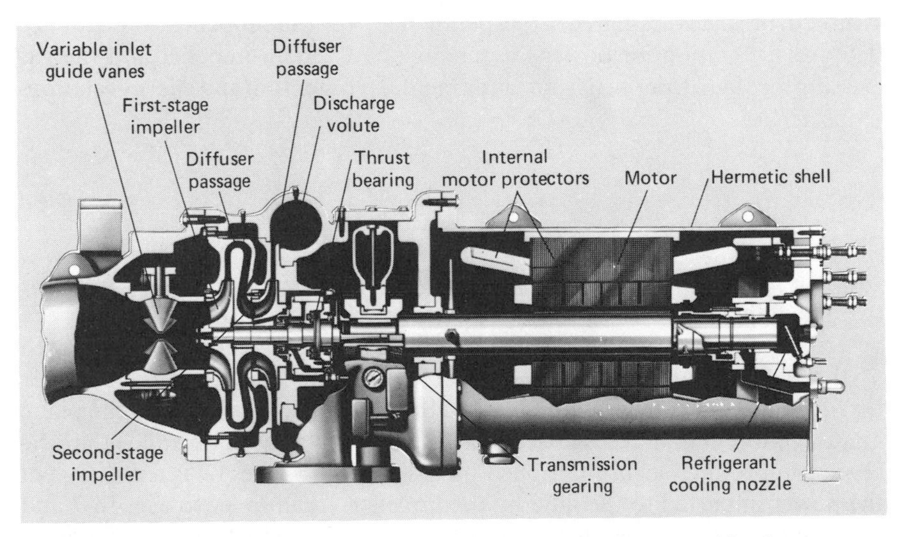

FIGURE 16-6. Sectional view of a Model-19EA Carrier hermetic-centrifugal compressor. (Courtesy Carrier Corporation.)

performance under diverse rates of flow. The vapor then enters the suction area or eye of the first stage. Then in passing through the flow passage of the rotating impeller the vapor greatly increases in velocity and a slight pressure rise also occurs. The vapor then passes into the diffusion passages in the casing where, as it decreases in velocity in moving through these passages of expanding cross section, a significant pressure rise occurs. The partially compressed gas then enters the second stage where a second increase in velocity followed by diffusion takes place in the diffusion passages and outlet volute. At this point the vapor has arrived at the maximum pressure that the compressor can create and the vapor passes to the condenser. To the right of the second stage are seals followed by a thrust bearing, then a main bearing and the gearing step-up from motor speed to the higher compressor speed. The encased motor which follows is cooled by a stream of refrigerant which vaporizes and, after passing around the windings, is led into the second-stage inlet area for recompression.

The centrifugal compressor is a dynamic device and as such, it can only impart energy by means of the impeller, which accelerates the fluid passing through. In the impeller the pressure is increased, and further pressure rise takes place in the fixed diffusing vanes (or in the volute casing of the compressor). Figure 16-7 is a diagrammatic front view of an impeller, such as either of the impellers shown from the side in Fig. 16-6. The inlet section of the impeller is known as the eye, and here the gas enters the impeller in a direction essentially parallel to the axis of the machine. Thus the inlet velocity triangle shown in Fig. 16-7 would really be completely visible only if viewed at 90 deg from the plane of the figure. This figure shows inducer guide vanes into the impeller. Such vanes are optional in use, depending upon the design. However, whether or not they are used, the absolute velocity (V_1) of the gas on inlet into the eye is essentially parallel to the axis (shaft) of the compressor and consequently lies essentially perpendicular to the impeller velocity u_1. The basic equation for power interchange between an impeller and the gas passing through it, derived from fundamental momentum and energy relationships, is

$$P = \frac{G}{g}(V_{u2}u_2 - V_{u1}u_1) \tag{16-1}$$

where P is the power interchange, in foot-pounds per second; G is the fluid (gas) flow through the impeller, in pounds per second; g is 32.17; u_1, u_2 are velocities of points on the impeller at varying radii, in feet per second, with u_1 the velocity value at the eye section, and u_2 the velocity at the impeller tip (u_t); and V_{u2}, V_{u1} are components of the absolute gas velocity in the plane and direction of u, in feet per second. In words, this equation states that the change in product terms multiplied by the fluid flow rate (G) is a measure of the power imparted to the fluid by the impeller. Again refer to Fig. 16-7 and note that on exit the relative velocity w_2 is usually almost radial and consequently the component of absolute velocity (V_{u2}) in the direction of u_2 is

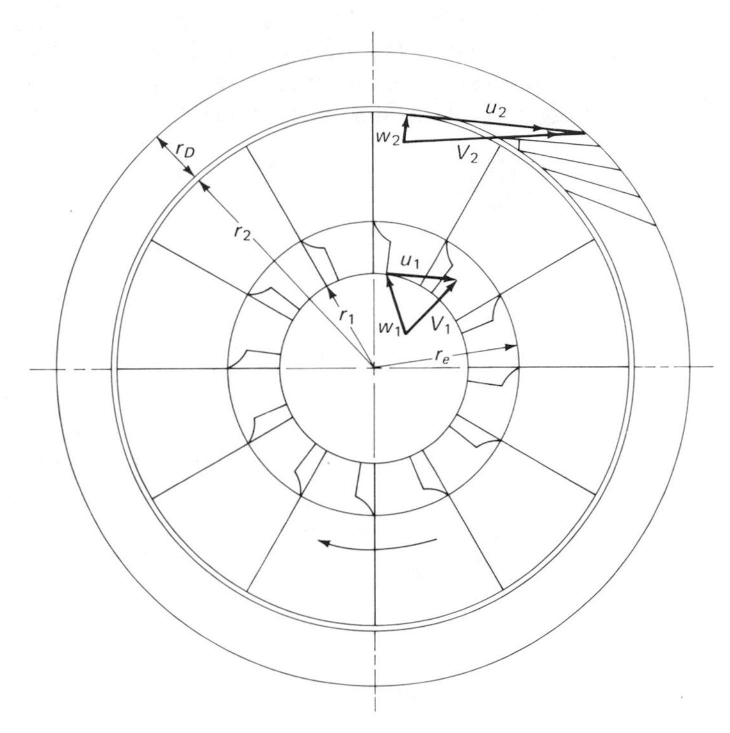

FIGURE 16-7. Inlet and exit velocity conditions for a centrifugal compressor impeller.

essentially equal to u_2 or $V_{u2} = \mu u_2$, where μ is a factor to account for any deviation in the exit direction of w_2 from being radial. With a radial-tipped impeller, μ can usually be considered as being 0.9 or higher. Figure 13-7 shows the flow patterns as these exist for similar fan impellers. On inlet, however, with the fluid flowing into the impeller in an almost axial direction, V_{u1} approaches zero in value. For these conditions, Eq. (16-1) becomes

$$P = \frac{G}{g}(V_{u2}u_2 - V_{u1}u_1) = \frac{G}{g}[\mu u_2 u_2 - (0)u_1]$$

$$= \frac{G}{g}(\mu u_2^2) \tag{16-2}$$

Since u_2 is the tip velocity of the impeller, it is desirable to write it thus, and Eq. (16-2) becomes

$$P = \frac{G}{g}(\mu u_t^2) \tag{16-3}$$

This equation is a basic measure of the power in foot-pounds per second absorbed by an impeller with a tip speed of u_t feet per second for a fluid (gas) flow of G pounds per second. However, the power imparted by the impeller is also measured by the increase in enthalpy (**h**) of the fluid plus the increase in kinetic energy in the fluid, and in equational form this appears as

$$P = G\left[778(\mathbf{h}_3 - \mathbf{h}_1) + \frac{V_3^2 - V_1^2}{2g}\right] \tag{16-4}$$

This becomes

$$P = G\left[778C_p(T_3 - T_1) + \frac{V_3^2 - V_1^2}{2g}\right] \tag{16-5}$$

since, for a gas, enthalpy change can be measured by temperature change (see Eq. 2-29). The velocity V_3 represents the velocity at outlet from the fixed diffuser section following the impeller, at the same point of measurement as T_3. The exit passage area and the passage area on inlet are usually so designed that V_3 and V_1 are essentially equal so that the kinetic energy term approaches zero and can be dropped. Thus Eq. (16-5) becomes

$$P = G(778)C_p(T_3 - T_1) = 778GC_pT_1\left(\frac{T_3}{T_1} - 1\right) \tag{16-6}$$

For a gas undergoing isentropic compression, Eq. (2-41) applies:

$$\frac{T_3}{T_1} = \left(\frac{p_3}{p_1}\right)^{(k-1)/k}$$

and substituting

$$P = 778GC_pT_1\left[\left(\frac{p_3}{p_1}\right)^{(k-1)/k} - 1\right] \tag{16-7}$$

Equations (16-3) and (16-7) represent the same process and are equalities; thus

$$\frac{G}{g}(\mu u_t^2) = 778GC_pT_1\left[\left(\frac{p_3}{p_1}\right)^{(k-1)/k} - 1\right]$$

$$\frac{p_3}{p_1} = \left(\frac{\mu u_t^2}{778C_pgT_1} + 1\right)^{k/(k-1)} \tag{16-8}$$

However, the compression process is not truly isentropic and if an efficiency (η_{ic}) is introduced to account for this fact, Eq. (16-8) then becomes usable. For general conditions, η_{ic} ranges from 0.70 to 0.80. Thus the performance of a centrifugal compressor can be expressed

$$\frac{p_3}{p_1} = \left(\frac{\eta_{ic}\mu u_t^2}{778C_pgT_1} + 1\right)^{k/(k-1)} \tag{16-9}$$

This equation is applicable to centrifugal air compressors, centrifugal steam compressors, or refrigerant compressors, provided the proper value of the polytropic coefficient k can be found, and the mean specific heat of the vapor during compression C_p is known. For design estimation, k can be taken as the ratio of specific heats.

Example 16-1. A two-stage centrifugal compressor has two impellers in series, each with a diameter of 34 in. The impellers are direct-connected to a motor turning at 3550 rpm and completely enclosed in a hermetic housing. Refrigerant-11 is used, and the suction temperature is to be 30 F in the evaporator. Compute the probable maximum pressure rise possible in this machine, and find a top condensing temperature.

Solution: The impeller tip speed is

$$u_t = \pi \frac{D}{12}\left(\frac{\text{rpm}}{60}\right) = \frac{\pi D}{720}(\text{rpm}) = \frac{(3.14)(34)}{720}(3550)$$

$$= 526 \text{ ft/s}$$

Using Eq. (16-9), assume $\eta_{ic} = 0.80$ and $\mu = 0.9$, and using $C_p = 0.136$ and $k = 1.13$ for refrigerant-11 (from Table 2-1),

$$\frac{p_3}{p_1} = \left[\frac{(0.8)(0.9)(526)^2}{(778)(0.136)(32.17)(460 + 30)} + 1\right]^{1.13/0.13} = 2.7$$

The pressure ratio for each stage is 2.7, and for two stages is thus

$$(2.7) \times (2.7) = 7.3$$

However, because of pressure loss between stages, perhaps allowance should be made for, say, 95% realization. From Table 15-7 the evaporator pressure is 5.56 psia; therefore the maximum condenser pressure would be

$$5.56\,(7.3 \times 0.95) = 38.5 \text{ psia}$$

$$\text{pressure rise} = 38.5 - 5.56 = 32.94 \text{ psia} \qquad \textit{Ans.}$$

This is beyond the range of values of Table 15-7, but more extended tables would show the saturation pressure to be 129 F. *Ans.*

Thus there is adequate pressure-ratio as well as pressure-range capacity in the machine running at this speed. The capacity design–pressure ratio would be set at a lower value, as shown in Fig. 16-8.

Figure 16-8 shows the general characteristic pressure-ratio capacity curves for a centrifugal compressor. At any fixed speed, such as 4 M in the figure, it will be observed that as the flow rate decreases, the pressure ratio which can be produced slowly increases but soon reaches a maximum value followed by instability. The region of instability is indicated as the region to the left of the dashed line. The design operating point should always be selected safely to the right of the instability region. It should be mentioned that the pressure ratio as computed by use of Eq. (16-9) represents essentially the maximum pressure ratio that could be produced by that machine.

A significant development in recent years has been to make centrifugal machines for small as well as large capacities, and one manufacturer has

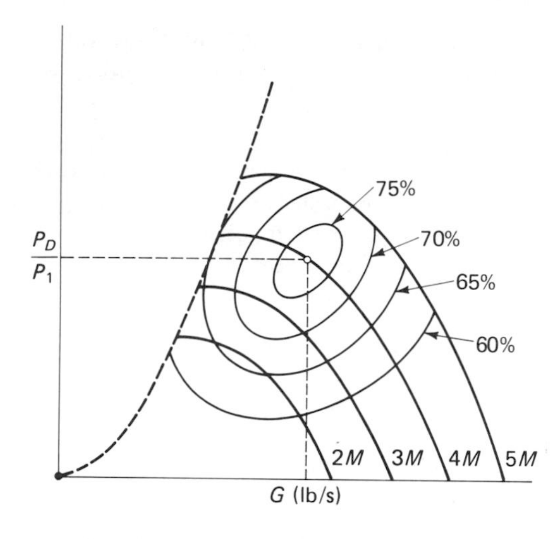

FIGURE 16-8. Characteristic pressure-ratio–capacity curves at various speeds for a centrifugal compressor.

brought out such a machine, which operates effectively at 50 tons of refrigeration. Another development has been to build hermetic units, in which the motor is completely enclosed in a shell inside the system with no shaft and packing gland leading to outside. The fact that some centrifugal machines employ low-pressure refrigerants with a total pressure range in the neighborhood of two atmospheres means that extremely heavy shells do not always have to be used, and the hermetic casing for the motor is not difficult to construct. The hermetic design reduces the possibility of leakage either in or out and minimizes the need of purging. Usually two impellers are used, which permit the pressure-ratio rise to be produced in two steps.

As was seen in Fig. 16-8, a characteristic of centrifugal machinery is the fact that the capacity falls off as the pressure ratio increases. In the case of a centrifugal machine running at constant speed and stable condensing pressure, as the external load falls off, the evaporator temperature and pressure start to drop. This in turn increases pressure ratio, and therefore decreases rate of refrigerant flow so that, before long, the machine stabilizes at a lower capacity point. Moreover, when it is necessary to decrease capacity at fixed evaporator temperature, a similar result can be produced by increasing the condenser temperature. This is readily accomplished merely by diminishing the amount of water flowing through the condenser. Most centrifugal machines being built at the present time have internal, hydraulically operated regulating vanes. These are installed at the inlet eye section to the low-pressure compressor and when operated to restrict this opening, the amount of gas entering the low-pressure impeller is reduced as closure takes place.

16-4. CONDENSERS

Condensers for refrigeration units are made in many forms and designs. The trend is toward horizontal or vertical *shell and tube* types. These condensers are similar in appearance to shell and tube evaporators, and the circulation within the tubes may be for multipass or single pass. The condenser heads should be easily removable for cleaning the sludge and debris which collect on the water side. A condenser can be seen at the top of Fig. 16-1.

Double-pipe condensers, in which the water passes through the inside pipe and the refrigerant vapor condenses in the annular space between the two pipes, were formerly used very extensively. They are very effective but require elaborate joints at the ends of each pipe for connecting the two tubes.

The so-called *atmospheric-type condenser*, in which water trickles down over condenser tubes placed outdoors, usually on the roof of a building, requires much surface for a given capacity (see Table 9-3), and installations are now seldom made.

The *evaporative condenser*, using finned surface, regulated water spray, and forced circulation of the water and air, is very effective. Figure 16-9 shows one type of evaporative condenser. In this condenser the (hot) vapor from the compressor is supplied to a bank of finned tubes enclosed in a metal cabinet. A circulating pump draws water from a basin in the bottom of the cabinet and sprays it over the tubes. Fans draw large quantities of air into the lower part of the cabinet. This air passes up around the wetted

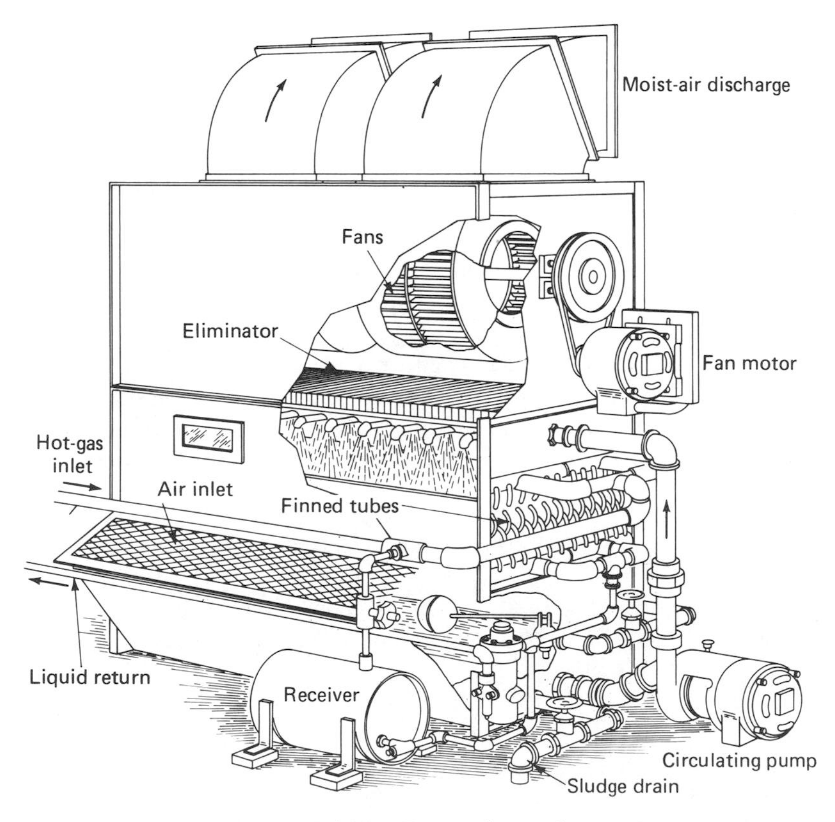

FIGURE 16-9. Evaporative condenser.

condenser tubes and a portion of the water evaporates into the air. The wetted surface and the rapidly moving air produce very effective heat transfer, permitting a large condensing capacity in a very compact space. In a cooling tower, with water flowing down and air rising up through it, the water cooling may approach the wet-bulb temperature of the entering air. In an evaporative condenser the recirculated spray water leaving the bottom tubes will have a temperature which lies between the original wet-bulb temperature of the air and the temperature at which the refrigerant condenses in the tubes. Tests on a certain evaporative condenser showed representative data as follows: temperature of refrigerant condensing, 100 F; temperature of spray water falling into recirculating basin, 89 F; wet-bulb temperature of air supply, 80 F; cfm of air circulated per ton of refrigeration, 265. Increasing the quantity of air circulated brings the water temperature closer to the wet-bulb temperature of the air, and increasing air quantities in the range from 110 to 300 cfm/ton of capacity shows justifiable gains; but above this point the rate of gain is so small that the fan power costs begin to offset gains from decreased water temperatures.

16-5. EVAPORATIVE CONDENSER CALCULATIONS

A basis for finding the load on an evaporative condenser can be had by starting back with the evaporator, where, by Eq. (14-9),

$$Q_R = \mathbf{h}_R - \mathbf{h}_{f1} \tag{14-9}$$

and in an evaporator producing **T** tons of refrigeration, the refrigerant flow (w'_R), in pounds per minute, is

$$w'_R = \frac{200\mathbf{T}}{Q_R} = \frac{200\mathbf{T}}{\mathbf{h}_R - \mathbf{h}_{f1}} \tag{16-10}$$

The heat output of the condenser is equivalent to the heat absorbed in the evaporator plus the work absorbed by the refrigerant in passing through the compressor. If we call W_k the Btu absorbed per pound as work, then the total work addition is $w'_R W_k$ and the condenser load, in Btu per minute, becomes

$$\begin{aligned} Q'_c &= w'_R(\mathbf{h}_R - \mathbf{h}_{f1} + W_k) \\ &= \frac{200\mathbf{T}}{\mathbf{h}_R - \mathbf{h}_{f1}}(\mathbf{h}_R - \mathbf{h}_{f1} + W_k) \\ &= 200\mathbf{T}\left(1 + \frac{W_k}{\mathbf{h}_R - \mathbf{h}_{f1}}\right) \end{aligned} \tag{16-11}$$

The ratio $W_k/(\mathbf{h}_R - \mathbf{h}_{f1})$ varies with the effectiveness of the compressor and with the pressure ratio through which the refrigerant is compressed. The value of W_k exceeds the isentropic work, and previously in this book it has been indicated that W_k is approximately 40% greater than the isentropic work, $(\Delta\mathbf{h})_s$. Equation (16-11) can thus be expressed

$$Q_c' = 200\mathbf{T}\left(1 + \frac{1.4(\Delta\mathbf{h})_s}{\mathbf{h}_R - \mathbf{h}_{f1}}\right) \qquad \text{Btu/min} \tag{16-12}$$

This can be written

$$Q_c' = 200\mathbf{T}K \qquad \text{Btu/min} \tag{16-13}$$

where K represents the composite term of Eq. (16-12). For representative systems the condensing load factor K will vary from 1.1 to 1.5 with a value of 1.25 being a value that might be used for estimation computations.

Thus an overall expression for the heat delivered per minute to the condenser in Btu per minute, for **T** tons produced, appears as

$$Q_c' = 200\mathbf{T}(1.25) \tag{16-14}$$

Consider now the evaporative-condenser aspects of this problem. Referring to Fig. 16-10, the following equation can be written in terms of the pounds of air flowing through the evaporator (w_a pounds per minute) and specific humidities (W_{s2} and W_{s1})

$$w_a\mathbf{h}_{a1} + w_a(W_{s2} - W_{s1})\mathbf{h}_{f1} + Q_c' = w_a\mathbf{h}_{a2}$$

$$w_a\mathbf{h}_{a1} + w_a(W_{s2} - W_{s1})\mathbf{h}_{f1} + 200(\mathbf{T})(1.25) = w_a\mathbf{h}_{a2}$$

$$w_a = \frac{200(\mathbf{T})(1.25)}{(\mathbf{h}_{a2} - \mathbf{h}_{a1}) + (W_{s1} - W_{s2})\mathbf{h}_{f1}} \tag{16-15}$$

In the above expression w_a represents the pounds of dry air flowing through the evaporative condenser per minute, $\mathbf{h}_{a1}$ and $\mathbf{h}_{a2}$ are the enthalpies in Btu per pound of dry air respectively on inlet and outlet from the evaporator, W_{s2} and W_{s1} are specific humidities, and $\mathbf{h}_{f1}$ is the enthalpy of the replacement water supply to the condenser sump. In general, it will be found that the airflow through an evaporative condenser, in pounds per minute, approximates 20**T**, or, expressed in volume terms, about 260 cfm/ton of refrigeration. However, this value ranges from perhaps 120 to 300 cfm/ton.

In operation, an evaporative condenser is not an adiabatic device, since heat is continuously being added from the refrigerant. Consequently, the final temperature of the water leaving the condenser coils will not reach the entering wet-bulb temperature as closely as would be the case were this a simple cooling tower. Expressed in equational form,

$$t_{\text{condensation}} > t_{\text{water in sump}} > t_{\text{wet bulb in}}$$

and

$$t_{\text{water in sump}} \text{ is 4 to 25 deg} > t_{\text{wet bulb in}}$$

and

$$t_{\text{condensation}} \text{ is 5 to 30 deg} > t_{\text{wet bulb in}}$$

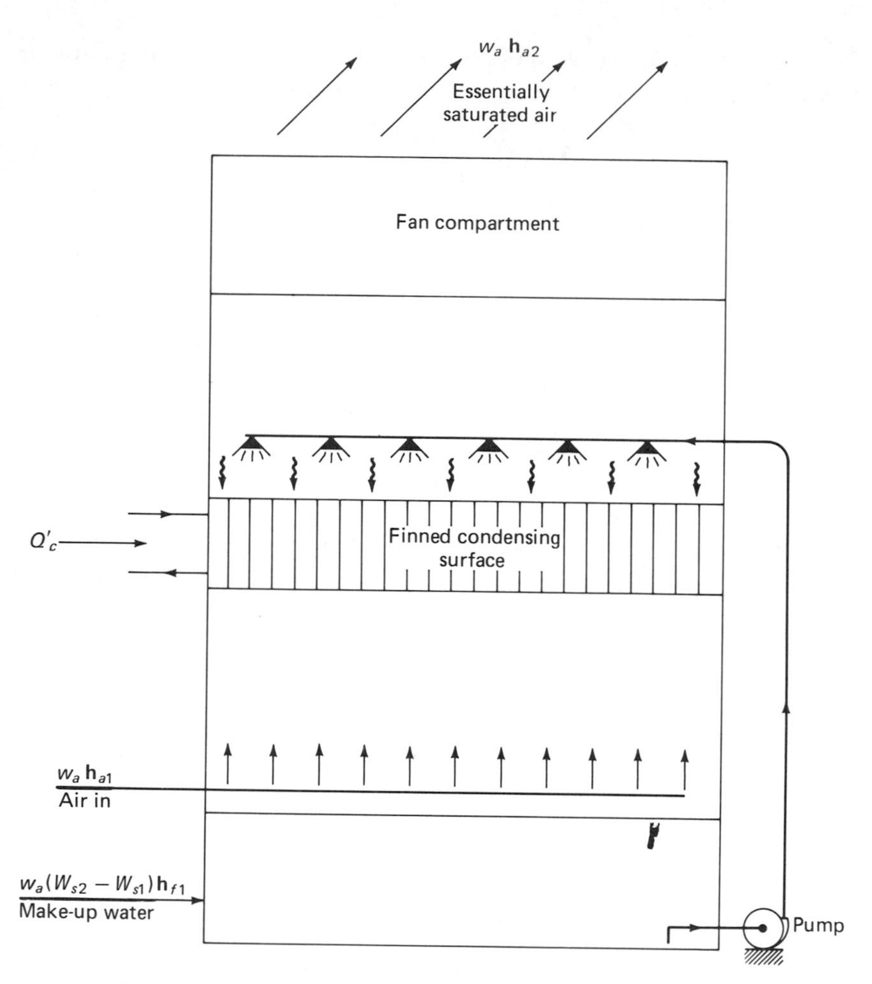

FIGURE 16-10. Diagrammatic arrangement of an evaporative condenser.

The rate of pumping must be sufficiently high to wet the coils at all times, and it has been found good practice to use pumping rates which range from 1 to 2 gpm/ton of refrigeration handled.

16-6. WATER COOLING FOR CONDENSERS

Recirculated water must frequently be employed for cooling condensers, since large quantities of water from city mains may be costly and the resultant overloading of mains and sewers makes it necessary to reduce, as much as possible, the quantity of condenser water required. By using evaporative condensers, cooling towers, and the like, only about 4 to 8% as much make-up water is required as when water is used directly. In some cases, water after passing through the condensers can be used for building service, but this arrangement seldom balances out perfectly. Heat dissipation from recirculated water is usually accomplished by *spray ponds*, *atmospheric cooling towers*, *mechanical-draft cooling towers*, or *evaporative condensers*.

In the case of a spray pond, the warm condenser water, under considerable pressure, is forced into the air by spray nozzles and allowed to fall into an open pond. This system requires a considerable area and is not usually adaptable to installations in cities.

In the case of an atmospheric cooling tower, the warm water is distributed by sprays or troughs at the top of the tower, allowed to trickle down over air-swept baffles, and then collected in a basin at the bottom. The sides of the tower are inclined louvres, to facilitate the circulation of air through the tower and to prevent the water from being blown out. The degree of cooling depends on the wet-bulb temperature of the air, the velocity of wind movement, and the intimacy of contact between the water and the air.

A mechanical-draft cooling tower is similar in construction to the atmospheric type, except that it is provided with large fans for moving the air through the tower and bringing it in contact with the water. The operation of a mechanical-draft tower is independent of the wind. Such a tower generally uses interior sprays and dispenses with the elaborate baffle system necessary with an atmospheric cooling tower.

With any of the foregoing methods the amount of make-up water required is only that sufficient to replace the moisture evaporated, plus loss by windage, and the small amount required for periodical flushing and cleaining. About 0.03 to 0.06 gpm of water per ton of refrigeration capacity is lost through evaporation into the air and carry-out as mist. Mechanical-draft cooling towers bring water to within 2 to 6 deg of the wet-bulb temperature of the air, with economy in operation (fan power, atomization, etc.) dictating 4 to 6 deg above the wet-bulb temperature as an economical range to meet.

16-7. ABSORPTION REFRIGERATION

In Fig. 16-11 is shown a diagrammatic arrangement of a conventional absorption system using ammonia and aqua ammonia. Water will absorb large quantities of ammonia vapor, the amount absorbed increasing with the external pressure and decreasing with rising temperature. The absorber, which operates at about evaporator pressure, is supplied with a cooled solution of water ammonia not saturated with ammonia. This so-called weak aqua absorbs ammonia gas from the evaporator suction line until the liquid becomes saturated at the evaporator (absorber) pressure. Heat is generated during the absorption process and is removed by cooling water. The saturated aqua is then pumped through a heat exchanger into the generator. The generator, which operates at about condenser pressure, is supplied with steam, or other heat supply, and ammonia is boiled off from the mixture until the aqua is reduced to a saturated condition at generator

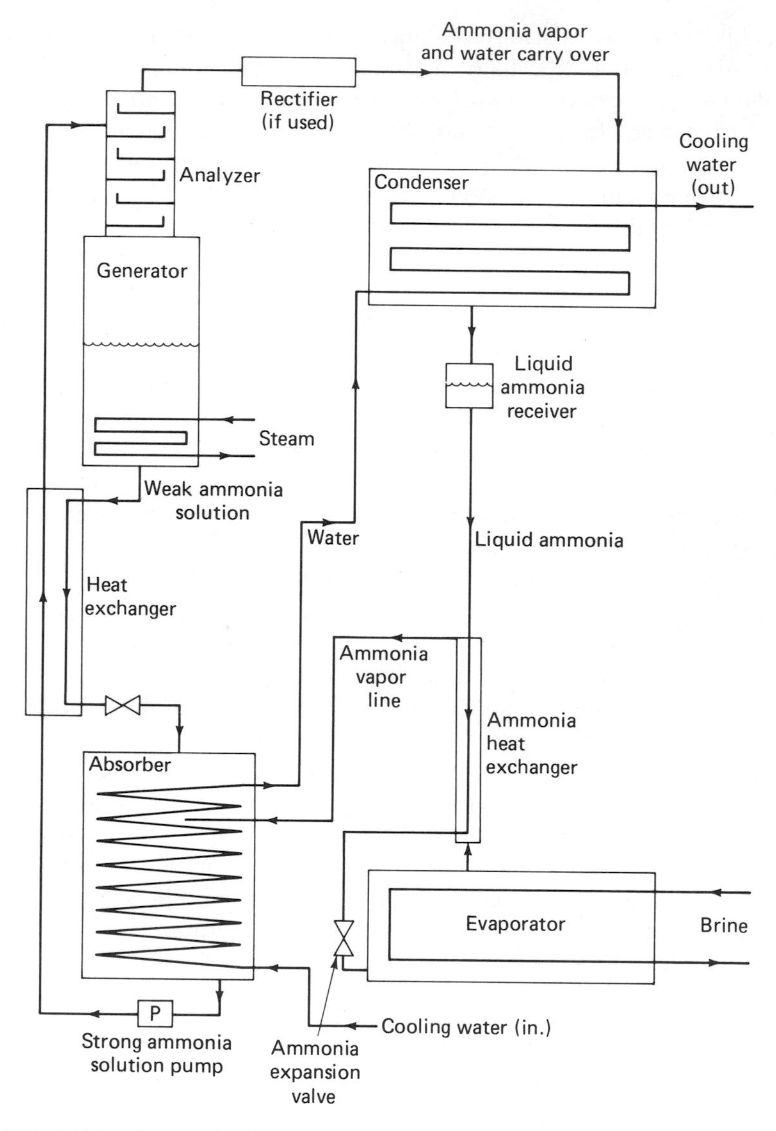

FIGURE 16-11. Conventional water-ammonia (aqua-ammonia) absorption-refrigeration system.

pressure and temperature. The ammonia vapor from the generator eventually passes to the condenser, where it is condensed, and as liquid passes through the expansion valve into the evaporator. The weak hot aqua from the generator passes through the heat exchanger where it is cooled in warming the strong liquid and is throttled as it passes into the absorber again to absorb a charge of ammonia vapor from the evaporator.

The absorption system resembles the compression system in several ways: the condenser, expansion valve, and evaporator are interchangeable with either system—the compressor is paralleled by the absorber for the suction stroke of the piston, by the aqua pump for the compression stroke, and by the generator for the delivery of the ammonia. The energy input to the system consists of a small amount of power consumed by the aqua pump and a large amount of thermal energy supplied by the heating medium to the generator. There are three major circuits of fluid throughout the system: the ammonia circuit from the generator eventually to the absorber; the strong aqua circuit from the absorber eventually to the generator; the weak aqua circuit from the generator finally to the absorber; and, in addition, steam and circulating water are required.

The vapor which rises from the solution in the generator consists of ammonia vapor along with small quantities of steam. As this vapor is cooled, the steam (saturated with ammonia) condenses out first. The analyzer performs this function of dehydration by bringing the vapor into contact with the aqua richest in ammonia and by cooling the vapor with this aqua. If the dehydration is not complete enough in the analyzer, an added water-cooled vessel called a rectifier may be used to complete the process for sending anhydrous (dry) ammonia to the condenser. Traces of moisture in the ammonia are not serious, although this moisture collects in the evaporator (if of flooded or partly flooded type) and must periodically be purged back to the absorber. If it is desired to fully dehydrate the ammonia vapor this can be done by feeding back to the rectifier a small amount of liquid ammonia from the condenser to serve as a reflux.

Tables of properties of aqua solutions[3] show the following values for a given pressure and liquid concentration: saturation temperature, enthalpy of the liquid, enthalpy of the vapor rising from such liquid, and the ammonia concentration in the vapor in equilibrium with the liquid. Typical values for saturated liquid leaving an absorber maintained at 93 F at 40 psia are $x_f = 42\%$ ammonia by weight in liquid; $t = 93.0$ F; $\mathbf{h}_f = -45.3$ Btu/lb of liquid (32 F datum for tables); $\mathbf{h}_v = 587.9$ Btu/lb of vapor (32 F datum for tables); and $x_v = 99.2\%$ of ammonia in vapor. For aqua leaving the generator saturated at 180 psia and at 239.5 F, values are $x_f = 28\%$, $\mathbf{h}_f =$ 135.9 Btu/lb, $\mathbf{h}_v = 710.2$ Btu/lb, $x_v = 88.8\%$.

The ammonia-water absorption system has not been used very extensively in recent years except where steam has been cheap or a by-product and where low refrigerant temperatures are required. However, the direct application of heat energy supply has certain advantages, and absorption systems using media other than ammonia-water are now in wide usage for air-conditioning applications.

16-8. SALT SOLUTION CYCLE ABSORPTION REFRIGERATION

For air conditioning in which refrigeration temperatures below 0°C (32 F) are not needed, an absorption refrigeration system using water as the refrigerant and lithium-bromide solution as the absorbent has been successfully developed and has achieved great commercial success. The vapor

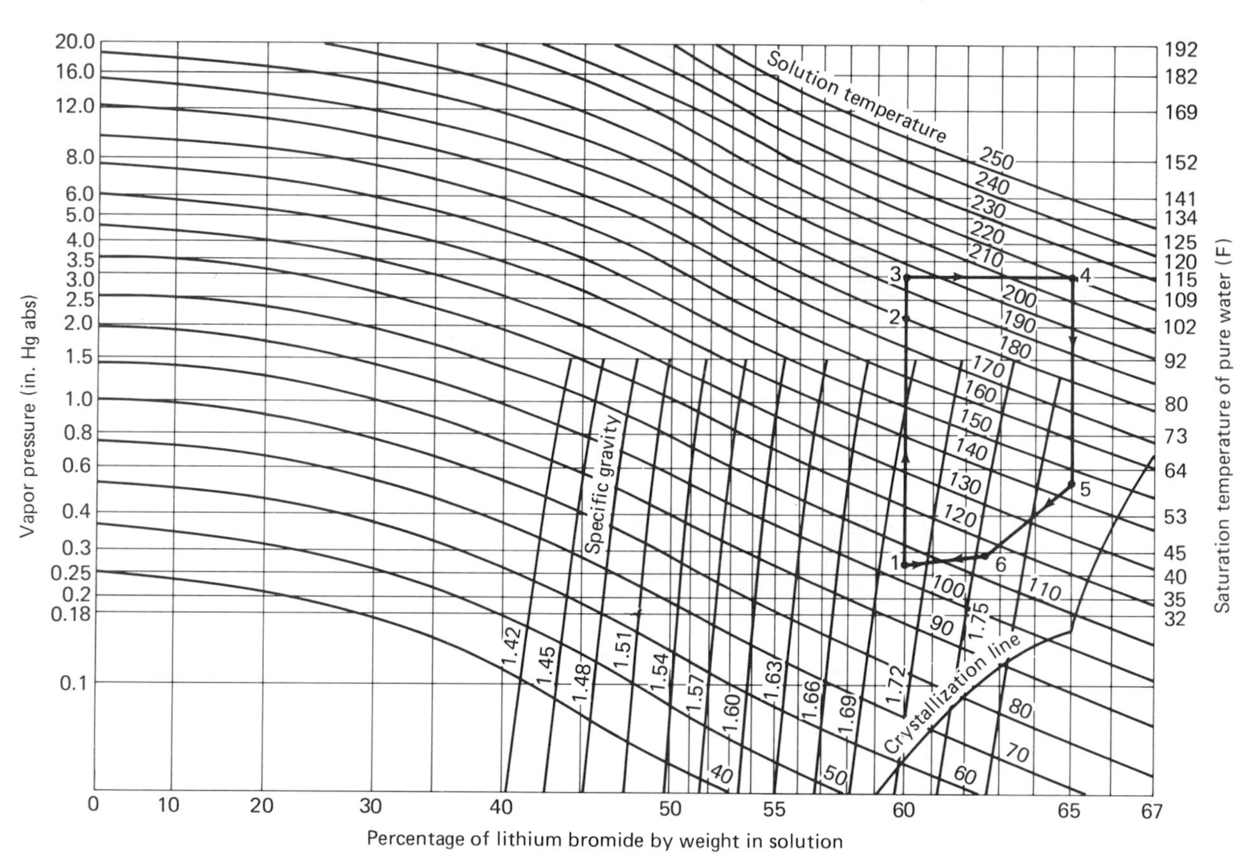

FIGURE 16-12. Equilibrium diagram for lithium-bromide-water solutions, with superposed refrigeration cycle. (Courtesy Carrier Corporation.)

pressure of an aqueous solution of lithium-bromide high in salt is very low, and if water and solution are placed adjacent to each other in a closed evacuated system, the water will evaporate. For example, a 60% lithium-bromide–water solution at 110 F has a vapor pressure of 0.27 in. Hg, which is sufficiently low to cause water at 43 F, or slightly lower, to boil. Figure 16-12 is an equilibrium diagram of lithium-bromide–water solutions, and use will be made of it and of the numbered cycle in describing the absorption system. Figures 16-13 and 16-14 show the system first in diagrammatic arrangement, and then with the vessels as actually placed in the machine.

Absorber. In the absorber, a circulating pump takes solution at roughly 60% from one end and recirculates it back into the absorber to mix with strong salt solution (point 5) at 65%, supplied to the absorber from the heat exchanger. The mixed solution at an average temperature of 115 F at 62.5% (point 6 in Fig. 16-12) has a sufficiently low vapor pressure that it can readily absorb water vapor from the evaporator, and the concentration further reduces with the addition of vapor toward point 1. As the water vapor enters the salt solution to enter the liquid phase, the heat of condensation must be removed as must also the heat of dilution (solution). This heat removal is accomplished by cooling water, which is circulated through the absorber. The main solution pump takes diluted salt solution at 60% and 105 F (point 1) and delivers it to the *heat exchanger*. In the heat exchanger the weak salt solution is warmed to 180 F (1 to 2 in Fig. 16-12). From the heat exchanger this weak solution then passes to the *generator*. In the generator it warms to 220 F (2 to 3 to 4). As the solution receives heat from the

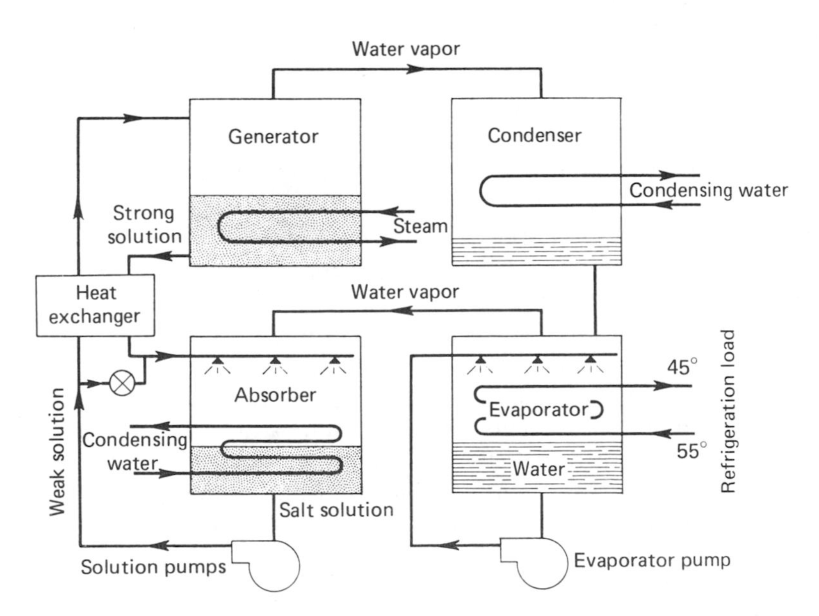

FIGURE 16-13. Operational arrangement of the water and lithium-bromide circuits of an absorption machine.

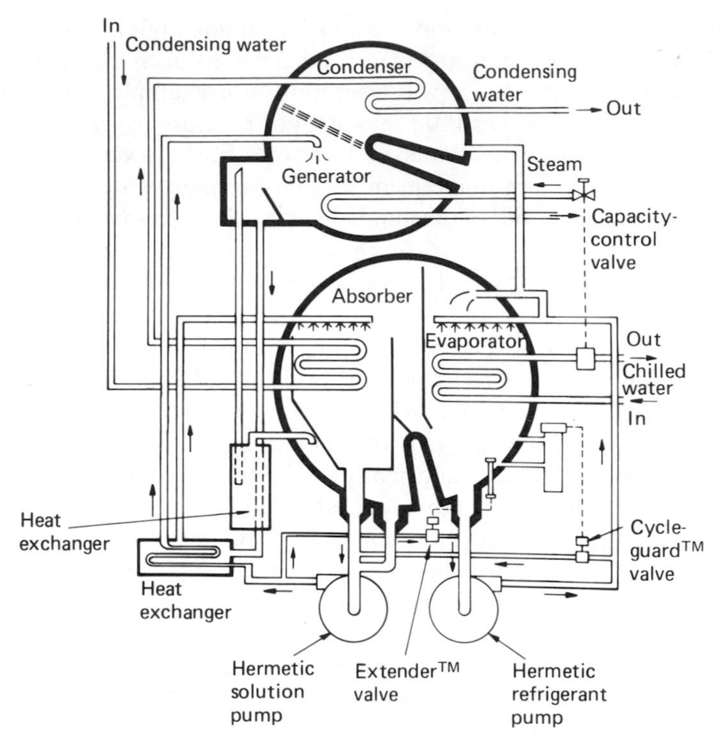

FIGURE 16-14. Placement of units and equipment in a Carrier absorption refrigeration unit. (Courtesy Carrier Corporation.)

steam coil in the generator, excess water is driven off from the solution to enrich it to a salt concentration of 65% (point 4). The strong solution, which is at 220 F and at the generator pressure of 3 in. Hg absolute, then enters the *heat exchanger*, where it cools to 145 F (points 4 to 5 in Fig. 16-12) as it warms the counterflowing weak solution. This strong solution then sprays into the *absorber*, dropping in pressure to 0.28 in. Hg absolute, where the previously described absorption process takes place.

The water vapor of the refrigerant circuit, when distilled from the solution in the generator, passes to the *condenser*, where circulating water removes heat and changes the water vapor into liquid at 115 F, 3 in. Hg absolute. From the condenser, water, the refrigerant of the system, then passes to the *evaporator*, where it drops in pressure to 0.28 in. Hg absolute, and a portion of it flashes into low-temperature steam. Water in the evaporator is continuously recirculated by a pump and sprayed over the refrigerant load coils. Because a portion of the water is continuously evaporated, the evaporator maintains a useful low temperature of approximately 43 F. The water vapor produced passes to the absorber.

The system, although apparently complex, is actually simple and compact. The evaporator and absorber are placed in a common lower shell (Figs. 16-14 and 16-15), which operates at the low pressure of the system at

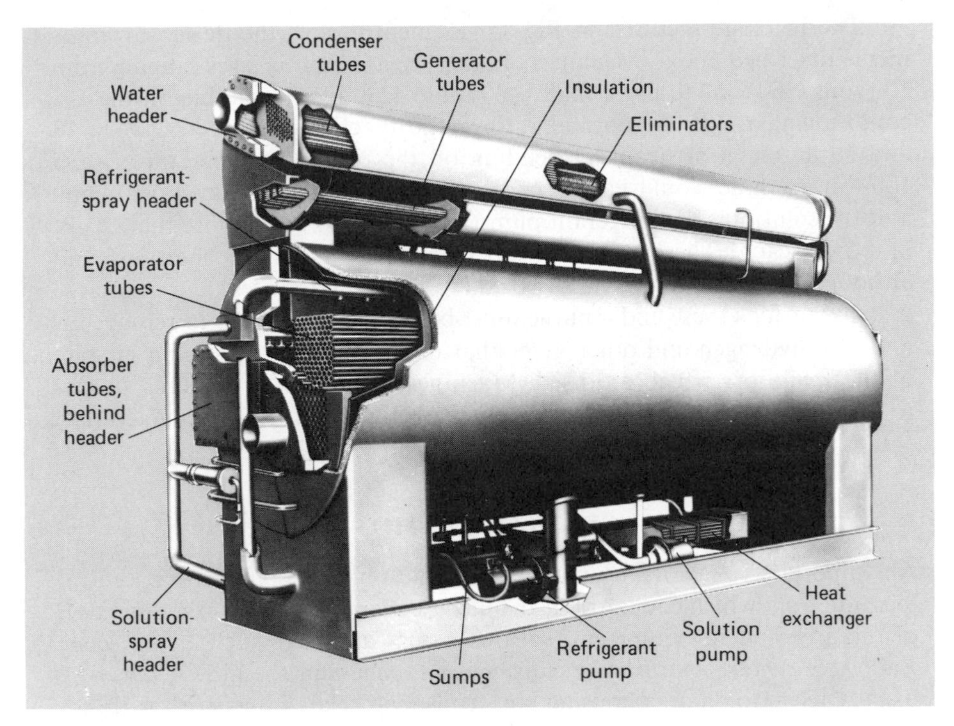

FIGURE 16-15. Cut-away view of a Model-19JB Carrier absorption liquid chiller. (Courtesy Carrier Corporation.)

around 0.26 to 0.36 psia. The generator and condenser are placed in a common upper shell and operate at the high pressure of the system in a usual range of 2.2 to 3.2 psia. The circulating (condensing) water passes through the absorber where it rises in temperature in picking up the solution load (heat of solution and condensation of the refrigerant from the evaporator), following which it moves to the condenser and warms an additional amount as it condenses the water vapor arriving from the generator. Steam is supplied at gage pressures of from 8 to 12 psia, producing temperatures of some 234 F (112.8°C) to 245 F (118.4°C). The major energy input into the system is from the steam since the pumps require relatively trivial horsepower.

One problem with water-salt machines is the possibility that under certain operating conditions the solutions can become too dry and partially crystallize, a situation which, if extreme, could made a machine inoperative. Two devices are provided on the Carrier unit to prevent this possibility. The cycle guard (Fig. 16-14) automatically transfers water (refrigerant) to the absorber whenever the refrigerant level in the evaporator reaches preset levels. These levels vary because of the wide variations in condenser water temperature which can occur. The extender valve control operates in an opposite sense to assure that an adequate supply of refrigerant (water) is always present in the evaporator.

The machine pictured in Fig. 16-15 incorporates the design arrangements described above. Machines of this type are built in sizes ranging from 100 tons (350 kW) to more than 1200 tons (4230 kW) of cooling in the air-conditioning range. At normal ratings the machines require between 18 and 19 lb/ton · h of steam. Depending on the planned temperature rise, a 100-ton machine would have a cooling-water flow rate of some 400 gpm, and the solution and refrigerant pumps would require not more than 2 kW of power each. These machines operate under vacuum conditions and although hermetic pumps are used, there is gradual air leakage into large systems such as these and a purge must be provided on them to remove air and also hydrogen and other gases that might form in the system in spite of the inhibitors which are used. Machines having many features similar to the one described here are now being made by a number of manufacturers.

16-9. CENTRAL STATION UNITS

An important area for any major air-conditioning system is that central location from which cooling is distributed. Depending on the exact function of the area and its equipment, various names are used, such as *fan room, equipment center, central station, machinery room*, and the like. A *fan room* usually refers to an area housing fans, coils, and related ductwork with hot water (or alternately steam) and chilled water made available to it from separate sources. An *equipment center* or *central station* normally has the same equipment as a fan room but is also provided with functional cooling equipment such as a compressor as well as a boiler or other heating source. The location of these service rooms or areas in a building is determined by many factors, but basements or rooftops are by far the most common since such locations remove these noise sources away from occupied and usually more valuable space. Very large buildings often need many central stations and/or fan rooms. Manufacturers now make available a variety of types of equipment to solve energy distribution problems. Some are completely self-contained package units while others are modular units which can be combined and arranged to serve different functions.

Figure 16-16 shows one such special-purpose air-distribution system. This is a completely self-contained, variable-volume, rooftop-mounted cooling system. Installation requires putting the roof curb support in place, cutting holes in the roof for the return- and supply-air ducting, and setting the weatherproofed package unit in place. After the unit is sealed against roof leakage and the ductwork and power lines have been connected, the unit is ready to run.

The right end of the figure shows the high side of the system, the compressors, the air-cooled condensers, and their fans. The evaporator shown in the left center of the figure is served by two fans which deliver the chilled supply air to the distribution system. Recirculated (return) air with proper amounts of outside (ventilation) air is supplied for the evaporator. Hermetic

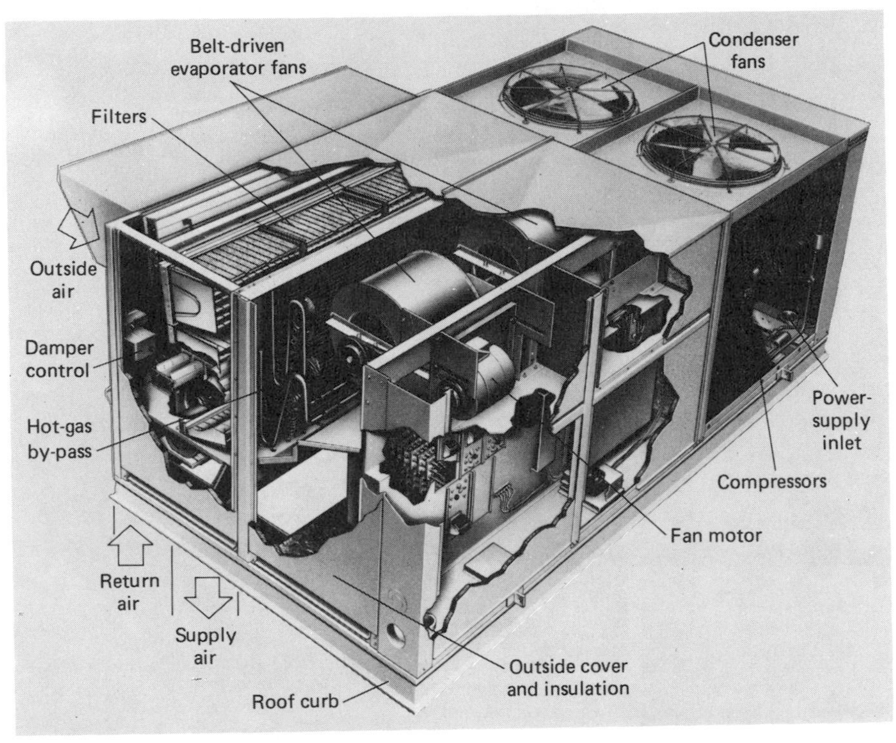

FIGURE 16-16. Carrier Modu-pac rooftop variable-volume cooling unit. (Courtesy Carrier Corporation.)

reciprocating compressors with four or six cylinders per compressor run at 1750 rpm. Each compressor has electric capacity-step unloaders to provide for space load variations. In addition, a hot gas by-pass can operate during extremely light space loads. Damper action can provide operation making use of maximum outside air when seasonal temperatures permit. This rooftop unit is designed to work effectively with the Carrier Moduline units described in Section 12-9. The rooftop units are built in 5 capacities ranging from approximately 280 000 Btuh (82 kW) to 650 000 Btuh (190 kW) of total cooling capacity. The largest unit requires a roof space of approximately 25 by 7.3 ft with a height approaching 5 ft.

16-10. UNIT AIR CONDITIONERS

In the absence of central air conditioning, self-contained units can be employed directly in the space being cooled, or immediately adjacent to it. Such units are particularly suitable in locations such as small stores or shops, restaurants, beauty parlors, offices, and waiting rooms. Noise can be a problem with some units if they are installed directly in the space to be

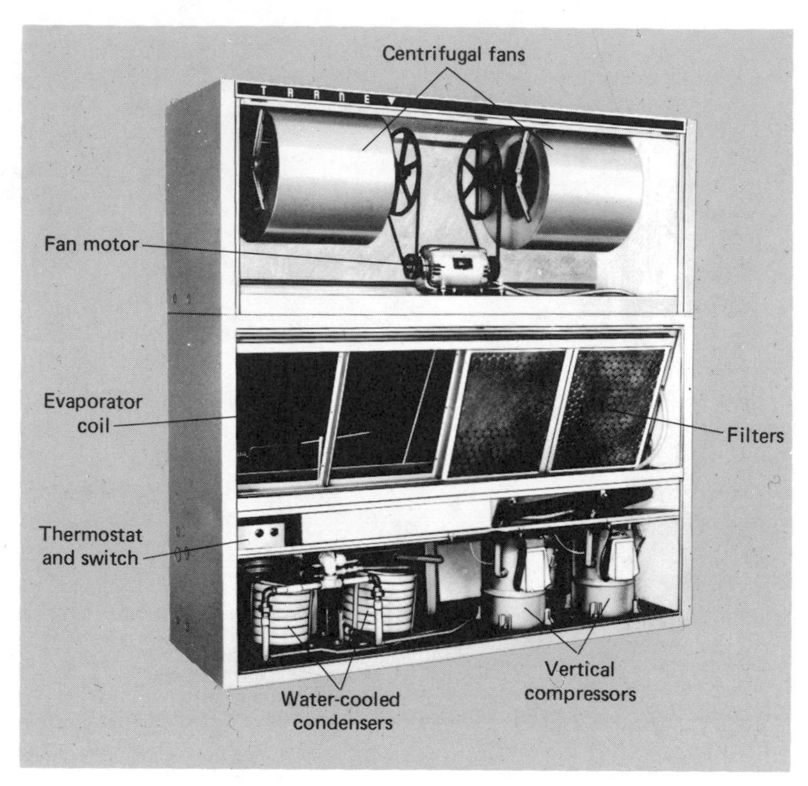

FIGURE 16-17. A vertical, self-contained, water-cooled, unit air conditioner, with covers removed to show operating elements. (Courtesy Trane Corporation.)

cooled, but in a number of recent designs noise has been reduced to acceptably low levels.

Figure 16-17 is a frontal view, with covers removed, of a water-cooled, vertical self-contained unit. Air, drawn in at the front center, passes through filters, over the evaporator coils, and then to fans for delivery into the cooled space. The air outlet can be selected for delivery rearward—as, for example, into an adjacent space—upward, or forward. To keep noise levels low, the fans are of large size and low speed, both the compressor suction and discharge lines have noise attenuation chambers, and the compressor is enclosed in a separate sound-insulated compartment. The compressor is a quiet-running, vertical, hermetic unit with the compressor in the lower part of the chamber and the motor above it. The weight of these is supported by a thrust bearing. These units are built in capacities from 30 000 to 200 000 Btuh of cooling (8.8 to 58 kW). Similar self-contained units can employ air cooling for the condenser. A modified condenser is required for this usage and the unit must be at an outside wall with suitable openings for air flow. All such units can be provided with a heating coil to make them adaptable for year-round operation.

PROBLEMS

16-1. The capacity of a cooling system under test can be found by measuring the chilled water flow into and from the evaporator. In one test, water entered the evaporator at 51.0 F (10.6°C), left at 42.6 F (5.9°C), and the flow rate was found to be 1250 lb/min. (a) Find the cooling capacity expressed in Btu per hour, tons of refrigeration, and kilowatts. (b) If refrigerant-12 is supplied to the evaporator at 80 F (26.7°C) and leaves at 34 F (1.1°C) in dry and saturated condition, compute the refrigerant flow in pounds per minute.

Ans. (a) 630 000 Btuh, 52.5 tons, 184.6 kW; (b) 192.8 lb/min

16-2. A cooling system was found to have a water flow through its evaporator of 567 kg/min while water entered at 10.55°C and left at 5.88°C. Compute the heat abstraction from the water expressed in kilocalories per hour, in kilowatts, and in tons of refrigeration.

Ans. 158 870 kcal/h, 184.6 kW, 52.5 tons

16-3. A calcium chloride brine with a specific gravity of 1.212 at 60 F and a freezing point of −10.6 F (−23.7°C) is cooled in an evaporator from 16 F to 5 F. This brine has a specific heat of 0.77 Btu/lb · F. In the evaporator the measured flow rate of the brine is 720 lb/min. Ammonia liquid supplied to this evaporator at 70 F (21.1°C) boils at 0 F (−17.8°C) and leaves dry and saturated. (a) Compute the cooling produced in Btu per hour, in tons, and in kilowatts. (b) Compute the refrigerant flow in pounds per minute.

Ans. (a) 365 900 Btu/h, 30.5 tons, 107.2 kW; (b) 12.4 lb/min

16-4. A two-cylinder single-acting compressor, with $2\frac{1}{4}$-in. bore and $1\frac{11}{16}$-in. stroke, runs at 1500 rpm. The condensing temperature is 82 F and the evaporating temperature is 30 F. A heat exchanger which warms to 65 F the 30 F dry-saturated vapor from the evaporator is used in subcooling the 82 F condensed liquid to a lower temperature before the liquid enters the evaporator. Refrigerant-12 is used. (a) Compute total piston displacement, in cubic feet per minute. (b) Compute the temperature of the liquid before it enters the expansion valve—after it has passed through the heat exchanger and has been cooled down by the vapor warming from 30 F to 65 F. Note that the decrease in enthalpy of the liquid equals the increase in enthalpy of the vapor. (c) Compute the pounds of refrigerant required per ton of refrigeration per minute. (d) For a volumetric efficiency of 88%, compute the pounds of refrigerant handled per minute by the compressor. (e) Find the refrigerating capacity, in tons and in units of 1000 Btuh. (f) Find the isentropic work of compression per pound. (g) Find the theoretical horsepower and specify a motor size for this unit

Ans. (a) 11.65 cfm; (b) 59 F; (c) 3.4 lb/ton · min; (d) 10.3 lb/min; (e) 3.0 tons, 36 500 Btuh; (f) 7.3 Btu/lb; (g) 1.77 hp and 2.5

16-5. A 100-ton refrigerant-12 compressor runs with a discharge pressure of 93.3 psig and a suction pressure of 11.79 psig. This liquid is subcooled to 70 F before expansion, and the vapor entering the compressor is superheated 15 deg. Find (a) the heat absorbed in the evaporator per pound of refrigerant circulated; (b) the weight of refrigerant flowing per minute per ton of refrigeration; (c) the theoretical piston displacement per minute per ton; (d) the isentropic horsepower; (e) the probable actual horsepower; and (f) the probable actual piston displacement per minute per ton.

Ans. (a) 56.0 Btu/lb; (b) 3.54 lb/ton · min; (c) 5.4 cfm/ton; (d) 0.9 hp/ton; (e) 1.26 shaft hp/ton; (f) 6.2 cfm/ton

16-6. In an evaporative condenser assume that 12 000 Btuh, plus 2500 Btuh in the form of work from the compressor, constitute the heat removed in the condenser each hour per ton of refrigeration. (a) If the latent heat of vaporization of water is taken as 1050 Btu/lb, compute, in pounds per hour, the water evaporated per ton of refrigeration in the evaporative condenser, and express the quantity of water evaporated also in gpm. (b) In a particular evaporative condenser used on a 5-ton unit, city water supplied has a hardness of 15 grains/gal. It has been found undesirable to operate evaporative condensers when the hardness exceeds 75 grains/gal. How much water should be bled (wasted) from this evaporative condenser every hour to keep the hardness at a safe level?

Ans. (a) 13.8 lb/ton · h, 0.0277 gpm/ton; (b) 2.08 gal/h

16-7. An evaporative condenser serving a 25-ton capacity unit is located on the roof of a building. Compute (a) the probable heat load in Btu per minute removed from the condenser and (b) the gpm evaporated if the latent heat of water is considered to be 1050 Btu/lb. If on a given day the outside air conditions are 80 Fdb and 70 Fwb and air leaves the condenser essentially saturated at 84 F, compute the air flow in (c) pounds per minute and (d) cubic feet per minute.

Ans. (a) 6250 Btu/min; (b) 0.717 gpm; (c) 488 lb/min; (d) 6750 cfm

16-8. Work Problem 16-7 for conditions of a 10-ton evaporative condenser and of outside air at 95 Fdb and 75 Fwb, inlet make-up water at 70 F, and air from the condenser at 100 Fdb, at 90% relative humidity.

Ans. (a) 2500 Btu/min; (b) 0.287 gpm; (c) 93.5 lb/min; (d) 1342 cfm

16-9. Consider that the two-stage centrifugal compressor described in Example 16-1 is recharged with refrigerant-113 (trichlorotrifluoroethane) for which $k = 1.09$ and the mean specific heat of its vapor can be taken as 0.15. Consider adiabatic compressor efficiency as 70%. For the same running conditions find the probable maximum pressure ratio and the resulting condensing pressure if this refrigerant when operating at 30 F (-1.11°C) has an evaporator pressure of 2.0 psia.

Ans. Total pressure ratio, 8.5; 17.0 psia

16-10. A two-stage centrifugal compressor, having two impellers each with a maximum tip diameter of 28 in., is driven by a steam turbine rotating at 4500 rpm. Refrigerant-12 is used, and the suction temperature is -60 F. The compressed refrigerant is liquefied in a condenser cooled by refrigerant from a separate system (cascade arrangement). The refrigerant flow under design conditions at 16 F condensation is 152 lb/min. (a) Compute the probable maximum pressure-rise ratio for this machine, and find the top condensing pressure for -60 F evaporation (consider $\eta_{ic} = 0.72$ and $\mu = 0.9$). (b) Compute the probable horsepower needed from the turbine to run the compressor under design conditions if the power factor compared to isentropic can be considered as 1.38. (The isentropic work can be found by considering the inlet enthalpy and entropy of dry-saturated vapor from Table 15-5 and then reading the compressed vapor enthalpy from Fig. 15-4 at the -60 F entropy value.)

Ans. (a) 7.14, 38.3 psia; (b) 65 shaft hp

16-11. The evaporator of a lithium-bromide absorption refrigeration system with water as a refrigerant operates at a 40 F (4.44°C) evaporating temperature. Refrigerant from the condenser is at 100 F (37.8°C). Water is chilled in the evaporator from 56 F (13.3°C) to 48 F (8.89°C) in sufficient amount to produce 250 tons (877 kW) of cooling.

(a) Compute the chilled water flow in pounds per hour, and in gpm, needed for 250 tons of cooling. (b) Compute the refrigerant flow in pounds per hour for 250 tons of cooling.

Ans. (a) 375 000 lb/h, 749 gpm; (b) 2966 lb/h

16-12. The absorber of the lithium-bromide absorption system of Problem 16-11 receives salt solution at 130 F (54.4°C) at concentration $x_a = 0.64$, namely, 64% LiBr, for absorbing the refrigerant vapor (water). The solution is diluted to $x_b = 0.60$ or 60% LiBr while the circulating water cools the resulting solution to 120 F (48.9°C) as it also removes the heat of condensation and heat of mixing. The specific heat of salt solutions in this range is about 0.44 Btu/lb · F and for estimation, use the heat of mixing as 20% of the latent heat of condensation. (a) Write two mass-balance equations, one for water-steam and one for salt, and solve to compute the ratio of solution to water refrigerant absorbed. Express your answer as the weight (w_b) of leaving (60%) solution per pound of water refrigerant absorbed. (b) For computational purposes, consider that all condensation and mixing take place at 120 F (48.9°C) instead of over the actual range of temperature and read the latent heat at this temperature from Table 2-2. Then compute the heat of condensation and mixing per pound of steam liquefied into the salt solution. Compute the additional heat removal required in cooling the solution from 130 F to 120 F, and then find the total heat load in the absorber. (c) For the 2966 lb/h of refrigerant flow of this 250-ton unit, compute the Btu per hour removed in the absorber. The cooling effect produced by the steam (refrigerant) as it warms from 40 F to 120 F final mixing temperature was not considered but would actually reduce the absorber load

Ans. (a) $w_b = x_a/(x_a - x_b) = 16.0$ lb solution/lb refrigerant; (b) 1297.0 Btu/lb of vapor; (c) 3 846 900 Btuh

16-13. The generator of the 250-ton absorption machine considered in Problems 16-11 and 16-12 receives solution strong in refrigerant at 195 F (90.6°C). Process steam at 20.0 psia boils out excess water refrigerant until the LiBr solution reaches a final equilbrium temperature of 210 F (98.9°C) on leaving the generator. Assume that 20% in excess of the latent heat of vaporization at 210 F is needed to break the salt-solution binding per pound of steam released. A specific heat of 0.44 Btu/lb · F is applicable for bringing the 16 lb of salt solution up to the final generator temperature. (a) Compute the generator load for final warming of the salt solution and also vaporizing and re-leasing each pound of refrigerant. (b) For each pound of refrigerant (steam) released, compute the process steam needed if this is provided dry and saturated and no sub-cooling takes place. (c) For the 250-ton unit with a refrigerant flow of 2966 lb/h. compute the process steam needed if it is estimated that 10% in excess is needed to take care of internal and external heat-transfer losses, incomplete equilibrium between water and salt, etc. (d) Express this process steam requirement in pounds per ton-hour.

Ans. (a) 1271.5 Btu/lb; (b) 1.32 lb/lb of refrigerant; (c) 4307 lb/h; (d) 17.2 lb/ton · h

16-14. Condensation of the steam refrigerant from the generator of the 250-ton absorption unit of Problems 16-11 to 16-13 takes place at 128 F (53.3°C) and the liquid subcools in the condenser to 110 F and in the exit piping and receiver to 100 F (37.8°C). Steam moves over from the generator at 210 F (98.9°C) in superheated state relative to the low condenser pressure of 2.11 psia (Table 2-2). [The enthalpy of superheated steam at 210 F and low pressure can be read from complete steam tables or computed by Eq. (3-12) as 1155.0 Btu/lb.] (a) Express the condenser pressure in millimeters of mercury and in pascals. (b) Using Table 2-2 or another source to find the enthalpy of

110 F (43.3°C) water, find the required heat removal for each pound of steam condensed. (c) Find the condenser load for the 250-ton unit in Btu per hour.

Ans. (a) 109.1 mm Hg, 14 548 Pa; (b) 1077.1, 3 194 700 Btuh

REFERENCES

B. H. Jennings and W. L. Rogers, *Gas Turbine Analysis and Practice* (New York: McGraw-Hill Book Co., 1953; Dover Press, 1969), "Centrifugal Compressors," pp. 223–41.

ASHRAE Handbook—Equipment—1975, "Centrifugal Compressors," pp. 12.16–12.25.

B. H. Jennings and F. P. Shannon, *Tables of Properties Aqua-Ammonia Solutions*, Lehigh University Studies No. 1 (July 1938).

ASHRAE Handbook of Fundamentals—1972, Chapter 1, "Absorption Refrigeration," pp. 19–24.

ASHRAE Handbook—Equipment—1975, "Absorption Refrigeration," pp. 14.1–14.8.

CHAPTER

17

AIR-CONDITIONING SYSTEM OPTIMIZATION AND HEAT-PUMP USAGE

17-1. BASIC AIR CONDITIONING SYSTEM DESIGNS

Air-conditioning objectives can often be met through a variety of alternate design patterns. Thus the designer should have some familiarity with all of the approaches so as to be in a position to choose the most satisfactory solution from the viewpoints of cost, reliability, energy usage, comfort performance, appearance, noise, type of control, and maintenance. It is obvious that the various objectives are not always compatible and thus each final design must represent, to some extent, a compromise between several alternative approaches.

Basic Single Duct Air Systems with Reheat. Let us refer back to Fig. 6-1 and review the explanation of it given at the start of Chapter 6. In such a system the air is heated or cooled, dehumidified or humidified at one central location and only air serves the working space to meet the needs of the occupants.

A variation of the system to make it applicable to several zones is shown in Fig. 17-1. Here we again have a central system for the air but separate

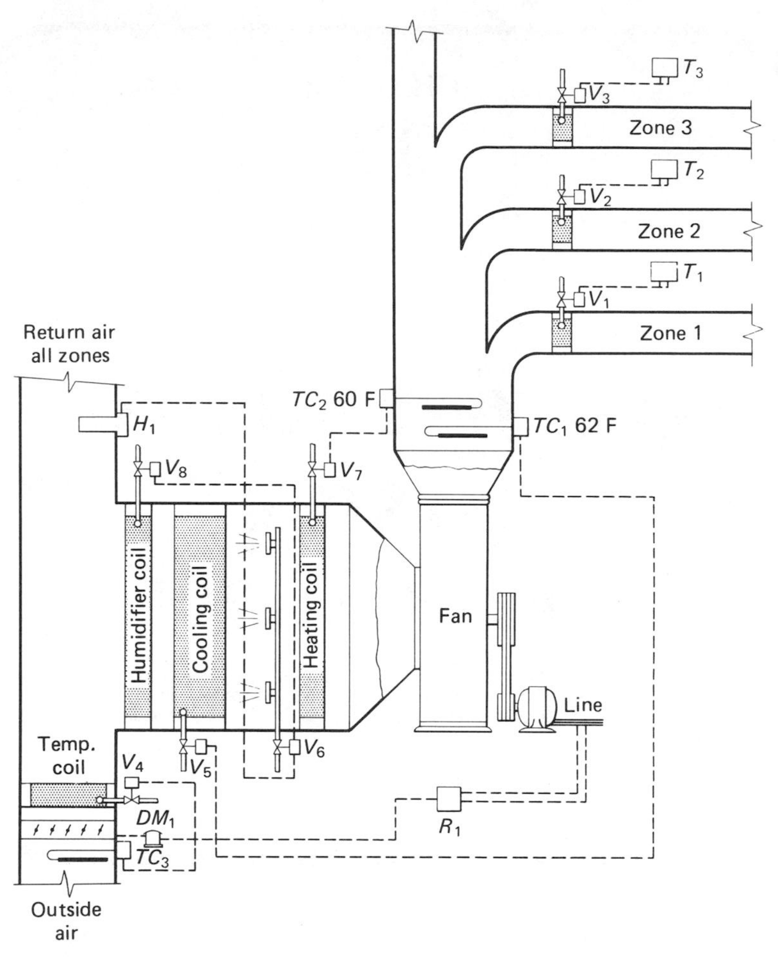

FIGURE 17-1. Single-duct year-round system using zone reheat.

reheat coils are provided for each zone. Figure 17-1 also shows in some detail how a complete humidification pattern could be operated in winter. The recirculated air from the different zones, after mixing with tempered ventilation air from outside, first passes through a humidifier warming coil, which operates in connection with the humidifier sprays. A heating coil follows the sprays. For summer operation the air is chilled by a cooling coil and then passes through the heating coil, which is under the action of the limit temperature control TC_2. Here, if the air temperature falls to less than 60 F, heat is supplied. If the temperature rises above 62 F, the limit temperature control TC_1 increases the cooling. The zone thermostats reheat as necessary. The tempering coil is employed under the action of TC_3 whenever the outside air temperature drops below a predetermined point (perhaps 45 F).

The reheat system works well in that it is easy to provide sensitive temperature control for each and every zone. However, when the latent heat loads in different zones are widely different, precise humidity control cannot be maintained because the central system does not produce at the same time the different dew points needed to serve the disparate needs of each space. Usually precise humidity control is not essential and a compromise dew point is chosen to provide the best overall performance. Reheating can be wasteful of energy because the air not only is cooled but then has to be reheated. If the reheat is extensive and electric units are used, the power costs can be substantial. However, the air by-pass around the cooling coils (see Fig. 6-1) can reduce or even eliminate reheating in many instances.

All-Air System with Precool and Reheat. Figure 17-2 shows one pattern of energy conservation applicable to a system with a substantial dehumidification load. Here return and ventilation air first pass through a precooling coil. The partly cooled air then enters the low-temperature cooling (dehumidifying) coil which condenses moisture from the air as required. The humidistat in the space acts to control the dehumidifier temperature. The water used in the precool coil, having been warmed by the air, is pumped back to the reheat coil, where it can rewarm the air which has been chilled to low temperature in the dehumidification coil. In this way the air is partly warmed and the actual load on the final heating coil is reduced to a minimum

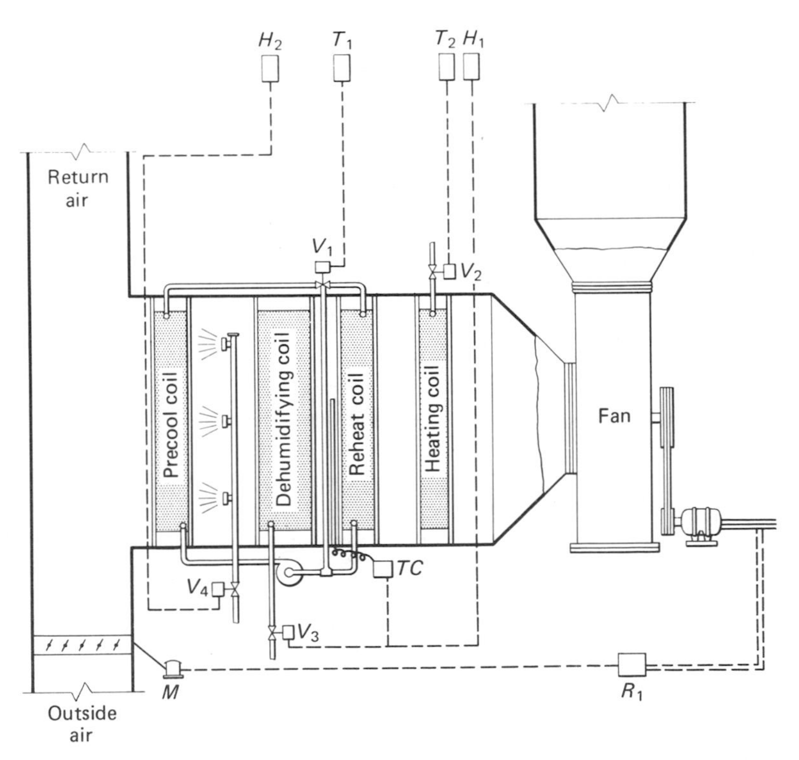

FIGURE 17-2. High-dehumidification system using precool energy for partial reheat in run-around circuit.

amount. This idea of using the same water to cool the air at one point of a cycle and to warm the air at another part of the cycle is known as a run-around system and can, in many cases, result in economy where the dehumidification load is particularly heavy. By close examination of the diagram, one can observe the places at which the thermostats, indicated as T_1 and T_2, and the humidistats, indicated as H_1 and H_2, act on different valves in the system. Notice that a damper motor (M) is placed so as to close the outside-air inlet in the event that the fan power fails. A limit temperature control (TC) operates should the dehumidifier temperature become too cold. For winter operation, a humistat (H_2) can call for humidification should this be necessary.

Example 17-1. Make appropriate calculations for the unit illustrated in Fig. 6-1 when the following operating conditions apply: conditioned space load, 90 000 Btuh sensible and 30 000 Btuh latent; recirculated (return) air at 78 Fdb and 66 Fwb; and minimum outside air for ventilation 1000 cfm, measured at outside conditions of 95 Fdb and 75 Fwb. The primary air supply can be as low as 58 Fdb because induction-type mixing outlets warm the air delivered into the space to acceptably higher levels. Reheating will be used sparingly, only when by-passing at the conditioner is not adequate or at light loads on the space. It is not desired to operate the conditioner coils below 46 F. (a) Find the sensible-total-heat ratio for space conditions and draw it on the psychrometric chart through 78 F, 66 Fwb with the proper slope. (b) Compute the space air flow needed for 58 F primary-air supply to the grilles and find the required wet-bulb temperature that the 58 F air should have. (c) Find the pounds per hour of ventilation air needed and the air conditions resulting from mixing this air with the return air. This represents the supply air to the conditioner. (d) Find the apparatus dew-point coil temperature to which some of this air must be cooled to produce the 58 F mixed air leaving the conditioner by-pass and its coils for supply to the conditioned space.

Solution: (a)

$$\text{ratio} = \frac{\Delta H_s}{\Delta H_T} = \frac{90\,000}{90\,000 + 30\,000} = 0.75$$

Using the Psychrometric Chart, Plate I, draw a line through 78 Fdb, 66 Fwb parallel to 0.75 on the protractor at the upper left center of the chart. Note that Fig. 6-7 shows the protractor and chart with a similar condition line drawn.

(b) Make use of Eq. (6-5) to compute the air flow needed for the sensible load.

$$90\,000 = 0.244\,(m_a)(78 - 58)$$

$$m_a = 18\,443 \text{ lb/h air flow needed} \qquad \textit{Ans.}$$

(c) Outside air at 95 Fdb, 75 Fwb has a specific volume of 14.27 ft^3/lb dry air (Plate I).

$$\text{ventilation air} = \frac{(60)(1000)}{14.27} = 4205 \text{ lb/h} \qquad \textit{Ans.}$$

This amount of air is of course exhausted from the conditioned space and the return air is reduced by that amount:

$$\text{return air} = 18\,443 - 4205 = 14\,238 \text{ lb/h}$$

From Plate I, enthalpy of outside air at 95 F, 75 F is 38.39 Btu; space air at 78 F, 66 F is 30.75 Btu. Mixed air to the conditioner is thus at

$$(14\,238)(30.75) + 4205(38.39) = 18\,443\,\mathbf{h}_m$$

$$\mathbf{h}_m = 32.49 \qquad \textit{Ans.}$$

and similarly its temperature is

$$(14\,238)(78) + (4205)(95) = 18\,443 t_m$$

$$t_m = 81.9 \text{ F} \qquad \textit{Ans.}$$

(d) For this solution, draw a line on Plate I, from the point $t_m = 81.9$ F at $\mathbf{h}_m = 32.49$ Btu/lb and passing through the necessary exit condition from the conditioner at 58 Fdb and 56.5 Fwb. Extending this line to the saturation line shows that the coil dew-point temperature must be closely 54.6 F. *Ans.*

The results found in Example 17-1 show that by-pass warming is adequate to bring the supply air to the inlet temperature at 58 F. However, at light or partial loads in the space, unless the coil temperature and by-pass conditions are changed, reheating will be necessary since no arrangement has been made for reduced volume of air flow.

Example 17-2. Consider the conditioned space of Example 17-1 under severe winter conditions of −23°C (−10 F). The conditioned space is heated exclusively by warmed air from the conditioner to maintain 24.4°C (76 F) at 30% relative humidity. The net sensible heating load loss (over and above lights and people) is 14.65 kW (50 000 Btuh); the occupants in the conditioned space provide the equivalent of 5.86 kW (20 000 Btuh) latent load. Ventilation air of 22.65 m^3/min (800 cfm) measured at outside conditions is needed. Primary air is supplied to the induction-type mixing grilles at 40°C (104 F). (a) Find the air flow through the conditioner needed to offset the net heat loss from the space under the indicated operating conditions. (b) Find the humidity ratio, specific volume, and density of the outside air at −10 F (−23.3°C) if its relative humidity is taken as 50%. Compute then the ventilation air flow in kilograms per hour. (c) Find the amount of water vapor removed from the conditioned space by the wasted air equivalent of the ventilation air supplied. (d) Over and above the water vapor supplied by the latent heat load, find the required water addition from the humidifier to satisfy the indicated operating conditions. (e) Find the humidity ratio of the humidified air as it leaves the conditioner at 40°C.

Solution: (a) Use can be made of Eq. (6-5). In SI units find the kilograms per hour of air flow needed. Note that the factor 0.244 Btu/lb · F = 0.244 kcal/kg · °C. For moist air:

$$C_p = 0.244\,\frac{\text{kcal}}{\text{kg}\cdot{}^\circ\text{C}} = 0.244 \times 1.162 \times 10^{-3}\,\frac{\text{kWh}}{\text{kg}\cdot{}^\circ\text{C}} = 2.84 \times 10^{-4}\,\frac{\text{kWh}}{\text{kg}\cdot{}^\circ\text{C}}$$

$$14.65\text{ kW} = 2.84 \times 10^{-4}\,\frac{\text{kWh}}{\text{kg}\cdot{}^\circ\text{C}} \times m\,\frac{\text{kg}}{\text{h}} \times (40^\circ\text{C} - 24.4^\circ\text{C})$$

$$m = 3307 \text{ kg/h} \qquad \textit{Ans.}$$

(b) Read from Plate II that air at -10 F (-23.3°C) has a specific volume of 11.33 ft^3/lb and a humidity ratio of 0.00023 lb/lb dry air, or 0.23 g/kg dry air.

$$11.33 \frac{ft^3}{lb} \times 0.062429 = 0.7073 \frac{m^3}{kg} \text{ or } 1.414 \frac{kg}{m^3} \qquad \textit{Ans.}$$

A ventilation air flow of 22.65 m^3/min is thus

$$22.64 \times 60 \times 1.414 = 1921 \text{ kg/h} \qquad \textit{Ans.}$$

(c) Ventilation air is wasted from the conditioned space at 30% relative humidity at 76 F (24.4°C). Read from Plate I a humidity ratio of 0.00572 lb/lb dry air or 0.00572 kg/kg dry air. The moisture carried away is thus

$$0.00572 \times 1921 = 10.99 \text{ kg/h} \qquad \textit{Ans.}$$

(d) The latent load of the space adds water vapor amounting to

$$\frac{20\,000 \text{ Btuh}}{1100 \text{ Btu/lb vapor}} = 18.18 \text{ lb/h} = 8.26 \text{ kg/h}$$

or by an alternative solution

$$\frac{5.86 \text{ kW}}{0.7107 \text{ kWh/kg vapor}} = 8.25 \text{ kg/h}$$

A moisture balance is required to find how much water vapor the humidifier must add.

Water vapor leaving the system (from part c) =		10.99 kg/h
Less water vapor entering the system with the ventilation air 1921 × 0.00023 kg/kg dry air =	0.44	
Less water vapor provided from the latent load =	8.25	
Total deductions =		−8.69 kg/h
Net water needed from the humidifier =		2.30 kg/h *Ans.*

(e) The conditioner must deliver a water vapor supply to the conditioned space adequate to balance:

water vapor in the recirculated air, (3307 − 1921)(0.00572) =	7.93kg/h
water vapor exhausted to waste =	10.99 kg/h
less the water vapor added by people from latent load =	8.25 kg/h
Total (net)	10.67 kg/h

This 10.67 kg/h of water vapor from the conditioner is carried by the 3307 kg/h of air flow; thus

$$\text{humidity ratio} = \frac{10.67}{3307} = 0.00323 \text{ kg/kg dry air} \qquad \textit{Ans.}$$

Single Duct System With Variable Air Volume. The systems just described adjust to meet changes in space cooling loads by raising or lowering the supply air temperature and also by varying the dew-point temperature as may be required. Variable-volume systems compensate for changing space cooling loads by varying the volume of cooled or heated air delivered into the space. Because the volume of air delivered is reduced with load reduction, refrigeration and fan power demands are also reduced. For the single room or area, reduced load requires merely throttling down the air supply into that room. However, for a fall off in load on the whole system the fan output must also be reduced by use of fan inlet dampers, by passing to the return side, or by other static control methods. Care must be exercised to see that air flow to a room or area never drops to such a low point as not to provide for adequate ventilation and moisture pickup. Noise can also be a factor if dampers are throttled down to the point of creating disturbing sounds at room outlets. With proper care in design and operation, variable air volume systems will have low operating costs and give satisfactory performance.

High Velocity Conduit Induction Systems. The previous sytems described all had one thing in common, a central unit where all of the air used in the cooling or warming process was conditioned to serve the occupied space. In some arrangements reheat coils also appeared adjacent to the occupied space, but these coils functioned merely to give a final tempering to the supply air and not as basic heater devices. The total heating or cooling load in the space was thus completely served by supply air and, because of this, relatively large quantities of air were required, necessitating large ducts for both supply and return. When induction systems are used only part of the heating or cooling load is provided by the supply air because each room unit contains a heating or cooling coil that provides the major energy source for the space. This coil in turn is served by a chilled or heated water supply.

Because so much less supply air is required it has been found advantageous to deliver the air through conduits at high velocity and at elevated static pressures, and usually no provision for return of the supply air to the central conditioner is made. In Section 12-7 several types of air-terminal induction units are described in detail.

Induction systems are particularly applicable when a large number of similar rooms need to be served by a central system. Such is the case in hotels and office buildings, where systems of this type have been used extensively. In office and commercial buildings which do not operate at night it is possible to shut off the air flow during such periods, and in winter the heating coil can act as a convector to keep the rooms tempered even with no primary air flow.

Example 17-3. An induction-air room cooler supplied with conduit air is installed in a small office having a sensible cooling load of 5200 Btuh. A primary air conduit supplies 55 cfm to the unit at a temperature of 56 F. Room temperature is maintained at 80 F and chilled water is supplied to the coil at 52 F. It is deemed inadvisable to deliver air from the top outlet of the module colder than 62 F even though this is directed upward and does not strike any occupants. In this unit inducted and not primary air passes over the coil. (a) Compute the cooling contribution of the primary air and by difference the cooling needed from the induced air passing over the coil. (b) If it is assumed that 4.5 cfm of inducted air flows for 1 cfm of primary air, compute the temperature change needed by this air to remove 3775 Btuh from initial room conditions of 80 F. (c) For the primary and induced air flow at their respective temperatures, find the delivery temperature from the unit.

Solution: (a) By Eq. (5-13), the 55 cfm of primary air ultimately warming from 56 F to 80 F room temperature produces cooling amounting to

$$q_s = (1.08)(55)(80 - 56) = 1425 \text{ Btuh} \qquad \textit{Ans.}$$

Remaining cooling load = 5200 − 1425 = 3775 Btuh, which must be provided by the coil of the unit as this chills the induced air passing over it.

(b) Let us assume that 4.5 cfm of air can be inducted over the coil for 1 cfm of primary air. Then for 3775 Btuh,

$$3775 = (1.08)(\text{cfm})(\Delta t) = (1.08)(4.5 \times 55)(\Delta t)$$

$$\Delta t = 14.2 \text{ deg} \qquad \textit{Ans.}$$

or

$$80 \text{ F} - \Delta t = 80 - 14.2 = 65.8 \text{ F}$$

This is the temperature to which the coil with 52 F inlet water should cool the inducted air.

(c) The temperature of the mixed air delivered from the unit is thus

$$(55 \text{ cfm})(56 \text{ F}) + (4.5 \times 55)(65.8 \text{ F}) = (302.5)t$$

$$t = 64 \text{ F} \qquad \textit{Ans.}$$

This is sufficiently warm for delivery into the space.

For induction units it is always necessary to made use of manufacturer's performance data to assure the adequacy of the unit.

17-2. DUAL DUCT SYSTEM

The dual-duct system deviates so greatly from the previously described systems that detailed discussion is required. This system has wide applicability but is particularly effective in structures having different zones or types of occupancy, such as office buildings, hospitals, hotels, schools, and stores. The system provides for central location of the heating and cooling plant, from which heating or cooling can be distributed to the multiple areas of a large building. Many variations in the arrangement of a dual-duct system are possible. Figure 17-3 uses more than the minimum number of elements but provides a degree of flexibility not possible with simpler dual-duct systems. All dual-duct systems use an air-supply duct carrying cool air and an air-supply duct carrying warm air. Air from these ducts enters a mixing and volume-control box which tempers and delivers the air-supply for a specific space or zone. Similar boxes serve other spaces to mix the cool and warm air to meet the desired conditions for each space.

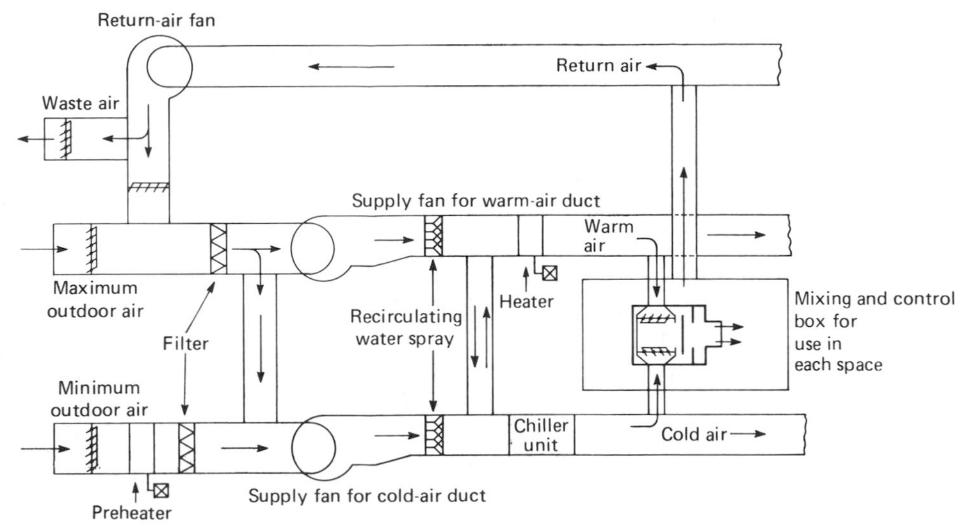

FIGURE 17-3. Dual-duct system using two supply fans.

Since both heating and cooling are available, the system can operate equally well in summer and winter, and between-season shutdown of part of the system is not necessary. The system is also flexible in that it is possible to use outdoor air to its fullest extent for between-season cooling when the outdoor temperature is suitable. This is made possible by having the system arranged with two outside air inlets—the one provides for maximum outdoor air supply and the other for minimum outdoor air supply (ventilation requirement only). Under summer conditions it is undesirable to use more than minimum outdoor air and the maximum outdoor air dampers, which are interlocked with the others, are closed. Each of two supply fans carries about 50% of the total air flow and at high loads the air flow in the two ducts is in this ratio. The cold air stream in summer usually provides air chilled to a 50 F to 55 F temperature range with the air almost saturated and at a low humidity ratio. When maximum cooling is required, a large fraction of this chilled air must be provided to each space and a smaller fraction of warm air is needed. Since the by-pass duct on the discharge side of the fans freely connect the two ducts, it follows that more air will flow in the cool duct. This occurs because the warm-air dampers at the mixing control boxes are largely in closed position. In contrast to this, if the cooling load in a space is light, less chilled air is required and a larger fraction of warm air from the duct is needed. This type of operation delivers more high-humidity warm air to the space than is justified from a moisture standpoint and the relative humidity in the space rises. To offset this happening, it is necessary to operate the heater coil in the warm-air duct. The warmer air makes it possible to provide a larger fraction of colder dryer air for the space and the relative humidity there lowers to a desirable range.

It is obvious that air cannot be stored in any space. Thus the return- or waste-air system must always exhaust an amount of air equivalent to the amount of outdoor air brought into the system. Moreover, with each of the controlled spaces requiring varying amounts of cool or warm air, the flow in the two ducts can be widely different. Where minimum outdoor air is provided, the return air can move freely into either the warm-air or the cold-air system by means of the by-pass connection preceding the fans.

The dual-duct system is also advantageous in that for a large building in which the outside areas need heating in winter while the inside areas need cooling, it is easily possible to control and supply the needs of both types of areas. Many systems are not nearly so complex as that shown in Fig. 17-3 because in many, a single supply fan, by means of control dampers, delivers air to both of the ducts and forces the air through appropriate heaters or coolers. Similarly, some systems do not employ a return-air fan, although this makes control of the system less flexible. In winter the temperature of the supply air in the warm duct often ranges between 110 F and 130 F. In summer it is seldom brought above 86 F and frequently is at return-air temperature. A dual-duct system, although it cannot provide exact control of humidity in every space, particularly when the sensible to total-heat ratio varies greatly in the different spaces, nevertheless does give approximate humidity control along with precise temperature control and it represents a very satisfactory method of operation.

17-3. DUAL DUCT SYSTEM CALCULATIONS

The air flow in the system is based on the sensible-heat load and sufficient capacity must be provided in each duct to serve the maximum summer sensible cooling load (Q_{sc}) and the maximum winter sensible heating load (Q_{sw}). Making use of an equation in the form of Eq. (5-13)

$$q_c = \frac{Q_{sc}}{1.08(t_r - t_c)} \tag{17-1}$$

$$q_w = \frac{Q_{sw}}{1.08(t_h - t_r)} \tag{17-2}$$

where q_c and q_w represent the capacity in cfm required for the cold and hot ducts, respectively; t_r is the temperature held in a space, that is, temperature of the recirculated air, F; t_c is the temperature in the cold-air duct, F; t_h is the temperature in the warm-air duct, F.

The expressions apply to either a whole system or to a room, space, or zone of the system as may be required, and the maximum flow ultimately determines the fan and duct sizes. In operation with either one or two supply fans, the total flow q is

$$q = q_c + q_w \tag{17-3}$$

It is, however, true that the solving of Eqs. (17-1) and (17-2) does not offer an optimum solution for the summer and winter air flows because of by-passing between the two ducts and also because, for each space, air flow from both ducts occurs and the air-inlet temperature will differ both from the very low value t_c and the high value t_h. In fact, the maximum flow in the warm duct usually occurs during light summer loads when the warm air temperature is held at a low level (80 F to 90 F) or during intermediate-season operation. As a basis for starting the computation, it is suggested that an appropriate summer cooling-air flow first be found and the warm-air flow then be expressed as an additional percentage of this flow; 30 to 40% is suggested. This will give a trial air flow which can be corroborated or if necessary revised after use is made of the following equations.

For summer

$$1.08q_c(t_r - t_c) = Q_{sc} + 1.08q_w(t_h - t_r) \tag{17-4}$$

For winter

$$1.08q_w(t_h - t_r) = Q_{sw} + 1.08q_c(t_r - t_c) \tag{17-5}$$

Observation of Eq. (17-4) will show that the left-hand term represents the cooling produced as the cold air (t_c) warms to space temperature (t_r). This cooling must serve two purposes, namely, handling the sensible cooling load for summer Q_{sc} in Btuh, and also cooling the warm-duct air at t_h to the space temperature (t_r). Equation (17-5) works in similar manner but in reversed sense to offset the sensible winter heat loss, Q_{sw} Btuh, and to heat the mixing air from the cold duct.

Example 17-4. For a dual-duct system, find the basic air-flow patterns if the system has to serve under the following conditions: in summer, a sensible cooling load of $Q_{sc} = 1\,600\,000$ Btuh, latent load 760 lb of water per hour, space temperature to be maintained $t_r = 78$ F (temperature of recirculated air), temperature of chilled air in cold-air duct, $t_c = 52$ F, temperature of warm air in warm duct $t_h = 86$ F. In winter, sensible heating load $Q_{sw} = 2\,400\,000$ Btuh, space temperature and temperature of recirculated air $t_r = 75$ F, temperature of chilled air in cold-air duct 55 F, temperature of hot air in warm duct $t_h = 125$ F. In summer chilled air cannot enter the space colder than 56.0 F without producing discomfort.

Solution: Make use of Eq. (17-1) to find a trial value of q_c

$$q_c = \frac{1\,600\,000}{1.08(78 - 52)} = 57\,000 \text{ cfm}$$

Increase this value by an assumed 35% for total air flow:

$$q = 57\,000(1.35) = 77\,000 \text{ cfm}$$

Use this value, 77 000 cfm, with Eqs. (17-3) and (17-4) for summer conditions:

$$77\,000 = q_c + q_w$$

$$1.08q_c(78 - 52) = 1\,600\,000 + 1.08q_w(86 - 78)$$

Solving these two equations simultaneously,

$$q_c = 61\,600 \text{ cfm} \qquad q_w = 15\,400 \text{ cfm}$$

Make a rough check for the inlet temperature to the space (t_i):

$$(61\,600)(t_c) + (15\,400)(t_h) = 77\,000\, t_i$$

$$t_i = 58.8 \text{ F}$$

Since this is higher than the 56 F minimum allowed on inlet, the flow ratios are suitable for summer operation.

For winter operation, substitute in Eqs. (17-3) and (17-5)

$$77\,000 = q_c + q_w$$

$$1.08q_w(125 - 75) = 2\,400\,000 + 1.08q_c(75 - 55)$$

$$q_w = 53\,700 \text{ cfm} \qquad q_c = 23\,300 \text{ cfm}$$

The problem is not complete until a check is made to determine whether the inlet air can be sufficiently dry at 52 F to absorb the 760 lb of water latent load without having the space humidity rise above an acceptable relative humidity. The situation is complicated by also having to consider the moisture brought in with the warm-air stream. The solution will not be worked out here but is presented at the end of the chapter as a problem.

Notice that this total air flow of 77 000 cfm can effectively serve the sensible load with different warm- and cool-air flows for summer and winter. Other air flows are also possible to made a satisfactory solution, and attention must also be given to part load operation.

Ceiling Induction Systems. Here the diffuser outlet is usually placed in the suspended-ceiling cavity. above the conditioned space. Primary air for the unit is supplied from either a conduit supply source or from conventional ductwork. Air from the space is inducted to mix with the primary air and the mixed air is then directed into the room. Although a reheat coil may be provided with a unit of this type, this is not usually done unless the internal load of the space is slight and the resultant mixing temperature of the air is too low for comfort. Often arrangements are provided to control the primary air flow so that the unit may operate as a variable air-flow system.

17-4. OTHER AIR CONDITIONING SYSTEMS

The basic central air-conditioning systems which have just been described are merely representative of numerous arrangements which can be envisaged by a designer. It should be noted that with the exception of the conduit-induction system, all of them produced their basic heating and cooling by means of convected air from a central point. The so-called *split system*, which deviates from this pattern, is a design that is extensively employed. All split systems supplement the heating or cooling needed by using water, steam, or sometimes direct electric heating to supplement the air control provided by the central conditioner. Split systems are particularly useful in the exposed or outer zones of large buildings where in winter so much heat loss occurs that it would be difficult to supply all the heating needed by air from the central system. Consequently, direct radiation is used to provide heat and the air-conditioning system provides tempered air, primarily for ventilation and humidity control.

Yet a different arrangemeent involves the use of units in a conditioned space which receive no conditioned air but instead are provided with fans for recirculating air from the space along with outside air for ventilation. This air is passed over coils supplied with only chilled or warmed water from a central unit. Since no air moves from the central unit, the cost of running ductwork is eliminated and the building space demand is reduced.

Finally, mention should be made of the complete air-conditioning units which are installed within spaces. These necessarily involve a refrigeration unit for providing the cooling, a means of dissipating the heat from the refrigeration unit either through an air circuit to outside or by means of circulating water, and, finally, a fan for recirculating air from the conditioned space along with sufficient outside air for ventilation. Frequently these are free-standing units that are placed along a wall of the space with suitable connections made for the various services required. A small version of a unit of this type is the familiar window unit which serves a similar function for an individual room or for a small space.

17-5. HEAT RECOVERY

A very usual problem which arises in larger buildings is the need for cooling year-round for the interior spaces while the peripheral areas of the building may require heating under winter conditions. It is also true that most computer rooms, as well as many laboratories and work spaces have such high internal heat loads that cooling is required during all seasons. Buildings with large ventilation requirements are particularly wasteful of energy unless heat-recovery or cooling-conservation practices are followed. Methods for accomplishing heat recovery are well known, but when energy was cheap and abundant there was little incentive to design for heat conservation because of the substantial increase in first cost, combined with additional maintenance expense and a more complicated operating cycle.

That substantial savings are possible can be realized from examining the heat-recovery system, diagrammed in Fig. 17-4. This system has a high ventilation-air demand (5000 cfm); thus as much of the waste air as can be collected (here 4000 cfm) is passed over the heat-exchange surface to absorb some of its energy before discharge to outside takes place. The liquid warmed in the heat exchanger is pumped to the supply-air heat exchanger where it in turn warms the ventilating air. For the winter conditions indicated, simple calculations can show that the heater-coil demand is reduced by 36% when this *run-around type* of liquid circuit heat exchanger is employed.

The arrangement shown in Fig. 17-4 is applicable to other temperature ranges in winter. The system is also applicable to the cooling (summer) cycle, in which situation the hot outside air is prechilled in the run-around circuit before entering the cooling coil of the conditioner. The magnitudes

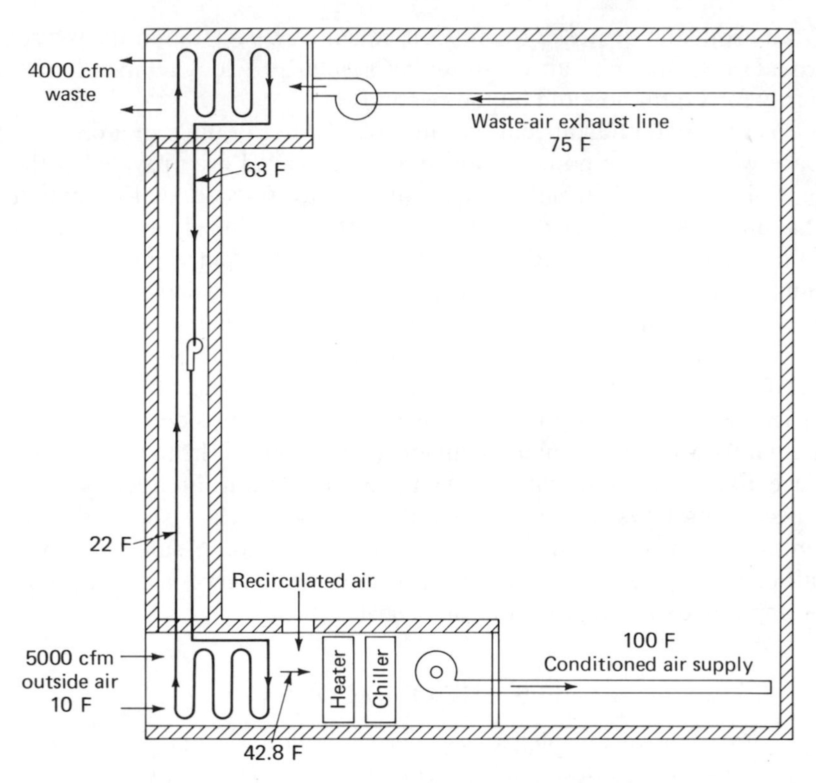

FIGURE 17-4. Representative energy conservation system for ventilation air using heat exchangers in a liquid run-around circuit.

of the savings diminish appreciably as the outside and inside temperatures approach each other, reaching zero when the temperatures are the same. In fact, the need of a significant temperature difference (from some 5 to 15 deg) for producing the effective heat flow is a deterrent to performance under less than optimal conditions.

However, the opportunities for energy conservation in complex structures are almost unlimited. Additional patterns will be discussed later in connection with the heat pump. Here it should be mentioned that besides liquid heat-transfer circuits and liquid storage of energy, use can be made of rotary mechanical devices and of heat pipes (tubes) extending from the warmer to cooler passages. In the heat pipe (tube), the heat is transferred by the passage of warmed vapor in a closed circuit with condensed liquid later returned by capillary action.

We must recognize that the core areas of larger buildings often generate so much heat from lights, machinery, and people that year-round cooling is needed there while the peripheral areas of the same building require heating only in winter. Conserving heat from energy otherwise wasted to outside and putting this heat to use in the perimeter areas represents a fertile area for design innovation. The problems are not simple ones, however, because often the heat exchanges must take place under conditions of very small temperature difference between warmer and cooler zones.

17-6. THE HEAT PUMP

The use of electric energy for heating has been growing at a rapid rate. Two basic factors account for this growth. The first is related to the fact that the peak capacity of most electric utility companies is required during the summer season, which makes excess capacity available for heating in winter. The second factor is that the cost of gas and oil fuels has greatly increased and this increase is apparently continuing. However, the fuel-cost impact for nuclear and coal-fired power plants has not been nearly so great, so that the rise in cost of electric energy has been relatively less than that of direct fuel. On balance, the cost of electric energy for heating in many instances can compete with the cost of heating using fossil fuels. When, in addition, the savings possible by using the heat pump are considered, year-round electric systems often present an optimum design arrangement.

Any mechanical refrigeration system has the necessary components to constitute a heat pump, because, whenever mechanical work is available for a refrigeration system, energy absorbed at a low temperature can be made available for use at a higher temperature. The key to success for a heat pump is having available a heat source that is not unreasonably cold. Regions where the winter temperatures are mild, seldom dropping below the 30 F to 40 F range (0°C to 5°C), are particularly suited to heating by means of the heat pump. In other areas, the heat pump is also applicable wherever there are suitable heat sources available, such as from well water or from industrial processes or energy from those interior areas of buildings which require cooling even in mid-winter. Direct electric or other heating may be required to supplement the heat pump of a system under unusually severe conditions. Where space heating alone is under consideration, the fixed investment in a heat-pump system is usually higher than for a direct-fired system, and thus the heat-pump system is most advantageous when it is also used for cooling at appropriate times.

Figure 17-5 shows in schematic arrangement how a heat-pump system can be substituted for the furnace of a residence or building which employs forced-warm-air circulation for heating. The liquid refrigerant from the condenser and receiver enters the expansion valve and expands into the cold evaporator. Here vaporization of the refrigerant is carried out as it absorbs heat from the cold outside air passing over the coils. The vaporized refrigerant then enters the compressor, where mechanical energy is added to raise the pressure and temperature of the refrigerant. The hot refrigerant then passes to the condenser, where it gives up superheat and enthalpy of vaporization in useful heating. For example, if 200 Btu are added in the evaporator and 100 Btu are supplied by the compressor, then 300 Btu are delivered to the condenser and the *heating coefficient of performance* (HCOP) is $\frac{300}{100}$, or 3. This figure is representative of what might actually be experienced

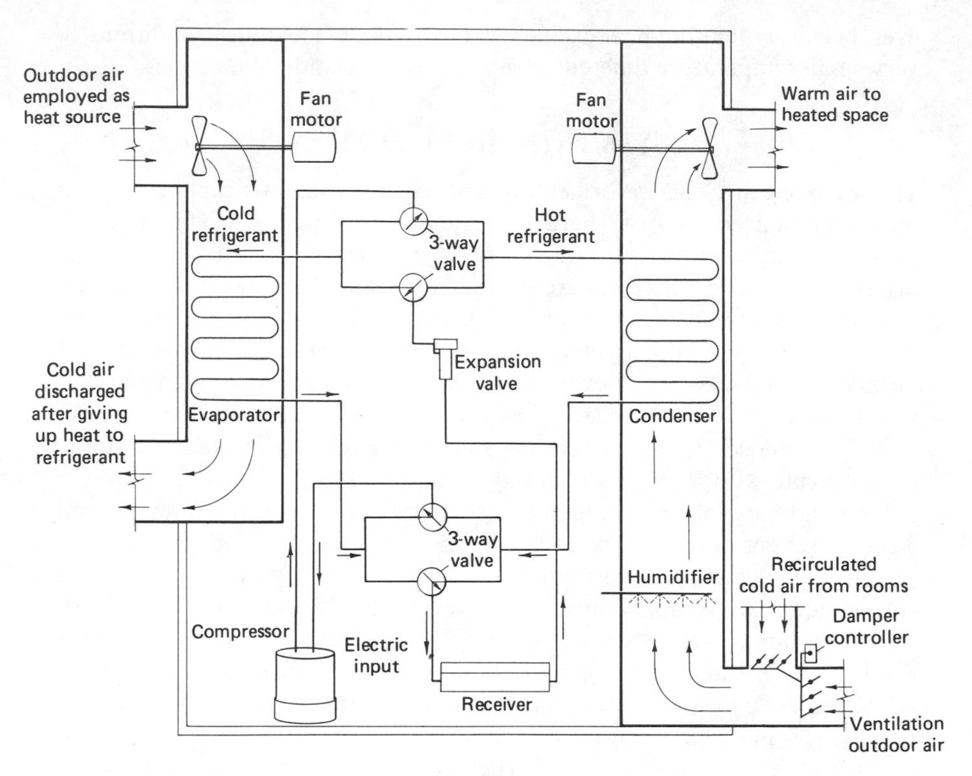

FIGURE 17-5. Diagram of heat pump used to heat residence.

with a temperature of 30 F outside. At 55 F outside temperature, the corresponding value for the HCOP would be 4. Because of the temperature difference required for heat transfer, the refrigerant is at least 10 deg colder than the air passing over it.

Notice that by changing the four three-way valves in Fig. 17-5 the same system can be used for cooling. The evaporator in this case becomes the condenser, and the former condenser serves as the evaporator for cooling the residence air.

Equation (14-3) showed that the coefficient of performance (CP) of a refrigeration system was expressed as the ratio of the refrigeration produced (Q_R) to the work (W) required to produce it. Equation (14-3) can also be written as

$$\mathrm{CP} = \frac{Q_R}{W} = \frac{Q_C - W}{W} \tag{17-6}$$

where Q_C represents the heat transfer at higher temperature during condensation. The useful output that could be derived from a heat pump (Q_C) is greater than the refrigeration effect Q_R to the extent of the work addition

W; that is, $Q_C = Q_R + W$. In equational form, the heating coefficient of performance is expressed as

$$\text{HCOP} = \frac{Q_C}{W} \tag{17-7}$$

Example 17-5. A certain residence with a design heat loss of 60 000 Btuh employs a heat pump. The pump has a refrigeration coefficient of performance of 3 when it is 50 F outside and when 70 F is maintained inside. Under these conditions the condenser and evaporator coil temperatures are, respectively, 80 F and 40 F. Disregarding power used by auxiliaries, compute (a) the kilowatts required from the power lines when the overall motor-compressor efficiency is taken at 0.82 and (b) the heating coefficient of performance.

Solution: (a) Substituting in Eq. (17-6), the shaft work is found as follows:

$$3 = \frac{60\,000 - W}{W}$$

$$W = 15\,000 \text{ Btuh}$$

Therefore, the number of kilowatts supplied is

$$\frac{15\,000}{(0.82)(3413)} = 5.35 \text{ kW} \qquad \textit{Ans.}$$

where 3413 represents the heat equivalent of a kilowatthour. (If power can be purchased at 2.5 cents per kilowatthour, the 5.35 kW represents an operating cost of 13.4 cents per hour.)

(b) The heating coefficient of performance, by Equation (17-7), is thus

$$\text{HCOP} = \frac{Q_c}{W} = \frac{60\,000}{15\,000} = 4 \qquad \textit{Ans.}$$

Table 17-1 gives specific data on two small heat-pump units of one manufacturer.

Table 17-1 Heat-Pump Specifications

Item	3 Hp	5 Hp
Winter heating output, Btuh*	32,500	54,200
Summer cooling rating, Btuh†	30,000	50,000
Winter: Heat delivered to	Building air	Building air
Summer: Heat dissipated to	Outdoor air	Outdoor air
Winter heat source	Outdoor air	Outdoor air
Winter supplementary heat	Electric strip heaters	Electric strip heaters
Indoor fan, cfm	1200	2000
Outdoor air fan, cfm	1200	2000
Compressor type	Hermetic	Hermetic
Size	50″ × 29″ and 75″ high	71″ × 29″ and 75″ high
Design for house having	5–6 rooms	6–8 rooms

* Based on 70 F inside, 35 F outside.
† Based on 80 F db and 67 F wb inside; 95 F outside.

17-7. HEAT SOURCES FOR HEAT PUMP

Air. The most obvious energy source for the heat pump is the air. Where the air temperature does not go to low extremes, this source is quite adequate. Certainly in climates where winter temperatures are in general above 32 F, air can be considered an effective and useful source of low-temperature energy. In regions where the temperature does drop to values around and below 32 F, not only is the temperature lift of the system increased, but the problem of frost formation on the heat-transfer coils may become serious and provision must be made for defrosting. The main disadvantage of air-to-coil transfer lies in the fact that the maximum heat-loss loading on the building occurs simultaneously with the lowest temperature available for supplying energy to the coils. This lowers the HCOP; and not only is more power needed per unit of heating output, but this condition occurs when the demand is greatest. The widespread use of the heat pumps within a given utility system might cause a very heavy loading on the system with every extreme drop in atmospheric temperature. If such a peak coincided with other peak loads of the system, generating capacity might be overextended or even inadequate.

Many designs of individual room-cooler, heat-pump units are now on the market. With the addition of supplementary electric heaters they can function under low outside-temperature conditions. Most are also provided with automatically cycling, electrically heated defroster units so they can operate under adverse atmospheric conditions. Most of the larger units are, however, centrally located, sending air or water, either chilled or warmed, to points of utilization.

Well Water. The temperature of water in wells does not deviate greatly from an annual minimum value; consequently, when such water is available, it forms an excellent source of energy for heat-pump coils. However, the cost of sinking a well and pumping the water must be considered, and unfortunately the water-level table in the United States has been dropping in almost all sections of the country, because of greater usage and because of less effective rain absorption into the soil (resulting from intensive land use and bare fields).

Some double wells have been drilled as a means of water conservation. Water is pumped from one of the wells, and the cooled water, after passing through the system, is forced back into the other well. The second well must be placed at a sufficiently great distance from the first well to permit the water, in percolating back to the source point, to be warmed by contact with the subsurface of the earth. The additional complications of this system, and the cost of the second well, are deterrents to its extensive use.

Underground Water. In the case of the double-well system, underground water is employed by bringing it to the surface of the earth and then sending it back. It is sometimes possible to accomplish the same result by placing a coil below ground in such manner that a moving underground current passes over the coil. It is desirable, of course, that the water be continuously moving; otherwise the coil will freeze a layer of ice around it as heat is abstracted, and the useful process of heat transfer will be retarded as the insulation effect of the ice becomes greater and greater. The problem of burying such a coil and finding a source of underground, water is, of course, difficult.

Earth Coils. Instead of depending on continuous movement of water below ground, it is possible to bury a network of pipe coils so arranged that they come into intimate contact with the earth itself. Heat can then flow from the earth into the colder coil. In turn, the temperature of the earth adjacent to the coil becomes reduced. It is necessary that the coil remove heat at a slow enough rate to prevent an excessive localized lowering of ground temperature. Earth has such widely different conductivities, ranging from 0.2 to 2.0 $Btu \cdot ft/h \cdot ft^2 \cdot F$, that designs of this nature are somewhat uncertain. If evaporator operation is carried out continuously at low temperatures, the moisture in the earth around the buried pipe freezes. This is not a serious disadvantage, since frozen ground with good moisture content has a higher thermal conductivity than unfrozen ground, namely, from 1 to 2.5 $Btu \cdot ft/h \cdot ft^2 \cdot F$. In the winter season the temperature of an underground area supplying coils would be greatly reduced, and therefore it might be desirable in summer to reverse the system and put back into the ground, with the former evaporator acting as the condenser of the system. The ultimate source of ground heat of this type is largely the sun, heat being obtained by conduction from the sun-warmed surface of the earth during summer periods. A relatively small amount of heat passes from the core of the earth outward. If sufficient investment is made in burying an adequate surface of pipe coils, the earth can be an effective source of heat for pumps.

With $\frac{3}{4}$-in. copper tubing buried at a depth of 6 ft, heat-absorption rates have ranged from 50 to 25 Btuh/lin ft over a winter season when the coil temperature was about 21 F and the minimum temperature of the uncooled ground not adjacent to the coil was 41 F.

Miscellaneous Sources. Natural lakes or even artificial water-storage reservoirs can be considered as possible heat-pump sources. Certainly such storage could be used to carry peak demands in the case of a normal air-to-air operating system. Even the possibility of using city water for peak demands may be feasible in some cases. The possibility of freezing water and utilizing this latent heat of fusion might also be considered, but the inflexibility of rigid ice, and the insulating effect of ice, make this scheme appear not particularly feasible.

Of increasingly great importance for large buildings are the various designs which make use of heat picked up in the inner (core) areas for use in heating the perimeter areas in winter. Such designs can be extremely flexible, even providing for heat storage in appropriate water tanks, when the necessary heat dissipation from the core areas exceeds the heat demands of the perimeter areas. When the imbalance is greater or when all areas require cooling, it is necessary to use cooling towers to dissipate the surplus heat.

Solar energy has also been employed as a heat-pump source. Here the usual arrangement is to have a solar collector arranged in such manner as to heat water during active insolation and store this heat in appropriate water tanks. When the temperature rise of the water is adequate, energy is available to serve as the heat-pump source at night and during cloudy periods. During long periods of solar inadequacy it becomes necessary to maintain proper water temperature by using fuel or electric heat.

17-8. HEAT PUMP ARRANGEMENTS

Methods of using the heat pump are many and varied. For example, Fig. 17-5 demonstrated an arrangement in which the refrigerant flow could be altered so that the evaporator could become the condenser and the condenser the evaporator or, by a simple reversal of the three-way valves, returned to the original arrangement. If a different arrangement appears preferable, it is possible by a change in the ducting layout and use of appropriate dampers to direct the air into the space so that it passes either over the condenser for space heating or over the evaporator for space cooling. Yet another arrangement, where water is used as the heat-transfer medium, is to direct the chilled water from the evaporator to the coils for chilling the space air when this is desired or, alternatively, to direct the warmed condenser water to the space coils when space heating is desired. Other arrangements of these design patterns are also possible, some of which are so ingenious that patents have been taken out on them and their exact utilization is subject to licensing by payment of a royalty to the patent holders, or one may avoid infringement by purchase of equipment from a licensed manufacturer.

One such arrangement, suitable for a complex building in which year-round cooling of the core area is required, is shown in Fig. 17-6. Here a large number of heat-pump units operate from a common hydronic circuit. This circuit always carries water in the range of approximately 70 F to 95 F. When heating of a particular space in the building is needed, this water acts as a heat source for the evaporator and the then air-cooled condenser of the unit delivers warmed air for heating that space. If cooling is needed elsewhere in the building, the refrigerant connections are reversed to the unit there, so that the evaporator provides chilled air for that space and water from the common hydronic circuit serves the condenser.

Figure 17-7 shows in detail the operational pattern for heat-pump units that might be employed in energy-conservation circuits. Each unit serves a room or zone of the building for either heating or cooling. When heating is needed, the reversing valve is set as shown in the left-hand diagram of Fig. 17-7. Water flow from the closed-loop system provides heat to the

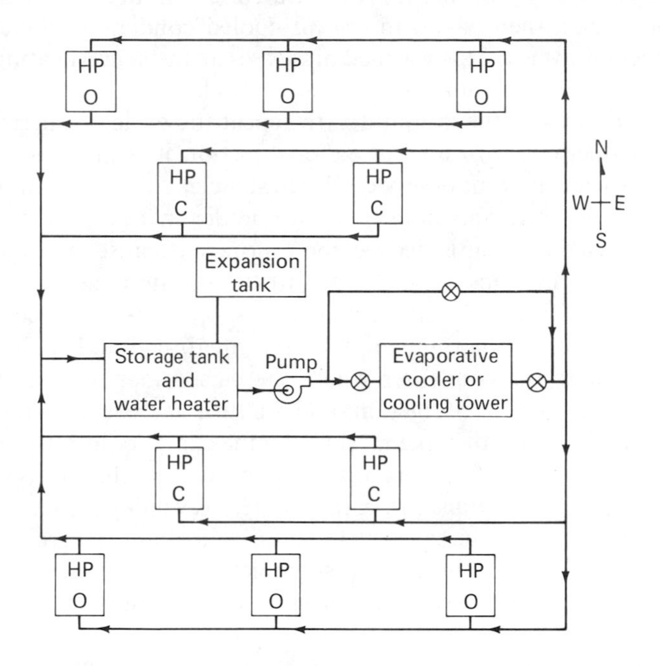

FIGURE 17-6. Diagrammatic layout of a closed-loop system showing six outside heat pumps and four core area heat pumps.

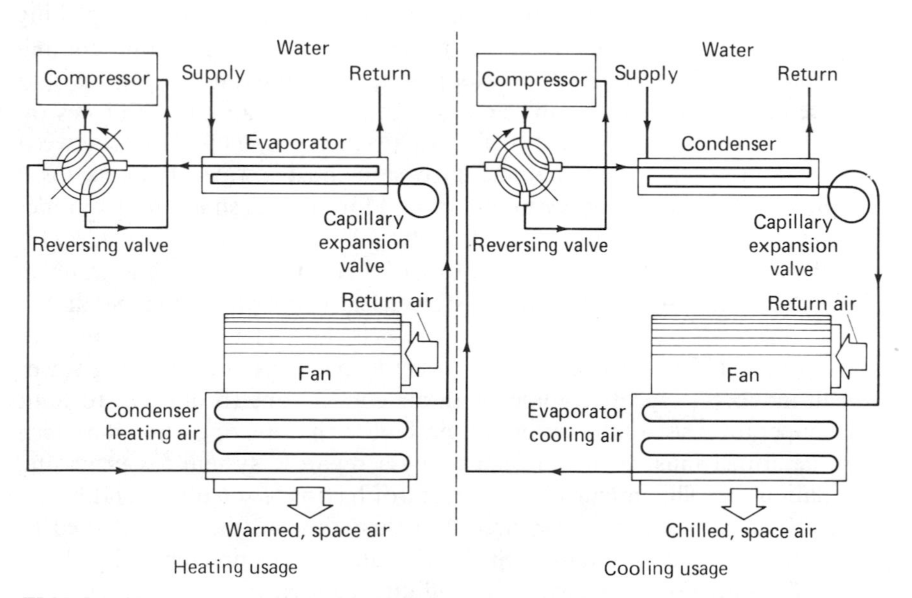

FIGURE 17-7. Heat pump and chiller unit, equipped with reversing valve, arranged to operate from common water source, for both heating and cooling.

evaporator. The vapor from the evaporator, after compression with resulting temperature rise, then passes to the air-cooled condenser where it warms the air passing over it. This warmed air serves as the space-heating medium. The condensed refrigerant then passes through a capillary (or expansion valve) and returns to the evaporator to repeat the cycle of operation. Since a capillary operates with flow in either direction it is preferable to an expansion valve for this type of service. When sufficient heating of the controlled space has taken place, the unit shuts down under thermostatic action.

When space cooling is needed the thermostat causes the internal mechanism of the reversing valve to turn to the position shown in the right-hand diagram of Fig. 17-7 and the compressor operates on the cooling cycle. Vapor flows through the system in the opposite direction. Leaving the compressor, it enters the condenser where the closed-loop system water now absorbs heat from the refrigerant and in turn increases in temperature. Condensed refrigerant then passes through the capillary into the evaporator where it cools the air for entry into the space. The thermostat stops the unit when space conditions are satisfied. The reversing valve, here shown diagrammatically, accomplishes its function by a simple 90° rotation. Commercial valves under magnetic or solenoid action accomplish the same result by opening and closing internal ports within the sealed body of the valve.

The system shown in Fig. 17-6 could represent a single-floor structure or be part of a multistory building. Usually in winter the perimeter outside rooms will be on heat-cycle operation and consequently act to reduce the temperature of the water in the loop. Inside or core-area rooms, picking up heat from lights, machinery, and people, will probably require cooling and consequently add heat to the loop water. In mild climates, this would also be the case for rooms subjected to high solar loading through fenestration. However, it is seldom possible to reach a perfect balance between heating and cooling demands and provision must always be made for a supplementary energy supply in winter and for heat dissipation in summer or even in fall and spring.

When water temperature drops to 70 F (21.1°C) or lower, the water in the storage tank must be heated either electrically or by fossil fuel. Similarly, in summer and even in other seasons, when the water temperature rises toward 95 F (35°C), it becomes necessary to make use of a cooling tower (or an indoor evaporative cooler with air connections to outside) to limit the temperature rise of the system water. Notice that the operational pattern here described must be supplemented by a separate system for providing ventilation air. This might take the form of having a separate conditioner heat or cool minimum air required and this air would then be ducted to each area in required amounts. Such ventilation air is not recirculated but is usually vented outside from toilet and kitchen areas.

Example 17-6. Refer to Fig. 17-6 and note that on a sunny spring day the 3 southerly outside heat pumps were on cooling cycle and delivering 15 000 Btuh each into the water circuit; the 4 core units at the same time were delivering 22 000 Btuh into the water circuit. The three outside heat pumps on the north side of the building were on heating cycle with each drawing 10 000 Btuh from the water circuit. (a) Compute the net heat addition to the water circuit from the indicated heat-pump operation. (b) If the storage tank and all connecting piping contain a total of 4000 gal of water, compute the temperature rise in the system water per hour under the operating conditions of part a. (c) Compute the flow in gpm through a core heat pump if the water passing through it rises 5 deg in temperature.

Solution: (a) The heat addition is

$$3 \times 15\,000 + 4 \times 22\,000 = 133\,000 \text{ Btuh}$$

$$\text{heat abstraction} = 3 \times 10\,000 = 30\,000 \text{ Btuh}$$

$$\text{net heat addition} = 103\,000 \text{ Btuh} \qquad \textit{Ans.}$$

(b) 4000 gal × 8.3 lb/gal = 33 200 lb in the system. This water must absorb the 103 000 Btuh released.

$$(M)(C_p)(\Delta t) = (33\,200)(1.0)\Delta t = 103\,000 \text{ Btuh}$$

$$\Delta t = 3.1 \text{ deg/h rise} \qquad \textit{Ans.}$$

(c) For a core unit:

$$\frac{22\,000}{60} = 366.67 \text{ Btu/min} = \left(\frac{\text{lb}}{\text{min}}\right)(C_P)(\Delta t) = \frac{\text{lb}}{\text{min}} \times 1.0 \times 5.0$$

$$\text{lb/min} = 75.33$$

$$\text{gpm} = \text{lb/min} \times \frac{1}{8.3} = 73.33 \times \frac{1}{8.3} = 8.8 \qquad \textit{Ans.}$$

Example 17-7. One of the heat-pump units of Fig. 17-6 on heating cycle absorbs 12 000 Btuh (3.52 kW) when supplied with 78 F (25.6°C) inlet water. Vaporization of the refrigerant-22 takes place at 67 F (19.4°C). Condensation of refrigerant takes place in the air-cooled condenser at 100 F (37.8°C) with trivial subcooling. A suction-line heat exchanger warms the vapor to 88 F (31.1°C). Assume that compression requirements exceed the isentropic by 50%. Motor efficiency can be taken as 80% for power estimation. (a) Find the refrigerant flow in pounds per minute per 12 000 Btuh absorbed in the evaporator with the heat exchanger in operation. (b) Find the enthalpy and temperature the refrigerant would have after compression under isentropic conditions. (c) Compute the probable enthalpy at entry to the condenser and find the heat delivery to air for space warming. (d) Compute the probable shaft horsepower needed and the kilowatt demand from the power lines. (e) Find the HCOP for this unit.

Solution: (a) Without the heat exchanger, liquid at 100 F showing $\mathbf{h}_f = 39.27$, and saturated vapor at 67 F (129.94 psia) showing $\mathbf{h}_g = 110.21$, would constitute values for

$$Q'_R = 110.21 - 39.27 = 70.94 \text{ Btu/lb}$$

The heat exchanger warms the vapor by the same amount that the liquid cools; thus we can read the enthalpy of the vapor at 88 F (129.94 psia) from Plate V as 114.6 Btu/lb and subcooling is thus

$$114.6 - 110.21 = 4.4 \text{ Btu/lb}$$

$$Q_R = 70.94 + 4.4 = 75.34 \text{ Btu/lb}$$

$$w_R = \frac{12\,000}{(60)(75.34)} = 2.65 \text{ lb/min}$$

(b) From Plate V, start with 88 F vapor at 129.94 psia, run isentropically to condenser pressure of 210.6 psia, and read $\mathbf{h}_{2s} = 120.0$ Btu/lb at about 134 F superheat temperature.

$$\Delta\mathbf{h})_s = 120.0 - 114.6 = 5.4 \text{ Btu/lb}$$

(c) Assume that all of the deviations from the isentropic appear as energy additions to the enthalpy of the vapor as this enters the compressor, and on this basis

$$\text{actual } \Delta\mathbf{h} = (1.5)(5.4) = 8.1 \text{ Btu/lb}$$

$$\mathbf{h} = 114.6 + 8.1 = 122.7 \text{ Btu/lb}$$

$$\text{heat delivery from condenser } Q_c = 122.7 - 39.27 = 83.4 \text{ Btu/lb}$$

$$\text{heat delivery} = (60)(w_R)(83.4) = 13\,261 \text{ Btuh}$$

(d) $$\text{shaft hp} = \frac{(60)(w_R)(\Delta\mathbf{h})}{2545} = \frac{(60)(2.65)(8.1)}{2545} = 0.51$$

$$\text{kW demand} = \frac{0.51}{0.80} \times 0.746 = 0.48$$

(e) Based on actual power demand to motor,

$$\text{HCOP} = \frac{13\,261}{(\text{kW})3412} = \frac{13\,261}{(0.48)(3412)} = 8.1$$

This HCOP is unusually high because of the very shallow temperature lift involved from a 67 F evaporator to a 100 F condenser. Moreover, the heat losses from the system would not all reach the condenser although they would still reappear in the heated space of the building.

PROBLEMS

17-1. In a system using face and by-pass dampers (see Fig. 6-1), it is found that with the face damper almost closed, the air delivered by the cooling coil is at a temperature of 55 Fdb and 53.2 Fwb, and that for every pound of air passing through the coil, 4 lb of air at 82 Fdb and 70 Fwb pass through the by-pass dampers. Compute the dry-bulb and wet-bulb temperature of the resultant air being delivered by the fan to the conditioned space.

Ans. 76.6 Fdb, 67.2 Fwb

17-2. For the face and by-pass system described in Problem 17-1, rework the problem under the assumption that half of the air passes over the coil and the other half moves through the by-pass dampers, and that the conditions of the air leaving the coil and dampers are the same as indicated in Problem 17-1.

Ans. 68.5 Fdb, 62 Fwb

17-3. In a precool and reheat system, air leaves the dehumidifying coil with a temperature of 53 F and a relative humidity of 90%. It then passes through the reheat coil where it is raised in temperature to 63 F. For each pound of water circulating through the precool and reheat system, 4 lb of air pass over the coils. (a) What temperature drop takes place in the water of the reheat coil? (b) If the mixture of outside and return air enters the precool coil at 84 Fdb and Fwb, how much is the air cooled in passing through the precool coil? Refer to Fig. 17-2 for a diagram of a typical system.

Ans. (a) 9.8 deg; (b) 10 deg (to 74 F)

17-4. In the zone control system of Fig. 17-1, it is found that for the air being supplied the temperature in a zone is 67 Fdb and 62 Fwb. Under these circumstances, the room thermostat calls for heat from the zone reheat coil and the supply temperature is brought to 70 F. If 5000 cfm of air at supply conditions of 67 Fdb and 62 Fwb enter the zone reheater, how much reheat is being employed?

Ans. 16 300 Btuh

17-5. For a unit of the type shown in Fig. 6-1, at a certain time the space loads were found to be 300 000 Btuh sensible and 190 000 Btuh latent. Space and recirculated air were at 78 Fdb and 66 Fwb; 5000 cfm of outside air at 95 Fdb and 75 Fwb were required for ventilation. Air at 58 Fdb is delivered into the space from induction mixing outlets. By-pass heating is used to avoid reheating whenever possible. Apparatus coils cannot operate at lower than 44 F. (a) Find the sensible to total-heat ratio for this space load. Use this value to draw the appropriate line on the psychrometric chart, making it pass through space conditions of 78 Fdb, 66 Fwb. On it, read the wet-bulb temperature corresponding to the 58 Fdb temperature of the supply air. (b) Compute the air flow needed based on sensible-heat requirements. (c) Find the ventilation air needed in pounds per hour and then compute the air condition which results when this air is mixed with the proper amount of return air. (d) Find the apparatus dew-point coil temperature required for the mixture of return and ventilation air of part c. Note that the apparatus dew-point can be found at the terminus of the line on the psychrometric chart, which passes through the supply-air condition of part a and the mixed-air condition of part c. (e) Compute the fraction of air which has to be by-passed.

Ans. (a) 0.612, 54.7 Fwb; (b) 61 475 lb/h; (c) 21 023 lb/h, $\mathbf{h}_m = 33.36$, 83.8 Fdb; (d) 45.3 F, 0.33

17-6. Rework Example 17-1 in the text when the same temperatures inside and outside are applicable and there is no change in the ventilation air. Conditioned space load is different at 150 000 Btuh sensible and 80 000 Btuh latent. Air at 58 Fdb is delivered into the conditioned space. (a) Find the sensible to total-heat ratio for the space loading and, using this value, draw an appropriate line on the psychrometric chart passing through space conditions of 78 Fdb, 66 Fwb. (b) For 58 Fdb temperature, read the proper wet-bulb temperature required for the supply air to the space. Compute the space air flow needed based on sensible-heat requirements. (c) Find the pounds per hour of ventilation air needed and the air condition which results when ventilation air is mixed with the proper amount of return air. (d) Find the apparatus dew-point coil temperature required when using the mixed air of part c. Note that the apparatus dew point appears at the saturation end of the line passing through the supply-air conditions as determined in part b and the mixed-air conditions of part c. (e) Compute the fraction of air which has to be by-passed.

Ans. (a) 0.652; (b) 55.3 Fwb and 30 737 lb/h; (c) 4205 lb/h, 31.8 Btu/lb at 80.3 Fdb; (d) 50.3 Fdb; (e) 0.257

17-7. Consider the energy-conservation system shown in Fig. 17-4 at a time when 5000 cfm of outside air at 100 F (37.8°C) enter and pass over the heat exchanger, while 4000 cfm of inside air at 75 F (23.9°C) are collected to pass over the heat exchanger for exhaust to outside. Because of the very adequate surface in this heat exchanger, the inside air warms to 88 F (31.1°C) before entry to the atmosphere as waste. (a) Compute the amount of heat added by the heat-exchanger coil to the waste air. (b) If it is assumed that all of the heat extracted from the water in the waste-air heat exchanger is used in cooling the incoming air, find the outside supply-air temperature on exit from the heat exchanger coil. (Note that 0.0180 $Btu/ft^3 \cdot F$ can be used with sufficient accuracy as the specific heat of air in parts a and b.) (c) The water in the heat-transfer coil system cools from 94 F (34.4°C) to 81 F (27.2°C) in passing through the waste-air coil. Compute the required pumping rate in pounds per minute and in gpm when the water cools to provide the indicated waste-air warming. (d) If all of the ventilation air entering the space ultimately is cooled to 65 F (18.3°C), what percentage of its total sensible cooling is provided by the energy conservation system?

Ans. (a) 936 Btu/min; (b) 89.6 F; (c) 72 lb/min, 8.6 gpm; (d) 29.7

17-8. Consider the energy-conservation system shown in Fig. 17-4 at a time when the following conditions are applicable: 5000 cfm of outside air at 10 F (−12.2°C) enter and pass over the heat exchanger; 4000 cfm of inside air at 75 F (23.9°C) are collected to pass over the heat exchanger for exhaust to outside. The 4000 cfm of waste air cool to 34 F (1.1°C) before entering the atmosphere. (a) Compute the amount of heat transferred to the heat-exchanger coil by the waste air. (b) If it is assumed that all of the waste heat transferred is utilized in warming the 5000 cfm of outside air, find the temperature rise in the supply air and its final temperature. (Note that 0.018 $Btu/ft^3 \cdot F$ can be used with sufficient accuracy as the specific heat of air in parts a and b.) (c) The ethylene glycol liquid in the heat-transfer coil system is warmed from 22 F to 63 F by the waste air. Its specific heat is 0.92 and it weighs 8.5 lb/gal. For the heat delivered by the air in part a and for the indicated temperature rise, find the required pumping rate for the glycol in pounds per minute and in gpm. (d) If the air entering the space must all be warmed to 100 F (37.8°C), what fraction of the total warming is provided by the energy-conservaton system?

Ans. (a) 2952 Btu/min; (b) 32.8 deg, 42.8 F; (c) 78.3 lb/min, 9.2 gpm; (d) 36.4

17-9. A 4-cylinder packaged hermetic unit using refrigerant-12 operates both as a cooler and as a heat pump. It was subjected to a test when the outside design temperature was 30 F (−1.11°C) and produced 24 000 Btuh of useful heating by heat-pump action. During the test run, it was found that the evaporator worked at 18 F (−7.78°C) and condensing took place at 105 F (40.6°C). The vapor supply warmed to 60 F (15.6°C) as it passed through a suction-line heat exchanger with the refrigerant liquid moving counterflow to the vapor. It can be assumed that compression requirements exceed the isentropic by 50% and the motor efficiency is 80% for this small unit. (a) Find the enthalpy and temperature of the vapor under isentropic conditions after compression from 60 F at evaporator pressure. (b) Assuming that the amount of loss deviation from isentropic compression all reappears in the leaving vapor, compute the enthalpy of the vapor entering the condenser. (c) Find the energy release to useful heating in the condenser and the refrigerant flow in pounds per hour required to realize 24 000 Btuh of useful heating. (d) Compute the shaft horsepower and the input kilowattage. (e) Find the HCOP.

Ans. (a) 85.6, 160 F; (b) 103.0 Btu/lb; (c) 70.7 Btu/lb, 339.5 lb/h; (d) 2.32 hp, 2.16 kW; (e) 3.25

17-10. A refrigeration unit in a factory building is required to produce a year-round supply of chilled water at 48 F (8.89°C). This water warms to 56 F during a processing operation and approximately 100 tons of refrigeration are needed for the process cooling. To conserve fuel energy, it was asked whether this ammonia system could be altered to serve as a heat pump and greatly reduce the fuel consumption of the winter heating load on the building, estimated at 1 000 000 Btuh. Water at 105 F (40.6°C) or higher at condenser outlet is needed for heating, which would necessitate condensing ammonia at approximately 115 F (46.1°C). The evaporator customarily operates at 41 F (5°C). Carry out computations as follows to obtain an answer to the question raised. (a) Compute the ideal (isentropic) work of compression per pound of ammonia circulating. (b) Estimate the total work energy added to each pound circulating if the actual to isentropic ratio is 1.4. (c) Find Q_R, the heat addition in the evaporator per pound of refrigerant, and the refrigerant flow per hour for 100 tons of cooling. (d) Compute the heat per pound of refrigerant available in the condenser with no liquid subcooling, considering the actual work of compression. (e) For the refrigerant flow of part c, compute the energy available for heating in the condenser if 90% of it can be considered usable. (f) Estimate the shaft horsepower needed and the line kilowattage if a motor efficiency of 88% applies. (g) Find the overall net HCOP (heating coefficient of performance).

Ans. (a) 77 Btu/lb; (b) 107.8 Btu/lb; (c) 450.2 Btu/lb, 2665.5 lb/h; (d) 558 Btu/lb; (e) 1 338 600 Btuh; (f) 112.9 hp, 95.7 kW; (g) 4.1

17-11. A heat-pump system must supply a design heat load of 50 000 Btuh. Refrigerant-12 is used with a minimum evaporator temperature of 30 F, and the condenser must work at not less than 106 F (143.2 psia). (a) Find the isentropic work of compression and the probable actual work input per pound of refrigerant (use a factor of 1.4). (b) Find the pounds of refrigerant which must be circulated per hour for the heating load. (c) Find the shaft horsepower required by the compressor. (d) What is the kilowatt input (assume a motor efficiency of 90%) and the hourly operating cost (assume a cost of 2.2 cents per kilowatthour)? (e) Find the overall HCOP in terms of input electric power.

Ans. (a) 9.6, 13.4 Btu/lb; (b) 817 lb/h; (c) 4.3 hp; (d) 3.6 kW, 7.7 cents; (e) 4.1

17-12. Work Problem 17-11 for a top pressure condition of 110 F (151.1 psia) and an evaporator at 38 F.

17-13. A heat-pump system in a large office building in Portland, Oregon, uses well water, being supplied with 150 gpm at 64.5 F from one well and with 450 gpm at 62.5 F from a second well. For heating, the water is chilled to 50.4 F and disposed of in a third well of greater depth. The heat-pump equipment consists of four centrifugal refrigerating units having a total rated capacity of 540 tons under summer cooling conditions. The water for warming the air in winter is heated to 101 F in the condenser and cools to 86 F. Various energy-saving and heat-saving features are employed in this system, but they will not be considered here. (a) Compute in Btu per hour the heat that is absorbed from the water when the full capacity of both wells is absorbed. (b) The evaporators operate at 46 F and condensation occurs at 105 F. Refrigerant-11 (Table 15-7) is used, for which the isentropic work of compression between evaporation at 46 F and condensation at 105 F is 8.9 Btu/lb. Making use of this value and a compressor efficiency of 77%, estimate the shaft horsepower that is required. (c) Assume the motor efficiency is 93% and find the power, in kilowatts, drawn from the power lines. (d) For these data find the overall HCOP.

Ans. (a) 3 764 000 Btuh (68.5 Btu/lb refrigerant, 916 lb/min); (b) 251 shaft hp, 193 isentropic hp; (c) 201 kW; (d) 6.4

17-14. A large building with perimeter and core areas is cooled and heated exclusively by multiple heat pumps which take energy for their evaporators from a common water loop. The water-loop temperature is maintained year-round between 70 F and 95 F. At a given period it was believed that ten heat pumps on cooling were picking up 600 000 Btuh with their average evaporator temperature 58 F (14.4°C). Condensing was taking place at 90 F (32.2°C). Subcooling to 86 F (30°C) took place but liquid-vapor heat exchangers were not used. Refrigerant-502 was used in the heat pumps. (a) Compute the probable refrigerant flow in each heat pump producing 60 000 Btuh of cooling. (b) Find the isentropic enthalpy increase for the operating conditions indicated. (c) Assume that a 50% increase over the isentropic value is needed to develop input shaft horsepower and find the necessary motor horsepower. (d) Find the condenser heat loading per hour added to the water loop by the ten units. (e) If 25 tons of water are stored in the loop piping and storage tank, estimate the hourly temperature rise in the loop water for these operating conditions.

Ans. (a) 1198 lb/h; (b) 3.5; (c) 2.5 hp per unit; (d) 663 700 Btuh; (e) 13.3 deg

Dual-Duct Problems

17-15. A dual-duct system is used in a complex building which has a summer sensible cooling load under maximum conditions of 2 100 000 Btuh with a latent load of 810 lb of water per hour. Spaces of the system must be maintained at 78 F or lower and with wet-bulb temperature never to exceed 69 F; temperature of chilled air in the cold-air duct is 54 F; temperature in the warm duct is 88 F. Outside summer design conditions are 95 Fdb and 69 Fwb. (a) Compute a trial cooling air flow or a total air flow. (b) Use these data and find a suitable design cold-air duct flow and warm-air duct flow. (c) Find for these data an approximate value of mixed inlet air temperature to a representative space.

Ans.(a) 110 000 cfm; (b) cold, 89 600 cfm, warm, 20 400 cfm; (c) 60.3 F

17-16. The building of the preceding problem has a design winter heat loss of 3 700 000 Btuh. A space air temperature of 75 F is to be maintained and the warm-air duct temperature can reach 130 F. The cold-air duct is 55 F. The total fan capacity, sized for summer, is 110 000 cfm. Compute the probable air flow in the warm- and cold-air ducts at winter design operating conditions.

17-17. Refer to problem 17-15 and for certain conditions assume that the total return-air flow from the conditioned space, 77 000 cfm at 78 Fdb and not over 69 Fwb, is recirculated except for 15% of its volume which is replaced by fresh (ventilation) air. The warm-air duct fan delivers 15 400 cfm of the recirculated air, and at the conditions under consideration there is no flow in the by-pass duct after the fans. Air enters the conditioned spaces at 58.8 F average temperature. It leaves the conditioner at 52 F and a relative humidity of 90%. The cold-air duct air is at 52 F and a relative humidity of 90%. The warm-air duct air is at 86 F. The latent load of the conditioned spaces is 760 lb of moisture/hour. (a) In line-diagram form, sketch the system to show the air flows indicated, adding additional data as these are computed in subsequent parts of the problem. Note that of the 77 000 cfm leaving the space, 15% or 11 550 cfm at 78 Fdb pass to waste and 65 450 cfm pass to the fans. (b) Find the weight flow rate in the cold-air duct. (c) Compute the humidity ratio of the air at exit from the conditioner. Then find how much additional moisture each pound of this air must absorb to handle the space moisture load. Note that the warm-duct air is at same humidity ratio as the space air and cannot pick up additional moisture load. (d) Find the humidity ratio, using the data of part c, and read the space wet-bulb temperature at 78 Fdb from the psychrometric chart.

Ans. (b) 61 000 cfm measured at 78 F or 268 000 lb/h; (c) $W = 0.00743$, $\Delta W = 0.00284$; (d) $W = 0.01027$, $t_{wb} = 65$ F

17-18. Refer to problem 17-15 and for certain conditions assume that 15% of the total air flow from the conditioned space, 77 000 cfm at 78 F, is replaced as ventilation (fresh) air with the remainder recirculated. Assume that the 78 F recirculated air should not exceed 68 Fwb. The latent load of the space is 760 lb of water per hour. Air enters the space at 58.8 F. Air leaves the conditioner at 52 F at $\varphi = 0.9$. Air by-passes up from the cold duct and is warmed to 86 F by the warm-air duct heater with 15 400 cfm at this temperature, entering the warm-air duct for delivery. (a) Sketch the system, in line-diagram form, to show the air flows as indicated, adding temperature values as these are found in the solution. Note that of the 77 000 cfm leaving the space, 15% or 11 550 cfm at 78 F pass to waste. Of the remaining 65 450 cfm, 53 900 cfm pass to the cold-duct fan and 11 550 cfm continue to the warm-duct fan. (b) Make a temperature correction to find the fresh-air flow in cfm measured at outside conditions of 95 Fdb and 78 Fwb. (c) Making use of Eq. (3-45) in consistent cfm, find the temperature of the outside and recirculated air after mixing and before entry to the conditioner. (d) Find the humidity ratio of the exit air from the conditioner. (e) For the 15 400 cfm at 86 F (15 160 cfm at 78 F) warm-air supply, find how much air has to be by-passed from the cold side before the chiller, if measured at 78 F. (f) Find the resulting cfm air flow into the chiller at 78 Fdb and convert this to a mass flow rate in pounds per hour. (g) Find the pickup per pound of chilled-air flow and the final humidity ratio of the space air at recirculation conditions.

Ans. (b) 11 920 cfm; (c) 81.0 F: (d) $W = 0.00743$; (e) 3 610 cfm at 78 F; (f) $m = 268\,200$ lb/h; (g) $\Delta W = 0.00283$, $W = 0.01026$ lb/lb air at 78 F and 65 Fwb

17-19. Refer to Problem 17-15 and check to see whether under the design conditions indicated the latent load can be properly served. All of the 78 F space air is recirculated except for 10 310 cfm of ventilation air at 95 Fdb 78 Fwb taken from outside. Assume that no bypassing occurs, after the fans, and that the cold- and warm-air duct flows are as computed for Problem 17-15. (a) Make a line-diagram flow chart and indicate on this known temperatures and other pertinent data for a 2-fan dual-duct system. (b) Find the approximate mass flow rate in the cold duct and use this to compute how much water vapor in pounds is absorbed by each pound of chilled air entering the space. (c) Assuming that the cold air leaves the chiller at 54 F at 90% relative humidity, compute the humidity ratio of the final space air and find its wet-bulb temperature at 78 Fdb.

REFERENCES

K. L. Bowler, "Energy Recovery from Exhaust Air," *ASHRAE Journal*, Vol. 16 (1974), pp. 49–56.

ASHRAE Handbook, SYSTEMS 1976, Chapter 11, "Applied Heat Pump Systems."

CHAPTER

18

SOLAR AND OTHER ENERGY SYSTEMS

18-1. PERSPECTIVE ON ENERGY

The United States, and in fact the whole world, is faced with an energy dilemma which already borders on being at crisis level. The most unfortunate aspect of this problem is that with the passage of time the overall situation can only worsen because the world is using up its energy resources at a much faster rate than these can be replaced. The seriousness of the problem was not appreciated by the public until the energy squeeze resulting from a Middle East oil embargo alerted the United States and the Western world to recognize their dependence on imported energy. In the United States, which is a relatively energy-rich nation, from 20 to 50% of its oil for energy needs came from foreign sources. Assuming that the international balance-of-payments issue for imported oil can be hurdled, the question still to be faced by many countries is how long it will take before import supplies approach exhaustion.

It must be realized that the fossil fuel resources resting in the outer crust of this earth are finite in amount. These reserves were created by the action of the sun on organic matter and then stored, when evolutionary changes in the crust of the earth locked the organic material underground. This process of creating energy reserves took place over a period of many millions of years. Now these fossil resources of liquid and gas fuel are being depleted by man in the short space of 100 to 150 years of intensive demand, with only 30 to 50 years of that period yet remaining. Until the last two decades, no serious alarm was experienced because as one field was depleted, one or two new fields were discovered and the everincreasing demands for energy were easily met. Unfortunately, this is no longer true and even the discovery of new fields will only postpone the day of reckoning on which so little oil will be available that all oil will be so costly as to prohibit its use.

We must also realize that the production of fuel and its delivery to the ultimate consumer is an energy-consuming process at all levels, from discovery through refining to the various transporting operations. Nor can we disregard the energy required to make the supplies and equipment needed for the various items used in getting the final product to the consumer—for steel, pipelines, tanks, ships, trucks, and refineries, to mention a few. It is often stated that oil shales can be developed to offset the diminishing output of present-day oil fields as these become depleted. There is indeed much to be hoped for from this potentially great resource, but at the same time we must not overlook the enormous energy demands which have to be met in mining the rock, transporting it to a production center, crushing and retorting it, as well as transporting back the residue for disposition, all before refining the distillate can even start. The amount of energy used in creating the final product is in fact so great that when labor and plant investment costs are added, shale oil is currently at a serious economic disadvantage.

Coal, in its many classifications, is available in relatively unlimited amounts, and this fuel must take over more and more of the energy requirements which are now met by oil. Particularly in power plants, coal and nuclear fuels will take a continuously increasing share in providing electric energy.

The high sulfur content of many coals serve as a deterrent to wide usage. However, research is under way to find means of gasifying coal with simultaneous sulfur removal, and SO_2 stack cleaners are being perfected which will make high sulfur coals usable directly. Conversion of coal to liquid fuel is possible and will be necessary as soon as the price of fossil oil becomes so high as to make the cost of synthetic oil from coal competitive. An immediate limitation on coal usage is set by the inability of operators to get new mines into operation rapidly, and strip mining is being held back by environmental litigation. Electric utilities too are having difficulty in getting new power plants constructed on schedule to meet the ever increasing demands of their customers.

What then is the energy picture faced by those concerned with serving the thermal environment as this relates to heating, cooling, and air conditioning in its many aspects?

1. Emphasis must be put on energy conservation by all possible means. For example, attention must be given to more careful design and to better and more complete use of insulation. Designs for maximum heat utilization should employ heat exchange between the naturally hotter and cooler parts of large buildings. Optimization of system selection, from an energy viewpoint as well as from a comfort-performance viewpoint must be given high priority.
2. With gas and even oil becoming less available for new installations, alternate energy sources must be given more consideration. Where electric-utility capacity permits, electric heating as well as cooling should be studied, and in this connection the energy savings resulting from heat-pump designs must be given careful study.
3. Older installations as well as new ones should be analyzed to see that only minimum overheating or minimum overcooling take place. This may require the installation of completely new control systems in some instances. In some older installations it may require merely a resetting of thermostats.
4. System operators and space users must be educated to the pressing need for conserving energy in every way possible. Not only should temperature be strictly controlled; wasteful air flow and ventilation practices also should be eliminated. More use of outside air must be made for cooling during periods of season change, particularly in spring and fall.
5. Solar energy as a supplement for building heating must now be seriously considered in all design planning, since the spectacular rise in the cost of fossil fuels has altered comparative price structures sufficiently to justify the high capital investment required for solar energy usage.

18-2. SOLAR ENERGY AVAILABILITY

A review of Section 6-3 would clearly show that the solar energy available for the use of man is almost unlimited. Table 18-1 also shows this availability for unit areas receiving radiation under different configurations. The table gives representative solar data for two latitudes, namely, 32°N and 40°N. At 32°N latitude, with lower heating loads needed for buildings, solar implementation can be very effective, while at 40°N latitude, with the colder winters which occur, larger collectors and more careful overall design are indeed necessary for effective solar implementation. The dates of June 21 and December 21 have been selected as representative of typical summer and winter conditions. For different hours of the day, the irradiation values are tabulated in the sixth column under conditions that might be expected for a normal atmosphere on a cloudless day. Note that these are not the maximum values that might occur. The remaining columns show for different orientations the solar transfer that could be expected to pass through double-strength glass at various times of the day on vertical or horizontal surfaces.

Example 18-1. Making use of data in Table 18-1 for 32°N latitude and winter conditions, read or compute, as may be necessary, the following: (a) for a south wall, the probable rate of heat gain between 11:30 A.M. and 12:30 P.M. (b) Estimate what this value would be if the fenestration were in such position as to lie normal to the sun. (c) Compare this answer with the probable normal irradiation exclusive of glass. (d) What is the total daily heat gain through the south-facing fenestration disregarding losses? (e) Express the previous answers in appropriate units of the foot-pound-second system.

Solution: (a) Use the average of the 11 A.M. to 12 noon and 12 noon to 1 P.M. values:

$$\frac{766 + 794}{2} = 780 \text{ W/m}^2$$

(b) Use the average solar altitude for the same period as (32.7 + 34.6)/2 = 33.65 degrees.

$$I \cos \beta = I \cos 33.65^\circ = 780$$

$$I = \frac{780}{0.8324} = 937 \text{ W/m}^2$$

(c) Read (946 + 958) = 952 W/m^2 from the table. This is higher than for part b because reflection and absorption in the glass is disregarded.

(d) Read the total insolation received per day through the glass as 19 156 kJ/m^2.

(e) $780 \text{ W/m}^2 \times 0.31721 = 247.4 \text{ Btu/h} \cdot \text{ft}^2$
$937 \text{ W/m}^2 \times 0.31721 = 297.2 \text{ Btu/h} \cdot \text{ft}^2$
$925 \text{ W/m}^2 \times 0.31721 = 302 \text{ Btu/h} \cdot \text{ft}^2$
$19\,156 \text{ kJ/m}^2 \times 0.088111 = 1688 \text{ Btu/ft}^2$ per solar day

Table 18-1 and Example 18-1, illustrating its usage, indicate many significant points relative to the entrapment of solar energy. For example, we must note that if substantial amounts of energy are needed, large collecting areas are required. Perhaps in some cases the roof area of a building is large enough to provide adequate collection area, even though this is not usually the case. In winter at 32°N latitude, a horizontal roof on a 20- by 10-m (65.6- by 32.8-ft) residence could collect 20 × 10 × 9988 = 1 997 600 kJ per day. If this energy could all be converted to useful heating, this would be the equivalent of some 14 gal of fuel oil, one-third of a 42 gal barrel. This value can be substantially increased if instead of horizontal placement of panels on the roof the collecting panels are inclined at a suitable angle to make their surfaces lie more normal to the rays of the sun.

The extreme differences in the solar heat-gain factors for different orientations at different times of the day makes it most obvious that the orientation most nearly in line or normal to the sun's rays receives maximum energy at any given hour. Of course the walls or fenestration of a building cannot change to follow the sun, as it appears to rise and fall in altitude while the earth turns through its daily orbit, but it is possible to design and build roof-or ground-placement collectors which can carry out such a pattern. Such collectors are usually too expensive to justify for installation but a compromise is usually reached by fixing the collectors at a vertical

Table 18-1 Instantaneous Solar Heat-Gain Factor for 32°N Latitude

	Sun Time		Solar Position		Normal Irradiation	Heat-Gain Factors through Double-Strength Glass (W/m^2)								
Date	A.M. (Read Down)	P.M. (Read Up)	Altitude, Deg.	Azimuth from South, Deg.	(W/m^2)	North	North-east NE–A.M. NW–P.M.	East	South-east SE–A.M. SW–P.M.	South	South-west SW–A.M. SE–P.M.	West	North-west NW–A.M. NE–P.M.	HOR.
June 21	6	6	12.2	110.2	410	139	388	400	173	32	32	32	32	88
	7	5	24.3	103.4	659	145	555	634	340	60	60	60	60	277
	8	4	36.9	96.8	769	113	539	675	426	88	82	82	82	476
	9	3	49.6	89.4	829	107	429	608	438	110	101	101	101	640
	10	2	62.2	79.7	861	120	271	460	385	142	113	113	113	769
	11	1	74.2	60.9	876	126	145	255	277	177	126	120	120	845
	12	12	81.5	0.0	883	126	129	132	164	189	164	132	129	870
	Total Insolation per Half Solar Day (kJ/m^2)				A.M. →	2860	8444	11032	7627	2520	2111	2043	2043	12700
					P.M. →	2860	2043	2043	2111	2520	7627	11032	8444	12700
December 21	8	4	10.3	53.8	555	22	57	426	523	303	22	22	22	69
	9	3	19.8	43.6	810	41	44	511	750	539	47	41	41	227
	10	2	27.6	31.2	905	57	57	400	776	684	164	57	57	375
	11	1	32.7	16.4	946	63	63	199	670	766	366	63	63	466
	12	12	34.6	0.0	958	66	66	73	558	794	558	73	66	498
	Total Insolation per Half Solar Day (kJ/m^2)				A.M. →	761	862	5471	10748	9578	3099	772	760	4994
					P.M. →	761	760	772	3099	9578	10748	5471	862	4994

Table 18-1 (*Continued*)
Instantaneous Solar Heat-Gain Factors for 40°N Latitude

	Sun Time		Solar Position		Normal Irradiation	Heat-Gain Factors through Double-Strength Glass (W/m^2)								
Date	A.M. (Read Down)	P.M. (Read Up)	Altitude, Deg.	Azimuth from South, Deg.	(W/m^2)	North	North-east NE–A.M. NW–P.M.	East	South-east SE–A.M. SW–P.M.	South	South-west SW–A.M. SE–P.M.	West	North-west NW–A.M. NE–P.M.	HOR.
June 21	5	7	4.2	117.3	66	32	66	63	19	3	3	3	3	6
	6	6	14.8	108.4	485	148	448	476	221	38	38	38	38	123
	7	5	26.0	99.7	678	117	542	653	385	66	63	63	63	306
	8	4	37.4	90.7	776	91	492	678	479	91	92	92	92	482
	9	3	48.8	80.2	826	104	356	605	508	142	98	98	98	634
	10	2	59.8	65.8	857	110	195	457	467	218	113	110	110	747
	11	1	69.2	41.9	870	117	126	252	366	277	129	117	117	820
	12	12	73.5	0.0	876	120	120	129	224	299	224	129	120	842
	Total Insolation per Half Solar Day (kJ/m^2)				A.M. →	2747	8104	11565	9193	3530	2236	2054	2043	12722
					P.M. →	2747	2043	2054	2236	3530	9193	11565	8104	12722
December 21	8	4	5.5	53.0	277	6	22	211	262	154	9	6	6	19
	9	3	14.0	41.9	684	28	31	426	646	476	38	28	28	123
	10	2	20.7	29.4	823	44	44	356	731	662	173	44	44	243
	11	1	25.0	15.2	880	50	50	176	684	763	378	50	50	324
	12	12	26.6	0.0	895	54	54	57	558	798	558	57	53	356
	Total Insolation per Half Solar Day (kJ/m^2)				A.M. →	556	613	4313	9431	8614	3099	567	556	3200
					P.M. →	556	556	567	3099	8614	9431	4313	613	3200

angle which permits them to lie in approximate normality to the rays of the sun during the hours of most intense insolation. No matter how important direct insolation is, we must never lose sight of the fact that diffuse radiation from the sky is always present and reflection and reradiation from the ground and other surfaces is a significant factor in the magnitude of the radiation for different orientations. The data of Table 18-1 have diffusion effects included in them.

Finally, recognition must be given to the fact that there will be cloudy days with little or no insolation and a similar situation exists at night. To meet these contingencies it is necessary to collect excess energy during insolation periods and store it for use at other times. The most available storage medium is, of course, water (or a water-antifreeze solution) but other things such as stones or coarse rocks have been also employed. The size of the storage system must be patterned to suit the period of nonsolar energy, matched to serve the degree of independence from fossil fuels or electric heat usage that might be desired. In parts of the country at 40°N latitude and higher, water collector temperatures in winter may be as low as 40°C (104 F), but in summer temperatures above 110°C (230 F) can be expected.

18-3. SOLAR DESIGN ARRANGEMENTS

Figures 18-1 and 18-2 show in diagrammatic form two arrangements for absorbing and utilzing solar energy to reduce consumption of fossil fuel or of electric energy. Figure 18-1 utilizes a loop circuit of water (antifreeze solution). The solution absorbs solar heat in the collector and then transfers this heat to water in the thermal storage tank. A pump must be used in the circuit whenever the storage tank is located below the collector panels. When the solar-heat input exceeds the rate at which energy is being withdrawn from storage the temperature in the storage tank rises. During night periods and on cloudy days the temperature of the stored water necessarily falls as heat is delivered to the system.

Here two kinds of utilization are shown, one for building heating and the other for warming the domestic water supply. In mild weather and on very sunny days, solar energy may be adequate to carry the full building load. However, systems are seldom designed with adequate collection surface and thermal storage to carry the load under cold winter conditions, so supplementary heat from a fossil-fuel boiler is required.

The building heating circuit takes energy both from the storage tank and from the fuel-fired boiler. The latter must necessarily insure an adequately high temperature for the building heater coils to maintain desired temperatures in the occupied spaces. A forced-warm-air distribution system is shown here for heating the building, but a hydronic (hot-water) system could be made to serve equally well. The boiler can be controlled to supply heat only when space temperature falls below a preset level.

The domestic water supply is warmed in passing through the thermal storage tank and if this heating is not sufficient, it is further warmed in a

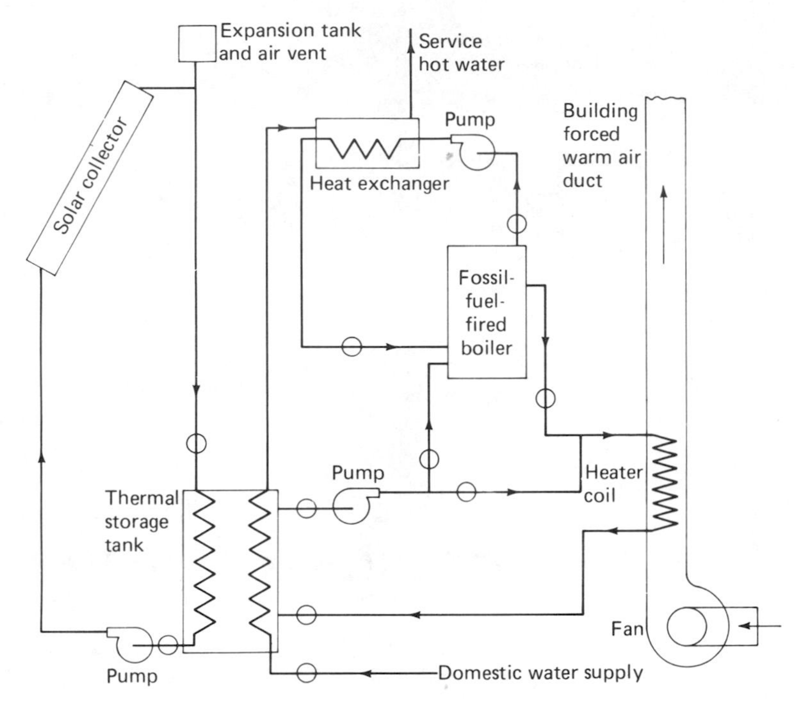

FIGURE 18-1. Schematic layout of solar-collection system, with storage, to provide energy, for space heating and domestic hot water.

second heater which is served by the boiler. In general it would be expected that in summer solar heat could suffice for complete heating of the domestic water supply but this condition would not apply in winter.

The installation of a dual system to serve both building heating and domestic water heating calls for a more complicated control system than is required for merely building heating, but simple thermostatically controlled cutoff valves for the two boiler circuits could serve in this connection. Circulation in the solar-collector circuit must be stopped at night and, in fact, whenever the temperature in this circuit is less than the temperature of the thermal storage tank.

Numerous variations from the design shown in Fig. 18-1 are possible, not the least of which concerns choosing the amount of collector surface to be installed and the volume (weight) of storage medium required to carry the system during inactive solar periods. Among the variations, mention should be made of the possibility of direct circulation of water (antifreeze solution) between the solar collector and the storage tank. Also the pump for this system may be omitted when the storage tank can physically be located above the solar collectors, making use of thermal gravity circulation. Many variations are also possible in the interconnections for the supplementary boiler in relation to the heater coils and the domestic water circuits.

In designing solar circuitry as much care for safety must be exercised as is employed for any hydronic circuit. Reference to Chapter 8 will show

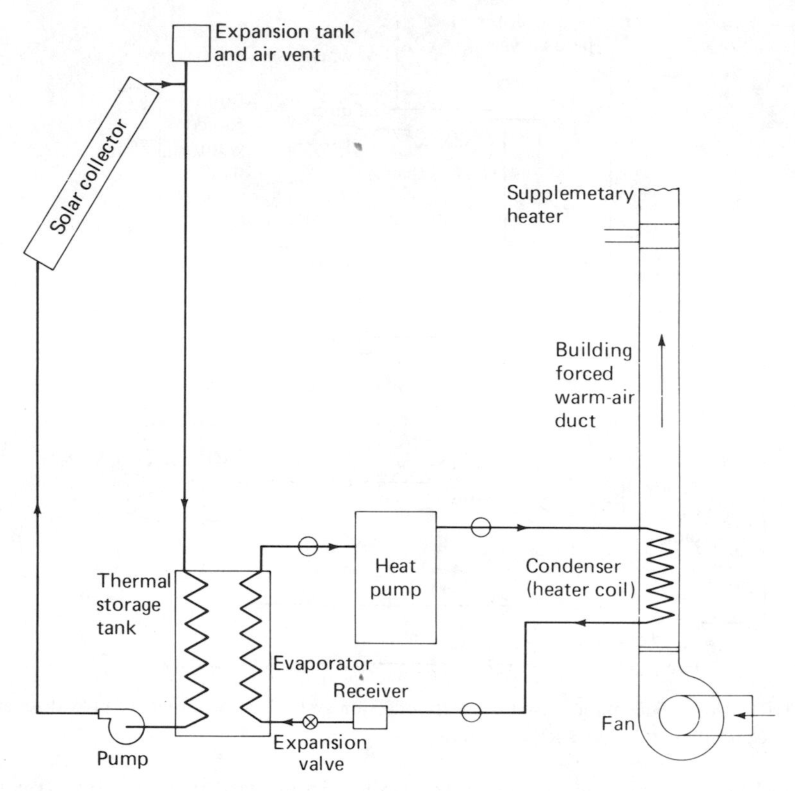

FIGURE 18-2. Schematic layout of solar-collection system with storage to operate in connection with a heat pump.

that variations in the volume of the water of the system make necessary the use of one or more expansion tanks in the solar circuits. Relief or safety valves and air venting for the system must be provided. Valves, in addition to those shown, are needed for various operating conditions, shutdown, and drainage.

The possibility of using solar energy for cooling in summer should not be overlooked because it is possible to produce solar collector temperatures in the range 35°C (203 F) to 45°C (235.4 F). Such temperatures are adequate to energize certain types of absorption refrigeration equipment to provide summer cooling. For this purpose, in Fig. 18-1 the heater coils would be valved off and connections from the supply circuit then would be made to serve the refrigeration unit.

Figure 18-2 shows a solar collection system employing a heat pump. Here the collector system, works as before, collecting solar energy when available, to provide energy for the storage tank. The heat pump, described more fully in Sections 17-7 to 17-9, is a complete refrigeration system. For its operation, heat flow to the evaporator is supplied from the warm water of the storage tank. The compressor of the heat pump raises the slightly warm evaporated refrigerant to a sufficiently higher pressure and temperature that the refrigerant, liquefying in the condenser coils, can provide the heat to warm the building under most load conditions.

Because of the warm evaporating conditions the pressure and temperature lift to the necessary condenser temperature is small and compressor power consumption is minimized. Thus some 2 to 8 times as much energy can be delivered at the condenser as is taken from the power system to drive the compressor motor. Effective use of the heat pump in this way can make the cost of electric energy for heating more competitive with that for fossil fuel usage. The key to success for such a system is providing enough energy into the thermal storage source to keep it operating for a major fraction of the time in the winter season. When sufficient energy is not available, as in very cold weather or in extended cloudy periods, supplementary direct electric heat without benefit of heat-pump action is required. Of course the supplementary heat could be provided from a fossil-fired unit, but since this would necessitate installation of a duplicate system, this practice is not usually followed.

The heat pump is particularly desirable when the same compressor is employed to produce refrigeration for summer cooling. The connections are not shown in Fig. 18-2, but for this purpose the condenser would have its operation reversed to serve as the evaporator or chiller of the building air supply and instead of the former evaporator in the thermal storage tank, a separate heat-dissipating condenser would have to be provided.

18-4. SOLAR COLLECTORS

The flat-plate collector is most widely in use at the present time. The plates are made in modules of convenient size for handling (see Figs. 18-3, 18-4, and 18-5. As many modules are installed as may be needed to collect the amount of solar energy desired. They are usually placed on roofs, in ground arrays adjacent to the building, or even on a sun-facing wall. The

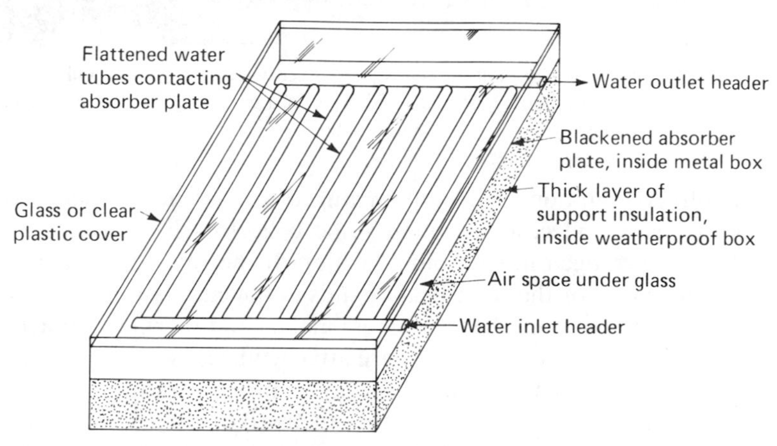

FIGURE 18-3. Flat-plate solar collector (metal encased, weather-proofed, with glass cover) for heating water (antifreeze solution).

FIGURE 18-4. Revere Tube-in-Strip Plate for collecting solar energy (Courtesy Revere Copper and Brass Inc.)

FIGURE 18-5. Solar modular collectors installed over windows of Sun Oil Company Research Center, Newtown Square, Pennsylvania. As located, the collectors also serve to provide shading, particularly in summer. (Courtesy Revere Copper and Brass Inc.)

water (antifreeze solution) is usually arranged to pass in parallel through the tubes of each collector, although some collectors have a continuous loop of tubing arranged in sinusoidal fashion. If the solution enters at the bottom of the collector the solar heating aids in the circulation, as the solution is warmed by the sun. The tubes must make good thermal contact with the blackened flat plate to which they are attached because it is the flat plate area which largely collects the solar energy in turn to transfer the heat to the tubes and into the solution. The tubes themselves are also blackened to aid in the heat absorption process. Figure 18-4 shows the tube-in-strip collector plate made by one manufacturer. Here the tube and strip are integrally the same metal and heat transfer into the water is greatly enhanced

Heat conservation is extremely important, so behind and under the plate and tubes, about 2 in. (5 cm) of solid insulation is usually employed. Above the plate and tubing and facing the sun it is customary to place one or two glass (or clear plastic) sheets to provide frontal insulation. The solar energy passes through the glass with little absorption and its energy then remains largely trapped in the collector. Side insulation at the edges is also provided, and the whole collector is mounted in a weatherproof enclosure.

When the collectors are mounted on the roof it is desirable to set them at an angle that will optimize the solar collection for winter, if building heating is the objective, or for summer, if cooling or domestic water service is involved. It is difficult, if not impossible, to change the angle on most installations and yet there are extreme differences between the best angle for the two seasons. In winter the usual angle is often taken as 10° to 15° greater than the latitude, measured from the horizontal; in summer it is taken as 10° to 15° less than the latitude, also measured from the horizontal. For year-round operation it is not possible to find an ideal solution, but a compromise angle at 4° less than the latitude, measured from the horizontal, is sometimes used. The collectors should face as much in a southerly direction as possible.

For the solar heating of air, similarly constructed flat collectors are used. In these, blackened steel or copper sheets, placed in the collector, are used in staggered sections. Between these the air is forced to flow, in turn picking up heat from the front and back surface of the plates. Back insulation and front glazing are very important for collectors in this type of service.

The performance or efficiency of these flat-plate collectors is good under most circumstances. It is best when the plate is normal to the sun and when the plate is not working at high temperature because then the losses from the plate rapidly increase. For example, a plate with two glass covers, heating water to 40°C (104 F), would utilize some 72% of the solar energy impinging on it; at 65°C (149 F) the efficiency drops to some 60% and at 110°C (230 F) it would be at only 40%. At low outside temperatures the efficiency drops even more markedly because of higher losses. For general design estimates, in summer employ an efficiency of 60 to 50%, in winter 50 to 35%.

18-5. ECONOMIC CONSIDERATIONS RELATIVE TO SOLAR ENERGY

The solar energy falling on the land area of the United States and in fact on most countries of the world is several hundred times greater than the total energy needed for vegetative growth and for the occupants of the area. Thus at first glance it is surprising why more use has not been made of this large source of free energy. There are, however, several reasons why solar usage has been so restricted. The most significant is the very substantial cost of equipment and plant, even to utilize energy at relatively low temperatures. For solar absorption and storage at high temperatures, plant costs have been and still are almost prohibitive. The lack of continuity of the solar energy supply as at night and on cloudy days has also interfered with widespread usage and necessitated the extra cost of storage facilities. Finally, until recent years, fossil fuels have been available in unlimited amounts at reasonable cost. With the availability of certain fossil-fuel supplies threatened, and with substantial increases in the prices of the available fuels a continuing threat, the picture has changed to the point that serious consideration must be given to solar-energy usage.

Let us put economic considerations into some perspective by means of simple illustrations. First, consider a small residence for which oil-fuel costs average \$80 per month for 8 months of the year and \$30 per month for 4 months when the major demand is domestic hot-water service. These figures indicate a total annual fuel cost of \$760. Suppose that a solar-collection system, with storage, could be installed that could provide 50% of the winter heat demand and 100% of the warm-month demand. This would reduce fuel cost and provide savings of \$440 per year. Suppose further that the cost of the solar system as a supplement to the present system would cost \$7000 installed. With interest at $7\frac{1}{2}$% and allowing 1% for annual maintenance, the annual investment and maintenance cost is \$7000 × 0.085, or \$595. As can be seen, this annual cost is more than the savings before any payoff on the system is made. The owner would therefore have to decide that the investment was unwise. However, if he had to face the possibility of the cost of fuel doubling over the next five years along with a possible reduction in his allocation of fuel, he might find it desirable to reconsider his decision. In such a case a more complete calculation including payoff over the useful life of the equipment should be made.

In making a substantial investment in equipment the factors of prime importance are the cost of money (interest rate) and the period of payoff. The latter in turn will often bear some relation to the useful life of the equipment. Equation (18-1), the proof of which can readily be found in textbooks of finance, covers this situation.

$$N = S\frac{i(1+i)^n}{(1+)^n - 1} = S\frac{i}{1-(1+i)^{-n}} \tag{18-1}$$

where N is the periodic payment, made at the end of each year or other chosen period, in dollars; S is the equipment cost, in dollars; i is the current interest rate, expressed as a decimal; and n represents the elapsed years to complete payoff.

Example 18-2. A residential supplementary solar installation can be bought and installed for \$7000. The current interest rate is $7\frac{1}{2}$% and 1% annually should be allocated for maintenance and upkeep. Assume that a 12-year payoff period is chosen. Compute the total annual cost of the installation.

Solution: Use Eq. (18-1). Note that the exponential term can readily be solved using a scientific hand calculator or a log-log slide rule or with logarithms if neither of the former is available.

$$N = 7000 \frac{0.075(1 + 0.075)^{12}}{(1 + 0.075)^{12} - 1} = 7000 \frac{0.075(2.382)}{2.382 - 1}$$

$$= \$905/\text{year}$$

To this the 1% maintenance at \$70 should be added, giving \$975 annual cost

Ans.

Usually, instead of one payment at the end of the year, semi-annual, quarterly, or monthly payments are made. To use Eq. (18-1) in this situation, i should be divided by the number of payments per year and n should be multiplied by the same factor. Let us rework on the basis of four (quarterly) payments per year.

$$N = 7000 \frac{(0.075/4)(1 + 0.075/4)^{48}}{(1 + 0.075/4)^{48} - 1} = 7000 \frac{(0.01875)(2.439)}{2.439 - 1} = \$222$$

annual cost = 4 × \$222 + \$70 = \$958 *Ans.*

Although on first examination these results appear unpromising for solar energy, no such generalization should be made. Here a 12-year payoff was assumed. This is longer than the 5 to 6 years one usually considers to amortize a project but is not a long period for a well-built residence. In fact, for new construction the payoff can be part of a morgage to run for 20 years or more. After the payoff is completed it must be realized that future solar energy is really "free" except for system maintenance, repairs, and the electric power needed for pumps and controls. Moreover, with a more expensive energy source such as electric power, the comparison would be more favorable for the use of supplementary solar energy.

Supplementary solar energy installations can be advantageously incorporated in new designs at less cost per million Btu (1.05 million kJ) than is involved for the same capacity in the renovation of older buildings with adaptation to an existing heating system. Many successful installations are now in use in residences, schools, and factory buildings. However, the designs almost always consider the solar energy on a supplementary basis with backup provided by a fossil-fuel-fired heater or by use of electric energy.

It is impossible to give accurate cost-design figures at this time because many design patterns and manufacturing methods are still in transition. A few generalizations, however, can be made. The major cost of the system rests in the solar collectors and here prices range from some \$5 to \$10 per square foot of collector area when simple flat plates are used. Water storage in inside tanks ranges from \$2 to \$7, based on each square foot of collector area. Pumps, exchangers, controls, and system adaption can add another \$3 to \$8 per square foot of collector area. In total, then, a system might cost from some \$10 to \$35 per square foot of collector area fully installed. The cost of storage using rocks in a warm-air system is slightly higher than for a water system of the same storage capacity.

Example 18-3. For a small residence located at 40°N latitude, it was found that there is sufficient room on its roof to place 20 flat-plate solar collectors, facing south, at an angle of 45° from the horizontal. Each proposed collector has a 0.91- by 1.98-m double-glazed surface, 12.7 cm deep including back insulation. The owner wishes to know how much energy can be collected if these units are installed and also wishes to explore the feasibility of installing them. The residence has a computed design heat loss of 7.5 kW, based on the lowest outside temperatures of significant duration which might occur during the heating season.

Solution: In Table 18-1, read for a south vertical wall in December that 8614 + 8614 or 17 228 kJ/m^2 of insolation can be expected for a glazed surface. Because the surface at 45° is more nearly normal to the rays of the sun than is the vertical, it is advisable to increase the insolation by a factor, say, 10%. Thus

$$17\,228 \times 1.10 = 18\,951 \text{ kJ/m}^2$$

For the whole panel array the insolation is

$$20 \times 0.91 \times 1.98 \times 18\,951 = 682\,920 \text{ kJ/day}$$

However, to run this system for heating, hot water from the collectors at, say, 55°C would be desirable, possibly cooling to 45°C on return. For collectors, under winter conditions at such temperatures, losses are substantial and the net realization of energy for inside usage is reduced to the point that the collector-system efficiency would probably not exceed 45%. Thus the solar energy actually delivered would approximate:

$$682\,920 \times 0.45 = 307\,310 \text{ kJ/day}$$

For the residence with its design heat loss of 7.5 kW (25 610 Btuh), the energy demand per day for heating is

$$7.5 \times 24 = 180 \text{ kWh}$$

$$180 \times 3600 = 648\,000 \text{ kW} \cdot \text{s} = 648\,000 \text{ kJ}$$

On a design heat-loss day, solar energy could thus supply

$$\frac{307\,310}{648\,000} \times 100 = 47.4\% \text{ of the fuel needed}$$

Even for this energy saving, a storage tank is necessary, and it must be realized that on cloudy days little or no solar energy can be entrapped. On the other hand, only rarely does the full design heat loss have to be supplied, and on warm days the solar contribution would represent a larger fraction of the total loss. In mild weather it may even be sufficient to provide for the whole heating load.

When the owner considers the cost of the solar equipment it may appear doubtful that the savings in fuel costs can justify a solar installation as described. If, however, fuel is very expensive or if electric heating is involved the owner may want to explore the problem in greater depth. In fact, he might also wish to consider the advisability of providing more insulation for his residence and look into ways of reducing infiltration, both for the purpose of reducing the building heat loss. Although winter solar heating in the northerly latitudes shows a lesser return on investment than is the case in the more southerly regions, careful studies have shown justification for its usage as a supplement to other energy sources.

18-6. SOLAR COOLING

The need for cooling in summer produces extreme demands on electric utility systems, in many instances causing peak loads in summer, higher than those met at any other period of the year. This tends to add to the ever rising cost of power, and large amounts of additional fuel are required to meet the cooling-load requirements. Widespread usage of solar cooling could greatly alleviate the summer peak-load problem faced by the electric utilities.

A number of successful solar cooling installations are in operation using absorption refrigerating units (see Sections 16-7 and 16-8). These units can operate effectively using solar energy provided the heating medium can be supplied at temperatures of 190 F (88°C) or higher and heat dissipation can take place at some 90 F (32°C) or lower. In general, fully satisfactory operation of the chiller at reasonable capacity is not possible at working temperature differences of less than some 56°C. With conventional flat-plate collectors, hardly able to attain 210 F (110°C), and with a 5 to 10 deg temperature differential (Celsius scale) needed for heat transfer at both the generator and the absorber-condenser, the working temperature difference is lower than desired. Moreover, the heat dissipation, from air-cooled condensers under customary 95 F (35°C) summer ambients, would necessitate 110 F (40.6°C) or higher condensing temperatures. Thus a dependable temperature difference of some 50 Celsius degrees or even less appears the best that can be expected. The absorption units will operate at this level but low capacity and poor machine performance can be expected.

Water-cooled or evaporative condensers can provide some improvement, but the real solution to the use of such equipment is to increase the temperature of the water coming from the solar heater to 230 F (110°C) or more. This can be done but it may involve concentrating (focusing) the solar energy, collected over a large area, into a smaller zone where the concentrated energy can produce higher temperatures. The principle is the same as that seen when an ordinary magnifying glass (lens), placed in the sun's rays, can concentrate the energy falling on the whole lens down to a point sufficiently hot to ignite a piece of paper. Lenses, as such are not used in solar heating, but many ingenious designs have appeared, most of which use parabolic shapes or paraboloids.

One of the most practical concentration devices is the parabolic trough shown as an end view in Fig. 18-6. These collectors can be made in appropriate lengths of perhaps 2 to 4 m. A tracking device is desirable so that the unit is faced as closely as possible into the rays of the sun. In the case of a parabolic-shaped surface, each of the parallel rays of the sun on striking the mirrorlike surface is reflected back to the blackened tube, located at the

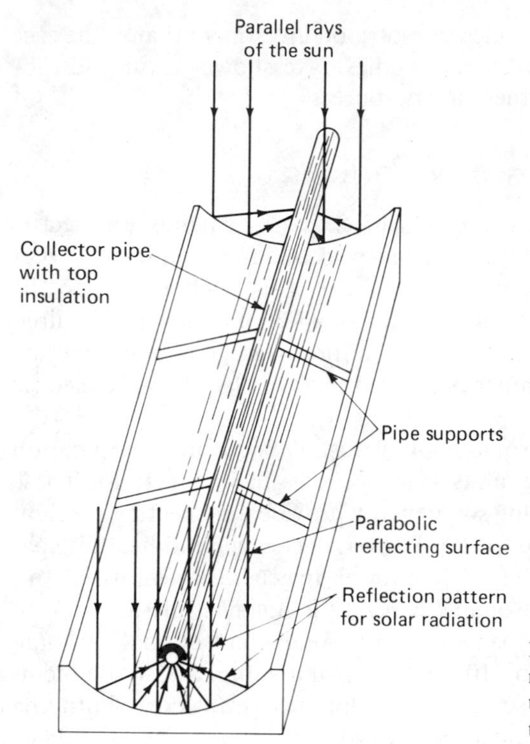

FIGURE 18-6. Diagrammatic arrangement for solar heating using a trough-type parabolic reflector with top-insulated collector plate.

focal point of the unit. The tube runs the full length of the trough and carries the water, which enters at its lower end and leaves at its upper end. Frequently they are set at a fixed horizontal angle, say, 40° and merely follow the orbit of the sun while being rotated about a vertical axis by the tracking mechanism. Theoretically, the energy falling on the whole reflecting surface is redirected onto the water tube except for the amount blocked out by the tube and its back insulation.

The performance of such units is good although it should be obvious that all of the energy impinging on the reflecting surface does not reach the tube. For example, imperfect geometry, poor solar alignment, irregularities of the surface, a dirty reflector, imperfect insulation, reradiation, and the glass cover for the unit (when used) all act against ideal performance. The tube, depending on its adjusted flow rate, can reach as high a temperature as needed although it continuously loses heat to its surroundings. Nevertheless, temperatures of 300 F (148.8°C) and even higher can be reached without difficulty. It is also possible to operate these units without tracking equipment, but their effectiveness is greatly reduced.

Many other arrangements have been contemplated. One of these involves a paraboloid-of-revolution shaped device with a pot-shaped boiler at its focus to receive the redirected solar rays. Consideration has also been given to evacuated bottom-silvered glass tubes which house suitably located absorbing tubes to pick up the reflected rays and impinging radiation. With heat loss limited by the insulating effect of the vacuum combined with the greenhouse effect of the glass tube, the liquid flowing through the core tube could reach temperatures in excess of 300 F (148.8°C).

Where land area or space for a solar installation is not at a premium, one method that has been followed is to carry out the heating in two or more stages. In the first stage, the heating panels are of cheap construction, that is, with limited insulation and glazing, but sufficiently well built to raise the water (solution) temperature to a substantially elevated temperature. This preheated water then passes into a more elaborately designed battery of solar heaters where the water can be brought to the final desired temperature. Primarily this arrangement is used for cost reduction since there is no need for the first stage to track the sun nor should it require more than minimal attention.

In summary, it should be noted that solar cooling is feasible and successful designs can be engineered readily. Higher temperatures approaching 300 F (148.8°C) are needed for good performance of the absorption unit with air-cooled condensers. To reach such temperatures flat-plate collectors are inadequate and more elaborate collector designs are required. These can be made but, even if mass produced, will be expensive, so that the investment required for a solar-cooling installation will necessarily be high and not as competitive with other forms of energy as is the case for solar heating.

18-7. ENERGY SOURCES

The discussions in this text have all shown that to provide heating or cooling, energy in one form or another is required. To the individual, it makes little or no difference what source of energy is used. Such, however, is not the case relative to the cost of providing this energy. Usually space heating can be provided most economically by burning the fossil fuels, oil or gas (whether supplemented by solar energy or not). However, with both oil and gas becoming even more expensive, all energy possibilities must be analyzed before a decision can be reached on an ultimate energy source.

Formerly electric heating was not competitive, but with the expanding use of the heat pump and with power being produced from coal and nuclear sources, electric heating is becoming more widespread. Where power is produced in oil- or gas-fired power plants and oil or gas is also available to heating users, it is difficult to justify electric heating. This is because in the production of power from fuel not more than 40% of the energy in the fuel can be converted to electric power with the remaining 60% necessarily dissipated to cooling water or to the atmosphere. With the combustion of fuel used directly for heating, 70 to 90% of the energy in the fuel can be expected as useful heat. There is thus a factor in the range of 2 to 1 in favor of direct use of fuel for heating; or in the case of less efficient power plants, the factor may be as high as 3 to 1. The heat pump can alter these figures because, with heating coefficients of performance of 2 to 4 available in mild climates, more useful heating may derive from each kilogram of fuel used electrically than from fuel burned directly.

The decisions that must be made in choosing the energy source for a major building development or merely for a residence are thus many and varied. Not only should the planner consider immediate cost to the owner; he must also give thought to the desirability of decreasing the nation's dependence on imported oil, to anticipated changes in the future price of fuel, and to the energy sources used by the electric utility. In addition, some consideration should be given to whether solar energy supplementation is desirable.

In broadest terms, relative to energy, we must ask ourselves whether all possible energy sources have been or are being explored. Unfortunately our horizons are limited since we cannot create energy and our only sources are (1) energy stored in the earth as coal, oil, gas, and geothermal; (2) nuclear potential; or (3) the continuously flowing energy from the sun. We are using our stored energy far too rapidly, so that all future development must necessarily involve effective adaptation of solar manifestations.

The enormous daily outpouring of solar energy to the earth is controlled by man to only a trivial extent. Evaporation from the oceans falling as rain is entrapped by dams and used for hydroelectric power and irrigation. Wind action arising from differential solar radiation and the earth's rotation is an almost untapped field of endeavor, and there is no question that enormous amounts of power could be generated if a gridwork of wind-powered generators were distributed over the landscape at strategic locations. Photovoltaic units for power generation need to be readied for mass production. Wave and tidal action have potential for the production of power. Finally, photosynthesis of vegetable and wood crops for energy, as well as for food, offers a continuing method of harnessing the sun directly for man's use. Investment capital is needed and must be found for all such projects or severe energy shortage will soon be felt by the present and following generations.

PROBLEMS

18-1. For a residence in architectural planning stages it is estimated that a conventional heating system can be installed for $6000, while the same system but with solar supplementation can be installed for $10 000. The solar system, it is believed, can reduce year-round fuel costs by $460. Money can be borrowed at 8% on a 25-year mortgage. Disregarding any additional maintenance and power costs for the solar part of the integrated system, is the extra cost of the solar supplementation justified? Monthly payments on the mortgage are required.

Ans. $46.31 and $77.18 per month for the two systems; solar system appears justifiable

18-2. Refer to Example 18-2 in the text where the payoff over a 12-year period was computed on the basis of both annual and quarterly payments. Compute annual cost using the same data except that payoff is made on a monthly basis (12 payments per year).

Ans. $887, disregarding maintenance

18-3. Refer to Example 18-2 in the text and recompute the required annual payments if these are made quarterly but extended over a 25-year period.

Ans. $622, disregarding maintenance

18-4. Double-glazed panels 910 by 1980 mm and 127 mm thick are available for placement on the roof of a public school to face in a southerly direction. The building is located at 32°N latitude and the panels would be set at an angle of 45° above the horizontal. Assume that at this angle the absorption is increased by 10% over a vertical south-faced collector. Water temperatures in the range of 35°C to 45°C are needed in winter, under which conditions a collection efficiency of 45% may be assumed. Make use of Table 18-1 and find the probable daily output per square meter of insolation surface. How many panels would be needed to serve a building with a design hourly heat loss of 50 kW under design conditions? For average heat-loss conditions of 30 kW, how many panels are needed, assuming storage capacity is increased?

Ans. 253; 152

18-5. Double-glazed panels 3 by 6.5 ft and 5 in. deep can be obtained for placement on the roof of a building to face in a southerly direction. The building is located at 40°N latitude and the panels will be set at an angle of 45° above the horizontal, for winter solar heating. In operation it is desired that the water warm to at least 131 F (55°C) and return from the storage reservoir at 113 F (45°C) or higher. The heat loss from the house is 30 000 Btuh or 720 000 Btu/day. Make use of Table 18-1 to find the probable daily output per square foot of panel insolation surface. Increase this value by 10% to allow for the benefit of the 45° panel slope over conditions of a vertical south wall. Considering losses under winter conditions it is doubtful that solar collection efficiency will exceed 45% on average. (a) Assume that a solar-heat system can be installed which will carry one-third of the design heat load on a clear solar day and compute the number of panels that would be needed for this purpose. (b) For serving this load, find the water (antifreeze) flow rate through the collectors when the average water temperatures are 131 F entering the storage tank and 113 F leaving, during the 8-h solar collection period.

Ans. (a) 17 p; (b) 1667 lb/h

18-6. Rework Problem 18-4 if it is found that the panels have to be placed in a southwesterly direction, all other conditions of the problem remaining the same.

18-7. Rework Problem 18-5 if it is found that the panels have to be placed in a southeasterly direction, all other conditions of the problem remaining the same.

18-8. A seaboard power plant uses oil as fuel with a heating value of 144 000 Btu/gal. Its performance is such that its heat rate is 9200 Btu/kWh; that is, 15.7 kWh are produced per gallon of fuel burned. A residence served by this utility is being renovated and the owner must decide whether to use fuel oil for heating at 36 cents per 144 000 Btu/gal in a furnace with an efficiency of 75%, or to use electric energy available on a sliding scale which averages 3 cents per kilowatthour. (a) Find energy cost to the user for each 100 00 Btu provided to the living space of his residence. (b) Find cost to the user for each 100 000 Btu of electric energy for heating if this is provided by a heat pump with a HCOP (heating coefficient of performance) of 3.

Ans. (a) $0.33 oil, $0.879 power; (b) $0.293 power

18-9. A seaboard power plant can produce power with an overall heat rate of 9200 Btu/kWh based on the fuel used. It formerly used residual fuel oil (144 000 Btu/gal) which it could buy at 20 cents. Changeover to coal at 13 000 Btu/lb was made and coal is available at \$22.00/ton delivered. The heat rate per kilowatthour was not affected by the change. In terms of fuel cost alone, compute the cost of a kilowatthour before and after the change.

Ans. Oil, 1.28 cents; coal 0.78 cent

18-10. It is desired to design a rock heat-storage chamber for a warm-air heating system. A pulldown storage capacity of 360 000 Btu is needed with lowest usable temperature 120 F (48.9°C) and probable top temperature 170 F (76.7°C). Crushed limestone fragments 1 to 2.5 in. in size have a density of 90 lb/ft^3 and the specific heat of limestone is approximately 0.22 Btu/lb · F. Find the weight of limestone needed for operation through 50 Fahrenheit degrees and the approximate volume required. If used in an 8ft wide by 6 ft high rectangular section container, what length is needed?

Ans. 32 730 lb, 363.6 ft^3, 7.6 ft long

18-11. A crushed-rock heat-storage chamber is needed to store 759 100 kJ when operating through 27.9 Celsius degrees, to provide reserve capacity for a warm-air solar-energy system. Limestone rock randomly sized at 25 to 64 mm is available with a specific heat of 0.92 kJ/kg · °C and at a density of 1442 kg/m^3. Find the weight and volume of limestone needed.

Ans. 29 574 kg, 20.5 m^3

18-12. At one time use was made of a unit for solar radiation called the Langley, which had units of calories per square centimeter, and tabular information is still found expressed in this unit. For example, in one published table we find that the average solar radiation on a horizontal surface in Philadelphia for each month, expressed in calories per square centimeter per day, is as follows:

Jan.	Feb.	Mar.	Apr.	May	June
175	242	347	425	493	554
July	Aug.	Sept.	Oct.	Nov.	Dec.
538	465	388	293	191	152

(a) Find the multiplying factor to convert Langleys to Btu per square foot.
(b) Find a similar multiplying factor to convert to kilojoules per square meter.
(c) Compare the corresponding June and December values above with those in Table 18-1 for 40°N latitude.

Ans. (a) 3.682; (b) 41.84; (c) reasonable check since value above is monthly average and Table 18-1 is for a given day

REFERENCES

E. J. Beck and R. L. Field, "Solar Heating of Buildings and Domestic Hot Water," *Technical Report R835 Naval Facilities Engineering Command* (April, 1976). Port Hueneme, Calif: Civil Engineering Laboratory, Dept. of the Navy.

F. de Winter, "Solar Energy and the Flat-plate Collector," *ASHRAE Journal*, Vol. 17 (Nov., 1975), pp. 56–59.

H. G. Larsch, "Thermal-Energy Storage for Solar Heating," *ASHRAE Journal*, Vol. 17 (Nov., 1975), pp. 47–52.

F. S. Dubin, "Solar-energy Design for Existing Buildings," *ASHRAE Journal*, Vol. 17 (Nov., 1975), pp. 53–55.

N. V. Suryanarayana and R. C. Bosio, "Solar Energy for Space Heating with Heat Pump," *Mechanical Engineering*, Vol. 98 (Oct., 1976), pp. 33–37.

R. T. Ruegg, "Life-Cycle Costs and Solar Energy," *ASHRAE Journal*, Vol. 18 (Nov., 1976), pp. 22–25.

A. Weinstein, R. T. Duncan, W. C. Sherbin, "Atlanta (Towns-School) Solar Experiment," *ASHRAE Journal*, Vol. 18 (Nov., 1976), pp. 32–35.

A. B. Newton, "Optimizing Solar Cooling Systems," *ASHRAE Journal*, Vol. 18 (Nov., 1976), pp. 26–31.

ASHRAE, Handbook and Product Directory, Applications, 1974, Chap. 59.

APPENDIX

Table A-1 Defining SI Equivalents

1 ft = 0.3048 m = 3.048 E − 01
1 lb · mass = 0.45359237 kg = 4.535 923 7 E − 01
1 lbf = 4.448221615 N = 4.448 221 6 E + 00
1 Btu (thermochemical) = 1054.350 J = 1.054 350 E + 03
1 Btu (IST 1956) = 1055.056 J = 1.055 056 E + 03
1 kcal (thermochemical) = 4184 J = 4.184 E + 03
1 kcal (IST 1956) = 4186.8 J = 4.1868 E + 03
1 kWh = 3 600 000 J = 3.6 E + 06

Table A-2 Equivalence of Miscellaneous Units

Lengths

1 ft	= 0.3048 m	= 12 in.	= 0.3333 yd
1 m	= 3.28084 ft	= 39.37008 in.	= 10^6 μ (microns)
1 mi	= 5,280 ft	= 1,760 yd	= 1,609.34 m
1 mi	= 0.86898 (nautical) mi	= 1.60934 km	= 320 rd

Areas

1 sq ft	= 0.09290 sq m	= 144 sq in.	= 0.11111 sq yd
1 sq m	= 1549.99 sq in.	= 10.7639 sq ft	= 1.19599 sq yd
1 Acre	= 43,560 sq ft	= 4,840 sq yd	= 0.40469 ha (hectare)
1 Acre	= 4046.87 sq m	= 0.001563 sq mi	
1 sq mi	= 640 A	= 3,097,600 sq yd	= 2,589,999 sq m
1 sq mi	= 2.59000 sq km	= 259.0 ha	
1 sq km	= 0.38610 sq mi	= 247.104 Acre	= 10^6 sq m = 100 ha

Masses and Weights

1 lb	= 0.45359 kg	= 16 oz	= 14.5833 oz (troy)	= 0.0005 ton
1 lb	= 7000 grains	= 0.000464 long ton		
1 kg	= 2.2046 lb av	= 2.2692 lb tr	= 35.274 oz av	
1 kg	= 15,432.4 grains	= 0.00110 ton	= 0.001 m ton	
1 ton	= 2,000 lb	= 907.185 kg	= 32,000 oz	= 0.90722 m ton

Volume and Capacity

1 cu ft	= 1728 cu in.	= 0.03704 cu yd	= 0.028317 cu m
1 cu ft	= 29.9221 qt (liq)	= 7.4806 gal (liq)	
1 cu ft	= 6.229 Imp, gal (Br)	= 0.80356 bu	
1 cu yd	= 46,656 cu in.	= 27 cu ft	= 0.76456 cu m
1 cu yd	= 807.896 qt (liq)	= 201.974 gal (liq)	
			= 21.6962 bu
1 gal (liq)	= 231 cu in.	= 0.13368 cu ft	= 4 qt
1 gal (liq)	= 0.83268 Imp. gal	= 0.00378543 cu m	
1 cu m	= 61,023 cu in.	= 35.314 cu ft	= 1056.7 qt (liq)
	= 264.18 gal (liq)	= 28.38 bu	= 1.308 cu yd

Table A-3 Conversion Factors for Pressure Units
(Multiply units in left column by appropriate factor to obtain result in units designated at the top of vertical column.)

	atm	lbf/in.^{2}	Pa $\equiv$ N/m^2	in. Hg at 32 F (0°C)	mm Hg at 0°C (32 F)
Atmosphere	1.0	14.6960	101 325	29.9212	760
lbf/in.^{2} (psi)	0.068 046	1.0	6894.8	2.036	51.715
Pa (N/m^2)	9.8692 × E−06	1.450 38 × E−04	1.0	2.953 × E−04	7.500 × E−3
in. Hg at 32 F (0°C)	0.033 421	0.491 16	3 386.39	1.0	25.40
m Hg at 0° (32 F)	1.315 79	19.337	133 322	39.370	1 000
m Hg at 20°C (68 F)	1.311 00	19.267	132 843	39.23	996.38
mm water at 20°C (68 F)	0.000 096 6	0.001 419 9	9.7894	0.002 890 7	0.073 4
in. water at 68 F (20°C)	0.002 454	0.036 065	248.652	0.073 423	1.865

Table A-4 Conversion Factors for Energy
(Multiply units of left column by appropriate factor* in table to obtain result in units designated at the top of vertical column.)

	Btu	J	kWh	kcal	ft · lbf	Int. Steam Table kcal
Btu	1.0000	1054.350	2.92875 × E−04	2.519957 × E−01	777.649	2.52164 × E−01
J ≡ W · s	9.4845 × E−04	1.0000	2.77778 × E−07	2.39006 × E−04	7.3756 × E−01	2.38845 × E−04
kWh	3414.43	3.6 E+06	1.000	860.420	2.6552 × E+06	859.845
kcal	3.96835	4184.0†	1.16222 × E−03	1.0000	3085.96	0.99933
hp · h	2547.16	2.6855 × E−06	0.7460†	641.87	1.9808 × E+06	641.44
ft · lbf	1.28592 × E−03	1.355818	3.76616 × E−07	3.2405 × E−04	1.0000	3.2383 × E−04
Int. Steam Table kcal	3.9656	4186.8‡	1.163 × E−03	1.00067	3088.04	1.0000

* Factors based on SI-unit definitions and the thermochemical calorie taken as 4184.0 J.
†Represents a defined SI unit.
‡Represents the International Steam Table kilocalorie defining equivalent.

Table A-5 Conversion Factors for Energy in Relation to Time and Area (Multiply units of left column by appropriate factor in table to obtain result in units designated at top of each other column.)

	Btu/h · ft^2	Btu/h · m^2	W/ft^2	W/m^2	kcal/h · m^2	Btu/s · ft^2
Btu/h · ft^2	1.0000	10.7639	2.92875 ×E−01	3.15248	2.712459	2.77778 ×E−04
Btu/h · m^2	9.29030 ×E−02	1.0000	2.72089 ×E−02	2.92875 ×E−01	2.51996 ×E−01	2.58064 ×E−05
W/ft^2	3.41443	36.7526	1.0000	10.76391	9.26142	9.48453 ×E−04
W/m^2	3.17210 ×E−01	3.41442	9.29030 ×E−2	1.0000	8.6042 ×E−01	8.81138 ×E−05
kcal/h · m^2	3.68669 ×E−01	3.96832	1.0797 ×E−01	1.16222	1.0000	1.02408 ×E−04
Btu/s · ft^2	3600	38750	1054.35	11348.9	9764.85	1.0000

Table A-6 Conversion Factors for Thermal Conductivity
(Multiply units of left column by appropriate factor* in table to obtain result in units designated at the top of vertical column.)

	Btu · ft/h · ft^2 · F	Btu · in./h · ft^2 · F	W/m · °C	W/cm · °C	cal/s · cm · °C	kcal/h · m · °C
Btu · ft/h · ft^2 · F	1.000	12	1.72958	1.72958 ×E−02	4.13378 ×E−03	1.48816
Btu · in./h · ft^2 · F	8.33333 ×E−02	1.000	1.441314 ×E−01	1.441314 ×E−03	3.44481 ×E−04	1.24013 ×E−01
W · m/m^2 · °C	5.78175 ×E−01	6.93811	1.000	1.0000 ×E−02	2.39006 ×E−03	8.60422 ×E−01
W· cm/cm^2 · C	57.8175	6.93811 ×E+02	100.00	1.000	2.39006 ×E−01	8.60422 ×E+01
cal · cm/s · cm^2 · °C	241.909	2.90291 ×E+03	4.18400 ×E+02	4.18400	1.000	360.00
kcal · m/h · m^2 · °C	6.71971 ×E−01	8.06365	1.16222	1.16222 ×E−02	2.77778 ×E−03	1.000

* Factors based on SI-unit definitions and the thermochemical calorie taken as 4184.0 J.

Table-A 7 Conversion Factors for Coefficients of Heat Transfer (Multiply units of left column by appropriate factor* in table to obtain results in units designated at top of vertical column.)

	Btu/h · ft² · F	W/m² · °C	W/cm² · °C	kcal/h · m² · °C	cal/s · cm² · °C
Btu/h · ft² · F	1.000	5.67446	5.67446 ×E−04	4.88243	1.35623 ×E−04
W/m² · °C	1.76228 ×E−01	1.000	1.0E−04	8.6042 ×E−01	2.3900 ×E−05
W/cm² · °C	1.76228 ×E+03	1.0 E+04	1.000	8.6042 ×E+03	2.3900 ×E−01
kcal/h · m² · °C	2.04816 ×E−01	1.16222	1.16222 ×E−04	1.000	2.77778 ×E−05
cal/s · cm² · °C	7.37338 ×E+03	4.1840 ×E+04	4.1840	3.6000 ×E+04	1.000

*Factors based on SI units and the thermochemical kilocalorie taken as 4184 J.

Table A-8 Miscellaneous Conversion Equivalents

Density						
(lb/ft³)		(g/cm³)		(kg/m³)		(lb/gal)
1.000	=	0.0160185	=	16.01846	=	0.133680
62.4280	=	1.000	=	1,000.0	=	8.34538
0.062428	=	0.001	=	1.000	=	0.008345
7.48055	=	0.119827	=	119.827	=	1.000
Enthalpy and Energy per Unit Mass						
(Btu/lb)		(kcal/kg)		(J/g)		(W · h/kg)
1.000	=	0.555556	=	2.32444	=	0.645679
1.8	=	1.000	=	4.184	=	1.16222
0.430210	=	0.239006	=	1.000	=	0.277778
1.54876	=	0.860422	=	3.600	=	1.000
Specific Heat and Entropy						
(Btu/lb · R)		(kcal/kg · K)		(kJ/kg · K)		(W · h/kg · K)
1.000	=	1.000	=	4.184	=	1.16222
0.239006	=	0.239006	=	1.000	=	0.277778
0.860422	=	0.860422	=	3.600	=	1.000

Table A-9 Temperature Conversion

The temperature in the center column in either Fahrenheit or Celsius (Centigrade) degrees can be read in equivalent degrees in the appropriate left or right column.

°C	°C or F	F	°C	°C or F	F	°C	°C or F	F	°C	°C or F	F	°C	°C or F	F
−40.0	−40	−40.0	−9.44	15	59.0	12.8	55	131.0	35.0	95	203.0	57.2	135	275.0
−37.2	−35	−31.0	−8.89	16	60.8	13.3	56	132.8	35.6	96	204.8	57.8	136	276.8
−34.4	−30	−22.0	−8.33	17	62.6	13.9	57	134.6	36.1	97	206.6	58.3	137	278.6
−31.7	−25	−13.0	−7.73	18	64.4	14.4	58	136.4	36.7	98	208.4	58.9	138	280.4
−29.4	−21	−5.8	−7.22	19	66.2	15.0	59	138.2	37.2	99	210.2	59.4	139	282.2
−28.9	−20	−4.0	−6.67	20	68.0	15.6	60	140.0	37.8	100	212.0	60.0	140	284.0
−28.3	−19	−2.2	−6.11	21	69.8	16.1	61	141.8	38.3	101	213.8	60.6	141	285.8
−27.8	−18	−0.4	−5.56	22	71.6	16.7	62	143.6	38.9	102	215.6	61.1	142	287.6
−27.2	−17	1.4	−5.00	23	73.4	17.2	63	145.4	39.4	103	217.4	61.7	143	289.4
−26.7	−16	3.2	−4.44	24	75.2	17.8	64	147.2	40.0	104	219.2	62.2	144	291.2
−26.1	−15	5.0	−3.89	25	77.0	18.3	65	149.0	40.6	105	221.0	62.8	145	293.0
−25.6	−14	6.8	−3.33	26	78.8	18.9	66	150.8	41.1	106	222.8	63.3	146	294.8
−25.0	−13	8.6	−2.78	27	80.6	19.4	67	152.6	41.7	107	224.6	63.9	147	296.6
−24.4	−12	10.4	−2.22	28	82.4	20.0	68	154.4	42.2	108	226.4	64.4	148	298.4
−23.9	−11	12.2	−1.67	29	84.2	20.6	69	156.2	42.8	109	228.2	65.0	149	300.2
−23.3	−10	14.0	−1.11	30	86.0	21.1	70	158.0	43.3	110	230.0	65.6	150	302.0
−22.8	−9	15.8	−0.56	31	87.8	21.7	71	159.8	43.9	111	231.8	71.1	160	320.0
−22.2	−8	−17.6	0	32	89.6	22.2	72	161.6	44.4	112	233.6	76.7	170	338.0
−21.7	−7	19.4	0.56	33	91.4	22.8	73	163.4	45.0	113	235.4	82.2	180	356.0
−21.1	−6	21.2	1.11	34	93.2	23.3	74	165.2	45.6	114	237.2	87.8	190	274.0

−20.6	−5	23.0	1.67	35	95.0	23.9	75	167.0	46.1	115	239.0	93.3	200	392.0
−20.0	−4	24.8	2.22	36	96.8	24.4	76	168.8	46.7	116	240.8	98.9	210	410.0
−19.4	−3	26.6	2.78	37	98.6	25.0	77	170.6	47.2	117	242.6	100.0	212	413.6
−18.9	−2	28.4	3.33	38	100.4	25.6	78	172.4	47.8	118	244.4	104.4	220	428.0
−18.3	−1	30.2	3.89	39	102.2	26.1	79	174.2	48.3	119	246.2	110.0	230	446.0
−17.8	0	32	4.44	40	104.0	26.7	80	176.0	48.9	120	248.0	115.6	240	461.0
−17.2	1	33.8	5.00	41	105.8	27.2	81	177.8	49.4	121	249.8	121.1	250	482.0
−16.7	2	35.6	5.56	42	107.6	27.8	82	179.6	50.0	122	251.6	126.7	260	500.0
−16.1	3	37.4	6.11	43	109.4	28.3	83	181.4	50.6	123	253.4	132.2	270	518.0
−15.6	4	39.2	6.67	44	111.2	28.9	84	183.2	51.1	124	255.2	137.8	280	536.0
−15.0	5	41.0	7.22	45	113.0	29.4	85	185.0	51.7	125	257.0	143.3	290	554.0
−14.4	6	42.8	7.78	46	114.8	30.0	86	186.8	52.2	126	258.8	148.9	300	572.0
−13.9	7	44.6	8.33	47	116.6	30.6	87	188.6	52.8	127	260.6	154.4	310	590.0
−13.3	8	46.4	8.89	48	118.4	31.1	88	190.4	53.3	128	262.4	159.9	320	608.0
−12.8	9	48.2	9.44	49	120.2	31.7	89	192.2	53.9	129	264.2	165.5	330	626.0
−12.2	10	50.0	10.0	50	122.0	32.2	90	194.0	54.4	130	266.0	171.3	340	644.0
−11.7	11	51.8	10.6	51	123.8	32.8	91	195.8	55.0	131	267.8	176.9	350	662.0
−11.1	12	53.6	11.1	52	125.6	33.3	92	197.6	55.6	132	269.6	182.5	360	680.0
−10.6	13	55.4	11.7	53	127.4	33.9	93	199.4	56.1	133	271.4	188.1	370	698.0
−10.0	14	57.2	12.2	54	129.2	34.4	94	201.2	56.7	134	273.2	193.6	380	716.0

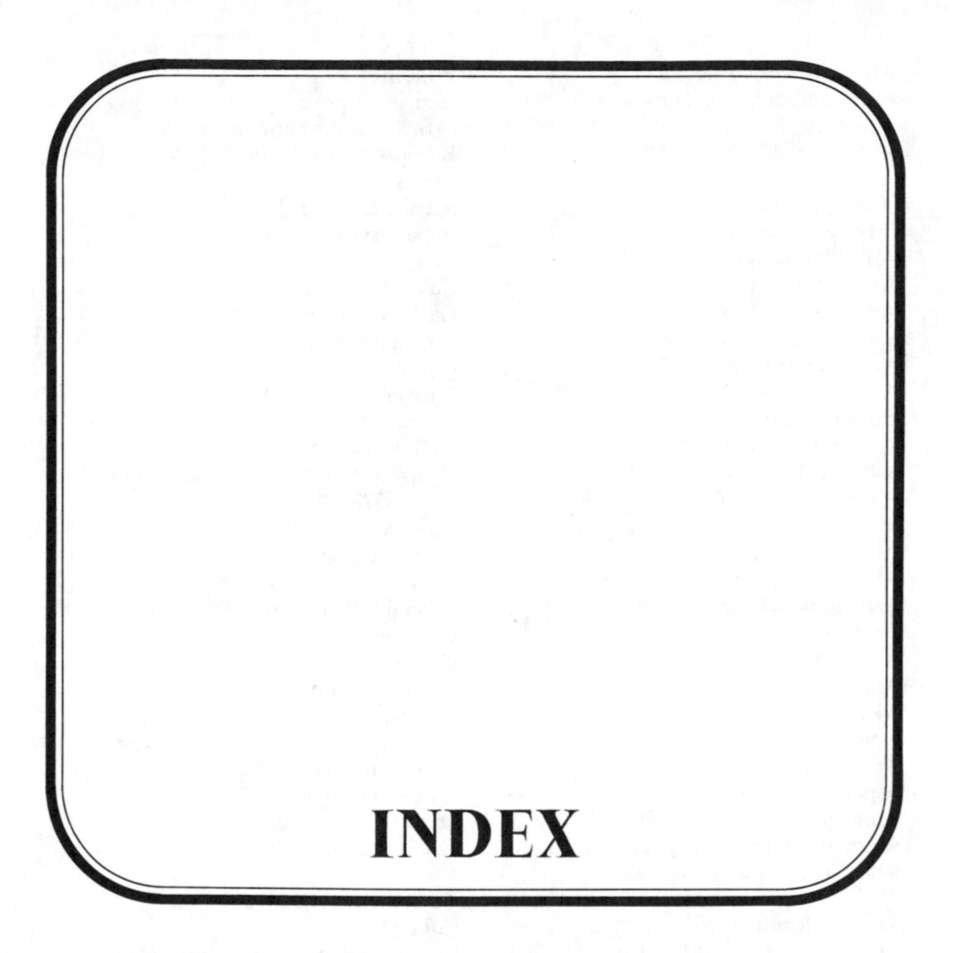

INDEX

78 79 80 81 82 9 8 7 6 5 4 3 2 1